HEAT TRANSFER PHYSICS

Heat Transfer Physics is a graduate-level textbook describing the atomic-level kinetics (mechanisms and rates) of thermal energy storage, transport (conduction, convection, and radiation), and transformation (various energy conversions) by principal energy carriers. These carriers are called *phonons* (lattice vibration waves, also treated as quasi-particles), *electrons* (classical or quantum entities), *fluid particles* (classical particles with quantum features), and *photons* (classical electromagnetic waves, also treated as quasi-particles), as shown in the cover figure. This approach combines fundamentals (through survey and summaries) of the following fields: molecular orbitals/potentials, statistical thermodynamics, computational molecular dynamics (including lattice dynamics), quantum energy states, transport theories (e.g., Boltzmann and stochastic transport and Maxwell equations), solid-state (including semiconductors) and fluid-state (including surface interactions) physics, and quantum optics (e.g., spontaneous and stimulated emission, photon-electron-phonon couplings). These are rationally connected to atomic-level heat transfer (e.g., heat capacity, thermal conductivity, photon absorption coefficient) and thermal energy conversion (e.g., ultrasonic heating, thermoelectric and laser cooling). This book presents a unified theory, over fine-structure/molecular-dynamics/Boltzmann/macroscopic length and time scales, of heat transfer kinetics in terms of transition rates and relaxation times and relates it to modern applications (including nanoscale and microscale size effects).

Massoud Kaviany is professor of mechanical engineering and applied physics at the University of Michigan. His research-education integrations include the monographs *Principles of Heat Transfer in Porous Media* (1991) and *Principles of Convective Heat Transfer* (1994) and the textbook *Principles of Heat Transfer* (2001).

Heat Transfer Physics

Massoud Kaviany
University of Michigan

CAMBRIDGE
UNIVERSITY PRESS

CAMBRIDGE UNIVERSITY PRESS
Cambridge, New York, Melbourne, Madrid, Cape Town, Singapore, São Paulo, Delhi

Cambridge University Press
32 Avenue of the Americas, New York, NY 10013-2473, USA

www.cambridge.org
Information on this title: www.cambridge.org/9780521898973

First published 2008

Printed in the United States of America

A catalog record for this publication is available from the British Library.

Library of Congress Cataloging in Publication data

Kaviany, M. (Massoud)
Heat transfer physics / Massoud Kaviany.
p. cm.
Includes bibliographical references and index.
ISBN 978-0-521-89897-3 (hardback)
1. Nuclear reactor kinetics. 2. Change of state (Physics) 3. Heat storage. 4. Heat – Transmission. I. Title.
QC787.N8K39 2008
536′.2 – dc22 2008009486

ISBN 978-0-521-89897-3 hardback

To curiosity, reason, doubt,
dialogue, understanding, tolerance,
and humility.

Contents

Preface

Heat transfer physics describes the thermodynamics and kinetics (mechanisms and rates) of energy storage, transport, and transformation by means of principal energy carriers. Heat is energy that is stored in the temperature-dependent motion and within the various particles that make up all matter in all of its phases, including electrons, atomic nuclei, individual atoms, and molecules. Heat can be transferred to and from matter by combinations of one or more of the principal energy carriers: electrons† (either as classical or quantum entities), fluid particles (classical particles with quantum features), phonons (lattice-vibration waves), and photons‡ (quasi-particles). The state of the energy stored within matter, or transported by the carriers, can be described by a combination of classical and quantum statistical mechanics. The energy is also transformed (converted) between the various carriers. All processes that act on this energy are ultimately governed by the rates at which various physical phenomena occur, such as the rate of particle collisions in classical mechanics. It is the combination of these various processes (and their governing rates) within a particular system that determines the overall system behavior, such as the net rate of energy storage or transport. Controlling every process, from the atomic level (studied here) to the macroscale (covered in an introductory heat transfer course), are the laws of thermodynamics, including conservation of energy.

The focus of this text is on the heat transfer behavior (the storage, transport, and transformation of thermal energy) of the aforementioned principal energy carriers at the atomic scale. The specific mechanisms will be described in detail, including elastic–inelastic collisions–scattering among particles, quasi-particles, and waves. Particular attention will be given to the various time scales over which energy transport or transformation processes occur, so that the reader will be given some sense of how they compare with one another, as well as how they combine to produce overall system energy storage–transport–transformation rates. The ap-

† For semiconductors, the holes are included as energy carriers. For electrolytes, ion transport is treated similarly.

‡ Here, *photon* refers to both the classical (Maxwell) and the quantum (quasi-particle, Schrödinger) descriptions of the electromagnetic waves.

proach taken here is to begin with a survey of fundamental concepts of atomic-level physics. This includes looking at the energy within the electronic states of atoms, as well as interatomic forces and potentials. Various theories of molecular dynamics and transport will also be described. Following this overview, in-depth, quantitative analyses will be performed for each of the principal energy carriers, including analysis of how they interact with each other. This combination should allow for the teaching of a thorough introduction of heat transfer physics within one semester, without prolonged preparation or significant prerequisites. In general, several areas of physics are relevant to the study of heat transfer: (a) atomic–molecular dynamics, (b) solid state (condensed matter), (c) electromagnetism, and (d) quantum optics. No prior knowledge of these is necessary to appreciate the material of this text (a knowledge of introductory heat transfer is assumed).

Crystalline solids and their vibrational and electronic energies are treated first. This is followed by energies of fluid particles and their interactions with solid surfaces. Then the interactions of photons with matter are posed with photons as EM waves, or as particles, or as quasi-particles.

The text is divided into seven chapters, starting with the introduction and preliminaries of Chapter 1, in which the microscale carriers are introduced and the scope of the heat transfer physics is defined. Chapter 2 is on molecular electronic orbitals, interatomic and intermolecular potentials, molecular dynamics, and an introduction to quantum energy states. Chapter 3 is on microscale energy transport and transition kinetics theories, including the Boltzmann transport equation, the Maxwell equations, the Langevin stochastic transport equation, the Onsager coupled transport relation, and the Green–Kubo fluctuation–dissipation transport coefficients and relations. Following these, Chapters 4, 5, 6, and 7 cover the transport and interactions of phonons, electrons, fluid particles, and photons, respectively.

The size effects (where the system size affects the atomic-level behavior) on transport and energy conversion, for each principal carrier, are considered at the ends of Chapters 4 to 7. This allows for reference to applications in nanostructured and microstructured systems.

Some of the essential derivations are given as appendices. Appendix B gives the Green–Kubo relation, Appendix C gives the minimum phonon conductivity relations, Appendix D gives the phonon boundary resistance, Appendix E gives the Fermi golden rule, and Appendix F gives the particle energy distribution (occupancy) functions for bosons (phonons and photons), fermions (electrons), and Maxwell–Boltzmann (fluid) particles.

Some end-of-chapter problems are provided to assist in further understanding and familiarity, and to allow for specific calculations. When needed, computer programs are also used.

In general, vectors (lowercase) and tensors (uppercase) are in bold symbols. A nomenclature and an abbreviation list are given at the end of the text. Numbers in parenthesis indicate equation numbers. A glossary of relevant terminology is also given at the end. The periodic table of elements, with the macroscopic (bulk) and atomic properties, is given in Appendix A (in Tables A.1 and A.2), along with the tables of the universal and derived constants and unit prefixes.

It is hoped that this treatment provides an idea of the scope and some of the fundamentals of heat transfer physics, along with some of the most recent findings in the field.

Massoud Kaviany
Ann Arbor
kaviany@umich.edu

Acknowledgments

Many doctoral students and postdoctoral Fellows working with me have contributed to this book. Among them are Jae Dong Chung, Luciana da Silva, Baoling Huang, Gi Suk Hwang, Dan Johnson, Ankur Kapoor, Jedo Kim, Scott Gayton Liter, Alan McGaughey, Da Hye Min, Brendan O'Connor, Xiulin Ruan, and Xiangchun Zhang. Baoling Huang, Gi Suk Hwang, Ankur Kapoor, Jedo Kim, Scott Gayton Liter, Alan McGaughey, and Xiulin Ruan have provided many ideas and have been constant sources of inspiration (the last three have also carefully read the manuscript and have made valuable comments). I am indebted to all of them; without them this task could not have been completed. I would also like to thank the National Science Foundation and Department of Energy (Basic Energy Sciences) for sponsoring the research leading to some of the materials presented here.

1

Introduction and Preliminaries

The macroscopic heat transfer rates use thermal-energy related properties, such as the thermal conductivity, and in turn these properties are related to the atomic-level properties and processes. Heat transfer physics addresses these atomic-level processes (e.g., kinetics). We begin with the macroscopic energy equation used in heat transfer analysis to describe the rates of thermal energy storage, transport (by means of conduction k, convection u, and radiation r), and conversion to and from other forms of energy. The volumetric macroscopic energy conservation (rate) equation is listed in Table 1.1.† The sensible heat storage is the product of density and specific heat capacity ρc_p, and the time rate of change of local temperature $\partial T/\partial t$. The heat flux vector $\boldsymbol{q}$ is the sum of the conductive, convective, and radiative heat flux vectors. The conductive heat flux vector $\boldsymbol{q}_k$ is the negative of the product of the thermal conductivity k, and the gradient of temperature ∇T, i.e., the Fourier law of conduction. The convective heat flux vector‡ $\boldsymbol{q}_u$ (assuming net local motion) is the product of ρc_p, the local velocity vector $\boldsymbol{u}$, and temperature. For laminar flow, in contrast to turbulent flow that contains chaotic velocity fluctuations, molecular conduction of the fluid is unaltered, whereas in turbulent flow this is augmented (phenomenologically) by turbulent mixing transport (turbulent eddy conductivity). The radiative heat flux vector $\boldsymbol{q}_r = \boldsymbol{q}_{ph}$ (ph stands for photon) is the spatial (angular) and spectral integrals of the product of the unit vector $\boldsymbol{s}$ and the electromagnetic (EM) spectral (and directional) radiation intensity $I_{ph,\omega}$, where ω is the angular frequency of EM radiation (made up of photons). This intensity is influenced by the emission, absorption, and scattering of photons by matter, i.e., its radiative properties. Among these properties are the photon spectral, absorption coefficient $\sigma_{ph,\omega}$, which results from the interaction of an EM waves with electric entities in its traveling medium (e.g., vibrating electric dipoles, free electrons).

† In deriving this equation, it is assumed that c_p is constant. The more general form is also given in [174].

‡ The surface-convection heat flux vector $\boldsymbol{q}_{ku}$ is the special case of conduction occurring on an interface separating a moving fluid from a generally stationary solid. Then, assuming no fluid slip on this interface, the heat transfer through the fluid is only by conduction, but the interface temperature gradient is influenced by the fluid motion [174].

Table 1.1. *Macroscopic energy conservation equation and the heat flux vector [174].*

Energy equation

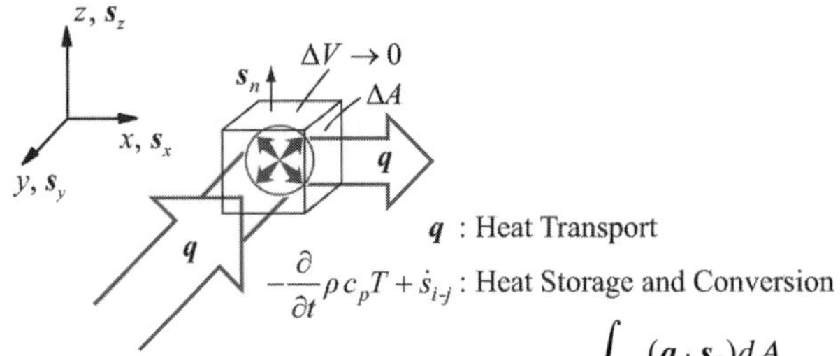

$$\nabla \cdot \boldsymbol{q} = -\frac{\partial}{\partial t}\rho c_p T + \sum_{i,j} \dot{s}_{i\text{-}j}, \qquad \nabla \cdot \boldsymbol{q} = \frac{\int_{\Delta \mathrm{A}} (\boldsymbol{q} \cdot \boldsymbol{s}_n) dA}{\Delta V \to 0}$$

Heat flux vector

$$\boldsymbol{q} = \boldsymbol{q}_k + \boldsymbol{q}_u + \boldsymbol{q}_r = -k\nabla T + \rho c_p \boldsymbol{u} T + 2\pi \int_0^\infty \int_{-1}^1 \boldsymbol{s} I_{ph,\omega} d\mu d\omega$$

A	area, m^2
$I_{ph,\omega}$	spectral, directional radiation intensity, W/(m^2-sr-rad/s)
c_p	specific heat capacity, J/kg-K
k	thermal conductivity, W/m-K
$\boldsymbol{q}$	heat flux vector, W/m^2
$\boldsymbol{s}_n$	surface normal, unit vector
$\boldsymbol{s}$	unit vector
$\dot{s}_{i\text{-}j}$	energy conversion rate between principal carrier i and carrier j, W/m^3
T	temperature, K
$\boldsymbol{u}$	velocity, m/s
V	volume, m^3
ρ	density, kg/m^3
μ	$\cos\theta$, θ is polar angle
ω	angular frequency, rad/s

In Table 1.1, it is noted that the divergence of $\boldsymbol{q}$ is integral to its surface normal component on the differential surface area ΔA of a vanishing deferential volume ΔV. The rate of energy conversions to and from thermal energy $\dot{s}_{i\text{-}j}$ is determined by the nature and frequency of the interactions between the principal energy carriers i and j. This rate describe various bond (chemical and physical), electromagnetic, and mechanical energy conversions. The rate is related to the available energy transitions as well as the contributions from promoting/limiting mechanisms (e.g., energy distribution probabilities, kinetics, transport, and temperature). It is this interplay among storage, transport, and transformation rates that allow for the behaviors exhibited by energy conversion phenomena and devices.

The motion of atoms, because of vibration, rotation, and translation, creates kinetic energy within matter. Electrons (valence electrons as they orbit the nuclei or conduction electrons as they move among atoms) are also central to the energy contained within matter. Electronic energies (including bond energy between atoms), kinetic energy (created by the motion of atoms), and even the annihilation/

creation of mass (i.e., relativistic effects), can all be converted to electromagnetic energy (photons). The energy contained in photons can then, in turn, interact with other matter (through electric entities such as dipole moments) resulting in energy conversion.

The state of an electron (which defines its energy level), as well as its coupling with atomic nuclei (in a free atom, or in molecules), is central to the energy of a system, its interactions (including conversion to other forms of energy), and its transport (both within the system and across its boundaries), especially when the energy per unit mass is much larger than $k_\mathrm{B}T$ (which is 0.026 eV for $T = 300$ K), where k_B is the Boltzmann constant. $k_\mathrm{B}T$ is the energy of thermal fluctuations (Section 2.3.2). In ideal electrical insulators, no conduction (free) electrons exist, metals have a large number (over 10^{21} cm^{-3}), and intrinsic (undoped) semiconductors have a small, temperature-dependent number (less than 10^{15} cm^{-3}) of conduction electrons. Mobile and stationary ions, as intrinsic constituents or as dopants, also have their own particular electronic energy states.

Thus the microscopic model of thermal energy storage, transport, and interactions is assembled through the study of each of the principal energy carriers, namely: phonons (p), electrons (e), fluid particles (f), and photons (ph).†

This introduction–preliminaries chapter continues with attributes of the four principal heat carries, including their combinatorial energy-state occupational probabilities and their wave, particle, and quasi-particle treatments. Then we present a brief history of contributions to heat transfer physics, and introduce the universal constants and atomic-level (fine-structure) time, length, and energy scales. Then we give examples of atomic-level kinetics controlling energy storage, transport, and conversion, and define the scope of heat transfer physics and this text.

1.1 Principal Carriers: Phonon, Electron, Fluid Particle and Photon

The energy of matter can be divided into potential and kinetic energies E_p and E_k. Each of the principal energy carriers can have potential and kinetic energy (for photon it is electric and magnetic field energies). Here we discuss the energy attributes of carriers, including their equilibrium occupancy probabilities, which allow for inclusion of these attributes into statistical presentation of energy of matter.

1.1.1 Phonon

A phonon is a quantized mode of vibration occurring in rigid atomic lattices, such as those in crystalline solids.‡ The properties of long-wavelength phonons give rise

† The suffix "on" indicates having properties of particles.

‡ In lattice dynamics, there are a finite number of vibrational modes, and the energy of each mode is quantized. So, phonons are also these normal modes. Also, although phonons are exclusively a property of periodic media, vibrations exist in all solids.

to sound in solids (hence the name phonon). Phonons participate in many of the physical properties of materials, including heat capacity and thermal/electrical conductivities (the propagation of phonons is responsible for the conduction of heat in electrical insulators).

In classical mechanics, any vibration of a lattice can be decomposed into a superposition of nonlocalized normal modes of vibration. When these modes are analyzed by use of quantum mechanics, they are found to possess some particlelike properties (wave–particle duality). Thus a phonon is an indistinguishable quasi-particle (see Glossary). When treated as particles, phonons are called bosons (see Table 1.2, and the derivation given in Appendix F) and are said to possess zero spin. The thermal equilibrium particle probability distribution (occupation) function $f_p^{\rm o}$ indicates, that at thermal equilibrium (the superscript $^{\rm o}$ indicates equilibrium), phonons are distributed based on their scaled energy $E_p/k_{\rm B}T$. This is also shown in Figure 1.1, which shows the phonon energy is the sum of its potential and kinetic energy. Here, $E_p = \hbar\omega_p$ is the phonon energy, and $\hbar = h_{\rm P}/2\pi$, where $h_{\rm P}$ is the Planck[†] constant (fundamental constants are discussed in Section 1.5.1).

Spin is one of the properties of elementary particles, which can be thought of as rotating tops. Based on spin, the particles are either fermions or bosons. The spin (quantum angular momentum) of a particle is given by $S = \hbar[s(s+1)]^{1/2}$ (Section 2.6.6), where s is 0, 1/2, 3/2, 2, ... If s is 1, 2, ..., then the particle is called a boson, and if s is 1/2, 3/2, ..., then it is called a fermion. No two fermions can be found in the same quantum state (because of the Pauli exclusion principle), whereas bosons tend to accumulate in certain favored states.

There are two types of phonons: acoustic phonons, denoted with the subscript A, and optical phonons, denoted with the subscript O. Acoustic phonons have frequencies that become small for long wavelengths and correspond to sound waves in the lattice (this is a property of phonon dispersion relation). These long wavelengths correspond to bulk translations. Longitudinal- and transverse-acoustic phonons are often abbreviated as LA and TA phonons, respectively. Optical phonons, which arise in lattices with more than one atom per unit cell, always have some minimum frequency of vibration, even when their wavelength is infinite. They are called optical, because in ionic crystals (such as NaCl) they are excited very easily by light (such as infrared radiation). This is because they correspond to a mode of vibration where positive and negative ions at adjacent lattice sites move, thus creating a time-varying electric dipole moment. Optical phonons that interact in this way with light are called infrared active. Optical phonons, which are known as Raman active, can also interact indirectly with light, through Raman scattering (an inelastic scattering of a photon that creates or annihilates an optical phonon). Optical phonons are often abbreviated as LO and TO for the longitudinal and transverse varieties, respectively (although they are readily distinguishable for low-symmetry crystals).

[†] The subscript P is for consistency with other fundamental constants (Table 1.4 and also listed in Table A.3) and also is used to avoid confusion with the specific enthalpy h.

Table 1.2. *Thermal equilibrium particle (energy occupancy) distribution (statistical) function $f_i^{\circ}(E_i)$, $i = p, e, f, ph$ and its temperature dependence for principal energy carriers.*

Attributes	Phonon	Electron and Hole[a]	Fluid Particle	Photon
Iconic presentation	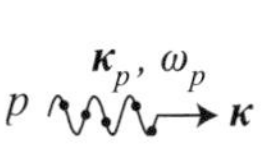	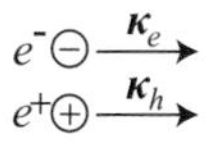	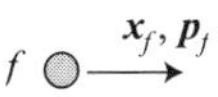	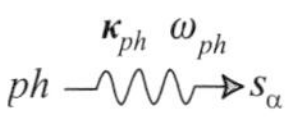
Energy presentation	wave vector $\boldsymbol{\kappa}_p$, dispersion $\boldsymbol{\kappa}_p(\omega_p)$, and polarizations in reciprocal lattice space $\boldsymbol{\kappa}$	wave vector $\boldsymbol{\kappa}_e$, band structure (conduction and valence bands), and spins	momentum $\boldsymbol{p}_f$, kinetic, potential, and electronic energy states	frequency ω_{ph}, dispersion $\boldsymbol{\kappa}_{ph}(\omega_{ph})$, and polarization $\boldsymbol{s}_\alpha$
Particle type	Bose–Einstein (boson)	Fermi–Dirac (fermion)	Maxwell-Boltzmann (M–B)	Bose–Einstein (boson)
Nature of particle	particles are indistinguishable, integer spin (angular momentum), and any number of particles may occupy a given eigenstate	particles are indistinguishable, odd, half-integer spin (angular momentum) and obey the Pauli exclusion principle (only one particle may be found in a given quantum state)	particles are distinguishable, and any number of particles may occupy a given eigenstate (classical particle or nondegenerate limit)	particles are indistinguishable, integer spin (angular momentum), and any number of particles may occupy a given eigenstate
Equilibrium distribution (occupancy) function, $f_i^{\circ}(E_i)$	$\dfrac{1}{\exp(\dfrac{E_p}{k_{\rm B}T})-1}$	$\dfrac{1}{\exp(\dfrac{E_e-\mu}{k_{\rm B}T})+1}$	$\dfrac{1}{\exp(\dfrac{E_f}{k_{\rm B}T})}$	$\dfrac{1}{\exp(\dfrac{E_{ph}}{k_{\rm B}T})-1}$
Energy	$E_p = E_{p,p} + E_{p,k}$ $= \hbar\omega_p$	$E_e = E_{e,p} + E_{e,k}$ $E_{e,k} = \dfrac{\hbar\kappa_e^2}{2m_e}$ μ is chemical potential $\mu = E_{\rm F}[1- \dfrac{1}{3}(\dfrac{\pi k_{\rm B}T}{2E_{\rm F}})^2]$	$E_f = E_{f,t} + E_{f,v} + E_{f,r} + E_{f,e} + E_{f,p}$ $E_{f,t} = \dfrac{p_f^2}{2m_f}$, $E_{f,v} = (l+\dfrac{1}{2})\hbar\omega_f$	$E_{ph} = E_{ph,e} + E_{ph,m}$ $E_{ph} = \hbar\omega_{ph}$

$\hbar = h_{\rm P}/2\pi$ and $k_{\rm B}$ are the reduced Planck and the Boltzmann constants

$k_{\rm B}T = 0.02585$ eV for $T = 300$ K

[a] Holes are represented by e^+

A general relation can be used for all particles as $f_i^{\circ} = \dfrac{1}{e^{(E_i-\mu)/k_{\rm B}T}+\gamma}$,

where $\gamma = 1$ (fermion), 0 (M–B), -1 (boson).

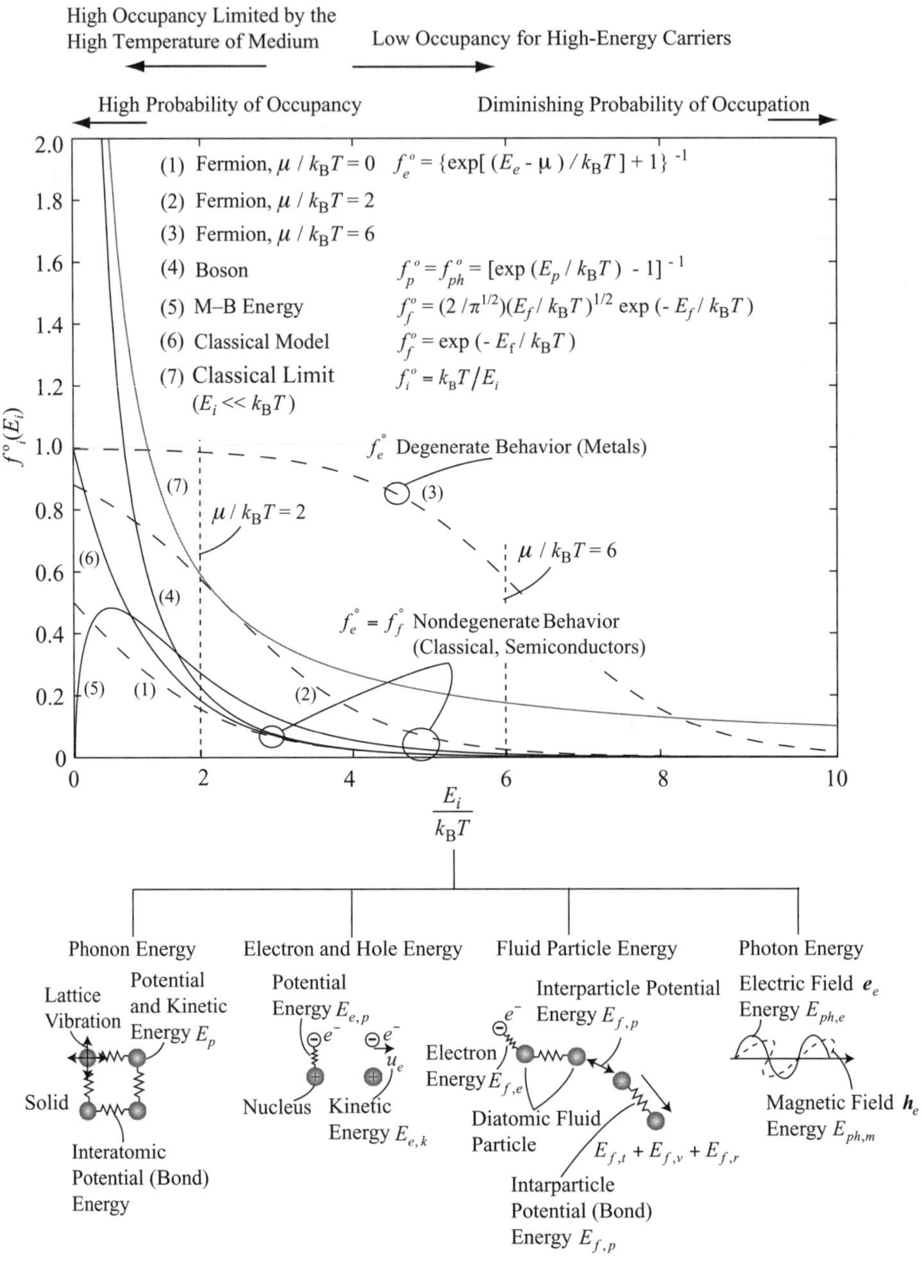

Figure 1.1. Variation of equilibrium particle (energy-state occupation) distribution function with respect to energy (scaled with respect to thermal fluctuation energy $k_B T$) for different energy carriers. For fermions, three different reduced chemical potentials $\mu / k_B T = E_F / k_B T$ (Fermi energy) are used. Electrons in metals have a highly degenerate behavior. The high-energy (low population) approximation (nondegenerate behavior) for electrons and holes in semiconductors is also shown (discussed in Section F.5 in Appendix F). The classical limit distribution function is also shown. Various contributions (e.g., kinetic and potential) to the total energy of each principal carrier, are also shown. The photon energy is divided between electric and magnetic energies.

1.1.2 Electron (and Hole)

The electron is a charged subatomic particle. In an atom, electrons surround the nucleus, made of protons and neutrons, in a manner termed the electronic configuration (or structure). The electron is among a class of subatomic particles called leptons, believed to be fundamental particles (i.e., they cannot be divided into smaller constituents). The electron has a spin of 1/2 (which makes it a fermion), and follows the Fermi–Dirac statistical energy distribution $f_i^{\rm o}$ (Table 1.2; derivation is given in Appendix F, these distributions are dictated by thermodynamics laws and relations). Electrons can exhibit properties of both particles and waves, and thus can be treated as quasi-particles. An electron bound to a nucleus behaves as a standing wave (due to the periodic boundary condition).

For matter in the solid state, electrons are responsible for bonding within crystals; they hold the nuclei together. These bonds belong to one of four types: van der Waals, ionic, covalent, or metallic. The Fermi surface is the set of loci in electron momentum space with zero excitation energy. The topology of the Fermi surface is important in understanding the electronic properties of materials. Electrons in solids are divided into core and outer electrons. Core electrons do not participate in bonding and are assumed to move with the nucleus at all times. Outer electrons reside farther away from the nucleus, and are in turn divided into conduction (or free) electrons and valence electrons.

In quantum mechanics, the electron is described by the Dirac equation. In the Standard Model of particle physics, it forms a doublet in SU(2) with the electron neutrino, as they interact through the weak force. The electron has two variations with the same charge but different masses: the muon and the tauon. The antimatter counterpart of the electron, its antiparticle, is the positron. The positron has the same amount of electrical charge, mass and spin as the electron, except that the charge is positive.

The variation of the three equilibrium distribution functions [boson, fermion, and Maxwell–Boltzman (M–B)] with respect to dimensionless (scaled) energy are shown in Figure 1.1. For electrons, the chemical potential μ is the datum of energy, and the results for a few values of the scaled (also called reduced) electron chemical potentials $\mu/k_{\rm B}T$ are also shown. All distributions become similar to the classical distributions for large energies (for electrons also in the case small μ). As will be discussed in Section 2.6.5 (metals) and 5.7 (semiconductors), the number of conduction electrons is related to the chemical potential (Fermi energy), and because semiconductors have smaller conduction electron density, their chemical potential is smaller than that of metals. As shown in Figure 1.1, whereas the occupation probability of conduction electrons in metals is high and subject to exclusion principle (degenerate), that of semiconductors is low and is generally treated with the nondegenerate, classical distribution.

For a general treatment, the energy of electron (and hole) are divided into potential (representing its bond energy) and kinetic (representing its velocity) energy.

1.1.3 Fluid Particle

Gases and liquids are composed of single atoms or molecules (here broadly termed fluid particles), which can be neutral or charged, in constant random motion. The fluid particle (see Glossary) energy is divided into potential, electronic, and kinetic energy. The fluid specific volume is fixed under constant temperature and pressure, and its shape may be determined by the container it fills or through cohesive forces such as surface tension. The moving particles constantly collide with each other (and, possibly, with the container wall). In ideal gas behavior, the collisions between the gas particles are elastic (energy is conserved among colliding particles) and the forces of attraction between the particles are negligible. For ideal gases, the M–B distribution f_f° (Table 1.2 and Figure 1.1; derivation is given in Appendix F) can be derived by use of statistical mechanics (and the concepts of energy partitioning and symmetry). Figure 1.1 also shows various contributions to the total energy of the fluid particle (translational, vibrational, rotational, electronic, intraatomic, and interatomic potential). For these ideal, noninteracting particles in the ground state, all energy is only in the form of kinetic energy, and (by use of the equipartition energy principle) each mode has a kinetic energy equal to $k_{\rm B}T/2$. This corresponds to the most probable energy distribution, in a collision-dominated system consisting of a large number of noninteracting particles with no net motion, but with a nonzero root-mean-square (RMS) thermal fluctuation velocity $\langle u_f^2 \rangle^{1/2}$. This velocity is related to the speed of sound in fluids a_s, and is temperature dependent. So, the probability of occurrence of very-high velocity fluid particles decreases as the temperature decreases.

A liquid is considered to be a substance that's particles have enough kinetic energy to stretch the intermolecular forces of attraction, but not completely overcome them (so their densities are close to solids). Collisions thus occur between the particles more often than in gases (ideal or otherwise). As the temperature of a liquid is raised, the velocity of the particles increases. The kinetic energy eventually becomes so great that the particles overcome all intermolecular forces and move freely, thus becoming a gas.

1.1.4 Photon

A photon (Greek for light) is a quantum of excitation of an electromagnetic field, and is also an elementary particle in quantum electrodynamics (QED), which is part of the Standard Model of particle physics. Photons are thus the building blocks of EM radiation, some of which we observe as light. According to quantum mechanics, all particles, including the photon (a quasi-particle), have some of the properties of a wave. Photons have zero invariant mass, but a definite, finite energy. Because they have energy, the theory of general relativity states that they are affected by gravity, something that has been confirmed by experiment.

Photons have a spin of 1, which makes them bosons (f_{ph}° in Table 1.2). Photons act as mediators to the EM spectrum; they are the particles that enable other particles to interact with each other electromagnetically and with an electromagnetic field. Individual photons are circularly polarized (as compared to electrons that have a spin up or spin down) because of their unit spin. They travel at the speed of light, u_{ph} (or c), and their lifetime is infinite, although they can be created and destroyed.

In general, an EM field consists of plane, monochromatic waves, of frequency f_{ph} (angular frequency $\omega_{ph} = 2\pi f_{ph}$), wavelength λ_{ph}, and speed u_{ph}, with $\omega_{ph} = 2\pi u_{ph}/\lambda_{ph}$. The quantum property of an EM waves is given by its energy $E_{ph} = \hbar\omega_{ph}$. The photon also has a quantum momentum $\boldsymbol{p}_{ph} = \hbar\boldsymbol{\kappa}_{ph}$, where $\boldsymbol{\kappa}_{ph}$ is the wave vector and the wave number $\kappa_{ph} = 2\pi/\lambda_{ph}$. Note that for a classical particle, the energy is $p^2/2m$, i.e., it is proportional to p^2.

In a vacuum, the dispersion relation of photons (the relation between angular frequency or energy and wave vector or momentum) is linear, and the constant of this proportionality is the Planck constant.

In matter, photons couple to the excitations of the material. These excitations can often be described as quasi-particles (such as phonons and excitons defined in Glossary) with quantized wavelike or particlelike entities propagating through the material. Photons can transform into these excitations (that is, a photon is absorbed and the medium is excited, creating a quasi-particle) and vice versa (the quasi-particle transforms back into a photon, i.e., the medium relaxes by reemitting the energy as a photon). These transformations are subject to probability rates and are presented as a polariton (Glossary). This is a quantum-mechanical superposition of the energy quantum being a photon as well as being one of the quasi-particle, matter excitations (such as photons). According to the rules of quantum mechanics, a measurement breaks this superposition; that is, the quantum is either absorbed into the medium and stays there, or it reemerges as a photon.

Matter excitations have a nonlinear dispersion relation, and their momentum is not proportional to their energy. Hence, these particles propagate more slowly than the speed of light in vacuum (Section 3.3.1). The propagation speed is the derivative of the dispersion relation with respect to momentum (angular frequency with respect to wave vector). A photon, by coupling with the excitation of matter and forming a polariton, acquires an effective mass, which means that it cannot travel at speed of light in vacuum.

1.2 Combinatorial Probabilities and Energy Distribution Functions

Starting with the four principal energy carriers first discussed, we note that our measurements and observations are at relatively large length scales, which include many such principal particles in each observation. Also, it is possible for each observed macrostate to correspond to many microstates, which is addressed by use of statistical mechanics [119]. For our purposes, we are interested in both the microstates (and what governs them) individually, and in their collective behavior.

In Chapter 3, we will discuss the Boltzmann transport equation that governs the dynamics of the probability distribution function $f_i (i = p, e, f, ph)$ under applied forces. In turn, the dynamics involved in the deviation of f_i from the equilibrium value $f_i^{\rm o}$, influences the transport properties. As will be shown, the function f_i is central to principal transport phenomena.

In a system made of a number of particles, the observed macrostate (ensemble averaged over all particles in the system) is related to the microstate of each particle (i.e., the position and momentum of each particle) by the probability of existence of each microstate, as described by the probability distribution function f_j, where j denotes the microstate. As shown in Figure 1.1, the higher energy states, in general, have lower probabilities of existence at a given temperature. As the temperature increases, the higher-energy states are more probable (because the energy is scaled with $k_{\rm B}T$).

An ensemble averaged quantity, denoted by the angle brackets $\langle\ \rangle$, is related to a corresponding microstate quantity, which we denote here as ϕ_j, through

$$\langle\phi\rangle = \sum_j \phi_j f_j. \qquad (1.1)$$

The summation can be changed to an integral when the probability can be given as a continuous function (as is given in Table 1.2).

The probability distribution (or occupancy) function is used in determining the carrier energy and its transport properties. For example, the conduction electron density, its current density vector, and its energy density are related to the moments of $f_e(\boldsymbol{x}, \boldsymbol{\kappa}, t)$, where $\boldsymbol{\kappa}$ is the wave vector, as

$$n_{e,c} = \langle n_{e,c}\rangle = \sum_{\kappa} f_e(\boldsymbol{x}, \boldsymbol{\kappa}, t) \qquad \text{carrier density} \qquad (1.2)$$

$$\boldsymbol{j}_e = \langle \boldsymbol{j}_e\rangle = -e_c \sum_{\kappa} \nabla_{\kappa} E_e(\boldsymbol{\kappa}) f_e(\boldsymbol{x}, \boldsymbol{\kappa}, t) \quad \text{carrier current density vector} \qquad (1.3)$$

$$\frac{E_e}{V} = \langle \frac{E_e}{V} \rangle = \sum_{\kappa} E_e(\boldsymbol{\kappa}) f_e(\boldsymbol{x}, \boldsymbol{\kappa}, t) \qquad \text{carrier energy density,} \qquad (1.4)$$

where e_c is charge of an electron.

For conduction electrons, these relations will be derived in Sections 5.7, 5.11, and 5.18. Each principal energy carrier has its own particular features. For example, the energy (or velocity) distribution of fluid particles in a dilute-gas phase may be described using classical mechanics with symmetry constraints. Electrons, by contrast, are treated with quantum mechanics, which includes an antisymmetry constraint (regarding electron spin).

The equilibrium probability distribution function $f_i^{\rm o}$ gives the most probable distribution of microstates under zero disturbing force or potential [including zero-temperature gradient, i.e., for constant (time invariant) uniform (space-invariant) temperature]. This would mean allowing sufficient elapsed time for the particles to thermalize, meaning they reach the equilibrium distributions given in Table 1.2. The thermalization time is actually rather small, generally of the order of 10^{-12} s, or 1 picosecond (ps), (see Table A.4 in Appendix A for unit prefixes) for most of the principal carriers.

These particle probability distributions functions enable us to describe the temperature dependence of the lattice (phonon) and electron specific heat capacities, the relation between the temperature and the kinetic energy of gases, and the blackbody thermal emission of photons. Many other principal heat carrier behaviors and properties can also be described and predicted by use of f_i.

Table 1.2 and Figure 1.1 summarize some of the characteristics of the four principal heat carriers and their equilibrium distribution functions. The derivations of f_p° and f_{ph}° are given in Appendix F and are also given in [9]. The derivation of f_e° is given in [183], and that of f_f° is given in [55]. For $E_i/k_{\mathrm{B}}T \gg 1$, all particles follow the classical (M–B) distribution. The deviation from the equilibrium distribution is used in the particle treatment of the transport of these carriers, i.e., Boltzmann transport theory. We start with a general treatment given in Chapter 3, and then specifics are given for each of the carriers in the subsequent chapters. A derivation of the Bose–Einstein (boson) distribution function will also be given in Section 2.6.4 (for a harmonic, quantum oscillator). The classical distribution function (M–B) will be discussed in Section 6.4.2.

1.3 Particles, Waves, Wave Packets and Quasi-Particles

Quasi-particles (including phonons, electrons, and photons) have properties of both particles and waves, and can be described by use of wave packets, i.e., localizations of energy that are due to the superposition of many plane waves of different wavelengths. These are, in a sense, both a particle and a wave at the same time, a concept termed particle–wave duality. Below, we discuss particle, wave, and particle–wave behaviors.

Particles, which are sometimes called corpuscles, are discrete; their energy is concentrated into what appears to be a finite space with definite boundaries, and its contents are considered to be homogenous (the same at any point within the particle). Particles exist at a specific location and can never exist in more than one place at once. To travel to a different point in space, a particle must move according to the laws of kinematics. Interactions between particles follow simple laws, such as the laws of conservation of energy and momentum in the case of elastic collisions. Such laws are fundamental to Newtonian mechanics and provide relatively simple, yet powerful, tools for predicting particle behavior. When there is no hindering interaction, the particle is said to be ballistic.

Waves, unlike particles, cannot be considered finite entities. Their energy is distributed in space and in time. Unlike a particle, a wave can propagate until it exists in all locations and at all times. Like particles, a part (or phase) of the wave can be analyzed to determine its velocity in space. Waves are specified by frequency and wavelength. An EM wave is a propagating oscillation of a perpendicular electric field and a magnetic field in space.

For quasi-particles, the classical distinctions between particles and waves can become blurred. They behave partly according to wave theory and partly according to particle theory. For example, in blackbody thermal radiation, by analysis of

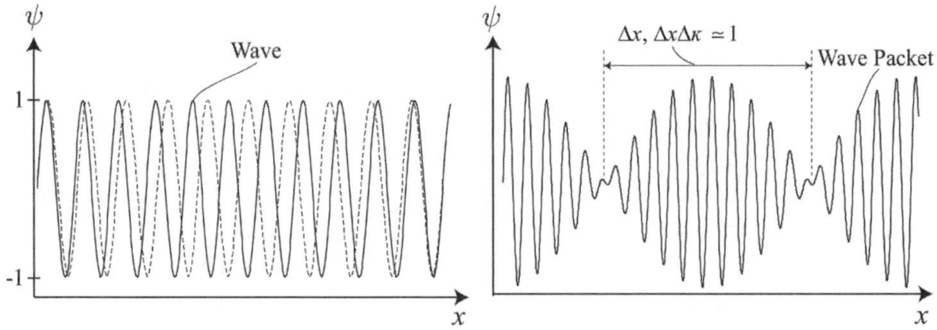

(a) $\psi = \sin \kappa x = \sin x$, and $\psi = \sin 1.1x$ (b) $\psi = \sin x + \sin 1.1x$

Figure 1.2. (a) Two simple (pure sine) waves, (b) beat phenomenon, demonstrating a wave packet created by superposition of the two simple waves.

the spectral intensity of the EM radiation for various temperatures, it is noted that the total radiation intensity (integrated over the entire spectrum) is a function of temperature to the fourth power. Also, a wavelength corresponding to a maximum radiant intensity is observed for a given temperature. As the temperature of the body increases, the wavelength of maximum intensity decreases. The wave treatment of a blackbody cavity shows that the cavity is full of EM standing waves with nodes at the walls, and the number of these standing waves can be determined as function of the size of the cavity. Each wave contributes $k_B T$ joules of energy to the system, giving the classical Rayleigh–Jeans formula. Compared with the measured blackbody spectral intensity (which is quasi-particle treatment, Section 7.1), this relation, based on wave treatment, overestimates significantly the intensity at short wavelengths (an error called the UV catastrophe).

The concept of wave packets addresses the overestimation problem by allowing us to almost determine the location of a wave. This in turn allows us to measure energy, location, and momentum at a specific time. A wave packet is a quasi-particle because it exhibits some particle-like behavior. The principle of superposition describes the addition of waves, in which the overlapping of two pure sine waves of similar amplitude and frequency produces areas of constructive and destructive interference. This produces a beat phenomenon, as shown in Figure 1.2. The first two waves ($\sin \kappa x$), shown in Figure 1.2(a), have slightly different frequencies ($1/\kappa$), leading to the result shown in Figure 1.2(b). The energy of the waves is located in an area that is approximated with a distance Δx. By adding together many waves of slightly different amplitudes and frequencies (with a range of wave numbers $\Delta \kappa$) it is possible to create a beat of any shape. Increasing the number of waves used (and thus $\Delta \kappa$), decreases Δx. This means the range of wave numbers and the size of the beats are inversely proportional, i.e., $\Delta x \Delta \kappa \simeq 1$. The smaller the wave packet (the more localized), the larger the spread of wavelengths needed to construct the packet.

Using the quasi-particle concept of a wave packet, the Planck distribution (Chapter 7) can be developed that describes the thermal equilibrium distribution of these packets of energy (photons).

1.4 A History of Contributions Toward Heat Transfer Physics

Heat is a form of energy that manifests itself as the motion of the molecules of a substance and is capable of being transmitted from one body to another by conduction (by means of phonons, electrons, and fluid particles), convection (by means of fluid particles), and radiation (by means of photons). It can also be transformed into other forms of energy, e.g., EM energy.

The areas of physics most relevant to heat transfer physics are (a) atomic/molecular dynamics, (b) solid-state (condensed-matter) and fluid-state physics, (c) electromagnetism, and (d) quantum optics.

Table 1.3 gives a chronological overview of the contributions to the physics of energy and its transport, as related to heat transfer. These include thermoelectricity, the theory of thermal motion (fluctuation), the theory of harmonic oscillation and photon emission, and the theory of phonons and phonon thermal conductivity. We will encounter and discuss these principles and contributions in the following chapters. A review of recent advances in heat transfer physics is given in [56].

Kelvin helped in the creation of a new kind of mathematical and experimental physics based on the concept of energy. In describing rules for the conservation of energy, he introduced the term energy dissipation/degradation (in terms of the kinetic and potential energy). These rules later became the first and second laws of thermodynamics.

Boltzmann developed the idea that heat content, entropy (a measure of the disorder of a system), and other thermodynamic properties were the result of the behavior of large numbers of atoms, and could be treated by use of mechanics and statistics. He introduced the equation for the relation of entropy to probability, not to mention the Boltzmann constant k_B. He made important contributions to the kinetic theory of gases and developed the law of equipartition of energy with Maxwell (the M–B law), stating that the total energy of an atom or molecule is, on average, equally distributed over the degrees of freedom.

Maxwell created the EM theory of light. He formulated a group of equations summarizing the relation of electric and magnetic fields to the charges and currents producing them. He also contributed to the kinetic theory of gases (the M–B law just mentioned), molecular physics, and thermodynamics.

Planck discovered that energy exists in discrete units, which came to be called quanta, a Latin word translated as how much?. He assumed that nature was being selective in the amount of energy it would allow a body to accept or emit, allowing only amounts (or quanta) that were multiples of $h_P f$, i.e., quanta, where f is the photon frequency.

Bohr laid the foundations for one of the most important scientific achievements of the time: the model of the atom. While considering the simplest atom, hydrogen, and studying its atomic line spectrum, he postulated that its electron radiated energy only when it dropped from one allowed level to a lower energy level. The atom can absorb and emit energy only in quanta, which correspond to the energy differences

Table 1.3. *A chronological list of historical contributions to the physics (atomic-level description) of energy storage, transport, and transformation.*

1773	Coulomb's and Navier's (1820) theory of elasticity (mechanics of solids) [343]
1811	Avagadro's number and Fourier's law of heat conduction [25]
1821	Seebeck's discovery of electromotive force of heated junctions of two conductors [123]
1827	Navier's and Stoke's (1845) fluid momentum conservation [25]
1828	Brown's observation of pollen grains' motion in water [25]
1834	Peltier's thermoelectric cooling/heating at junction of two current-carrying conductors [123]
1845	Waterston's suggestion of what later became the kinetic theory of gases [25]
1859	Kirchhoff's law of equality of spectral, directional absorptivity and emissivity [306]
1865	Loschmidt's estimation of the diameter of gases [212], also Clausius' introduction of entropy [212]
1866	Maxwell's dynamical theory of gases [212], also Boltzmann's first attempt to relate entropy with average property of gas in thermal motion [212]
1872	Boltzmann's minimum (or *H*-) theorem and M–B distribution function [212]
1873	Gibb's first publication of thermodynamic properties and their relations [212], van der Waals' equation of state for liquids (and dense gases) including the attractive intermolecular forces [156]
1877	Boltzmann's introduction of his proportionality constant [212]
1879	Stefan's empirical relation for total emissive power of blackbody [25, 261]
1897	Thomson's discovery of electrons [9]
1900	Drude's theory of electron and phonon thermal conductivity [9]
1901	Planck's spectral distribution of blackbody emissive power [306]
1905	Einstein's molecular theory of heat capacity of solids [25, 261]
1908	Langevin's stochastic particle dynamics equation [73]
1909	Knudsen's experiment on viscous, transitional, and molecular-flow regimes [25]
1911	Eucken's T^{-1} relation for thermal conductivity of dielectrics [9]
1912	Debye's theory of crystal specific heat capacity based on lattice vibration [85]
1914	Millikan's verification of photoelectric effect (quantization of energy and Planck constant)
1916	Chapman–Enskog's expansion of distribution using the Knudsen number in derivation of Navier–Stokes equations from Boltzmann transport equation [339]
1918	Langmuir's theory of adsorbed monolayers and isotherms [205]
1924	de Broglie wave–matter theory [25, 261], Güneisen's lattice dynamics-based solid equation of states [132]
1925	Heisenberg's and Schrödinger's wave function-equation for matter wave [327], Pauli's exclusion principle [25, 261], Prandtl's phenomenological turbulent mixing-length theory [148]
1926	Schrödinger's equation using de Broglie's hypothesis that each particle has a wavelength [25], Fermi's quantum state exclusion principle for ideal gas [25, 261], Dirac's similar treatment [25, 261]
1928	Bloch's theory of electron energy surfaces in crystals [327]
1929	Peierl's inclusion of lattice anharmonicity and introduction of phonons as wave packets and U–processes [265], Pringsheim's radiation cooling by anti-Stokes fluorescence [269]
1931	Onsager's reciprocal relations for coupled irreversible thermoelectric processes [257]
1932	Sommerfeld's model of electron gas in solids [25, 261], Dirac's positron experimentally confirmed [25, 261]
1934	Fermi's (contributions from Dirac) golden rule for electronic transition probability rate [261]
1935	Taylor's introduction of velocity correlation and statistical theory of turbulence [19]
1937	Wheeler's quantum-mechanical description of particle scattering [25, 261]
1938	Kapitsa's (also by Allen and Misener) discovery of superfluid liquid helium [25, 261]

1939	Landau's theory on superthermal conductor liquid helium II [25, 261]
1949	Feynman's quantum electrodynamics diagram for charged-particle scattering [25, 261]
1954	Green's and Kubo's (1957) transport coefficient determined from current autocorrelation integral [142], Bhatagar–Gross–Krook's model for restoration of equilibrium distribution due to particle collision [339]
1959	Ziman's variational treatment of nonequilibrium phonon in transport properties [360], Callaway's inclusion of N– and U–processes in the relaxation-time model of lattice thermal conductivity [123]
1967	Tien's radiation tunneling analysis [79] and contribution to photon (infrared) gaseous absorption [325], Slack's high-temperature phonon conductivity relation T^{-1} [313]
1997	Chu/Cohen–Tannoudji/Phillips's Bose-Eienstein condensate experiment using laser cooling [269]

between allowed levels. The electrons in the atom begin to occupy the orbit nearest the nucleus, but not all the electrons can fill the lowest orbit.

Pauli articulated a rule for atomic structure (commonly known as the Pauli exclusion principle) in which the state of each electron in an atom is defined by a unique set of four quantum numbers, and no two electrons in a single atom may have the same set of quantum numbers or the same energy state.

Schrödinger introduced the equation that describes the form of the probability waves (or wave functions) that govern the motion of small particles and that specify how these waves are altered by external influence. These functions, and the idea that particles have wavelike properties, are the foundation of quantum wave mechanics.

Fermi devised a method for calculating the behavior of particle systems obeying the Pauli exclusion principle (i.e., having nonequal quantum numbers), later termed Fermi statistics. Dirac independently developed an equivalent theory. Fermi also developed a method for calculating the quantum electron transition probability rates.

Green and Kubo developed the fluctuation–dissipation theory of transport coefficients, and Callaway and Holland formulated (and solved) the single-mode relaxation-time model of lattice thermal conductivity.

Chu/Cohen–Tannoudj/Phillips laser cooled Na gas atoms to very close to 0 K to form the quantum, Bose–Einstein condensate.

In the history of mathematical physics the most notable milestone between Newton and Einstein is the work of Hamilton, who brought the theory of classical mechanics to a level of great sophistication.

1.5 Fundamental Constants and Fine-Structure Scales

1.5.1 Boltzmann and Planck Constants

The Boltzmann constant is $k_B = 1.3806503(24) \times 10^{-23}$ J/K and is defined in the relation between the average thermal energy (related to the potential and kinetic energy) of a principal energy carrier (phonon, electron, photon, or fluid particle) to its absolute temperature T (The digits in parentheses correspond to uncertainty in the last two digits.[†]). This thermal energy $k_B T$ is used to scale (normalize) the energy of principal energy carriers when they are treated as particles. In the classical systems, thermal energy is assumed to vary continuously with temperature (in contrast to quantum systems).

The Boltzmann constant is a classical quantity. In statistical mechanics, the entropy S (a macroscopic property) of a system of N particles is defined as

$$S = k_B N \ln Z. \tag{1.5}$$

Here, Z is called the partition function, which is the probability function describing the distribution of energy states available to a system given its macroscopic constraints (such as a fixed total energy E). This will be further discussed in Section 2.4.1 and in Chapter 6. In principle, the Boltzmann constant is a derived physical constant, as its value is determined by other physical constants. Its independent derivation has yet to be completed due to its complexity.

In kinetic theory, from the equipartition of energy, for each degree of motion an energy equal $k_B T/2$ is assigned (Section 2.3.1).

The Planck constant h or $h_P = 6.62606896(33) \times 10^{-34}$ J-s arises in quantum mechanics, such that the energy within a body obeying quantum mechanics is the product of its frequency f and h_P. The reduced Planck constant (also called the Dirac constant) is $\hbar = h_P/2\pi$.

The Planck constant is used to describe quantization, a phenomenon for principal energy carriers, in which certain physical properties occur in fixed amounts rather than a continuous range of possible values.

The Planck constant also occurs in statements of the Heisenberg uncertainty principle, in which the uncertainty (the standard deviation) in any position measurement Δx and the uncertainty in a momentum measurement along the same direction Δp_x, obeys (and also energy/time) the relations (Chapter 2 problem)

$$\Delta p_x \Delta x \geq \frac{\hbar}{2}, \quad \Delta E \Delta t \geq \frac{\hbar}{2}. \tag{1.6}$$

Table 1.4 lists the fundamental physical constants used in the text. This table is also repeated as Table A.3 in Appendix A, for easy reference. The electron charge e_c, the Newton gravitational constant G_N, electron and nuclear particle masses, and EM constants are also listed. Some derived constants, such as the Bohr radius r_B, the universal gas constant R_g, and the Stefan–Boltzmann constant σ_{SB} are also listed.

1.5.2 Atomic Units and Fine-Structure Scales

Four fundamental constants, the reduced Planck constant $\hbar$, the electron mass m_e, the Coulomb (electrostatic) constant $1/4\pi\epsilon_o$ (where ϵ_o is the free-space electric permittivity), and the electron charge e_c, are used to define atomic units. Based

† In this text, we use four significant figures, for consistency. Whenever appropriate, more significant figures are shown. Also, when accuracy is limited, only the available significant digits are given.

Table 1.4. *Fundamental constants, derived quantities, and units conversions [209, 210].*

Symbol	Magnitude and units
Fundamental constants:	
c_o	speed of light in vacuum, 2.99792458×10^8 m/s
e_c	electron charge, 1.602×10^{-19} C (1 eV = 1.602×10^{-19} J where V = J/C)
G_N	Newton (gravitational) constant, 6.673×10^{-11} m^3/kg-s^2
h_P	Planck constant, 6.626×10^{-34} J-s = 4.136×10^{-15} eV-s
$\hbar$	$h_P/2\pi = 1.055 \times 10^{-34}$ J-s = 6.583×10^{-16} eV-s
k_B	Boltzmann constant, 1.381×10^{-23} J/K = 8.618×10^{-5} eV/K
m_e	electron mass, 9.109×10^{-31} kg
m_n	neutron mass, 1.675×10^{-27} kg
m_p	proton mass, 1.673×10^{-27} kg
N_A	Avogadro number, 6.022×10^{23} molecule/mole = 6.022×10^{26} molecule/kmole
μ_o	free-space magnetic permeability, $4\pi \times 10^{-7} = 1.257\times10^{-6}$ N-s^2/C^2
Derived quantities:	
$\epsilon_o = \dfrac{1}{c_o^2\mu_o} = 8.854 \times 10^{-12}$ C^2/N-m^2	free-space electric permittivity
$D = e_c r_B/2.54 = 3.3356 \times 10^{-30}$ C-m	Debye (atomic dielectric dipole moment)
$N_L = \dfrac{\pi^2}{3}\dfrac{k_B T^2}{e_c^2} = 2.442 \times 10^{-8}$ W-Ω/K^2	Lorenz number
$r_C = \dfrac{e_c^2}{4\pi\epsilon_o m_e c_o^2} = 2.8179 \times 10^{-15}$ m	classical (Compton) electron radius
$r_B = \dfrac{4\pi\epsilon_o\hbar^2}{m_e e_c^2} = 5.292 \times 10^{-11}$ m	Bohr radius (atomic length unit)
$\dfrac{e_c^2}{4\pi\epsilon_o r_B} = 27.2114\,\text{eV} = 4.35975 \times 10^{-18}$ J	hartree (atomic energy unit)
$\dfrac{e_c^2}{4\pi\epsilon_o r_B^2} = 8.2378 \times 10^{-8}$ N	atomic force unit
$R_g \equiv k_B N_A = 8.3145 \times 10^3$ J/kmole-K	universal gas constant
$\text{Ry} = \dfrac{m_e e_c^4}{8\epsilon_o^2 c_o \hbar^3} = 1.0974 \times 10^5$ cm^{-1}	Rydberg ground-state energy of H
$\dfrac{1}{4\pi\epsilon_o} = 8.9876 \times 10^9$, Nm^2/C^2	Coulomb constant
$\alpha = \dfrac{e_c^2}{4\pi\epsilon_o\hbar c_o} = 7.29735 \times 10^{-3}$	fine-structure constant
$\sigma_{SB} \equiv \dfrac{\pi^2 k_B^4}{60\hbar^3 c_o^2} = 5.670 \times 10^{-8}$ W/m^2-K^4	Stefan–Boltzmann constant
Units conversions:	
$1\text{eV} = 8.0655 \times 10^4$ cm^{-1} $= 2.418 \times 10^{14}$ Hz $= 11,600$ K $= 1.602 \times 10^{-19}$ J	
1 cm^{-1} $= 0.12398$ meV $= 2.998 \times 10^{10}$ Hz	

on these, the atomic length unit (the Bohr radius for electron in hydrogen atom) is $4\pi\epsilon_o\hbar^2/m_e e_c^2 = 5.2917725\times10^{-11}$ m = 0.52917725 Å, the atomic energy unit (1 hartree = $e_c^2/4\pi\epsilon_o r_B$) is $e_c^4 m_e/(4\pi\epsilon_o\hbar)^2 = 4.3597482\times10^{-18}$ J = 27.211396 eV, the atomic time unit $\tau_a = m_e r_B^2/\hbar$ is 2.4188843×10^{-17} s, the atomic (electron) velocity $u_e = r_B/\tau_a = \hbar/m_e r_B$ is 2.1876914×10^6 m/s, and the atomic electric dipole moment $e_c r_B$ is 8.4783579×10^{-30} C-m = 2.54 D, where D = 3.3356×10^{-30} C-m is one Debye. The atomic (*ab initio*) scales are summarized in Table 1.5.

Table 1.5. *Atomic scales (*ab initio *or fine-structure), based on hydrogen atom.*

Scale	Relation
Length	$r_B = \dfrac{4\pi\epsilon_o\hbar^2}{m_e e_c^2} = 5.2918 \times 10^{-11}$ m
Time	$\tau_a = \dfrac{m_e r_B^2}{\hbar} = 2.4189 \times 10^{-17}$ s
Energy	$\dfrac{e_c^2}{4\pi\epsilon_o r_B} = 27.211$ eV $= 4.3597 \times 10^{-18}$ J
Velocity	$\dfrac{r_B}{\tau_a} = 2.1877 \times 10^6$ m/s
Dipole moment	$e_c r_B = 8.4783 \times 10^{-30}$ C-m
Fine-structure constant	$\alpha = \dfrac{e_c^2}{4\pi\epsilon_o\hbar c_o} = 7.29735 \times 10^{-3}$

More commonly, the electron-volt (eV) is used as the unit of atomic-level energy and is defined as the amount of kinetic energy gained by a single unbound electron (of charge e_c) when it passes through an electrostatic potential difference $\Delta\varphi_e$ of one volt (V) in vacuum.

The fine-structure constant

$$\alpha = \frac{e_c^2}{4\pi\epsilon_o\hbar c_o} = 7.29735 \times 10^{-3} \tag{1.7}$$

is the ratio of electron velocity in the Bohr model of an atom to the speed of light. It is also an indicative of scaled EM interaction (energy needed to bring two electrons from infinity to a close distance, divided by the energy of a photon having a wavelength equal to the same separation distance).
We will discuss this electrostatic force in Section 2.1.

The hyperfine structure is a small perturbation in the energy levels of atoms and molecules that is due to the magnetic dipole–dipole interactions (nuclear–electron magnetic moment interactions). The hyperfine energies are not in the optical regime, but in the radio waves and microwaves regimes.

1.6 Principal Carriers: Concentration, Energy, Kinetics and Speed

The energy storage, transport, and transformation involving the four principal carriers are related to their concentration n_i, energy E_i, kinetics (represented by time constants, relaxation times, and rates) τ_i and speed u_i.

In this section these properties of the principal carriers are briefly examined, and examples of their magnitudes are given. Full treatments are given in Chapters 4 to 7.

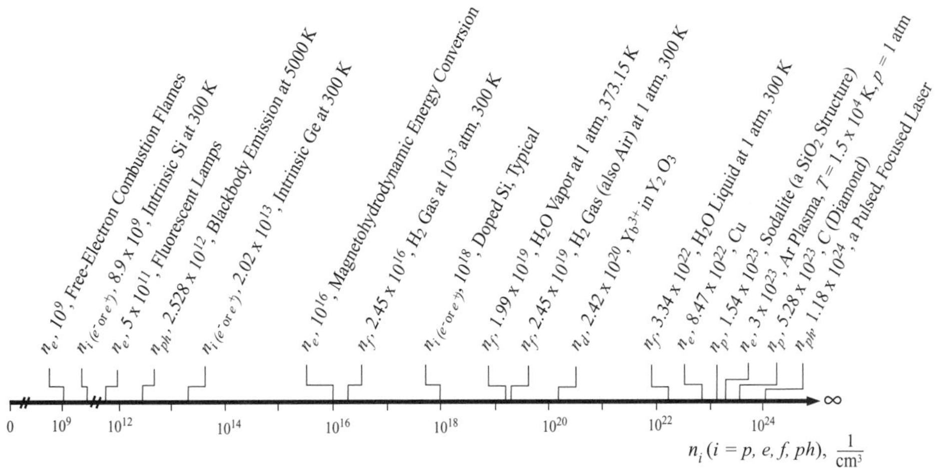

Figure 1.3. Typical number densities for phonons, electrons (conduction), fluid particles, and photons. For Yb^{+3} in Y_2O_3, n_d is the optimum ion dopant concentration for the laser cooling of an Y_2O_3 crystal.

1.6.1 Principal Energy Carriers Concentration

Figure 1.3 lists several typical carrier densities, in units of carriers per cubic centimeter. In the Debye model of the lattice specific heat of monotonic crystals, the number of phonon modes is three times (for dimension of motion) the number density of the atoms. For Al [face-centered-cubic (FCC) structure] this phonon density is

$$n_p = 3\frac{\rho N_A}{M} \quad \text{monatomic FCC crystals,} \tag{1.8}$$

(ρ is density and M is molecular weight), which gives 5.28×10^{23} 1/cm^3 and is shown in Figure 1.3. For the polyatomic crystalline sodalite, a silica (SiO_2) structure, the number of phonon modes is 1.54×10^{23} 1/cm^3.

We concentrate on free (conduction) electrons, because of their role in transport, with metals having one or more conduction electrons per atom, such as Cu having one (found from the outmost electron orbital). The number density of free (conduction) electrons in metals is given by

$$n_{e,c} = \frac{\rho N_A}{M} z_e \quad \text{metals,} \tag{1.9}$$

where M is the molecular weight, ρ is the density, and z_e is the number of free electrons per atom. (Table 5.2 gives listing of n_e for other metals.) There are no conduction electrons in semiconductors at $T = 0$ K, and the number of conduction electrons per unit volume increases as the temperature increases. Semiconductors can also have electron-donating dopants, which add to their overall electron density. Typical conduction electron densities for metals are of the same order as that for Cu, 8.47×10^{22} 1/cm^3. For semiconductors, the value for intrinsic Ge at 300 K of 2.5×10^{13} 1/cm^3 is typical. These are listed in Figure 1.3. For semiconductors, the density value reported includes both electrons and holes. Intrinsic (not doped) Si, for example, at $T = 300$ K has equal concentrations of electrons and holes of 8.72×10^9 1/cm^3 each. An ion dopant (Yb^{3+}) in the Y_2O_3 crystal, with applications in laser cooling of solids, is also shown ($n_d = 2.42 \times 10^{20}$ 1/cm^3). The number of free electrons in a typical combustion flame and in an Ar plasma ($T = 1.5 \times 10^4$ K) is also shown in Figure 1.3. Thermal plasmas reach the free-electron density of metals.

The atomic and molecular gases and liquids, which we call fluids for simplicity, have carrier densities n_f that depend on their pressures and temperatures. The molecular number density of H_2 (gas) at 10^{-3} atm and 300 K is found from

$$n_f = \frac{\rho_f N_{\mathrm{A}}}{M} = \frac{p}{k_{\mathrm{B}} T} \quad \text{ideal gas,} \tag{1.10}$$

and is 2.45×10^{16} 1/cm^3. As an example of liquids, H_2O liquid at 1 atm and 300 K has $n_f = 3.34 \times 10^{22}$ 1/cm^3.

In the blackbody emission of photons, the number of photons per unit volume is given by (Chapter 7 problem)

$$n_{ph} = 2.404 \times 8\pi \left(\frac{k_{\mathrm{B}}}{h_{\mathrm{P}} u_{ph,\mathrm{o}}}\right)^3 T^3 \quad \text{blackbody emission.} \tag{1.11}$$

From the constants in Table 1.4, this gives 2.023×10^{10} photons/cm^3 for $T = 1000$ K. For $T = 5000$ K this is 2.528×10^{12} 1/cm^3 and is marked in Figure 1.3. In lasers and other amplified, pulsed, and focused radiation, the number of photons per unit volume can be extremely large. For example, a 10^{-3} J laser pulse at a wavelength of 5×10^{-7} m (green light) released over a period of 10^{-15} s (femtoseconds, fs) with a beam cross-sectional area of 10^{-10} m^2, and using $\dot{s}_{ph} = \hbar\omega_{ph} n_{ph} u_{ph} A$, gives $n_{ph} = 1.18 \times 10^{22}$ 1/cm^3. (Table 1.4 is repeated in Appendix A, along with a table of unit prefixes.)

1.6.2 Principal Carrier Energy

As was listed in Table 1.2, wave or particle characteristics are used in presenting the carrier energy. The energy of a phonon is generally given in terms of its wave vector κ_p, and by use of the phase velocity, it can also be given in terms of its angular frequency ω_p and direction $\boldsymbol{s}_\alpha$. The energy of an electron is also given in terms of $\boldsymbol{\kappa}_e$. For fluid particles (and classical particles in general), the particle momentum $\boldsymbol{p}_f$ in its kinetic energy is used. For photons, generally the frequency ω_{ph} (or wavelength λ) is used. Regardless of their specific presentations, these carriers contain continuous and discrete energies and exchange these energies through various interactions.

Figure 1.4 shows examples of typical energies for phonons, fluid particles, electrons, and photons. Carrier energies larger than 10 eV (of the order of 1 hartree = 27.21 eV) are considered high energy here, as most of the phonon, electron, fluid particle, and photon energies in heat transfer phenomena are less than 1 hartree. As shown, the very high energies (10^8 eV) are associated with nuclear forces (strong and weak nuclear electronic energy) and associated with γ-rays (made up of photons). The corresponding thermal (or mechanical) energy represented by the

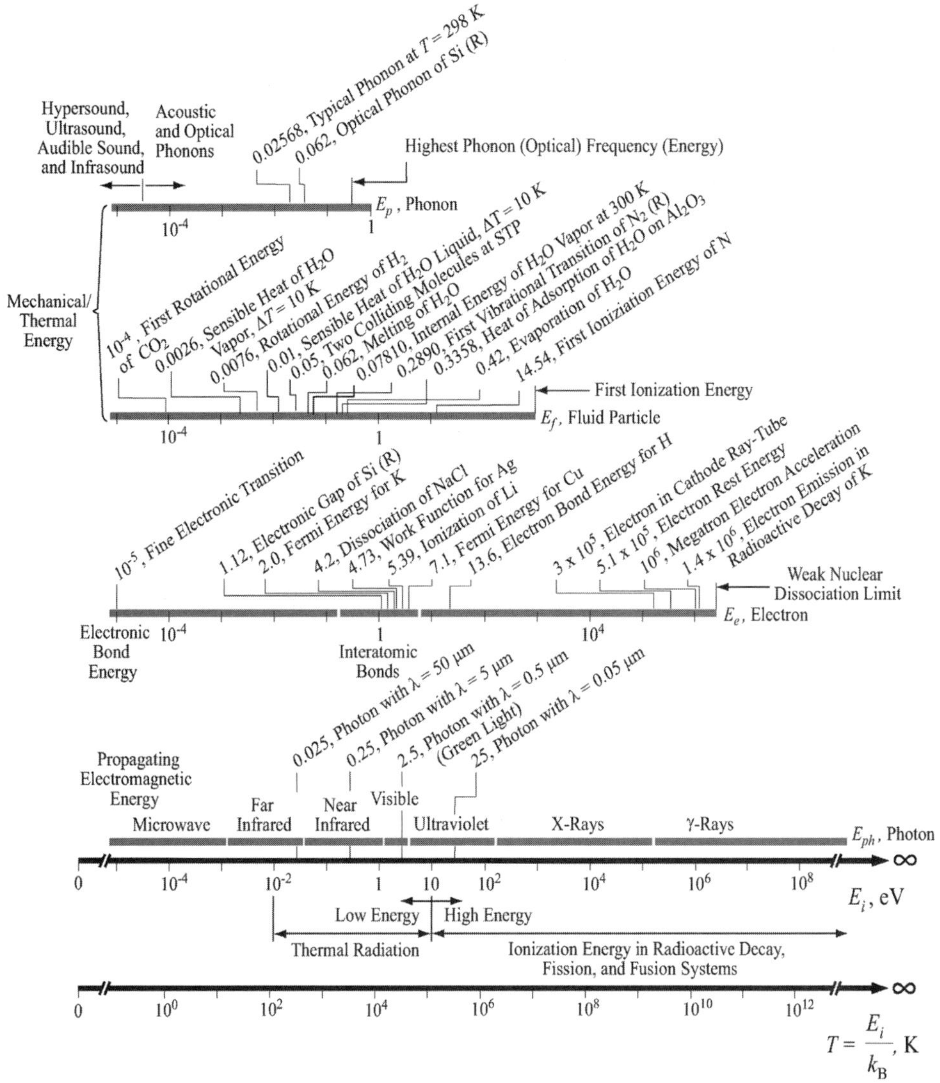

Figure 1.4. Examples of typical energies for the four principal energy carriers. The highest phonon energies are in the optical phonons, the highest fluid particle energies are the electronic energy, the highest electron energies are in electron emissions and in accelerated electrons, and the highest photon energies are in γ-rays and cosmic radiations. Symbol R stands for a resonance phenomenon.

product of the Boltzmann constant and absolute temperature is exceptionally high. For fluid particles, high-temperature thermal plasmas may have ionized atoms with energies of the order of 10 eV. Phonon energies are limited by the temperature of the crystalline solid they travel through, which is limited to around 4000 K. The lattice vibrations with the highest energies are associated with optical phonons, which typically contain less than 1 eV. The highest electron energies are associated with accelerated electrons and some electron emissions. The highest photon energies are in the γ-ray and cosmic radiation range.

In Figure 1.4, the symbol R stands for a resonance phenomenon, and indicates an electronic transition energy, or bond energy, or mechanical vibration energy. Resonance phenomena require atomic-level energy-matching metrics and allow for atomic-level design of energy conversion devices.

1.6.3 Principal Carrier Energy Transport/Transformation Kinetics

The smallest time unit typically used is the Planck time $\tau_P = (h_P G_N/c_o^5)^{1/2}$, where G_N is the Newton gravitational constant (Table 1.4), yielding an atomic time unit equal to 2.42×10^{-17} s (Table 1.5). Figure 1.5 shows typical times for various microscale energy interactions (transitions) $\tau_{i\text{-}j}$. The inverse of $\tau_{i\text{-}j}$ is the interaction (transition) rate $\dot{\gamma}_{i\text{-}j}$, which signifies the rate of hindering or enabling atomic-level energy transfers, i.e., microscale energy kinetics. The examples in Figure 1.5 are for $T = 300$ K. As shown in Figure 1.5, the fastest processes are associated with photons and their interactions with electrons (or electric fields). An electron orbiting a hydrogen atom, for example, has an orbital period of about 2.42×10^{-17} s. The time it takes for particles to thermalize to their equilibrium distribution is around 10^{-12} s, which also corresponds to phonon–phonon interaction relaxation times. Collision time intervals (relaxation times) for gas molecules are around 10^{-10} s. There are also much slower times, such as the lifetime of excited electronic states of Yb^{+3} (a typical rare-earth ion in dielectric crystals), which is about 10^{-6} s.

As also emphasized in Figure 1.5, in transport, the process (scattering) rate that is the fastest (with the smallest $\tau_{i\text{-}j}$) dominates, because the resistances are additive (for independent scattering mechanisms). There are also cases in which the conductances are additive, in which the slowest process dominates. For example, in solids both phonons and free electrons contribute to thermal conductivity and the transport with least scattering (as well as other desired attributes such as number density) dominates. Also, for energy conversions (transitions) involving multiple processes, the slowest process dominates.

As an example, in the thermoelectric material Bi_2Te_3 (a semiconductor of the p-type, described in Section 5.17), electrons are scattered most rapidly by the polar optical phonons ($\tau_{e\text{-}p} = 5\times10^{-14}$ s), and least rapidly by Coulombic scattering from the nucleus ($\tau_{e\text{-}C} = 8\times10^{-13}$ s).

1.6.4 Principal Carrier Speed

Three of the four carriers have speeds associated with the motion of atoms and free (conduction) electrons. For electrons, quantum effects persist to high temperatures, whereas for fluid particles and phonons, room temperature classical descriptions generally suffice. The speed of light in vacuum c_o is a physical constant, although it decreases during interactions with matter. Figure 1.6 gives examples of thermal speeds, starting with the Brownian motion of copper particles in water (with a dependence on the mass of the particle, Section 6.10), which is rather low. The lighter particles in thermal motion have higher thermal speeds. Electrons, traveling at the

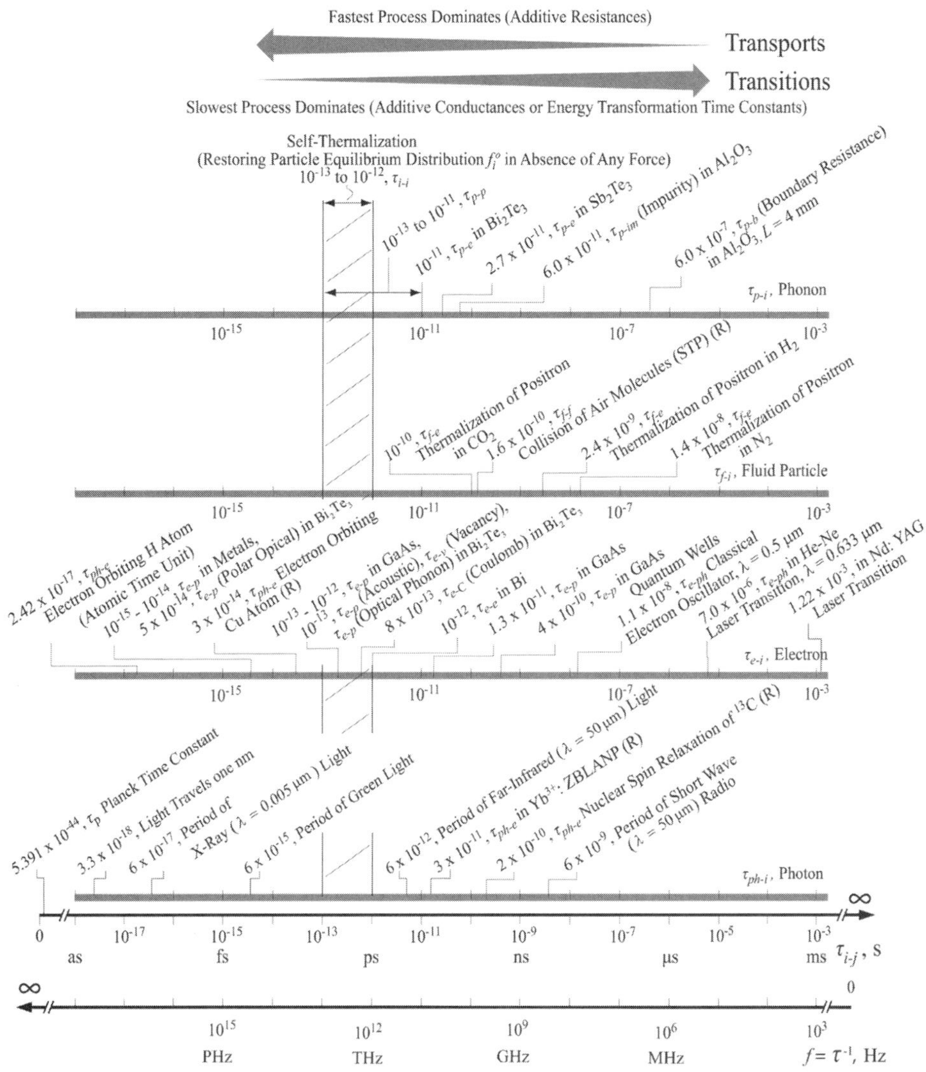

Figure 1.5. Kinetics of atomic-level energy transport and transition interaction. Examples of typical energy interaction (transition) times for various principal energy carrier pairs. The frequency (here defined as $1/\tau$) is also shown. Unless specified, $T = 300$ K. Symbol R indicates resonance.

Fermi speed, have the highest speed of such particles. The highest speed achievable is the speed of light in a vacuum $u_{ph,o} = c_o$, which is reached by photons (quasi-particles).

1.7 Periodic Table of Elements

The periodic table of elements, with their bulk and atomic properties, is given in Appendix A, Tables A.1 and A.2. The principal carrier energies and interactions determine their energy-storage and energy-transport properties, and some of these properties for elements are also listed in Tables A.1 and A.2. The properties listed

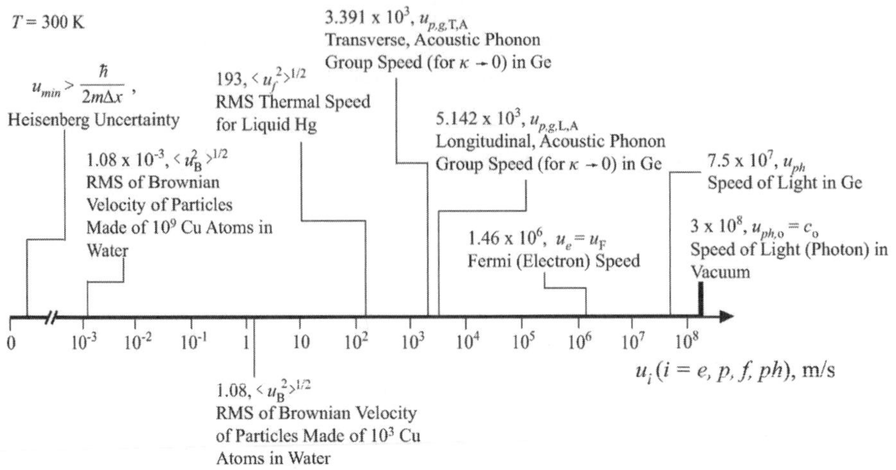

Figure 1.6. Typical principal carrier speeds (when applicable, $T = 300$ K). For the electrons, the Fermi speed is used, whereas for the phonon and fluid particles, the thermal speed is used. For the photons, the speed of light is used.

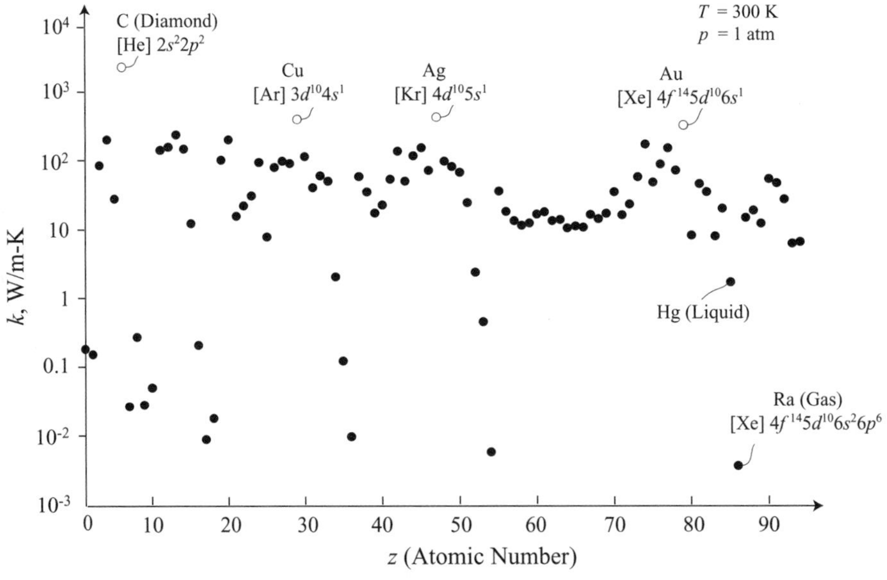

Figure 1.7. Variation of thermal conductivity k of elements (Appendix A, and Table A.1) with respected to atomic number z. The experimental results are for $T = 300$ K (and $p = 1$ atm, when relevant). Most elements are in the crystalline phase, but a few are in the gas phase and Hg is in the liquid phase. High k solids are also marked along with their electronic orbitals.

are for the state at $T = 300$ K and $p = 1$ atm, with the most common phase or a stated phase (the diamond phase of carbon, for example). An example is given in Figure 1.7, where the thermal conductivity of elements (Table A.1) is plotted versus the atomic number. Among the high thermal conductivity elemental solids, the thermal conductivity of diamond (a crystalline solid phase of C) is almost entirely due to phonon transport k_p, while those of metals such as Cu, Ag, and Au are dominated by electron transport k_e. There are marked in Figure 1.7. Those elements in the gaseous phase have the lowest thermal conductivities k_f.

The transport of thermal energy in various phases and by various principal heat carriers is discussed in Chapters 3 to 7. The bulk properties of elements (Table A.1) include the molecular weight, specific heat capacity, density, thermal conductivity, melting temperature, phase state, electrical conductivity, heat of fusion, heat of evaporation, and melting temperature. The atomic properties (Table A.2) include the electronic orbital structure, first ionization energy, crystal structure, electronegativity, cohesive energy, atomic radius, oxidation states, isotopes, lattice constant, and mean Debye temperature. Electronegativity is the potential of an atom to attract electrons to itself (fluorine is the most electronegative).

The group number (1 through 18) is the number assigned to the vertical columns of the periodic table, and elements within the same group all have the same number of electrons in their outer electron shells (Table A.2), as well as similar chemical properties. The group 18 elements have a full outer electronic shell and no tendency to lose, gain, or share electrons. The elements in that group are thus chemically inert and gaseous under atmospheric conditions (0.1013 MPa and 288.15 K), leading to their being termed the noble gases.

Group 1 metals on the left-hand side are followed by a column of group 2 alkaline earth metals. These columns are followed by a block of 40 elements divided into ten columns of four elements each. This group is collectively called the transition metals. Groups 13 through 18 make up the right-hand side of the table. A diagonal dividing line separates the nonmetals in the upper-right-hand portion of this block, such as oxygen, carbon, and nitrogen, from the metals such as tin and lead in the lower-left portion.

There is an additional block of 28 elements, divided into two rows of 14 elements each, that is usually placed beneath the main table. These are the rare-earth elements, whose properties are all remarkably similar. This additional block belongs between the first block, consisting of groups 1 and 2, and the transition-metals-block. The chemical properties of the lanthanides (elements 57–71) and the actinides (elements 89–103) are also very similar to each other. Semiconductors are defined in the Glossary and have no conduction electrons at $T = 0$ K, but at higher temperatures become conductors. The elemental and compound semiconductors are made from elements in groups 12-16. Elemental semiconductors (group 12) have increasing room-temperature electrical conductivity σ_e with increasing atomic weight.

The filling order of the quantum states (wave functions) of electrons, commonly known as orbitals, is

$$
\begin{array}{llll}
1s & & & \\
2s & 2p & & \\
3s & 3p & 3d & \\
4s & 4p & 4d & 4f \\
5s & 5p & 5d & 5f \\
6s & 6p & 6d & \\
7s. & & &
\end{array}
$$

We will discuss quantum numbers and electron wave functions in connection to the solution of the Schrödinger equation in Section 2.6. In the periodic table of the elements we also use the letter and number of the highest energy orbital to designate a group of elements, as shown in Table A.2. The f block is divided into a lanthanide (La) series and an actinide (Ac) series.

As the nuclear charge and the number of electrons increase, the electrons are pulled closer to the nucleus and the atomic diameter decreases. There is a screening effect by the inner electrons, so the outer electrons experience an effective nuclear charge.

1.8 Heat Transfer Physics: Atomic-Level Energy Kinetics

Heat transfer physics describes the transport, storage, and transitions of principal energy carriers appearing in the macroscopic energy equation (Table 1.1), which is central to heat transfer analysis.

Energy is stored in principal carriers both as kinetic (translational) and potential (positional) energy. Depending on the degrees of freedom of motion and the extent of influence of other carriers (or other entities in the system), these energies vary among the different carriers. Table 1.6 shows how the specific heat capacity (at constant volume or pressure) describes energy storage (sensible heat) in each carrier.

Internal motion enters into the carrier specific heat capacity through, for example, the frequency of vibration of atoms in a crystal (lattice). In the Debye model of solid specific heat capacity, the maximum frequency of vibration is designated as ω_D, called the Debye cut-off frequency or Debye frequency (a filtering parameter). We will show that, for monatomic crystals, ω_D is close to the maximum frequency obtained from the interatomic force constant and the atomic mass (Section 2.5.3). The temperature-dependent specific heat capacity of monatomic, crystalline solids is related to this frequency through (Section 4.7)

$$c_{v,p}(T) = 9\frac{k_\mathrm{B} T}{m}\left(\frac{k_\mathrm{B} T}{\hbar\omega_\mathrm{D}}\right)^3 n\int_0^{\hbar\omega_\mathrm{D}/k_\mathrm{B} T}\frac{x^4 e^x}{e^x - 1}\,\mathrm{d}x, \quad \frac{\hbar\omega_\mathrm{D}}{k_\mathrm{B} T} \equiv \frac{T_\mathrm{D}}{T}$$

phonon specific heat of monatomic, crystalline solids, (1.12)

where m is the mass of atom, n is the number of atoms per unit volume, ω is the phonon angular frequency, and the scaled energy $x = \hbar\omega/k_\mathrm{B} T$ is integrated up to the Debye energy $\hbar\omega_\mathrm{D}/k_\mathrm{B} T \equiv T_\mathrm{D}/T$, which defines the Debye temperature T_D. The higher the cut-off frequency, the higher the temperature needed to reach a constant specific heat (or the more temperature sensitive the heat capacity is at moderate temperatures). Here, the kinetics of phonon energy storage are contained in ω_D.

In (1.12),[†] the Debye frequency is related to the Debye temperature T_D, and for $T \gg T_\mathrm{D}$ we will show that per atom and per mass $c_{v,p}$ is equal to $3k_\mathrm{B}$, which represents 3 degrees of freedom for both the kinetic and potential energies.

[†] (x.y) denotes it means Chapter x and Equation number y.

Table 1.6. *Thermal energy storage (kinetic and potential) in principal carriers. The specific heat capacities are defined as,* $\partial e/\partial T|_v \equiv c_v$, $\partial h/\partial T|_p \equiv c_p$, $c_p - c_v = T\beta^2/\rho\kappa$, *where the volumetric thermal expansion coefficient and isothermal compressibility are* $\beta \equiv \partial v/\partial T|_p/v$, *and* $\kappa \equiv -\partial v/\partial p|_T/v$.

Carrier	Kinetic energy	Potential energy storage	Temperature dependence of energy occupancy	Quantum effects
Phonon (lattice vibration)	atoms vibrate about their equilibrium position, governed by interatomic potential; motion is anharmonic and three dimensional	weak and strong interatomic forces, and potentials, for harmonic motion the kinetic and potential energies are equal	boson	at high temperatures (marked by the Debye temperature), anharmonic classical behavior is reached, but at low temperatures quantum effects and harmonic behavior are dominant
Electron (free or bound)	electrons move around nuclei in three-dimensional motion and depending on their orbits are divided into core, valence, and conduction electrons	electric charge potential between electrons and nuclei	fermion	strong quantum behavior at all temperatures; heat capacity per electron is small at moderate temperatures
Fluid particle (gas or liquid phase)	nucleus–electron collective translational, vibrational and rotational motions (polyatomic) as well as electron motion around nuclei	for gases the potential energy vanishes for ideal gas and increases as the thermodynamic critical point is approached (dense gas); for liquids the potential energy can reach the kinetic energy	in additional to M–B distribution, vibrational and rotational energies are activated at certain transition temperatures	strong quantum behavior, at high temperatures (above transition temperatures) heat capacity of polyatomic gas approaches the classical behavior
Photon	under diffusive photon transport (Table 1.7), specific heat capacity can be assigned to photons	photon trapping	boson	significant quantum characteristics, although classical treatment (EM wave) is also used

Transport of thermal energy by phonons and electrons is presented by the Fourier law and phonon and electron conductivities k_p and k_e. The magnitude of these conductivities is limited by carrier scattering (the diffusion limit), and when these resistances are not present, ballistic transport occurs. Fluid particles experiencing thermal motion while under an external force (i.e., pressure gradient) undergo a net motion. The thermal motion of fluid particles results in a similar Fourier conduction and conductivity k_f. The kinetic theory of gases (Chapter 6) leads to the following relation for the thermal conductivity, and we generalize it for all carriers as

$$\begin{aligned} k_i &= \frac{1}{3} n_i c_{v,i} u_i \lambda_i \\ &= \frac{1}{3} n_i c_{v,i} u_i^2 \tau_i, \end{aligned} \tag{1.13}$$

where n_i is the number of carriers per unit volume, u_i is the carrier-average speed, λ_i is the carrier-average mean free path, and τ_i is the carrier-average scattering relaxation time ($\tau_i = \lambda_i / u_i$).

In general, the carrier population will have a range of transport properties and here averages are used. For a carrier traveling in a medium of linear dimension L, ballistic motion refers to $\lambda_i \gg L$ and localization refers to $\lambda_i \ll L$. When fluid particle collisions are not present, collisions with the bounding surface determine energy transport (this happens in the molecular-flow regime, or Knudsen regime). The Knudsen number $\mathrm{Kn}_L = \lambda_i / L$, defines the molecular-flow regimes as $\mathrm{Kn}_L \gg 1$. In the viscous-flow regime ($\mathrm{Kn}_L \ll 1$, where the molecular collisions dominate), the net (macroscopic) motion can be laminar (ordered) or turbulent (chaotic fluctuations). Heat transfer by net motion is referred to here as convective heat transfer.

Photons carry energy without any resistance in a vacuum and experience scattering, absorption, and emission when interacting with matter. When the absorption is rather large, such that distant effects are negligible, the system exhibits the diffuse limit of photon transport. Table 1.7 summarizes some of these attributes of microscale transport. The radiative properties of matter include its spectral (wavelength-dependent) absorption and emission, which depend on its fine electronic structure and their motions/displacements.

Similar to the phonon specific heat capacity previously given, the Callaway model (single-mode relaxation time) of the phonon thermal conductivity of crystalline solids is (Section 4.9.2)

$$k_p = \left(48\pi^2\right)^{1/3} \frac{1}{a} \frac{k_\mathrm{B}^3}{h_\mathrm{P}^2} \frac{T^3}{T_\mathrm{D}} \int_0^{T_\mathrm{D}/T} \tau_p(x) \frac{x^4 e^x}{(e^x - 1)^2} \mathrm{d}x \quad \text{phonon thermal conductivity,} \tag{1.14}$$

where a is the lattice constant. Here τ_p is the effective phonon relaxation time. The longer the relaxation time, the larger the phonon thermal conductivity. The relaxation time is related to the mean free path, through the carrier speed as, $\lambda_i = u_i \tau_i$. The kinetics of phonon transport is represented in τ_p and ω_D.

Table 1.7. *Transport kinetics of thermal energy by principal carriers (e.g., k_i thermal conductivity, $i = p, e, f, ph$).*

Carrier	Transport regime	Transport mechanism	Controlling kinetics
Phonon	diffusive ($\lambda_p \ll L$, controlled by scattering) or ballistic ($\lambda_p \gg L$, no resistance, except at boundaries)	phonons of energy $\hbar\omega_p$ (or bulk heat capacity $\rho c_{v,p}$) travel at speed u_p and encounter resistance (as represented by mean free path λ_p) by conduction electrons, other phonons, impurities, and grain boundaries	interphonon scattering rate, phonon-impurity, and phonon-grain boundary scattering rates
Electron	diffusive (controlled by scattering), tunneling, or by electric field (drift)	electrons of heat capacity $n_e c_{v,e}$ travel at Fermi speed $\langle u_F \rangle$ and encounter resistance (mean free path λ_e) by phonons and impurities	electron–phonon and electron–impurity scattering rates
Fluid particle	convective	fluid particle transit, fluid flows with a net speed $\overline{u}_f$ and travels distance L in time $\tau_f = L/\overline{u}_f$, in single-phase flow the fluid particle will have change in its sensible heat during this transit	in laminar flow the fluid thermal conductivity is not altered, but in turbulent flow (containing chaotic velocity fluctuations) it is augmented by turbulent mixing (eddy) conductivity
	diffusive/viscous (controlled by collision), ballistic (molecular flow), transitional (moderate Knudsen number $\mathrm{Kn}_L = \lambda_f/L$), and net (bulk) motion called convection	fluid particles of energy $\rho_f c_{v,f}$ travel at thermal speed $\langle u_f \rangle$ and collide (mean free path λ_f) with each other or with surfaces, when there is no collision the transport is ballistic	interfluid particle collisions or fluid particle surface scattering rates
Photon	transparent (vacuum, no absorption or emission), semi-transparent, or diffusive (dominated by strong absorption/scattering, also called optically thick)	photons of energy $\hbar\omega_{ph}$ travel at speed u_{ph} and are absorbed, scattered, and emitted (mean free path λ_{ph}); the interaction is mostly through coupling of the photon as an electromagnetic wave with electric entities such as dipoles	photon scattering and absorption (EM wave–electric entity interactions) rates, photon emission rate

The transport of photon through media is hindered by absorption and for the case of a medium allowing for absorption of photons/excitation of free electrons (metals), we have (Section 7.8)

$$\sigma_{ph,\omega} = \frac{1}{\lambda_{ph,\omega}} = \frac{1}{u_{ph}\tau_{ph,a,\omega}} = \left(\frac{2n_{e,c}e_c^2\langle\langle\tau_e\rangle\rangle\omega}{\epsilon_o m_e u_{ph}^2}\right)^{1/2}, \qquad (1.15)$$

Table 1.8. *Examples of volumetric $\dot{S}_{i\text{-}j}/V$ and surface $\dot{S}_{i\text{-}j}/A$ thermal energy conversions (resulting in heating/cooling) and their kinetics relations. R stands for a resonance phenomenon.*

Thermal energy conversion	Relation
Phonon	
Ultrasonic heating (Section 4.18)	$\frac{\dot{S}_{p\text{-}p}}{V} = 2\sigma_p I_{ac} = 2\gamma_G^2 \frac{c_{v,p}}{2u_{p,\mathrm{A}}^2} \frac{\omega\tau_p}{1+2\omega^2\tau_p^2} I_{ac}$
Laser cooling of solids (R) (Section 7.12)	$\frac{\dot{S}_{ph\text{-}e\text{-}p}}{A} = N_d \frac{\tau_{ph,tr}}{\tau_{ph\text{-}e\text{-}p}}(1 - \frac{\omega_{ph,e}}{\omega_{ph,i}} \frac{\tau_{e\text{-}ph}^{-1}}{\tau_{e\text{-}ph}^{-1} + \tau_{e\text{-}p}^{-1}}) q_{ph,i}$
Electron	
Joule heating (Section 5.13)	$\frac{\dot{S}_{e\text{-}p,\mathrm{J}}}{V} = \rho_e j_e^2 = \frac{j_e^2}{\sigma_e}$,
	$\sigma_e = \frac{8\pi}{3}(\frac{1}{2\pi^2\hbar^2})^{3/2} e_c^2 m_e^{1/2} \tau_{e,o} (\frac{E_\mathrm{F}}{k_\mathrm{B}T})^{s+3/2}$ metals
Peltier cooling/heating (Section 5.16)	$\frac{\dot{S}_{e\text{-}p,\mathrm{P}}}{A} = \mp\alpha_S T j_e,\ \alpha_S = \mp\frac{\pi^3}{3}\frac{k_\mathrm{B}}{3}\frac{s+3/2}{E_\mathrm{F}/k_\mathrm{B}T}$ metals
Fluid particle	
Exothermic or endothermic reaction (Section 6.5)	$\frac{\dot{S}_{f\text{-}f,r}}{V} = -\rho_\mathrm{F}\Delta h_{r,\mathrm{F}} \frac{n_{f,\mathrm{F}}}{\tau_{f\text{-}f}} e^{-\Delta E_a/k_\mathrm{B}T}$
Surface evaporation (R) (condensation) (Section 6.8.1)	$\frac{\dot{S}_{f\text{-}f,lg}}{A} = -\Delta h_{lg}\dot{m}_{f,lg},\ \dot{m}_{f,lg,max} = \frac{1}{4} m n_f (\frac{8k_\mathrm{B}T}{\pi m})^{1/2}$
Photon	
Photovoltaic power generation (R) (Section 7.14)	$\frac{\dot{S}_{ph\text{-}e}}{A} = \int_0^\infty (\Delta\varphi_e j_e)_{max} \mathrm{d}\omega$
Dielectric heating (Section 3.4)	$\frac{\dot{S}_{ph\text{-}p,\epsilon_e}}{V} = \epsilon_o \epsilon_{e,c} \omega \overline{\boldsymbol{e}_e \cdot \boldsymbol{e}_e}$

where $\sigma_{ph,\omega}$(1/m) is the spectral photon absorption coefficient (and is related to $\dot{s}_{ph\text{-}e}$, as listed in Table 1.8), $\tau_{ph,a,\omega}$ is the spectral photon relaxation (absorption) time, $\langle\langle\tau_e\rangle\rangle$ is the energy-averaged electron relaxation time (Section 5.7), and ω is the angular frequency of the photon. The natural and engineered transformation of energy among principal energy carriers continues to be an important subject of study. In microscopic heat transfer there are many complex heating and cooling processes, and as our understanding of phenomena occurring at the atomic scale increases, we expand the possibilities for energy conversion. Table 1.8 lists some relations of energy transformations involving the four principal carriers as well as their heating or cooling effects. Table 1.9 gives examples of mechanisms of energy conversion with explanations of kinetics.

Some of the more novel energy-conversion methods being studied include the absorption of ultrasound in solids (which involves the interaction of forced external vibrations with phonons), thermoelectric cooling/heating, dielectric and Joule heating, endothermic/exothermic phase changes and chemical reactions, laser cooling of gases, and various other photon–electron interaction-mediated heating–cooling processes in solids and fluids.

Cooperative processes (Glossary) are those in which more than one carrier assists in promotion (and decay) of an energy transition. This involves couplings among various carriers, these coupling kinetics are critical in such energy

Table 1.9. *Examples of microscale energy transformation (conversion) kinetics (from/to thermal energy), involving principal carriers.*

Carrier	Energy conversion mechanism	Cooling/heating	Controlling kinetics
Phonon	phonon-assisted fluorescence emission in electronic transitions (R)	phonon emission or absorption	relaxation rate of thermal phonons, coupling between photon (electromagnetic wave) and electric entity, coupling between electron and phonon, number density of these three carriers
	ultrasonic heating	absorption of ultrasound waves	rate of inelastic relaxation of external sound waves by thermal phonons and inter-phonon relaxation
Electron	metals: electrical conductivity (conduction electron)	Joule resistive heating	scattering of electrons by phonons and impurities
	dielectrics: coupling of photon (electromagnetic wave) with electric dipoles (R)	dielectric (microwave) heating	dielectric loss factor (imaginary part of electric permittivity) and frequency of EM wave
	semiconductors: thermoelectric coupling (R)	Peltier cooling/heating	bandgap, Fermi energy, electron relaxation rate, and temperature
Fluid particle	interparticle collision rate in gases	endothermic and exothermic thermal reactions	collision rate, particle concentration, and temperature
	physical adsorption and desorption of gas molecules on solid surfaces (R)	heating (adsorption) and cooling (desorption)	attraction force between the gas and surface atoms, number of adsorbed atomic layers, temperature
Photon	conversion of valence electrons in semiconductors to conduction electrons by absorbed photon (R)	heating when some photon absorption leads to high kinetic energy (hot electrons)	absorption rate of photon, lack of multiple-bandgap semiconductors, electron–phonon coupling
	photon emission by conversion of kinetic energy of excited gases (mixture of molecules in translational, vibrational, and rotational motion) (R)	heating by nonradiative decays	rate of collision between species, relaxation time for radiative and nonradiative decays

conversions. R stands for resonance, indicating an electronic for mechanical resonance energy match is needed or involved. For example, an electronic bandgap, or a bond energy, or a mechanical vibration frequency.

The expressions given in Table 1.8, will be described in the related sections noted, and various variables will be discussed, and the relations developed.

For example, similar to the preceding phonon specific heat and thermal conductivity examples, the acoustic absorption rate of ultrasound in crystals can be described as a volumetric heat source $\dot{s}_p$(W/m^3) as (Section 4.18)

$$\frac{\dot{S}_{p\text{-}p}}{V} = \dot{s}_{p\text{-}p} = 2\sigma_p I_{ac} = 2\gamma_{\rm G}^2 \frac{c_{v,p}}{2u_{p,{\rm A}}^2} \frac{\omega\tau_p}{1+2\omega^2\tau_p^2} I_{ac}$$

$$\text{acoustic wave absorption rate,} \qquad (1.16)$$

where $\gamma_{\rm G}$ is the Grüneisen parameter (see Glossary), $u_{p,{\rm A}}$ is the average acoustic-phonon speed, ω is the frequency of the external ultrasound wave, and I_{ac} is the ultrasonic intensity of this wave. Again, the kinetics of the energy conversion rate involves the phonon relaxation time τ_p and frequency ω.

Heat transfer physics (the kinetics of microscale energy storage, transport, and transitions) addresses energy dynamics at the atomic scale. Examples of the time scales of these atomic-level dynamic interactions are given in Figure 1.5. In general, although shorter time scales are desirable in energy-storage and conversion rates, longer time scales are desirable for high transport rates. When multiple mechanisms are involved in the kinetics, the mechanism with the longest response time dominates the conversion rate, whereas the mechanism with the shortest response (scattering) time dominates the transport. This is evident in the acoustic wave absorption rate in (1.16) and in the phonon thermal conductivity in (1.14). We will discuss the evaluation of the various interaction rates, time constants, and relaxation times throughout the text. These are the central governors of the atomic-level heat transfer.

These kinetics are associated with some atomic-level resonances, which allow for the creation of desired spectral behavior in materials and devices. For example, crystalline Si has a relatively narrow optical-phonon bandwidth, as shown in Figure 1.8(a). This material also has an electronic energy bandgap, which is shown in Figure 1.8(b). Both figures show the density of states (number of carriers per unit energy level of the carriers). The resonant optical phonon (marked with R) has an energy much less than the electron bandgap energy, but does play a role in the absorption and emission of photons from Si (doping can charge the electronic bandgap). Heat transfer physics examines and exploits resonances in these and other carrier interactions. In Figures 1.4 and 1.5, some carrier kinetics resonances are marked with (R).

1.9 *Ab Initio*/MD/BTE/Macroscopic Treatments

Figure 1.9 shows the regimes of application of the Schrödinger equation (Table 2.8), computational molecular dynamics (MD) simulations (this classical treatment is

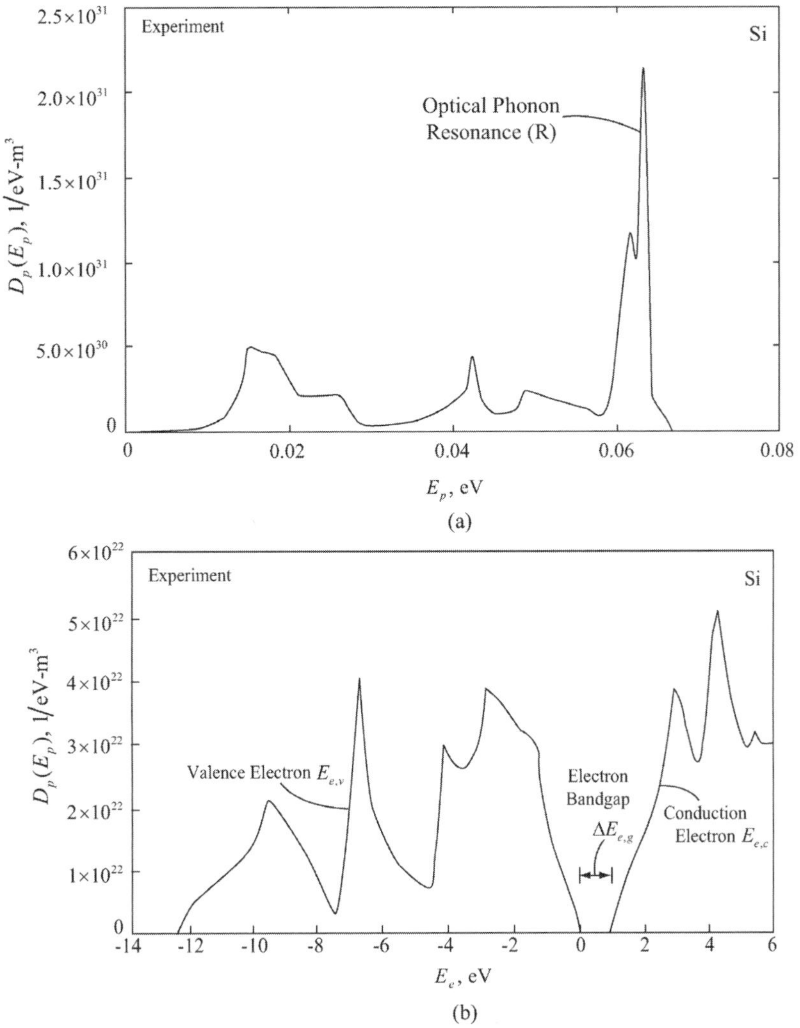

Figure 1.8. (a) Phonon density of states for Si (described in Section 4.6), and (b) electron density of states for Si (described in Section 5.7). Here R stands for resonance. $E_{e,c}$ and $E_{e,v}$ indicate conduction and valence electrons.

given in Table 2.5), the Boltzmann transport equation (BET) (Table 3.1), and the macroscopic energy equation (including all carriers) (Table 1.1).

The Schrödinger equation uses the atomic scale units of Table 1.5, Section 1.5.2 ($r_{\rm B}$ for length and $\tau_a = m_e r_{\rm B}^2/\hbar$ for time).

MD uses interatomic interaction units that are r_{nn} the distance between the nearest-neighbor atoms, and $\tau_n = \pi/\omega_n$ where ω_n is the natural frequency of atomic vibration, (Section 2.5.3 and Table 2.7).

The BTE uses the many (N-) atoms length scale $N^{1/3} r_{nn}$ and the particle collision (scattering) relaxation time $\tau_{i\text{-}j}$ (Figure 1.5). The Boltzmann transport scales (Table 3.2) are discussed in Section 3.1.6.

The macroscopic energy equation (Table 1.1) assumes that the system dimension and time resolutions are large enough that local and temporal variations in

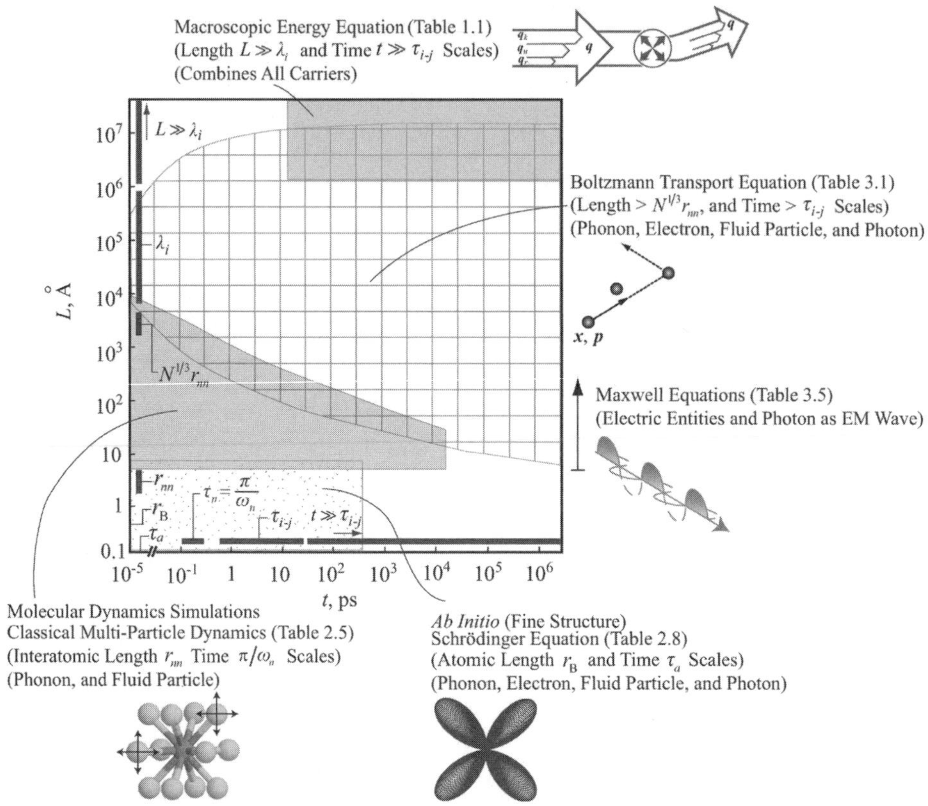

Figure 1.9. Length–time scale regimes for *ab initio*, MD, Boltzmann transport, and macroscopic treatments of heat transfer.

density, temperature, velocity, etc., can be represented by smooth (continuous) functions. This is marked by $L \gg \lambda_i$, where λ_i is the mean free path of carrier i. The mean free path signifies how far a carrier travels before it undergoes an encounter with another carrier that significantly alters its momentum (Section 3.1.3). Similarly, the macroscopic time scale is $t \gg \tau_{i-j}$. The macroscopic length and time scales are given in Table 3.12.

Note that the t axis in Figure 1.9 can be further marked by other kinetic times listed in Figure 1.5 for each of the carriers. For photons treated as EM waves, the Maxwell equations (Table 3.5) are used at length scales for which quantum effects are not important. The Maxwell equations are also used for describing electric quantities. The Schrödinger-equation-based calculations are called *ab initio* (Latin for from the beginning) calculations and describe what is referred to as the fine-structure of matter. *Ab initio* calculations require the least amount of empiricism.

1.10 Scope

The transport and transitions of principal thermal energy carriers are rendered in Figure 1.10, and are discussed in the following chapters. As shown in Figure 1.10,

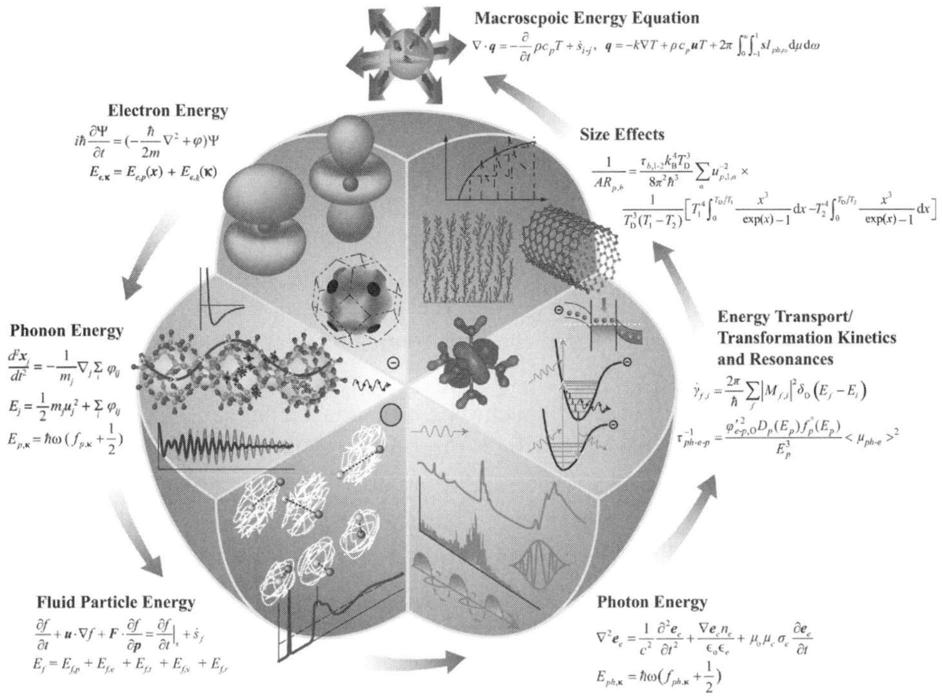

Figure 1.10. A summary rendering of the four principal thermal energy carriers and their transport and interaction. Also shown are the size effects and the macroscopic energy equation.

size effects and the inclusion of various phonon, electron, fluid particle, and photon energy transport/transition kinetics in the macroscopic energy equation, connect heat transfer physics to its applications at various length scales.

In Chapter 2, atoms, their interactions (such as forming bonds), and their collective properties lead to the definition of temperature and other properties. The Schrödinger equation Newtonian and Hamiltonian mechanics (and energy relations) are discussed in relation to molecular interactions and molecular dynamics.

In Chapter 3, the principal energy carrier transport and kinetics theories are reviewed. These form the foundations for the treatments of each carrier in the subsequent four chapters. The Boltzmann transport equation used here for all principal carriers (phonons, electrons, fluid particles, and photons), the probability distribution functions f_i, the stochastic transport equation, and the G–K autocorrelation equation are reviewed. There are also equations that are derived from these, such as the macroscopic energy equation, the equation of radiative transport, the Navier–Stokes equation, and the macroscopic elastic fluid mechanics equation. The unified treatment of transport phenomena by fluid particles, electrons, and phonons has been made in [315], and for all four principal carriers in [66].

Chapter 4 describes phonons, Chapter 5 covers electrons, Chapter 6 details fluid particles, and Chapter 7 explains photons. The unique energy transport (including various inter- and intra-carrier interactions) and storage characteristics of each carrier are described in each of these chapters. The common element unifying the coverage of all of these carriers is the use of the Boltzmann transport equation.

1.11 Problems

Problem 1.1

(a) Using Table A.1, plot the scaled specific heat capacity ($c_v m/k_\mathrm{B}$) versus atomic number z for the elements (at 300 K), where m is the mass of the atom (M/N_A). Note that here c_v is per atom or molecule. Also, note that for a monatomic solid $c_v m/k_\mathrm{B} = 3$ (Section 4.7), and for monatomic ideal gas $c_v m/k_\mathrm{B} = 3/2$ (Section 6.2).

(b) Similarly, plot the scaled specific heat of phase change, $\Delta h_{sl} m/k_\mathrm{B} T_{sl}$ and $\Delta h_{lg} m/k_\mathrm{B} T_{lg}$, versus z, where T_{sl} is the saturation solid–liquid temperature and T_{lg} is the saturation liquid–gas (vapor) temperature at $p = 1$ atm.

(c) Comment on the trends (Figure 1.7 displays similar results for the thermal conductivity k). Note that Δh_{ij} is scaled with respect to the specific energy $k_\mathrm{B} T_{ij}$, which represents the kinetic energy.

Problem 1.2

(a) Express the conductive heat flux vector as $|\boldsymbol{q}_k| = k_s dT_s/dx = k_s \Delta T/\Delta x$. Determine $|\boldsymbol{q}_k|$ for (i) a high-conductivity elemental solid (Table A.1) along with $\Delta T = T_s - T_{ls}$, where T_{ls} is the melting temperature, and $\Delta x = l$, where l is the thickness along which $T_s - T_{lg}$ is applied (this is an approximation for a thin slab as part of a structure). (ii) Repeat using a high-melting-temperature elemental solid (Table A.1). Use $T_s = 300$ K.

(b) For the convective heat flux vector, determine $|\boldsymbol{q}_u| = (\rho c_p)_f u_f T_f$, using (i) a gas, and (ii) a liquid, moving at velocities equal to the speeds of sound and at $T_f = 300$ K. (iii) For the surface-convection heat flux vector, $|\boldsymbol{q}_{ku}| = \langle \mathrm{Nu} \rangle_D k_f (T_s - T_{f,\infty})/D$, for Nusselt number $\langle \mathrm{Nu} \rangle_D = 100$, $D = 10^{-2}$ m, $T_s - T_{f,\infty} = 100$ K, and k_f of nitrogen at $T_f = 300$ K, at 1 atm pressure (Table A.1).

(c) Determine the radiative heat flux vector, $|\boldsymbol{q}_r| = n_{ph} u_{ph,c} \hbar\omega_{ph}$, for a typical (i) continuous laser and (ii) pulsed laser, for $\lambda = 0.5$ μm. (iii) Also use the blackbody (thermal) radiation $|\boldsymbol{q}_r| = \sigma_\mathrm{SB} T^4$, for $T = 300$ K, where σ_SB is the Stefan–Boltzmann constant (Table 1.4).

Problem 1.3

(a) For Si, typical maximum E_p for acoustic and optical phonons are 0.019 and 0.062 eV, respectively. Using $T = 300$ and 1000 K, determine f_p^o (boson) for these two energies.

(b) Consider a conduction electron in Cu and use $\mu = E_\mathrm{F}$ (Table 5.2), and for T = 300 and 1000 K, determine f_e^o (fermion) for $E_e = 1$ and 10 eV.

(c) For an Ar gas, with the average thermal speed given by (6.63), determine the M–B energy distribution f_f^o, given in Figure 1.1, for $T = 300$ K.

(d) For a photon of energy $E_{ph} = \hbar\omega_{ph}$, and for (i) $\lambda = 20$ μm (near infrared), (ii) $\lambda = 0.63$ μm (red), (iii) $\lambda = 0.50$ μm (blue), (iv) $\lambda = 0.30$ μm (ultraviolet), and (v) $\lambda = 0.001$ μm (X-rays), determine $f_{ph}^{\rm o}$ (boson) at $T = 300$ K.

(e) Compare the results of part (a) with those of part (d).

Note that these are equilibrium (thermalized or allowed to reach thermal equilibrium) distributions. For lasers, for example, a selective amplification of a narrow band of energy is used, so the photon distribution does not follow the equilibrium boson. The Fermi energy of semiconductors is temperature dependent.

Problem 1.4

(a) Show that M–B energy probability distribution function (6.65), given by

$$f_f^{\rm o}(E_f^*) = \frac{2}{\pi^{1/2}} E_f^{*\,1/2} \exp(-E_f^*), \quad E_f^* = \frac{E_f}{k_{\rm B}T},$$

satisfies

$$\int_0^\infty f_f^{\rm o}(E_f^*) {\rm d}E_f^* = 1.$$

Superscript * indicates dimensionless quantity.

(b) Show that the classical distribution

$$f_f^{\rm o}(E_f^*) = \exp(-E_f^*),$$

also satisfies the preceding integration.

(c) Show that

$$\langle E_f \rangle = \int_0^\infty f_f^{\rm o}(E_f^*) E_f^* {\rm d}E_f = \frac{3}{2} k_{\rm B} T,$$

i.e., the average energy (in a M–B system) is equal to $3k_{\rm B}T/2$. Use gamma functions given by (5.113).

The integral of the other probability distribution function is also physically meaningful. For example for quasi-particles, the integral of $f_p^{\rm o}$ ($f_e^{\rm o}$ and $f_{ph}^{\rm o}$) leads to the number densities. For example, from (1.2) for p and e, we have

$$n_i = \sum_\kappa f_i^{\rm o} = \frac{1}{(2\pi)^3} \int_0^\infty f_i^{\rm o} 2\pi\kappa^2 {\rm d}\kappa$$

$$= \frac{1}{4\pi^2} \int_0^\infty \frac{\kappa^2 {\rm d}\kappa}{\exp(\dfrac{\hbar^2\kappa^2}{2m_i k_{\rm B}T} - \mu) \pm 1}, \quad i = \begin{cases} +\, e\ (\mu \neq 0) \\ -\, p\ (\mu = 0) \end{cases}.$$

Problem 1.5

(a) Plot $f_e^{\rm o}$ listed in Table 1.2, versus E_e for copper ($E_{\rm F}$ listed in Table 5.2). Use (i) $T = 0$, (ii) $T = 300$ K, (iii) $T = 1000$ K, and (iv) $T = 10{,}000$ K, and use $0 \leq E_e \leq 1.5E_{\rm F}$. Note the rather abrupt change in $f_{\rm o}^e$ for metals.

(b) Indicate in the graph the region $\Delta E_e = k_B T$ on both sides of $E_e = E_F$, where f_o^e changes the most.

(c) For metals a constant Fermi energy is used, and the $T = 0$ K approximations of f_e^o, i.e., $f_e^o = 1$ for $E_e \leq E_F$, and $f_e^o = 0$ for $E_e > E_F$, are used. Based on the results of parts (a) and (b), is this a good approximation?

Problem 1.6

Similar to Figure 1.7, plot electronegativity of elements versus the atomic number. Electronegativity is a measure of atom attracting electrons (forming bonds). What trends are observed?

Problem 1.7

(a) Verify the magnitude and unit of the atomic length, time, velocity, energy, and dipole moment given in Table 1.5.

(b) Show that the atomic unit of energy is the product of the electron mass (at rest) and the square of the atomic velocity.

(c) Compare the atomic energy unit with $k_B T$ ($T = 300$, 10^3 and 10^5 K), in eV.

2

Molecular Orbitals/Potentials/Dynamics, and Quantum Energy States

Energy is stored within atoms (electrons and relativistic effects) and in the bonds among them when they form molecules, and also between these molecules when they form clusters and bulk materials. In this chapter we review the intraatomic and interatomic forces and energies (potentials). We examine electron orbitals of atoms and molecules and review *ab initio* computations of interatomic potentials and potential models. Then we discuss the dynamics (and scales) of molecular (or particle) interactions in classical (Newtonian and Hamiltonian) treatments of multi-body dynamics and the relationship between the microscopic and macroscopic states (particle ensembles). We also begin examining the quantum energy states of the simple harmonic oscillator, free-electron gas, and electrons in hydrogen atom orbitals, by seeking exact solutions to the Schrödinger equation.

2.1 Interatomic Forces and Potential Wells

Atoms and molecules, at sufficiently low temperatures, can aggregate into linked structures that form what we know as the condensed phases of matter. What causes this aggregation behavior are the attractive forces generated by the opposite electrical charges of each particle (be it individual atom, ion, or polar molecule) in the system. These forces are in turn caused by the innate charges of the constituent subatomic particles of each atom, ion, or molecule. All physical and chemical bonding is the result of such molecular forces that are electrical in origin (the Coulomb forces). Intramolecular (or interatomic) forces keep atoms in a molecule (a compound of atoms) together, such as the H–O bond in a H_2O molecule, whereas intermolecular forces keep neighboring molecules together, such as H_2O molecules in the gas, liquid, or solid phase. The intramolecular forces are stronger than the intermolecular forces. The difference among solids, liquids, and gases is the degree to which the thermal energy of the particles $k_B T$ in the system can overcome the attractive forces holding them together. At a given temperature, substances that contain strong interatomic bonding are more likely to be solids (or liquids), whereas those that contain weak or no bonding will be gases. The interatomic potential φ, a form of energy

Interatomic and Intermolecular Forces

Force	Structure	Electric Force	φ, eV/molecule	Example	r_e, Å
Interatomic (δ Denotes Charge Fraction)					
Ionic bond	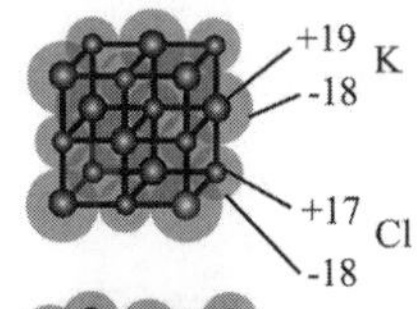	Opposite charges	4 to 40	K–Cl	2.36
Covalent bond	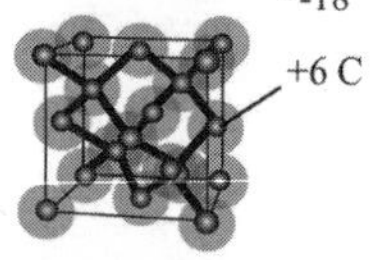	Nuclei shared electron pair	1.5 to 11	C–C (Diamond)	1.54
Metallic bond	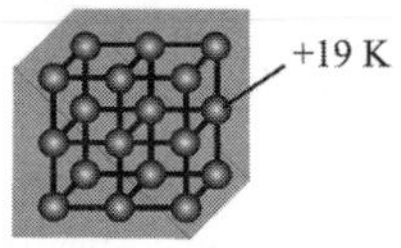	Metal cations and delocalized electrons	0.75 to 10	K	4.54
London dispersion (weak) bond	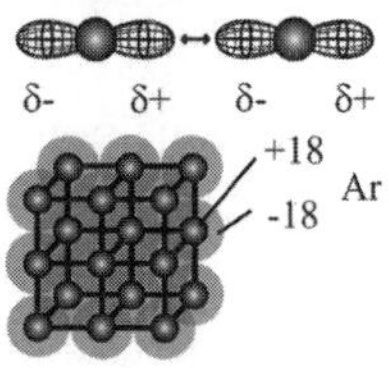	Induced dipoles of polarizable mole-cules	0.005 to 0.4	Ar–Ar	3.76
Intermolecular (δ Denotes Charge Fraction)					
Ion-dipole bond	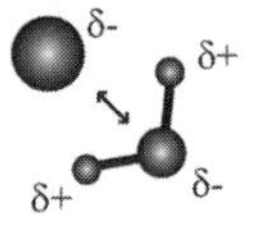	Ion and polar molecule	0.4 to 6	Na^+–H_2O	
Dipole-dipole bond	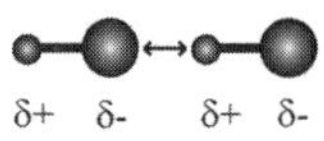	Partial charges of polar molecules	0.05 to 0.25	HCl–HCl	
Hydrogen bond	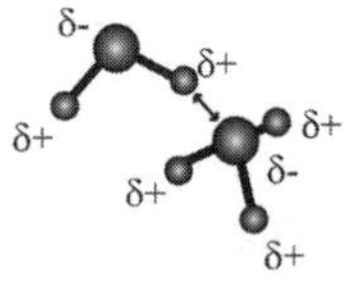	H bonded to N, O, or F, and another N, O, or F	0.1 to 0.4	NH_3–H_2O	

Figure 2.1. A classification of the interatomic and intermolecular forces, atomic structures, electric forces, ranges of their bond energy (potential), examples, and interatomic bond lengths.

(which explains/describes interatomic and intermolecular integrations) and a measure of bond strength, is generally expressed in electron volts (eV), which is the work required to displace an electron through an electric potential of 1 volt (1 V is 1 J/C). Figure 2.1 gives a summary of the interatomic and intermolecular forces and bond length r_e and they are subsequently described.

2.1.1 Interatomic Forces

These forces include ionic bonding forces and bond lengths r_e, covalent bonding forces, and metallic bonding forces, and London dispersion (van der Waals) bonding forces. Ionic bonding forces are generated between oppositely charged ions. Ions themselves can arise from electron transfer that can occur when atoms of elements that have a large difference in electronegativity (the tendency of an atom to draw electrons in a bond toward itself) come in close proximity to one another. Such bonding is frequently seen between a metal and a nonmetal (cation and anion), i.e., NaCl, MgF_2. In ionic crystals, an array of ions are arranged at regular positions in a crystal lattice.

Covalent bonding forces are formed by two atoms coming together and sharing one of more electrons. Neither atom is able to overcome the attractive forces of the nucleus of the other and completely take its electron(s) for itself. This results in a balance between the forces of attraction and repulsion that act between the nuclei and electrons of the atoms. In many cases, this electron sharing enables each atom to reach a noble gas configuration (i.e., a filled outer electron shell).

Metallic bonding forces arise when metallic atoms (typically with two valence electrons each, Table 5.2) bond to one another. In the free-electron model of this type of bonding, the valence electrons of each atom are no longer restricted to a single atom but are delocalized, forming a sea of electrons that hold the positively charged nuclei together. Metals are easily deformed, due to the layers of positive ions (the nuclei) being able to slide over one another without disrupting the sea of electrons between them. It is also due to the delocalized behavior of the electrons in these bonds that metals are good electrical conductors. The greater the number of valence electrons (and the larger the positive charge of the nuclei), the stronger the metallic bonding forces.

2.1.2 Intermolecular Forces

The attractive forces between neutral molecules are usually ion–dipole forces, dipole–dipole forces, London dispersion forces, or hydrogen bonding forces. The dipole–dipole and dispersion varieties are grouped together and termed van der Waals forces (sometimes hydrogen bonding is also included with this group). The attractive forces between neutral and charged (ionic) molecules are called the ion–dipole forces.

Ion–dipole forces arise between charged ions and polar molecules (molecules with an asymmetric charge distribution), where cations are attracted to the negative end of a dipole, and anions are attracted to the positive end. The magnitude of the interaction potential (with the force being the negative gradient of the potential) is directly proportional to the charge of the ion and the dipole moment of the molecule, and inversely proportional to the distance from the center of the ion to the midpoint of the dipole. Ion–dipole forces are important in solutions of ionic substances in polar solvents (e.g. a salt in an aqueous solvent).

Dipole–dipole forces exist between neutral or polar molecules and polar molecules. They attract one another when the partial positive charge (δ_+ in Figure 2.1) on one molecule is near the partial negative charge (δ_-) on the other. The molecules must be in relatively close proximity (a few Å) for dipole–dipole forces to be significant. These forces are characteristically weaker than ion–dipole forces, and increase with an increase in the polarity of the molecules.

Hydrogen bonds are special dipole–dipole forces between electronegative atoms and hydrogen atoms bonded to adjacent electronegative atoms. Their bond strength is weak covalent. The intermolecular hydrogen bond results in relatively high boiling temperature of H_2O.

2.1.3 Kinetic and Potential Energies and Potential Wells

A system in which intramolecular and intermolecular forces act can possess two types of energy: kinetic energy E_k and potential energy $E_p = \varphi$. The magnitude of the kinetic energy is a function of temperature only. Potential energy arises from the action of the Coulombic forces previously mentioned. Although there are differences among the various forces, they all give rise to potential energy between the interacting particles. It is the relative difference between the kinetic and potential energies of a system of particles that determines many of the properties of the substance.

Kinetic energy is energy of motion. A body of mass m, moving with velocity u, has kinetic energy given by

$$E_k = \frac{mu^2}{2}. \tag{2.1}$$

The kinetic energy of atoms and molecules has a thermal origin and, from equipartition of energy [119], is related to the absolute temperature as follows:

$$E_k = \frac{\alpha k_{\rm B} T}{2}, \tag{2.2}$$

where $k_{\rm B}$ is the Boltzmann constant and α is related to the degrees of freedom of the atomic motion of the system. Potential energy φ is the energy of position. When a particle begins to move, potential energy is converted to kinetic energy. The opposite happens as a moving particle is affected by forces generated as it approaches another body. The total energy of any mechanical system is the sum of its kinetic and potential energies, and on a per-atom/molecule basis we have

$$E = e = E_k + E_p \equiv E_k + \varphi, \tag{2.3}$$

where e is the specific (per-atom/molecule) internal energy. In addition to the gravitational potential, there are three other potentials: the strong nuclear, electrostatic (or Coulombic), and weak nuclear potentials. As discussed, the Coulombic force is the force by which electrons are attracted to and held by the nucleus of an atom, and this force is responsible for all interactions between atoms and molecules. It is far stronger than the gravitational force, which may be neglected in discussing atomic and molecular interactions (the electromagnetic force of repulsion between two electrons is 10^{42} times stronger than the gravitational force of attraction between them). The magnitude of the electrical force is given by the Coulomb law [215]

$$F_{12} = \frac{1}{4\pi\epsilon_o}\frac{q_1 q_2}{r^2} \quad \text{Coulomb law,} \tag{2.4}$$

where q_1 and q_2 are the magnitudes of the two interacting electrical charges, in coulombs (C), r is the distance between them, in meters and ϵ_o is the free-space electric permittivity and has a value of 8.854×10^{-12} $C^2/N\text{-}m^2$ (also listed in Table 1.4). Here $1/4\pi\epsilon_o = 8.9876 \times 10^9$ $N\text{-}m^2/C^2$ is called the Coulomb constant. The potential energy associated with the Coulombic force is called the Coulomb potential energy. Because force is the negative derivative of potential energy with respect to distance ($F_{12} = -\partial\varphi_{12}/\partial r_{12}$), the Coulombic potential energy can be obtained by the integration of the Coulomb law

$$\varphi_{12} = -\frac{1}{4\pi\epsilon_o}\frac{q_1 q_2}{r}, \quad \boldsymbol{F}_{12} = -\frac{\partial\varphi}{\partial r}\boldsymbol{s}_{12} \quad \text{force–potential relation,} \tag{2.5}$$

where $\boldsymbol{s}_{12}$ is the unit vector along $\boldsymbol{r}$, connecting 1 and 2. Bold characters signify vectors (lowercase) or tensors (uppercase).† When one of the charges is positive and the other negative, the potential energy is negative at all finite values of r.‡

For a system of two or more particles (atoms, molecules, and/or ions), we assume that the interparticle forces are electrical and are attractive, except at very close distances, and described by the Coulomb law. The magnitude (strength) of the force between two particles varies with their distance of separation. The force is zero for an infinite separation, increases as the particles get closer, reaches a maximum when the particles just touch, and becomes strongly repulsive when the centers of the particles are separated by less than one particle diameter because of

† The screened Coulomb potential is

$$\varphi = -\frac{q_1 q_2}{4\pi\epsilon_o r}e^{-r/r_o},$$

where r_o is the screening distance.

‡ In the Bohr model of a hydrogen atom, the single electron is in the ground state has $r_B = 4\pi\epsilon_o\hbar^2/m_e e_c^2 = 5.292\times10^{-11}$ m (Bohr radius, Table 1.4), and the potential energy [for charge of unity on proton and electron, using (2.5)] is twice the kinetic energy $m_c u_e^2/2$ (but negative), and this gives the total energy as

$$\begin{aligned} E_k + \varphi &= \frac{1}{2}\frac{1}{4\pi\epsilon_o}\frac{e_c^2}{r_B} - \frac{1}{4\pi\epsilon_o}\frac{e_c^2}{r_B} \\ &= -\frac{1}{2}\frac{1}{4\pi\epsilon_o}\frac{e_c^2}{r_B} = -13.606 \text{ eV} \quad \text{H ground state for hydrogen.} \end{aligned}$$

Thus, to remove the electron from hydrogen, an ionization energy of 13.606 eV is needed. Note that this is for $n = 1$ (the ground state). For other principal quantum numbers, we have $r_B n^2$ in place of r_B.

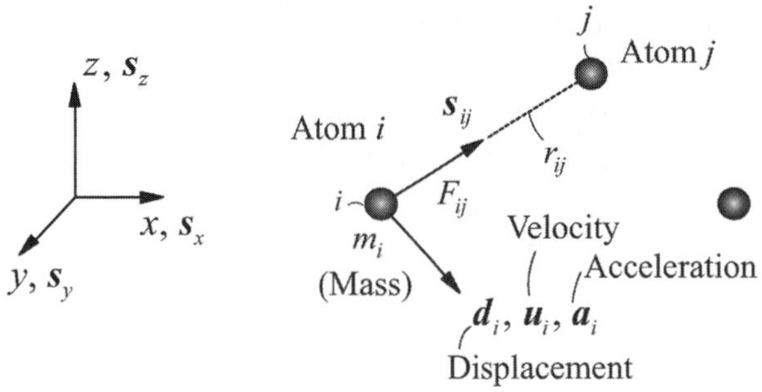

$$m_i \frac{\mathrm{d}\boldsymbol{u}_i}{\mathrm{d}t} = m_i \boldsymbol{a}_i = \sum_j F_{ij} \boldsymbol{s}_{ij} = -\sum_j \Gamma_{ij} \boldsymbol{d}_i \quad \text{Newton Law}$$

$$F_{ij} = -\frac{\partial \varphi_{ij}}{\partial r_{ij}} \quad \text{Interatomic Force and Potential}$$

$$\Gamma_{ij} = \left.\frac{\partial^2 \varphi_{ij}}{\partial r_{ij}^2}\right|_e \quad \text{Spring Force Constant (Harmonic Approximation)}, \quad F_{ij} = -\Gamma_{ij} d_{ij}$$

$$\frac{d^2 \boldsymbol{x}_i}{dt^2} = -\frac{1}{m_i} \sum_j \nabla \varphi_{ij} \quad \text{Position-Potential Relation}$$

φ_{ij} Interatomic Potential from Solution to Schrödinger Equation (*ab initio* Calculations) or from Potential Models

$$N\langle \varphi \rangle = \frac{1}{2} \sum_{i,j} \varphi_{ij} \quad \text{Total Potential Energy}$$

Figure 2.2. Interatomic forces F_{ij} acting on atom i and presentation of these forces by the interatomic potentials. The net effect of these forces determines the classical (Newton) dynamics of the atom (particle) i, while the electronic structure of atom i and its neighbors determine the interatomic potential (fine-structures from the Schrödinger equation). $\langle \varphi \rangle$ is the effective potential and $N\langle \varphi \rangle$ is the total potential energy.

interpenetration and repulsion of electron clouds (Pauli exclusion principle). The force has two components, one repulsive and dominant at only very small distances, and the other attractive and acting at all distances. The sum of these forces between two particles is thus

$$F_{12} = \frac{1}{4\pi\epsilon_o} \frac{q_1 q_2}{r_{12}^2} + b \quad \text{attractive and repulsive forces,} \tag{2.6}$$

where b is a positive, repulsive force that noticeably acts at only very small distances.

At distances greater than one particle diameter, intermolecular forces draw particles together as part of the natural tendency for the system to seek the lowest potential energy state possible for the pair (or collection) of particles. As mentioned, the attractive force becomes stronger until the particles nearly touch, which means the potential energy also keeps decreasing until this point. Figure 2.2 shows the interatomic forces acting on atom i and the resulting motion from the net force on this atom (particle). When the distance between atom/particle i and atom/particle

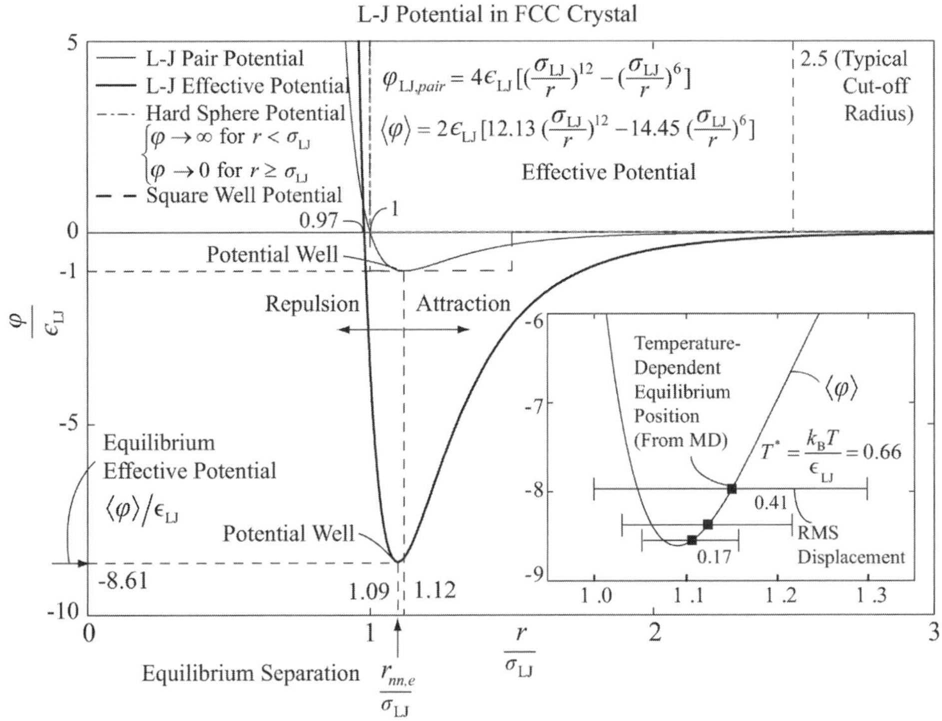

Figure 2.3. Variation of the scaled L–J interatomic potential with respect to scaled separation distance for Ar. The effective $\langle\varphi\rangle$ (per atom) potential indicating the cumulative potential that is due to presence of all neighboring atoms (FCC arrangement) is also shown. The hard-sphere and square-well approximations are also shown. The RMS displacement of each atom in the crystalline solid phase, at three different reduced temperatures $T^* = k_B T/\epsilon_{LJ}$ (scaled with $\epsilon_{LJ}/k_B = 120.7$ K for Ar), are shown around the equilibrium (e) nearest-neighbor spacing $r_{nn,e}$, for solid Ar [237]. The constants ϵ_{LJ} and σ_{LJ} for noble gases are listed in Table 2.2 in Section 2.2.

j, r_{ij} [called the normal coordinate (Glossary)], is less than one particle diameter, the force becomes repulsive and the potential energy begins to increase, meaning the system is again deviating from its ideal, lowest energy state. The interatomic potential is determined by solving the Schrödinger equation (Section 2.2.2). The force (spring) constant is related to the second derivative of the interatomic potential along the normal coordinate (at equilibrium location), and the interatomic force is related to the first derivative.

For all types of interatomic forces, the potential energy varies with distance between the interacting bodies r. An example of a typical potential energy $\varphi(r)$ curve (often called the potential energy well) is shown in Figure 2.3 for a FCC crystal, using the Lennard–Jones (L–J) potential model. The depth of the well is the energy that must be added to increase the distance of separation between particles until the force between them is nearly zero. Most interatomic potentials are characterized by a potential well of similar shape. However, the depth of the well increases with the strength of the force. We discuss interatomic potentials in Section 2.2.3. The effective, or net, potential $\langle\varphi\rangle$ that a particle experiences when surrounded by many others is discussed in Section 2.5.3. Because the potential is shared by adjacent particles, $\langle \varphi \rangle$ is the sum of the potentials between all interacting pairs divided by the total number of particles in the system N.

For a system in which there is a very large separation between atoms, molecules, or ions, there is a state of deaggregation. The gas phase of a pure substance is an example of a deaggregated state. At the minimum level of separation distance, the particles nearly touch, and the attractive forces are maximized while potential energy is at a minimum. This is an aggregated state (e.g., the solid phase of a pure substance). In an ionic compound, such as KBr, this corresponds to a situation in which positive ions are surrounded by equally spaced negative ions. In a covalent compound, such as H_2O, this state occurs when the hydrogen atoms are covalently bonded to the oxygen atom at a distance corresponding to the lowest point of the well and are separated from each other by an angle of 104°.

A collection of atoms or molecules seeks to minimize potential energy by aggregating in such a way that a particular atom/molecule nearly touches as many of its neighbors as possible. This organization is referred to as close packing, and characterizes the solid state of a pure substance (not all solids are chose packed, there is also an amorphous phase that has no long-range order). Note that the potential energy increases when an aggregated state is converted to a deaggregated state. The phases (solid, liquid, and gas) of a pure substance fit the aggregation/deaggregation model very well. In the solid phase, the fundamental units of a substance are packed together in an orderly, often repeating, arrangement (or lattice) that minimizes potential energy. In the gaseous phase, the fundamental units are widely separated and experience only very weak forces from one another. The liquid phase has distances of molecular separation and potential energy only somewhat greater than those of the solid, but no long-range order.

The inset in Figure 2.3 shows equilibrium separation (interatomic) distance at three different dimensionless temperatures. Note that the equilibrium interatomic distance increases with temperature. Also note that in the limit of $T \to 0$, the displacement becomes nearly symmetric. This is the harmonic limit of the displacements. In many analytical treatments (Sections 2.5.3, 2.6.4, and 4.4) this harmonic approximation is used.

Figure 2.4 gives examples of intraatomic and interatomic forces and distances. Refining to Figure 1.9, the L range here covers *Ab initio*, and MD treatments. For all cases shown, the gravitational force is less than 10^{-33} N. For electron–proton interactions, the weak nuclear force is 10^{-37} N, and the strong nuclear force is negligible compared with that. Note that atomic force microscopy techniques generate forces smaller than those holding together some solids, allowing them to be used without damaging those materials. For interatomic interactions, the length given is the nearest neighbor distance. The data for quartz and argon are obtained from MD simulations. The calculation of potential energy as FL (force times separation) is an estimate of the potential energy scale for a pair of particles [based on (2.5)]. A calculation exhibiting better accuracy would need to consider how the force changes as a function of the separation [such as (2.6)]. For example, the energy required to

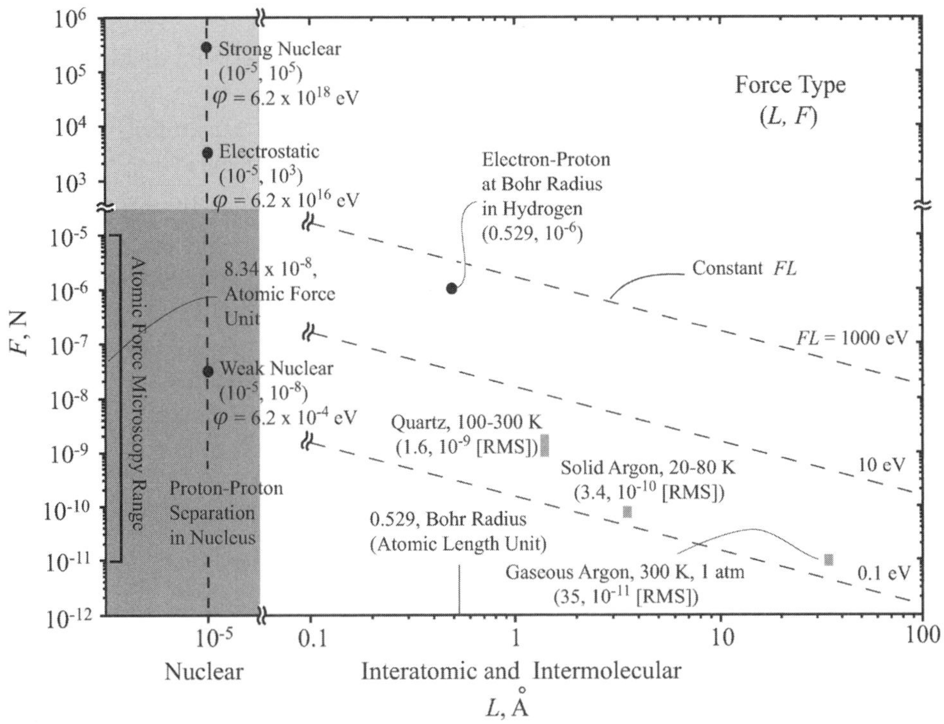

Figure 2.4. Forces at nuclear, interatomic, and intermolecular length scales. For demonstration of the potential energy, a linear relation FL is used to show equipotential lines.

remove the electron from a hydrogen atom is 13.6 eV, compared with the value of FL, which is 312 eV.

2.2 Orbitals and Interatomic Potential Models

2.2.1 Atomic and Molecular Electron Orbitals

The electronic orbitals are the wave-function solutions obtained from the Schrödinger equation (as will be discussed in Section 2.6.6) of the nucleus–electron electrodynamic (i.e., in motion) system. The arrangement of electrons in the orbitals of an atom (known as the electronic configuration) follows the Pauli exclusion principle. The configuration is specified by a combination of a letter and number for each sublevel, along with a superscript number to represent the number of electrons contained in the sublevel. The periodic table (Section 1.7 and Appendix A) is arranged based on these electronic configurations within orbitals. For example, $1s^1$ describes the state of the single electron in a hydrogen atom, which resides in the first s-orbital (a spherical shape). In geometric space, an orbital is the space having the highest probability of containing an electron.

Figures 2.5(a) to (e) show typical s, p, d, and f (two) orbitals for hydrogenlike atoms, for principal quantum numbers $n = 1, 2, 3$, and 4. The four quantum numbers are defined in Section 2.6.6. Every energy level has an s-sublevel (momentum

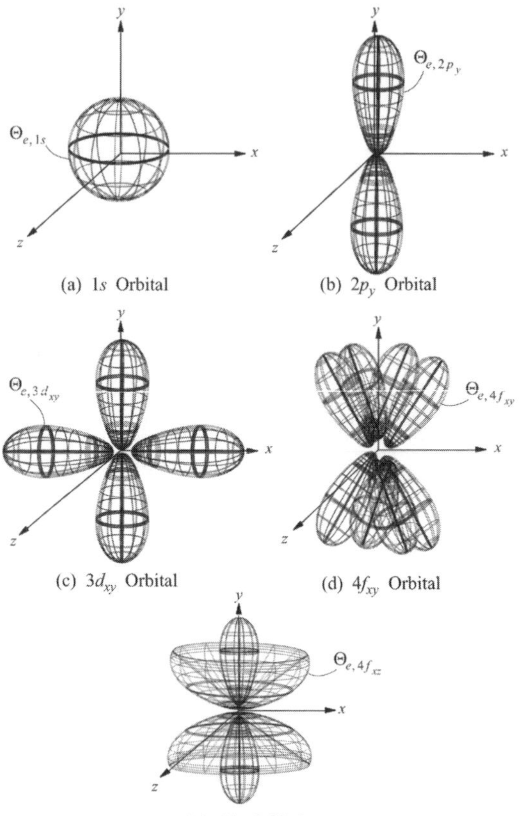

Figure 2.5. Three-dimensional angular part of probability (orbital) of the hydrogenlike atom electron wave function represented as $\psi_e = R_e(r)\Theta_e(\theta, \phi)$, for (a) 1$s$-, (b) 2$p$-, (c) 3$d$-, and (d), and (e) 4$f$- orbitals around the nucleous.

or azimuthal quantum number $l = 0$) containing one s orbital (magnetic quantum number $m = 0$). The fourth quantum number is the spin quantum number. The first energy level has a 1s-orbital, the second a 2s orbital, etc. The geometry of s orbitals is spherical, and the higher the energy level, the larger the radius. The charge density of the electron cloud generally increases the closer it is to the nucleus. Above the first energy level, s orbitals have radial nodes (regions inside the orbital with zero charge density). A 2s orbital has one radial node, a 3s orbital has two, etc. Each orbital has one less radial node than the number of energy levels. The second energy level and higher have p sublevels (azimuthal quantum number $l = 1$), each containing three p orbitals. There are two lobes for each p orbital, separated from each other by a planar node passing through the nucleus. The three p orbitals are mutually perpendicular and are denoted as p_x, p_y, and p_z. The third and higher energy levels have d sublevels (quantum number $l = 2$), each containing five d orbitals ($m = 0, +1, -1, +2$, and -2), and four of the five d orbitals are identical in shape, having four lobes. These orbitals are designated as d_{xy}, d_{yz}, d_{xz}, and $d_{x^2-y^2}$. The fifth orbital consists of a torus with two lobes perpendicular to the plane of the torus. The fourth and higher energy level have f sublevels (magnetic quantum number

$l = 3$), each containing seven f orbitals ($m = 0, +1, -1, +2, -2, +3$, and -3). The geometries of the f orbitals are more complex, as shown in Figures 2.5(d) and (e). These figures in general represent the angular part of the hydrogen wave function Θ_e, where $\psi_e = R_e(r)\Theta_e(\theta, \phi)$. They show a probability for the specific orbital. We will discuss this further in Section 2.6.6. Table A.2 (periodic table) in Appendix A gives the electron orbitals of the elements.

Molecular orbitals represent electron positions in molecules (which contain more than one nucleus). The probability function shows that the electron density is the highest in the space between adjacent nuclei.

When two atoms are brought close together, an electron can simultaneously experience a strong attraction to both nuclei, leading it to exist in a molecular orbital that is distributed between the two atoms, producing the chemical bond. A molecular orbital is modeled as a linear combination (overlap) of the wave functions for two (or more) atomic orbitals. If the two wave functions have the same sign in the region of overlap, a constructive interaction between the two wave functions occurs and a bonding orbital is produced. If the two wave functions have opposite signs in the region of overlap, a destructive interaction between the two wave functions occurs and an antibonding orbital is produced. For example, in H_2, the sigma bond (i.e., a single lobe of one electron orbital overlaps a single lobe of the other electron orbital) is formed by the overlap of two $1s$ orbitals. Pi bonds are formed where two lobes of one electron orbital overlap two lobes of the other electron orbital (for example, B_2H_2). Pi bonds are weaker than sigma bonds. A single bond is always a sigma bond. A double bond (e.g., C_2H_4) is made up of a sigma bond and a pi bond. A triple bond (e.g., HCN) contains a sigma bond and two pi bonds.

2.2.2 *Ab Initio* Computation of Interatomic Potentials

The interaction between electrons, as atoms are brought in close contact, is treated by use of various approximations of the potential force appearing in the Schrödinger equation (electronic structure theory) [9]. As will be discussed in Section 2.6.4, the Schrödinger equation contains a potential energy that here is the Coulombic repulsion/attraction (2.6). For an electron the charge is $-e_c$ and for the nucleus it is ze_c, where z is the atomic number.†

The *ab initio* calculations, such as those described in the Gaussian series of computational chemistry computer programs [108], determine the potential and electron cloud densities by using the Hartree–Fock self-consistent field algorithm. The

† The system potential energy (N_e electrons and N_n nuclei) electron–nuclear attraction, the electron–electron repulsion, and the nuclear–nuclear repulsion are given by [108]

$$\varphi_{e\text{-}n} + \varphi_{e\text{-}e} + \varphi_{n\text{-}n} = \frac{1}{4\pi\epsilon_o}\Big(\underbrace{-\sum_i^{N_e}\sum_I^{N_n}\frac{z_I e_c^2}{\Delta r_{iI}}}_{\text{electron–nuclear}} + \underbrace{\sum_i^{N_e}\sum_{j<i}^{N_e}\frac{e_c^2}{\Delta r_{ij}}}_{\text{electron–electron}} + \underbrace{\sum_I^{N_n}\sum_{J<I}^{N_n}\frac{z_I z_J e_c^2}{\Delta r_{IJ}}}_{\text{nuclear–nuclear}}\Big),$$

where z is the atomic number and Δr is the absolute value of separation distance between the two charged entities.

Hartree–Fock method is an approximation to the electronic Schrödinger equation. In this approximation, the exact Hamiltonian (the sum of kinetic and potential energies) is replaced with the Fock operator. This is an approximation describing the interaction of each electron with the average electronic field (rather than with every other electron separately). We will discuss the solutions to the Schrödinger equation, including the density functional theory (DFT), in relation to electronic band structures in Section 5.5.

Figure 2.6(a) shows an O-O bond calculated with the Gaussian.‡ As discussed in Section 1.5.2, we use the atomic length unit, the Bohr radius $r_B = h_P^2/4\pi^2 m_e e_c^2 =$ 0.52917725 Å, and energy unit e_c^2/r_B (a hartree), 27.2114 eV = 4.35975×10^{-18} J. Atomic oxygen has the electronic configuration $1s^2 2s^2 2p^4$ (Table A.2).

After obtaining the *ab initio*, results for $\varphi(r)$ are obtained by use of several existing potential models. These contain constants that are fitted by the use of least-square fit. The models (from Table 2.1) considered here are the Morse, expanded harmonic, and L–J potentials, i.e.,

$$\text{Morse potential} \quad \varphi = \varphi_o\{[1 - e^{-a_o(r-r_o)}]^2 - 1\},$$

$$\text{expanded harmonic potential} \quad \varphi = \frac{1}{2}\Gamma_a(r - r_o)^2 + \frac{1}{6}k_1(r - r_o)^3 + \frac{1}{12}k_2(r - r_o)^4,$$

$$\text{L–J potential} \quad \varphi = \frac{A}{r^{12}} - \frac{B}{r^6}. \tag{2.7}$$

Here we have used Γ instead of k as the spring constant. Note that, compared with the L–J model given in Figure 2.2, here $\sigma_{LJ} = (B/A)^{1/6}$, and $\epsilon_{LJ} = A^2/2B$.

For the O–O bond considered here, we have

$$\text{Morse potential } \varphi_o = 5.14228 \text{ eV},\ a_o = 2.87104 \text{ Å}^{-1},\ r_o = 1.17618 \text{ Å},$$

$$\text{expanded harmonic potential } \Gamma_a = -1.1695\times10^{-6} \text{ eV-Å}^{-2},\ k_1 = 276.678 \text{ eV-Å}^{-3},$$

$$k_2 = 518.141 \text{ eV-Å}^{-4},\ r_o = 2.02355 \text{ Å},$$

$$\text{L–J potential } A = 2.7602 \text{ eV-Å}^{-12},\ B = 5.79201 \text{ eV-Å}^{-6}.$$

(

‡ The input begins with the commend line scan requested followed by a file title line. The third input line begins specification of particular species, beginning with the number of charge on the molecule (zero in this case) and then the spin multiplicity (one here). The fourth line sets oxygen as the first atom. In the fifth, the second atom is set as oxygen and is bonded to the first atom in the molecule "1" at a distance r from the first. This distance is variable as defined by the sixth (r as variable) and seventh (range of r starting from 0.8 Å, with 37 steps of size 0.03 Å) lines. The self energy is -150.12056 hartree. Note that 1 hartree = 27.2114 eV. The input file ⟨⟩ to Gaussian is listed subsequently (no entry implies blank line).

1. ♯ T B3LYP/6-311+=g(3df,3pd) scan test
2.
3. OO scan
4.
5. 0 1
6. O
7. O 1 r
8. Variable
9. r 0.8 37 0.03
10. add a blank line

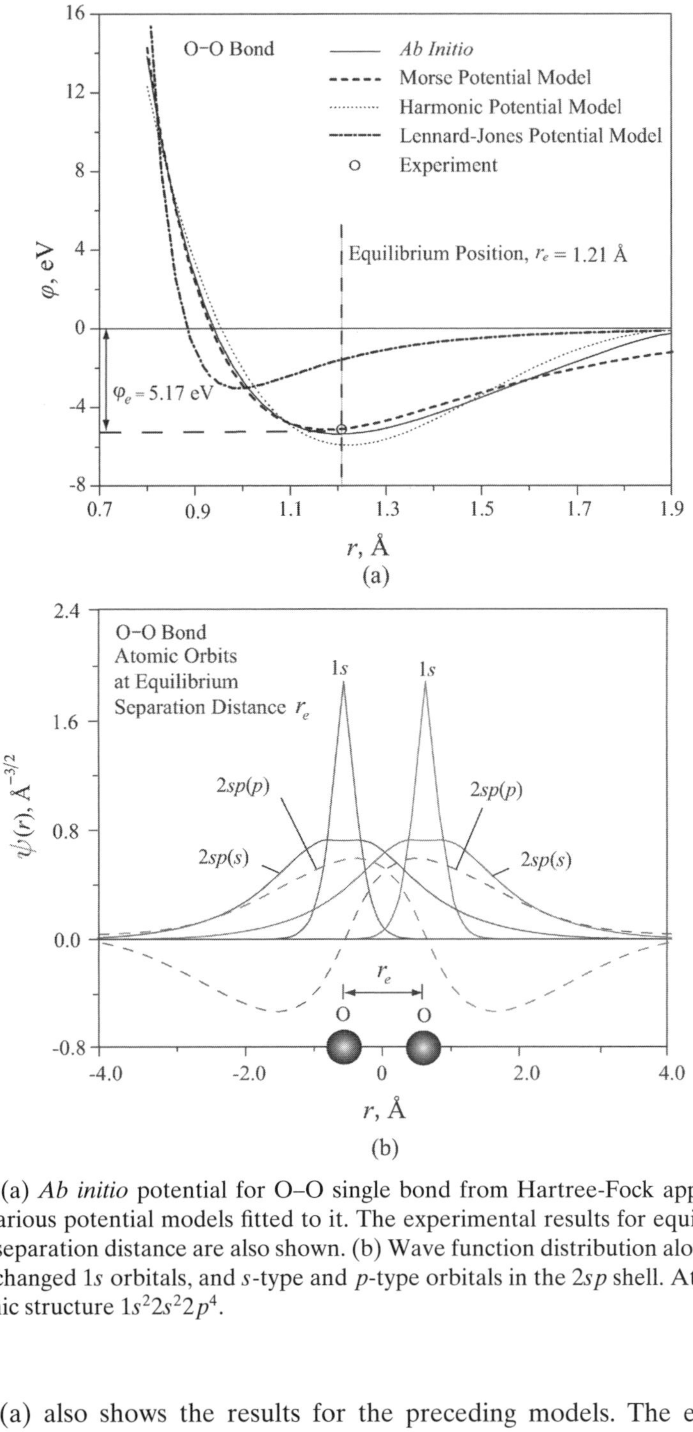

Figure 2.6. (a) *Ab initio* potential for O–O single bond from Hartree-Fock approximation [108] and various potential models fitted to it. The experimental results for equilibrium potential and separation distance are also shown. (b) Wave function distribution along the O–O axis, for unchanged 1*s* orbitals, and *s*-type and *p*-type orbitals in the 2*sp* shell. Atomic O has the electronic structure $1s^2 2s^2 2p^4$.

Figure 2.6(a) also shows the results for the preceding models. The experimental result for the equilibrium bond length r_e (1.21 Å), and bond energy φ_e (5.17 eV) are also shown and are in good agreement with the *ab initio* results (some other bonds are listed in Table 2.3 and will be discussed in Section 2.2.4). The closest fit to the *ab initio* data, over the entire range, is the Morse potential.

In addition, this potential matches the magnitude and location of the potential well.

Figure 2.6(b) shows the wave function distributions of the oxygen atomic orbits along the O–O axis.† The results are for the equilibrium separation distance. The $1s$ orbital is a core orbital and the overlap between the two $1s$ orbitals is small. The $2s$ orbital has symmetry around the nucleus centers. However, the $2p$ has an anti-symmetry around the nucleus centers. Thus the possibility of having an electron located between the two nuclei is high.

† In molecular physics, the atomic orbits are often described by a set of Gaussian-type orbitals (GTOs) for fast calculations, i.e.,

$$\psi_\mu = \sum_i^N d_{\mu,i} g_i,$$

where the coefficients $d_{\mu,i}$ are called the contraction coefficients, and g_i is a GTO.
The GTO (also called the Cartesian Gaussian or Gaussian primitive function) is expressed as

$$g(\alpha, l, m, n, f; x, y, z) = c\, e^{-\alpha f^2 r^2} x^l y^m z^n.$$

Here α is called the exponent, f is a scale factor (determined by the basis), and c is a normalization constant determined by f and α. Also x, y, z are Cartesian coordinates, and the summation of their exponents $L = l + m + n$ is used to mark functions as s-type ($L = 0$), p-type ($L = 1$), d-type ($L = 2$), and so on. For example, the following GTOs are normally used

$$\begin{aligned}
1s &= c\, e^{-\alpha f^2 r^2} \\
2p_x &= c\, e^{-\alpha f^2 r^2} x \\
2p_y &= c\, e^{-\alpha f^2 r^2} y \\
2p_z &= c\, e^{-\alpha f^2 r^2} z.
\end{aligned}$$

Gaussian03 will output the components and contraction coefficients when the keyword "GFPrint" appears in the command line. The output of Gaussian with the basis of 6-31G is as follows

```
 1. Atom O1 Shell 1 S 6
 2.        0.5484671660D+04    0.1831074430D-02
 3.        0.8252349460D+03    0.1395017220D-01
 4.        0.1880469580D+03    0.6844507810D-01
 5.        0.5296450000D+02    0.2327143360D+00
 6.        0.1689757040D+02    0.4701928980D+00
 7.        0.5799635340D+01    0.3585208530D+00
 8. Atom O1 Shell 2 SP 3
 9.        0.1553961625D+02   -0.1107775495D+00    0.7087426823D-01
10.        0.3599933586D+01   -0.1480262627D+00    0.3397528391D+00
11.        0.1013761750D+01    0.1130767015D+01    0.7271585773D+00
12. Atom O1 Shell 3 SP 1
13.        0.2700058226D+00    0.1000000000D+01    0.1000000000D+01
```

Here, lines 1, 8, and 12 start with the atom index, followed by the index of the electronic shell, shell type, and the number of primitive functions. In the lines following them, the first column shows the exponents, the second column shows the contraction coefficients for s-type GTOs, and the third column shows the contraction coefficients for p-type GTOs. The shell type "SP" means that the s-type GTOs and p-type GTOs share the same exponents. The summation of the GTOs for an orbit gives the wave function of the orbit (note that the wave function should be normalized to make sure the integral of the square of it over the space is equal to 1.0)

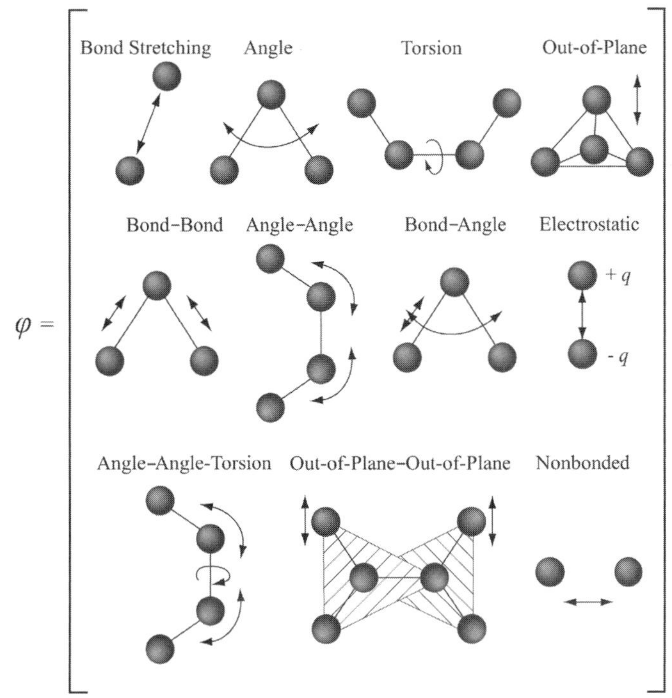

Figure 2.7. Various atomic motions contributing to the interatomic potential.

2.2.3 Potential Models

Table 2.1 lists some of the most commonly used interatomic potentials [116]. A review of various potentials is given in [70]. In addition to the motion along the bond axis shown in Figure 2.6, for a diatomic molecule, there are torsional and various other in-phase and out-of-plane motions. Figure 2.7 lists 11 such motions. These are bond stretching, angle, torsion, out-of-plane, bond–bond, angle–angle, bond–angle, electrostatic, angle–angle-torsion, out-of-plane–out-of-plane, and nonbonded. The complete interatomic potential (not the form always used) would contain the contribution from all of these motions, when applicable. The L–J potential is an effective potential that describes the interaction between two uncharged molecules or atoms (Figure 2.4). This potential is mildly attractive as two uncharged molecules or atoms approach one another from a distance, but strongly repulsive when they come too close to each other. At equilibrium, the pair of atoms or molecules tends to go toward a separation distance corresponding to the minimum of the L–J potential.

London dispersion forces exist between nonpolar molecules that would seem to have no basis for attractive interactions. However, gases of nonpolar molecules can be liquefied, indicating that, if the thermal (kinetic) energy is reduced to a sufficiently low level, an attractive force can dominate. The London forces arise from the motions of electrons within an atom or nonpolar molecule that can produce a fluctuating dipole moment. This causes neighboring atoms to be attracted to one another, but is significant only when the atoms are close together. The ease with which an

Table 2.1. *Some interatomic potential models (two, three, and four-body, including bond angles) [116].*

Potential designation	Model	Units for parameters
Buckingham	$A\exp(-r/r_o) - Cr^{-6}$	A in eV, r_o in Å, C in eV-Å^6
L–J (combination rules permitted)	$Ar^{-m} - Br^{-n}$ or $\epsilon_{LJ}[c_1(\sigma/r)^m - c_2(\sigma_{LJ}/r)^n]$ $c_1 = [n/(m-n)](m/n)^{[m/(m-n)]}$ $c_2 = [m/(m-n)](m/n)^{[m/(m-n)]}$	A in eV-Å^m, B in eV-Å^n ϵ_{LJ} in eV, σ_{LJ} in Å
Expanded harmonic (k_3 and k_4 optional)	$(1/2)\Gamma_a(r-r_o)^2 + (1/6)k_1(r-r_o)^3 + (1/12)k_2(r-r_o)^4$	Γ_a in eV Å^{-2}, r_o in Å, k_1 in eV-Å^{-3}, k_2 in eV-Å^{-4}
Born harmonic	$(1/2)\Gamma(r-r_o)^2$	Γ in eV-Å^{-2}
Morse	$\varphi_o(\{1-\exp[-a_o(r-r_0)]\}^2 - 1)$	φ_o in eV, a_o in Å^{-1}, r_0 in Å
Spring (core-shell)	$(1/2)\Gamma_a r^2 + (1/24)k_1 r^4$	Γ_a in eV-Å^{-2}, k_1 in eV-Å^{-4}
General	$A\exp(-r/r_o)r^{-m} - Cr^{-n}$	A in eV-Å^m, r_o in Å, C in eV-Å^n
Two-body Stillinger–Weber	$A\exp[r_o/(r-r_{max})](Br^4 - 1)$	A in eV-Å, r_o in Å, B in eV-Å^4
Three-body Stillinger–Weber	$\varphi_o\exp[r_o/(r_{12}-r_{max}) + r_o/(r_{13} - r_{max})][\cos(\theta_{213}) - \cos(\theta_0)]^2$	φ_o in eV, r_o in Å
Expanded three-body harmonic	$(1/2)k_1(\theta-\theta_o)^2 + (1/6)k_2(\theta-\theta_o)^3 + (1/12)k_3(\theta-\theta_o)^4$	k_1 in eV-rad^{-2}, θ_o in degrees, k_2 in eV-rad^{-3}, k_3 in eV-rad^{-4}
Expanded three-body harmonic + exponential	$(1/2)k_1(\theta_{213}-\theta_0)^2 \times \exp(-r_{12}/r_o)\exp(-r_{13}/r_o)$	k_1 in eV-rad^{-2}, θ_0 in degrees, r_o in Å
Axilrod–Teller	$k[1 + 3\cos(\theta_{213})\cos(\theta_{123})\cos(\theta_{132})] /(r_{12}r_{13}r_{23})^3$	k in eV-Å^9
Three-body exponential	$\varphi_o\exp(-r_{12}/r_o)\exp(-r_{13}/r_o)\exp(-r_{23}/r_o)$	φ_o in eV, r_o in Å
Urey–Bradley	$(1/2)\Gamma(r_{23}-r_0)^2$	Γ in eV-Å^{-2}, r_o in Å
Four-body	$\varphi_o[1+\cos(n\phi - \phi_o)]$	φ_o eV, ϕ_o in degrees
Ryckaert–Bellemans	$\sum_{n=1}^{N} \varphi_n(\cos\phi)^n$	φ_n in eV

θ_{ijk} represents the angle between the two interatomic vectors i-j and j-k and ϕ_{ijk} is the torsional angle between the planes ijk and jkl. We will use Γ instead of k for the spring constant, since k is used for the thermal conductivity.

Table 2.2. *L–J potential model constants for noble-gas element atomic pairs [9, 183].*

Noble-gas pair	ϵ_{LJ}, eV	σ_{LJ}, Å	$r_{nn,e}$, Å
Ne–Ne	0.0031	2.74	3.13
Ar–Ar	0.0104	3.40	3.76
Kr–Kr	0.0140	3.65	4.01
Xe–Xe	0.0200	3.98	4.35

external electric field can induce a dipole (alter the electron and charge distribution) within a molecule is referred to as the polarizability of that molecule. The greater the polarizability of a molecule, the easier it is to induce a momentary dipole and the stronger the dispersion forces. The larger molecules tend to have greater polarizability because their electrons are farther away from the nucleus (any asymmetric distribution produces a larger dipole due to larger charge separation), and their number of electrons is greater (higher probability of asymmetric distribution).

Hydrogen bonding forces arise when a hydrogen atom in a polar bond (e.g., H–F, H–O or H–N) experiences an attractive force with a neighboring electronegative molecule or ion that has an unshared pair of electrons (usually a F, O or N atom on another molecule). Hydrogen bonds are considered to be dipole–dipole-type interactions. The hydrogen atom has no inner core of electrons, so the side of the atom facing away from the bond represents a virtually naked nucleus (a proton). The positive charge of this electron-deficient side is attracted to the negative charge of an electronegative atom in a nearby molecule. Water is unusual in its ability to form an extensive hydrogen bonding network.

The simple interatomic potential (for van der Waals forces) of a pair of atoms/mole-cules following the L–J potential model is given by [similar to that given in (2.7)]

$$\varphi_{LJ} = 4\epsilon_{LJ}[(\underbrace{\frac{\sigma_{LJ}}{r})^{12}}_{\text{repulsion}} - (\underbrace{\frac{\sigma_{LJ}}{r})^{6}}_{\text{attaction}}] \quad \text{L–J potential.} \tag{2.9}$$

In Figure 2.3, the effective L–J potential is that which results when interactions with all the atoms in a FCC lattice are taken into account. The inset of Figure 2.3 shows the equilibrium position and the RMS displacement obtained by MD simulations for three temperatures. The maximum temperature is close to the melting point (89.8 K for molecular dynamics). Note that, as the temperature increases, the solids generally expand (increasing the atomic spacing). The cutoff used is $r/\sigma_{LJ} = 2.5$, as shown in the figure. Values of the L–J scales for some noble gas element pairs are given in Table 2.2.

The temperature scale is ϵ_{LJ}/k_B. More information on the L–J potential can be found in [9]. The hard-sphere and square-well potentials are not appropriate in MD simulations because they are not differentiable, as needed to calculate forces. They

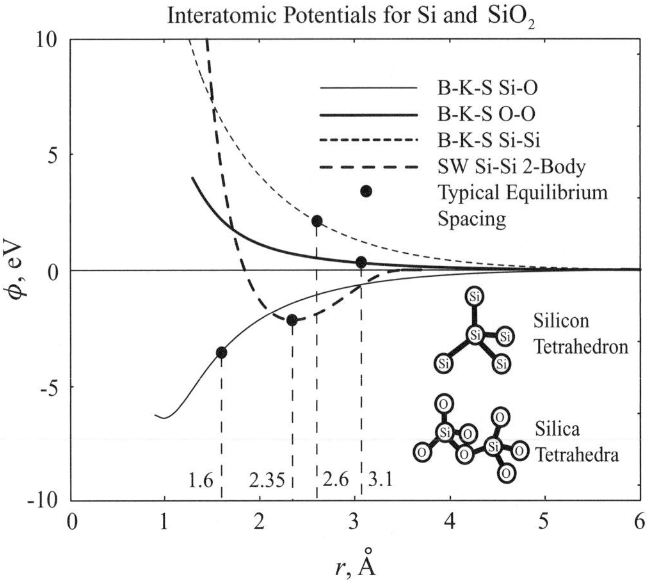

Figure 2.8. Variation of the interatomic potential for silicon and silica (SiO_2) with respect to separation distance.The B–K–S potential model for Si–O, O–O, Si–Si, and the S–W two-body Si–Si potential model are shown. Typical equilibrium spacings are also shown [235].

are more commonly used in Monte Carlo simulations, for which only energy calculations are required. The choice of the transition lengths in these potentials may depend on the system of interest. Here, they are chosen at $r/\sigma_{\rm LJ}$ values of 1 (hard sphere and square well) and 1.5 (square well). The distance to nearest neighbors in the FCC solid state is also listed.

As another example of interatomic potential, consider a polyatomic molecule such as SiO_2. The van Beest–Krammer–van Staten (B–K–S) interatomic potential between atom i and j separated by a distance r is [334]

$$\varphi = \frac{q_i q_j}{r} + Ae^{-r/r_{\rm o}} - \frac{C}{r^6} \quad \text{B-K-S potential,} \tag{2.10}$$

shown in Figure 2.8. To match the dynamic behavior and potential of SiO_2 crystals, further refinement to this potential is made to arrive at [235]

$$\varphi = \frac{q_i q_j}{r}\text{erf}(\alpha r) + Ae^{-r/r_{\rm o}} - \frac{C}{r^6} + 4\epsilon[(\frac{\sigma}{r})^{24} - (\frac{\sigma}{r})^6] \quad \text{modified B-K-S potential.} \tag{2.11}$$

This is a combined Wolf [350] and modified L–J (2.9). Only the two-body term is shown in Figure 2.8 for the Stillenger–Weber (S–W) model (Table 2.1) for silicon [320]. For the B–K–S silica curves [192, 334], the Wolf method [350] has been used to model the electrostatic interactions with an α value of 0.431 Å^{-1}. The cutoff radius used is 6.44 Å. The Si–O and O–O curves show nonphysical behavior below the minimum value of the distance plotted. However, this atomic separation falls below anything encountered in the solid state. If liquid silica is modeled, an L–J

24–6 potential must be added to the B–K–S potential to prevent the system from blowing up [133]. Note the different behaviors of the Si–Si curves. This is because the Si–Si interactions in silicon are covalent, whereas in silica they are ionic. The energies associated with the silicon and silica potentials are on the order of 100 times greater than those encountered in the L–J potential. This is an indication of the large strength of covalent and ionic bonds compared with bonds that result from dipole fluctuations (i.e., London forces).

Various silica structures can be constructed from the preceding potentials, and Figure 2.9(a) shows some of these [235]. Quartz is an anisotropic crystal (a and c directions) and has an ordered tetrahedral building block. Amorphous silica (glass) has a random structure made of such tetrahedrons. Zeolites have cages (crystal pores), and some silica zeolites have sodalite (SOD) cage building blocks.

A combination of the formation process (from gas, liquid, or solid phase), atomic stoichiometry of the ingredients temperature and pressure, as well a presence of mediating chemical and physical agents (e.g., catalysts), determines the particular atomic arrangement. For example, Figure 2.9(b) shows the various solid-state structures of carbon (e.g., diamond, graphite, carbon nanotube, graphene, fullerene carbon, and soot). The diameters d of a single-walled nanotube, Buckyball (e.g., carbon cage or fullerene carbon), and soot are of the order of one nanometer and soots can conglomorate and become larger. Figure 2.9(c) shows the variation of phonon (equal to the total thermal conductivity, i.e., no electron contribution) thermal conductivity of carbon in diamond, graphite (which is an anisotropic crystal), Buckminister fullerene solid, and amorphous solid phases, with respect to temperature [237]. Diamond has the overall tightest (smallest average interatomic spacing) packing among the crystalline phases shown, and has the highest thermal conductivity. These show the dominant role of molecular arrangement on the properties (including transport properties), and the ability to synthesize the desired structure (for function) from constituent atoms.

The properties of a nanotube depend on the wall (single-, double-, or multiple layer), chirality (Glossary), and diameter. We will discuss the temperature dependence of the crystalline and amorphous phase phonon conductivities and the crystalline atomic structures in Chapter 4.

Figures 2.10(a) and (b) show the atomic structure of some gases. Figure 2.10(a) shows the linear, trigonal, tetrahedral, trigonal bipyramidal and octahedral molecular geometries. Figure 2.10(b) shows the electron pairing and the bond angles for some gas molecules.

2.2.4 Examples of Atomic Bond Length and Energy

Table 2.3 lists the bond length r_e and energy φ_e for some atom pairs appearing in compounds of common chemical reactions of interest [298]. The bond lengths range from 1 to 3 Å, and the bond energies range from 1 to 10 eV. The nitrogen triple bond in N_2 is among the tightest and most energetic. Note that some of the bonds,

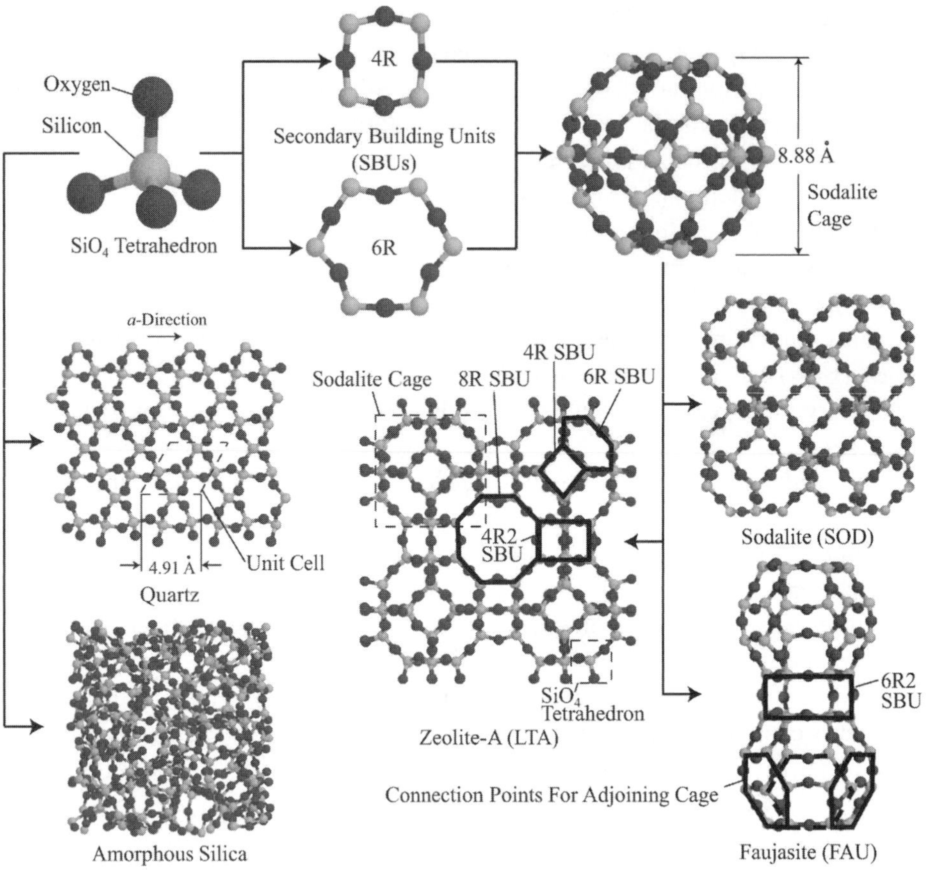

(a) Various SiO_2 (Silica) Structures

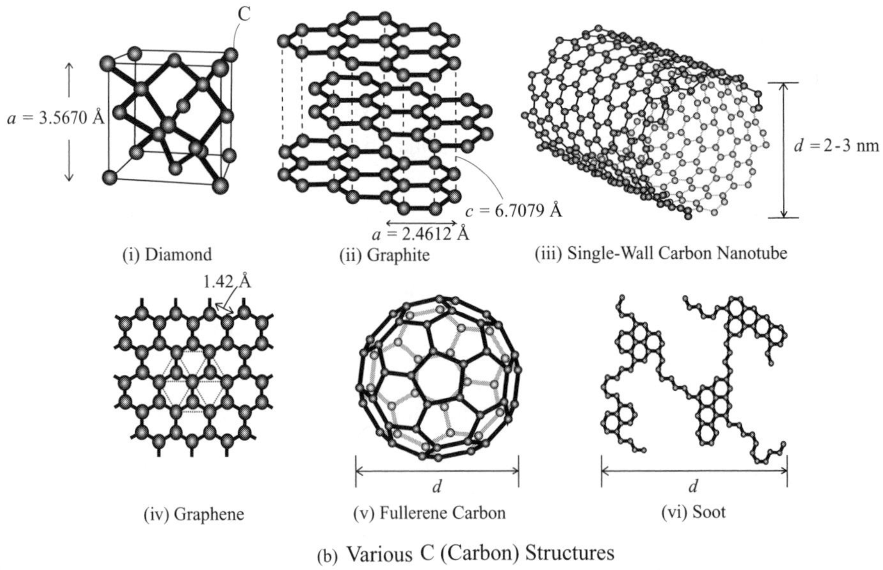

(b) Various C (Carbon) Structures

Figure 2.9. (a) Various solid-state phases (structures) of silica, and (b) various solid-state phases of carbon [237]. These include crystalline and amorphous phases.

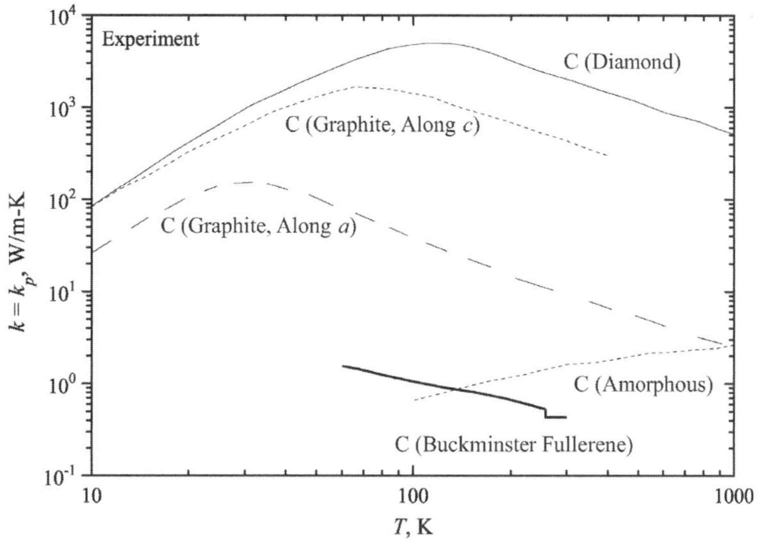

(c) $k_p(T)$ for Solid Carbon

Figure 2.9. (c) Variations of thermal conductivity of carbon solids with respect to temperature. The experimental results for diamond, graphite (anisotropic), fullerene, and amorphous phase are shown.

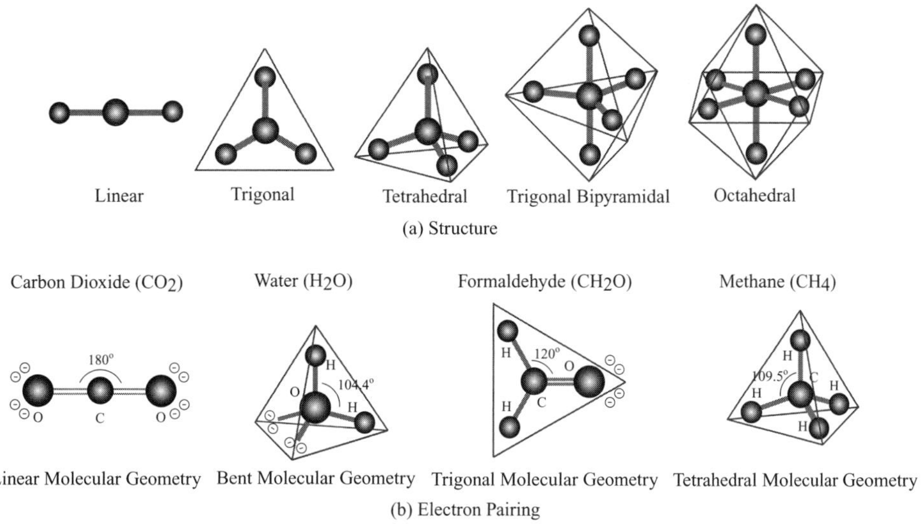

Figure 2.10. (a) Examples of classifications of gaseous molecular structures, and (b) electron pairing and bond angles for the same gas molecules in (a).

for example C=C, can have different lengths and energy levels when they appear in different compounds. Examples of weaker bonds are the noble-gas elements, for which the pair bonding energies (same as ϵ_{LJ}) are listed in Table 2.2, such as 0.0031 eV for Ne–Ne.

Table 2.3. *Atomic pair bond length r_e and bond energy φ_e for some typical strong bonds and their compounds, as encountered in common chemical reactions [298].*

Atom pairs	Compound	r_e, Å	φ_e, eV	Atom pairs	Compound	r_e, Å	φ_e, eV
Al–F	AlF	1.65	6.86	K–Br	KBr	2.82	3.95
Al–H	AlH	1.65	2.95	K–Cl	KCl	2.95	4.41
Al–O	AlO	1.62	5.02	K–I	KI	3.29	3.37
As–F	AsF_3	1.71	5.04	Mg–Cl	MgCl	2.18	4.22
B=O	BOF	1.23	9.00	Mg–F	MgF	1.75	5.72
B–Br	BBr	1.87	4.32	N≡N	N_2	1.1	9.80
B–Cl	BCl	1.76	5.22	N=O	NO	1.15	6.55
B–I	BI	2.1	3.54	Na–Br	NaBr	2.68	3.85
Br–Cl	BrCl	2.14	2.22	Na–Cl	NaCl	2.53	4.29
Br–F	BrF	1.76	2.58	N–F	NF	1.37	2.99
C≡C	C2H2	1.2	8.20	N–F	NOF	1.52	2.61
C≡O	CO	1.13	11.34	N–S	NS	1.5	4.99
C=C	CH_2CO	1.31	6.11	O=O	O_2	1.21	5.17
C=C	C_2	1.31	6.53	O–F	OF	1.42	2.00
C=O	CO_2	1.16	8.34	P≡P	P2	1.89	5.42
Ca–F	CaF	2.02	5.51	Pb–F	PbF	2.11	3.24
Ca–O	CaO	1.82	4.03	P–Br	PBr	2.23	2.75
C–C	CCl_3CHO	1.52	3.11	P–Cl	PCl	2.04	3.26
Cl–F	ClF	1.63	2.60	P–F	PF	1.59	4.87
Cl–O	ClO	1.55	2.80	P–H	PH	1.42	3.14
Cs–I	CsI	3.32	3.25	P–O	PO	1.45	5.22
Ga–H	GaH	1.58	2.84	P–S	PS	1.92	5.36
Ge–Br	GeBr	2.3	2.62	Rb–Br	RbBr	2.95	3.92
Ge–Cl	GeCl	2.08	3.56	Rb–F	RbF	2.87	5.02
Ge–F	GeF	1.68	5.07	Sb–Cl	$SbCl_3$	2.33	9.76
H–As	AsH_3	1.51	3.07	Si≡S	SiS	1.93	6.66
H–C	CH_2CO	1.08	4.28	Si=Si	Si_2	2.25	3.29
H–H	H_2	0.74	4.52	Si–Cl	SiCl	2.0	4.02
H–O	H_2O	0.96	4.80	Si–Cl	SiC	1.65	4.48
H–O	OH	1.03	4.44	Si–F	SiF	1.6	5.72
H–S	HS	1.36	3.78	Si–N	SiN	1.57	4.58
I–Br	IBr	2.45	1.84	Sn–H	SnH	1.79	3.21
I–Cl	ICl	2.3	2.18	Te–Se	TeSe	2.5	2.30
I–F	IF	1.91	2.91	Tl–F	TlF	2.08	4.60
In–H	InH	1.84	1.99	Xe–O	XeO_3	1.76	1.31

2.2.5 Radial Distribution of Atoms in Dense Phase

The local density of a collection of particles (molecules) is given in terms of the radial distribution (RDF) function $g(r)$ as

$$\rho(r) = \langle\rho\rangle g(r), \quad \langle g(r)\rangle = \frac{1}{V}\int_0^r g(r)4\pi r^2 \mathrm{d}r, \quad \langle g(r)\rangle \to 1 \text{ for } r \to \infty. \tag{2.12}$$

Here the radial location r is measured from the center of a reference particle. For ideal gases, $g(r) = 1$, everywhere, whereas for nonideal gases, liquids, and solids

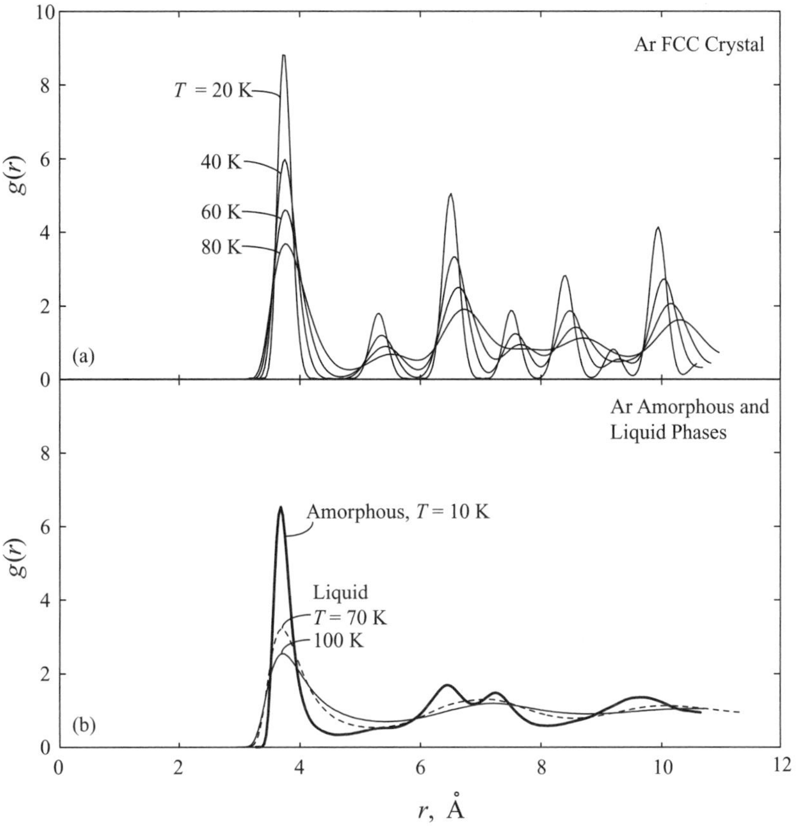

Figure 2.11. L–J radial distribution functions for (a) Ar FCC crystal at T = 20, 40, 60, and 80 K, (b) the amorphous phase at T = 10 K, and the liquid phase at T = 70 and 100 K [232].

$g(r)$ varies greatly over several intermolecular lengths (measured from the central particle).

The RDF describes the distribution of the atoms in a system from the standpoint of a fixed, central atom. Its numerical value at a position r is the probability of finding an atom in the region $r - \mathrm{d}r/2 < r < r + \mathrm{d}r/2$. The Fourier transform of the RDF is called the structure factor, which can be determined from scattering experiments. The RDFs for the Ar FCC crystal from MD simulations temperatures of 20, 40, 60, and 80 K, the amorphous phase at a temperature of 10 K, and the liquid phase at temperatures of 70 and 100 K are shown in Figure 2.11. The RDF can be determined only up to one half of the simulation cell size, which here is about 3.25 σ_{LJ} for the solid phases and slightly larger for the liquid.

The RDF of the Ar FCC crystal phase shows well-defined peaks that broaden as the temperature increases and the atomic displacements increase. Each peak can be associated with a particular set of nearest-neighbor atoms. The locations of the peaks shift to higher values of r as the temperature increases and the crystal expands. In the amorphous phase, the first peak is well-defined, but after that, the disordered nature of the system leads to a much flatter RDF. There is no order beyond a certain point. The splitting of the second peak is typical of amorphous phases, and consistent with the results for L–J Ar [207]. The presence of only short-range order is also seen in the liquid phase, in which only the first neighbor peak is well-defined. This is expected, because the physical size of the atoms defines the minimum distance over which they may be separated.

2.3 Molecular Ensembles, Temperature and Thermodynamic Relations

2.3.1 Ensembles and Computational Molecular Dynamics

The ensemble-average behavior of a collection of particles is used in statistical mechanics (Section 1.3) and allows for the determination of macroscopic properties of the system. Each ensemble includes a collection of positions and velocities of a given number of particles at a given energy and volume (for NVE ensemble, discussed in Section 2.5.1). It is possible for these same particles to have different positions and velocities, but the same macroscopic properties. This means the particles are allowed to differ microscopically, while retaining the same macroscopic, independent properties when looked at as a whole.

Because of the way an ensemble is constructed, if snapshots of all the possible configurations (different particle positions and velocities) is taken at some instant in time, we will find that they differ in the instantaneous values of their bulk properties. This phenomenon is called fluctuation. Thus the true value of any particular bulk property must be calculated as an average over all the possible microscopic arrangements of the ensemble. This is what is meant by an ensemble average, and the instantaneous values are said to fluctuate about the average value.

MD proceeds by numerically integrating the equations of motion of these particles. Each time step generates a new arrangement (or configuration) of the atoms and new instantaneous values for macroscopic properties such as temperature, pressure, and energy. Determining the thermodynamic values of these variables requires an ensemble average. In MD, this is achieved by finding the average over successive configurations generated by the simulation. It is assumed that an ensemble average (which relates to many configurations of the system) is the same as an average over time of one configuration (the system simulated). This assumption is known as the ergodic hypothesis and is generally true, provided a long enough time behavior is included in the average. The specifics of this hypothesis are given in Section 2.3.2. Other examples of properties that can be calculated as ensemble averages include the elastic properties of solids and time correlations.

2.3.2 Thermodynamic Relations

From the microscopic (statistical) perspective, temperature is the measure of the average kinetic energy of the molecules (or atoms) of a substance, i.e., the greater the kinetic energy, the higher the temperature. Note that this energy is solely due to thermal fluctuations of the constituent particles (the particular phase these particles assume is determined in part by their temperature). The absolute zero of temperature (0 K) is where the molecules have no motion in classical systems (there is a zero point motion in quantum mechanics). This definition of temperature, for a system of N particles, with individual particles having mass m_i and instantaneous speed u_i, as shown in Figure 2.2, is [326]

$$T \equiv \frac{\langle \sum_{i=1}^{N} m_i u_i^2 \rangle}{N_f k_\mathrm{B}}, \tag{2.13}$$

where N_f is the number of degree of freedom of the system of particles. This is based on the equipartition of energy allowing one $k_\mathrm{B} T/2$ for each degree of motion (freedom). This kinetic definition of temperature is an integral part of its of relation to the entropy S (a measure of multiplicity) of the system.

The equilibrium thermodynamic state of N particles in a volume V is given in terms of $E = E(S, V, N)$, and T is also related to the derivatives of these functions. We defined S in terms of the partition function Z in (1.5), and Z in turn is related to the energy (Table 2.4), so we have

$$S \equiv k_\mathrm{B} N \ln Z = k_\mathrm{B} N \ln \sum_{i=0}^{\infty} e^{E_i/k_\mathrm{B} T} + \frac{E}{T}, \tag{2.14}$$

where each E_i is an individual micro energy state of the system and E is the total internal energy. These are repeated in Table 2.4, along with the definition of N_f. In fluctuation theory, allowance is made for the anharmonic motions, and then the temperature is defined in terms of the anharmonicity parameter $\alpha(T)$.

Thermal equilibrium in a system of particles corresponds to its maximum entropy, and the rate of approach to that state is proportional to the difference in temperature between parts of the system. The equilibrium state of a collection of particles then becomes the state of greatest multiplicity, and the temperature defined in terms of this multiplicity (entropy) is [54]

$$T \equiv \frac{1}{\left.\dfrac{\partial S}{\partial E}\right|_{V,N}}, \quad p \equiv -\left.\frac{\partial E}{\partial V}\right|_{S,N}, \quad \mu \equiv \left.\frac{\partial E}{\partial N}\right|_{S,V}, \tag{2.15}$$

where we have also included the definitions of pressure and chemical potential μ. The macroscopic thermodynamic definition of temperature is expressed as the inverse of the rate of change of entropy with respect to internal energy, with the volume V and the number of particles N held constant. A few other useful thermodynamics relations are given in Table 2.4.

Table 2.4. *Microscopic (statistical) and macroscopic thermodynamic relations [326, 54].*

Microscopic (statistical) thermodynamic relations include the following:

Temperature, pressure and potential, and kinetic energy are related to individual particle (atom or molecule in motion)

$$T = \frac{\langle \sum_{i=1}^{N} m_i u_i^2 \rangle}{N_f k_B} = \frac{2E_k}{3(N-1)k_B} = \frac{\sum_{i=1}^{N} \frac{|\boldsymbol{p}_i|^2}{m_i}}{3(N-1)k_B}$$

$$p = \frac{Nk_B T}{V} + \frac{1}{3V}\left(\sum_{i=1}^{N}\sum_{j\neq i}^{N} \boldsymbol{x}_{ij} \cdot \boldsymbol{F}_{ij}\right)$$

$$E = E_k + E_p$$

$$E_k = \sum_{i=1}^{N} \frac{1}{2} m_i u_i^2, \quad E_p = \langle \varphi \rangle = \frac{1}{2}\sum_{i=1}^{N}\sum_{j\neq i}^{N} \varphi(\boldsymbol{x}_{ij}) \quad \text{for pair potential}$$

where N_f is number of degrees of freedom in the system [in MD, $N_f = 3(N-1)$, where N is the number of particles] for systems with $N \gg 1$, and $N-1$ is required for having zero momentum for the system. Entropy, internal energy, and pressure are related to the partition function Z (and Hamiltonian H)

$$Z = \iint \exp\left[\frac{-\mathrm{H}(\boldsymbol{x}, \boldsymbol{p})}{k_B T}\right] \mathrm{d}\boldsymbol{x}\, \mathrm{d}\boldsymbol{p}, \quad \mathrm{H} = \sum_l \mathrm{H}_l \quad \text{partition of energies and partition function}$$

$$Z = \frac{1}{h_{\mathrm{P}}^{3N} N!} \int \exp\left[\frac{-\mathrm{H}(1, \ldots, N)}{k_B T}\right] \mathrm{d}^{3N}x\, \mathrm{d}^{3N}p \quad \text{integral grand (canonical) partition function}$$

$$S = k_B N \ln Z = k_B N \ln \sum_i^{\infty} e^{E_i/k_B T} + \frac{E}{T}$$

$$E = \frac{\sum_i E_i e^{-E_i/k_B T}}{Z} = k_B T^2 N \frac{\partial \ln Z}{\partial T}\Big|_{N,V}$$

$$p = k_B T \frac{\partial \ln Z}{\partial V}\Big|_{T,N}.$$

Thermal fluctuation theory uses energy deviations, and temperature is defined as

$$T = \left[\frac{\langle (E - \langle E \rangle)^2 \rangle}{k_B c_v}\right]^{1/2}, \qquad c_v = \alpha(T) N_f k_B$$

where $\alpha(T)$ represents anharmonicity in the system.

Macroscopic thermodynamic relations include:

$$\mathrm{d}E = \mathrm{d}Q + \mu \mathrm{d}N - p\mathrm{d}V, \quad \mu - T\frac{\partial \mu}{\partial T}\Big|_p = h, \quad T = \frac{1}{\frac{\partial S}{\partial E}\Big|_{V,N}}, \quad -\frac{\mu}{T} = \frac{\partial S}{\partial N}\Big|_{E,V}, \quad c_v \equiv \frac{\partial e}{\partial T}\Big|_v$$

E	internal energy, J, e specific internal energy, J/kg
h	specific enthalpy, J/moleculae
H	Hamiltonian, J
S	entropy, J/K
V	volume (m^3), v specific volume, m^3/kg
Z	partition function
μ	chemical potential, J/molecule
φ	potential energy, J

The quantum mechanical expectation value of temperature is based on the quasi-particle momentum and is [131] (derivation left as Chapter 5 problem)

$$T = \langle T \rangle = -\frac{\hbar^2}{2m} \int \Psi^{\dagger} \nabla^2 \Psi \mathrm{d}\boldsymbol{x}.$$

2.4 Hamiltonian Mechanics

2.4.1 Classical and Quantum Hamiltonians

In Newtonian mechanics, the equation of motion of particle k, in a multibody system (subject to two- and three-body forces) is

$$\frac{\mathrm{d}}{\mathrm{d}t}(m_k \boldsymbol{u}_k) = \boldsymbol{F}_k, \tag{2.16}$$

where m_k is the mass, $\boldsymbol{u}_k$ is the velocity vector of particle k, and $\boldsymbol{F}_k$ is the sum of all the external forces acting on it (Table 2.4).

In classical mechanics, the Hamiltonian $\mathrm{H}(\boldsymbol{x}, \boldsymbol{p})$ is a function describing the energy state of a mechanical system. The Hamiltonian equations of motion are the relations between H, $\boldsymbol{x}$, and $\boldsymbol{p}$ (i.e., a generalized coordinate $\boldsymbol{x}$ and a generalized momentum $\boldsymbol{p}$) are

$$\frac{\partial \boldsymbol{x}}{\partial t} = \frac{\partial \mathrm{H}}{\partial \boldsymbol{p}}, \quad \frac{\partial \boldsymbol{p}}{\partial t} = -\frac{\partial \mathrm{H}}{\partial \boldsymbol{x}}, \tag{2.17}$$

which is a set of first-order differential equations. The dynamical states of a system of N particles is completely determined by $2N$ vectors $\boldsymbol{x}$ and $\boldsymbol{p}$, and the energy of the system is given by the Hamiltonian function (for 3 degrees of freedom)

$$E = \mathrm{H}(\boldsymbol{x}, \boldsymbol{p}) = \sum_l \mathrm{H}_l = \sum_{k=1}^{3N} \frac{p_k^2}{2m} + \sum_{i=1}^{N} [\varphi_e(\boldsymbol{x}_i) + \sum_{i<j}^{N} \varphi_{ij}(\boldsymbol{x}_i - \boldsymbol{x}_j)] \quad \text{classical Hamiltonian}, \tag{2.18}$$

where φ_e is the external potential and φ is the interparticle pair potential (Section 2.2 and depends on particle position only). The Hamiltonian is the central function in the theory of statical mechanics. The preceding relations are also listed in Table 2.5.

As indicated in (2.18), the total Hamiltonian of a system contains the various energies (e.g., kinetic, electronic). Each of these can be further divided; the kinetic energy can be divided into translational, rotational, and vibrational, through the partition of energies.

In (2.18) we have partitioned the energy into kinetic and potential, and further partitioned each one. Now using (2.13) we can relate the kinetic energy and temperature, and also by defining the partition function Z we relate thermodynamics properties to Z. The partition function is used to define the probability of a system (equilibrium) of particles to be found in a particular energy state P_n, where n designates the energy eigenstate [119]. The grand canonical (for systems that can exchange both energy and particles with a much larger reservoir) partition function is also listed in Table 2.4.

In quantum mechanics, the Hamiltonian operator (or observable) of the system is the sum of its kinetic and potential energy φ, but in an operator form also

Table 2.5. *Classical particle (molecular) dynamics equations (Newton and Hamiltonion equations).*

Newton particle momentum conservation equation:

$$\frac{\mathrm{d}}{\mathrm{d}t}(m_k \boldsymbol{u}_k) = \frac{\mathrm{d}}{\mathrm{d}t}(m_k \frac{\mathrm{d}\boldsymbol{x}_k}{\mathrm{d}t}) = \boldsymbol{F}_k \equiv \sum_j \boldsymbol{F}_{kj} + \sum_{i,j} \boldsymbol{F}_{kij} + \cdots$$

$$\boldsymbol{F}_{kj} = -\frac{\partial \varphi}{\partial \boldsymbol{x}_{kj}} = -\frac{\partial \varphi}{\partial r}\boldsymbol{s}_{kj} \quad \text{interatomic force}$$

$$\boldsymbol{F}_{kj} = -\Gamma_{kj}\boldsymbol{d}_{kj},\ \Gamma_{kj} = \left.\frac{\partial^2 \varphi_{kj}}{\partial r_{kj}^2}\right|_e \quad \text{force constant}$$

Γ_{kj} is force constant (harmonic approximation)

In terms of displacement vector $\boldsymbol{d}$,

$$m_k \frac{\mathrm{d}\boldsymbol{d}}{\mathrm{d}t^2} = \boldsymbol{F}_k$$

Hamiltonian (energy) function:

$$\mathrm{H} = \mathrm{H}(\boldsymbol{x}, \boldsymbol{p}) = \sum_l \mathrm{H}_l \quad \text{Hamiltonian, and partition of energies}$$

$$= \frac{1}{2}\frac{|\boldsymbol{p}|^2}{m} + \varphi(\boldsymbol{x}) = E_k + E_p = E(\boldsymbol{x}, \boldsymbol{p}) \quad \text{classical Hamiltonian}$$

Hamilton equations of motion:

$$\frac{\partial \boldsymbol{x}}{\partial t} = \frac{\partial \mathrm{H}(\boldsymbol{x}, \boldsymbol{p})}{\partial \boldsymbol{p}} = \frac{\partial E(\boldsymbol{x}, \boldsymbol{p})}{\partial \boldsymbol{p}}$$

$$\frac{\partial \boldsymbol{p}}{\partial t} = -\frac{\partial \mathrm{H}(\boldsymbol{x}, \boldsymbol{p})}{\partial \boldsymbol{x}} = -\frac{\partial E(\boldsymbol{x}, \boldsymbol{p})}{\partial \boldsymbol{x}}$$

Hamiltonian function for a system of N particles:

$$E = \mathrm{H} = \sum_{k=1}^{3N} \frac{p_k^2}{2m} + \sum_{i=1}^{N} [\varphi_e(\boldsymbol{x}_i) + \sum_{i<j}^{N} \varphi_{ij}(\boldsymbol{x}_i - \boldsymbol{x}_j)] + \ldots$$

E	energy, J
$\boldsymbol{d}_{kj}$	displacement vector, m
$\boldsymbol{F}_k$	applied external force on particle k, N
H	Hamiltonian, J
$\boldsymbol{p}$	particle momentum, N-s
φ_e	external potential energy, J
φ_{ij}	interparticle potential energy, J
Γ_{kj}	spring (force) constant, N/m

(operating on a quantum wave function):

$$\mathrm{H} = -\frac{\hbar^2}{2m}\nabla^2 + \varphi(\boldsymbol{x}) \quad \text{quantum Hamiltonian operator.} \tag{2.19}$$

We discuss this in relation to the Schrödinger equation in Section 2.6.

2.4.2 Probability and Partition Function

The energy probability density of state n of the ensemble (assuming zero chemical potential, $\mu = 0$) is given by

$$P_n(\boldsymbol{x}, \boldsymbol{p}) = \frac{1}{Z}\exp[-\mathrm{H}_n(\boldsymbol{x}, \boldsymbol{p})/k_\mathrm{B}T] \quad \sum_n P_n = 1, \tag{2.20}$$

where $\mathrm{H}_n(\boldsymbol{x}, \boldsymbol{p}) = E_n(\boldsymbol{x}, \boldsymbol{p})$ is the Hamiltonian of state n, and Z is the partition function (dimensionless), introduced in (1.5), and is defined in the general form as

$$Z \equiv \int\int \exp[\frac{-\mathrm{H}(\boldsymbol{x}, \boldsymbol{p})}{k_B T}]\mathrm{d}\boldsymbol{x}\mathrm{d}\boldsymbol{p} = \frac{1}{h_\mathrm{P}^{3N} N!}\int \exp[\frac{-\mathrm{H}(1, \dots, N)}{k_\mathrm{B} T}]\mathrm{d}^{3N}x\, \mathrm{d}^{3N}p, \tag{2.21}$$

where the grand canonical ensemble (Section 2.5.1) is a statistical ensemble in equilibrium with a external bath (particle and energy exchange) at fixed T, V, and μ. This integral is in turn represented as a summation, as given in Table 2.4, and is generally extremely difficult to calculate because one must calculate all possible states of the system. In MD simulation, the points in the ensemble are calculated sequentially in time, so to calculate an ensemble average; the MD simulations must pass through states corresponding to the particular thermodynamic constraints. To cover all possible states, simulations with different initial conditions are used. As shown in Appendix F, the energy occupancy distribution functions f_i^{o} ($i = p, e, f, ph$) are derived based on the partition function (and the scaled energy $E_i/k_\mathrm{B}T$).

From (2.21), we note that the contribution from each component of the Hamiltonian to the partition function is multiplicative, i.e., $Z = \Pi_i Z_i$, where each Z_i is for a Hamiltonian. We will use this in Chapter 6 for the fluid particles and its various energies.

2.4.3 Ergodic Hypothesis in Theoretical Statistical Mechanics

In statistical mechanics, average values are defined as ensemble averages. The ensemble average is given by

$$\langle\phi\rangle_{x,p} = \int_p\int_x \phi(\boldsymbol{x}, \boldsymbol{p})P(\boldsymbol{x}, \boldsymbol{p})\mathrm{d}\boldsymbol{x}\,\mathrm{d}\boldsymbol{p}, \tag{2.22}$$

where ϕ is the observable of interest and it is expressed as a function of the positions $\boldsymbol{x}$ and momenta $\boldsymbol{p}$ (or velocity $\boldsymbol{u}$) of the system. The integration is over all possible values of $\boldsymbol{x}$ and $\boldsymbol{p}$.

Another way, as done in a MD simulations, is to determine a time average of ϕ, which is expressed as

$$\langle\phi\rangle_t = \lim_{\tau\to\infty}\frac{1}{\tau}\int_{t=0}^{\tau}\phi[\boldsymbol{x}(t), \boldsymbol{p}(t)]\mathrm{d}t \simeq \frac{1}{K}\sum_{t=1}^{K}\phi[\boldsymbol{x}(t), \boldsymbol{p}(t)], \tag{2.23}$$

where τ is the simulation time, K is the number of time steps in the simulation and $\phi[\boldsymbol{x}(t), \boldsymbol{p}(t)]$ is the instantaneous value of ϕ.

The ergodic hypothesis states that the phase-space ($\boldsymbol{x}$, $\boldsymbol{p}$) ensembles and time ensemble (t) are equal for ergodic systems, i.e., ergodic hypothesis:

$$\langle \phi \rangle_{\boldsymbol{x},\boldsymbol{p}} = \langle \phi \rangle_t \quad \text{ergodic hypothesis.} \tag{2.24}$$

In an ergodic system, all points in phase space ($\boldsymbol{x}$, $\boldsymbol{p}$) are accessible from any starting point.

2.5 Molecular Dynamics Simulations

In a MD simulation [3, 112], the position and momentum space trajectories of a system of classical particles are predicted by the Newtonian laws of motion and an appropriate interatomic potential. By use of the positions and momenta, it is possible to investigate a variety of thermal transport problems at the atomic level. Below, we summarize key aspects of MD simulations.

2.5.1 Ensemble and Descretization of Governing Equations

MD [112] is concerned with molecular motion, i.e., conformational transitions (fluids) and local vibrations (solids), which are represented by changes in the positions and velocities of molecules (or atoms/particles) with respect to time. The Newtonian equations of motion are used in the MD formalism to simulate these particle motions as classical, multibody systems. The particles are generally considered hard spheres, but elastic particles can also be studied. The rate and direction of motion (velocity) are governed by the forces that the particles of the system exert on each other as described by the Newtonian momentum conservation equation. In practice, the atoms are assigned initial velocities that conform to the total kinetic energy of the system (but this may not remain constant during simulation), which in turn is dictated by the desired simulation temperature. This can be carried out by slowly heating the system (initially at absolute zero) and then allowing the energy to equilibrate among the constituent particles. The basic elements of molecular dynamics are the calculation of the force on and (from that information) position of each particle throughout a specified period of time Δt (typically of the order of fs, 10^{-15} s). The force on a particle can be calculated from the change in potential energy between its current position and its position a small distance away. This can be recognized as the derivative of the potential energy with respect to the change in position (Figure 2.2).

The potential energies can be calculated by either molecular mechanics or quantum mechanical methods. Molecular-mechanic energies are limited to applications that do not involve drastic changes in electronic structure, such as bond making/breaking. Quantum-mechanical energies can be used to study dynamic processes involving chemical changes, but computation time becomes intensive. Knowledge of the atomic forces and masses can then be used to solve for the positions of each atom along a series of extremely small time steps Δt. The resulting series of snapshots of structural changes over time is called a trajectory.

These particle motions, which are microscopic, are used in ensemble averages (statistical mechanics) to calculate macroscopic observable properties, including both static (e.g., pressure, energy, heat capacity) and kinetic-transport (e.g., thermal conductivity) properties. The thermodynamic state of a system is usually defined by a small set of parameters, for example, the temperature T, the pressure p, and the number of particles N. Other thermodynamic properties may be derived from the equations of state and other fundamental thermodynamic relations.

(A) *Ensembles*

The mechanical or microscopic state of a system is defined by the particle positions $\boldsymbol{x}$ and momenta $\boldsymbol{p}$ (or velocities $\boldsymbol{u}$), and these can also be considered as coordinates in a multidimensional space called the phase space. For an N particle system, this space has $6N$ dimensions. A single point in the phase space describes the state of the system. An ensemble is a collection of points in phase space satisfying the conditions of a particular thermodynamic state. A MD simulation generates a sequence of points in phase space as a function of time. These points belong to the same ensemble, and they correspond to the different conformations of the system and their respective momenta.

Several different ensembles are used, including the microcanonical ensemble (NVE), where the thermodynamic state is characterized by a fixed number of particles N, a fixed volume V, and a fixed energy E, which corresponds to an isolated system. The canonical ensemble (NVT) is a collection of all systems whose thermodynamic state is characterized by a fixed N, a fixed V, and a fixed temperature T (the system can exchange heat with an ambient reservoir). The isobaric–isothermal ensemble (NpT) is characterized by a fixed N, a fixed pressure p, and a fixed T. The grand canonical ensemble (μVT) is characterized by a fixed chemical potential μ, a fixed V, and a fixed T (the system can exchange heat and particle with an ambient reservoir).

(B) *Ensemble-Averaged Energy*

The average potential energy is

$$\langle \varphi \rangle_K = \frac{1}{K} \sum_{i=1}^{K} \varphi_i, \tag{2.25}$$

where K is the number of configurations in the MD trajectory and φ_i is the potential energy of each configuration.

The average kinetic energy is

$$\langle E_k \rangle_K = \frac{1}{K} \sum_{j=1}^{K} \Big(\sum_{i=1}^{N} \frac{m_i}{2} \boldsymbol{u}_i \cdot \boldsymbol{u}_i \Big)_j, \tag{2.26}$$

where K is the number of configurations in the simulation, N is the number of atoms in the system, m_i is the mass of the particle i, and $\boldsymbol{u}_i$ is the velocity of particle i.

Using the degeneracy g_n (microstates having the same energy E_n), discrete canonical partition function (2.21) is defined as [55]

$$Z = \sum_{n=0}^{\infty} g_n \exp(-\frac{E_n}{k_B T}) \quad \text{discrete canonical partition function,} \tag{2.27}$$

where the summation is over all the energy levels.

A degenerate state is that for which $g_n \geq 2$ for an energy state E_n. For the non-degenerate states, $g_n = 1$.

Here we first proceed with the computational MD simulations, and in the following chapters (in particular Chapters 4 and 6) we will make use of the preceding theoretical statistical mechanics (thermodynamics). A MD simulation must be performed over a sufficiently long elapsed time so sufficient representative configurations have been sampled (Section 2.4.3).

(C) *Computational Classical Particle Dynamics*

The MD simulation method is based on the Newton second law, $\boldsymbol{F} = m\boldsymbol{a}$, where $\boldsymbol{F}$ is the force exerted on the particle, m is its mass, and $\boldsymbol{a}$ is its acceleration. From a knowledge of the force on each atom/particle, it is possible to determine the acceleration of each atom/particle in the system. Integration of the equations of motion then yields a trajectory that describes the positions, velocities, and accelerations of the particles as they vary with time. From this trajectory, the average values of properties can be determined. The method is deterministic; once the positions and velocities of each atom/particle are known, the state of the system can be predicted at any time in the future or the past. MD simulations can become computationally extensive.

The Newton equation of motion is given by

$$\boldsymbol{F}_i = m\boldsymbol{a}_i, \tag{2.28}$$

where F_i is the force exerted on particle i, m_i is the mass of particle i and $\boldsymbol{a}_i$ is the acceleration of particle i. The force can also be expressed as the gradient of the potential energy φ, i.e.,

$$\boldsymbol{F}_i = -\nabla_i \varphi. \tag{2.29}$$

Combining the preceding two equations, and for a one-dimensional translation, we have

$$-\frac{\mathrm{d}\varphi}{\mathrm{d}r_i} = m_i \frac{\mathrm{d}^2 x_i}{\mathrm{d}t^2}. \tag{2.30}$$

Note that φ is the collective (or effective/average) potential energy that a particle experiences. The Newtonian equation of motion can then relate the derivative of the potential energy to changes in position as a function of time.

The Newton second law of motion is

$$F_i = m_i a_i = m_i \frac{\mathrm{d}u_i}{\mathrm{d}t} = m_i \frac{\mathrm{d}^2 x_i}{\mathrm{d}t^2}. \tag{2.31}$$

Taking the simple case in which the acceleration is constant, we have

$$a_i = \frac{\mathrm{d}u_i}{\mathrm{d}t}, \tag{2.32}$$

we obtain an expression for the velocity after integration,

$$u_i = a_i t + u_{i,o}, \tag{2.33}$$

and because

$$u_i = \frac{\mathrm{d}x_i}{\mathrm{d}t}, \tag{2.34}$$

we can once again integrate to obtain

$$x_i = u_i t + x_{i,o}. \tag{2.35}$$

Combining this equation with the expression for the velocity, we obtain the following relation that gives the value of x at time t as a function of the acceleration a, the initial position, x_{o}, and the initial velocity u_{o}:

$$x_i = a_i t^2 + u_{i,o} t + x_{i,o}. \tag{2.36}$$

The acceleration is given as the derivative of the potential energy with respect to the position x, i.e.,

$$a_i = -\frac{1}{m_i}\frac{\mathrm{d}\varphi_i}{\mathrm{d}x}. \tag{2.37}$$

Therefore, to calculate a trajectory, one needs only the initial positions of the atoms, an initial distribution of velocities and the acceleration, which is determined by the gradient of the potential energy function that is dependent on positions. The equations of motion are deterministic, e.g., the positions and the velocities at time zero determine the positions and velocities at all other times t. The initial positions can be obtained from experimental structures, such as the X-ray crystal structure of the protein or the solution structure determined by nuclear magnetic resonance (NMR) spectroscopy (or quantum calculations).

The initial distributions of velocities are usually determined from a random distribution with the magnitudes conforming to the required temperature and corrected so there is no overall momentum, i.e.,

$$\langle \boldsymbol{p} \rangle = \sum_{i=1}^{N} m_i \boldsymbol{u}_i = 0. \tag{2.38}$$

This operation removes 3 degrees of freedom from the system and leads to $3(N-1)$ degrees of freedom for the particles (Table 2.4).

(D) *Descretization Verlet Algorithm for MD Computation*

The potential energy of a system is a function of the positions of all of the particles in the system. Because of the complicated nature of this function, there is no analytical solution to the equations of motion; they must be solved numerically. Numerous numerical algorithms have been developed for integrating the equations of motion. Among them are the leap-frog, position-Verlet, velocity-Verlet, and Beeman algorithms [112]. These algorithms conserve energy and momentum, are computationally efficient, and permit a large time step for integration. All the integration algorithms assume the position, velocity, and acceleration are expandable in Taylor series as

$$x(t+\Delta t) = x(t) + u(t)\Delta t + \frac{1}{2}a(t)\Delta t^2 + \cdots \tag{2.39}$$

$$u(t+\Delta t) = u(t) + a(t)\Delta t + \cdots \tag{2.40}$$

$$a(t+\Delta t) = \frac{F(t+\Delta t)}{m} + \cdots \tag{2.41}$$

For example, to derive the position-Verlet algorithm, consider $t \pm \Delta t$, so we have

$$x(t+\Delta t) = x(t) + u(t)\Delta t + \frac{1}{2}a(t)\Delta t^2 \tag{2.42}$$

$$x(t-\Delta t) = x(t) - u(t)\Delta t + \frac{1}{2}a(t)\Delta t^2, \tag{2.43}$$

and by summing these two equations, we have

$$x(t+\Delta t) = 2x(t) - x(t-\Delta t) + a(t)\Delta t^2 \quad \text{position-Verlet algorithm.} \tag{2.44}$$

The position-Verlet algorithm uses position and acceleration at time t and the position from time $t - \Delta t$ to calculate the new position at time $t + \Delta t$. The Verlet algorithm uses no explicit velocities. The advantages of the Verlet algorithm are that it is straightforward and the storage requirements are modest. The disadvantage is that the algorithm is of only moderate precision.

Another approach is the velocity-Verlet algorithm, which yields position, velocity, and acceleration at time $t + \Delta t$, so there is no compromise on precision. This uses (2.39) and an average acceleration, i.e.,

$$x(t+\Delta t) = x(t) + u(t)\Delta t + \frac{1}{2m}F(t)\Delta t^2 \tag{2.45}$$

$$u(t+\Delta t) = u(t) + \frac{1}{2m}[F(t) + F(t+\Delta t)]\Delta t \quad \text{velocity-Verlet algorithm,} \tag{2.46}$$

where $a(t + \Delta t)$ uses $x(t + \Delta t)$. So, the new positions and forces (accelerations) are computed before the new velocities are calculated [112].

2.5.2 A Molecular Dynamics Simulation Case Study

Choosing a simple system allows for the elucidation of results that may be difficult to resolve in more complex materials, in which multi-atom unit cells (and thus optical phonons) can generate additional effects. The L–J atomic interactions are described by the pair potential (2.9), i.e.,

$$\varphi_{\mathrm{LJ},ij}(r_{ij}) = 4\epsilon_{\mathrm{LJ}}[(\frac{\sigma_{\mathrm{LJ}}}{r_{ij}})^{12} - (\frac{\sigma_{\mathrm{LJ}}}{r_{ij}})^{6}], \tag{2.47}$$

where $\varphi_{\mathrm{LJ},ij}(r_{ij})$ is the potential energy associated with particles i and j (i is not equal to j), which are separated by a distance r_{ij}. The depth of the potential energy well is ϵ_{LJ}, and corresponds to an equilibrium particle separation of $2^{1/6}\sigma_{\mathrm{LJ}}$.

The force acting between i and j is

$$F_{\mathrm{LJ},ij}(r_{ij}) = -\frac{\partial \varphi_{\mathrm{LJ},ij}}{\partial r_{ij}} = 24\epsilon_{\mathrm{LJ}}(2\frac{\sigma_{\mathrm{LJ}}^{12}}{r_{ij}^{13}} - \frac{\sigma_{\mathrm{LJ}}^{6}}{r_{ij}^{7}}). \tag{2.48}$$

Ar and its FCC crystal (Section 4.3 describes this crystal structure), amorphous, and liquid phases are considered. The 12 nearest neighbors of an atom in a FCC crystal and of a sample atom in an amorphous structure are shown in Figures 2.12(a) and 2.12(b). In the FCC crystal, all atoms are at equivalent positions, and the atomic displacements are isotropic. In the amorphous phase, each atom has a unique environment, with a range of neighbor orientations and bond lengths. The resulting atomic displacements are anisotropic. The plane formed by the [100] and [010] directions in the FCC crystal is shown in Figure 2.12(c). Both the lattice constant a and the size of the simulation cell L are indicated. Note that the conventional unit cell, which is cubic, contains four atoms. The true FCC unit cell, containing one atom, is rhombohedral, and not as suitable for analysis.

The simulations are run in the NVE ensemble at zero pressure with a time step $\Delta t = 4.285$ fs. The simulation cell is cubic and contains 256 atoms for the crystal and liquid phases, and 250 atoms for the amorphous phase. In similar simulations [168], good agreement has been found between the FCC crystal thermal conductivities predicted from cells containing 256 and 500 atoms [236]. This result indicates that 256 atoms are sufficient to eliminate size effects. Periodic boundary conditions are imposed in all directions. The equations of motion are integrated with a Verlet leapfrog algorithm described above. Atomic interactions are truncated and shifted at a cutoff radius equal to $2.5\sigma_{\mathrm{LJ}}$ (for computational economy).

For the FCC crystal, temperatures between 10 and 80 K are considered in 10 K increments. Melting occurs at a temperature of about 87 K. The amorphous phase is generated by starting with the desired number of atoms placed in a FCC lattice, running at a temperature of 300 K for 10^4 time steps to eliminate any memory of the initial configuration, and then quenching at 8.5×10^{11} K/s back to a temperature of 10 K. Table 2.6 lists the results of the simulations. The amorphous phase is stable up to a temperature of 20 K. Above this point, the equilibrium thermal fluctuations in the system are large enough to return the atoms to a FCC crystal structure. Temperatures of 10, 15, and 20 K are considered. Three different amorphous phases (each with 250 atoms) have been formed to check if the systems are truly disordered, and cells with 500 and 1000 atoms have been created to investigate

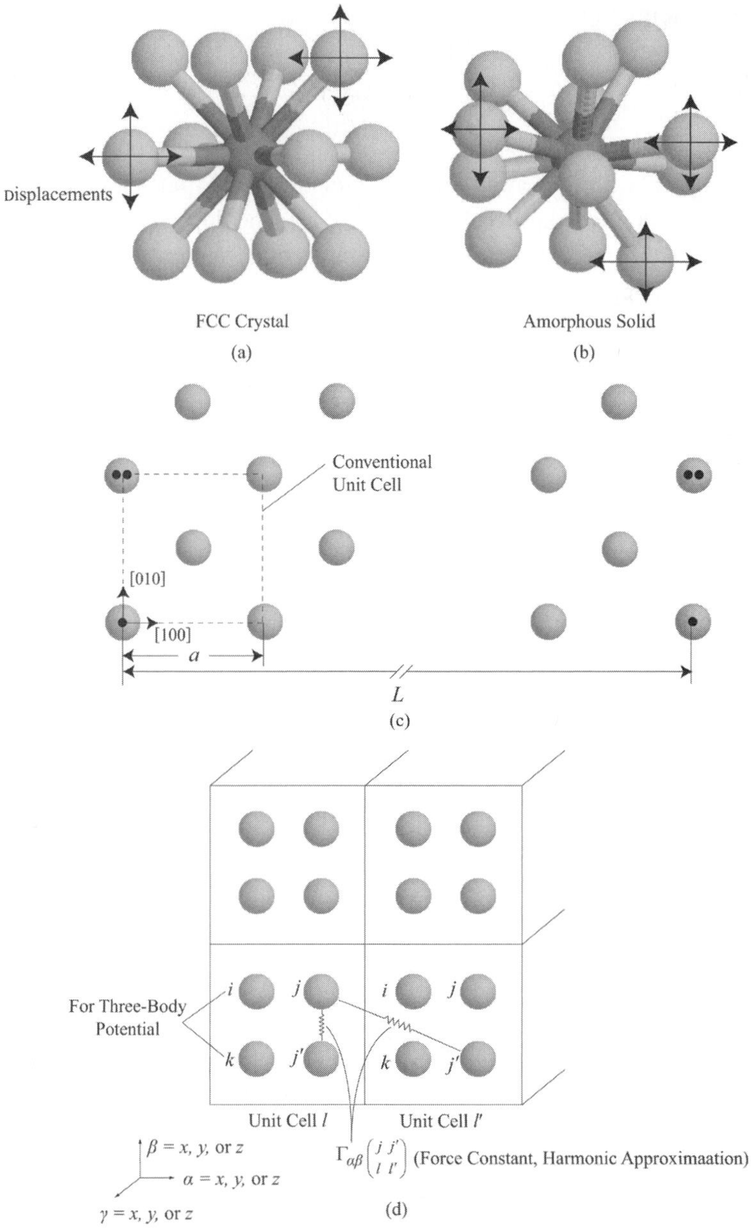

Figure 2.12. Local environment for an atom in (a) the FCC crystal, and (b) the amorphous structure. In the crystal, there are 12 nearest neighbors. For the amorphous phase, the 12 nearest atoms are shown. Although the color of the center atom is a darker gray than the neighbors, all the atoms are the same. Also shown are representative atomic displacements (not to scale) for some of the neighbor atoms for each case. The motions in the FCC crystal are isotropic and equivalent between atoms, whereas those in the amorphous structure are not. (c) Indicals in the FCC crystal. The atoms with black dots are equivalent through the use of periodic boundary conditions [236]. (d) Shows coordinate and atom designations used a dynamical analysis of lattice vibration (will be further discussed in Section 4.4).

Table 2.6. *Simulation cell parameters and equilibrium properties of the Ar FCC L–J phases. Note that the specific heat capacity is per atom and per degree of freedom [236].*

Structure	T, K	a, Å	ρ, kg/m^3	$\tau_{nn,e}$, ps	$\langle\varphi\rangle_e$, eV	$\langle\lvert\Delta_i\rvert^2\rangle^{1/2}$, Å	c_v/k_B
FCC crystal	0	5.269	1813	–	−0.0778	0	1
	10	5.290	1791	0.240	−0.0765	0.117	0.988
	20	5.315	1766	0.270	−0.0751	0.180	0.976
	30	5.341	1740	0.278	−0.0736	0.244	0.959
	40	5.370	1713	0.278	−0.0720	0.309	0.953
	50	5.401	1684	0.278	−0.0703	0.383	0.944
	60	5.436	1651	0.283	−0.0684	0.464	0.937
	70	5.476	1615	0.291	−0.0663	0.560	0.922
	80	5.527	1571	0.296	−0.0638	0.682	0.914
Amorphous	0	–	1717	–	−0.0719	0	1
	10	–	1694	0.291	−0.0706	–	0.976
	15	–	1682	0.291	−0.0699	–	0.979
	20	–	1669	0.283	−0.0691	–	0.970
Liquid	70	–	1432	–	−0.0562	–	0.857
	80	–	1363	–	−0.0527	–	0.826
	90	–	1287	–	−0.0491	–	0.803
	100	–	1202	–	−0.0453	–	0.758

size effects. The liquid phase is obtained by first heating the crystal phase to a temperature of 100 K to induce melting, then lowering the temperature. When this approach is used, a stable liquid is found to exist at temperatures as low as 70 K. Because of the small length and time scales used (necessary for reasonable computational times), the melting/solidifying temperature is not well-defined, and it is possible to have stable FCC crystal and liquid phases at the same temperature and pressure, although the densities differ. Temperatures of 70, 80, 90, and 100 K are considered for the liquid simulations. The time scale $\tau_{nn,e} = \pi/\omega_n$, where ω_n is the natural frequency evaluated at equilibrium interatomic distance (next section).

2.5.3 L–J MD Scales in Classical Harmonic Oscillator

When relaxed to zero temperature, the MD FCC crystal unit-cell parameter, a (the lattice constant), is 5.2686 Å. The experimental value for Ar is 5.3033 Å [9].

The lattice constant can also be predicted from the analytical form of the L–J potential. To do this, one must consider the total potential energy associated with one atom $\langle\varphi\rangle$. If the energy in each pair interaction is assumed to be equally distributed between the two atoms, $\langle\varphi\rangle$ will be given by [183]

$$\langle\varphi\rangle = \langle\varphi_i\rangle = \frac{1}{2}\sum_{i\neq j}\varphi_{\text{LJ},ij}. \tag{2.49}$$

The equilibrium position-effective potential (per atom) for the FCC crystal lattice can be expressed as

$$\langle\varphi\rangle_e = 2\epsilon_{\rm LJ}[A_{12}(\frac{\sigma_{\rm LJ}}{r_{nn,e}})^{12} - A_6(\frac{\sigma_{\rm LJ}}{r_{nn,e}})^6] = -8.607\epsilon_{\rm LJ} \quad \text{L–J FCC crystal}, \tag{2.50}$$

where A_{12} and A_6 have values of 12.13 and 14.45, respectively (for FCC, other lattices are also reported in [9]), and r_{nn} is the nearest neighbor (nn) separation. This effective L–J potential is plotted in Figure 2.3 alongside the pair potential, given by (2.47). By setting

$$\frac{\partial\langle\varphi\rangle}{\partial r_{nn}}|_{r_{nn,e}} = 0, \tag{2.51}$$

the equilibrium value of $r_{nn,e}$ is found to be

$$r_{nn,e} = (\frac{2A_{12}}{A_6})^{1/6}\sigma_{\rm LJ} = 1.090\sigma_{\rm LJ} \quad \text{L–J FCC crystal}. \tag{2.52}$$

The location of the minimum is slightly shifted from that in the pair potential, and the energy well is deeper and steeper. The value of $r_{nn,e}$ for noble-gas elements are listed in Table 2.2. For Ar, the equilibrium separation (2.52) gives $r_{nn,e} = 3.706$ Å, and corresponds to a unit-cell parameter (lattice constant $a = 2^{1/2}r_{nn,e}$) of $a = 5.2411$ Å, which agrees with the zero-temperature MD result to within 0.6%.

In a simplified real-space model of atomic-level behavior, the energy transfer between neighboring atoms can be assumed to occur over one half of the period of oscillation of an atom [51, 52, 100]. The associated time constant $\tau_{\rm D}$ can be estimated from the Debye temperature (listed in Table A.2 for elements and discussed in Sections 4.2 and 4.7) $\tau_{\rm D}$ as

$$\tau_{\rm D} = \frac{2\pi\hbar}{2k_{\rm B}T_{\rm D}}. \tag{2.53}$$

The factor of 2 in the denominator is included because of the half period of oscillation that is used. By fitting the specific heat (as predicted by the MD zero-temperature phonon density of states) to the Debye model using a least squares method, the Debye temperature, is found to be 81.2 K. This compares well with the experimental value of 85 K [9] listed in Table A.2 (Appendix A). The MD result is used in subsequent calculations and gives $\tau_{\rm D} = 0.296$ ps ($\sim$ 69 time steps).

An estimate of this time constant can also be made with the L–J potential. The time constant is related to the curvature of the potential well that an atom experiences at its minimum energy. Assuming that the potential is harmonic at the minimum, then we can write the interatomic forces as $-\Gamma d_x$ (Table 2.5), where d_x is the displacement in a one-dimensional motion along x. The natural angular frequency (for sinusoidal motion) ω_n of the atom is given by †

$$\omega_n = \omega_{\rm LJ} = (\frac{1}{m}\frac{\partial^2\langle\varphi\rangle}{\partial r_{nn}^2}|_{r_{nn}=r_{nn,e}})^{1/2} = (\frac{\Gamma}{m})^{1/2} = 22.88(\frac{\epsilon_{\rm LJ}}{m\sigma_{\rm LJ}^2})^{1/2},$$

$$\Gamma_{\rm LJ} = \left.\frac{\partial^2\langle\varphi\rangle}{\partial r_{nn}^2}\right|_{r_{nn}=r_{nn,e}} = (22.88)^2\frac{\epsilon_{\rm LJ}}{\sigma_{\rm LJ}^2}, \qquad (2.54)$$

where m is the mass of one atom, which for Ar, $m = 6.634\times10^{-26}$ kg, and $\langle\varphi\rangle$ and $r_{nn,e}$ are taken from (2.50) and (2.52). Note that the larger the spring (or force) constant Γ (which in turn requires larger $\epsilon_{\rm LJ}$) and the smaller the mass, the larger is ω_n. For FCC Ar which $\Gamma = 7.545$ N/m, and $\omega_n = 1.068\times10^{13}$ rad/s or 10.68 Trad/s.

The second derivative of the potential is designated as Γ and has the role of a spring constant in that it relates force to displacement in the mechanical system. For an isotropic, cubic lattice, the relation between the spring constant and the bulk modulus E_p (and the lattice constant a which is equal to $2^{1/2}r_e$ for FCC) is

$$\Gamma = E_p a = E_p(2^{1/2}r_{nn,e}),\ E_p = \frac{1}{\kappa_p},\ \kappa_p = -\frac{1}{\rho}\frac{\partial\rho}{\partial p}|_T. \qquad (2.55)$$

The bulk modulus in a L–J FCC crystal is $E_p = 75.14\epsilon_{\rm LJ}/\sigma_{\rm LJ}^3$ (end-of-chapter problem). We further discuss this in Section 4.7.2. One half of the period of oscillation is then

$$\tau_{\rm LJ} = \frac{1}{2}\frac{2\pi}{\omega_n} = 0.137(\frac{\epsilon_{\rm LJ}}{m\sigma_{\rm LJ}^2})^{-1/2}, \qquad (2.56)$$

which gives $\tau_{\rm LJ} = 0.294$ ps, or within 1% of τ_D. The temperature scale here is $\epsilon_{\rm LJ}/k_{\rm B}T$ and for Ar crystal gives $T_{\rm LJ} = 120.7$ K. The mean Debye temperature for Ar FCC is (Table A.2) 85 K.

The physical significance of this time constant can be further investigated by considerations of the flow of energy between atoms in the MD simulation cell. As the atomic separation increases, it takes longer to transfer energy between two atoms. The time constants $\tau_{\rm D}$, $\tau_{\rm LJ}$, and τ_{nn} agree to within 10%, which supports the assumed link between the period of atomic oscillation and the time scale of the atom to atom energy transfer.

Table 2.7 list the various scales derived from the L–J potential for these FCC crystals. Other crystal geometries are explained in [9]. For gases for which the potential energy vanishes, $\epsilon_{\rm LJ}$ and $\sigma_{\rm LJ}$ are used for scaling with similar results. It is also common to use m, $\sigma_{\rm LJ}$, and $\epsilon_{\rm LJ}$ to scale the quantities, then numerical constants appearing in Table 2.7 will be changed to unity.

† In the simple classical harmonic oscillator $F = -\Gamma x$ (Table 2.5) or
$m\frac{{\rm d}^2x}{{\rm d}t^2} = -\Gamma x,\ x = x_{\rm o}\cos(\omega_n t + \phi),\ \omega_n = (\frac{\Gamma}{m})^{1/2}$.
The instantaneous kinetic and potential energies are
$E_k(t) = \frac{1}{2}\Gamma x_{\rm o}^2\sin^2(\omega_n t + \phi),\ E_p(t) = \varphi(t) = \frac{1}{2}\Gamma x_{\rm o}^2\cos^2(\omega_n t + \phi),\ E_k + \varphi = \frac{1}{2}\Gamma x_{\rm o}^2$.
Also, in harmonic lattice vibration, $\langle E_k\rangle = \langle\varphi\rangle = 3k_{\rm B}T/2$, which is the expectation in kinetic and potential energy here. The lattice energy for a linear chain of atoms is discussed in Section 4.1.

Table 2.7. *MD scales, for L–J effective potential in FCC crystal [9].*

Scale	Relation
Natural frequency	$\omega_{\rm LJ} = \omega_n = 22.88(\frac{\epsilon_{\rm LJ}}{\sigma_{\rm LJ}^2 m})^{1/2} = (\frac{\Gamma_{\rm LJ}}{m})^{1/2}$
Effective force constant	$\Gamma_{\rm LJ} = (22.88)^2 \frac{\epsilon_{\rm LJ}}{\sigma_{\rm LJ}^2}$
Length	$L_{\rm LJ} = r_{nn,e} = 1.09\sigma_{\rm LJ}$
Effective potential	$\langle\varphi\rangle_{e,\rm LJ} = \langle\varphi\rangle_{\rm LJ} = 2\epsilon_{\rm LJ}[12.13(\frac{\sigma_{\rm LJ}}{r_{nn,e}})^{12} - 14.45(\frac{\sigma_{\rm LJ}}{r_{nn,e}})^6] = -8.61\epsilon_{\rm LJ}$
Temperature	$T_{\rm LJ} = \frac{\epsilon_{\rm LJ}}{k_{\rm B}}$
Time	$\tau_{\rm LJ} = \frac{\pi}{\omega_n} = 0.137(\frac{\sigma_{\rm LJ}^2 m}{\epsilon_{\rm LJ}})^{1/2} = 0.137(\frac{m}{\Gamma_{\rm LJ}})^{1/2}$

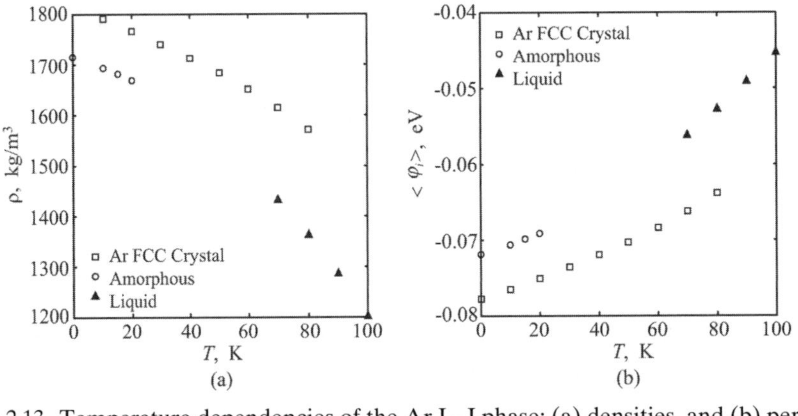

Figure 2.13. Temperature dependencies of the Ar L–J phase: (a) densities, and (b) per particle potential energies [237].

2.5.4 L–J Potential Phase Transformations

The densities and potential energies per particle of the zero pressure L–J FCC crystal, liquid, and amorphous phases are plotted as functions of temperature in Figures 2.13(a) and 2.13(b), and are also given in Table 2.6. The energies correspond to the shifted potential that is used as a result of the application of the potential cutoff. As would be expected, the FCC phase has the lowest potential energy at a given temperature. Note the consistent trend between the amorphous and liquid phases in both density and potential energy. This is consistent with the idea of an amorphous phase being a fluid with a very high viscosity.

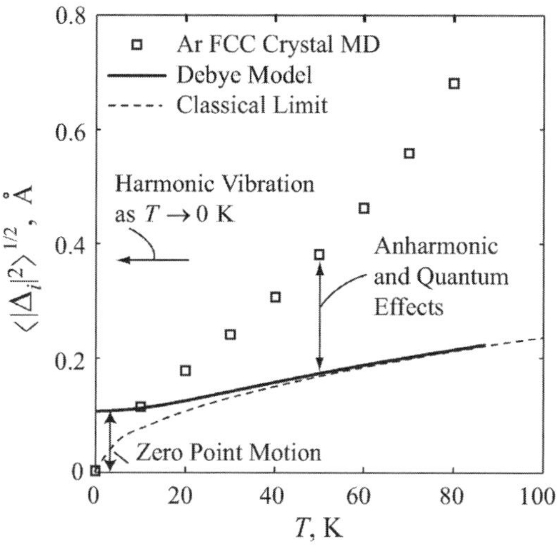

Figure 2.14. MD predicted RMS atomic mean-square directional displacement displacement for Ar FCC and comparison with Debye model, a quantum-mechanical prediction. The zero-point RMS atomic displacement value is $(3\hbar/4m\omega_D)^{1/2}$ [237].

2.5.5 Atomic Displacement in Solids and Quantum Effects

The RMS displacement, $\langle|\boldsymbol{\Delta}_i|^2\rangle^{1/2}$ where $\boldsymbol{\Delta}_i$ is the displacement of atom indirection i from its equilibrium position, of the atoms in the FCC crystal is shown as a function of temperature in Figure 2.14. The numerical values are given in Table 2.6. The results presented are based on 10^5 time steps of NVE simulation, with data extracted every five time steps. The mean-square displacement (MSD) atomic displacement can be predicted from a quantum-mechanical description of the system under the Debye approximation (for isotropic lattice) in [23] (the more general Beni–Platzman–Debye model is derived in Section 4.8)

$$\langle|\boldsymbol{\Delta}_i|^2\rangle^{1/2} = (\boldsymbol{d}_j - \boldsymbol{d}_o) \cdot \boldsymbol{s}_j = \{\frac{3\hbar}{m\omega_D}[\frac{1}{4} + (\frac{T}{T_D})^2 \int_0^{T_D/T} \frac{x\mathrm{d}x}{e^x - 1}]\}^{1/2}$$

Debye model for mean-square directional displacement, (2.57)

where $\boldsymbol{d}_j$ and $\boldsymbol{d}_o$ are the displacement vector of the atom j and the central atom, $\boldsymbol{s}_j$ is the equilibrium position unit vector of the atom j, T_D is given in Table A.2 (Appendix A), and $\omega_D = \pi/\tau_D$ and τ_D is given by (2.53).

This relation is also shown in Figure 2.14. Considering the minimal input required in the theoretical model (only the atomic mass and the Debye temperature), the agreement between the two curves is fair because the anharmonicity is not included in the model (the model uses the dynamical matrix, which assumes a harmonic behavior and is valid as $T \to 0$ K). Note that although the quantum model predicts the finite zero-point motion $(3\hbar/4m\omega_D)^{1/2}$, the MD results show a trend toward no motion at zero temperature. This is what one expects in a classical system, as the phase space approaches a single point when motion ceases. The results for the classical harmonic oscillator† is also shown in Figure 2.14.

Also, from the Heisenberg uncertainty principles (1.6), we have (end-of-chapter problem)

$$\Delta p_x \Delta x \geq \frac{\hbar}{2} \quad \text{or} \quad \Delta E \Delta t \geq \frac{\hbar}{2}, \quad \frac{\Delta x}{\Delta t} = u_g = \frac{\hbar \kappa}{m}, \tag{2.58}$$

where u_g is the group velocity. After substitutions (including $\Delta E = \hbar\omega$), we have

$$\Delta x_{min} = (\frac{\hbar}{2m\omega})^{1/2}, \tag{2.59}$$

where ω is the maximum frequency (here ω_D). As expected, this is slightly smaller than the zero-point motion displacement $(3\hbar/4m\omega_D)$ given in (2.57).

The displacement (2.59) is used in the definition of quantum normal coordinate (Glossary), for atomic displacement and in relation to phonons.

The specific heat is defined thermodynamically as the rate of change of the total system energy (kinetic and potential) as a function of temperature at constant volume [9]

$$c_v = \left.\frac{\partial \langle E \rangle}{\partial T}\right|_V . \tag{2.60}$$

Such a calculation can be explicitly performed by use of the results of the MD simulations. The predicted, scaled specific heats for the FCC crystal, amorphous, and liquid phases are plotted in Figure 2.15 and given in Table 2.6. The values given correspond to the specific heat per degree of freedom [there are $3(N-1)$, Table 2.4]. The calculation is performed by varying the temperature in 0.1 K increments over a ± 0.2 K range around the temperature of interest.

The specific heat predicted from the MD simulations is a classical-anharmonic value. Also shown in Figure 2.15 are the classical-harmonic and the quantum-harmonic specific heats for the crystal phase. The classical-harmonic value, k_B, is based on an assumption of equipartition of kinetic and potential energy between normal modes. The equipartition assumption is always valid for the kinetic energy (i.e., it contributes $k_B/2$ to c_v, which has been verified). For the potential energy, however, it is true only under the harmonic approximation, which itself is valid only at zero temperature. The deviations of the classical-anharmonic results from the classical-harmonic model (the Dulong–Petit limit) are significant. The quantum-harmonic specific heat that is based on the zero-temperature phonon density of

† The high-temperature (classical) limit of (2.57) is found from the simple, classical harmonic oscillator of Section 2.5.3, as is
$\langle |\boldsymbol{\Delta}_i|^2 \rangle^{1/2} = (\boldsymbol{d}_j - \boldsymbol{d}_o) \cdot \boldsymbol{s}_j = x_o = \left(3k_B T / m\omega_n^2\right)^{1/2}$,
where $\boldsymbol{d}_j$ and $\boldsymbol{d}_o$ are the j atom and the central atom displacements and $\boldsymbol{s}_j$ is the equilibrium position unit vector for atom j. $\langle |\boldsymbol{\Delta}_i|^2 \rangle$ shows a linear increase with temperature.

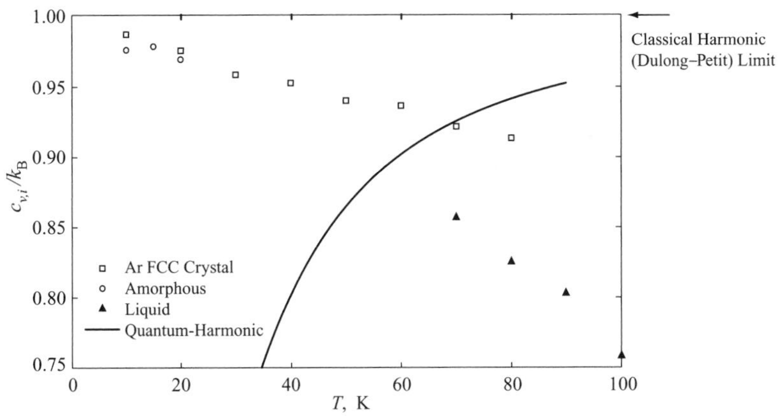

Figure 2.15. The classical-anharmonic specific heat per degree of freedom predicted from the MD simulations, and the classical-harmonic and quantum-harmonic curves for the crystal phase (all scaled by k_B), for Ar FCC. The theoretical predictions are stopped at $T = 87$ K, the melting point of the MD system [237].

states is discussed in Section 4.7.1 (the Dulong–Petit limit) and is given by

$$c_{v,p,quant-harm} = k_B \sum_{i}^{N_p} \frac{x_i^2 e^{x_i}}{(e^{x_i} - 1)^2}, \tag{2.61}$$

where x_i is $\hbar\omega_i / k_B T$ and the summation is over the normal modes (N_p) of the system. As expected, the classical and harmonic specific heats are significantly different at low temperatures, where quantum effects are important. Prediction of the quantum-anharmonic specific heat (as would be measured experimentally) would require taking into account the temperature dependence and coupling of the normal modes. The results would be expected to converge with the classical-anharmonic value at high temperatures (i.e., of the order of the Debye temperature).

Not surprisingly, the FCC crystal and amorphous data are not significantly different. The crystal structure should not have a significant effect on the specific heat, especially at low temperatures for which the harmonic approximation is still reasonable. There is a definite drop in the liquid values, and the specific heat would continue to drop as the temperature is increased. The lower limit for the specific heat is $0.5k_B$, when potential energy effects have been completely eliminated (i.e., an ideal gas).

It should be noted that the MD simulations are clearly not harmonic in nature. It is sometimes assumed that the mode specific heat of solids in MD is equal to k_B, which is not the case and can lead to errors at high temperatures.

2.6 Schrödinger Equation and Quantum Energy States

Experiments show that atomic particles are also wavelike in nature (e.g., electrons give diffraction patterns when passed through a double slit in a similar way to light waves). The Schrödinger equation is a wave equation that explains the behavior of atomic particles. We consider the electromagnetic wave in Section 3.3.1 and discuss some of its similarities and differences with the Schrödinger wave. The eigenvalues of the wave equation are the energy levels of the quantum mechanical system. This is used to find the energy levels of the hydrogen atom, and the energy levels are in good agreement with the Rydberg law.

In quantum mechanics, the Schrödinger equation plays the role of the Newton laws of motion and the energy conservation in classical mechanics. This is a wave equation in terms of the wave function that predicts analytically (although limited to a few cases) and precisely the probability of events or outcome. The detailed outcome is not strictly determined, but given a large number of events, the Schrödinger equation will predict the distribution of the results. In this equation, the kinetic and potential energies are transformed into the Hamiltonian H that acts upon the wave function Ψ to generate the evolution of the wave function in time and space.

In the Born–Oppenheimer approximation, the nuclear and electronic motions are separated (due to large mass for nuclear compared with electron). The electron distribution depends on the fixed position of nuclei.[†]

The Schrödinger equation gives the quantized energies of the system and determines everything that can be known about the system. The wave function (or wavefunction) is a single-valued function of position and time, because that is sufficient to guarantee an unambiguous value of probability of finding the particle at a particular position and time. The wave function may be a complex function, because it is a product with its complex conjugate that specifies the real physical probability of finding the particle in a particular state. Also, the wave number ($\kappa = 2\pi/\lambda$, where λ is wavelength) is related to the wave momentum by equation $p = \hbar\kappa$. The Schrödinger equation and its variables are listed in Table 2.8. We will discuss this equation next. Then we solve this equation for the simple quantum-harmonic oscillator, the electron gas, and the electron in hydrogenlike atoms.

2.6.1 Time-Dependent Schrödinger Equation and Wave Vector

The time-dependent Schrödinger equation (also listed in Table 2.8) is

$$\begin{aligned} i\hbar\frac{\partial\Psi}{\partial t} &= (-\frac{\hbar^2}{2m}\nabla^2 + \varphi)\Psi \\ &\equiv \mathrm{H}\Psi, \quad \varphi = \varphi_a + \varphi_c + \varphi_s, \end{aligned} \tag{2.62}$$

and describes the $\boldsymbol{x}$- and t-dependence of a quantum mechanical system described by H and φ. Here the potential energy φ has been decomposed to applied potential

[†] The full Hamiltonian for the molecular system is (footnote of Section 2.2.2)

$$\mathrm{H} = \mathrm{H}_e + \varphi_{n\text{-}e} + \varphi_{e\text{-}e} + \varphi_{n\text{-}n},$$

where e and n stand for electron and nucleus.

Table 2.8. *Quantum (wave–particle) wave function dynamics equation (Schrödinger equation).*

$$i\hbar\frac{\partial\Psi}{\partial t} = (-\frac{\hbar^2}{2m}\nabla^2 + \varphi)\Psi \equiv \mathrm{H}\Psi \quad \text{time dependent}$$

$$\mathrm{H} = -\frac{\hbar^2}{2m}\nabla^2 + \varphi(\boldsymbol{x}) \quad \text{quantum Hamiltonian operator}$$

$$E\psi = \mathrm{H}\psi \quad \text{time independent}$$

$$\varphi = \varphi_a + \varphi_c + \varphi_s \quad \text{various potentials}$$

$$\int \Psi^\dagger \Psi \mathrm{d}V = \int |\Psi|^2 \mathrm{d}V = 1 \quad \text{normalization condition } (V \text{ contains the particle})$$

† indicates complex conjugate (or in general Hermitian conjugate)

$$P_V = \int_V |\Psi|^2 \mathrm{d}V \quad \text{probability of finding the particle in } V$$

$$\langle\phi\rangle = \int_V \Psi^\dagger \phi \Psi \mathrm{d}V \quad \text{expectation value of quantity } \phi \text{ in } V$$

$$\int \Psi_m^\dagger \mathrm{H} \Psi_n \mathrm{d}V = \langle\Psi_m|\mathrm{H}|\Psi_n\rangle \quad \text{matrix element}$$

E	energy eigenstates, J
H	quantum Hamiltonian operator, J
Ψ, ψ	wave function (for matter wave), $1/\mathrm{m}^{3/2}$
φ	total potential energy, J
φ_a	applied potential energy, J
φ_c	crystal potential energy, J
φ_s	scattering potential energy, J

Note: The quantum electronic transition rates are determined based on a time-dependent perturbation theory as transition probabilities. This is referred to as the Fermi golden rule and is written as (derivation is given in Appendix E)

$$\dot{\gamma}_{1\text{-}2} = \frac{2\pi}{\hbar}|M_{1\text{-}2}|^2\delta_{\mathrm{D}}(E_1 - E_2 \pm \hbar\omega),$$

for transition between energy levels 1 and 2. Here $M_{1\text{-}2}$ is called the interaction matrix element and δ_{D} is the Dirac delta function (has units of J^{-1} as defined in Glossary). We use this for phonon absorption (–) and emission (+), in electron scattering (Section 5.10), and in photon absorption or emission in semiconductor (Section 7.8).

φ_a, crystal potential φ_c, and scattering potential φ_s (there may be others). For example, in a quantum harmonic oscillator (Section 2.6.4) $\varphi_c = \Gamma x^2/2$, while for electron orbiting H atom we will use (Section 2.6.6) the Coulomb potential (2.5). The quantum Hamiltonian operator was also defined in (2.19), but can have a more general form when multiple interactions (such as electronic-vibronic coupling) are considered.

Compared to the Helmholtz electromagnetic wave equation [given in Chapter 3, as (3.35)], the Schrödinger equation is first-order differential in time. Here $\Psi(\boldsymbol{x}, t)$

is the wave function and H is the Hamiltonian.† Using the separation of variables technique, this is written as a product of time-dependent and spatial-dependent parts as,

$$\Psi = \psi(\boldsymbol{x})\theta(t). \tag{2.63}$$

Then,

$$i\hbar\frac{d\theta}{dt}\frac{1}{\theta} = (-\frac{\hbar^2}{2m}\nabla^2 + \varphi)\psi = \mathrm{H}\psi = E\psi. \tag{2.64}$$

From these, we find,

$$\mathrm{H}\psi = (-\frac{\hbar^2}{2m}\nabla^2 + \varphi)\psi = E\psi, \tag{2.65}$$

$$\theta(t) = \exp(-iEt/\hbar), \text{ exponential time dependence.} \tag{2.66}$$

Because $\Psi(\boldsymbol{x})$ has infinite solutions, a linear combination of the solutions gives [131]

$$\Psi = \sum_{n=1}^{\infty} a_n\psi_n(\boldsymbol{x})\exp(-iE_nt/\hbar), \tag{2.67}$$

where E_n are the eigenstates.

The molecular Hamiltonian operator determines the dependence of the wave function on positions of electrons and the nuclei within the molecule. The potential energy φ portions of the Hamiltonian can for example be the Coulomb interaction between a pair of charged entities (2.5), where for electron $q = -e_c$ and for the nucleus it is ze_c, where z is the atomic number. In Section 2.2.2 (footnote), examples of electrons–electrons, electrons–nuclei, and nuclei–nuclei potential energies were given.‡

† The de Broglie quantum wave with $\boldsymbol{p} = \hbar\boldsymbol{\kappa}$, is

$$\Psi = A\exp[i(\boldsymbol{\kappa}\cdot\boldsymbol{x} - \omega t)] = A\exp[\frac{i}{\hbar}(\boldsymbol{p}\cdot\boldsymbol{x} - Et)].$$

Then the momentum is extracted from

$$\frac{\hbar}{i}\nabla\Psi = \boldsymbol{p}\Psi, \quad \frac{\hbar^2\nabla^2}{i^2} = -\hbar^2\nabla^2.$$

Similarly, the energy is extracted as

$$-\frac{\hbar}{i}\frac{\partial}{\partial t}\Psi = E\Psi.$$

Then, the classical Hamiltonian (Table 2.5) becomes

$$\mathrm{H} = \frac{p^2}{2m} + \varphi = E = -\frac{\hbar^2}{2m}\nabla^2\Psi + \varphi\Psi = i\hbar\frac{\partial\Psi}{\partial t},$$

which is (2.62).

‡ If we neglect the kinetic energy term for the nuclei, the electronic Hamiltonian in atomic units becomes (adding H_e to potential energies of expression in footnote of Section 2.2.2)

$$\mathrm{H} = -\frac{\hbar^2}{2m}\sum_i \nabla_e^2 - \frac{e_c^2}{4\pi\epsilon_o}(\sum_i^{N_e}\sum_I^{N_n}\frac{3I}{|\boldsymbol{R}_I - \boldsymbol{r}_i|} + \sum_i^{N_e}\sum_{j<i}^{N_e}\frac{1}{|\boldsymbol{r}_i - \boldsymbol{r}_j|} + \sum_I^{N_n}\sum_{J<I}^{N_n}\frac{z_I z_J}{|\boldsymbol{R}_I - \boldsymbol{R}_J|}),$$

where ∇_e^2 and $\boldsymbol{r}$ indicate position of electrons and $\boldsymbol{R}$ indicates the position of nuclei.

2.6.2 Bloch Wave Form

For a free particle ($\varphi = 0$) we have $E = p^2/2m = \hbar^2\kappa^2/2m$, and for a free particle in periodic arrangement the spatial part of ψ is expressed as $\exp(i\boldsymbol{\kappa}\cdot\boldsymbol{x})$, and $\boldsymbol{\kappa}$ varies continuously [131], so we can use an integral presentation. Then the wave function of a continuous wave packet takes the form (called the Bloch wave form)

$$\Psi = \frac{1}{(2\pi)^{1/2}}\int_{-\infty}^{\infty} g(\boldsymbol{\kappa})\exp[i(\boldsymbol{\kappa}\cdot\boldsymbol{x} - \frac{\hbar\kappa^2 t}{2m})]\mathrm{d}\boldsymbol{\kappa} \quad \text{Bloch wave form,} \qquad (2.68)$$

where $\boldsymbol{\kappa}$ is the wave number vector (or wave vector), and $(2\pi)^{-1/2}g(\boldsymbol{\kappa})\mathrm{d}\boldsymbol{\kappa}$ normalizes the wave packet. Here $g(\boldsymbol{\kappa})$ is an inverse Fourier transform [131] that allows for a short-form presentation of (2.68). Using (2.68), time and spatial operators in the time-dependent Schrödinger equation (2.62), become

$$\frac{\partial\Psi}{\partial t} = -i\frac{\hbar\kappa^2}{2m}\frac{1}{(2\pi)^{1/2}}\int_{-\infty}^{\infty} g(\boldsymbol{\kappa})\exp[i(\boldsymbol{\kappa}\cdot\boldsymbol{x} - \frac{\hbar\kappa^2 t}{2m})]\mathrm{d}\boldsymbol{\kappa}, \qquad (2.69)$$

$$\nabla^2\Psi = -\kappa^2\frac{1}{(2\pi)^{1/2}}\int_{-\infty}^{\infty} g(\boldsymbol{\kappa})\exp[i(\boldsymbol{\kappa}\cdot\boldsymbol{x} - \frac{\hbar\kappa^2 t}{2m})]\mathrm{d}\boldsymbol{\kappa}. \qquad (2.70)$$

2.6.3 Quantum-Mechanics Formalism, Bra–Ket and Matrix Element

In quantum mechanics, the state of the physical system is identified with a vector designated by $|\phi\rangle$, in a Hilbert space H. Hilbert space is a complete (including all limits) linear product space [131].[‡] The inner product of two functions $\psi(\boldsymbol{x})$ and $\phi(\boldsymbol{x})$ is defined as $\langle\psi|\phi\rangle = \int \psi^*(\boldsymbol{x})\phi(\boldsymbol{x})\mathrm{d}\boldsymbol{x}$. The inner product of two vectors $|\psi\rangle$ and $|\phi\rangle$ is written as $\langle\psi|\phi\rangle$. Each vector is called a ket, and written as $|\psi\rangle$. The bra–ket inner product, $\langle\psi|\phi\rangle$ is a complex quantity.

‡ For distinguishable (separable) particles we have for particles A and B, using the joint wave function Ψ^{AB} and individual states Ψ^{A} and Ψ^{B}

$$|\Psi\rangle^{\mathrm{A}} \in H^{\mathrm{A}}, \quad |\Psi\rangle^{\mathrm{B}} \in H^{\mathrm{B}},$$
$$|\Psi^{\mathrm{AB}}\rangle = |\Psi\rangle^{\mathrm{A}}|\Psi\rangle^{\mathrm{B}}.$$

For indistinguishable (tangled) particles, we have

$$|\Psi^{\mathrm{AB}}\rangle \neq |\Psi\rangle^{\mathrm{A}}|\Psi\rangle^{\mathrm{B}}.$$

Every ket has a dual bra, written as $\langle\psi|$, a continuous linear function on H defined as [along with its extension to include additional operator $A(\boldsymbol{x})$]

$$\langle\psi|\phi\rangle \equiv \int \psi^{\dagger}(\boldsymbol{x})\phi(\boldsymbol{x})\mathrm{d}\boldsymbol{x} \equiv (|\phi\rangle, |\psi\rangle),$$

$$\langle\psi|A|\phi\rangle = \int \psi^{\dagger}(\boldsymbol{x})A(\boldsymbol{x})\phi(\boldsymbol{x})\mathrm{d}\boldsymbol{x}. \tag{2.71}$$

The bra–ket (Dirac notation) is the standard notation for the quantum-mechanical state. The name comes from the linear product of two states denoted by a bracket $\langle\psi|\phi\rangle$ consisting of a left part $\langle\psi|$ bra and a right part $|\phi\rangle$ ket.

The commutator operator, operating on a function $f(\boldsymbol{x})$, is defined as

$$[A, B]f(\boldsymbol{x}) = (AB - BA)f(\boldsymbol{x}), \tag{2.72}$$

and if $[A, B] = 0$, then A and B commute, and otherwise they do not.

This is used in the uncertainty principle, for example,

$$[\boldsymbol{x}, \boldsymbol{p}]\Psi = -i\hbar, \tag{2.73}$$

(end-of-chapter problem), which states that position and momentum cannot be given precisely and simultaneously at a given time.

The matrix element i, f of the operator H (Hamiltonian) is defined as $\langle\psi_i|\mathrm{H}|\psi_f\rangle$. Here the operator represents the physical interaction that couples the initial and final states of the same carrier. This is a linear transformation and i and f are the matrix elements of operator H [131]. Examples are given in Sections 3.2.3, 4.9.4, 4.15, 5.10, 5.15, 7.8.1, and 7.12, and in Appendix E. We now discuss solutions to (2.62) for the simple harmonic oscillator, electron gas, and electron in hydrogenlike atoms. We also discuss the perturbation and numerical techniques and for solving the Schrödinger equation.

2.6.4 Quantum Mechanical, Harmonic Oscillator

We begin with the quantum-harmonic oscillator to demonstrate the eigenvalue nature of quantum energy and the probability nature of the quantum-particle location. Consider a particle of mass m moving in one direction along x (taken to be the same as displacement d) and connected to a fixed point with a spring constant Γ, with its potential energy ($F = -\mathrm{d}\varphi/\mathrm{d}x = -\Gamma d = -\Gamma x$, when we place $x = 0$ at the equilibrium position) given by

$$\varphi = \frac{1}{2}\Gamma x^2 = \frac{1}{2}m\omega^2 x^2, \tag{2.74}$$

where $\omega = (\Gamma/m)^{1/2}$ is its natural angular frequency given by (2.54).

In the classical treatment based on the Newton law described in Section 2.5.3, this oscillator experienced sinusoidal motion with natural frequency given by (2.54). For this simple (potential is proportional to displacement squared) harmonic system, the maximum kinetic and potential energies are equal and each is equal to $3k_{\mathrm{B}}T/2$.

The quantum-mechanical description of a simple harmonic oscillation begins with the Schrödinger equation giving the probability of finding the atom at a given location $\boldsymbol{x}$. The solution gives the eigenvalue E_n and $\psi_n(\boldsymbol{x})$ for the energy and the wave function, where n is the quantum number. Here we review the solution to the Schrödinger equation for this simple case.

Using the time-independent Schrödinger equation (Table 2.8), in one dimension, we have

$$(-\frac{\hbar^2}{2m}\frac{\mathrm{d}^2}{\mathrm{d}x^2} + \frac{1}{2}m\omega^2 x^2)\psi = E\psi \ \text{ for } \varphi = \frac{1}{2}m\omega^2 x^2. \tag{2.75}$$

We write this in dimensionless form as

$$(-\frac{\mathrm{d}^2}{\mathrm{d}x^2} - \frac{m^2\omega^2}{\hbar^2}x^2)\psi = -\frac{2mE}{\hbar^2}\psi. \tag{2.76}$$

Two solution methods are available [131], the power-series expansion and the ladder operator.†

Here we use the power series solution, and begin by writing (2.76) as

$$\frac{\mathrm{d}^2\psi}{\mathrm{d}x^{*2}} + (E^* - x^{*2})\psi = 0, \tag{2.77}$$

where the dimensionless energy and coordinate are defined as

$$E^* = \frac{2E}{\hbar\omega}, \quad x^* = x(\frac{m\omega}{\hbar})^{1/2}. \tag{2.78}$$

Now, consider a solution to (2.77) in the form

$$\psi = X(x^*)\exp(-\frac{x^{*2}}{2}). \tag{2.79}$$

† The ladder operators ($a^\dagger$ is the raising or creation operator and a is the lowering or annihilation operator), are defined as (in one dimension x) and the Hamiltonian operator is related to these as

$$a^\dagger = \frac{1}{2^{1/2}m}(\frac{\hbar}{i}\frac{\mathrm{d}}{\mathrm{d}x} + im\omega x),$$

$$a = \frac{1}{2^{1/2}m}(\frac{\hbar}{i}\frac{\mathrm{d}}{\mathrm{d}x} - im\omega x)$$

$$\mathrm{H} = \hbar\omega(a^\dagger a + \frac{1}{2}).$$

The commentator operator (2.72) is used to find H from [131]

$$a^\dagger a = \mathrm{H} - \frac{1}{2}\hbar\omega, \ \ \mathrm{H} = \frac{1}{2m}[(\frac{\hbar}{i}\frac{\mathrm{d}}{\mathrm{d}x})^2 + (m\omega^2 x)^2].$$

Using these, Schrödinger equation (2.75) becomes

$$\mathrm{H}\psi = \hbar\omega(a^\dagger a + \frac{1}{2})\psi = E\psi.$$

Using the ket notation, we have from the property of ladder operators,

$$a^\dagger|\psi_n\rangle \rightarrow |\psi_{n+1}\rangle$$

$$a|\psi_n\rangle \rightarrow |\psi_{n-1}\rangle.$$

Here we use $(a, a^\dagger)$ for electron, $(b, b^\dagger)$ for phonon, and $(c, c^\dagger)$ for photon.

Substituting this solution into (2.77), we have

$$X'' + x^* X' + (E^* - 1)X = 0. \tag{2.80}$$

This is the harmonic differential equation and with $\lambda \equiv E^* - 1$, and the solution is

$$X(x^*) = A\, H_{\lambda/2}(x^*) + B\, {}_1F_1(-\frac{1}{4}\lambda;\ \frac{1}{2};\ x^{*2}), \tag{2.81}$$

where A and B are constants, $H_n(x^*)$ is a Hermite polynomial [1], and

${}_1F_1(-\lambda/4; 1/2; x^{*2})$ is a confluent hypergeometric function [1] of the first kind and is given as a series in x^* with the leading term being unity and the second term being $(-\lambda/4)x^*/(1/2)$. The first few terms of $H_n(x^*)$ are given in [1, 131] (for example, $H_0 = 1$, $H_1 = 2x^*$, $H_2 = 4x^* - 2$, $H_3 = 8x^{*3} - 12x^*$).

The solution to the original differential equation is

$$\psi(x^*) = e^{-x^{*2}/2}[A\, H_{\lambda/2}(x^*) + B\, {}_1F_1(-\frac{1}{4}\lambda;\ \frac{1}{2};\ x^{*2})]. \tag{2.82}$$

Because for $x^* \to \infty$ the wave function should diminish, the coefficient B should then be set to zero. Then we have

$$\psi(x^*) = Ae^{-x^{*2}/2} H_{\lambda/2}(x^*). \tag{2.83}$$

This solution always converges, but in order for the wave function to be square integrable, λ must be an even integer, i.e.,

$$\lambda_n = \frac{2E}{\hbar\omega} - 1 = 2n, \quad n = 0, 1, 2, \ldots, \tag{2.84}$$

which gives allowed energies as

$$E_n = \hbar\omega(n + \frac{1}{2}), \quad n = 0, 1, 2 \ldots \quad \text{fundamental quantization condition.} \tag{2.85}$$

The ground state is designated as $n = 0$ and is populated at $T = 0$ K. Then $\hbar\omega/2$ is called the zero-point energy (motion) and is a quantum-mechanical phenomenon. The eigenfunctions (wave functions) are

$$\psi_n = C H_n(x^*)e^{-x^{*2}/2}. \tag{2.86}$$

Here, C is the normalization constant that is determined by the orthogonality of ψ_n and the final solution (dimensionless and dimensional) is

$$\begin{aligned}\psi_n &= (\frac{m\omega}{\pi\hbar})^{1/4}(\frac{1}{2^n n!})^{1/2} H_n(x^*)e^{-x^{*2}/2}\\ &= (\frac{m\omega}{\pi\hbar})^{1/4}(\frac{1}{2^n n!})^{1/2} H_n[(\frac{m\omega}{\hbar})^{1/2}x]\exp(-\frac{m\omega x^2}{2\hbar}).\end{aligned} \tag{2.87}$$

The first four dimensionless wave functions ψ_n^*, are listed in Table 2.9, where $\psi^* = \psi/(m\omega/\hbar)^{1/4}$.

These are plotted (qualitatively) in Figure 2.16. For higher quantum numbers, the wave function extends to higher values of x^*, which allow for the probability

Table 2.9. *The first four dimensionless wave functions for the quantum-harmonic oscillator,* $x^* = x(m\omega/\hbar)^{1/2}$ *[240].*

Quantum Number	Wave Function
0	$\psi_0^* = \pi^{-1/4}\exp(-x^{*2}/2)$
1	$\psi_1^* = (\frac{4}{\pi})^{1/4}x^*\exp(-x^{*2}/2)$
2	$\psi_2^* = (4\pi)^{-1/4}(2x^{*2}-1)\exp(-x^{*2}/2)$
3	$\psi_3^* = (9\pi)^{-1/4}(2x^{*3}-3x^*)\exp(-x^{*2}/2)$

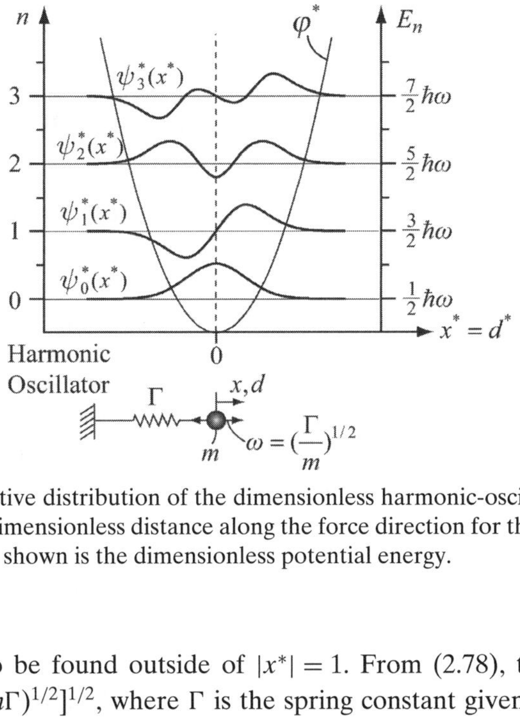

Figure 2.16. Qualitative distribution of the dimensionless harmonic-oscillator wave function with respect to the dimensionless distance along the force direction for the first four quantum numbers [240]. Also shown is the dimensionless potential energy.

of the particles to be found outside of $|x^*| = 1$. From (2.78), this would be $x > (\hbar/m\omega)^{1/2} = [\hbar/(m\Gamma)^{1/2}]^{1/2}$, where Γ is the spring constant given in (2.54). The dimensionless potential energy φ^* is also shown. Further graphical presentation of the wave function can be found in [41]. For the classical oscillator, $|x^*|$ is limited to ≤ 1 (end-of-chapter problem).

Although the calculation of $\psi_n(x)$ requires evaluation of H_n (examples given in [131]), a general result can be derived based on (2.87). The partition function (2.27) for this simple harmonic oscillator (with degeneracy $g_n = 1$) is

$$Z = \sum_{n=0}^{\infty}\exp[-\frac{(n+1/2)\hbar\omega}{k_\mathrm{B}T}] = \exp(-\frac{\hbar\omega}{2k_\mathrm{B}T})\sum_{n=0}^{\infty}[\exp(-\frac{\hbar\omega}{k_\mathrm{B}T})]^n$$

$$= \frac{\exp(-\frac{\hbar\omega}{2k_\mathrm{B}T})}{1-\exp(-\frac{\hbar\omega}{k_\mathrm{B}T})} = \frac{\exp(-\frac{\hbar\omega}{k_\mathrm{B}T})}{\exp(-\frac{\hbar\omega}{k_\mathrm{B}T})-1}, \quad (2.88)$$

so

$$\ln Z = -\frac{\hbar\omega}{2k_B T} - \ln[1 - \exp(-\frac{\hbar\omega}{k_B T})]. \tag{2.89}$$

Relation (2.88) gives the Bose–Einstein (boson) distribution function given in Table 1.2 (a more detailed derivation of f_p^o is given in Appendix F). This is characteristic of phonons and photons. Finally, using Table 2.4, we have for the mean energy

$$\frac{\langle E \rangle}{N} = k_B T^2 \frac{\partial \ln Z}{\partial T} = \hbar\omega[\frac{1}{2} + \frac{1}{\exp(\frac{\hbar\omega}{k_B T}) - 1}]$$

$$\text{mean energy of quantum oscillator.} \tag{2.90}$$

Note that in comparison the mean energy of the classical oscillator is $3k_B T/2$.

2.6.5 Periodic, Free Electron (Gas) Model for Metals

The simplest electron system is the periodic, free (conduction) electron gas, where there is no potential, i.e., $\varphi = 0$. This is applicable to solid metals and is called the Drude–Sommerfeld electron gas model. If there are $n_{e,c}$ electrons per volume and the electrons do not interact (are prescribed within a periodic, three-dimensional space), then the ground state of the electron energy is found by examining a single electron in a volume V and the additional free electrons, while following the Pauli exclusion principle (two electrons per energy level). This volume is found from the number of free electrons per unit volume $n_{e,c}$, which in turn is found from the properties of metallic elements (valence electrons that are the outmost orbitals in Table A.2) and their densities in solid state. Then from the solution of the Schrödinger equation for a free electron in a box, we will also related this number density $n_{e,c}$ with the number of quantum states (and through this define the Fermi energy).

The wave function $\psi_e(\boldsymbol{x})$ with two possible spins and energy E_e satisfies the time-independent Schrödinger equation (Table 2.8), i.e.,

$$-\frac{\hbar^2}{2m_e}\nabla^2\psi_e(\boldsymbol{x}) = E_e\psi_e(\boldsymbol{x}) \text{ for } \varphi = 0. \tag{2.91}$$

The periodic boundary conditions on a cube $L^3 = V$ is

$$\psi_e(x, y, z) = \psi_e(x, y, z + L), \tag{2.92}$$

etc. The solution† is

$$\psi_{e,\kappa}(\boldsymbol{x}) = L^{-3/2} e^{i(\boldsymbol{\kappa}\cdot\boldsymbol{x})} \quad \text{standing plane (Bloch) wave,} \tag{2.93}$$

with the electron energy (kinetic energy) given by

$$E_e = E_{e,k}(\boldsymbol{\kappa}) = \frac{\hbar^2\kappa^2}{2m_e}. \tag{2.94}$$

This relation between E_e and κ is called the parabolic relation.

The electron wave function $\psi_{e,\kappa}$ satisfies the normalization condition (Table 2.8). Applying the boundary condition (at L and at 0 for all three directions) gives (this is called the particle-in-box solution)

$$e^{i\kappa_x L} = e^{i\kappa_y L} = e^{i\kappa_z L} = e^0 = 1, \tag{2.95}$$

which shows that κ_i has discrete (quantized) values

$$\kappa_x = \frac{2\pi n_x}{L},\ \kappa_y = \frac{2\pi n_y}{L},\ \kappa_z = \frac{2\pi n_z}{L},\ n_i = 1, 2, 3, \cdots . \tag{2.96}$$

Figure 2.17(a) shows the one-dimensional (x direction) wave function (sine function) that satisfies $\psi_{e,n}(0) = \psi_{e,n}(L) = 0$ for $n = 1,\ 2,$ and 3. The corresponding one-dimensional energy $E_{e,n}$ is also shown.

In these dimensions, we have $\kappa^2 = \kappa_x^2 + \kappa_y^2 + \kappa_z^2$. An electron in the level $\psi_{e,\kappa}(\boldsymbol{x})$ has a momentum, velocity, and energy given by

$$\boldsymbol{p}_e = \hbar\boldsymbol{\kappa}, \quad \boldsymbol{u}_e = \frac{\hbar\boldsymbol{\kappa}}{m_e}, \quad E_e = \frac{p_e^2}{2m_e} = \frac{1}{2}m_e u_e^2 = \frac{\hbar^2\kappa^2}{2m_e}. \tag{2.97}$$

This constructs the $\boldsymbol{\kappa}$-space or wave-vector space, noting that the wave number is $2\pi/\lambda$, where λ is the wavelength. Figure 2.17(b) shows the two-dimensional Cartesian κ_x–κ_y space, with the coordinates $\kappa_x = 2\pi n_x/L$ and $\kappa_y = 2\pi n_y/L$. Then in the three-dimensional $\boldsymbol{\kappa}$-space of unit $(2\pi/L)^3$, the number of allowed $\boldsymbol{\kappa}$ values per unit κ space is $(L/2\pi)^3 = V/8\pi^3$. This also means that there is one distinct triplet quantum-numbers set $(\kappa_x, \kappa_y, \kappa_z)$ for volume $(2\pi/L^3)$ in the $\boldsymbol{\kappa}$-space. Now, for very large N_e, we can consider the volume to be spherical and referring to the radius of this sphere in the $\boldsymbol{\kappa}$-space as $\kappa_{\rm F}$ (having a volume $4\pi\kappa_{\rm F}^3/3$), the number of allowed values of κ (or orbitals) within this sphere is [Figure 2.17(b)]

† Using (2.93) in (2.91), we have

$$-(\frac{\partial^2}{\partial x^2} + \frac{\partial^2}{\partial y^2} + \frac{\partial^2}{\partial z^2})\psi_\kappa(\boldsymbol{x}) = \kappa^2\psi_\kappa(\boldsymbol{x})$$

$$\kappa^2 = \kappa_x^2 + \kappa_y^2 + \kappa_z^2.$$

Then separating the variables, we have

$$\frac{\partial^2\psi_{\kappa,x}(x)}{\partial x^2} + \kappa_x^2\psi_{\kappa,x}(x) = 0, \ \text{etc.}$$

The solution is $\psi_\kappa(\boldsymbol{x}) = A\exp[i(\kappa_x x + \kappa_y y + \kappa_z z)] = A\exp[i(\boldsymbol{\kappa}\cdot\boldsymbol{x})]$. Then A is determined from the normalization condition (Table 2.8), which gives $A = V^{-1/2} = L^{-3/2}$ (end-of-chapter problem).

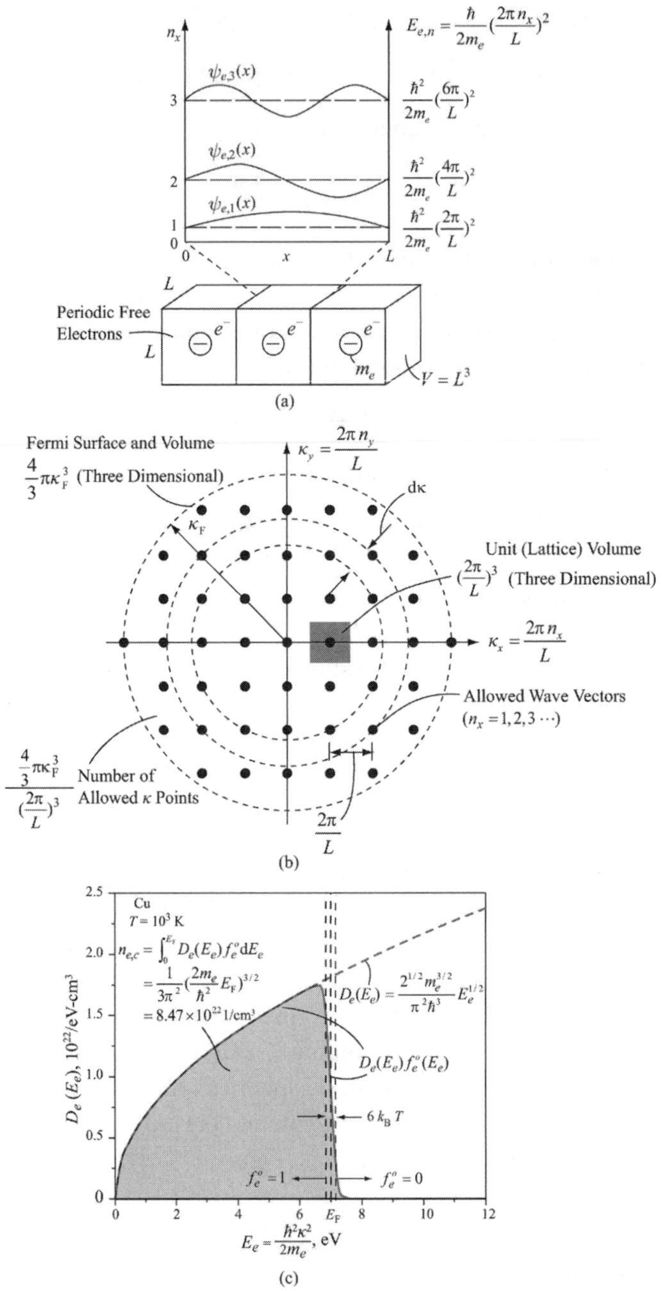

Figure 2.17. (a) Distribution of the one-dimensional (x direction) free-electron wave function $\psi_{e,\kappa_x}(x)$ and its energy, for $n_x = 1,\ 2,$ and 3. (b) κ-space shown in two dimensions, where each κ point occupies a space $(2\pi/L)^3 = 8\pi^3/V$ [9, 183]. (c) Variation of electron density of states, as a function of electron energy for Cu at $T = 10^3$ K.

$$\frac{4\pi\kappa_F^3/3}{(2\pi/L^3)} = \frac{4\pi\kappa_F^3}{3}\frac{V}{8\pi^3} = \frac{\kappa_F^3}{6\pi^2}V \text{ number of allowed } \kappa \text{ points in } \kappa_F \text{ volume.} \quad (2.98)$$

Below, we will show that V depends on the number of free electrons in metallic solids. Note that the wave number and energy of electron density of states (Glossary) are related based on Figure 2.17(b) as

$$D_e(\kappa)\mathrm{d}\kappa = \frac{4\pi\kappa^2\mathrm{d}\kappa}{(2\pi)^3},\ D_e(\kappa) = \int_0^{\kappa_F} D_e(\kappa)\mathrm{d}\kappa \text{ electron density of states,} \quad (2.99)$$

$$D_e(E_e)f_e^{\mathrm{o}}(E_e)\mathrm{d}E_e = \mathrm{d}n_{e,c},\ D_e(E_e) = \frac{2^{1/2}m_e^{3/2}}{\pi^2\hbar^3}E_e^{1/2},$$

and D_e is the number of electronic states (allowed $\boldsymbol{\kappa}$ values) in the increment $\mathrm{d}\kappa$ of a sphere of radius κ. Figure 2.17(c) shows the variation of energy density of states $D_e(E_e)$ for copper electron gas, at $T = 10^3$ K. As shown in the figure, it is common to assume that $f_e^{\mathrm{o}}(E_e) = 1$ for $E_e \le E_F$, and zero beyond that.

Because each allowed κ is for two one-electron levels (for each spin), then using (2.98) or performing the integration over $\mathrm{d}E_e$ in (2.99), up to E_F, we have

$$N_{e,c} = 2\frac{\kappa_F^3}{6\pi^2}V, \quad (2.100)$$

$$n_{e,c} = \frac{N_e}{V} = \frac{\kappa_F^3}{3\pi^2} = \frac{1}{3\pi^2}(\frac{2m_e}{\hbar^2}E_F)^{3/2}, \quad (2.101)$$

where the Fermi energy is found from (2.97) as

$$E_F = \frac{\hbar^2\kappa_F^2}{2m_e}. \quad (2.102)$$

For conduction electrons, $n_{e,c}$ is also given by

$$n_{e,c} = \frac{N_A}{M}\rho z_e, \quad (2.103)$$

where ρ is the density, and z_e is the number of free electrons (i.e., valence electrons, outermost orbitals in Table A.2, in metallic bond) per atom. This is also listed as one of the oxidation states in Table A.2. Then

$$\kappa_F = (3\pi^2\frac{N_A}{M}\rho z_e)^{1/3}. \quad (2.104)$$

Equations (2.103) and (2.104) show that the Fermi energy and the number density of free electrons are related. The tabulated results for solid metals will be given in Table 5.2; for example, for copper, $z_e = 1$, $n_{e,c} = 8.47 \times 10^{22}$ 1/cm^3, and $E_F = 7.00$ eV. Metals $n_{e,c}$ and E_F (which is also closely related to the chemical potential μ) are directly related and are independent of temperature, and $n_{e,c}$ in semiconductors is temperature dependent (zero at $T = 0$ K, and increasing with increase in T). We will discuss these in Sections 5.6 and 5.7.

By defining a radius r_s containing a conduction electron, i.e.,

$$\frac{1}{n_{e,c}} = \frac{4\pi r_s^3}{3}, \quad (2.105)$$

we have

$$\kappa_{\rm F} = \frac{(\frac{9\pi}{4})^{1/3}}{r_s} = \frac{1.92}{r_s}. \tag{2.106}$$

The Fermi velocity is found from the momentum, i.e.,

$$u_{\rm F} = \frac{\hbar}{m_e}\kappa_{\rm F}. \tag{2.107}$$

The energy density of the electron gas is (for two spins) [9]

$$\begin{aligned}\frac{\langle E_e\rangle}{V} &= \frac{2}{(2\pi)^3}\int E_e \mathrm{d}\kappa \\ &= \frac{1}{4\pi^3}\int_0^{\kappa_{\rm F}} \mathrm{d}\kappa \frac{\hbar^2\kappa^2}{2m_e} = \frac{1}{4\pi^3}\frac{\hbar^2}{2m_e}\int_0^{\kappa_{\rm F}} 4\pi\kappa^4 \mathrm{d}\kappa \\ &= \frac{1}{\pi^2}\frac{\hbar^2\kappa_{\rm F}^5}{10m_e}.\end{aligned} \tag{2.108}$$

We will further discuss this integral over $\mathrm{d}\kappa$ is further discussed in Chapters 3 and 4.

Using (2.101), $N_e/V = \kappa_{\rm F}^3/3\pi^2$, the energy per electron, in ground state, is

$$\frac{\langle E_e\rangle}{N_{e,c}} = \frac{3}{10}\frac{\hbar^2\kappa_{\rm F}^2}{2m_e} = \frac{3}{5}E_{\rm F} \equiv \frac{3}{5}k_{\rm B}T_{\rm F}, \quad T_{\rm F} = \frac{E_{\rm F}}{k_{\rm B}} \quad \text{Fermi temperature.} \tag{2.109}$$

This constant $\langle E_e\rangle$ is one of this limitations of the Drude–Sommerfeld model. We will further discuss the free-electron gas model for metals in Section 5.6. This simple model again allows for an analytical solution to the Schrödinger equation and the resulting discretization of energy. We will now do another analytical solution.

2.6.6 Electron Orbitals in Hydrogenlike Atoms

In the hydrogen atom model, the electron interacts with a stationary proton (nucleus), through the Coulombic potential (2.5), i.e.,

$$\varphi_e = -\frac{1}{4\pi\epsilon_{\rm o}}\frac{e_c^2}{r}. \tag{2.110}$$

Then, using the spherical coordinates (r, θ, ϕ), the time-independent Schrödinger equation (Table 2.8) becomes [240]

$$\begin{aligned}(-\frac{\hbar^2}{2m_e}\nabla^2 + \varphi_e)\psi_e &= -\frac{\hbar^2}{2m_e}[\frac{1}{r^2}\frac{\partial}{\partial r}(r^2\frac{\partial}{\partial r}) + \frac{1}{r^2\sin\theta}\frac{\partial}{\partial\theta}(\sin\theta\frac{\partial}{\partial\theta}) \\ &\quad + \frac{1}{r^2}\frac{1}{\sin^2\theta}\frac{\partial^2}{\partial\phi^2} - \frac{e_e^2}{4\pi\epsilon_{\rm o}r}]\psi_e \\ &= E_e\psi_e.\end{aligned} \tag{2.111}$$

Then variables are separated by use of

$$\psi_e(r, \theta, \phi) = R_e(r)\Theta_e(\theta, \phi) \quad \text{radial angular decomposition.} \tag{2.112}$$

Then when this is substituted, the equation for $R_e(r)$ is called the radial equation. The equation for P_e is called the angular equation. The general form of the wave function (in terms of the first three quantum numbers) is [solution to (2.111)] [131, 240]

$$\psi_{e,nlm} = \{(\frac{2}{nr_B})^3 \frac{(n-l-1)!}{2n[(n+1)!]}\}^{1/2} e^{-r^*/n} (\frac{2r^*}{n})^l \mathrm{L}_{n-l-1}^{2l+1}(\frac{r^*}{n}) \mathrm{Y}_{l,m}(\theta, \phi), \qquad (2.113)$$

where L is the associated Laguerre (orthogonal confluent hypergeometric) polynomial, and Y is a spherical harmonic function [1].

There are four quantum numbers, n, l, m, and s. The first one n, is called the principal quantum number such that (it does not include the fine-structure that is due to spin)

$$E_{e,n} = -\frac{e_c^2}{8\pi \epsilon_o r_B n^2}, \quad n = 1,\ 2,\ 3, \ldots, \qquad (2.114)$$

where r_B is the Bohr radius defined in Section 1.5.2 and Table 1.4. This is the kinetic energy given in Section 2.1.3 (footnote). The total energy (kinetic and potential, which is $-e_c^2/4\pi\epsilon_o r_B n^2$, is also given there.

The second one, l, is the azimuthal quantum number, and $l = 0,\ 1, \ldots,\ n-1$. Also,

$$\begin{aligned} &l = 0 \text{ is designated by } s,\\ &l = 1 \text{ is designated by } p,\\ &l = 2 \text{ is designated by } d,\\ &l = 3 \text{ is designated by } f, \end{aligned} \qquad (2.115)$$

followed by g, h, i, etc. Here s stands for sharp, p for principal, d for diffuse, and f for fundamental. Then for $n = 1$ and $l = 0$, we have 1s, and for $n = 2, l = 0$, we have 2s, etc.

The third quantum number, m, is called the magnetic quantum number, and it takes on values of $2l + 1$, i.e., $m = 0,\ \pm 1,\ \pm 2,\ \ldots, \pm l$.

The fourth quantum number is the spin quantum number (Section 1.1.1). The first electron is designated $s = +1/2$ and the next are $-1/2$, and then repeated.

The lowest energy state for H is 1s and the radial function is

$$R_{e,1s}(r) = \frac{2}{r_B^{3/2}} e^{-r/r_B}, \qquad (2.116)$$

and the complete 1s wave function is

$$\psi_{e,1s}(r) = \frac{1}{\pi^{1/2} r_B^{3/2}} e^{-r/r_B}. \qquad (2.117)$$

The local probability of having a 1s electron between r and r+dr is (Table 2.8)

$$P(r) = 4\pi \int_r^{r+dr} \psi_{e,1s}^2(r) r^2 dr = \frac{4}{r_B^3} r^2 e^{-2r/r_B} dr \quad \text{local probability}. \qquad (2.118)$$

Then local probability $P(r = 0) = P(r \to \infty) = 0$, with a peak near $r/r_B = 1.5$. Note that in Table 2.8 the probability P_V is integrated over the entire volume V.

The electron wave functions for $n = 1$, 2, and 3 for hydrogenlike atomic wave functions (where z is atomic number) and $r^* = zr/r_B$ are listed in Table 2.10 [240].

The possibility distribution P_e of some of the hydrogen atom electron orbitals shown in Figures 2.5(a)–(e). Note that in Figure 2.5(c), the x–y plane is used showing $3d_{xy}$ instead of $3d_{xz}$ shown in Table 2.10 (the choice of plane is arbitrary).

Figures 2.18(a) and (b) show the variation of the projected angular function of the $3d_{xy}$ wave function and the electron orbital (probability). Figure 2.19 shows radial variation of angular integral of the $\psi^{\dagger}_{3d_{xy}}\psi_{3d_{xy}}$ with respect to r^*. For the latter, the probability P_V (Table 2.8) of finding the $3d_{xy}$ electron within some radial locations are also shown (the 99% probability is for $r^* = 21.9$).

In Sections 2.6.4, 2.6.5, and 2.6.6, we reviewed three exact, analytical solutions to the Schrödinger equation; we now proceed to discuss perturbation and numerical solution methods.

2.6.7 Perturbation and Numerical Solutions to Schrödinger Equation

In addition to the preceding three exact solutions, we review the exact solutions to the Schrödinger equation for the case of an electron in a one-dimensional, periodic, ionic potential in Section 5.1 and for the translational motion of an ideal gas in a box in Section 6.1. These are other exact (analytic) solutions as discussed in [41, 131]. Because the electron–nuclei interactions are strong and the practical ones are many-body systems, the exact solutions are not possible and practical. There are also approximate (including numerical) solutions, including the perturbation theory, the variational principle, quantum Monte Carlo, DFT, Wentzel–Kramers–Brillouin (W-K-B) approximation, discrete delta-potential method, pseudo-potential method, etc. [308].

The development of the DFT and the accuracy of the local density approximation (LDA) and generalized gradient approximation (GGA) have allowed for useful quantum mechanics calculations.

The perturbation theory approximates the solution to the Schrödinger equation assuming the external disturbance to be a small perturbation, then by using solutions to simpler systems (i.e., simpler or weaker Hamiltonians). In the time-independent perturbation theory, the perturbation Hamiltonian is static. The solution begins with an unperturbed Hamiltonian $\mathrm{H_o}$ giving a set of discrete energy levels $E_{\mathrm{o},n}$ from the Schrödinger equation written as

$$\mathrm{H_o}|n_\mathrm{o}\rangle = E_{\mathrm{o},n}|n_\mathrm{o}\rangle, \quad n = 1,\ 2,\ 3, \ldots. \tag{2.119}$$

Then the perturbation Hamiltonian H' is introduced, which represents a weak disturbance, along with a perturbation parameter ϵ $(0 \le \epsilon \le 1)$, such that

$$\mathrm{H} = \mathrm{H_o} + \mathrm{H}' = \mathrm{H_o} + \epsilon \mathrm{H_1} + \epsilon^2 \mathrm{H_2} + \cdots$$

$$(\mathrm{H_o} + \epsilon\mathrm{H}')|n\rangle = E_n|n\rangle, \quad \text{first-order perturbation in } \mathrm{H}_{1,n}. \tag{2.120}$$

Table 2.10. *Hydrogenlike atom, electron wave functions for the principal quantum numbers, 1, 2, and 3, where z is the atomic number, and the dimensionless radial position is* $r^* = zr/r_B$ *[240]. The spin number s is* $+1/2$ *for first electron and* $-1/2$ *for the next, and then repeated.*

Principal	Azimuthal	Magnetic	Wave function
$n=1$	$l=0$	$m=0$	$\psi_{1s} = \frac{1}{\pi^{1/2}}(\frac{z}{r_B})^{3/2}e^{-r^*}$
$n=2$	$l=0$	$m=0$	$\psi_{2s} = \frac{1}{4(2\pi)^{1/2}}(\frac{z}{r_B})^{3/2}(2-r^*)e^{-r^*/2}$
	$l=1$	$m=0$	$\psi_{2p_z} = \frac{1}{4(2\pi)^{1/2}}(\frac{z}{r_B})^{3/2}r^*e^{-r^*/2}\cos\theta$
	$l=1$	$m\pm 1$	$\psi_{2p_x} = \frac{1}{4(2\pi)^{1/2}}(\frac{z}{r_B})^{3/2}r^*e^{-r^*/2}\sin\theta\cos\phi$
			$\psi_{2p_y} = \frac{1}{4(2\pi)^{1/2}}(\frac{z}{r_B})^{3/2}r^*e^{-r^*/2}\sin\theta\sin\phi$
$n=3$	$l=0$	$m=0$	$\psi_{3s} = \frac{1}{81(3\pi)^{1/2}}(\frac{z}{r_B})^{3/2}(27-18r^*+2r^*)e^{-r^*/3}$
	$l=1$	$m=0$	$\psi_{3p_z} = \frac{2^{1/2}}{81(\pi)^{1/2}}(\frac{z}{r_B})^{3/2}r^*(6-r^*)e^{-r^*/3}\cos\theta$
	$l=1$	$m=\pm 1$	$\psi_{3p_x} = \frac{2^{1/2}}{81(\pi)^{1/2}}(\frac{z}{r_B})^{3/2}r^*(6-r^*)e^{-r^*/3}\sin\theta\cos\theta$
			$\psi_{3p_y} = \frac{2^{1/2}}{81(\pi)^{1/2}}(\frac{z}{r_B})^{3/2}r^*(6-r^*)e^{-r^*/3}\sin\theta\sin\theta$
	$l=2$	$m=0$	$\psi_{3d_{z^2}} = \frac{1}{81(6\pi)^{1/2}}(\frac{z}{r_B})^{3/2}r^{*2}e^{-r^*/3}(3\cos^2\theta-1)$
	$l=2$	$m=\pm 1$	$\psi_{3d_{xz}} = \frac{2^{1/2}}{81(\pi)^{1/2}}(\frac{z}{r_B})^{3/2}r^{*2}e^{-r^*/3}\sin\theta\cos\theta\cos\phi$
			$\psi_{3d_{yz}} = \frac{2^{1/2}}{81(\pi)^{1/2}}(\frac{z}{r_B})^{3/2}r^{*2}e^{-r^*/3}\sin\theta\cos\theta\sin\phi$
	$l=2$	$m=\pm 2$	$\psi_{3d_{x^2-y^2}} = \frac{1}{81(2\pi)^{1/2}}(\frac{z}{r_B})^{3/2}r^{*2}e^{-r^*/3}\sin^2\theta\cos 2\phi$
			$\psi_{3d_{xy}} = \frac{1}{81(2\pi)^{1/2}}(\frac{z}{r_B})^{3/2}r^{*2}e^{-r^*/3}\sin^2\theta\sin 2\phi$

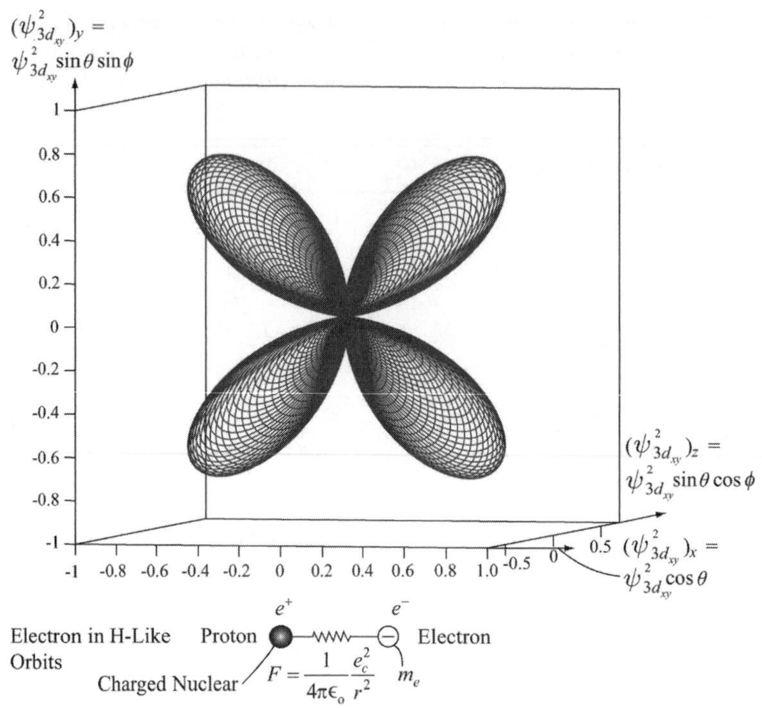

(a) Projected Variation of $3d_{xy}$ Wave Function $\psi = \cos^2\theta \sin 2\phi$

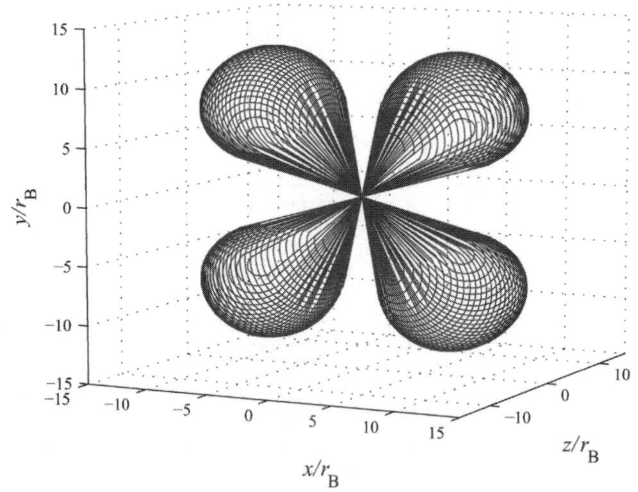

(b) Constant Probability Surface of $3d_{xy}$ Electron Orbital

Figure 2.18. (a) Projected, three-dimensional variation of $3d_{xy}$ wave function, and (b) three-dimensional electron orbital showing the surface of a constant probability.

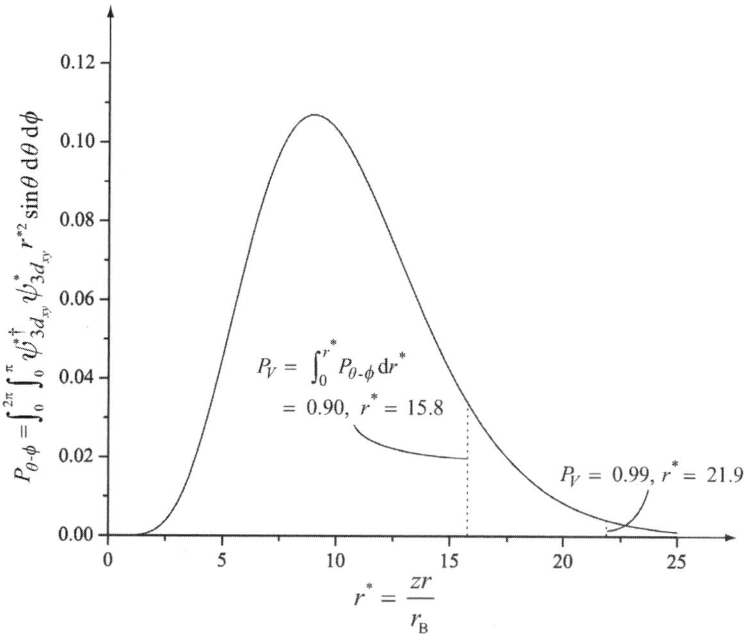

Figure 2.19. Distribution of the probability-related function with respect to dimensionless radial position for $3d_{xy}$ orbital. z is the atomic number.

In the perturbation approximation, instead of solving the preceding Schrödinger equation, a power-series solution is used for E_n, i.e.,

$$E_n = E_{o,n} + \epsilon \mathrm{H}_{1,n} + \epsilon^2 \mathrm{H}_{2,n} + \cdots ,$$

$$|n\rangle = |n_o\rangle + \epsilon |n_1\rangle + \epsilon^2 |n_2\rangle + \cdots . \quad (2.121)$$

Now using these expansions in (2.119) and sorting the terms with the same power of ϵ, a series of equations emerges. The first one is the first-order energy shift:

$$E_{1,n} = \langle n_o | \mathrm{H}_{1,n} | n_o \rangle \quad \text{expectation value of } \mathrm{H}'. \quad (2.122)$$

The perturbation changes the average energy of the state by $E_{1,n}$. There are higher-order terms that can be indicated.

We have discussed and will discuss again numerical solutions to the Schrödinger equation in Section 2.2.2 (Hartree–Fock approximation), Section 5.3 (tight-binding approximation), Section 5.5 (DFT), and Section 7.12.6 (time-dependent DFT).

The time-dependent perturbation theory (given in Appendix E) includes $\mathrm{H}'(t)$. This leads to the Fermi golden rule, which we use often (Sections 3.2.3, 4.9.4, 5.10, 5.15, 5.17.3, 7.8, and 7.12) in determining the various transition rates (atomic-level kinetics).

2.7 Problems

Problem 2.1

Using an available computer program for *ab initio* calculation of the interatomic potential (e.g., Gaussian), reproduce the O–O bond potential shown in Figure 2.6(a). After obtaining the *ab initio* potential, using a least-squares method, find the constants in the three interatomic potential models shown in Figure 2.6 and discussed in Section 2.2.2.

Problem 2.2

Using the L–J potential given by (2.9), the L–J constants for Xe given in (Table 2.2), and the Debye temperature given in Table A.2, determine the equilibrium nearest-neighbor location $r_{nn,e}$, the total potential energy $\langle\varphi\rangle$ at $r_{nn,e}$, the natural angular frequency ω_n, the Debye time constant $\tau_{\rm D}$, and the L–J time constant $\tau_{\rm LJ}$. Note that MD scales are given in Table 2.7. Comment on their magnitudes. Compare with any similar length, energy, frequency, and time scales you have experience with. Use SI units.

Problem 2.3

(a) Plot the MSD atomic displacement $\langle|\boldsymbol{\Delta}_i|^2\rangle^{1/2}$ for Xe as a function of temperature for $0 < T < 150$ K. This is similar to Figure 2.14 given for Ar. Comment on the zero Kelvin motion predicted by the quantum mechanical treatment of the atomic motion.

(b) In the same graph, plot the classical atomic displacement $x_{\rm o} = (3k_{\rm B}T/m\omega_n^2)^{1/2}$, using ω_n given by (2.54).

Problem 2.4

Using the two-dimensional MD Matlab code, choose a simulation unit-cell size (dimension) and the number of particles in it. Use the dimensionless form of the variables as they are in the code. Since the L–J potential is used, it can apply to noble-gas elements (Table 2.2), but computation is done for dimensionless variable.

(a) Then plot, as a function of time, the variation in the kinetic energy, potential energy, total energy, temperature, and pressure by taking the proper ensemble averages over this simulation (Table 2.4, in dimensionless forms). Note that for this simulation volume is replaced with area. (Add code lines to compute these quantities.)

(b) Comment on your observations about fluctuations. Would these fluctuations disappear as the simulation size (both the dimension and the number of particles, but in the same proportion) is increased? (Try doubling these values a few times.) Note that using $m_{\rm o}$, $\epsilon_{\rm LJ}$, and $\sigma_{\rm LJ}$, the dimensionless quantities become, $x^* - x/\sigma_{\rm LJ}$, $m = m/m_{\rm o} = 1$, $E^* = E/\epsilon_{\rm LJ}$, $u^* = u/(\epsilon_{\rm LJ}/m_{\rm o})^{1/2}$, $t^* = t/(m_{\rm o}\sigma_{\rm LJ}^2/\epsilon_{\rm LJ})^{1/2}$, etc., similar to those in Table 2.7 (where we use the effective FCC potential).

Problem 2.5

Show that the wave function solution to the time-independent Schrödinger equation for free-electron gas in a volume $V = L^3$ is (Section 2.6.5 footnote)

$$\psi_{e,\kappa} = L^{-3/2} \exp[i(\boldsymbol{\kappa} \cdot \boldsymbol{x})].$$

This is for a free quasi-particle in a box. Show the separation of variables and the exponential solutions, and evaluate the single proportionality constant of integration (which becomes $L^{-3/2}$ when the normalization condition is used).

Problem 2.6

For the free-electron-gas model, determine the electron number density, equivalent radius, the Fermi wave number, Fermi energy, Fermi velocity, and Fermi temperature for Li, Cu, and Bi. Check with the values available in Table 5.2.

Problem 2.7

Complete the steps in deriving the expression for $\Delta x_{\rm min}$ from the Heisenberg uncertainty principle (2.58).

Problem 2.8

In a polymer membrane fuel cell, the interatomic bond energy from hydrogen fuel is converted to electric potential energy. The chemical reactions that occur throughout the membrane are

hydrogen is oxidized	$2H_2 \rightarrow 4H^+ + 4e^-$,
oxygen is reduced	$O_2 + 4e^- \rightarrow 2O^{2-}$,
net reaction	$2H_2 + O_2 \rightarrow 2H_2O$.

In this reaction, the oxidation of two hydrogen gas molecules releases four electrons, which later recombine to reduce the oxygen in the formation of two H_2O molecules. Use Table 2.3 for the bond energies and show that the maximum electric potential for this reaction is 1.248 V.

Problem 2.9

(a) Using the dimensionless form of the solution for the wave function of quantum harmonic oscillator (2.87), plot $\psi_n/(m\omega/\pi\hbar)^{1/4}$ versus x^* $(-5 \le x^* \le +5)$, for $n = 0,\ 1,\ 2$ and 3. Compare the results with those shown in Figure 2.16.

(b) Comment on the trend with increasing n.

(c) What does ψ_n indicate about the probability of finding the particle at a given location x^*?

Problem 2.10

Compare the behavior of the classical and quantum simple harmonic oscillators.

(a) Show that the classical harmonic oscillator (Section 2.5.3) extends to $x_o = (\hbar/m\omega)^{1/2}$ only (use $m\omega^2 x_o^2/2 = E_o = \hbar\omega/2$).

(b) Using the definition for finding a particle in $x_o \le x < \infty$ given in Table 2.8, show that this probability for a one-dimensional motion and for the ground state ($n = 0$) is 0.157 (need to determine the error function for the interval 1 to ∞).

(c) Repeat (b) for $n = 1$.

Problem 2.11

Plot

(a) P_V versus x^*,

(b) dimensionless potential energy φ_n^* versus dimensionless distance x^* (quantitative), for the first four quantum numbers of the harmonic oscillator.

(c) Comment on the trends.

Problem 2.12

(a) Show that the effective equilibrium potential for the L–J FCC atomic structure is $\langle\varphi\rangle_e = -8.6067\epsilon_{\rm LJ}$, and the equilibrium separation distance is $r_{nn,e} = 1.090\sigma_{\rm LJ}$.

(b) Show that the isothermal bulk modulus $E_p = -v(\partial p/\partial v)_T = v\partial^2 e/\partial^2 v = v\partial/\partial v(\partial\langle\varphi\rangle/\partial v)$ [where $p = -(\partial e/\partial v)_{T=0}$ is used] is equal to $75.14\epsilon_{\rm LJ}/\sigma_{\rm LJ}^3$, for the L–J FCC atomic structure. Note that per particle, $v = a^3/4 = r^3/2^{1/2}$ (r is the interatomic spacing and a is the lattice constant) and use this to relate $\partial/\partial v$ to $\partial/\partial r$. Also note that at $r_{nn,e}$, $\partial\langle\varphi\rangle/\partial r = 0$.

Problem 2.13

(a) Energy broadening occurs in high scattering rates or atomic vibration phenomena. Calculate the energy uncertainty associated with vibration of Ar–Ar [use $\tau_{\rm LJ} = \Delta t$, in (2.58)].

(b) Using Ar again, determine Δx_{min} (2.59) for interatomic vibration displacement uncertainty and the associated uncertainty in x-direction momentum Δp_x. Use the natural frequency ω_n.

(c) Compare Δx_{min} in (b) with the zero-point atomic displacement, which is the leading term in (2.57).

Problem 2.14

(a) For the $3d_{xy}$ orbital, reproduce the graph angular distribution of the wave function, Figure 2.18(a). Note that the 3 projections are $x = \psi_{3d_{xy}}^2 \sin\theta\cos\phi$, $y = \psi_{3d_{xy}}^2 \sin\theta\sin\phi$, and $z = \psi_{3d_{xy}}^2 \cos\theta$, the standard transformation from spherical to Cartesian coordinate. Here $\psi_{3d_{xy}} = \sin^2\theta\sin 2\phi$.

(b) Also, reproduce Figure 2.19 showing the function related to the probability distribution (as a function of the dimensionless radial position) for finding the $3d_{xy}$ electron with the volume of radius r^*.

(c) What is the volume of r^* at which the probability $P_V = 0.5$?

(d) What is the energy of the hydrogen $3d_{xy}$ electron in eV?

Problem 2.15

The commutator operator $[A, B]$ has the property $[A, B] = AB - BA$.

Using momentum and Hamiltonian operators (footnote of Section 2.6.1), we have

$$\boldsymbol{p} \equiv \frac{\hbar}{i}\nabla \quad \text{and} \quad \mathrm{H} \equiv i\hbar\frac{\partial}{\partial t}.$$

Show that

(a) $[\boldsymbol{x}, \ \boldsymbol{p}] = i\hbar$, and
(b) $[\boldsymbol{x}, \ \mathrm{H}] = -\frac{i\hbar}{m_e}\boldsymbol{p}$.

3

Carrier Energy Transport and Transformation Theories

The Boltzmann transport equation (BTE) is based on classical Hamiltonian-statistical mechanics, as described in Section 2.4. The BTE gives the particle energy (in a system of particles) in terms of its position and momentum ($\boldsymbol{x}$, $\boldsymbol{p}$), and also allows for the determination of a nonequilibrium probability distribution of particles f_i under an applied force (on a return to equilibrium, after an initial nonequilibrium state). These distributions are used in determining transport coefficients under the influence of driving forces in cases of local nonequilibria.

The Maxwell equations describe the propagation of EM waves and their interactions with electronic entities. These are among the most useful fundamental equations (including the laws/relations of Gauss, Faraday, Ampere, and Ohm).

In treating carriers as particles, the fluctuation–dissipation transport theory associated with Green and Kubo [239] gives general expressions for the transport coefficients, valid at all times and densities, in terms of correlation or autocorrelation functions calculated from a system at equilibrium. The kinetic-theory-based transport coefficients require, as observed in experiments, imposition of an external gradient. These gradients lead to a breakdown of the correlation. Computer simulations (i.e., many-body problems, such as MD simulations) have provided equilibrium results, which are used in the G–K fluctuation dissipation theory, as well as in nonequilibrium theory. The nonequilibrium (imposed gradient) simulations require extrapolation to large systems, as well as having the complications of steep gradients and boundary conditions.

Stochastic transport processes can be described by stochastic particle dynamics equations, which are used for Brownian motion/diffusion and other transport phenomena.

In this chapter, these transport theories are reviewed to predict properties of the microscale carriers and their interaction rates. Macroscopic (continuum) equations for Newtonian fluid mechanics and elastic solid mechanics are also reviewed.

3.1 Boltzmann Transport Equation

3.1.1 Particle Probability Distribution (Occupancy) Function

In Sections 2.4 and 2.5 (also Appendix F) on classical particle dynamics, we discussed describing particles in a system by using particle position and momentum space (i.e., phase space). In a dilute gas, the fluid particles (atoms or molecules) are in free linear flight most of the time, and only in rare instances will their flight will be interrupted by a collision with another particle. The result of such an event, which in general may be treated as a classical elastic collision, is a change in the speeds and directions of both particles. Boltzmann derived, in the kinetic theory of gases [37], a statistical treatment of such binary collisions, as well as predicted the macroscopic properties of gases. He defined a particle probability distribution function f_f that denotes the fraction of particles that, at time t, are situated at a location $\boldsymbol{x}$ and have a given momentum $\boldsymbol{p}$ (in phase space). This BTE is also used for phonons, electrons, and photons, as well as their associated probability distribution functions f_p, f_e, and f_{ph}. We begin by noting that, in classical mechanics (Section 2.4), particles are described by their position and momentum, to which we now add time. Thus we have $\boldsymbol{x}$, $\boldsymbol{p}$, and t as the independent variables (i.e., seven-dimensional space).

3.1.2 A Simple Derivation of BTE

The equation governing the evolution of f is the BTE and its derivation starts with $f(\boldsymbol{x}, \boldsymbol{p}, t)$ being the fraction of particles whose positions and momenta are $\boldsymbol{x}$ and $\boldsymbol{p}$ at time t. If there were no collisions, then a short time Δt later, each particle would move from $\boldsymbol{x}$ to $\boldsymbol{x} + \boldsymbol{u}\Delta t$, and each particle momentum would change from $\boldsymbol{p}$ to $\boldsymbol{p} + \boldsymbol{F}\Delta t$, where $\boldsymbol{F}$ is the sum of the external forces on a particle at time t. Any difference between $f(\boldsymbol{x}, \boldsymbol{u}, t)$ and $f(\boldsymbol{x} + \boldsymbol{u}\Delta t, \boldsymbol{p} + \boldsymbol{F}\Delta t, t + \Delta t)$ is thus due to collisions. The collision term is set with respect to a collision rate $\partial f/\partial t|_s$ (this is not a formal derivative, but represents change) involving particles of positions $\boldsymbol{x}'$ and momenta $\boldsymbol{p}'$ over the time interval Δt entering $(\boldsymbol{x}, \boldsymbol{p})$. Figure 3.1(a) gives a one-dimensional presentation of the balance on f in the x–p_x space. Putting these terms together, we have [37, 216]

$$[f(\boldsymbol{x} + \boldsymbol{u}\Delta t, \boldsymbol{p} + \boldsymbol{F}\Delta t, t + \Delta t) - f(\boldsymbol{x}, \boldsymbol{p}, t)]\mathrm{d}\boldsymbol{x}\mathrm{d}\boldsymbol{p} = \frac{\partial f}{\partial t}|_s \mathrm{d}\boldsymbol{x}'\mathrm{d}\boldsymbol{p}'\Delta t, \tag{3.1}$$

where $\partial f/\partial t|_s$ is the time rate of change of f that is due to collisions. Expanding the first term on the left as a Taylor series about $f(\boldsymbol{x}, \boldsymbol{p}, t)$, we have

$$\begin{aligned} f(\boldsymbol{x} + \boldsymbol{u}\Delta t, \boldsymbol{p} + \mathbf{F}\Delta t, t + \Delta t) &= f(\boldsymbol{x}, \boldsymbol{p}, t) + (\frac{\partial f}{\partial x_j}u_j + \frac{\partial f}{\partial p_j}F_j + \frac{\partial f}{\partial t})\Delta t \\ &= f(\boldsymbol{x}, \boldsymbol{p}, t) + [(\nabla_x f) \cdot \boldsymbol{u} + (\nabla_p f) \cdot \boldsymbol{F} + \frac{\partial f}{\partial t}]\Delta t. \end{aligned} \tag{3.2}$$

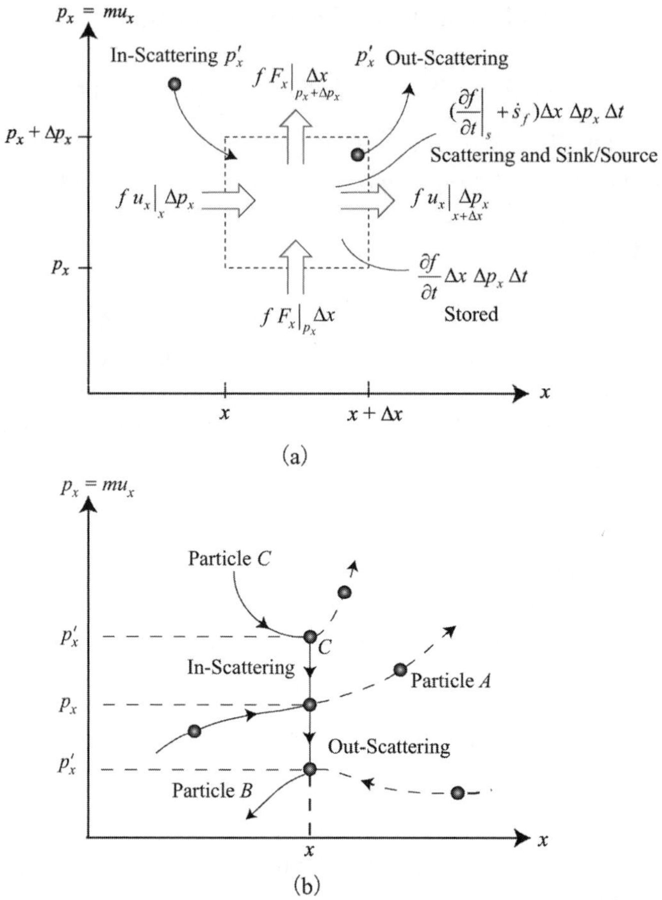

Figure 3.1. (a) Simple, infinitesimal ($\Delta x \to 0$, $\Delta p_x \to 0$) balance on conserved property f, in single-space x and single-momentum p_x coordinates. The storage, scattering, and source terms are also shown. (b) Two-dimensional (x, p_x) rendering of particle scattering that results in a change in the particle momentum. In-scattering adds particles to state (x, p_x), whereas out-scattering removes particles from it.

Taking the limit as $\Delta t \to 0$, and by adding a sink/source term $\dot{s}_f$, we have

$$\frac{\partial f}{\partial t} + u_j \frac{\partial f}{\partial x_j} + F_j \frac{\partial f}{\partial p_j} = \frac{\partial f}{\partial t}|_s + \dot{s}_f$$

$$\frac{\partial f}{\partial t} + \boldsymbol{u} \cdot (\nabla_{\boldsymbol{x}} f) + \boldsymbol{F} \cdot (\nabla_{\boldsymbol{p}} f) = \frac{\partial f}{\partial t}|_s + \dot{s}_f \quad \text{BTE}. \tag{3.3}$$

The velocity term in BTE is the group velocity (the speed of propagation of the energy of the carrier), which is further defined in Section 4.1.

In (3.1), we have used unity for the determinant of the Jacobian matrix $\mathbf{J}$ for the transformation $\mathrm{d}\boldsymbol{x}'\mathrm{d}\boldsymbol{p}' = |\mathbf{J}|\mathrm{d}\boldsymbol{x}\mathrm{d}\boldsymbol{p}$, where $\mathbf{J}$ is the 6×6 matrix written as [37]

$$|\mathbf{J}| = \left| \frac{\partial(x, y, z, p_x, p_y, p_z)}{\partial(x', y', z', p'_x, p'_y, p'_z)} \right| = 1, \tag{3.4}$$

Table 3.1. *Probability particle transport equation (BTE) for particle treatment of principal energy carrier i, $i = p, e, f, ph$ [219].*

$$\frac{\partial f_i}{\partial t} + \boldsymbol{u}_i \cdot \nabla_x f_i + \boldsymbol{F}_i \cdot \nabla_p f_i = \frac{\partial f_i}{\partial t}|_s + \dot{s}_{f,i}, \quad i = p, e, f, ph$$

$\boldsymbol{F}_i$	applied force, N
f_i	i particle probability distribution function
$\dot{s}_{f,i}$	i carrier source rate , 1/s
∇_x	spatial gradient, 1/m
∇_p	momentum gradient, 1/N-s

to indicate no change in the coordinates. To extend (3.3) to phonons, electrons, and photons, we have also introduced a source/sink rate term $\dot{s}_f$, and (3.3) is also repeated in Table 3.1 for carrier i.

Examples of the force $\boldsymbol{F}$ include gravitational $m\boldsymbol{g}$ for fluid particles and Coulombic $\boldsymbol{e}_c\boldsymbol{e}_e$ for electrons. Derivation of f_i^o include thermodynamics consideration and is given in Appendix E. For f_f^o, we will also use the equilibrium BTE discussed below, and use symmetry to arrive at the M–B distribution Section 6.4.2.

3.1.3 In- and Out-Scattering

The particles represented by a probability distribution (occupancy) function f_i (governed by the BTE) can have their positions and momenta $(\boldsymbol{x}, \boldsymbol{p})$ altered after collisions (scattering) with each other or with other microscale energy carriers. The instantaneous position of a colliding particle at time t_0 does not change, but its momentum does. This is shown in Figure 3.1(b).

For no scattering, where $\partial f_i/\partial t|_s = 0$, the particle continues on its trajectory (dashed curve) and the probability of occupying an energy state f_i^o does not change, i.e.,

$$\begin{aligned} f_i^o \boldsymbol{x} - \boldsymbol{u}\Delta t, \boldsymbol{p} - \boldsymbol{F}\Delta t, t - \Delta t) &= f_i^o(\boldsymbol{x}, \boldsymbol{p}, t) \\ &= f_i^o(\boldsymbol{x} + \boldsymbol{u}\Delta t, \boldsymbol{p} + \boldsymbol{F}\Delta t, t + \Delta t), \end{aligned} \tag{3.5}$$

or

$$\frac{\mathrm{D} f_i^o}{\mathrm{D}t} = (\frac{\partial}{\partial t} + \boldsymbol{u} \cdot \nabla_x + \boldsymbol{F} \cdot \nabla_p) f_i^o = 0 \text{ equlibrium distribution.} \tag{3.6}$$

This means that, if a given energy state is occupied, it remains occupied (and vice versa).

As shown in Figure 3.1(b) for particle B, as a result of collision (scattering), a particle can move out of its state $(\boldsymbol{x}, \boldsymbol{p})$ into $(\boldsymbol{x}, \boldsymbol{p}')$, thus decreasing the probability of occupation of this state. This is called out-scattering. Scattering can also cause a particle $(\boldsymbol{x}, \boldsymbol{p}')$ to enter a state $(\boldsymbol{x}, \boldsymbol{p})$, as shown in Figure 3.1(b) for particle C, a phenomenon called in-scattering (joining the state shared with particle A).

The right-hand side of the BTE allows for this change in f_i through

$$\frac{\mathrm{D} f_i}{\mathrm{D}t} = \frac{\partial f_i}{\partial t}|_s + \dot{s}_{f,i} \quad \text{nonequilibrium distribution,} \tag{3.7}$$

where $\dot{s}_{f,i}$ would account for any other source/sink not included in $\partial f_i/\partial t|_s$. Again, note that $\partial f_i/\partial t|_s$ is not a formal derivative, while $\partial f_i/\partial t$ is.

Deviation from equilibrium can be due to a force, a field, or a source. For example, an electric force on an electron for f_e, or a temperature field in a solid, fluid, or gas for f_p and f_f, or emission of photons for f_{ph}. The return to equilibrium is through collisions (scattering) $\dot{\gamma}_i$.

The sum of in- and out-scattering from ($\boldsymbol{p}'$ to $\boldsymbol{p}$ and from $\boldsymbol{p}$ to $\boldsymbol{p}'$) is represented in $\partial f/\partial t|_s$ as

$$\frac{\partial f_i}{\partial t}|_s = \underbrace{\sum_{\boldsymbol{p}'} f_i(\boldsymbol{p}')[1 - f_i(\boldsymbol{p})]\dot{\gamma}_i(\boldsymbol{p}', \boldsymbol{p})}_{\text{in-scattering}} - \underbrace{\sum_{\boldsymbol{p}'} f_i(\boldsymbol{p})[1 - f_i(\boldsymbol{p}')]\dot{\gamma}_i(\boldsymbol{p}, \boldsymbol{p}')}_{\text{out-scattering}}, \tag{3.8}$$

where the first term is for in-scattering, $f_i(\boldsymbol{p}')$ gives the probability that $\boldsymbol{p}$ is occupied, and $[1 - f_i(\boldsymbol{p})]$ is the probability that $\boldsymbol{p}$ is empty (so it can be occupied). In the second term, $f_i(\boldsymbol{p})$ and $[1 - f_i(\boldsymbol{p}')]$ are the probabilities that $\boldsymbol{p}$ is occupied and $\boldsymbol{p}'$ is empty, respectively.

Here $\dot{\gamma}_i(\boldsymbol{p}', \boldsymbol{p})$ is the transition-probability rate, or the probability of transition per unit time (1/s) that particle at state $\boldsymbol{p}'$ scatters to state $\boldsymbol{p}$. Note that we have also used $\dot{\gamma}_i$ as the scattering rate. This is the generalized symbol used throughout the book, which will be further discussed in Section 3.2.

For nondegenerate states (low occupancy, as shown in Figure 1.1) $f_i(\boldsymbol{p}) \ll 1$, and (3.8) becomes

$$\frac{\partial f_i}{\partial t}|_s = \sum_{\boldsymbol{p}'} f_i(\boldsymbol{p}')\dot{\gamma}_i(\boldsymbol{p}', \boldsymbol{p}) - \sum_{\boldsymbol{p}'} f_i(\boldsymbol{p})\dot{\gamma}_i(\boldsymbol{p}, \boldsymbol{p}'), \quad \text{nondegenerate particles.} \tag{3.9}$$

The sum of the collisions presented in $\partial f/\partial t|_s$ does not create or destroy carriers (particles), leading to

$$\sum_{\boldsymbol{p}} \frac{\partial f_i}{\partial t}|_s = 0 \quad \text{overall balance equation.} \tag{3.10}$$

This indicates that in-scattering originates from out-scattering of another state, and when adding all exchanges the net result is zero. The summation implies that integration for quasi-particles (including electrons) is taken over the momentum space (for isotropic behavior) by use of a spherical volume of radius $\boldsymbol{p}$.

Because we will apply the BTE to quasi-particles, here we make a preparation. We note that the integral (of any quantity φ) over momentum space $\mathrm{d}\boldsymbol{p}$ follows (2.108) for $\mathrm{d}\boldsymbol{\kappa}$, i.e., by using (2.97) $p_i = \hbar\kappa_i$ (the de Broglie relation, describing the parabolic energy dependence on κ, Glossary) and for isotropic behavior (spherical symmetry) noting that $d\boldsymbol{p} = 4\pi p^2 dp$, we have

$$\langle \varphi \rangle = \sum_{\kappa} \varphi = \frac{1}{8\pi^3} \int \varphi d\boldsymbol{\kappa} = \frac{1}{8\pi^3\hbar^3} \int \varphi d\boldsymbol{p} = \frac{1}{8\pi^3\hbar^3} \int_p \varphi \, 4\pi p^2 dp. \tag{3.11}$$

For electrons, allowing for two spins, we use $4\pi^3$ instead of $8\pi^3$. Thus, assuming isotropic behavior, the integral is over a radius p in momentum space.

Also, note that by allowing for the azimuthal angle θ and polar angle ϕ dependence (nonspherical symmetry), the integral can be written as

$$\langle \varphi \rangle = \sum_{p} \varphi = \frac{1}{8\pi^3\hbar^3} \int \varphi d\boldsymbol{p} = \frac{1}{8\pi^3\hbar^3} \int_0^{\pi} \int_0^{2\pi} \int_p \varphi \, p^2 dp \sin\theta d\phi d\theta, \tag{3.12}$$

where for isotropic behavior the integral over the angles gives 4π.

3.1.4 Relaxation-Time Approximation and Transport Properties

The scattering term in the BTE describes the elastic- (in which the energy of principal carrier groups is conserved) or inelastic- (energy exchange between principal carriers) scattering rate of particles as they collide with each other or different carriers. In Section 3.2, examples are given of such collisions involving phonons, electrons, fluid particles, and photons.

The phonon scattering is mostly with other phonons. In a three-phonon-scattering event, a phonon of wave number $\boldsymbol{\kappa}_p$ can split its energy and create two other phonons (which keeps the energy and momentum conserved). These are generally inelastic scattering events.

The electron scattering is mostly by phonons, and again, it conserves total energy and momentum. Most electron collisions (scattering) are assumed to be elastic.

The fluid particle collision is with mostly another fluid particle, and while the total momentum and energy of the two particles are conserved (an elastic collision), the particle trajectories undergo changes.

The inelastic photon interaction can for example be for a photon interacting with dipole moment of charge-unbalanced, vibrating atoms in a lattice. These can lead to absorption, emission, and/or elastic scattering of photons by lattice atoms.

Examples of expressions for the scattering rates is given in Section 3.2. These are generally in integral form, with the integral taken over the energy of the final (post-collision) state and the energy of the other participating particle. We will discuss these later in their appropriate chapters.

A more general treatment of f_i° (the equilibrium distribution) uses symmetric and antisymmetric decomposition, both with respect to $\boldsymbol{p}$. The integrals of the $\boldsymbol{p}$-moments of the symmetric and antisymmetric portions of f_i and f_i/τ_i may then vanish, depending on the even or odd integrand [219].

As will be discussed in the following chapters, it is also customary to represent the collision rate by using a relaxation-time τ_i approximation, i.e.,[†]

$$\frac{\partial f_i}{\partial t}|_s = -\frac{f_i - f_i^{\rm o}}{\tau_i} = -\frac{f_i'}{\tau_i} = -\dot{\gamma}_i f_i'. \qquad (3.13)$$

This approximation is generally valid for a small force field and under elastic or isotropic scattering (collisions).

Equation (3.13) represents a return to the equilibrium distribution $f_i^{\rm o}$ at a rate proportional to the deviation $f_i' = f_i - f_i^{\rm o}$ and to the inverse of the collision rate τ_i (τ_i is in turn equal to $1/\dot{\gamma}_i$, where $\dot{\gamma}_i$ is the transition probability rate).

As an example, for conduction electrons in semiconductors, under a steady-state and uniform electric field along the x direction $e_{e,x}$, we have $\partial f_e/\partial t = 0$ and $\nabla_{\boldsymbol{x}} f_e = 0$. Then, using the relaxation-time approximation ($F_x = -e_c e_{e,x}$), we find that (3.3) becomes

$$-e_c e_{e,x}\frac{\partial f_e}{\partial p_x} = -\frac{f_e - f_e^{\rm o}}{\tau_e} = -\frac{f_e'}{\tau_e}. \qquad (3.14)$$

Furthermore, under the low-field approximation it is assumed that $\partial f_e/\partial p_x = \partial f_e^{\rm o}/\partial p_x$. Then

$$-\frac{e_c e_{e,x}}{\hbar}\frac{\partial f_e^{\rm o}}{\partial \kappa_x} = -\frac{f_e'}{\tau_e}, \qquad (3.15)$$

where from (2.97) $p_x = \hbar\kappa_x$. Then

$$f_e' = \frac{e_c}{\hbar}e_{e,x}\tau_e\frac{\partial f_e^{\rm o}}{\partial \kappa_x} \quad \text{or} \quad f_e = f_e^{\rm o} + \frac{e_c}{\hbar}e_{e,x}\tau_e\frac{\partial f_e^{\rm o}}{\partial \kappa_x}, \qquad (3.16)$$

where in general $\tau_e = \tau_e(\kappa)$. Then, with $f_e^{\rm o}(\kappa)$ from Table 1.1, f_e is determined, and $e_{e,x}$ is shown to cause departure from the equilibrium.

For example, consider a nondegenerate (so we can drop the unity from the denominator of $f_e^{\rm o}$) electron energy distribution, which is used for semiconductors. Assume a one-dimensional electron motion, in which the total energy (represented as $E_e - \mu$ in Table 1.1) can be divided into the potential and kinetic energy as

$$E_e = E_{e,p}(x) + E_{e,k}(p_x). \qquad (3.17)$$

Then (this will be further explained in Section 5.7)

$$f_e^{\rm o} = \exp[-\frac{E_{e,p}(x)}{k_{\rm B}T}]\exp[-\frac{E_{e,k}(p_x)}{k_{\rm B}T}] \quad \text{nondegenerate } f_e^{\rm o}. \qquad (3.18)$$

[†] Considering time dependence only, and with $\dot{s}_{f,i} = 0$, (3.7) and (3.13) give

$$\frac{{\rm d}f_i(t)}{{\rm d}t} = \frac{\partial f_i}{\partial t}|_s = -\frac{f_i(t) - f_i^{\rm o}}{\tau_f}.$$

Using $f_i(t=0) = f_i(0)$, the solution to this equation is

$$f_i(t) = f_i^{\rm o} + [f_i(0) - f_i^{\rm o}]e^{-t/\tau_i}.$$

This shows that within a few time constants τ_i, the equilibrium distribution $f_i^{\rm o}$ is restored.

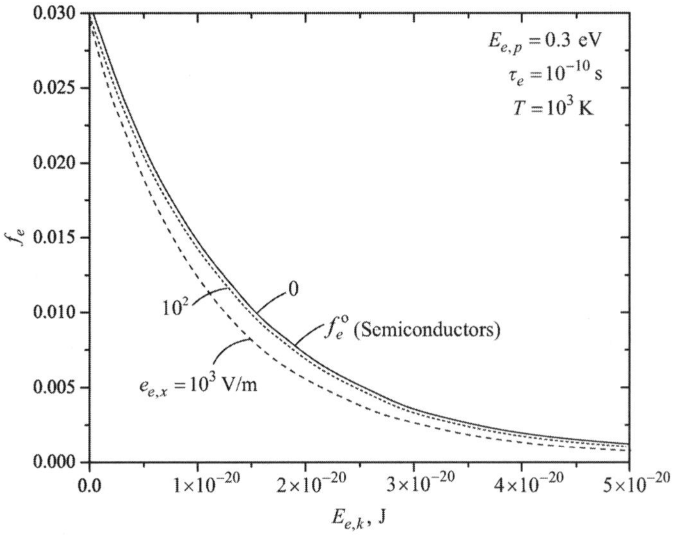

Figure 3.2. Variation of f_e with respect to the kinetic energy for nondegenerate electrons in a one-dimensional electric field.

Using (2.97) again, $E_{e,k} = p_x^2/2m_e = \hbar^2\kappa_x^2/2m_e$, and we have

$$\frac{\partial f_e^{\text{o}}}{\partial \kappa_x} = -\frac{\hbar^2\kappa_x}{m_e k_B T}\exp[-\frac{E_{e,p}(x)}{k_B T}]\exp(-\frac{\hbar^2\kappa_x^2}{2m_e k_B T}). \tag{3.19}$$

Finally,

$$\begin{aligned} f_e &= f_e^{\text{o}}(1 - \frac{\hbar e_c e_{e,x}\tau_e}{m_e k_B T}\kappa_x) \\ &= f_e^{\text{o}}(1 - \frac{e_c e_{e,x}\tau_e}{m_e k_B T}p_x) \\ &= f_e^{\text{o}}(1 - \frac{2^{1/2}e_c e_{e,x}\tau_e}{m_e^{1/2}k_B T}E_{e,k}^{1/2}). \end{aligned} \tag{3.20}$$

This shows that f_e is shifted on the $E_{e,k}$ axis when compared with the equilibrium distribution f_e^{o}.

Figure 3.2 shows the variation f_e with respect to $E_{e,k}$ for $E_{e,p} = 0.3$ eV, $\tau_e = 10^{-10}$ s, $T = 1000$ K, and a range of e_e (0, 10^2, and 10^3 V/m). Note that for the smaller electric fields (e.g., $e_{e,x} = 0$), the deviation from equilibrium is not noticeable. This response to an electric field determines the transport properties. In Section 5.11 we will use (3.20) to find the directional electron drift velocity $u_{e,x}$, the electron mobility $\mu_{e,x}$, and electrical conductivity $\sigma_{e,x}$, which are defined as

$$\mu_{e,x} = \frac{\langle u_{e,x}\rangle}{-e_{e,x}} \equiv \frac{1}{-e_{e,x}}\frac{\sum_p u_{e,x} f_e}{\sum_p f_e}, \quad \sigma_{e,x} = n_{e,c}e_c\mu_{e,x}, \tag{3.21}$$

Table 3.2. *Boltzmann transport scales (time, length, momentum, and force), based on relaxation time.*

Scale	Relation
Time (relaxation time)	τ_i
Length (mean free path)	$\lambda_i = u_i \tau_i$
Momentum	$\boldsymbol{p}_i = m\boldsymbol{u}_i = \boldsymbol{F}_i \tau_i$
Force	$\boldsymbol{F}_i$

where $n_{e,c}$ is the conduction electron density, and e_c is the electron charge. Through (3.20), τ_e and other properties enter into $\mu_{e,x}$.

The relaxation time τ_e represents the inverse of the transition rate $\dot{\gamma}_e$. This indicates that for a very large transition rate, in which the scattering drives the distribution toward equilibrium, the relaxation time tends to zero. This was also previously found for $\tau_e \rightarrow 0$.

Similarly, as will be shown in Section 4.9.2, the thermal conductivity tensor $\boldsymbol{K}_i$ and the conduction heat flux $\boldsymbol{q}_{k,i}$ for carrier i are given by

$$\boldsymbol{q}_{k,i} = \sum_{\alpha} \frac{1}{8\pi^3} \int E_i \boldsymbol{u}_i f_i' \mathrm{d}\boldsymbol{\kappa}_\alpha \equiv -\boldsymbol{K}_i \cdot \nabla T, \quad \boldsymbol{K}_i = \sum_{\alpha} \frac{1}{8\pi^3} \int c_{v,i} \tau_i \boldsymbol{u}_i \boldsymbol{u}_i \mathrm{d}\boldsymbol{\kappa}_\alpha, \qquad (3.22)$$

where α is the polarization of the carrier i (for electrons $8\pi^3$ is replaced with $4\pi^3$ to account for the two spins). Similar to (3.16), the deviation f_i' is related to the driving potential ∇T by use of BTE (as was previously done for electrons in an electric field).

As just demonstrated, the BTE is also used to determine macroscopic transport properties. We will do this in Chapters 4 to 7 for all four microscale energy carriers.

3.1.5 Boltzmann Transport Scales

The relaxation time of principal carrier i, τ_i, signifies the time between encounter (scattering) events that alter the particle momentum $\boldsymbol{p}_i$. Examples are given in Figure 1.5, ranging from fs to ms. The particle momentum is in turn affected by the external force $\boldsymbol{F}_i$. The velocity of the particle $\boldsymbol{u}_i$ is related to its momentum. Then the time, length, and momentum scales of Boltzmann transport, τ_i, λ_i, and $\boldsymbol{p}_i$, respectively signify how long and how far the particle can retain its identity before the next scattering event. This length scale is the mean free path λ_i as introduced in (1.13).

Table 3.2 summarizes the Boltzmann transport scales. These can be compared with Table 1.5 which shows the *ab initio* (fine-structure) scales, and Table 2.7 which shows the MD scales. We also note that the macroscopic treatment scales are $L \gg \lambda_i$ and $t \gg \tau_i$, where the continuum theories are valid. The four levels of length and time scales were presented together in Figure 1.9. When λ_i is larger than the considered system length, i.e., $\lambda_i \gg L$, the transport is said to be ballistic (no momentum altering event during the transport).

3.1.6 Momentum, Energy and Average Relaxation Times

Various approximations are used for the relaxation time. The simplest treatment assumes a constant τ_i. However, in most cases, this does not lead to accurate predictions.

The scattering rate $\dot{\gamma}(\boldsymbol{p}, \boldsymbol{p}')$ is the rate at which carriers with specific momentum $\boldsymbol{p}$ scatter to $\boldsymbol{p}'$. The relaxation time is the average time between such collisions, or the lifetime of state $\boldsymbol{p}$. Some scattering processes are not isotropic, such as small angle-momentum deflections, and using the angle θ between $\boldsymbol{p}$ and $\boldsymbol{p}'$, a momentum relaxation time τ_i(momentum) is defined as [219]

$$\tau_i^{-1}(\text{momentum}) = \sum_{\boldsymbol{p}'} \dot{\gamma}_i(\boldsymbol{p}, \boldsymbol{p}')(1 - \frac{|\boldsymbol{p}'|}{|\boldsymbol{p}|}\cos\theta), \tag{3.23}$$

where θ is the polar angle between the incident and the scattered (initial and final) momenta. A few relaxation times are needed to randomize the momentum.

In Chapter 5, for electron–phonon scattering, we will also be interested in the rate of electron energy scattering by phonons. The energy relaxation time weighs each collision by the fractional change in the energy [219]

$$\tau_i^{-1}(\text{energy}) = \sum_{\boldsymbol{p}} \dot{\gamma}_i(\boldsymbol{p}, \boldsymbol{p}')[1 - \frac{E(\boldsymbol{p}')}{E(\boldsymbol{p})}]. \tag{3.24}$$

This is the time required to dissipate or exchange carrier energy. For elastic scattering, $E(\boldsymbol{p}') = E(\boldsymbol{p})$, and $\tau_i(\text{energy}) \rightarrow \infty$ (Section 5.10).

There are also simple models for energy-dependent relaxation times. Among them is the power-law energy-dependent relaxation time, which is used for electron scattering [219] and is given by

$$\tau_e(E_e) = \tau_{e,\text{o}}(\frac{E_e}{k_\text{B}T})^s \quad \text{energy-dependent electron relaxation time,} \tag{3.25}$$

where s is the power-law constant. For example, for electron–phonon (acoustic) scattering $s = -1/2$ (Table 5.3).

Using $\tau_e(E_e)$, the energy-averaged relaxation time, we find the energy-averaged $\langle\langle\tau_e\rangle\rangle$ (end-of-chapter problem). The double brackets indicates that the average is not a simple average, but actually a particularly weighted average (Section 5.11). We will apply this model to the electron scattering by phonons and impurities.

3.1.7 Moments of BTE

Within the particle treatment model, each microscale BTE can be used for deriving conservation equations involving the moments of f_i with respect to the power n of its momentum $\boldsymbol{p}_i^n$. This moment is then averaged over the entire momentum (or wave vector) space to arrive at the average conservation equations. When this procedure is applied to fluid particles, it gives the conservation of mass (continuity) equation for $\boldsymbol{p}_f^\text{o}$, the momentum conservation equation (the Navier–Stokes equations,

Table 3.10, Section 3.7) for $\boldsymbol{p}_f^1$, and the mechanical energy equation for $|\boldsymbol{p}_f|^2$. Similarly, we will arrive at the electron conservation (drift–diffusion) equation for $\boldsymbol{p}_e^{\rm o}$ and the energy conservation equation for $\boldsymbol{p}_e^2$. A similar set of equations can be developed for phonons. We will discuss the electron and phonon conservation equations in Section 5.18. For photons, the energy equation is developed from the BTE by use of the photon energy $\hbar\omega$, which is called the equation of radiative transfer (Section 7.6).

As an example, following (1.2) to (1.4), the nth moment $\boldsymbol{p}_e^n$ of the electron distribution function f_e averaged over the momentum space (allowing for two spins) (3.12) is

$$\langle n_{e,c}\rangle = \frac{1}{\hbar^3}\sum_{\boldsymbol{p}_e} \boldsymbol{p}_e^0 f_e = \frac{1}{\hbar^3}\sum_{\boldsymbol{p}_e} f_e$$

$$\langle n_{e,c} e_e \boldsymbol{u}_e\rangle = \langle j_e\rangle = \frac{e_c}{\hbar^3}\sum_{\boldsymbol{p}_e} \frac{\boldsymbol{p}_e f_e}{m_e}$$

$$\langle E_e\rangle = \frac{1}{\hbar^3}\sum_{\boldsymbol{p}_e} \frac{1}{2m_e}\boldsymbol{p}_e^2 f_e = \frac{1}{\hbar^3}\sum_{\boldsymbol{p}_e} \frac{p_e^2}{2m_e} f_e, \qquad (3.26)$$

where we have used $\boldsymbol{p}_e = \hbar\boldsymbol{\kappa}_e$ to turn summation that over $\boldsymbol{\kappa}_e$ to that over $\boldsymbol{p}_e$. The integral of the zeroth-moment gives the carrier density, that of the first-moment gives the current density vector, and that of the second-moment gives the energy density (these lead to the carrier continuity, momentum, and energy equations. An example is given in Section 5.18). Then, BTE (3.3) is accordingly multiplied by $\boldsymbol{p}_i^2$ and averaged. The wave number vector $\boldsymbol{\kappa}_i$ and momentum vector $\boldsymbol{p}_i$ are related through (2.98).

When the relaxation-time approximation is used, the momentum and energy equations will contain the moment-averaged relaxation times. As mentioned in Section 3.1.6, these are designated as τ_i(momentum) and τ_i(energy), and examples are given in Section 5.18.

3.1.8 Numerical Solution to BTE

There are two classes of computational techniques that are used to solve the BTE: the deterministic method and the Monte Carlo method [306].† In the first method, the BTE is discretized by use of a variety of methods and then solved directly or iteratively. The discretization techniques include the discrete ordinates S_N, spherical harmonics P_N, collision probabilities, nodal methods, and others. The Monte Carlo method constructs a stochastic model in which the expected value of a certain random variable is equivalent to the value of a physical quantity to be determined. The expected value is estimated by the average of many independent samples

† Note that the equation of radiative transfer (for photon transport), given in Section 7.6.1, is derived from the BTE, and therefore, in its most general form is similar to the BTE. Many numerical methods used for the solution of the equation of radiative transfer, as described in [306], apply in general to solving the BTE.

representing the random variable. Random numbers, following the distributions of the variable to be estimated, are used to construct these independent samples. There are two different options to choose from in constructing a stochastic model for the Monte Carlo computations. In the first case, the physical process is stochastic and the Monte Carlo computation involves a simulation of the real physical process. In the other case, a stochastic model is constructed artificially, such as the solution of the deterministic equations, by the Monte Carlo method. Both the deterministic and the Monte Carlo methods have errors, but their source is different for each method. In the deterministic computational method, the computing errors are systematic. They arise from the discretization of the time–space–angle-energy phase space and the approximate geometry. The errors in the Monte Carlo method are from stochastic uncertainties.

3.2 Energy Transition Kinetics and Fermi Golden Rule

Central to heat transfer is the rate (time) of change of energy to and from thermal energy $\dot{s}_{i\text{-}j}$, as given in Table 1.1. These rates (energy transformation kinetics) are governed by the slowest of the processes involved (in a multiprocess energy conversion). For example, in laser cooling of solids, the photon-electron interactions (absorption), the electronic transition, the coupling of this transition with phonons (absorption of phonons), and luminescence (photon emission), are all encountered in the processes. Each of these processes, in which a carrier undergoes interaction with the same or different carrier, has an interaction/transition time.

Energy conversion is an inelastic carrier interaction. In Figure 1.5, examples of various interaction times $\tau_{i\text{-}j} = 1/\gamma_{i\text{-}j}$ were listed. In semiconductors, for example, the presence of holes results in electron–hole radiative recombination, which depends strongly on dopant concentration.

We also consider the rate of return to equilibrium in particle interactions. An isolated, many-particle system will eventually reach equilibrium, irrespective of its initial state. The typical time for this is called the thermalization time, and depends on the nature of the interparticle interactions.

Because carrier energy distribution is temperature dependent (Table 1.2), all interaction and transition rates are temperature dependent. This gives rise to change in the dominant (slowest) process as temperature changes, in addition to a strong dependence of the rates.

A general qualitative discussion of the various interactions encountered by the carriers is given in subsequent Sections. Then, in the following chapters, quantitative treatments of these rates are given while considering a central (e.g., source of energy) carrier.

3.2.1 Elastic and Inelastic Scattering

In classical particle collisions, an ideally elastic collision is one in which there is no loss of kinetic energy in the collisions; in an inelastic collision, part of the kinetic

Table 3.3. *Some carrier scatterings of significance in heat transferred and thermal energy conversion.*

Carrier	Scattered by	Elastic or inelastic scattering
Phonon	phonon	elastic or inelastic
	impurity	elastic
	vacancy	elastic
	electron	elastic or inelastic
	grain boundary	elastic
Electron	phonon	elastic or inelastic
	ionized impurity	elastic
	piezoelectric	elastic
	photon	elastic or inelastic
Fluid particle	electron and ion	inelastic
	fluid particle	mostly elastic
	solid surface	inelastic
Photon	electron and ion	elastic or inelastic
	oscillating dipole	elastic or inelastic
	fluid particle	elastic or inelastic
	solid particles	elastic or inelastic
	gas plasma	inelastic

energy is changed to some other form of energy (e.g., internal energy). Collisions between hard spheres are considered elastic (the other extreme is that in which the colliding particles coalesce). In addition to energy, momentum conservation is also imposed in some treatments.

Therefore, an inelastic event is one in which part of the energy is "lost" in the scatter, giving rise to internal energy, and thus requiring additional independent variables.

In scattering among the same principal carriers (e.g., phonon–phonon), while energy (and momentum) is exchanged between them, such scattering is considered elastic. In scattering between different carriers (e.g., photons and electrons), although total energy (and in some analysis the momentum) conservation holds among them, the energy exchange between these carriers is considered inelastic. Table 3.3 gives examples of intracarrier and intercarrier scatterings. In many cases, idealized elastic scattering can be assumed.

In addition to energy, particle momentum conservation may also be included in the interaction analysis. In electron–phonon and phonon–phonon scattering events, this conservation is also imposed, with the added complexity of the analysis. In binary fluid particle collision, momentum conservation is readily imposed.

3.2.2 Phonon Interaction and Transition Rates

Phonons interact with each other, electrons, impurity molecule sites, grain boundaries, and dislocations. A phonon's energy, manifested as lattice vibration, is proportional to $k_B T$, and its distribution also depends on temperature. This temperature dependence also creates a very strong temperature dependence for the specific heat capacity and interactions involving phonons. At high temperatures, interphonon interactions dominate others, whereas close to absolute zero, impurity and boundary interactions dominate. Also, as the melting temperature is reached and large atomic displacements occur, the ordered behavior of lattice vibration breaks down, leading to the disorder of the amorphous and liquid phases.

Liquid crystals are intermediates between liquids and crystals. The molecules are typically rod-shaped organic moieties (about 25 Å) and their order is temperature dependent. The nematic phase, for example, is characterized by the orientational order of the constituent molecules and can be controlled with an applied electric field. The smectic phases are found at lower temperatures and form well-defined layers that can slide over one another. Among such molecules are the chiral molecules that possess permanent polarizations, are ferroelectric, and whose time constant (relaxation time) in an applied field can be very small.

Figure 3.3 entry (a) and Table 3.4 entry (a), show interphonon scattering $\tau_{p\text{-}p}^{-1} = \partial f_p/\partial t|_s$ and these will be discussed in Chapter 4, including Section 4.13.

In some cases, the relaxation time is either directly or indirectly measured and is used as an empirical constant in the BTE.

3.2.3 Electron (and Hole) Interaction and Transition Rates

Quantum theory gives an explanation for the probability of electronic transitions (interactions with other carriers, such as phonons and photons) in terms of the wave functions; this applies for transition probabilities (rates) that are time invariant, and it is generally expressed in a relationship referred to as the Fermi golden rule (FGR).

The transition rate depends on the strength of the coupling between the initial and the final state of a system and on the number of ways the transition may occur (i.e., the density of the final states). This transition probability is discussed in Section 5.10, and the derivation of the FGR is given in Appendix E. When the wave vector $\boldsymbol{\kappa}$ (which is more suitable for electrons) is used, the modal ($\boldsymbol{\kappa}'$) form of (E.24) is

$$\dot{\gamma}_{e\text{-}i}(\boldsymbol{\kappa}', \boldsymbol{\kappa}) = \frac{1}{\tau_{e\text{-}i}(\boldsymbol{\kappa}', \boldsymbol{\kappa})} = \frac{2\pi}{\hbar}|M_{\boldsymbol{\kappa}',\boldsymbol{\kappa}}|^2 \delta_{\mathrm{D}}(E_{\boldsymbol{\kappa}'} - E_{\boldsymbol{\kappa}} \mp \hbar\omega_i)$$

FGR for transition probability rate, (3.27)

where $\dot{\gamma}_{e\text{-}i}(\boldsymbol{\kappa}', \boldsymbol{\kappa})$ is transition probability rate, $M_{\boldsymbol{\kappa}',\boldsymbol{\kappa}}$ is the interaction matrix element for the interaction (for transition from state $\boldsymbol{\kappa}$ to state $\boldsymbol{\kappa}'$), and the energy Dirac delta function $\delta_{\mathrm{D}}(E)$ [216] has a unit of J^{-1} (see Glossary). Here ω_i is the angular frequency of carrier i, which causes the transition. The integral of δ_{D} over $\boldsymbol{\kappa}'$-space (Section 5.7) gives the energy density of states, (2.99) written in a more general form i.e.,

$$D_e(E) = \sum_{\boldsymbol{\kappa}'} \delta_{\mathrm{D}}(E_{\boldsymbol{\kappa}'} - E_{\boldsymbol{\kappa}} \mp \hbar\omega_i) = \frac{1}{(2\pi)^3}\int_{\boldsymbol{\kappa}'} \delta_{\mathrm{D}}(E_{\boldsymbol{\kappa}'} - E_{\boldsymbol{\kappa}} \mp h\omega_i)\mathrm{d}\boldsymbol{\kappa}', \qquad (3.28)$$

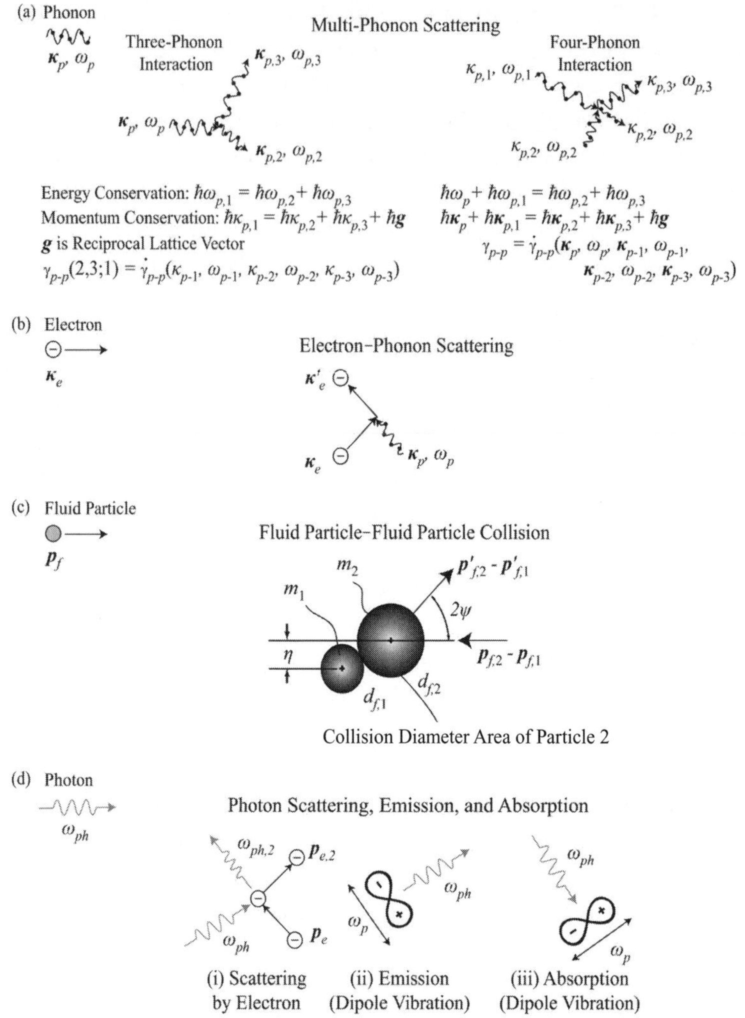

Figure 3.3. Carrier presentation icons, and inter- and intraparticle (carrier) interactions. (a) Three- and four-phonon, (b) electron–phonon, (c) binary fluid particle, and (d) various photon interactions.

where from (3.11), $\mathrm{d}\boldsymbol{\kappa}' = 4\pi\kappa'\mathrm{d}\kappa$ (bold designation indicate three-dimensional integration, or here a spherical volume).†

In some cases, (3.27) is expanded to include momentum conservation, and in that case, the right-hand side is also multiplied by $\delta_{\mathrm{D}}(\boldsymbol{\kappa}' - \boldsymbol{\kappa} \pm \boldsymbol{\kappa}_i)$. Because the FGR is a regular perturbation derivation, it applies to weak interactions. The transition-probability rate, also called the decay-probability rate, is related to the mean

† Alternatively, (3.27) is written as

$$\dot{\gamma}_{e\text{-}i} = \frac{2\pi}{\hbar}|M_{\kappa',\kappa}|^2 D_e(E),$$

where $D_e(E)$ is per unit volume or per unit-cell volume.

Table 3.4. *Examples of intercarrier and intracarrier scattering rates and relaxation times (in BTE) [297, 315, 318].*

(a) Phonon $\tau_{p\text{-}p}$:

$$\left.\frac{\partial \langle f_{p,1}\rangle}{\partial t}\right|_s = \sum_{\kappa_2}\sum_{\alpha_2}[\sum_{\alpha_3}\dot{\gamma}_{\alpha_1,\alpha_2,\alpha_3}(\kappa_1,\kappa_2,-\kappa_3)\langle f_{p,3}(f_{p,2}+1)(f_{p,1}+1)$$
$$-(f_{p,3}+1)f_{p,2}f_{p,1}\rangle\delta_D(\omega_1+\omega_2-\omega_3)$$
$$+\frac{1}{2}\sum_{\alpha_3}\dot{\gamma}_{\alpha_1,\alpha_2,\alpha_3}(-\kappa_1,\kappa_2,\kappa_3)\langle(f_{p,1}+1)f_{p,2}f_{p,3}-$$
$$f_{p,1}(f_{p,2}+1)(f_{p,3}+1)\rangle\delta_D(\omega_1-\omega_2-\omega_3)]$$

(three-phonon scattering, Section 4.15)

(b) Electron $\tau_{e\text{-}p}$:

$$\left.\frac{\partial f_e(\kappa)}{\partial t}\right|_s = -\int\frac{d\kappa'}{4\pi^3}\{\dot{\gamma}_e(\kappa,\kappa)f_e(\kappa')[1-f_e(\kappa)]-$$
$$\dot{\gamma}_e(\kappa,\kappa')f_e(\kappa')[1-f_e(\kappa')]\}$$

(conduction-electron binary scattering, including electron-impurity and electron–phonon scattering, Section 5.11)

(c) Fluid particle

$$\tau_{f\text{-}f}:\ \left.\frac{\partial f_f(p)}{\partial t}\right|_s = -\int\frac{dp'}{(2\pi\hbar)^3}[\dot{\gamma}_f(p',p)f_f(p)-\dot{\gamma}_f(p,p')f_f(p')]$$

(fluid particle classical binary collision, Section 6.5)

(d) Photon

$$\tau_{ph\text{-}e\,\mathrm{or}\,p}:\ \left.\frac{\partial f_{ph,\omega}(p,s)}{\partial t}\right|_s = -\sum_{i,j,k}\int\int\frac{u_{ph}dp_{i,j}}{(2\pi\hbar)^3}dA_{ph,\omega}(p_{i,j},\omega,s\rightarrow p_{i,k},\omega_2,s_2)\times$$
$$f_{i,j}(p_{i,j})[1\pm f_{i,k}(p_{i,k})]\{f_{ph,\omega}(s)[1+f_{ph,\omega_2}(s_2)]-$$
$$\{f_{ph,\omega}(s)[1+f_{ph,\omega}(s)]\times\frac{[1\pm f_{i,j}(p_{i,j})]f_{i,k}(p_{i,k})}{[1\pm f_{i,k}(p_{i,k})]f_{i,j}(p_{i,j})}\}$$

[photon binary scattering by boson (+) and fermion (-) carrier i, Section 7.5]

$A_{ph,\omega}$	phonon-scattering cross-section area with initial state j and final state k, m^2
p	particle momentum, N-s
α_i	polarization of particle i
κ	wave vector, 1/m
s	unit vector
i	type of particle
j	initial state
f	final state
ω	angular frequency, rad/s
u_{ph}	speed of light, m/s
$\dot{\gamma}_i(p',p)$	scattering transition rate from p' to p, 1/s
$\tau_{i\text{-}j}$	relaxation time in scattering between carrier i and j, s

lifetime $\tau_{e\text{-}i}$ of the state by $\dot{\gamma}_{e\text{-}i} = 1/\tau_{e\text{-}i}$. The general form of the FGR can apply to atomic transitions, nuclear decay, scattering, and some other transitions.

A transition will proceed more rapidly if the coupling between the initial and the final states is stronger. The coupling term is the matrix element $M_{\kappa',\kappa}$ for the transition (emphasizing the formulation of quantum mechanics in terms of matrices rather than the differential equations of the Schrödinger equation). The matrix element is an integral, in which the interaction that causes the transition is expressed as an interaction Hamiltonian (for example, a scattering potential φ_s) that operates on the initial-state wave function. The transition-probability rate is proportional to the square of the integral of this interaction over all of the appropriate space, i.e.,

$$M_{\kappa',\kappa} = \int \psi^{\dagger}_{\kappa'} \mathrm{H}' \psi_{\kappa} \mathrm{d}V \quad \text{interaction matrix element,} \tag{3.29}$$

where $\psi^{\dagger}_{\kappa'}$ is the conjugate wave function for the final state, ψ_{κ} is the wave function for the initial state, and H' is the operator (perturbation Hamiltonian) for the physical interaction that couples the initial and final states of the system (for example, the scattering potential). Here $M_{\kappa',\kappa}$ has units of J, $D_e(E)$ has units of 1/J-m^3, and δ_{D} has units of 1/J. In Section E.4 (Appendix E), we give examples of the perturbation Hamiltonian for some scatterings and couplings, involving the principal carriers.

The transition-probability rate is also proportional to the final density of states, which may be composed of several states with the same energy (i.e., degenerate states). This degeneracy is expressed as a statistical weight in the transition probability. For continuum final states, this final density of states is expressed as a function of energy.

Figure 3.3 entry (b) and Table 3.4 entry (b) show an electron–phonon scattering event, which will be discussed further in Chapter 5, including Sections 5.10, 5.11, and 5.15.

An example of nonradiative (no photon emission) coupling of transitions (decay) of excited electrons to the ground state and the vibrational/rotational (vibronic) energy of molecules is given in [138] and also in Chapter 7. These couplings (energy conversions) are referred to as internal conversions. The interaction matrix thus contains both the electronic and vibronic wave functions (from the Schrödinger equation). This involves the interaction Hamiltonian, as will be discussed in Appendix E.

A more general description of the FGR will be given in Section 5.7.

3.2.4 Fluid Particle Interaction and Transition Rates

Fluid particles in thermal motion constantly collide (in their translational motion) with each other and with their container wall, and this collision rate is central in gas kinetic theory. In addition, fluid particles can rotate and vibrate, and these have their own respective time constants. These are all temperature dependent.

Solids interact with fluid particles through surface forces, which cause physical phenomena such as gas adsorption (a thin layer of the condensed fluid phase) or physiochemical adsorption, as in surface-mediated (catalytic) reactions. These adsorbed fluid particles move about, and in and out of the adsorbed layer, because of thermal motion. Under equilibrium, these in and out flow rates balance, and as their kinetic energy (and temperature) increases beyond a threshold, they desorb.

Figure 3.3 entry (c) and Table 3.4 entry (c) show a binary, interfluid particle collision $\tau_{f\text{-}f}$. This will be further discussed in Section 6.5.

3.2.5 Photon Interaction and Transition Rates

As propagating electromagnetic waves, photons interact with electric fields (static and dynamic) in matter. This has been previously mentioned in regard to electron interactions with photons (radiation). Both classical (Maxwell) and quantum theory treatments of these interactions have been addressed. Photons can also participate in a mechanical interaction with fluid particles, as in photophoresis of nanoparticle and microparticles in rarefied gases.

Figure 3.3 entry (d) and Table 3.4 entry (d) show a phonon–electron scattering $\tau_{ph\text{-}e}$. This will be discussed in Chapter 7, including Sections 7.3, 7.4 and 7.10.

The photon interaction with electric entities (such as oscillating dipoles) can be assisted by phonons. The most fundamental approach uses the FGR (3.27). In general, photonic interactions with matter are the central phenomena in many energy-conversion processes (including photovoltaic, discussed in Chapter 7).

3.3 Maxwell Equations and Electromagnetic Waves

3.3.1 Maxwell Equations

Two of the microscale energy carriers, electrons (along with holes) and photons, can (in some cases) be treated by the Maxwell equations, which represent the fundamentals of classical electricity and magnetism. In their dynamical form, the Maxwell equations describe the propagation of EM waves in a medium, which is called the wave treatment of photons. Here, the differential (pointwise, as compared to integral) form of these equations is presented. These equations are compilations of various electrical and magnetic laws (including those of Ampere, Faraday, Gauss, and Ohm), and are listed in Table 3.5 [215].

The continuum properties of a medium that influence the propagation of electromagnetic waves are the electrical conductivity σ_e, the relative complex electric permittivity (or relative dielectric function) $\epsilon_e = \epsilon_{e,r} - i\epsilon_{e,c}$, and the complex relative magnetic permeability $\mu_e = \mu_{e,r} - i\mu_{e,c}$ (which are wavelength λ and temperature dependent, and can be complex variables).‡ The electric susceptivity is defined

‡ The quantum-mechanics definition of the imaginary part of the dielectric function is [22]

$$\epsilon_{e,c} = \frac{4\pi^2}{\omega^2}\sum_f \langle i|\boldsymbol{s}_\alpha \cdot \boldsymbol{j}_e| f\rangle \delta_{\rm D}(|\hbar\omega| - E_f),$$

where $\boldsymbol{s}_\alpha$ is polarization direction unit vector and $\boldsymbol{j}_e$ is the current density vector.

Table 3.5. *Continuum (classical) EM wave-propagation equations (the Maxwell equations), in nonhomogeneous, nonlinear anisotropic media, and some relationships with photons [215].*

$$\nabla \cdot \boldsymbol{e}_e = \frac{e_c n_e}{\epsilon_o} \quad \text{Gauss law for electricity, } \boldsymbol{d}_e = \epsilon_o \epsilon_e \boldsymbol{e}_e$$

$$\nabla \cdot \boldsymbol{b}_e = 0 \quad \text{Gauss law for magnetism, } \boldsymbol{b}_e = \mu_o \mu_e \boldsymbol{h}_e$$

$$\nabla \times \boldsymbol{e}_e + \frac{\partial \boldsymbol{b}_e}{\partial t} = 0 \quad \text{Faraday law}$$

$$\nabla \times \boldsymbol{b}_e - \frac{1}{c^2}\frac{\partial \boldsymbol{e}_e}{\partial t} = \mu_o \mu_e \boldsymbol{j}_e \quad \text{Ampere law}$$

$$\boldsymbol{j}_e = \boldsymbol{j}_{e,f} + \boldsymbol{j}_{e,p} + \boldsymbol{j}_{e,m} = \sigma_e \boldsymbol{e}_e + \frac{\partial \boldsymbol{p}_e}{\partial t} + \nabla \times \boldsymbol{M} \quad \text{Ohm law, with field and polarization currents}$$

$$n_e = n_{e,c} + n_{e,b} \quad \text{total (conduction and bound) electron density}$$

$$E_{ph,\omega} = E_{ph,e} + E_{ph,m} = \frac{1}{2}\int_V \epsilon_o \epsilon_e e_{e,o}^2 \mathrm{d}V \equiv V D_{ph,\omega}(f_{ph} + \frac{1}{2})\hbar\omega,$$

$$\boldsymbol{e}_e = e_{e,o}\boldsymbol{s}_\alpha e^{i(\omega t - \boldsymbol{\kappa}\cdot\boldsymbol{x})}$$

quantum spectral energy of electromagnetic field [for harmonic (EM) field]

$$I_{ph,\omega} = |\boldsymbol{s}_e(\boldsymbol{x})| g(\omega) = \frac{1}{2}|\mathrm{Re}(\boldsymbol{e}_e(\boldsymbol{x}) \times \boldsymbol{h}_e^\dagger(\boldsymbol{x}))| g(\omega) \quad \text{spectral intensity of radiation}$$

$$\int_{-\infty}^{\infty} g(\omega)\mathrm{d}\omega = 1, \; g(\omega) \text{ is line-shape function (broadening), } g(\omega) = \frac{1}{\pi}\frac{\frac{1}{2}\Gamma}{(\omega - \omega_o)^2 + (\frac{1}{2}\Gamma)^2}$$

(Lorentzian function)

$$\nabla^2 \boldsymbol{e}_e = \frac{1}{c^2}\frac{\partial^2 \boldsymbol{e}_e}{\partial t^2} + \nabla \frac{e_c n_e}{\epsilon_o \epsilon_e} + \mu_o \mu_e \frac{\partial \boldsymbol{j}_{e,f}}{\partial t} \quad \text{electromagnetic wave equation for constant } \epsilon_e \text{ and } \mu_e$$

$$I_{ph} = I_{ph,o} e^{i(\boldsymbol{\kappa}\cdot\boldsymbol{x} - \omega t)}$$

$$m_\lambda = n_\lambda - i\kappa_\lambda = [\mu_o \mu_e c_o^2(\epsilon_o \epsilon_e - \frac{i\sigma_e}{\omega})]^{1/2}, \quad c = u_{ph} = (\mu_o \mu_e \epsilon_o \epsilon_e)^{-1/2},$$

$$\kappa = \kappa_o m_\lambda = m_\lambda \frac{\omega}{c_o} \quad \text{dispersion relation,} \quad \sigma_{ph,\omega} = \frac{2\omega\kappa_\lambda}{c} = \frac{4\pi\kappa_\lambda}{\lambda} \quad \text{absorption coefficient}$$

Irrotational electric field, the scalar electric potential φ_e is defined through $\boldsymbol{e}_e = -\nabla\varphi_e$, replacing this in the Gauss law, gives the Poisson equation, $-\nabla^2\varphi_e = e_c n_e/\epsilon_o\epsilon_e$, also the magnetic induction vector is represented by magnetic vector potential $\boldsymbol{a}_e$ through $\boldsymbol{b}_e = \nabla \times \boldsymbol{a}_e$

Electromagnetic force (Lorentz) equation is $\boldsymbol{F}_e = e_c\boldsymbol{e}_e + e_c\boldsymbol{u} \times \boldsymbol{b}_e$

For optical fields, the spectral optical properties $m_\lambda = n_\lambda - i\kappa_\lambda$ are related to spectral ϵ_e, μ_e and σ_e, and the spectral absorption coefficient $\sigma_{ph,\omega}$

$\boldsymbol{a}_e$	vector potential, N/A-m^2
$\boldsymbol{b}_e$	magnetic induction vector, T = N/A-m = V-s/m^2
D_{ph}	photon density of states, 1/m^3-rad/s
$\boldsymbol{d}_e$	electric displacement vector, C/m^2
$\boldsymbol{e}_e$	electric field intensity vector, V/m
$\boldsymbol{h}_e$	magnetic field intensity, A/m, † indicates complex conjugate
$\boldsymbol{j}_e$	current density vector, A/m^2 = C/m^2-s
m_λ	spectral complex index of refraction, $m_\lambda = n_\lambda - i\kappa_\lambda$
$\boldsymbol{M}$	magnetization vector, A/m
n_e	total (free and bond) electron density, 1/m^3
n_λ	spectral index of refraction
$\boldsymbol{p}_e$	electric polarization vector (total dipole moment) per unit volume, C/m^2
$\boldsymbol{s}_e$	Poynting vector, W/m^2
ϵ_e	complex relative electric permittivity (or relative dielectric function)
ϵ_o	free-space permittivity, 8.854×10^{-12} C^2/N-m^2
κ_λ	spectral index of refraction and extinction
μ_e	complex relative magnetic permeability
μ_o	free-space permeability, $4\pi \times 10^{-7}$ N/A^2
σ_e	electrical conductivity, 1/Ω-m

as $1-\epsilon_e$, and we will discuss ϵ_e for different materials in Section 3.3.5. For vacuum, $\epsilon_e = \mu_e = 1$, and the electric susceptivity χ_e is zero. The speed of light is $c = (\mu_o \mu_e \epsilon_o \epsilon_e)^{-1/2}$, where ϵ_o and μ_o are the free-space permittivity and permeability (Table 1.4). For ideal insulators, $\sigma_e = 0$, and we will discuss σ_e for metals and semiconductors in Chapter 5 (experimental data in Section 5.16).

In the quantum treatment, these in turn depend on the molecular properties of the medium. In solids, for example, these depend on the electronic band structure, the scattering rates (electron by phonon, etc.), the conduction electron density, etc.

We will discuss these in Chapter 5. Figure 3.4 shows the progression of representing EM-related properties of matter, starting from quantum mechanics, to Maxwell equations, to optical properties, and finally the surface properties.

The optical properties presented by the spectral complex index of refraction m_λ and is introduced through the photon dispersion relation (for phase velocity) $u_{ph} = u_{ph,p} = \omega/\kappa = u_{ph,o}/m_\lambda$. In turn m_λ is given in terms of these three fundamental, classical properties (σ_e, ϵ_e and μ_e), and has two components, i.e.,

$$m_\lambda = n_\lambda - i\kappa_\lambda = \mu_o \mu_e c_o^2 (\epsilon_o \epsilon_e - i\frac{\sigma_e}{\omega})^{1/2}, \quad \lambda f = c = u_{ph}, \quad \omega = 2\pi f, \tag{3.30}$$

where n_λ is the spectral index of refraction, κ_λ is the spectral index of extinction, and f is the frequency (ω is the angular frequency). Derivation of (3.30), from the Maxwell equation, is an end-of-chapter problem.

The surface-radiation properties are represented by the spectral reflectivity $\rho_{ph,\lambda}$, and spectral emissivity $\epsilon_{ph,\lambda}$, which can be related to EM or optical properties (Sections 7.10 and 7.15.1). For semiconductors, the drift–diffusion equation is used for the electric current and is listed in Table 3.6. The electron (and hole) conservation (continuity) equation is also listed. These are derived and discussed in Section 5.18.

In addition to electric potential φ_e ($\boldsymbol{e}_e \equiv -\nabla\varphi_e$), we will also be the magnetic vector potential $\boldsymbol{a}_e$ ($\boldsymbol{b}_e \equiv \nabla \times \boldsymbol{a}_e$), when treating the dipole radiation (end-of-chapter problem), and in quantization of EM wave in Section 7.3.

3.3.2 Electromagnetic Wave Equation

The EM wave equation for a constant property medium is obtained by taking the curl of the Faraday law, then substituting the Ampere law, followed by use of the Ohm and Gauss laws (for constant ϵ_e and μ_e), i.e.,

$$\nabla \times \nabla \times \boldsymbol{e}_e = -\nabla^2 \boldsymbol{e}_e + \nabla(\nabla \cdot \boldsymbol{e}_e) \tag{3.31}$$

$$-\frac{\partial}{\partial t}\nabla \times \boldsymbol{b}_e = -\nabla^2 \boldsymbol{e}_e + \nabla(\nabla \cdot \boldsymbol{e}_e) \tag{3.32}$$

$$\nabla^2 \boldsymbol{e}_e - \nabla(\nabla \cdot \boldsymbol{e}_e) = \frac{\partial}{\partial t}(\frac{1}{c^2}\frac{\partial \boldsymbol{e}_e}{\partial t} + \mu_o \mu_e \boldsymbol{j}_{e,f}), \quad c \equiv u_{ph} \tag{3.33}$$

$$\nabla^2 \boldsymbol{e}_e = \frac{1}{c^2}\frac{\partial^2 \boldsymbol{e}_e}{\partial t^2} + \nabla \frac{e_c n_e}{\epsilon_o \epsilon_e} + \mu_o \mu_e \frac{\partial \boldsymbol{j}_{e,f}}{\partial t}. \tag{3.34}$$

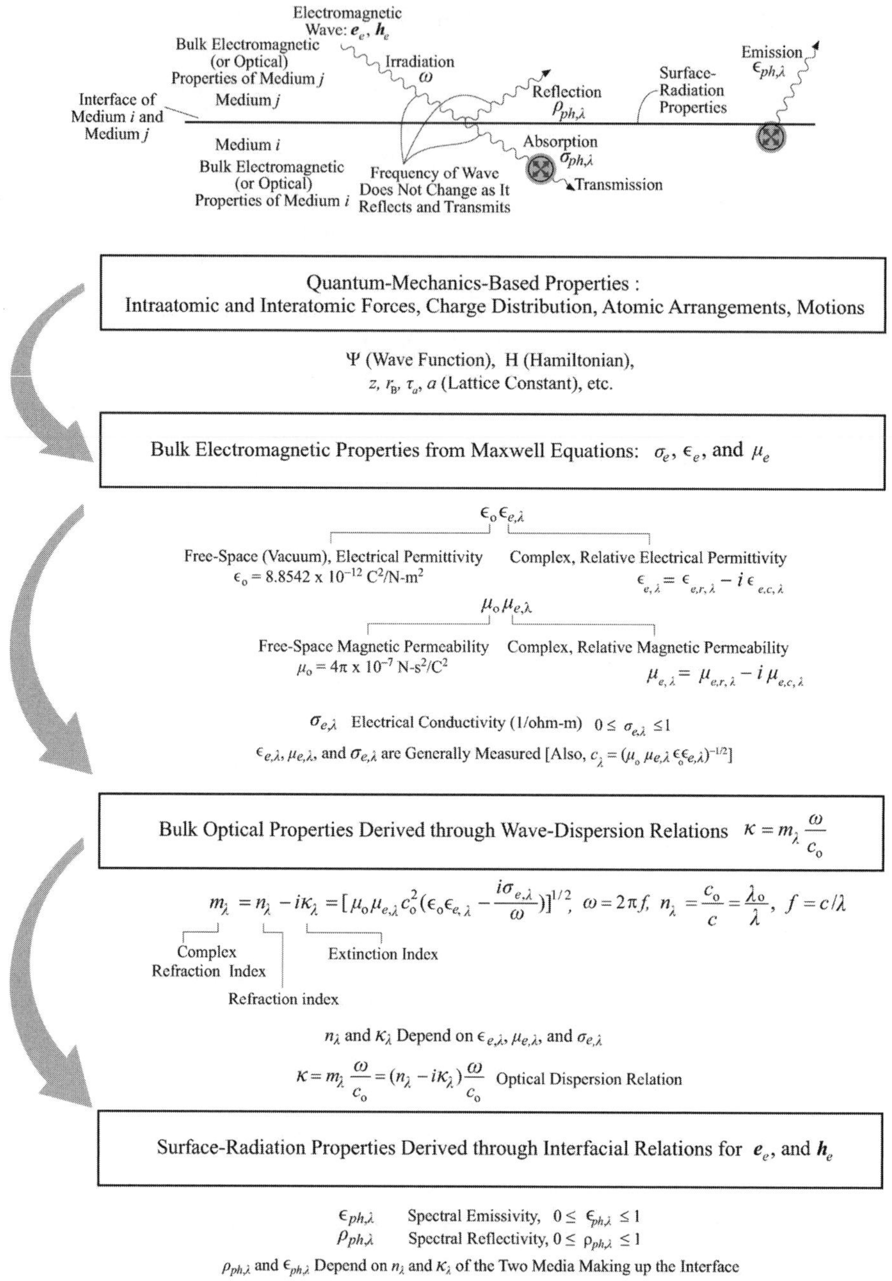

Figure 3.4. The classical (Maxwell) interaction of electromagnetic radiation with an interface marking a change in radiation properties. The quantum-mechanical, electromagnetic, optical (derived), and surface-radiation (derived) properties are also shown.

Table 3.6. *Electron (or hole) drift–diffusion equation, and continuity equation for semiconductors [219].*

Current is given by the drift–diffusion equation (for semiconductor heterostructures)

$$\boldsymbol{j}_e = \sigma_e \boldsymbol{e}_e + e_c D_e \nabla n_e + e_c \mathrm{S}_e \nabla T \quad \text{drift–diffusion equation,}$$

where the drift is represented by conductivity and the diffusion is represented by the carrier density and temperature gradients, D_e is the electron diffusivity, and S_e is the Soret coefficient

For electrons

$$D_e = \frac{k_\mathrm{B} T}{e_c}\mu_e, \quad \mathrm{S}_e = n_e \mu_e \frac{k_\mathrm{B}}{e_c}.$$

Carrier concentration is given by the electron (or hole) continuity equation (spatial variation of carrier distribution)

$$\frac{\partial n_e}{\partial t} = -\frac{1}{e_c}\nabla \cdot \boldsymbol{j}_e + \sum_i \dot{n}_{e,i} \quad \text{carrier continuity equation,}$$

where $\dot{n}_{e,i}$ stands for various generations and recombinations.

D_e	carrier (electron or hole) diffusivity, $\mathrm{m^2/s}$
n_e	carrier (electron or hole) density, $\mathrm{1/m^3}$
S_e	Soret coefficient, 1/K-m-s
σ_e	electrical conductivity, $1/\Omega$-m
μ_e	electron mobility, $\mathrm{m^2/V\text{-}s}$

This EM wave propagation equation is used to derive (3.30),† using the optical dispersion relation $\kappa = m_\lambda \omega / c_\mathrm{o} = (n_\lambda - i\kappa_\lambda)\omega / c_\mathrm{o}$, where $n_\lambda = c_\mathrm{o}/c$, c is the phase velocity (will be discussed in Section 4.2). When there is no charge gradient or current ($\sigma_e = 0$), then from (3.34) we have the simple wave (Helmholtz) equation

$$\nabla^2 \boldsymbol{e}_e = \frac{1}{c^2}\frac{\partial^2 \boldsymbol{e}_e}{\partial t^2} \quad \text{EM (classical) wave equation for } \nabla n_e = \boldsymbol{j}_{e,f} = 0. \tag{3.35}$$

The electric $\boldsymbol{e}_e$ and magnetic $\boldsymbol{h}_e$ field vectors are oriented perpendicular to each other, while propagating along the same direction. In Section 7.7.1, we discuss the continuity of the surface tangent components of these vectors as they cross interfaces with discontinuity in properties (e.g., change in m_λ). When compared with the time-dependent Schrödinger equation (2.62), which is first-order in the time derivative and has $i\hbar$ as the coefficient of this first derivative, we can expect similar solution forms for Ψ and $\boldsymbol{e}_e$, although $\boldsymbol{e}_e$ is a vector. Also note that the quantum Hamiltonian has a Laplacian (kinetic energy) term as well as a potential term.

† This begins by introducing n_λ and κ_λ through (for propagation along x direction)

$$e_{e,y} = e_{e,\mathrm{o}} \exp\{i[\omega t - \frac{(n_\lambda - i\kappa_\lambda)}{c_\mathrm{o}}\omega x]\},$$

etc.

The classical (hyperbolic, linear), propagating EM wave equation (3.35) has a solution of the type [215]†

$$\boldsymbol{e}_e = e_{e,\mathrm{o}}\boldsymbol{s}_\alpha e^{i(\omega t-\boldsymbol{\kappa}\cdot\boldsymbol{x})} = e_{e,\mathrm{o}}\boldsymbol{s}_\alpha e^{i(\omega t-\kappa_\mathrm{o}m_\lambda\cdot\boldsymbol{x})}, \quad (3.36)$$

where $\boldsymbol{s}_\alpha$ is the polarization vector, and $\boldsymbol{\kappa}$ is the wave vector.

Note the similarity to the time-dependent wave function (2.68) used for the Schrödinger equation in (2.69), and also note that $E = \hbar\omega$ or $E/\hbar = \hbar\kappa^2/2m$ in quantum mechanics. The Schrödinger equation has a first-order time derivative, but this term contains i.

We will use the Helmholtz equation (3.35) in treatment of radiation propagation in Section 7.7, where we show local interference of waves results in local electric field (photon energy) enhancement.

3.3.3 EM Wave and Photon Energy

The Poynting vector $\boldsymbol{s}_e$, when integrated over a closed surface, gives the total, outward flow of energy per unit time. The time-average magnitude (propagating in one direction) in a nonconducting medium‡ is

$$|\overline{\boldsymbol{s}}_e| = \frac{1}{2}c\,\epsilon_\mathrm{o}\epsilon_e e_{e,\mathrm{o}}^2 \quad \text{plane wave}, \quad (3.37)$$

and is equal to the phase velocity c times the average energy density $E_{ph,\omega}/V$. This uses the cycle-averaged fields $\overline{\boldsymbol{e}}_e$ and $\overline{\boldsymbol{h}}_e$, in which electric and magnetic contributions to E_{ph} are equal.

Figure 3.5 shows the EM wave and photon presentations of energy contained in a volume V. The polarization $\boldsymbol{s}_\alpha$ is along the y direction, and the wave travels along the x direction.

The relation among energy density of EM wave with field intensity vector $\boldsymbol{e}_e$, i.e., (3.37), and frequency ω, the quantum occupancy number of photons with the

† Compared with de Broglie quantum wave, which is

$$\Psi = Ae^{i(\boldsymbol{\kappa}\cdot\boldsymbol{x}-\omega t)},$$

where $\omega = E/\hbar = \hbar\kappa^2/2m$ as given in (2.68).

‡ Plane-polarized electric field (using $\kappa = \omega/c$)

$$\boldsymbol{e}_e = e_{e,\mathrm{o}}e^{i\omega(t-\frac{x}{c})}\boldsymbol{s}_y,$$

has a counterpart magnetic induction field (from Faraday law in Table 3.5), given as

$$\boldsymbol{h}_e = \epsilon_\mathrm{o}\epsilon_e c_\mathrm{o} e_{e,\mathrm{o}}e^{i\omega(t-\frac{\omega x}{c})}\boldsymbol{s}_y,$$

so the Poynting vector is

$$\boldsymbol{s}_e = \boldsymbol{e}_e \times \boldsymbol{h}_e = \frac{1}{\mu_\mathrm{o}\mu_e c}e_{e,\mathrm{o}}^2\boldsymbol{s}_z,$$

and its time average is (end-of-chapter problem) [215]

$$|\overline{\boldsymbol{s}}_e| = \frac{1}{2}c\epsilon_\mathrm{o}\epsilon_e e_{e,\mathrm{o}}^2,$$

which is half because of $\boldsymbol{e}_e$ and half because of $\boldsymbol{h}_e$.

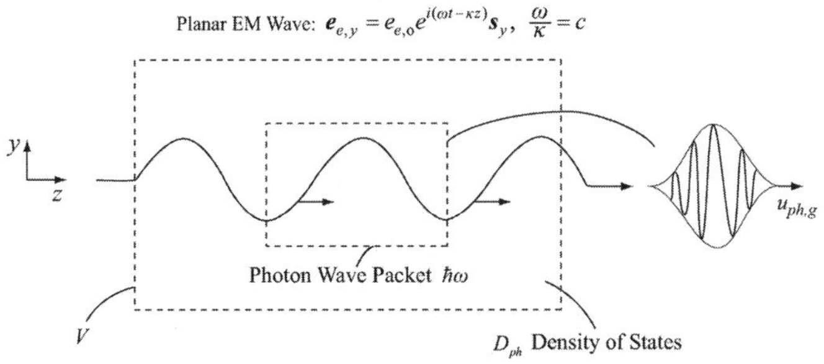

Figure 3.5. The planar EM wave and its photon presentation (wave packet of energy $\hbar\omega$, Figure 1.2). The number of photons in the volume V is such that it matches the EM energy in the same volume.

same energy and frequency f_{ph}, and number of phonons of energy $\hbar\omega$ (Section 7.1) per volume (photon density of state) $D_{ph,\omega}$, using (2.85) for quantized photon energy density, is (Section 7.3)

$$\frac{E_{ph,\omega}}{V} = \frac{1}{2V}\int_V \epsilon_o\epsilon_e e_{e,o}^2 \mathrm{d}V = D_{ph,\omega}(\omega)(f_{ph} + \frac{1}{2})\hbar\omega \quad \text{plane wave.} \tag{3.38}$$

For a plane wave, the radiative intensity $I_{ph,\omega}$ given by the average Poynting vector $|\bar{\boldsymbol{s}}_e|$ (3.37) becomes

$$I_{ph,\omega} = \frac{1}{2}c\,\epsilon_o\epsilon_e e_{e,o}^2 \quad \text{plane wave.} \tag{3.39}$$

In general, a line-shape function is also added to (3.39), as shown in Table 3.5. These classical, electromagnetic waves can travel in vacuum ($\sigma_e = 0$, $\mu_e = 1$, $\epsilon_e = 1$).

The limits of the classical treatments are in the description of the electronic energy bandgaps, and in quasi-particle features of photons (including blackbody radiation). We will address these in Chapters 5 (electrons) and Chapter 7 (photons).

3.3.4 Electric Dipole Emission, Absorption and Scattering of EM Waves

Molecules with nonuniform charge distributions (Table 2.1) have permanent electric dipole moments. This strictly applies to two charges, one positive and one negative, of the same magnitude and separated by a distance x. Water vapor, for example, has a dipole moment of 6.20×10^{-30} C-m.† The presence of an electric field can also induce a dipole moment. The permanent dipole moment (vector) $\boldsymbol{p}_{e,o}$ is the product of the charge q and the separation distance and points toward the positive charge, i.e.,

$$\boldsymbol{p}_{e,o} = \frac{1}{V}\int \boldsymbol{x}q(\boldsymbol{x})\mathrm{d}V \quad \text{electric dipole moment.} \tag{3.40}$$

These dipoles can be oscillating at the atomic frequencies related to their vibrations.

† One Debye D is 3.3356×10^{-30} C-m, Table 1.4. Then for water vapor, $p_{e,o} = 1.85$ D.

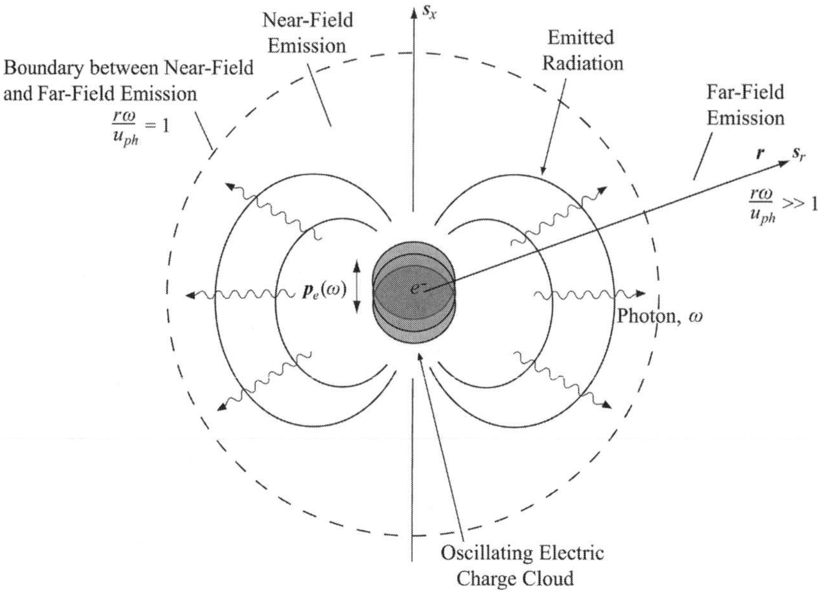

Figure 3.6. Emission of EM radiation by sinusoidally oscillating electronic charge [307].

The Debye model of polar molecules is an electric dipole contained in a spherical molecule, free to rotate into alignment with the electric field, but subject to disorientation by collision with the neighboring molecules (by thermal motion). The net dipole moment (per molecule) is expressed as [215]

$$\boldsymbol{p}_e = N\alpha_e(\boldsymbol{e}_e + a\frac{\boldsymbol{p}_e}{\epsilon_o}) \quad \text{induced linear dipole}$$

$$\boldsymbol{p}_e = \frac{N\alpha_e}{1 - \dfrac{N\alpha_e a}{\epsilon_o}}\boldsymbol{e}_e = \epsilon_o \chi_o \boldsymbol{e}_e, \tag{3.41}$$

where $\boldsymbol{p}_e$ is the induced dipole moment, α_e is the polarizibility (induced moment) produced by the applied field, a is a constant, N is the number of molecules per unit volume, and χ_e is electric susceptibility.

The emission of EM radiation by a sinusoidally oscillating charge is shown in Figure 3.6. The purely radiative decay rate or spontaneous decay rate of a classical electron oscillator $\dot{\gamma}_{ph,e,sp,o}$ is determined from the classical electromagnetic theory [215]. For vacuum ($\epsilon_e = 1$), this is given as (end-of-chapter problem) [307]

$$\dot{\gamma}_{ph,e,sp,o} = \tau^{-1}_{e\text{-}p,o} = \frac{e_c^2\omega^2}{6\pi\epsilon_o m_e u_{ph}^3}$$

spontaneous emission (decay rate) by classical oscillating electron. (3.42)

The emission rate by transition dipole moment $\boldsymbol{\mu}_e$ (quantum-mechanics quantity) causing a energy transition $\hbar\omega_{e,g}$ in an excited electric entity (e.g., electron in an ion) is [147] (derivation is Chapter 7 problem)

$$\dot{\gamma}_{e\text{-}p}\,(\text{spontaneous}) = \dot{\gamma}_{ph,e,sp} = \frac{2\omega_{e,g}^3}{3\epsilon_o \hbar_P u_{ph}^3}|\boldsymbol{\mu}_e|^2$$

spontaneous emission by transition dipole. (3.43)

In this relation it is assumed that $\dot{\gamma}_{e\text{-}p} \ll \omega_{e,g}/\pi$. This is the photon spontaneous emission and is part of the Einstein spontaneous and stimulated emission rate equation (Section 7.4.1). The transition dipole moment (a quantum-electrodynamic quantity) will be discussed in Section 7.12.1. The energy generation $\dot{s}_{ph,e,sp}$ is $\hbar\omega_{e,g}\dot{\gamma}_{ph,e,sp}$, and in thermal radiation the thermal fluctuations in matter (including solids) results in thermal radiation emission (sum of stimulated and spontaneous emission, Section 7.4.3).

The far-field ($r\omega/u_{ph} \gg 1$), i.e., spherical-wave limit,† energy of interaction of EM with a molecular dipole is given by $\boldsymbol{p}_e \cdot \boldsymbol{e}_e$. EM waves are also absorbed by oscillating dipoles. There is also scattering of EM waves by dipoles, such as in Rayleigh scattering [38, 178, 244]. In Section 5.15.3(B) we will discuss the polarization (dipole moment created by polar optical-phonon lattice displacement). In quantum mechanics, the transition dipole moment is represented by an operator $\boldsymbol{\mu}_e$ (Section 7.12.1).

3.3.5 Dielectric Function and Dielectric Heating

Dielectrics are materials with electrical conductivities lower than those of metals (Section 5.16). Dielectric materials are generally categorized as materials having electrical resistivity larger than 10^4 Ω-m. These materials can absorb EM energy from an oscillating electric field $\boldsymbol{e}_e(t)$ by molecular rotation, especially at lower frequencies. This molecular rotation occurs in polar molecules. A material capable of being heated with radiowave or microwave energy is said to be polar or dipolar. An oscillating electric field is applied to the material, causing its molecules to rotate at the frequency of the field and line up their corresponding fields. The molecular friction generated by the molecules moving relative to each other generates heat. This method of generating heat within a material is termed dipole rotation and can be used to heat solids, liquids, or gases. Electric permittivity is a measure of ability of a material to resist the formation of an electric field within it. Dielectric materials have polar molecules (see Glossary) that can store charge and rotate under an

† The far-field ($r \gg \lambda$) electric and magnetic field intensities are (Figure 3.6)

$$\boldsymbol{p}_e = \boldsymbol{p}_{e,o} e^{i\omega t}$$

$$\boldsymbol{e}_e = \left(\frac{\mu_o}{\epsilon_o}\right)^{1/2} \boldsymbol{h}_e \times \boldsymbol{s}_r$$

$$\boldsymbol{h}_e = \frac{\omega^2}{4\pi u_{ph,o}} (\boldsymbol{s}_r \times \boldsymbol{p}_e) \frac{e^{\frac{-i\omega r}{u_{ph,o}}}}{r}.$$

More detail derivation is guided through end-of-chapter problem.

applied alternating electric field of frequency f. The complex electric permittivity (also called the dielectric function) is defined as

$$\epsilon_e(\omega) = \epsilon_{e,r}(\omega) - i\epsilon_{e,c}(\omega) \quad \text{dielectric function (electric permittivity)}, \qquad (3.44)$$

where $\epsilon_{e,r}$ is called the dielectric constant and $\epsilon_{e,c}$ is called the dielectric loss (or loss factor).

The dielectric constant $\epsilon_{e,r}$ is a measure of the capacitance of the material and the dielectric loss $\epsilon_{e,c}$ is a measure of the electrical energy dissipated as heat. Both $\epsilon_{e,r}$ and $\epsilon_{e,c}$ are functions of frequency and temperature.

Figures 3.7(a) and (b) show variations of $\epsilon_{e,r}$ and $\epsilon_{e,c}$ as functions of angular frequency, for five materials representing dielectrics (acrylic and quartz), semiconductors (GaAs), metals (Al), and liquids (H_2O), all at $T = 300$ K.

Although $\epsilon_{e,r}$ affects refraction and reflection, $\epsilon_{e,c}$ affects absorption and emission. It is possible for $\epsilon_{e,r}$ to have a negative value (shown for Al). H_2O as a liquid has strong absorption (large $\epsilon_{e,c}$) in the microwaves, far-infrared, and ultraviolet regimes. Acrylic has a small absorption coefficient (related to $\epsilon_{e,c}$), whereas GaAs has a strong absorption near its bandgap energy (1.42–1.90 eV, ultraviolet regime). Quartz (crystalline SiO_2) has a high absorption because of its lattice vibration in the near-infrared regime. The free electrons of Al are excited by photons over a large range of energy.

The Kramers–Kronig dispersion relation describes the interdependency of $\epsilon_{e,r}$ and $\epsilon_{e,c}$. The classical electron oscillator model for the dielectric function (the Drude model, Glossary) is based on the damping motion of the electron and an oscillating EM field. This is shown in Figure 3.8. For an electron as a particle (Table 2.5) the momentum conservation is

$$m_e \frac{\mathrm{d}^2\boldsymbol{x}}{\mathrm{d}t^2} = -m_e\omega_{n,e}^2\boldsymbol{x} - \frac{m_e}{\tau_e}\frac{\mathrm{d}\boldsymbol{x}}{\mathrm{d}t} - e_e\boldsymbol{e}_e \quad \omega_{n,e} = 0$$

$$\text{for as unbound electron (Drude damped oscillator model)}, \qquad (3.45)$$

where $\omega_{n,e}$ is the natural frequency related to the force constant through (2.54), and τ_e is the relaxation time for the damping mechanism (e.g., scattering by nuclei). For a time-periodic EM field $\boldsymbol{e}_e = \boldsymbol{e}_{e,\mathrm{o}}e^{i\omega t}$ (3.36), we have the form [357]

$$\frac{\mathrm{d}\boldsymbol{x}}{\mathrm{d}t} = \frac{e_c/m_e}{i\omega - 1/\tau_e}\boldsymbol{e}_{e,\mathrm{o}}, \qquad (3.46)$$

which on using $\boldsymbol{j}_e = -n_{e,c}e_c\mathrm{d}\boldsymbol{x}/\mathrm{d}t = \sigma_e(\omega)\boldsymbol{e}_{e,\mathrm{o}}$, where σ_e is the complex electric conductivity, and $n_{e,c}$ is the free-electron density, gives

$$\sigma_e(\omega) = \frac{n_{e,c}e_e^2/m_e}{\tau_e - i/\omega} = \frac{\sigma_{e,\mathrm{o}}}{1 - i\omega\tau_e} \quad \text{Drude free-electron model}, \qquad (3.47)$$

where $\sigma_{e,\mathrm{o}} = n_{e,c}e_c^2\tau_e/m_e$ is the DC ($\omega = 0$) electrical conductivity.

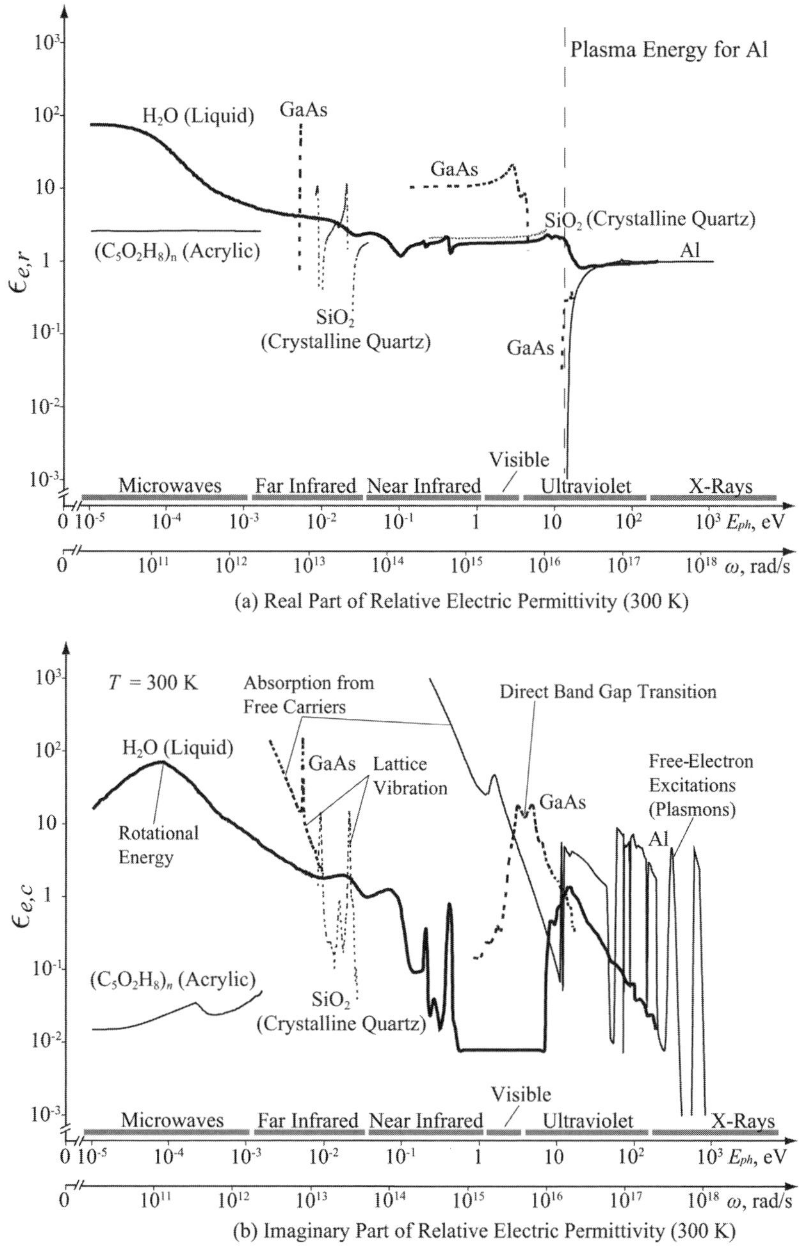

Figure 3.7. Variation of (a) the real and (b) imaginary parts of the electric permittivity (dielectric function) with respect to EM wave energy (angular frequency), for typical insulator (including an organic compound), conductor, and semiconductor solids and for liquid water. The data are for $T = 300$ K.

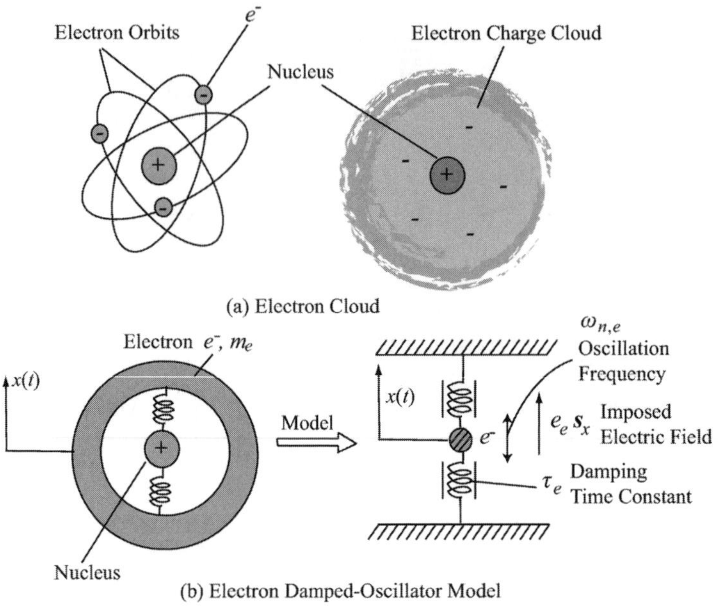

Figure 3.8. (a) Electronic models for an atom, and (b) the classical electron damped-oscillator model [307].

The components of the dielectric function are also related to $\sigma_e(\omega)$ through [357]

$$\sigma_{e,c} = -\omega\epsilon_o\epsilon_{e,r}, \quad \epsilon_{e,c} = \frac{\sigma_{e,r}}{\omega\epsilon_o}. \tag{3.48}$$

Then

$$\epsilon_e(\omega) = \epsilon_{e,\infty} - \frac{\sigma_{e,o}\tau_e}{\epsilon_o(\omega^2 + i\omega/\tau_e)} = \epsilon_{e,\infty} - \frac{\omega_{pl}^2}{\omega(\omega + i/\tau_e)}, \tag{3.49}$$

where $\epsilon_{e,\infty}$ is the high-frequency asymptote and ω_{pl} is the plasma frequency [quantum of plasma energy $\hbar\omega_{pl}$ is called plasmon, Figure 3.7(b) for Al] $(n_{e,c}e_c^2/\epsilon_o m_e)^{1/2}$. Generally, τ_e is the inverse of ω at which $\epsilon_{e,c}$ has a maximum.†

Using the low- and high-frequency limits, the modified Debye model is listed in Table 3.5, which is

$$\frac{\epsilon_e - \epsilon_{e,\infty}}{\epsilon_{e,s} - \epsilon_{e,\infty}} = \frac{1}{1 + i\omega\tau_e}, \tag{3.50}$$

where $\epsilon_{e,s}$ and $\epsilon_{e,\infty}$ are the low- and high-frequency emissions (real quantities). The Debye model is used in the electron-optical-phonon scattering (Section 5.15.3).

† In Debye model of a molecule of diameter d rotating in a viscous medium is,

$$\tau_e = \frac{4\pi(\frac{d}{2})^3\mu_f}{k_B T},$$

where μ_f is the dynamic viscosity.

Similarly, the Lorentz model treats the oscillatory motion of a band carrier in an oscillating EM field [357], and is used for dielectrics (e.g., SiO_2).

For semiconductors (direct and indirect bands, Chapter 5), the Drude model is extended to include the bandgap (we will discuss this in Section 7.15.1).

In general, $\epsilon_e(\omega)$ is related to the electronic band structure and is given in terms of ω_{ij}, which is used along with the Brillouin zone average transition probability (between states i and j) and the joint density of states [276], so we have

$$\epsilon_e(\omega) = 1 + \frac{8\pi e_e^2}{m_{e,e}^2} \sum_{i,j} \int \omega_{ij}[g(\hbar\omega - E_e) - g(\hbar\omega + E_e)]\mathrm{d}\boldsymbol{x} \quad \text{optical dielectric,} \tag{3.51}$$

where $m_{e,e}$ is the effective electron (or hole) mass, g is the line-shape function, and E_e is the electronic energy difference $E_{e,ij}$. We will apply this to photon absorption by semiconductors in Section 7.8.1.

The Adachi model [2] for semiconductor ϵ_e is based on the Kramers–Kronig transformation and is a summation over active bands. Other models are discussed in [258].

The dielectric heating (commonly called microwave heating) rate is

$$\dot{s}_{ph\text{-}e} = \dot{s}_{e,m} = \sigma_{e,r}|\overline{\boldsymbol{e}_e}|^2 = \epsilon_{\mathrm{o}}\epsilon_{e,c}\omega\overline{e_e^2} = \frac{1}{2}\epsilon_{\mathrm{o}}\epsilon_{e,c}\omega e_{e,\mathrm{o}}^2 \quad \text{volumetric dielectric heating,} \tag{3.52}$$

where $\sigma_{e,r}$ is given by (3.48), and $\overline{e_e^2}$ is the time average of the square of the electric field (square of RMS). In general, $\epsilon_{e,c}(\omega, T)$ is found from absorption data.

Note the high magnitude of $\epsilon_{e,c}$ for H_2O in Figure 3.7(b) in the microwave regime making for high heating rates (thus the name microwave heating).

3.3.6 Electrical Resistance and Joule Heating

Electrical resistivity $\rho_e = 1/\sigma_e$ also measures the loss of electric power that is due to the inelastic scattering of electrons with the lattice, impurities, etc. These scattering mechanisms are discussed in Sections 5.13 and 5.15. According to the Matthiessen rule, the resistances that are due to various scattering mechanisms are additive in series [174], i.e.,

$$\rho_e = \sum_i \rho_{e,i} \quad \text{Matthiessen rule (series transport).} \tag{3.53}$$

Using this ρ_e, the volumetric Joule heating rate for conductors (metals) is (for direct current)

$$\dot{s}_{e\text{-}p} = \dot{s}_{e,\mathrm{J}} = \boldsymbol{j}_e \cdot \boldsymbol{e}_e = \sigma_e|\boldsymbol{e}_e|^2 = \rho_e j_e^2 \quad \text{volumetric Joule heating,} \tag{3.54}$$

where j_e is the current density, and we have used the Ohm law (Table 3.5). In Section 5.18 we will derive this, showing that the Joule heating represents relaxation of the electron kinetic energy.

Electron-lattice (phonon) scattering is dominant in inelastic scattering mechanisms. Joule heating results in phonon emission and, as governed by the energy equation (Table 1.1), may lead to sensible heating.

For semiconductors, Joule heating is given as [in addition to that due to net current of conduction electrons presented in (3.54)]

$$\dot{s}_{e\text{-}p} = \dot{s}_{e,\mathrm{J}} = \boldsymbol{j}_e \cdot \boldsymbol{e}_e + (\dot{n}_{e,rec} - \dot{n}_{e,gen})(\Delta E_{e,g} + 3k_{\mathrm{B}}T), \tag{3.55}$$

where $\dot{n}_{e,rec}$ and $\dot{n}_{e,gen}$ are volumetric, nonradiative recombination and generation rates (1/m^3-s). The electron–hole carrier energy equation will be derived in Section 5.18.

3.4 Onsager Transport Coefficients

The Gibbs fundamental thermodynamic relations hold between the state variables and involve derivatives of these variables. The entropy balance equation is derived from the continuity equation, with proper substitutions. From these, the first postulate of irreversible thermodynamics allows for defining the conjugated thermodynamic fluxes and forces. In the second postulate, a linear relation is assumed between the two and the phenomenological transport equations are introduced.

The Onsager relationship for transport, represented for flux j_i (where i can be e for electrical transport of t for thermal current), is [257]

$$j_i = \sum L_{ik} F_j \quad \text{Onsager relation}, \tag{3.56}$$

where for each driving force F_i there is a corresponding conjugate primary flux j_i. Here L_{ik} is a phenomenological transport coefficient relating these two. These phenomena are called cross-coupling transport.

For example, for a system with two fluxes caused by two driving forces (i.e., two simultaneous irreversible processes), we have

$$j_1 = L_{11}F_1 + L_{12}F_2, \tag{3.57}$$

$$j_2 = L_{21}F_2 + L_{22}F_2, \tag{3.58}$$

where j_1 and j_2 denote the fluxes and F_1 and F_2 denote the forces causing these flows. The coefficients L_{ik} (with $i \neq k$) describe the interference (coupling) of the two irreversible processes i and k.

The Onsager reciprocity relation between the cross coefficients L_{ik} and L_{ki} is expressed as

$$L_{ik} = L_{ki}(i \neq k) \quad \text{Onsager reciprocity relation}. \tag{3.59}$$

This Onsager reciprocity relations allows a reduction in the number of transport coefficients and provides for the following constraints

$$L_{11} \geq 0, \quad L_{22} \geq 0, \quad (L_{12} + L_{21})^2 < 4L_{11}L_{22}$$
$$\text{constraints on transport coefficients}. \tag{3.60}$$

The conjugate coefficients (L_{11} and L_{22}) must be positive, whereas the crossed (coupling or interference) coefficients (L_{12} and L_{21}) have no definite sign.

The last condition in (3.60) is based on the position derivative of the entropy with respect to temperature [257].

As an example, consider the coupled electric and thermal current vectors $\boldsymbol{j}_e$ and $\boldsymbol{q}$ in solids. The choice of driving force is generally made by the examination of the transport equations, such as the BTE. As will be shown in Section 5.9, the driving forces suitable for this analysis are $\nabla(E_\mathrm{F}/e_c)$ and $\nabla(1/T)$, where E_F is the Fermi energy. Then, following (3.57) and (3.58), the general form of these coupled currents is

$$\boldsymbol{j}_e = \boldsymbol{L}_{ee} \cdot \nabla \frac{E_\mathrm{F}}{e_c} + \boldsymbol{L}_{et} \cdot \nabla \frac{1}{T} \tag{3.61}$$

$$\boldsymbol{q} = \boldsymbol{L}_{te} \cdot \nabla \frac{E_\mathrm{F}}{e_c} + \boldsymbol{L}_{tt} \cdot \nabla \frac{1}{T}. \tag{3.62}$$

The coupling tensors $\boldsymbol{L}_{ee}$, $\boldsymbol{L}_{et}$, $\boldsymbol{L}_{te}$, and $\boldsymbol{L}_{tt}$ are the electrothermal transport properties of the solid. The Onsager reciprocity relation (3.59) then requires that $L_{et,ij} = L_{te,ij}$, where ij designates the component of the tensor.

The more common form of the electrothermal transport equations (which are more suitable for experiments), and for the simple case of isotropic properties, is given in terms of the electric field intensity vector $\boldsymbol{e}_e$, i.e.,

$$\boldsymbol{e}_e = \rho_e \boldsymbol{j}_e + \alpha_\mathrm{S} \nabla T, \quad \rho_e = \frac{1}{\sigma_e} \tag{3.63}$$

$$\boldsymbol{q} = \alpha_\mathrm{P} \boldsymbol{j}_e - k_e \nabla T, \quad \alpha_\mathrm{P} = \alpha_\mathrm{S} T. \tag{3.64}$$

Here, $\rho_e(\Omega\text{-m})$ is the electrical resistivity (the inverse of electrical conductivity), α_S(V/K) is the Seebeck coefficient, α_P(V) is the Peltier coefficient, and k_e is the electronic thermal conductivity. Here, the Onsager reciprocity relation leads to $\alpha_\mathrm{P} = \alpha_\mathrm{S} T$, as will be discussed in Section 5.13. Here ρ_e and k_e have positive values.

3.5 Stochastic Particle Dynamics and Transport

3.5.1 Langevin Particle Dynamics Equation

The Langevin stochastic particle dynamics equation governs the motion of a particle through a medium by a combination of a mean force $\overline{\boldsymbol{F}}_k$ and a time-dependent random force $\boldsymbol{F}_k(t)$, with friction (e.g., viscous) from the medium $-\gamma \boldsymbol{u}_k$. This equation is listed in Table 3.7. The fluctuation-dissipation theorems also require some constraints on the random force, and these are also stated in the table. The brackets $\langle\ \rangle$ indicate integration over the volume. We will use the Langevin equation in Section 6.10 for Brownian motion of solid particles suspended in a liquid. In that case, the random force is from thermal fluctuations of fluid particles colliding with the solid particles of much larger size and mass.

Table 3.7. *Stochastic (Brownian) particle dynamics equation (Langevin stochastic equation) [73].*

Langevin equation for particle k

$$\frac{\mathrm{d}}{\mathrm{d}t}(m_k \boldsymbol{u}_k) = \overline{\boldsymbol{F}}_k - \gamma \boldsymbol{u}_k + \boldsymbol{F}_k(t)$$

Fluctuation-dissipation theorems

$$\langle \boldsymbol{F}_k(t) \rangle = 0 \quad \text{zero-mean } \overline{\boldsymbol{F}}_k(t) \text{ force}$$

$$\int \langle \boldsymbol{F}_k(t) \cdot \boldsymbol{F}_k(0) \rangle \mathrm{d}t = 6k_\mathrm{B} T \gamma \quad \text{correlation of random force (defines } T)$$

$\overline{\boldsymbol{F}}_k$	steady, deterministic force on particle k, N
$\boldsymbol{F}_k(t)$	random (stochastic), time-dependent force on particle k, N
m_k/γ	Langevin relaxation time, s
$-\gamma \boldsymbol{u}_k$	frictional force on particle k, N

The ratio m_k/γ is called the collision (viscous or Langevin) relaxation time τ_μ and under no mean (steady) force $\overline{\boldsymbol{F}}_k = 0$, the MSD $\langle [x(t) - x_\mathrm{o}]^2 \rangle$ is (note that $\langle m u_k^2/2 \rangle = k_\mathrm{B} T/2$, Section 6.10)

$$\langle x^2 \rangle - \langle x_\mathrm{o}^2 \rangle = -\frac{2k_\mathrm{B} T}{\gamma}[\tau_\mu (e^{-t/\tau_\mu} - 1) - t], \tag{3.65}$$

where x_o is the initial position. The short-time solution $t/\tau_\mu \ll 1$ gives the ballistic regime

$$\langle x^2 \rangle - \langle x_\mathrm{o}^2 \rangle = \frac{k_\mathrm{B} T}{m_k} t^2. \tag{3.66}$$

The long-time solution $t/\tau_\mu \gg 1$ gives the viscous regime

$$\langle x^2 \rangle - \langle x_\mathrm{o}^2 \rangle = -\frac{2k_\mathrm{B} T}{\gamma} t = \frac{2k_\mathrm{B} T}{m_k} t \tau_\mu, \quad \tau_\mu = \frac{m_k}{\gamma}. \tag{3.67}$$

This is the Brownian motion of particle in a viscous fluid. We will use these in derivation of the Brownian diffusion in Section 6.10.

3.5.2 Fokker–Planck Particle Conservation Equation

The Fokker–Planck equation is a particle drift–diffusion equation and represents the probability of finding a particle in a drift vector field $\boldsymbol{u}$, and a diffusion tensor D is used in the probability density function (pdf) P. This time-evolution equation for stochastic processes [239] is listed in Table 3.8.

The Fokker–Planck equation is a powerful tool for solving stochastic problems. It usually appears for variables (represented by P) describing a macroscopic but small subsystem in the field of a stochastic force. The diffusion term is the result of the stochastic force [similar to $R_k(t)$ in the Langevin equation].

Note that the pdf must satisfy the normalization constraint, as was the case case for the other distribution functions and for the wave function in the Schrödinger equation (Table 2.8).

Table 3.8. *Stochastic differential equation for evolution of the pdf (Fokker–Planck stochastic equation) [239].*

$$\frac{\partial P}{\partial t} + \nabla \cdot \boldsymbol{u} P = \frac{1}{2} \sum_{i,j} \frac{\partial^2}{\partial x_i \partial x_j} D_{ij} P$$

$$\equiv L_{FP} P$$

$$\int_o^\infty P \mathrm{d}V = 1 \text{ normalization constraint}$$

D_{ij}	diffusion tensor, m²/s
L_{FP}	Fokker–Planck operator, 1/s
P	probability density function (pdf)
$\boldsymbol{u}$	drift coefficient vector

3.5.3 Mean-Field Theory

In statistical mechanics, the interactions among the particles in a many-body system are generally very difficult to solve exactly without using those numerical approximations in MD, except for a few simple cases. The great difficulty in determining the partition function of the system results from the combining of the interaction terms in the Hamiltonian when summing over all states. The goal of mean-field theory (also known as self-consistent field theory) is to resolve these combinatorial problems.

In mean-field theory, all interactions to any one body are replaced with an average (or effective) interaction. This reduces any multibody problem to an effective, one-body problem. This allows for predicting the behavior of the system without extensive inclusion of all interactions. In mean-field theory, the Hamiltonian may be expanded in terms of the magnitude of fluctuations around the mean of the field. In this context, mean-field theory can be viewed as the zeroth-order expansion of the Hamiltonian in fluctuations. Physically, this means a mean-field system has no fluctuations, but this coincides with the idea that one is replacing all interactions with a mean field. In the formalism of fluctuations, mean-field theory provides a convenient starting point for studying the first- or second-order fluctuations.

The stronger the role of the long-range forces, and the larger the dimensionality of the system (one versus two or three dimensional), the more appropriate the use of mean-field theory becomes. We will have an example of a mean-field potential in Section 6.11.6.

3.6 Fluctuation–Dissipation and Green–Kubo Transport Theory

The G–K relations state that the space–time integral of the flux–flux equilibrium correlation function divided by $k_B T$ gives the corresponding steady-state transport coefficient. An example for thermal conductivity is shown in Table 3.9, and the derivation is given in Appendix B.

The complexity of many-particle systems is frequently the motivation for introducing mean-field approximations so as to simplify and solve the complicated

Table 3.9. *G–K heat current autocorrelation decay relation for phonon or fluid particle thermal conductivity (linear response theory) [239].*

G–K autocorrelation for thermal conductivity (for cubic isotropy, derivation is given in Appendix B)

$$k = \frac{1}{3Vk_BT^2}\int_0^\infty \langle \dot{\boldsymbol{w}}(t)\cdot\dot{\boldsymbol{w}}(0)\rangle_o \mathrm{d}t,$$

where the volume integral (over V) of the heat current autocorrelation function (based on equilibrium fluctuation) is $\langle \dot{\boldsymbol{w}}(t)\dot{\boldsymbol{w}}(0)\rangle_o$, and $\dot{\boldsymbol{w}}$ is defined as

$$\dot{\boldsymbol{w}} = \frac{\mathrm{d}}{\mathrm{d}t}\sum_i E_i\boldsymbol{x}_i = \frac{\mathrm{d}}{\mathrm{d}t}\sum_i (E_{k,i}+E_{p,i})\boldsymbol{x}_i$$

is the heat current vector ($E_{k,i}$ is the kinetic energy and $E_{p,i}$ is the potential energy of particle i) $\dot{\boldsymbol{w}}$ is also called the energy flux (or current) vector (when divided by V)

For a two-body potential, we have

$$\dot{\boldsymbol{w}} = \sum_i E_i\boldsymbol{u}_i + \frac{1}{2}\sum_{i,j}(\boldsymbol{F}_{ij}\cdot\boldsymbol{u}_i)\boldsymbol{x}_{ij}$$

based on equilibrium distributions of $\boldsymbol{x}_i$ and $\boldsymbol{u}_i$

k	thermal conductivity, W/m-K
E	energy, J
$\boldsymbol{F}_{ij}$	force vector, N
$\boldsymbol{u}_i$	velocity vector, m/s
$\boldsymbol{w}$	energy flux (current) vector, W-m
$\boldsymbol{x}$	position vector, m

equations of motion or the time-evolution equation for the pdf in configuration space. We are interested in the linear response of such a system to a small external influence, starting from the equilibrium situation. The usual procedure is to calculate the first-order deviation of the distribution function from its equilibrium form. This can be done if the statistical model equation is linear in the distribution function, as is the case, for example, in the Liouville equation (describes time evolution of phase space distribution function, i.e., conservation applied to evolution of ensemble in the phase space, $\boldsymbol{x}$, $\boldsymbol{p}$). The linear response to an external influence is around a distribution function and is valid for the Liouville equation with reversible dynamics and the dissipative Fokker–Planck equation.

Let $\mathrm{H_o}$ be the Hamiltonian of the given system when it is isolated. If we apply a weak disturbing field $\boldsymbol{F}(t)$ that couples to a property $\boldsymbol{b}(t)$, with $\langle \boldsymbol{b}\rangle = 0$, the Hamiltonian of the perturbed system is given by [338]

$$\mathrm{H} = \mathrm{H_o} - \boldsymbol{b}\cdot\boldsymbol{F}(t). \tag{3.68}$$

Linear response theory then leads to the following first-order expression for the mean temporal change rate of $\boldsymbol{b}$, i.e., $\langle \dot{\boldsymbol{b}}(t)\rangle$ in terms of the autocorrelation of equilibrium

$$\langle \dot{\boldsymbol{b}}(t) \rangle = \frac{1}{k_{\rm B}T}\int_{-\infty}^{t} \boldsymbol{F}(t')\langle \dot{\boldsymbol{b}}(t)\dot{\boldsymbol{b}}(t-t')\rangle_{\rm o}{\rm d}t'$$

relation between mean rate and equilibrium autocorrelation, (3.69)

where the ensemble average of equilibrium function (no gradients present) $\langle\ \rangle_{\rm o}$ is for the unperturbed system.

Assuming a constant force field $\boldsymbol{F}$ starting at $t<0$, we have

$$\langle \dot{\boldsymbol{b}} \rangle = \frac{\boldsymbol{F}}{k_{\rm B}T}\int_{-\infty}^{t} \langle \dot{\boldsymbol{b}}(0)\dot{\boldsymbol{b}}(-t')\rangle_{\rm o}{\rm d}t'. \tag{3.70}$$

Independently, the fundamental G–K relation states that, for any conserved quantity $\boldsymbol{b}$, the appropriate isotropic Onsager transport coefficient $L_{ii} = L$ (3.56) is given by the equilibrium average

$$L_{ii} = L \equiv \int_{0}^{\infty} \langle \dot{\boldsymbol{b}}(0)\cdot\dot{\boldsymbol{b}}(t)\rangle_{\rm o}{\rm d}t. \tag{3.71}$$

The integral is over the equilibrium flux autocorrelation (also called self-correlation) function $\dot{\boldsymbol{b}}(0)\cdot\dot{\boldsymbol{b}}(t)$. At $t=0$, the autocorrelation function is positive, as it is the mean-square value of the flux at equilibrium. At large t, $\dot{\boldsymbol{b}}(t)$ is uncorrelated with $\dot{\boldsymbol{b}}(0)$ and the autocorrelation decays to zero. This behavior is used in computational MD simulations (an example is given in Section 4.12).

Combining this with (3.70) and (3.71), we find that $\langle \dot{\boldsymbol{b}}(t)\rangle = \boldsymbol{F}\nu/k_{\rm B}T$, or

$$L_{ii} = L = \frac{k_{\rm B}T}{\boldsymbol{F}}\langle \dot{\boldsymbol{b}}(t)\rangle. \tag{3.72}$$

Thus we may determine the transport coefficient L either from an equilibrium simulation by using (3.71), or from a nonequilibrium simulation with applied field by using (3.72). Generally, the second method yields better statistics, but it is more prone to nonlinearity problems, such as large fields. Also, systems responding to an external field must be thermostated.

Consider a fluid sample of ions in an electric field $\boldsymbol{F}(t) = \boldsymbol{e}_e(t)$. The charge density is given by $\boldsymbol{b}(t) \equiv \sum_i q_i \boldsymbol{x}_i(t)$, where q_i is the charge of particle i. Then (3.68) becomes

$$\mathrm{H} - \mathrm{H}_{\rm o} = -\boldsymbol{b}\cdot\boldsymbol{F}(t) = -[\sum_i q_i \boldsymbol{x}_i(t)]\cdot\boldsymbol{e}_e(t). \tag{3.73}$$

The electric current density vector $\boldsymbol{j}_e$ is defined as

$$\dot{\boldsymbol{b}} = \sum_i q_i \dot{\boldsymbol{x}}_i(t) \equiv V\boldsymbol{j}_e(t), \tag{3.74}$$

where V is the volume.

Then the electrical conductivity σ_e is determined from an equilibrium simulation, using the isotropic G–K relation (3.71), i.e.,

$$\sigma_{e,xx} = \sigma_e = \frac{V}{k_{\rm B}T}\int_{0}^{\infty}\langle j_{e,x}(0)j_{e,x}(t)\rangle_{\rm o}{\rm d}t. \tag{3.75}$$

Alternatively, in a nonequilibrium simulation, using the measured response to an applied field $\boldsymbol{e}_e = \{e_{e,x}, 0, 0\} = -\{\mathrm{d}\varphi_e/\mathrm{d}x, 0, 0\}$, we have from the Ohm law (Table 3.5)

$$\sigma_{e,xx} = \sigma_e = \frac{\langle j_{e,x}\rangle}{e_{e,x}} = -\frac{\langle j_{e,x}\rangle}{\mathrm{d}\varphi_e/\mathrm{d}x}. \tag{3.76}$$

Table 3.9 gives a similar expression for the thermal conductivity k, with the corresponding nonequilibrium relation given by the Fourier law

$$k = -\frac{\langle q_x\rangle}{\mathrm{d}T/\mathrm{d}x}, \tag{3.77}$$

for $\nabla T = \{\mathrm{d}T/\mathrm{d}x, 0, 0\}$, and for heat flux vector $\boldsymbol{q} = \{q_x, 0, 0\}$.

For cubic isotropy, the thermal conductivity (for phonons and fluid particles) is given in terms of the so-called heat current vector (also called the energy flux, when divided by V) or the correlation function (or vector) $\dot{\boldsymbol{w}}(t)$, which replaces $\dot{\boldsymbol{b}}$ in (3.69), i.e.,

$$k_{xx} = k = \frac{1}{3Vk_\mathrm{B}T^2}\int_0^\infty \langle \dot{\boldsymbol{w}}(t)\cdot\dot{\boldsymbol{w}}(0)\rangle \mathrm{d}t,\quad \dot{\boldsymbol{w}} = \frac{\mathrm{d}}{\mathrm{d}t}\sum_i (E_{k,i} + E_{p,i})\boldsymbol{x}_i. \tag{3.78}$$

Note that the unit of heat current vector is $\dot{\boldsymbol{w}}$(W-m).

The G–K relation derivation in Appendix B is for the thermal conductivity, and uses the Fourier transform of the energy equation in the absence of net motion but subject to thermal fluctuation.

Section 4.12 presents the results of using (3.78) along with MD to predict the phonon conductivity of some crystals. We will note there that the heat current autocorrelation decays in an exponential envelope (can have oscillations), and that the decay is presented by time constants indicating the strength of the thermal transport.

3.7 Macroscopic Fluid Dynamics Equations

The Navier–Stokes equations are the macroscopic ($L \gg \lambda_f$), fundamental partial differential equations that describe the flow of incompressible fluids. Using the rate-of-stress and rate-of-strain tensors, the components of a viscous force in a nonrotating frame are derived. Table 3.10 gives the Navier–Stokes equations in vectorial form. Similar to thermal conductivity, the dynamic viscosity μ is a transport property related to atomic and dynamic behavior of the fluid. In Chapter 6, we will discuss various flow regimes, including molecular- and turbulent-flow regimes.

Also listed in Table 3.10 are the normal and tangential force balances at the liquid–gas interface with surface tension σ_{lg}, which is allowed to change along the interface.

A wave equation for a thermobuoyant disturbance (no mean velocity, $\overline{\boldsymbol{u}} = 0$) in a fluid layer with a mean temperature gradient $\overline{T} = \overline{T}(x)$ along the direction gravity acts is also listed. This wave can grow under favorable conditions to form a cellular motion with wave number κ [59, 163, 171, 173].

Table 3.10. *Macroscopic Newtonian fluid mechanics (Navier–Stokes) equations (including thermomechanical wave propagation) [163, 173].*

$$\rho\frac{\partial \boldsymbol{u}}{\partial t} + \rho(\boldsymbol{u}\cdot\nabla)\boldsymbol{u} = -\nabla p + \nabla\cdot \boldsymbol{S}_\mu + \rho\boldsymbol{g} + \boldsymbol{f}_e \quad \text{Navier–Stokes momentum equation}$$

$$\tau_{ij} = \mu(\frac{\partial u_i}{\partial x_j} + \frac{\partial u_j}{\partial x_i}) + \delta_{ij}\lambda_\mu\frac{\partial u_k}{\partial x_k} \quad \text{Newtonian viscous fluid}$$

$$\frac{\partial \rho}{\partial t} + \nabla\cdot\rho\boldsymbol{u} = 0 \quad \text{continuity equation}$$

Liquid–gas interfacial force balance:

$$\text{normal to interface: } p_e - p_g + \sigma_{lg}(\frac{1}{r_1} + \frac{1}{r_2}) = 2\mu_l\frac{\partial u_{l,n}}{\partial x_n} - 2\mu_l\frac{\mu_{g,n}}{\partial x_n} \quad \text{on } A_{lg}$$

$$\text{tangent to interface: } \mu_l(\frac{\partial u_{l,n}}{\partial x_t} + \frac{\partial u_{l,t}}{\partial x_n}) - \mu_g(\frac{\partial u_{g,n}}{\partial x_t} + \frac{\partial u_{g,t}}{\partial x_n}) = \frac{\partial \sigma_{lg}}{\partial x_t} \quad \text{on } A_{lg}$$

r_1 and r_2 are principal radii of interface curvature

Wave equation for a sinusoidal disturbance $\boldsymbol{u}'$ in the mean velocity $\overline{\boldsymbol{u}}$, i.e., $\boldsymbol{u} = \overline{\boldsymbol{u}} + \boldsymbol{u}'$, of wave number κ caused by thermobuoyancy ($\overline{\boldsymbol{u}} = 0$), with the temperature field similarly given by $T = \overline{T}(x) + T'$, for the component u'_x along the gravity is

$$(\frac{\partial}{\partial t} + \mu\nabla^2)\nabla^2 u'_x = -\rho_o\beta g(\frac{\partial^2}{\partial y^2} + \frac{\partial^2}{\partial z^2})T', \quad \overline{\boldsymbol{u}} = 0$$

$$\rho c_p\frac{\partial T'}{\partial t} + \rho c_p u'_x\frac{\partial \overline{T}}{\partial x} - k\nabla^2 T' = 0$$

$$\rho c_p\frac{\partial \overline{T}}{\partial t} - k\frac{\partial^2 \overline{T}}{\partial x^2} = 0$$

or

$$[\frac{\partial}{\partial t} + \mu(\kappa^2 - \frac{\partial^2}{\partial x^2})](\frac{\partial^2}{\partial x^2} - \kappa^2)u'_x = \kappa^2\rho_o\beta g T'$$

$$\rho c_p(\frac{\partial T'}{\partial t} + u'_x\frac{\partial \overline{T}}{\partial x}) - k(\frac{\partial^2}{\partial x^2} - \kappa^2)T' = 0$$

$$\rho c_p\frac{\partial \overline{T}}{\partial t} - k\frac{\partial^2 T^2}{\partial x^2} = 0$$

$\boldsymbol{f}_e$	volumetric electromagnetic force vector, N/m^3
$\boldsymbol{g}$	gravitational acceleration vector, m/s^2
p	pressure, Pa
β	volumetric thermal expansion coefficient, 1/K
κ	wave vector, 1/m
σ_{lg}	liquid–gas surface tension, N/m
τ_{ij}	component of viscous shear stress tensor $\boldsymbol{S}_\mu$, Pa
μ	dynamic viscosity, Pa-s
λ_μ	second viscosity, Pa-s

Table 3.11. *Macroscopic elastic solid-mechanics equations (including mechanical wave propagation) [27].*

$$\rho \frac{\partial^2 \boldsymbol{d}}{\partial t^2} = \frac{E_Y}{2(1+\nu_P)} \nabla \cdot (\nabla \boldsymbol{d}) + \frac{E_Y}{2(1+\nu_P)(1-\nu_P)} \nabla^2 \boldsymbol{d} + \boldsymbol{f} \quad \text{Navier equation}$$

$$\epsilon_{ij} = \frac{1}{2}(\frac{\partial d_i}{\partial x_j} + \frac{\partial d_j}{\partial x_i}) \quad \text{strain-displacement, or compatibility equation}$$

$$\tau_{ij} = c_{ijkl}\epsilon_{kl} \quad \text{(stress-strain relation, generalized Hooke law)}$$

c_{ijkl} elastic stiffness constants (elastic constants)

$$\epsilon_{ij} = \frac{1+\nu_P}{E_Y}\tau_{ij} - \frac{\nu_P \tau_{mm}\delta_{ij}}{E_Y} + \beta_s(T - T_o) \quad \text{isotropic (cubic) crystal}$$

(only three independent elastic constants), or Hooke law

Shear (or isovoluminous) wave (also called S or secondary wave) for isotropic solid is

$$\rho \frac{\partial^2 \boldsymbol{d}}{\partial t^2} = G \nabla^2 \boldsymbol{d},$$

moving at the transverse wave speed, $u_{p,\mathrm{T}} = \omega/\kappa = (G/\rho)^{1/2}$
For the compression wave (also called P or primary wave), $u_{p,\mathrm{L}} = (E_Y/\rho)^{1/2}$
Both $u_{p,\mathrm{L}}$ and $u_{p,\mathrm{T}}$ are independent of the frequency (no dispersion, large phonon wavelength limit)
The displacement plane wave is of the form

$$\boldsymbol{d} = \boldsymbol{d}_o \boldsymbol{s}_\alpha e^{i(\omega t - \boldsymbol{\kappa}\cdot\boldsymbol{x})}.$$

Relation between the isothermal compressibility κ_p, the Young modulus E_Y, and the shear modulus G, for cubic crystals, are

$$E_p = \frac{E_Y}{3(1-2\nu_P)} = \frac{2G(1+\nu_P)}{3(1-2\nu_P)}, \quad E_p^{-1} = \kappa_p = \frac{1}{\rho}\frac{\partial \rho}{\partial p}|_T,$$

c_{ijkl}	elastic constant, Pa
$\boldsymbol{d}$	displacement vector, m
E_p	isothermal bulk modulus, Pa
E_Y	Young modulus, Pa
$\boldsymbol{f}$	volumetric external force, N/m^3
β_s	coefficient of linear thermal expansion, 1/K
ϵ_{ij}	strain tensor component
ν_P	Poisson ratio
G	shear modulus, Pa
κ_p	isothermal compressibility, Pa^{-1}
τ_{ij}	component of stress tensor $\boldsymbol{S}$, Pa

Table 3.12. *Macroscopic scales, based on thermal energy transport by conduction, convection, and radiation.*

Scale	Relation
Length	L
Time:	
conduction (diffusion time)	$\tau_\alpha = \frac{L^2}{\alpha^2}, \alpha = \frac{k}{\rho c_p}$ (thermal diffusivity)
convection (transit time)	$\tau_u = \frac{L}{\lvert \boldsymbol{u}_o \rvert}$, $\boldsymbol{u}_o$ characteristic net velocity
radiation (absorption time)	$\tau_\sigma = \frac{1}{\sigma_{ph} u_{ph}}$, σ_{ph} absorption coefficient $\sigma_{ph} L \gg 1$ optically thick media (radiation diffusion limit) $\sigma_{ph} L \ll 1$ optically thin media
Temperature:	
single phase	0 K, or T_o characteristic temperature
phase change	T_{lg}, and T_{sl} (phase-change temperatures), or T_c (critical-point temperature)

3.8 Macroscopic Elastic Solid-Mechanics Equations

Elastic solid mechanics describes the macroscopic ($L \gg r_{nn}$) elastic deformation of a solid that is due to applied stress. For cubic (isotropic) crystals, there are only two independent elastic constants. The resistance to deformation is represented by the Young modulus E_Y. The Hooke law relates the deformation (strain) to the stress (including thermal stress), through the Young modulus and the Poisson ratio ν_P.

Table 3.11 lists the macroscopic elastic solid mechanics equation. The Young modulus is related to the bulk modulus E_p (and in turn this is related to the isothermal compressibility κ_p). We will discuss the relation between E_p and the atomic structure of the solid in Chapter 4, along with the speed of sound in solids.

Also listed in Table 3.11 is the displacement wave equation for the transverse wave (shear wave with no change in volume) [27]. There is also a longitudinal (compressional or density) wave. The shear waves are slower than the compressional waves.

3.9 Macroscopic Scales

Similar to Table 1.5 for atomic scales, Table 2.7 for MD scales, and Table 3.2 for BTE scales, Table 3.12 lists the macroscopic scales for heat transfer. Here L is the system (macroscopic) length scale. Then depending on presence/dominance of conduction, convection, and radiation, the time scale is diffusion time ($\tau_\alpha = L^2/\alpha$, and thermal diffusivity $\alpha = k/\rho c_p$), transit time ($\tau_u = L/\lvert \boldsymbol{u}_o \rvert$, $\boldsymbol{u}_o$ is the characteristic velocity), or photon absorption time ($\tau_\sigma = 1/\sigma_{ph} u_{ph}$, σ_{ph} is the absorption coefficient). The temperature scale is generally absolute 0 K, or a characteristic temperature T_o

such as the initial or undisturbed (far-field) temperature, or the phase change temperature T_{ij}.

In Figure 1.9, we marked possible ranges of the length and time scales for atomic, MD, BTE, and macroscopic treatments.

3.10 Problems

Problem 3.1

Show that the heat current vector, in the G–K transport relation, i.e.,

$$\dot{\boldsymbol{w}} = \frac{d}{dt}\sum_i (E_{k,i} + E_{p,i})\boldsymbol{x}_i,$$

can be written as

$$\dot{\boldsymbol{w}} = \sum_i E_i \boldsymbol{u}_i + \frac{1}{2}\sum_{i,j}(\boldsymbol{F}_{ij}\cdot\boldsymbol{u}_i)\boldsymbol{x}_{ij},$$

for two-body interatomic potentials.

Note that by using the Newton second law, we have

$$\frac{dE_{k,i}}{dt} = \sum_j \boldsymbol{F}_{ij}\cdot\boldsymbol{u}_i,$$

and also

$$\frac{dE_{p,i}}{dt} = \sum_j \frac{1}{2}\boldsymbol{F}_{ij}\cdot(\boldsymbol{u}_j - \boldsymbol{u}_i),$$

and that

$$\begin{aligned}\sum_i[\sum_j \frac{1}{2}\boldsymbol{F}_{ij}(\boldsymbol{u}_j + \boldsymbol{u}_i)]\boldsymbol{x}_i &= \frac{1}{2}\sum_{i,j}[\frac{1}{2}\boldsymbol{F}_{ij}(\boldsymbol{u}_i + \boldsymbol{u}_j)\boldsymbol{x}_i - \frac{1}{2}\boldsymbol{F}_{ij}(\boldsymbol{u}_i + \boldsymbol{u}_j)\boldsymbol{x}_j]\\ &= \frac{1}{2}\sum_{i,j}\frac{1}{2}\boldsymbol{F}_{ij}(\boldsymbol{u}_i + \boldsymbol{u}_j)\boldsymbol{x}_{ij},\end{aligned}$$

because $\boldsymbol{F}_{ij} = -\boldsymbol{F}_{ji}$.

Problem 3.2

The nondegenerate behavior leads to classical distribution and simplifies the analysis when justified. The conduction electron density $n_{e,c}$ is temperature-dependent in the semiconductors.

(a) For electrons in semiconductors, the equilibrium probability distribution function is

$$f_e^{\rm o} = \frac{1}{\exp\dfrac{E_{e,p}+E_{e,k}-E_{\rm F}}{k_{\rm B}T}+1}$$

$$= \frac{1}{\exp(-\eta_c + \dfrac{p^2}{2m_{e,e}k_{\rm B}T})+1} \quad \text{degenerate electrons}$$

$$f_e^{\rm o} \simeq \exp(\eta_c - \frac{p^2}{2m_{e,e}k_{\rm B}T}) \text{ for nondegenerate electrons for } -\eta_c k_{\rm B}T + \frac{p^2}{2m_{e,e}} \gg k_{\rm B}T,$$

where $\eta_c = (E_{\rm F} - E_{e,p})/k_{\rm B}T$, and $m_{e,e}$ is the effective electron mass (5.44).

The equilibrium electron density is given by

$$n_{e,c} = \frac{1}{4\pi^3\hbar^3}\int f_e^{\rm o}\mathrm{d}\boldsymbol{p}$$

$$= \frac{1}{4\pi^3\hbar^3}\int_0^\infty \exp(-\frac{p^2}{2m_{e,e}k_{\rm B}T})e^{\eta_c}4\pi p^2\mathrm{d}p.$$

Show that

$$n_{e,c} = 2(\frac{m_{e,e}k_{\rm B}T}{2\pi\hbar^2})^{3/2}e^{\eta_c}.$$

Note that the value of the definite integral can be found from mathematical tables, or from the gamma functions (5.113).

(b) Repeat (a) assuming degenerate electrons and show that

$$n_{e,c} = 2(\frac{m_{e,e}k_{\rm B}T}{2\pi\hbar^2})^{3/2}\mathrm{F}_{1/2}(\eta_c),$$

where $\mathrm{F}_{1/2}$ is the Fermi–Dirac integral of the order 1/2:

$$\mathrm{F}_{1/2}(\eta_c) = \frac{2}{\pi^{1/2}}\int_0^\infty \frac{x^{1/2}}{1+\exp(x-\eta_c)}\mathrm{d}x.$$

The general form for Fermi–Dirac integral of the order of j is

$$\mathrm{F}_j(\eta_c) = \frac{1}{\Gamma(j+1)}\int_0^\infty \frac{x^j}{1+\exp(x-\eta_c)}\mathrm{d}x,$$

where $\Gamma(j+1)$ is the gamma function, given by (5.113). The Fermi–Dirac integral indicates a quantum behavior.

Problem 3.3

Show that when the in- and out-scattering presentation of $\partial f/\partial t|_s$ given by (3.8) is used, the summation over all $\boldsymbol{p}$ vanishes, i.e.,

$$\sum_{\boldsymbol{p}} \frac{\partial f}{\partial t}|_s = 0.$$

(Note to interchange the order of summations over $\boldsymbol{p}$ and $\boldsymbol{p}'$.)

Problem 3.4

(a) Show that (B.4) is the Fourier transform of (B.1). Note that the Fourier transform has the property $F(\nabla_1 E') = i\kappa_1 F(E')$.

(b) Show that (B.10) is the Taylor series expansion of the left-hand side of (B.9).

(c) Explain the transformation of the time integral to the difference in (B.31).

(d) Explain in words the steps in derivation of the right-hand side of (B.33).

Problem 3.5

(a) Use the Matlab solid-phase MD code (three-dimensional) set for Ar FCC and change the potential constants to that for Xe (Tables 2.2 and A.1 give potential and atomic properties). Then calculate the thermal conductivity of Xe FCC with the G–K autocorrelation relation in the code. The decay of the heat current autocorrelation function (HCACF), and the lattice thermal conductivity are shown in Figure 4.31 for Ar.

(b) Show the results in the forms of dimensionless autocorrelation function versus time. Use temperatures of 20 and 40 K (both are solid phases, for Xe, $T_{\rm D} = 55$ K in Table A.2).

(c) Comment on the autocorrelation function decay and any trends as the temperature increases.

Note that this simulation includes only the phonon–phonon interactions (the other interactions are discussed in Section 4.9.4, including the impurity, grain boundary, and electron scattering).

Problem 3.6

The conduction heat flux by a phonon can be written in terms of the nonequilibrium phonon population f_p' as

$$\boldsymbol{q}_{k,p} = \sum_{\alpha} \frac{1}{8\pi^3} \int E_p \boldsymbol{u}_p f_p' \mathrm{d}\boldsymbol{\kappa},$$

where α is for the different polarizations (longitudinal transverse).

Starting with the BTE, for steady-state condition and $\boldsymbol{F} = \dot{s}_f = 0$, use $\nabla f_p = \nabla f_p^{\rm o}$, the spatial variation of temperature ∇T, and relaxation-time approximation (3.13)

(a) Show that

$$f_p' = -\tau_p \frac{\partial f_p^{\rm o}}{\partial T} \boldsymbol{u}_p \cdot \nabla T.$$

(b) Show that

$$\boldsymbol{q}_{k,p} = -\left(\frac{1}{8\pi^3} \sum_{\alpha} \int E_p \tau_p \frac{\partial f_p^{\rm o}}{\partial T} \boldsymbol{u}_p \boldsymbol{u}_p \mathrm{d}\boldsymbol{\kappa}\right) \cdot \nabla T$$

$$\equiv -\boldsymbol{K}_p \cdot \nabla T,$$

where $\boldsymbol{K}_p$ is the phonon thermal conductivity tensor. Note that $\boldsymbol{u}_p\boldsymbol{u}_p$ is a tensor (diadic product).

Problem 3.7

Consider the power-law, energy-dependent relaxation time given by (3.25). The average relaxation time $\langle\langle\tau_e\rangle\rangle$ is defined as

$$\langle\langle\tau_e\rangle\rangle \equiv \frac{\langle E_e\tau_e\rangle}{\langle E_e\rangle} = \frac{\displaystyle\int E_e\tau_e f_e^{\rm o}{\rm d}\boldsymbol{p}}{\displaystyle\int E_e f_e^{\rm o}{\rm d}\boldsymbol{p}}.$$

Using the approximation

$$E_e = E_{e,k} = \frac{p^2}{2m_e} \text{ and } f_e^{\rm o} = \exp(\frac{p^2}{2m_e k_{\rm B}T}),$$

show that

$$\langle\langle\tau_e\rangle\rangle = \tau_{e,{\rm o}}\frac{\Gamma(s+\frac{5}{2})}{\Gamma(\frac{5}{2})},$$

where Γ is the gamma function (5.113). This average relaxation time is related to the electron mobility and is a momentum relaxation time.

Problem 3.8

The EM wave equation follows from the Maxwell equations and for the electric field intensity the wave equation is given by (3.34) for nonconducting media ($\sigma_e = 0$) with zero gradient in the charge distribution. Consider a plane wave moving along the x direction.

(a) Show that (3.34), i.e., $e_{e,y} = e_{e,{\rm o}}\cos(\omega t - \kappa x)$ satisfies this equation, where $c = \omega/\kappa$ is the phase velocity.

(b) Show that the exponential notation presentation of this planar wave is $e_e = e_{e,{\rm o}}\exp[i(\omega t - \kappa x)]$.

(c) Explain that when this wave is attenuated, the decreasing amplitude can be shown by $e_{e,y} = e_{e,{\rm o}}\exp[i(\omega t - \kappa_r x) - \kappa_c x]$, where the real part of the wave vector κ_r remains as before, and κ_c is called the attenuation constant (has the unit of 1/m, and is related to the absorption coefficient σ_{ph}). Note that $\exp(-\kappa_c x)$ represents decay of the wave.

(d) Show that this can also be written as $e_{e,y} = e_{e,{\rm o}}\exp[i(\omega t - \kappa x)]$, where now $\kappa = \kappa_r - i\kappa_c$.

(e) Compare this solution with (2.68) for the Schrödinger equation.

Note that the time derivative term in Schrödinger equation (2.62) already has a coefficient $i = (-1)^{1/2}$, so the wave function for the Schrödinger equation has the similar form $\exp(-i\omega t)$ in (2.68), where the energy per particle is $E = \hbar\omega$.

The absorption coefficient, $\sigma_{ph} = 2\kappa_c$. The factor 2 is due to relation of radiation intensity to e_e^2, through (3.39).

Problem 3.9

(a) Using (3.18), plot $f_e^{\rm o}$ versus $E_{e,k}$ (up to 0.3 eV) for $T = 10^3$ K, and $E_{e,p} = 0.2$.

(b) On the same graph, plot (3.20) for f_e versus $E_{e,k}$ for $T = 10^3$ K, $E_{e,p} = 0.2$ eV, $\tau_e = 10^{-11}$ s, $e_{e,x} = 10^3$ and 10^4 V/m, using the rest electron mass m_e.

(c) Comment on the shift from the equilibrium particle energy distribution, under applied external force.

Problem 3.10

(a) Using the definition of $\langle E_e\rangle$ is (3.26), and $f_e^{\rm o}$ and $n_{e,c}$ in Problem 3.2(a), show that

$$\langle E_e\rangle = \frac{3}{2}k_{\rm B}Tn_{e,c},$$

which is the classical kinetic energy of conduction electrons.

(b) Using degenerate electrons, repeat part (a) and using $f_e^{\rm o}$ and $n_{e,c}^{\rm o}$ in Problem 3.2(b), to show that

$$\langle E_e\rangle = \frac{3}{2}n_{e,c}k_{\rm B}T\frac{{\rm F}_{3/2}(\eta_c)}{{\rm F}_{1/2}(\eta_c)}.$$

Problem 3.11

Use the current density $j_{e,x}$

$$j_{e,z} = \frac{e_c}{4\pi^3\hbar^3}\sum_{p_z>0} u_{e,z}f_e^{\rm o}, \quad m_{e,e}u_{e,z} = p_z\cos\theta,$$

[similar to (3.26)], where θ is the polar angle, and use (3.12) for the integration, along with $f_e^{\rm o}(\eta_c)$ in Problem 3.2(a), and show that

$$j_{e,z} = \frac{m_{e,e}(k_{\rm B}T)^2}{2\pi^2\hbar^3}e^{\eta_c} = (\frac{k_{\rm B}T}{2\pi m_{e,e}})^{1/2}n_{e,c}.$$

Note that for $p_z > 0$, the integral is over a hemisphere.

Problem 3.12

Show that the equilibrium fermion

$$f_e^{\rm o} = \frac{1}{\exp[(\varphi_{e,{\rm o}} + E_{e,k} - E_{\rm F})/k_{\rm B}T] + 1}, \quad E_{e,k} = \frac{p_e^2}{2m_{e,e}},$$

satisfies BTE (3.3).

Note that in general $\varphi_{e,{\rm o}}$, $E_{e,k}$, $E_{\rm F}$, and T are position dependent, and use

$$\boldsymbol{F}_e = -\nabla\varphi_{e,{\rm o}} \quad \text{built-in potential.}$$

At equilibrium, $\mathrm{D} f_e^{\mathrm{o}}/\mathrm{D}t = 0$, and

$$\nabla E_{\mathrm{F}} = \nabla T = 0.$$

Problem 3.13

The collision rate term in BTE (3.8) is zero under the condition of detailed balance states in equilibrium, giving for the electron systems

$$\dot{\gamma}_{e,\mathrm{o}}(\boldsymbol{p}', \boldsymbol{p}) f_e^{\mathrm{o}}(\boldsymbol{p}')[1 - f_e^{\mathrm{o}}(\boldsymbol{p})] = \dot{\gamma}_{e,\mathrm{o}}(\boldsymbol{p}, \boldsymbol{p}') f_e^{\mathrm{o}}(\boldsymbol{p})[1 - f_e^{\mathrm{o}}(\boldsymbol{p}')].$$

Show that

$$\frac{\dot{\gamma}_{e,\mathrm{o}}(\boldsymbol{p}', \boldsymbol{p})}{\dot{\gamma}_{e,\mathrm{o}}(\boldsymbol{p}, \boldsymbol{p}')} = e^{[E_{e,k}(\boldsymbol{p}') - E_{e,k}(\boldsymbol{p})]/k_{\mathrm{B}}T}.$$

Problem 3.14

(a) Using $\boldsymbol{F} = \partial \boldsymbol{p}/\partial t$, and for electrons $\boldsymbol{F}_e = -\nabla(\varphi_e + E_{e,k})$, show that the BTE becomes

$$\frac{\partial f_e}{\partial t} + \boldsymbol{u}_e \nabla f_e - \nabla \varphi_e \cdot \nabla_{\boldsymbol{p}} f_e - \nabla E_{e,k}(\boldsymbol{p}) \cdot \nabla_{\boldsymbol{p}} f_e = \frac{\partial f_e}{\partial t}|_s + \dot{s}_e.$$

(b) For $\nabla E_{e,k} = 0$, and $-\nabla \varphi_e = \boldsymbol{e}_e$, and using only the x component, show that (3.14) is obtained.

Problem 3.15

Consider electron scattering from a periodic potential (for example, scattering by acoustic phonons), along one dimension. The scattering potential and the confined wave function ($-L/2 \le x \le L/2$) are

$$\varphi_s = A_{\kappa_p} e^{\pm i \kappa_p x},$$

$$\psi_e = L^{-1/2} e^{i \kappa x}.$$

(a) Write the expression for the matrix element (3.29), and show that the modal transition rate is

$$M_{\kappa',\kappa} = \int_{-L/2}^{L/2} L^{-1} A_{\kappa_p} e^{i(\kappa - \kappa' \pm \kappa_p)x} \mathrm{d}x.$$

(b) For $\kappa' = \kappa \pm \kappa_p$ (momentum conservation, $\hbar\kappa' = \hbar\kappa + \hbar\kappa_p$), show that

$$M_{\kappa',\kappa} = A_{\kappa_p}.$$

(c) Using (3.27), show that

$$\dot{\gamma}_{e\text{-}p}(\kappa', \kappa) = \frac{2\pi}{\hbar} A_{\kappa_p}^2 \delta_{\mathrm{D}}[E_e(\kappa') - E_e(\kappa) \mp \hbar\omega_p],$$

where $E_e(\kappa') = E_e(\kappa) \pm \hbar\omega$ states energy conservation.

Note that the integrated $\dot{\gamma}_{e\text{-}p}$ over all κ' gives the overall transition rate. The integral of $\delta_{\mathrm{D}}(E_e)$ over all energy states ($\boldsymbol{\kappa}'$) leads to the density of states D_e, which is available in analytic form or as experimental data (e.g., Section 5.7). So $\dot{\gamma}_{e\text{-}p}$ integrated over all electron energy states ($\boldsymbol{\kappa}'$) becomes $2\pi A^2_{\kappa_p} D_e(E_e)/\hbar$.

Problem 3.16

(a) In the derivation of the FGR, substitute (E.16) into (E.14), using zero for the constant of integration.

(b) Show that for a perturbation H$'$ applied from $t = 0$ to $t = t_\mathrm{o}$, the first order solution gives

$$a_{1,\kappa'}(t \geq t_\mathrm{o}) = -M_{\mathrm{o},\kappa',\kappa}\{\frac{\exp\{i\dfrac{[E(\boldsymbol{\kappa}') - E(\boldsymbol{\kappa}) + \hbar\omega]}{\hbar}t_\mathrm{o}\} - 1}{E(\boldsymbol{\kappa}') - E(\boldsymbol{\kappa}) + \hbar\omega} - \frac{\exp\{i\dfrac{[E(\boldsymbol{\kappa}') - E(\boldsymbol{\kappa}) - \hbar\omega]}{\hbar}t_\mathrm{o}\} - 1}{E(\boldsymbol{\kappa}') - E(\boldsymbol{\kappa}) - \hbar\omega}\}.$$

(c) Show that for the case of $E(\boldsymbol{\kappa}') > E(\boldsymbol{\kappa})$, i.e., absorption, we have

$$|a_{1,\kappa'}(t \geq t_\mathrm{o})|^2 = 4M^2_{\mathrm{o},\kappa',\kappa}\frac{\sin^2(\omega' t_\mathrm{o}/2)}{\hbar^2\omega'^2},$$

where ω' is defined by (E.18).

(d) Reproduce Figure E.1.

Problem 3.17

Consider plane EM wave in vacuum (free space), presented by

$$\boldsymbol{e}_e = e_{e,\mathrm{o}}e^{i(\omega t - \omega\frac{z}{c_\mathrm{o}})}\boldsymbol{s}_x.$$

(a) Show that

$$\boldsymbol{h}_e = c_\mathrm{o}\epsilon_\mathrm{o}e_{e,\mathrm{o}}e^{i(\omega t - \omega\frac{z}{c_\mathrm{o}})}\boldsymbol{s}_y.$$

So,

$$\frac{\boldsymbol{e}_e}{\boldsymbol{h}_e} = \frac{1}{\epsilon_\mathrm{o}c_\mathrm{o}}, \quad \frac{\frac{1}{2}\epsilon_\mathrm{o}e_e^2}{\frac{1}{2}\mu_\mathrm{o}h_e^2} = 1.$$

(b) Show that

$$\nabla\cdot(\boldsymbol{e}_e \times \boldsymbol{h}_e) = -\frac{\partial}{\partial t}(\frac{\epsilon_\mathrm{o}e_e^2}{2} + \frac{\mu_\mathrm{o}h_e^2}{2}).$$

This shows that energy lost per unit time and volume is equal to the outward flux of energy through the surface.

(c) Using the result of part (a), show that the integral of the Poynting vector is

$$\boldsymbol{s}_e = \boldsymbol{e}_e \times \boldsymbol{b}_e = \epsilon_\mathrm{o}c_\mathrm{o}e_e^2\boldsymbol{s}_z.$$

(d) Show that the time-average energy flux can be decomposed as

$$|\bar{\boldsymbol{s}}_e| = \frac{1}{2}c_o\epsilon_o(\overline{e_e^2}) + \frac{1}{2}c_o\mu_o(\overline{h_e^2}) = c_o\epsilon_o(\overline{e_e^2}) = \frac{1}{2}c_o\epsilon_o e_{e,o}^2,$$

where $(\overline{e_e^2})^{1/2}$ and $(\overline{e_e^2})^{1/2}$ are the RMS values.

Problem 3.18

The EM waves attenuate over a short distance in metals (good conductor, where $\omega\epsilon_o\epsilon_e \le \sigma_e/50$).

(a) Use the Maxwell equations in Table 3.5 to arrive at expressions for $\nabla^2\boldsymbol{e}_e$ (3.34) and $\nabla^2\boldsymbol{h}_e$

$$\nabla^2\boldsymbol{h}_e = \frac{1}{c^2}\frac{\partial \boldsymbol{h}_e}{\partial t^2} - \nabla \times \boldsymbol{j}_{e,f}.$$

(b) Start with a plane EM wave with $\boldsymbol{e}_e = e_{e,o}e^{i(\omega t-\kappa z)}\boldsymbol{s}_x$, and show that

$$\boldsymbol{h}_e = e_{e,o}\frac{\kappa}{\omega\mu_o\mu_e}e^{i(\omega t-\kappa z)}\boldsymbol{s}_y,$$

and use this in the results of part (a) to show that

$$-\kappa^2 + \omega^2\epsilon_o\epsilon_e\mu_o\mu_e - i\omega\sigma_e\mu_o\mu_e = 0,$$

where κ is in general complex ($\kappa = \kappa_r - i\kappa_c$). Use constant charge density ρ_e.

(c) Using the definition of a good conductor, show that spectral skin depth (penetration distance) δ_ω is

$$\delta_\omega = \frac{1}{\kappa_c} = (\frac{2}{\sigma_e\mu_o\mu_e\omega})^{1/2}, \text{ or } \frac{e_e(z)}{e_e(0)} = e^{-\kappa_c z} = e^{-z/\delta_\omega}.$$

Note that it should be shown that $\kappa_r = \kappa_c$.

(d) The absorption coefficient $\sigma_{ph,\omega}$ is related to the decay of intensity that is the square of the field (3.39). Then find $\sigma_{ph,\omega}$ as twice the inverse of penetration distance with the expression for $\sigma_{ph,\omega}$. This is generally expressed as

$$\frac{I_{ph}(z)}{I_{ph}(0)} = e^{-\sigma_{ph,\omega}z}.$$

Problem 3.19

(a) Using EM wave-propagation equation (3.34) and $\boldsymbol{j}_e = \sigma_e\boldsymbol{e}_e$, show that for uniform charge density and for free charge only, we have for a transverse EM field $e_{e,y}$,

$$\mu_o\mu_e\epsilon_o\epsilon_e\frac{\partial^2 e_{e,y}}{\partial t^2} = \frac{\partial^2 e_{e,y}}{\partial x^2} + \frac{\mu_o\mu_e}{\rho_e}\frac{\partial e_{e,y}}{\partial t}.$$

(b) Then using $m_\omega = n_\omega - i\kappa_\omega$, as the complex index of refraction, and the dispersion relation

$$\kappa = m_\lambda\frac{\omega}{c_o} = (n_\lambda - i\kappa_\lambda)\frac{\omega}{c_o}$$

or

$$e_{e,y} = e_{e,o} \exp\{i[\omega t + (n_\omega - i\kappa_\omega)\frac{\omega x}{c_o}]\},$$

show that

$$c_o^2 \epsilon_o \epsilon_e \mu_o \mu_e = \epsilon_e \mu_e = (n_\lambda - i\kappa_\lambda)^2 + \frac{i\mu_o\mu_e\lambda_o c_o}{2\pi\rho_e}.$$

(c) Separating the real and imaginary parts, show that

$$n_\lambda^2 - \kappa_\lambda^2 = \mu_o\mu_e\epsilon_o\epsilon_e c_o^2 = \epsilon_e\mu_e, \quad n_\lambda\kappa_\lambda = \frac{\mu\lambda_o c_o}{4\pi\rho_e}$$

or

$$\begin{Bmatrix} n_\lambda^2 \\ \kappa_\lambda^2 \end{Bmatrix} = \frac{\epsilon_e\mu_e}{2}\left\{ \begin{matrix} + \\ - \end{matrix} 1 + [1 + (\frac{\lambda_o}{2\pi c_o \rho_e \epsilon_o \epsilon_e})^2]^{1/2} \right\}.$$

(d) Show that for metals (small ρ_e), and for long wavelengths, we have

$$n_\lambda = \kappa_\lambda = (\frac{\lambda_o\mu_o\mu_e c_o}{4\pi\rho_e})^{1/2}.$$

Problem 3.20

For an oscillating dipole moment $\boldsymbol{p}_{e,o} = q\boldsymbol{d}$ in vacuum with magnetic vector potential $\boldsymbol{a}_e$ and electric scalar potential φ_e [215],

$$\boldsymbol{a}_e = \frac{\mu_o}{4\pi r} I_o \exp[i\omega(t - \frac{r}{u_{ph}})]\boldsymbol{d},$$

$$\varphi_e = \frac{p_{e,o}}{2\epsilon_o r\lambda}(\frac{\lambda}{2\pi r} + i)\exp[i\omega(t - \frac{r}{u_{ph}})]\cos\theta,$$

where $\boldsymbol{d}$ is the separation distance between charges $+q$ and $-q$, $I_o d = i\omega p_{e,o}$.

(a) Write expressions $\boldsymbol{e}_e$ and $\boldsymbol{h}_e$ in terms of above parameters and variables.

(b) Show that

$$|\bar{\boldsymbol{s}}| = \frac{1}{2}\mathrm{Re}(\boldsymbol{e}_e \times \boldsymbol{h}_e^\dagger) = \frac{\mu_o u_{ph} I_o^2 d^2}{8r^2\lambda^2}\sin^2\theta \boldsymbol{s}_r.$$

(c) Show that the total irradiated power $|\bar{\boldsymbol{s}}|$ integrated over a sphere of radius r is

$$P_e = \frac{\omega^4|p_{e,o}|^2}{12\pi\epsilon_o u_{ph}^3}.$$

(d) Using the energy of the classical electric dipole as

$$\Delta E_d = \frac{1}{2}m_e|p_{e,o}|^2\frac{\omega^2}{e_c^2},$$

derive (3.42) by using

$$\tau_{ph,e,sp,o} = \frac{\Delta E_d}{P_e}.$$

Note that here $\boldsymbol{p}_{e,\mathrm{o}} = e_c \boldsymbol{d}$, so ΔE_d is the electric dipole energy.

Problem 3.21

Starting from the Gauss, Faraday, and Ampere laws (Table 3.5) and using the magnetic vector potential and the electric potential

$$\boldsymbol{b}_e \equiv \nabla \times \boldsymbol{a}_e, \quad \boldsymbol{e}_e \equiv -\nabla \varphi_e - \frac{\partial}{\partial t}\boldsymbol{a}_e,$$

and also using the Coulomb gauge

$$\nabla \cdot \boldsymbol{a}_e = 0,$$

show that for vacuum, we have

$$\nabla^2 \boldsymbol{a}_e - \frac{1}{u_{ph,\mathrm{o}}^2}\frac{\partial^2}{\partial t^2}\boldsymbol{a}_e = 0.$$

4

Phonon Energy Storage, Transport and Transformation Kinetics

Heat transfer by lattice (phonon) conduction is proportional to the lattice thermal conductivity tensor $\boldsymbol{K}_p$(W/m-K), i.e., $\mathbf{q}_k = -\boldsymbol{K}_p \cdot \nabla T$ (the Fourier law, Table 1.1), and sensible heat storage is determined by the phonon (lattice) specific heat capacity $c_{v,p}$(J/kg-K). The specific heat capacity is also given per unit volume(J/m^3-K), or per atom(J/K). Phonons participate in many thermal energy conversion phenomena, including laser cooling of solids, discussed in Chapter 7 [$\dot{s}_{i\text{-}j}$(W/m^3) in Table 1.1]. In this chapter, we examine how the atomic structure of a solid influences $c_{v,p}$, $\boldsymbol{K}_p$, and $\dot{s}_{i\text{-}j}$ involving phonons.

Phonons are lattice-thermal-vibration waves that propagate through a crystalline solid. Most lattice vibrations have higher frequencies than audible sound, ultrasound, and even hypersound. Figure 4.1 shows the various sound- and vibrational-wave regimes. A single, constant speed (dispersionless, i.e., having a linear frequency dependence on the wave number) of 10^3 m/s is used for the sake of illustration. As will be shown, the vibrational waves have different modes, and the propagation speed can be strongly frequency dependent. In this chapter, we begin with lattice vibration and the relation between frequency and wave number (the dispersion relation) for a simple, harmonic, one-dimensional lattice. Then we discuss the quantization of phonons and a general three-dimensional treatment of dispersion. We discuss lattice specific heat capacity and thermal conductivity (from the BTE for phonons), including quantum effects, and discuss the atomic structural metrics of the thermal conductivity at high temperatures. Then we discuss the phonon boundary resistance (interface of dissimilar materials) and phonon absorption of external ultrasound. Finally we discuss the size effects.

4.1 Phonon Dispersion in One-Dimensional Harmonic Lattice Vibration

In this section we derive the phonon dispersion relation and the phonon velocities, for an idealized chain of atoms. In a crystalline solid, because of the temperature of this system of atoms, the constituent atoms vibrate about their equilibrium positions, and the extent of these thermal vibrations is determined by the intermolecular forces they exert on each other. These vibrations are quantized

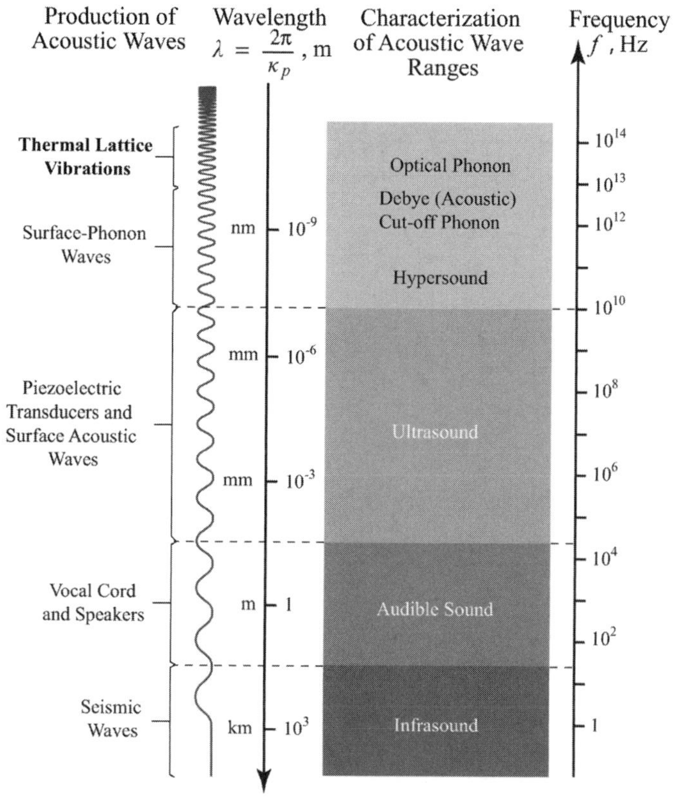

Figure 4.1. Spectra (wavelength and frequency) of acoustic- and vibrational- (mechanical- and thermal-) wave regimes. An average group speed of 10^3 m/s is used for the waves. The thermal lattice-vibration modes (phonons) have the highest frequency among the acoustic and vibrational waves.

elastic waves (phonons) and persist even at $T = 0$ K (the zero-point motion, Section 2.5.3).

The quantized treatment of waves is done in the normal coordinates (through the Schrödinger equation), and this introduces $\hbar$ (discussed in Section 4.5). This is also termed the harmonic-oscillator treatment. Here, to review some of the general features of these elastic waves, we begin with a classical dynamics (Newtonian) equation of the motion of two different atoms on a primitive basis (for example, SiC). We start with a linear (one-dimensional) lattice, as shown in Figure 4.2(a). A similar treatment is given in [91].

From Figure 2.2 and Table 2.5, and by expressing the external force with a spring force ($\boldsymbol{F}_{kj} = -\Gamma \boldsymbol{d}_{kj}$), where Γ is the force constant (2.54), we write a one-dimensional equation of motion. This is the harmonic approximation, as will be discussed in Section 4.4, and is valid near the bottom of the potential well as the zero-temperature asymptote (inset of Figure 2.3). We write the discretized form of this particle momentum (based on the nearest-neighbor interaction only) equation. We use the instantaneous locations of particle 1 and its relative position with respect

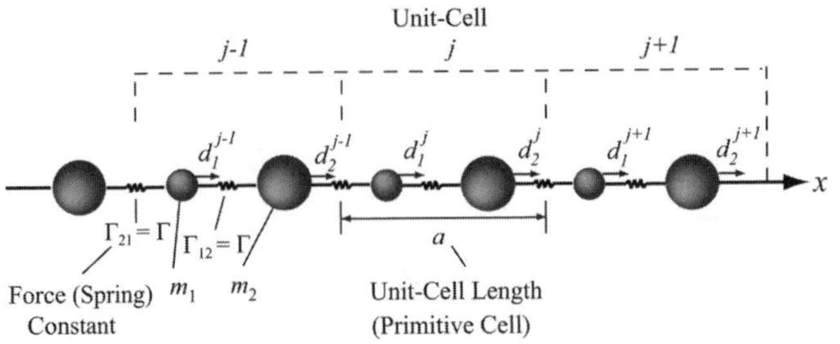

(a) Schematic of A Linear, Diatomic Chain

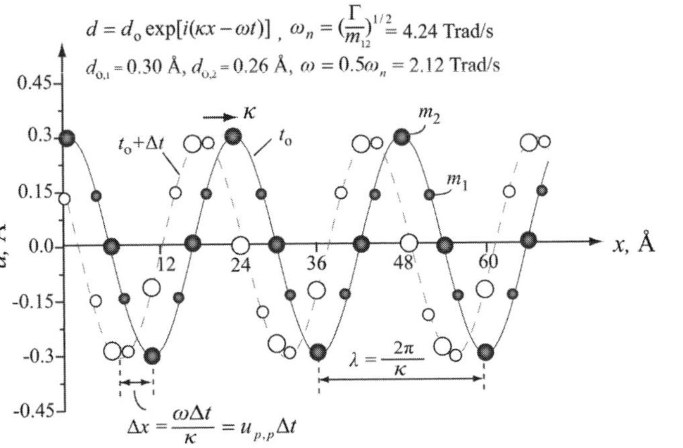

(b) Acoustic Phonon, $m_1 = 0.5m_2 = m_{\rm Ar}$, $a = 6$ Å, $\Gamma = 0.8$ N/m

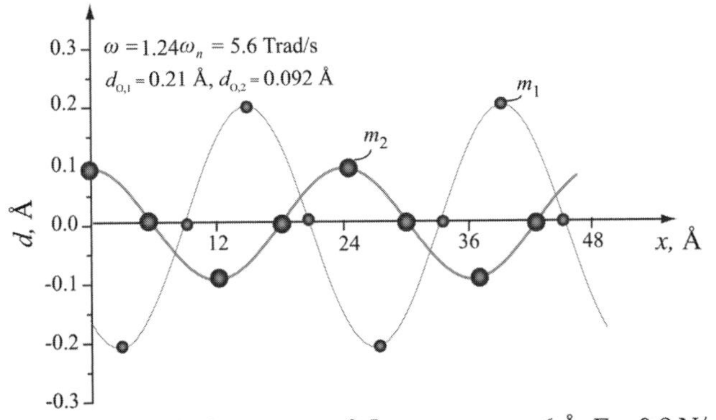

(c) Optical Phonon, $m_1 = 0.5m_2 = m_{\rm Ar}$, $a = 6$ Å, $\Gamma = 0.8$ N/m

Figure 4.2. (a) Designation of a diatomic, linear chain solid (lattice), having the force constant Γ, and its displacements around equilibrium positions. (b) Displacement of a linear, diatomic chain of Ar atoms, but every other atom has a mass of $1/2m_{\rm Ar}$. The acoustic phonon waves and their propagation is also shown. (c) Same as (b), but for optical phonon.

to particle 2 to its left and right in a unit cell designated with index j, i.e.,

$$m_1 \frac{\mathrm{d}^2 d_1^j}{\mathrm{d}t^2} = -\Gamma(d_1^j - d_2^j) - [-\Gamma(d_2^{j-1} - d_1^j)] = \Gamma(d_2^j + d_2^{j-1} - 2d_1^j) \qquad (4.1)$$

$$m_2 \frac{\mathrm{d}^2 d_2^j}{\mathrm{d}t^2} = \Gamma(d_1^{j+1} + d_1^j - 2d_2^j) \quad \text{harmonic motion of lattice atoms,} \qquad (4.2)$$

where the movement to the right creates a restoring force to the left, and movement to the left creates a restoring force to the right. Here Γ represents the second derivative of the effective interatomic potential evaluated at the equilibrium separation distance, as given by (2.54) and further discussed in Section 4.4. The proceeding are wave equations and admit plane-wave solutions such as those discussed in Sections 2.6 and 3.3.2 for the Schröndinger and the Maxwell equations, respectively.

Noting that the equilibrium locations of atoms take discrete values, $x = ja$, where j is the unit-cell designation, the solutions are of the form

$$d_1^j = d_{o,1}\exp[i(\kappa ja)]\exp[-i(\omega t)] \qquad \kappa = \frac{2\pi}{\lambda} \tag{4.3}$$

$$d_2^j = d_{o,2}\exp[i(\kappa ja)]\exp[-i(\omega t)], \tag{4.4}$$

where κ is the wave number, λ is the wavelength, and $\omega = 2\pi f$ (f is the frequency) is the angular frequency. In Section 4.4, we will write the general form of (4.1) and (4.2).

Substituting (4.3) and (4.4) into (4.1) and (4.2), and setting the determinant of $d_{o,1}$ and $d_{o,2}$ in the matrix given below equal to zero, we have the characteristic equation (dispersion relation), i.e.,

$$-m_1 d_{o,1}\omega^2\exp[i(\kappa ja)] = \Gamma\{d_{o,2}\exp[i(\kappa ja)] + d_{o,2}\exp\{i[\kappa(j-1)a]\} - 2d_{o,1}\exp[i(\kappa ja)]\}$$

$$-m_2 d_{o,2}\omega^2\exp[i(\kappa ja)] = \Gamma\{d_{o,1}\exp\{i[\kappa(j+1)a]\} + d_{o,1}\exp[i(\kappa ja)] - 2d_{o,2}\exp[i(\kappa ja)]\},$$

or

$$-m_1 d_{o,1}\omega^2 = \Gamma d_{o,2}[1+\exp(-i\kappa a)] - 2\Gamma d_{o,1}$$

$$-m_2 d_{o,2}\omega^2 = \Gamma d_{o,1}[1+\exp(i\kappa a)] - 2\Gamma d_{o,2},$$

or

$$\begin{vmatrix} 2\Gamma - m_1\omega^2 & -\Gamma[1+\exp(-i\kappa a)] \\ -\Gamma[1+\exp(i\kappa a)] & 2\Gamma - m_2\omega^2 \end{vmatrix} = 0,$$

or[†]

$$m_1 m_2\omega^4 - 2\Gamma(m_1+m_2)\omega^2 + 2\Gamma^2(1-\cos\kappa a) = 0 \text{ characteristic equation.} \tag{4.5}$$

[†] This can be written as (Section 4.4)

$$\begin{bmatrix} \frac{1}{m_1} & 0 \\ 0 & \frac{1}{m_2} \end{bmatrix}\begin{bmatrix} 2\Gamma & -\Gamma[1+\exp(-i\kappa a)] \\ -\Gamma[1+\exp(i\kappa a)] & 2\Gamma \end{bmatrix} \equiv \boldsymbol{M}^{-1}\boldsymbol{D}(\kappa) = \omega^2\boldsymbol{I},$$

where $\boldsymbol{I}$ is the identity matrix, $\boldsymbol{D}(\kappa)$ is the dynamical matrix and will be discussed in Section 4.4, and $\boldsymbol{M}$ is the mass matrix. For such harmonic displacements, the dynamical matrix facilitates the compact presentation of multidimensional, multiple-mass structures. The dimension of the dynamical matrix is the number of spatial dimensions times the number of atoms in a primitive cell, i.e., $3N_\circ$ for three-dimensional atomic structures.

The roots of this quartic equation are

$$\omega^2 = \frac{2\Gamma(m_1+m_2) \pm [4\Gamma^2(m_1+m_2)^2 - 8\Gamma^2 m_1 m_2(1-\cos\kappa a)]^{1/2}}{2m_1m_2}$$

dispersion relation for linear, diatomic lattice. (4.6)

In general, the total number of branches (here two) is the number of dimensions of the motion (here one) times the number of atoms per primitive cell $N_\circ$ (here two). We will discuss this further in Section 4.3.3.

The dimensionless form of this dispersion relation is

$$\omega^{*2} = \frac{1 \pm [1 - 2m_1^* m_2^*(1-\cos\pi\kappa^*)]^{1/2}}{2} \quad \text{dimensionless dispersion relation,} \quad (4.7)$$

where the dimensionless quantities are

$$\omega^* = \frac{\omega}{\omega_\mathrm{o}} = \frac{\omega}{[2\Gamma(\frac{1}{m_1}+\frac{1}{m_2})]^{1/2}}, \qquad \kappa^* = \frac{\kappa a}{\pi}, \qquad m_i^* = \frac{m_i}{m_1+m_2},$$

$$\omega_\mathrm{o} = [2\Gamma(\frac{1}{m_1}+\frac{1}{m_2})]^{1/2}, \qquad (4.8)$$

where ω_o is twice the natural frequency ω_n (vibration of a pair of atoms) and for diatomic cell the reduced mass is $m_{12} = m_1m_2/(m_1+m_2)$ and the natural frequency (2.54) is $\omega_n = (\Gamma/m_{12})^{1/2}$. Note that the dimensionless wave vector is $\kappa a/\pi$, but we could have also used $\kappa a/2\pi$, which is the more general usage. The (–) branch is acoustic (at $\kappa = 0$, $\omega = 0$ for the acoustic branch) and the (+) branch is optical.

As was mentioned in the last footnote, the dispersion relation (4.5) can be written in terms of the dynamical matrix $\boldsymbol{D}(\boldsymbol{\kappa})$ and this will be further discussed in Section 4.4.

Figure 4.3 shows the variation of scaled (dimensionless) angular frequency ω^* with respect to the scaled wave number κ^*, for $m_1^* = 0.75$ and 0.6, and 0.5 (same mass for all atoms). The solutions represent the two polarizations (branches of vibration modes). These are the acoustic (–) and optical (+) polarizations.

Some of the features of the dispersion relation can be explored for further insight. For $\kappa = 0$ (long-wavelength assumption, which is called the Brillouin zone center), we find by the expansion of $\cos\kappa a$, that

$$\kappa = 0: \qquad \omega_{\mathrm{LO}} = \omega_\mathrm{o} = [2\Gamma(\frac{1}{m_1}+\frac{1}{m_2})]^{1/2}$$

longitudinal-optical branch (LO) at Brillouin zone center, (4.9)

$$\omega_{\mathrm{LA}} = (\frac{1}{2}\frac{\Gamma}{m_1+m_2})^{1/2}\kappa a$$

longitudinal-acoustic branch (LA) at Brillouin zone center. (4.10)

The sound speed is found from (4.10).

The Brillouin zone center is designated as the Γ point (in the $\boldsymbol{\kappa}$ space), and the boundary is designated as X. For $\kappa = \pi/a$ (at the edge of the Brillouin zone), we

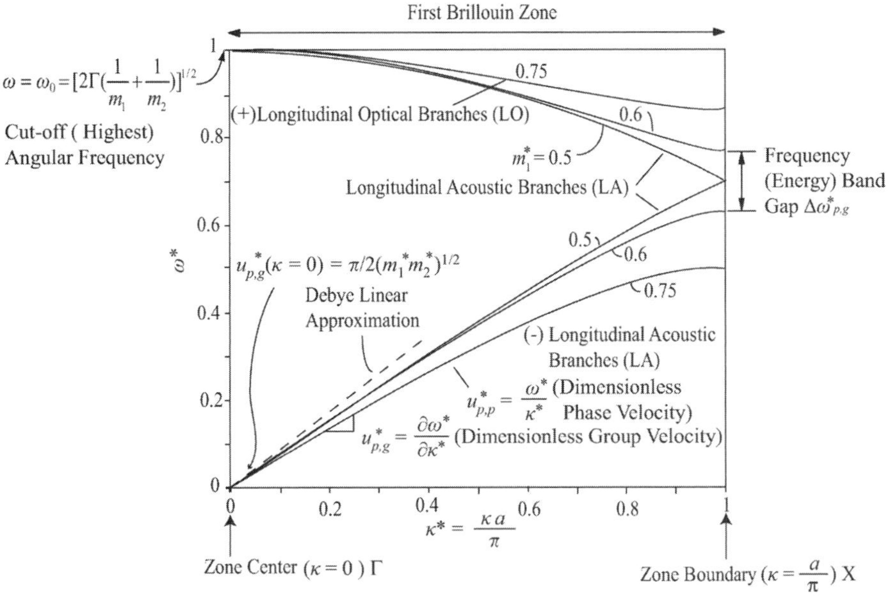

Figure 4.3. Dimensionless dispersion relation for a linear, diatomic chain. Both acoustic (–) and optical (+) branches are shown. The group velocity is also shown. Note that the maximum for ω^* is 1, and that for $m_1^* \neq 0.5$, at $\kappa = \pi/a$, there is a symmetry point referred to as the Brillouin-zone boundary. In the Debye approximation a linear dispersion relation (also called dispersionless, as the speed is independent of κ) is used.

have

$$\kappa = \frac{\pi}{a}: \qquad \omega_{\rm LO} = (\frac{2\Gamma}{m_1})^{1/2}$$

$$\text{LO at Brillouin zone edge (boundary X),} \qquad (4.11)$$

$$\omega_{\rm LA} = (\frac{2\Gamma}{m_2})^{1/2},$$

$$\text{LA at Brillouin zone edge (boundary X).} \qquad (4.12)$$

The difference between these is called the phonon energy gap $\Delta\omega_{p,g}$, i.e.,

$$\Delta\omega_{p,g} = \Delta\omega_{p,\rm X} = \omega_{\rm LO}(\kappa = \pi/a) - \omega_{\rm LA}(\kappa = \pi/a) \text{ at Brillouin zone boundary X.} \qquad (4.13)$$

The Brillouin zone that is in the space with coordinates that are in units of inverse of the lattice spacing or the wave vector (called the reciprocal lattice), will be discussed in Section 4.3. The Brillouin zone describes the symmetries of the dispersion relation.

The cut-off (highest) frequency is $\omega_{\rm o} = [2\Gamma(m_1^{-1} + m_2^{-1})]^{1/2} = 2^{1/2}\omega_n$ and is the largest frequency achievable. As an example, the spring constant Γ for the L–J potential in monatomic FCC, is given in Table 2.7 as

$$\Gamma_{\rm LJ} = \frac{\partial^2\langle\varphi\rangle}{\partial r_{nn}^2}|_{r_{nn,e}} = (22.88)^2\frac{\epsilon_{\rm LJ}}{\sigma_{\rm LJ}^2}. \qquad (4.14)$$

For Ar FCC, we have $\omega_n = 1.068 \times 10^{13}$ rad/s (10.68 Trad/s), which is larger than the atomic-pair value used in Figure 4.2.

The phonon group velocity is defined as

$$u_{p,g} \equiv \frac{\partial \omega}{\partial \kappa}, \quad \text{in general } \boldsymbol{u}_{p,g} \equiv \nabla_{\boldsymbol{\kappa}} \omega \quad \text{phonon group velocity.} \tag{4.15}$$

The phase speed (velocity) is defined as

$$u_{p,p} \equiv \frac{\omega}{\kappa} \quad \text{phase speed (velocity),} \tag{4.16}$$

and only for a nondispersive phonon are acoustic phonon the group and the phase speed the same.

For this example, we have the dimensionless (and dimensional) group velocity at $\kappa = 0$ (called the characteristic velocity):

$$u^*_{p,g}(\kappa^* = 0) = \frac{\pi}{2}(m_1^* m_2^*)^{1/2}, \; u_{p,g}(\kappa = 0) = \frac{\omega_n a}{2}, \tag{4.17}$$

and this is also shown in Figure 4.3. This is the long-wavelength, elastic, longitudinal wave (constant-sound-speed) asymptote (end-of-chapter problem) found from the macroscopic equation (Table 3.11).

Note that, for the optical branch at $\kappa = 0$, we have, by using (4.3) in (4.4),†

$$\frac{d_{o,1}}{d_{o,2}} = -\frac{m_2}{m_1}, \quad \kappa = 0 \quad \text{long-wavelength limit in optical branch,}$$

$$\frac{d_{o,1}}{d_{o,2}} = 1, \quad \kappa = 0 \quad \text{long-wavelength limit in acoustic branch,} \tag{4.18}$$

i.e., the neighboring masses (atoms) move in the opposite directions for the optical branch, whereas for the acoustic branch. For the long-wavelength-acoustic branch, the neighboring atoms move together and with the same magnitude. These long-wavelength-acoustic phonons move at a constant speed of $a\omega_n(m_1^* m_2^*)^{1/2}/2$. This is the elastic wave speed used in the macroscopic, elastic, solid mechanics (Navier) equation (Table 3.11).

For $m_1 = m_2$ ($m_1^* = 0.5$), the dispersion spectrum becomes that of a monatomic chain. The gap disappears, and the top branch can be unfolded to cover the region $1 \le \kappa^* \le 2$. Then there is only the acoustic branch. This is shown in Figure 4.4, where four Brillouin zones (for atomic chain) are shown. Note that, for a monatomic chain, the interatomic distances are $a/2$. This is the reason that the first symmetry point X is at $\kappa a/\pi = 2$. If we scale κ with $2\pi/a$, then for a monatomic chain, X will be at $\kappa a/2\pi = 1$.

We generalize the result to three dimensions and introduce the dynamical matrix, which allows for a more general and compact analysis, in Section 4.4.

† The general relation for the ratio of displacements is

$$\frac{d_{o,1}}{d_{o,2}} = -\frac{m_1\omega^2 - 2\Gamma}{2\Gamma \cos \frac{a}{2}\kappa},$$

for $m_1 < m_2$, and ω is taken for the A or O branch for the same κ.

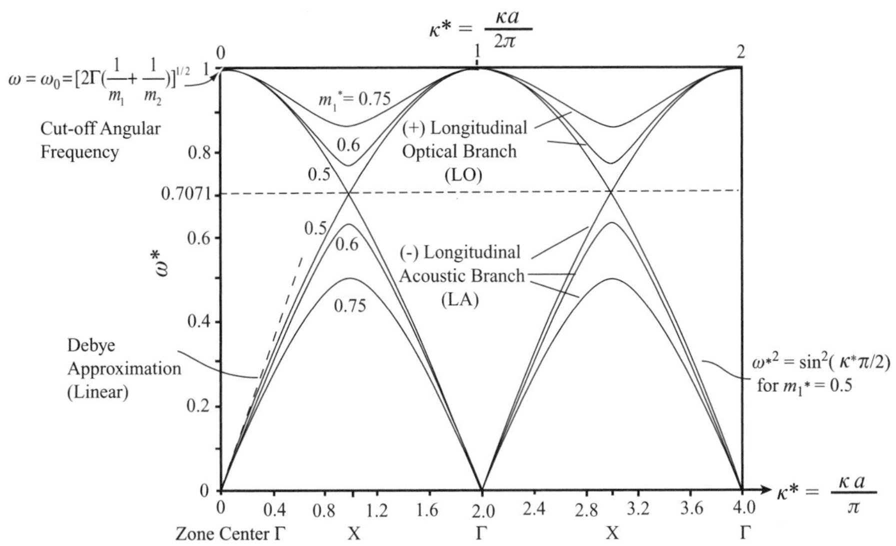

Figure 4.4. Dimensionless dispersion relation over four Brillouin zones. The results are for a linear, diatomic chain. The zone center is designated as Γ point (in the κ space), and the first symmetry point as X.

In the Debye approximation, a linear dispersion relation assumes $\kappa \rightarrow 0$ behavior, and this is shown in Figure 4.4.

The average lattice potential and kinetic energy are equal, and for a monatomic linear chain the total energy is $m\omega^2 a/2$ (end-of-chapter problem).

As an example, atomic displacements of a diatomic chain of Ar atoms (pair force constant in Table 4.4), with every other atom having a mass $0.5m_{\rm Ar}$, are shown in Figures 4.2(b) and (c). The acoustic wave propagation and phase velocity is shown in Figure 4.2(b). The wave at times $t_\circ$ and $t_\circ + \Delta t$ are shown, where the wave has travelled a distance $\omega\Delta t/\kappa$. Note that the displacement wave connects all adjacent atoms. In Figure 4.2(b), the optical waves are shown, with adjacent atoms moving in opposite directions. The results are frequencies near (but not at) the Γ point (Figure 4.3), so the amplitude ratio nearly follows (4.18). For both the acoustic and optical branches, κ is the same, while ω is taken along the appropriate branch ($\omega_{\rm A} < \omega_\circ$). In crystals, $d_{\circ,1}$ is not a constant, but a function of frequency (Section 4.8).

4.2 Phonon Density of States and Phonon Speeds

Wave number density of states (DOS) $D_p(\kappa)$ is defined in the Glossary (and also in Section 2.6.5). The DOS of phonons is used in determining the total number of phonons N_p as well as the phonon-related properties. The frequency (in linear ω space) density of the normal modes or phonon DOS $D_p(\omega)$ is the total number of modes in the frequency range ω to $\omega + d\omega$, divided by volume V. In (2.99), the integral is taken to include all κ corresponding to ω in this range. Then we use the integral defined in Section 3.1.3 to make the integration over $d\kappa$, as in (3.11), using

$\nabla_{\kappa}\omega_{\alpha}$ to relate $d\boldsymbol{\kappa}$ and $d\omega_{\alpha}$, where α is the polarization (or branch) of the dispersion relation. The density of normal modes (Section 4.4) for three-dimensional, cubic lattices (of cubic volume L^3) is

$$D_p(\omega) \equiv \frac{1}{L^3}(\frac{L}{2\pi})^3 \sum_{\alpha} \frac{1}{|\nabla_{\kappa}\omega_{\alpha}|} = (\frac{1}{2\pi})^3 \sum_{\alpha} 4\pi\kappa^2 \frac{d\kappa}{d\omega_{\alpha}}, \tag{4.19}$$

$$\alpha = \mathrm{A, O} \quad \text{for acoustic and optical branches,}$$

where α is the branch index (the acoustic branches has 3 branches, two transverse, and one longitudinal). We have used integration (3.11) over the $\boldsymbol{\kappa}$-space [or the $\boldsymbol{p}$-space, because $\boldsymbol{p} = \hbar\boldsymbol{\kappa}$, de Broglie relation for parabolic energy distribution, (2.97)].

The phonon DOS is used, similar to (2.99) to obtain integral of quantities which depend on the modes. For any quantity ϕ related to the modes, we have

$$\begin{aligned} \sum_{\alpha}\sum_{\kappa} \phi[\omega_{\alpha}(\boldsymbol{\kappa})] &= \sum_{\alpha} \int_{\kappa} \frac{1}{(2\pi)^3} \phi[\omega_{\alpha}(\boldsymbol{\kappa})] d\boldsymbol{\kappa} \\ &\equiv \sum_{\alpha} \int_{\omega_{\alpha}} D_p(\omega_{\alpha})\phi(\omega_{\alpha}) d\omega_{\alpha}. \end{aligned} \tag{4.20}$$

Here again $D_p(\omega)d\omega$ is the total number of modes in the interval ω to $\omega + d\omega$, divided by volume V. So $D_p(\omega)$ has the units of 1/m^3-rad/s. Then DOS can be viewed as the degeneracy of states with energy $\hbar\omega$.

The total DOS for normal modes can also be written in terms of the Dirac delta function δ_D. Then, by use of (4.19), $D_p(\omega)$ also given by (3.28), for a particular frequency ω can be given as [219]

$$D_p(\omega) = \sum_{\alpha} \int_{\kappa'} \frac{d\boldsymbol{\kappa}'}{(2\pi)^3} \delta_D[\omega - \omega_{\alpha}(\boldsymbol{\kappa}')], \; d\boldsymbol{\kappa}' = 4\pi\kappa'^2 d\kappa'. \tag{4.21}$$

The Dirac delta function $\delta_D[\omega - \omega_{\alpha}(\boldsymbol{\kappa}')]$ is defined in the Glossary, where the energy Dirac delta function $\delta_D[E - E_{\alpha}(\boldsymbol{\kappa}')]$ is used. This is an alternative presentation of (4.19). Note that $w_{\alpha}(\boldsymbol{\kappa})$ is the dispersion relation and the gradient (derivative) $|\nabla_{\kappa}\omega_{\alpha}|$, has the same role in (4.19). DOS is generally scaled, such that its integral gives the number of modes.

4.2.1 Phonon DOS for One-Dimensional Lattice and van Hove Singularities

The phonon frequency density of states $D_p(\omega)$ converts the number of phonons of wave number κ [similar to (2.99)] to those of frequency ω, using the dispersion relation (Figure 4.3). The conversion leads to singularities in $D_p(\omega)$ at Brillouin zone centers and edges.

As an example, for the one-dimensional (chain) lattice, using (4.19), $D_p(\omega)$ becomes

$$D_p(\omega) = \frac{1}{a}\frac{a}{\pi}\sum_\alpha \frac{1}{|\nabla_\kappa \omega_\alpha|}$$
$$= \frac{1}{a}\sum_\alpha \frac{1}{[2\Gamma(\frac{1}{m_1}+\frac{1}{m_2})]^{1/2}}\frac{1}{|\frac{d\omega^*_\alpha}{d\kappa^*}|}. \qquad (4.22)$$

Then using (4.7) for ω^*, we have

$$\frac{d\omega^*}{d\kappa^*} = \frac{1}{2}\{\frac{1 \pm [1-2m_1^*m_2^*(1-\cos\pi\kappa^*)]^{1/2}}{2}\}^{-1/2}$$
$$\times \frac{1}{4}[1-2m_1^*m_2^*(1-\cos\pi\kappa^*)]^{-1/2}(-2m_1^*m_2^*\pi\sin\pi\kappa^*), \qquad (4.23)$$

and

$$D_p^*(\omega^*) = \frac{D_p(\omega)}{a^{-1}[2\Gamma(\frac{1}{m_1}+\frac{1}{m_2})]^{-1/2}} = \sum_\alpha \frac{1}{u^*_{p,g}(\omega_\alpha)}$$
$$= 8|\{\frac{1 \pm [1-2m_1^*m_2^*(1-\cos\pi\kappa^*)]^{1/2}}{2}\}^{1/2}$$
$$\times [1-2m_1^*m_2^*(1-\cos\pi\kappa^*)]^{1/2}(2m_1^*m_2^*\pi\sin\pi\kappa^*)^{-1}|. \qquad (4.24)$$

Now with dispersion relation (4.7) again used [which is inverse of $\omega^*(\kappa^*)$] for $\kappa^*(\omega^*)$, (4.24) is graphed with D_p^* as a function of ω^*, in Figure 4.5, for $m_1^* = 0.5$ and 0.75. Note the lack of the optical branch for the case of $m_1^* = m_2^* = 0.5$. Also note the singularities (where the group velocity is zero) at the Brillouin zone points corresponding to $\omega_{\rm LA}$(X) and $\omega_{\rm LO}$(X), given by (4.11) and (4.12), for $m_1^* = 0.75$.

The DOS can also be given as $D_p(E_p)dE_p$, with $E_p = \hbar\omega$. Then the energy Dirac delta function $\delta_{\rm D}[E - E_\alpha(\boldsymbol{\kappa}')]$ is used.

In addition to the longitudinal displacement and vibration just discussed for a linear array of atoms, in three-dimensional lattice vibrations there are transverse displacements and vibrations. The transverse vibrations are presented by two mutually perpendicular planes and these vibrations also travel along the lattice. For each phonon acoustic and optical branch, there are three polarizations (one longitudinal and two transverse). Some of the characteristics of phonon modes are listed in Table 4.1. These include their external excitations. The total number of modes is three times the number of atoms per unit primitive cell. The number of optical modes is equal to this total minus 3 (for acoustic phonons).

4.2.2 Debye and Other Phonon DOS Models

For acoustic phonons, the case of linear dispersion relation at zone center ($\kappa \rightarrow 0$) corresponds to a constant phase velocity (which is also equal to the

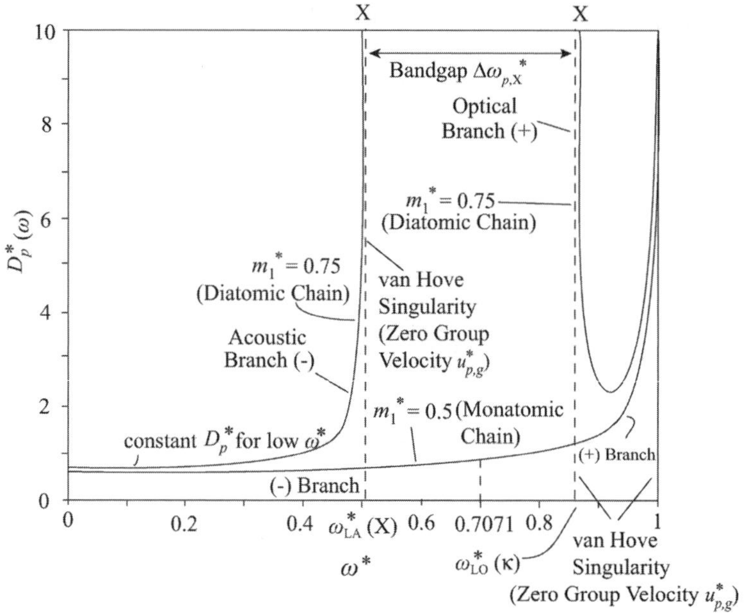

Figure 4.5. Dimensionless distribution of phonon DOS $D_p^*(\omega^*)$ as a function of dimensionless frequency for a linear, diatomic chain. Note that ω^* is the vertical axis in Figure 4.4 and the optical phonons are limited to high frequencies only.

Table 4.1. *Characteristics of lattice-vibrational acoustic and optical polarizations.*

Characteristic	Acoustic polarization	Optical polarization
Group velocity	sound waves, linear as $\kappa \rightarrow 0$	smaller than acoustic polarization
Displacement of adjacent atoms	in-phase	out-of-phase
Largest frequency	at the edge of Brillouin zone	higher than acoustic polarization
Number of polarizations	two transverse and one longitudinal (per unit-cell lattice)	number of atoms per primitive cell times 3, minus 3
f or ω versus λ or $1/\lambda$ (dispersion relation)	longer-wavelength modes have smaller frequency and for $\lambda \rightarrow \infty$, $f \rightarrow 0$	even long-wavelength modes have a finite frequency
External excitement	excited by microwave radiation	excited by infrared radiation (optical phonons cause time-varying electrical dipole moments)

constant group velocity). This leads to the parabolic DOS (proportional to E_p^2), or the Debye model. In the simple Debye model, a constant phonon speed is used for each polarization (two transverse and one longitudinal), with $E_p = \hbar\omega = \hbar u_{p,g}\kappa$.

From these, from (4.19) for three-dimensional $\boldsymbol{\kappa}$-space, we have the model form of the Debye DOS model

$$D_{p,\mathrm{D},\alpha}(\omega) = \frac{4\pi\kappa^2}{(2\pi)^3}\frac{1}{u_{p,g,\alpha}} = \frac{1}{2\pi^2}\frac{\omega^2}{u_{p,g,\alpha}^3}, \quad \frac{\mathrm{d}\omega}{\mathrm{d}\kappa} = u_{p,g,\alpha}$$
$$= \frac{1}{2\pi^2}\frac{E_p^2}{\hbar^2 u_{p,g,\alpha}^3}$$
$$\text{polarized (modal), Debye DOS model with cut-off frequency } \omega_\mathrm{D}. \quad (4.25)$$

Now assuming all modes (two transverse and one longitudinal-acoustic polarizations, Table 4.1) are the same, from the summation in (4.19), we have

$$D_{p,\mathrm{D}}(\omega) = \frac{3}{2\pi^2}\frac{\omega^2}{u_{p,\mathrm{A}}^3} \quad \text{simple Debye DOS model,} \quad (4.26)$$

where $u_{p,\mathrm{A}}$ is a single (average), modal acoustic-phonon speed.

This is similar to the DOS for photons in vacuum (which also has a linear dispersion relation, with equal phase and group velocity, $u_{ph} = \omega/\kappa = \mathrm{d}\omega/\mathrm{d}\kappa$, Section 7.1), except there are only two transverse photon polarizations.

The Debye model $D_{p,\mathrm{D}}(\omega)$ is valid near the zone center, where the dispersion relation is linear (also called no dispersion) (because $\mathrm{d}\omega/\mathrm{d}\kappa$ is constant). It also assumes a cut-off angular frequency ω_D. This is generally from the experimentally determined Debye temperature T_D (using for example, variation of the average specific heat capacity of crystal with respect to temperature, as will be discussed in Section 4.7). The mean Debye temperature, for elemental crystals, is listed in Table A.2.

The polarized Debye temperature $T_{\mathrm{D},\alpha}$ for a monatomic crystal is related to the polarized cut-off frequency and in turn the polarized phonon group speed $u_{p,g,\alpha}$ (further discussed in Section 4.7.1) by

$$T_{\mathrm{D},\alpha} = \frac{\hbar\omega_{\mathrm{D},\alpha}}{k_\mathrm{B}} = \frac{\hbar}{k_\mathrm{B}} u_{p,g,\alpha}(6\pi^2 n)^{1/3} \quad \text{Debye polarization temperature,} \quad (4.27)$$

where $u_{p,g,\alpha}$ is as previously defined and n is the atomic number density (1/m^3). We will use the Debye DOS and temperature in Section 4.7, and where $u_{p,\mathrm{A}}$ is also defined.

For the case of a simplified model of a single phonon speed, the Debye cut-off frequency is related to the average acoustic-phonon speed $u_{p,\mathrm{A}}$ through (as will be shown in Section 4.7.1)

$$\omega_\mathrm{D} = (6\pi^2 n u_{p,\mathrm{A}}^3)^{1/3} = (6\pi^2 n)^{1/3} u_{p,\mathrm{A}} \quad \text{Debye cut-off frequency.} \quad (4.28)$$

In Section 2.5.3, we compared the Debye and natural frequency ω_n for Ar FCC and found a good agreement. Examples of the Debye DOS model, and comparison with experimental results, will be given in Figures 4.11 and 4.41(a).

Other phonon DOS models are also available, one example is the Debye–Gaussian model, i.e.,

$$D_p(E_p) = CE_p^2 \exp[-(\frac{E_p - E_{p,p}}{\Delta E_p})^2] \quad \text{Debye–Gaussian phonon DOS,} \qquad (4.29)$$

where C is a normalization constant, $E_{p,p}$ is energy at the center of the DOS, and ΔE_p is the width of the distribution. At low E_p, this behaves similarly to the Debye, and near the center behaves similarly to a Gaussian distribution. This also avoids the appearance of a cut-off frequency. This model preserves the optical phonons and is used in applications such as the anti-Stokes luminescence (Section 7.12) and in examining the size effects on $D_p(E_p)$ in Section 4.19.3. An example is given in Section 4.6.

4.3 Reciprocal Lattice, Brillouin Zone, and Primitive Cell and Its Basis

In the discussion of phonon energy $\hbar\omega$ and its dispersion relation $\omega(\boldsymbol{\kappa})$, we defined the lattice structure by using a lattice constant, and we noted that the dispersion relation exposed special symmetries in the $\boldsymbol{\kappa}$-space, which has a relation with the inverse (or reciprocal) of the lattice. We now proceed to generalize these symmetries to three-dimensional lattices.

An ideal crystal is constructed by the infinite repetition of identical structural units. In the simplest crystals the structural unit is a single atom. The structure of a crystal is defined in terms of a lattice with the structural unit or basis attached to each lattice point. The lattice points form a set such that the structure is the same as seen from each point. An ideal crystal is described by three fundamental translation vectors, $\boldsymbol{l}_1$, $\boldsymbol{l}_2$, and $\boldsymbol{l}_3$ (also called primitive vectors of the crystal lattice). If there is a lattice point represented by the position vector $\boldsymbol{x}$, there is then also a lattice point represented by the position vector $\boldsymbol{x}'$, i.e.,

$$\boldsymbol{x}' = \boldsymbol{x} + i\boldsymbol{l}_1 + j\boldsymbol{l}_2 + k\boldsymbol{l}_3 \quad \text{primitive lattice vectors,} \qquad (4.30)$$

where i, j, and k are arbitrary integers. If all pairs of lattice points $\boldsymbol{x}'$ and $\boldsymbol{x}$ are given by (4.30), then the lattice is called primitive.

Figures 4.6(a), (c) and (e) show the primitive lattice vectors $\boldsymbol{l}_1$, $\boldsymbol{l}_2$, and $\boldsymbol{l}_3$, for the three cubic lattice structures, simple cubic (SC), face-centered cubic (FCC), and body-centered cubic (BCC).

4.3.1 Reciprocal Lattice

As we noted from Figure 4.4, the periodicity of the dispersion relation (and other associated symmetries) is conveniently presented when the wave number is scaled with the inverse of the interatomic (or interplanar) spacing. For a cubic lattice of lattice constant a, this scale is $2\pi/a$ and is called the reciprocal lattice constant, and, when specified along a direction, it is the reciprocal lattice vector $\boldsymbol{g}$. The primitive

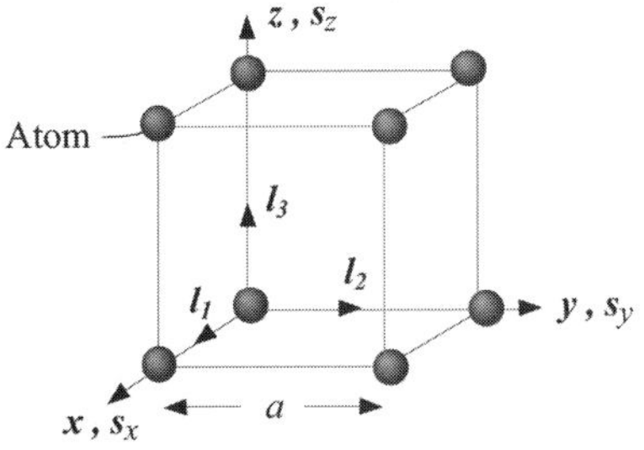

(a) Primitive Lattice Vectors in Simple Cubic (SC) Lattice

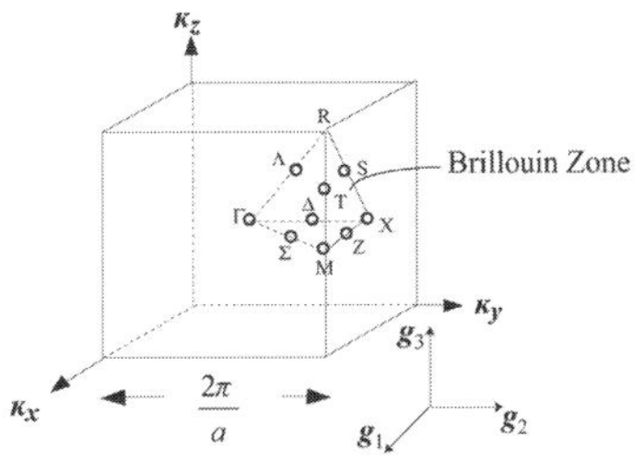

(b) Reciprocal Lattice Space and Brillouin Zones in SC

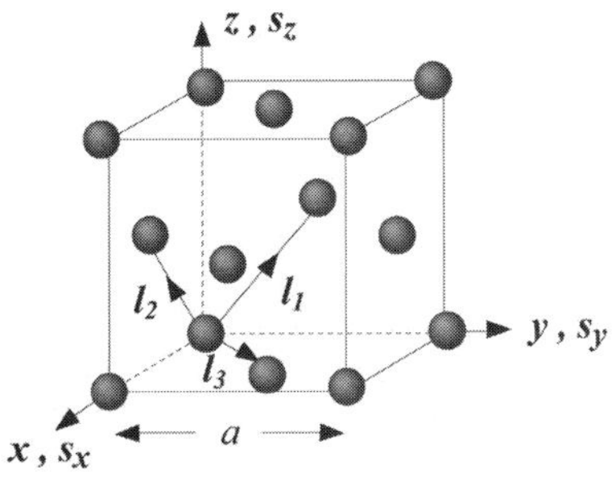

(c) Primitive Lattice Vectors in Face-Centered Cubic (FCC) Lattice

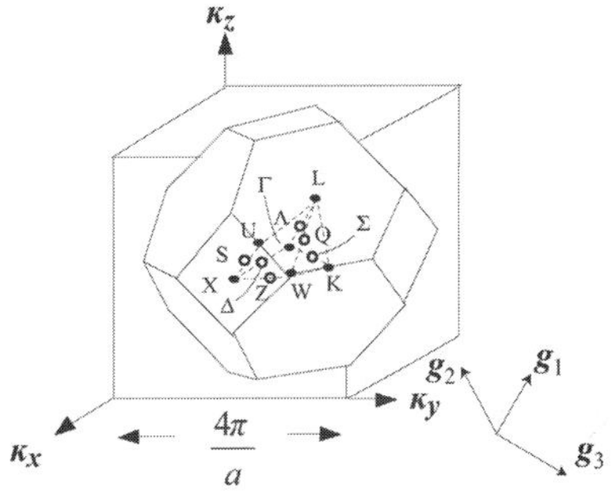

(d) Reciprocal Lattice Space and Brillouin Zones in FCC

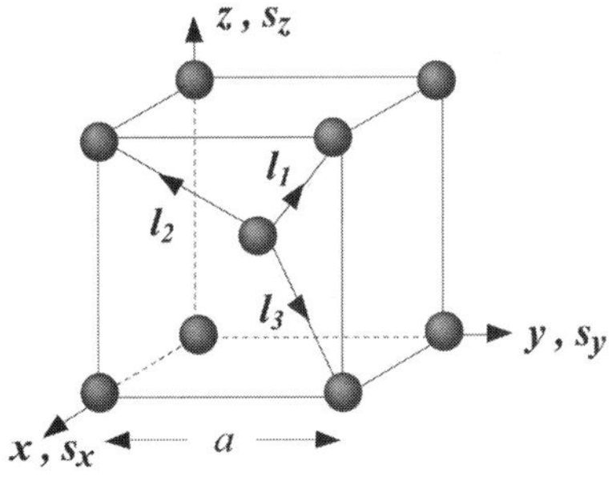

(e) Primitive Lattice Vectors in Body-Centered Cubic (BCC) Lattice

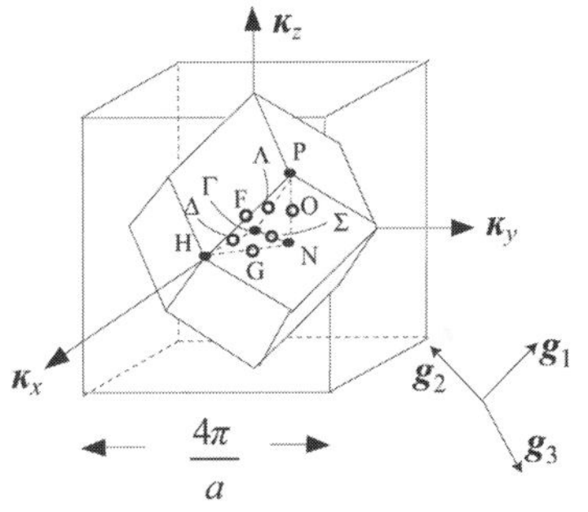

(f) Reciprocal Lattice Space and Brillouin Zones in BCC

Figure 4.6. The cubic lattices. (a) SC, (c) FCC, and (e) BCC The primitive translation vectors $\boldsymbol{l}_1$, $\boldsymbol{l}_2$, and $\boldsymbol{l}_3$ are also shown for each lattice. The Brillouin zone for cubic lattices are shown by (b) SC, (d) FCC, (f) BCC. The irreducible part of the zone is also shown. The Brillouin zone of BCC is identical to that for the diamond/zinc-blende structure crystals. $\boldsymbol{g}$ is the reciprocal lattice vector.

(the actual atomic locations) and reciprocal (wave vector or momentum vector) space lattices are shown in Figure 4.6 for a two-atom lattice (SC, FCC, and BCC). Because a Fourier transform is used to convert the physical space results (i.e., MD simulations) to the reciprocal lattice space, the latter is also called the Fourier space. From the solution to the lattice atom displacement of (4.3), and noting the symmetries in the κ-space, we note that the reciprocal lattice vectors satisfy $\exp[i(\boldsymbol{\kappa} \cdot \boldsymbol{g})] = 1$. For the translational vectors in (4.30), the three reciprocal lattice vectors $\boldsymbol{g}_1$, $\boldsymbol{g}_2$, and $\boldsymbol{g}_3$ are defined as [183]

$$\boldsymbol{g}_1 = 2\pi \frac{\boldsymbol{l}_2 \times \boldsymbol{l}_3}{\boldsymbol{l}_1 \cdot \boldsymbol{l}_2 \times \boldsymbol{l}_3}$$

$$\boldsymbol{g}_2 = 2\pi \frac{\boldsymbol{l}_3 \times \boldsymbol{l}_1}{\boldsymbol{l}_1 \cdot \boldsymbol{l}_2 \times \boldsymbol{l}_3}$$

$$\boldsymbol{g}_3 = 2\pi \frac{\boldsymbol{l}_1 \times \boldsymbol{l}_2}{\boldsymbol{l}_1 \cdot \boldsymbol{l}_2 \times \boldsymbol{l}_3} \quad \text{primitive reciprocal-lattice vectors,} \tag{4.31}$$

in $\boldsymbol{g}$ space, the space defined by $\alpha \boldsymbol{g}_1 + \beta \boldsymbol{g}_2 + \gamma \boldsymbol{g}_3$, where α, β, and γ are arbitrary integers. The reciprocal lattice is a lattice in the Fourier space associated with the crystal. The diffraction pattern of a crystal maps the reciprocal lattice [183].

4.3.2 Brillouin Zone

Each of the $\boldsymbol{g}_i$ is orthogonal to two $\boldsymbol{l}_i$, such that $\boldsymbol{g}_i \cdot \boldsymbol{l}_i = 2\pi \delta_{ij}$, where δ_{ij} is the Kronecker delta. As noted in Section 4.1, in the propagation of lattice vibrations through a crystal lattice, the phonon frequency is a periodic function of the wave vector $\boldsymbol{\kappa}$, $\omega(\boldsymbol{\kappa}) = \omega(\boldsymbol{\kappa} + \boldsymbol{g})$, as shown in Figure 4.4. This function can be multivalued, i.e., it has more than one branch. Discontinuities may also occur. As we also noted, a zone in the $\boldsymbol{\kappa}$-space is defined that forms the fundamental periodic region, such that the frequency or energy for a $\boldsymbol{\kappa}$ outside this region may be determined based on those within it. This region is known as the Brillouin zone (Figure 4.3) (sometimes called the first or the central Brillouin zone). In general we consider only $\boldsymbol{\kappa}$ values inside the zone. Discontinuities occur only on its boundaries. If the zone is repeated indefinitely, all $\boldsymbol{\kappa}$-space will be filled. The first Brillouin zone is defined to be the primitive cell of the reciprocal lattice, and is the set of points in the $\boldsymbol{\kappa}$-space that can be reached from the origin without crossing any Bragg plane. A Bragg plane for two points in a lattice is the plane that is perpendicular to the line between the two points and passes through the bisector of that line. The second Brillouin zone is the set of points that can be reached from the first zone by crossing only one Bragg plane. For adjacent Brillouin zones, the numbered $(n+1)$th Brillouin zone is the set of points not in the $(n-1)$th zone that can be reached from the nth zone by crossing $n-1$ Bragg planes.

Figures 4.6(b), (d), and (f) show the reciprocal space lattices, and the Brillouin zone planes and points designations are shown for the three lattices.

For an SC lattice, the primitive lattice (transition) vectors are $\boldsymbol{l}_1 = a\boldsymbol{s}_x$, $\boldsymbol{l}_2 = a\boldsymbol{s}_y$, and $\boldsymbol{l}_3 = a\boldsymbol{s}_z$. The boundaries of the first Brillouin zones are the planes normal to the six reciprocal lattice vectors, $\pm\boldsymbol{g}_1$, $\pm\boldsymbol{g}_2$, and $\pm\boldsymbol{g}_3$, i.e.,

$$\pm\frac{1}{2}\boldsymbol{g}_1 = \pm\frac{\pi}{a}\boldsymbol{s}_x, \ \pm\frac{1}{2}\boldsymbol{g}_2 = \pm\frac{\pi}{a}\boldsymbol{s}_y, \ \pm\frac{1}{2}\boldsymbol{g}_3 = \pm\frac{\pi}{a}\boldsymbol{s}_z,$$

$$\boldsymbol{g}_1 = \frac{2\pi}{a}\boldsymbol{s}_x, \ \boldsymbol{g}_2 = \frac{2\pi}{a}\boldsymbol{s}_y, \ \boldsymbol{g}_3 = \frac{2\pi}{a}\boldsymbol{s}_z, \tag{4.32}$$

where $\boldsymbol{s}_i$ are the Cartesian unit vectors. These six Brillouin zone planes bound a cube of volume $(2\pi/a)^3$. This is shown in Figure 4.6(b).

For an FCC lattice, the primitive and reciprocal lattice vectors are

$$\boldsymbol{l}_1 = \frac{a}{2}(\boldsymbol{s}_y + \boldsymbol{s}_z),\ \boldsymbol{l}_2 = \frac{a}{2}(\boldsymbol{s}_x + \boldsymbol{s}_z),\ \boldsymbol{l}_3 = \frac{a}{2}(\boldsymbol{s}_x + \boldsymbol{s}_y)$$

$$\boldsymbol{g}_1 = \frac{2\pi}{a}(-\boldsymbol{s}_x + \boldsymbol{s}_y + \boldsymbol{s}_z)$$

$$\boldsymbol{g}_2 = \frac{2\pi}{a}(\boldsymbol{s}_x - \boldsymbol{s}_y + \boldsymbol{s}_z)$$

$$\boldsymbol{g}_3 = \frac{2\pi}{a}(\boldsymbol{s}_x + \boldsymbol{s}_y - \boldsymbol{s}_z). \quad (4.33)$$

The volume of the reciprocal lattice cell is $4(2\pi/a)^3$. The FCC Brillouin zone is a truncated octahedron bounded by

$$\boldsymbol{g} = \frac{2\pi}{a}(\pm\boldsymbol{s}_x \pm \boldsymbol{s}_y \pm \boldsymbol{s}_z), \quad (4.34)$$

which for the octahedron and its corners, can be presented as having been cut with six planes given by

$$\frac{2\pi}{a}(\pm 2\boldsymbol{s}_x),\ \frac{2\pi}{a}(\pm 2\boldsymbol{s}_y),\ \frac{2\pi}{a}(\pm 2\boldsymbol{s}_z). \quad (4.35)$$

These planes are shown in Figure 4.6(d).

The BCC lattice with cubic cell side a has an FCC reciprocal lattice with cubic cell side $4\pi/a$. This is shown in Figure 4.6(f).

The high symmetry points and lines in the $\boldsymbol{\kappa}$ space have been designated by symbols. For example, the Γ point is [0,0,0] in the $\boldsymbol{\kappa}$-space. The first boundary point is designated by X, and in an FCC crystal it is at $[2\pi/a,0,0]$ in the $\boldsymbol{\kappa}$-space and indicates cyclic permutation of the axes. Another symmetry point in FCC is L at $[\pm\pi/a, \pm\pi/a, \pm\pi/a]$, and this and X and Γ are shown in Figure 4.6(d). The points of symmetry in the $\boldsymbol{\kappa}$ space are also important for electrons in semiconductors, because similar to phonons having $E_p = \hbar\omega_p(\boldsymbol{\kappa})$, electrons are represented by $E_e = E_e(\boldsymbol{\kappa})$.

4.3.3 Primitive Cell and Its Basis: Number of Phonon Branches

The real-space primitive cell is a volume of real space that when translated fills all of the space without overlapping or leaving any voids. A primitive cell contains one lattice point and its atomic basis. The number of atoms in the basis determines the number of branches in the dispersion relation. The number of branches (three acoustic branches and the rest optical branches) are three times the number of atoms in the basis $N_\circ$, i.e., $3N_\circ$ branches. One of the properties of a lattice is that any lattice point is close to its nearest neighbor and has the same number of nearest neighbors, and this is the coordination number.

The total number of phonon modes for a crystal having n atoms per unit volume is $n_p = 3n$ (or $N_p = 3N_\circ$ on total volume basis).

Single-atom SC, FCC, and BCC primitive cell structures [Figure 2.12(a)] such as Al (FCC) have only one atom in the primitive cell and this results in only three branches in the dispersion relation (acoustic only). The zinc-blende structures [including a diamond structure, Figure 2.9(b)] such as C, Si and GaAs have two atoms (in C and Si, crystals, the two atoms are identical) in their basis, so there are total of six branches (three acoustic and three optical). SiO_2 crystal (quartz), shown in Figure 2.9(a), has nine atoms in the tetrahedra-based anisotropic structure, so it has 27 branches (3 acoustic and 24 optical).

Table A.2 gives the crystal structures of elements (e.g., FCC, tetrogonal, orthorhombic, rhombohedral, monoclinic), as well as the lattice constant(s).

4.4 Normal Modes and Dynamical Matrix

Along the lines developed in Section 4.1, in three-dimensional lattice-dynamics calculations, a frequency-space description is used for the motions of the atoms. Instead of the localized motions of individual atoms, the system is described by energy waves with given wave vector $\boldsymbol{\kappa}$, frequency ω, and polarization vector $\boldsymbol{e}_\alpha$. The formulation of lattice-dynamics theory is described in [9] and [360].

For an equilibrium potential energy, defined by (2.47), of a system with N atoms that is designated by $\langle\varphi\rangle_\text{o}$, atom i is moved by an amount $\boldsymbol{d}_i$, and the resulting energy of the system, $\langle\varphi\rangle$, the total potential energy, is found by a Taylor series expression around the equilibrium state (small displacement of all atoms) as

$$
\begin{aligned}
\langle\varphi\rangle &= \langle\varphi\rangle_\text{o} + \sum_i \sum_\alpha \frac{\partial\langle\varphi\rangle}{\partial d_{i\alpha}}\Big|_\text{o} d_{i\alpha} + \frac{1}{2}\sum_{i,j}\sum_{\alpha,\beta}\frac{\partial^2\langle\varphi\rangle}{\partial d_{i\alpha}\partial d_{j\beta}}\Big|_\text{o} d_{i\alpha}d_{j\beta} \\
&\quad + \frac{1}{6}\sum_{i,j,k}\sum_{\alpha,\beta,\gamma}\frac{\partial^3\langle\varphi\rangle}{\partial d_{i\alpha}\partial d_{j\beta}\partial d_{k\gamma}}\Big|_\text{o} d_{i\alpha}d_{j\beta}d_{k\gamma} + \cdots + \\
&\equiv \langle\varphi\rangle_\text{o} + \frac{1}{2}\sum_{i,j}\sum_{\alpha,\beta}\Gamma_{\alpha\beta}d_{i\alpha}d_{j\beta} \\
&\qquad \text{harmonic approximation and spring constant } \Gamma_{\alpha\beta}, \qquad (4.36)
\end{aligned}
$$

where the i, j, and k sum over the atoms in the system (when including anharmonicity, for third-order interactions and for pair potentials, $i = j = k$ or $i = j$ or $j = k$), and the α, β, and γ sums are over the x, y, and z directions, as shown in Figure 2.12(d). Both $\langle\varphi\rangle$ and $\langle\varphi\rangle_\text{o}$ are only functions of the atomic positions. The first derivative of the potential energy with respect to each of the atomic positions is the negative of the net force acting on that atom. Evaluated at equilibrium, this term is zero. The first nonnegligible term in the expansion is thus the second-order term. The harmonic approximation is made by truncating the Taylor series at the second-order term. For a given i and j, the nine elements of the form $\partial^2\langle\varphi\rangle/\partial d_{i\alpha}\partial d_{j\beta}$ make up the force-constant matrix. Γ was introduced in (2.54), when the natural frequency was defined.

The harmonic approximation is valid for small displacements ($d_{i\alpha} \ll r_{nn}$) about the zero-temperature minimum and corresponds to the well minimum (Figure 2.3). Raising the temperature will cause deviations, making it anharmonic.

Given the crystal structure of a material, the determination of the allowed wave vectors (whose extent in the wave vector-space makes up the first Brillouin zone) is now addressed.

Consider a general crystal with an n-atom unit cell, such that the displacement of the jth atom in the lth unit cell is denoted by $\boldsymbol{d}(jl,t)$. The force-constant matrix [made up of the second order derivatives in (4.36)] between the atom (jl) and the atom $(j'l')$ is denoted by $\boldsymbol{\Gamma}\binom{jj'}{ll'}$. The indices are shown in Figure 2.12(d). Note that this matrix is defined for all atom pairs, including the case of $j=j'$ and $l=l'$. Imagine that the atoms in the crystal are all joined by harmonic springs; the equation of motion (Table 2.5), written in terms of displacement, for the atom (jl) can be written as [the general form of (4.1)]

$$m_j \frac{\mathrm{d}^2 \boldsymbol{d}(jl,t)}{\mathrm{d}t^2} = -\sum_{j'l'} \boldsymbol{\Gamma}\begin{pmatrix} jj' \\ ll' \end{pmatrix} \cdot \boldsymbol{d}(j'l',t) \text{ harmonic approximation,} \qquad (4.37)$$

subject to a suitable cut-off radius (Figure 2.3)†.

In (4.3), the modal displacement was given as a function of $\boldsymbol{\kappa}$, and to account for all modes (which are called the normal modes of vibration), we will use summation over $\boldsymbol{\kappa}$ space and also on the branches. For a system made of N atoms, there are $3N$ normal modes ($3N$ oscillator system) represented by $\boldsymbol{\kappa}$ and α.

Now assume that the displacement of an atom can be written as a summation over the normal modes [using the form given by (4.3)] of the system, such that

$$\boldsymbol{d}(jl,t) = \sum_{\boldsymbol{\kappa},\alpha} \boldsymbol{s}_\alpha(j,\boldsymbol{\kappa},\alpha) \exp\{i[\boldsymbol{\kappa}\cdot \boldsymbol{x}(jl)]\} \exp[-i\omega(\boldsymbol{\kappa},\alpha)t] \text{ plane-wave displacement} \qquad (4.38)$$

where $\boldsymbol{s}_\alpha(\boldsymbol{\kappa})$ is the unit normal coordinate vector (such that $\sum_{\alpha,\boldsymbol{\kappa}} \boldsymbol{s}_\alpha(\boldsymbol{\kappa})\cdot \boldsymbol{s}_\alpha(\boldsymbol{\kappa}) = N$), and $\boldsymbol{x}(jl)$ is the equilibrium location of atom jl. This equation expresses how this atom moves under the influence of $(\boldsymbol{\kappa},\alpha)$.

At this point, the wave vector is known, but the frequency and polarization vector are not (these are given by the dispersion relation). Note that the index k introduced in (4.36) has been replaced with $(\boldsymbol{\kappa},\alpha)$. The polarization vector and frequency are both functions of the wave vector and the dispersion branch, denoted by α. Substituting (4.38) and its second derivative into the equation of motion leads to the eigenvalue equation [similar to (4.5)]

$$\boldsymbol{M}\omega^2(\boldsymbol{\kappa},\alpha)\boldsymbol{s}_\alpha(\boldsymbol{\kappa}) = \boldsymbol{D}(\boldsymbol{\kappa})\boldsymbol{s}_\alpha(\boldsymbol{\kappa})$$

$$\text{equation of motion for a plane wave using dynamical matrix,} \qquad (4.39)$$

† Similarly, the lattice potential energy is given as (Tables 2.4 and 2.5)

$$E_{p,p} = \langle\varphi\rangle = \frac{1}{2}\sum_{jj',ll'} \varphi\begin{pmatrix} jj' \\ ll' \end{pmatrix},$$

which is a form of writing $\varphi_{jl-j'l'}$, etc., (Figure 2.12).
The harmonic displacement (kinetic) energy is

$$E_{p,k} = \frac{1}{2}\sum_{jj',ll'}\sum_{\alpha,\beta} d_\alpha(jl)\Gamma_{\alpha\beta}d_\beta(j'l').$$

where $\boldsymbol{M}$ is the diagonal mass matrix $(m_1, m_2, \dots)$, the mode frequencies are the square roots of the eigenvalues, and the polarization vectors are the eigenmodes. They are obtained by diagonalizing the matrix $\boldsymbol{D}(\boldsymbol{\kappa})$, which is known as the dynamical matrix, and has size $3n \times 3n$ (n is the number of atoms per unit cell). It can be broken down into 3×3 blocks (each for a given jj' pair), which will have elements [91]

$$D_{\alpha\beta}(jj', \boldsymbol{\kappa}) = \sum_{l'} \Gamma_{\alpha\beta}\begin{pmatrix} jj' \\ ll' \end{pmatrix} \exp\{i\boldsymbol{\kappa} \cdot [\boldsymbol{x}(j'l') - \boldsymbol{x}(jl)]\} \text{ harmonic dynamical matrix,} \tag{4.40}$$

where we used the indices as in (4.36), and the sum is over the unit cells.

The dynamical matrix uses the dot product of the wave vector and the position vectors of the atoms in periodic primitive cells to express an infinite number of atoms by a set of linear, homogeneous equations. The nontrivial solution of the dispersion relation containing the dynamical matrix gives the eigenvalues. The dynamical matrix has the two symmetry properties discussed in [91], which ensure that its eigenvalues are real. The dynamical matrix is central to the lattice-dynamics analysis.

For the L–J monatomic crystal phase, the dynamical matrix has size 3×3. Each wave vector will therefore have three modes associated with it (Figure 4.7). Given the equilibrium atomic positions and the interatomic potential, the frequencies and polarization vectors can be found by substituting the wave vector into the dynamical matrix and diagonalizing. Although this calculation can be performed for any wave vector, it is important to remember that only certain values are relevant to the analysis of the MD simulation cell. The phonon dispersion curves are obtained by plotting the normal-mode frequencies as functions of the wave number in different directions.

The dispersion relation is further developed because of some of the properties of $\boldsymbol{D}(\boldsymbol{\kappa})$ [9]. $\boldsymbol{D}(\boldsymbol{\kappa})$ is an even function of $\boldsymbol{\kappa}$, a real matrix, and is symmetric. Thus, it has three eigenvectors, $\boldsymbol{s}_1$, $\boldsymbol{s}_2$, and $\boldsymbol{s}_3$, which satisfy

$$\boldsymbol{D}(\boldsymbol{\kappa})\boldsymbol{s}_\alpha(\boldsymbol{\kappa}) = \lambda_\alpha(\boldsymbol{\kappa})\boldsymbol{s}_\alpha(\boldsymbol{\kappa}) \quad \text{eigenvalues, } \alpha \text{ is polarization.} \tag{4.41}$$

The formulation of [9] does not include the mass matrix in $\boldsymbol{D}(\boldsymbol{\kappa})$, and the corresponding force-displacement equation will be discussed in Section 4.8. The three normal modes of wave-number vector $\boldsymbol{\kappa}$ will have polarization $\boldsymbol{s}_\alpha$ and frequency $\omega_\alpha(\boldsymbol{\kappa})$.

For a unit cell of $N_\circ$ atoms, the matrix $\boldsymbol{D}(\boldsymbol{\kappa})$ is $3N_\circ \times 3N_\circ$, and there are $3n$ frequencies for each discrete value of $\boldsymbol{\kappa}$. Then care must be made in distinction of the restricted phonon branches in the polyatomic lattices, such as SiO_2 in Figure 4.8.

Then the dispersion relation becomes [9]

$$\omega_\alpha(\boldsymbol{\kappa}) = [\frac{\lambda_\alpha(\boldsymbol{\kappa})}{M}]^{1/2} \text{ dispersion relation.} \tag{4.42}$$

In summary, in a monatomic cubic lattice of lattice constant a, the discrete values that $\boldsymbol{\kappa}$ (normal modes) take on are

$$|\boldsymbol{\kappa}| = \frac{2\pi n}{Na}, \quad n = 0, \pm 1, \pm 2, ..., \pm(N-1), \tag{4.43}$$

where N is the number of unit cells in the system. This leads to $3N$ vibration modes (N longitudinal and $2N$ transversal).

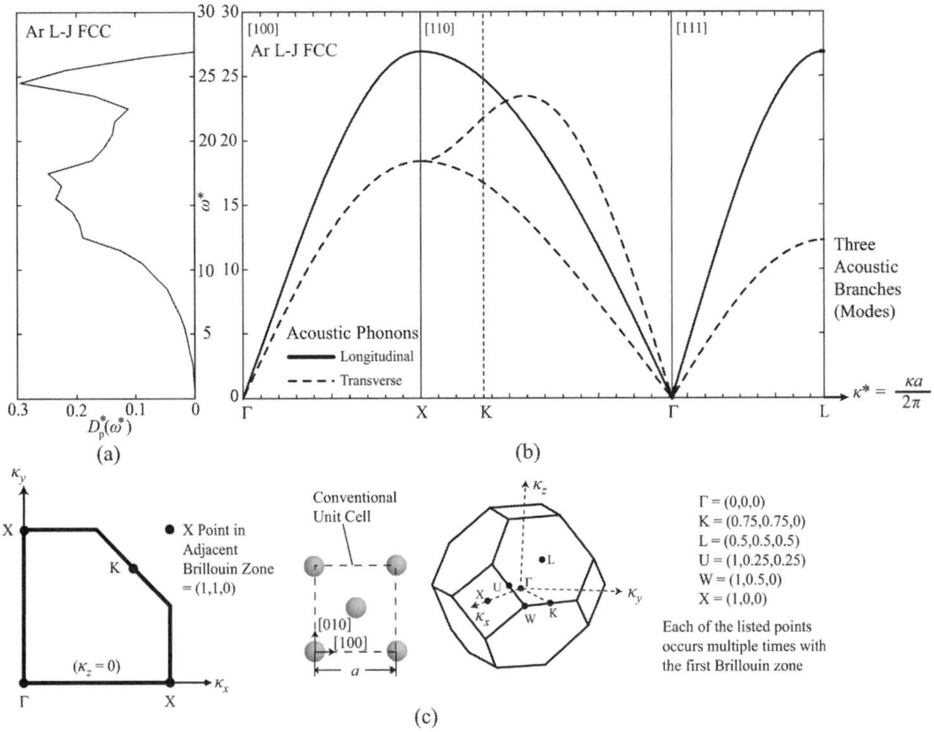

Figure 4.7. (a) DOS, (b) phonon dispersion for the Ar L–J FCC crystal [234], and (c) the directional designation and the Brillouin zone.

4.5 Quantum Theory of Lattice Vibration

It can be shown that the classical Hamiltonian of Table 2.5, which is applicable to each atom (center of mass) in a crystal lattice, has a general solution. This can be cast in terms of phonon annihilation and creation operators [9, 91, 127].

The Hamiltonian (Table 2.5) is given in terms of the dynamical matrix (4.40) as

$$\mathrm{H}_p = \sum_{\boldsymbol{x}} \frac{1}{2m} p^2(\boldsymbol{x}) + \frac{1}{2} \sum_{\boldsymbol{x},\boldsymbol{x}'} d_i(\boldsymbol{x}) D_{ij}(\boldsymbol{x} - \boldsymbol{x}') d_j(\boldsymbol{x}'). \tag{4.44}$$

Then with the solution to the dispersion relation, which gives $\omega_\alpha(\boldsymbol{\kappa})$, the phonon annihilation operator (footnote of Section 2.6.4 and the Glossary under creation and annihilation operators) is defined as

$$b_{\boldsymbol{\kappa},\alpha} = \frac{1}{N^{1/2}} \sum_{\boldsymbol{x}} e^{-i(\boldsymbol{\kappa}\cdot\boldsymbol{x})} \boldsymbol{s}_\alpha(\boldsymbol{\kappa}) \cdot [(\frac{m\omega_\alpha}{2\hbar})^{1/2} \boldsymbol{d}(\boldsymbol{x}) + i(\frac{1}{2\hbar m\omega_\alpha})^{1/2} \boldsymbol{p}(\boldsymbol{x})], \tag{4.45}$$

where N is the number of normal modes divided by α.

The annihilation operator is similar to the ladder operator defined in Section 2.6.5 footnote. Note that $\boldsymbol{s}_\alpha(\boldsymbol{\kappa})$ is the unit vector in the normal coordinates. One property of this is that $\sum_{\alpha,\boldsymbol{\kappa}} \boldsymbol{s}_\alpha(\boldsymbol{\kappa}) \cdot \boldsymbol{s}_\alpha(\boldsymbol{\kappa})$ is equal to N.

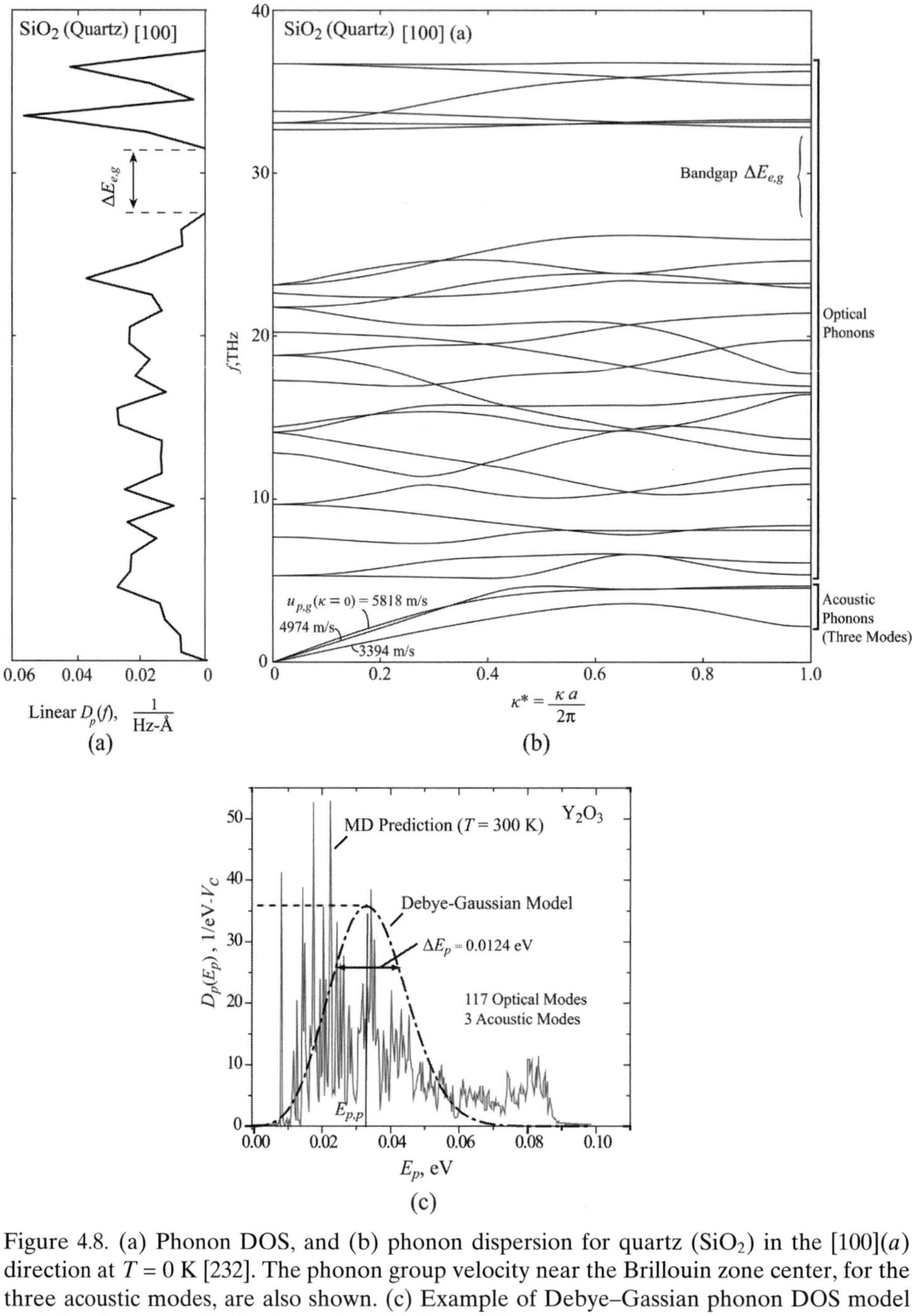

Figure 4.8. (a) Phonon DOS, and (b) phonon dispersion for quartz (SiO_2) in the [100](a) direction at $T = 0$ K [232]. The phonon group velocity near the Brillouin zone center, for the three acoustic modes, are also shown. (c) Example of Debye–Gassian phonon DOS model used for Y_2O_3 crystal. V_c is the unit-cell volume [180].

Similarly, the phonon creation operator is defined as

$$b^{\dagger}_{\kappa,\alpha} = \frac{1}{N^{1/2}} \sum_{x} e^{i(\kappa\cdot x)} s_\alpha(\kappa) \cdot [(\frac{m\omega_\alpha}{2\hbar})^{1/2} d(x) - i(\frac{1}{2\hbar m\omega_\alpha})^{1/2} p(x)]. \qquad (4.46)$$

Then the displacement $d(x)$ and momentum $p(x)$ are expressed by use of the properties of ladder operators $b_{\kappa,\alpha}$ and $b^{\dagger}_{\kappa,\alpha}$ (footnote of Section 2.6.4), as the new variables. This gives (the expression in terms of the normal coordinates will be given in Section 5.15.1)

$$d(x) = \frac{1}{N^{1/2}} \sum_{\kappa,\alpha} (\frac{\hbar}{2m\omega_\alpha})^{1/2} (b_{\kappa,\alpha} + b^{\dagger}_{\kappa,\alpha}) s_\alpha(\kappa) e^{i(\kappa\cdot x)} \qquad (4.47)$$

$$p(x) = -\frac{i^{1/2}}{N^{1/2}} \sum_{\kappa,\alpha} (\frac{\hbar m\omega_\alpha}{2})^{1/2} (b_{\kappa,\alpha} - b^{\dagger}_{\kappa,\alpha}) s_\alpha(\kappa) e^{i(\kappa\cdot x)}. \qquad (4.48)$$

Finally, the Hamiltonian (4.44) expressed in terms of these variables (i.e., for the momentum and potential energies) becomes†

$$\begin{aligned} \mathrm{H}_p &= \frac{1}{4} \sum_{\kappa,\alpha} \hbar\omega_\alpha (b_{\kappa,\alpha} - b^{\dagger}_{\kappa,\alpha})(b^{\dagger}_{\kappa,\alpha} - b_{\kappa,\alpha}) + \\ &\quad \frac{1}{4} \sum_{\kappa,\alpha} \hbar\omega_\alpha (b_{\kappa,\alpha} + b^{\dagger}_{\kappa,\alpha})(b^{\dagger}_{\kappa,\alpha} + b_{\kappa,\alpha}) \\ &= \frac{1}{2} \sum_{\kappa,\alpha} \hbar\omega_\alpha (b_{\kappa,\alpha} b^{\dagger}_{\kappa,\alpha} + b^{\dagger}_{\kappa,\alpha} b_{\kappa,\alpha}) \\ &= \sum_{\kappa,\alpha} \hbar\omega_\alpha (b^{\dagger}_{\kappa,\alpha} b_{\kappa,\alpha} + \frac{1}{2}), \end{aligned} \qquad (4.49)$$

where we have used the commutator operator (2.72) for the last step.

† The Hamiltonian is made of kinetic and potential energies that are

$$\text{kinetic energy} = \sum_{x} \frac{1}{2m} p^2(x) = \frac{1}{4} \sum_{\kappa,\alpha} \hbar\omega_\alpha(\kappa)(b^{\dagger}_{\kappa,\alpha} - b_{\kappa,\alpha})(b^{\dagger}_{\kappa,\alpha} - b_{\kappa,\alpha})$$

$$\text{potential energy} = \frac{1}{2} \sum_{x,x'} d_i(x) D_{ij}(x - x') d_j(x') = \frac{1}{4} \sum_{\kappa,\alpha} \hbar\omega_\alpha(\kappa)(b^{\dagger}_{\kappa,\alpha} + b_{\kappa,\alpha})(b^{\dagger}_{\kappa,\alpha} + b_{\kappa,\alpha})$$

$$\text{Hamiltonian } \mathrm{H}_p = \frac{1}{2} \sum_{\kappa,\alpha} \hbar\omega_\alpha(\kappa)(b_{\kappa,\alpha} b^{\dagger}_{\kappa,\alpha} + b^{\dagger}_{\kappa,\alpha} b_{\kappa,\alpha}).$$

Using commutator operator (2.72), we have

$$[b_{\kappa,\alpha}, b^{\dagger}_{\kappa,\alpha}] = 1 = b_{\kappa,\alpha} b^{\dagger}_{\kappa,\alpha} - b^{\dagger}_{\kappa,\alpha} b_{\kappa,\alpha}$$

$$b_{\kappa,\alpha} b^{\dagger}_{\kappa,\alpha} = b^{\dagger}_{\kappa,\alpha} b_{\kappa,\alpha} + 1$$

$$\begin{aligned} \mathrm{H}_p &= \frac{1}{2} \sum_{\kappa,\alpha} \hbar\omega_\alpha(\kappa)(b^{\dagger}_{\kappa,\alpha} b_{\kappa,\alpha} + 1 + b^{\dagger}_{\kappa,\alpha} b_{\kappa,\alpha}) \\ &= \sum_{\kappa,\alpha} \hbar\omega_\alpha(\kappa)(b^{\dagger}_{\kappa,\alpha} b_{\kappa,\alpha} + \frac{1}{2}). \end{aligned}$$

This is written in terms of energy (footnote of Section 2.6.4) by use of the probability distribution function $f_p(\boldsymbol{\kappa}, \alpha)^\dagger$ as [similar to (2.85) for the quantum-harmonic oscillator]

$$E_p = \sum_{\boldsymbol{\kappa},\alpha}[f_p(\boldsymbol{\kappa}, \alpha) + \frac{1}{2}]\hbar\omega_\alpha(\boldsymbol{\kappa}). \quad (4.50)$$

Therefore this is another solution path in arriving at the quantum energy of the harmonic oscillator (Section 2.6.4). The quanta of energy are $\hbar\omega$. We will make use of the quantum treatment of phonons in Sections 4.8, 5.15, and 7.12.

4.6 Examples of Phonon Dispersion and DOS

Figure 4.7 shows the MD-computed dimensionless phonon dispersion and DOS for the Ar FCC crystal. The atomic coordinates used in the generation of the dispersion data were obtained from MD simulations as discussed in Section 2.5 [232]. A structure at finite temperature is slowly quenched to a temperature of 0 K. The cut-off used is $2.5\sigma_{\rm LJ}$. As the crystal structure is monatomic (number of atoms per primitive cell, $N_\circ = 1$, total number of modes is $3N_\circ$), all modes correspond to acoustic phonons.

The plotted angular frequency is dimensionless and has been normalized by the L–J time scale (Table 2.7, except for the constant), $(\sigma_{\rm LJ}^2 m/\varepsilon_{\rm LJ})^{1/2}$. Note that from (2.54), ω_n is 22.88 time this ω scale, which is near the cut-off frequency in Figure 4.7(a). For Ar, the time scale has a value of 2.14 ps. The divisions on the horizontal axis (the wave number) are separated by $0.1 \times 2\pi/a$ (i.e., one-twentieth of the size of the first Brillouin zone in the [100] direction).

Note the degeneracies of the transverse branches in the [100] and [111] directions, but not in the [110] direction. Also, as seen in the [110] direction, the longitudinal branch does not always have the highest frequency of the three branches at a given point.

The volumetric density of states is based on a Brillouin zone with a grid spacing of $1/21 \times 2\pi/a$. This leads to 37,044 distinct points (each with three polarizations) covering the entire first Brillouin zone. The frequencies are sorted by a histogram with a bin size of 1, and the resulting data are plotted at the middle of each bin. The DOS axis is defined such that an integration over frequency gives $3(N-1)/V \simeq 12/a^3$ (for large V, where N is the number of atoms in volume V).

Figures 4.8(a) and (b) show the MD-computed phonon dispersion and DOS for silica (quartz). The predictions are made with the B-K-S potential (2.10). The Wolf method [350] is used to model the electrostatic interactions with an α value of 0.431 Å^{-1}. The cut-off radius used is 6.44 Å. The zero-temperature unit-cell parameters are a = 4.89 Å and c = 5.51 Å. The anisotropic structure of quartz is shown in

† The total number of phonons can also be written as sum of the normal modes as

$$N_p = \sum_{\boldsymbol{\kappa},\alpha} f_p.$$

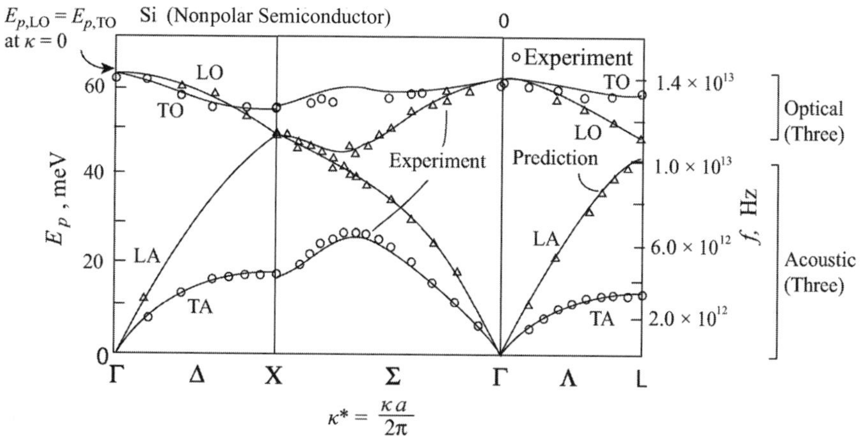

Figure 4.9. Predicted and measured phonon dispersion curves for Si (diamond structure). The solid curves represent the predicted results, and the open circles are experimental results [318]. The key Brillouin points are also shown. There are six branches, three acoustic and three optical; the transverse branches are degenerate.

Figure 2.9(a). The lattice-dynamics calculations are based on the description given in [91]. The dispersion curves are in reasonable agreement with experimental and theoretical predictions [15, 321].

The linear density of states is based on a grid spacing of (1/1000) $\times$ $2\pi/a$ in the [100] direction. This leads to 27,027 frequencies (there are $N_\circ = 9$ atoms per primitive cell and as discussed in Section 4.33, there are $3N_\circ = 27$ branches). The frequencies are sorted by a histogram with a bin size of 1 THz, and the resulting data are plotted at the middle of each bin. The DOS axis is defined such that an integration over frequency gives $3(N_c - 1)/L \simeq 3/a$ (for large L, where N_c is the number of unit cells in a length L in the [100] direction).

Note that the Debye model for D_p represents the energy of acoustic phonons. To include the high-energy phonons, the dimensionless Debye–Gaussian distribution function (4.29) can be used instead of the Debye DOS, which has the form

$$D_p^*(E_p) = CE_p^2 \exp[-(\frac{E_p - E_{p,p}}{\Delta E_p})^2], \quad E_p = \hbar\omega, \tag{4.51}$$

where C is the normalization constant making D_p dimensionless, $E_{p,p}$ is the energy at the center of the distribution, and ΔE_p is the width (at $1/e$ from the peak) of the distribution. Note that this model has only one peak, i.e., the positive root of $E_p^2 - E_p E_{p,p} - \Delta E_p^2 = 0$. Figure 4.8(c) shows (4.51) applied to the Y_2O_3 crystal D_p (where D_p is dimensional) from MD results [290]. These MD results will be discussed in Section 4.19.3. The agreement near the peak is reasonable.

As further results, typical dispersion relations are given in Figures 4.9 and 4.10, for Si and GaAs, respectively. Both predictions and measurements are shown and are in good agreement.

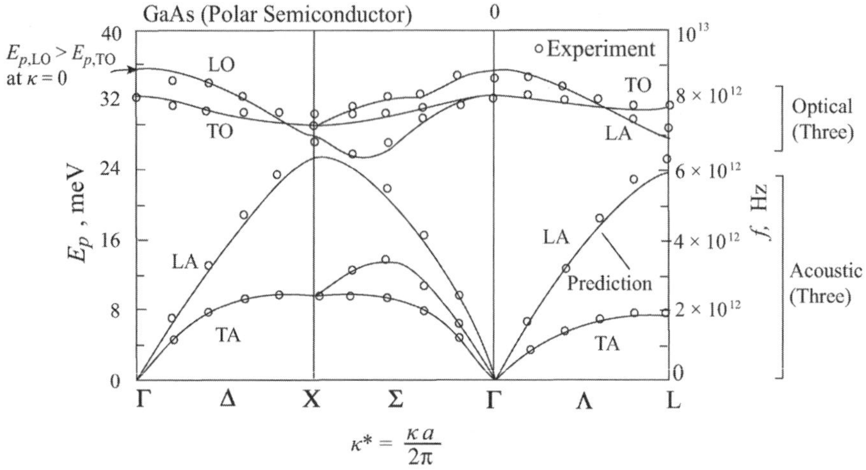

Figure 4.10. Predicted and measured phonon dispersion curves for GaAs (zinc-blende structure). The solid curves represent the predicted results, and the open circles are experimental results [318]. The key Brillouin points are also shown. There are six branches, three acoustic and three optical, the transverse branches are degenerate.

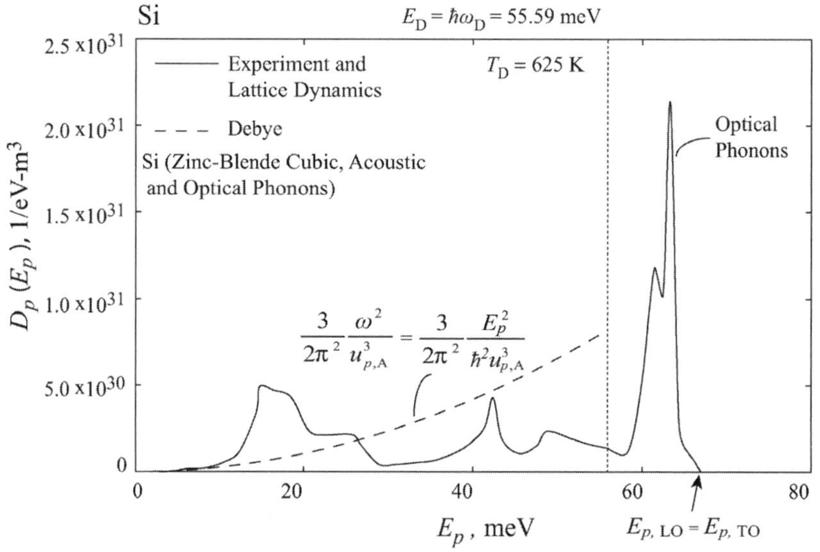

Figure 4.11. Measured and predicted phonon DOS for Si [346]. The Debye cut-off frequency (angular) is $\omega_D = T_D k_B/\hbar = 8.443 \times 10^{13}$ rad/s $= 84.43$ Trad/s.

The phonon DOS for Si is shown in Figure 4.11. Note that the maximum phonon energy is that $E_{p,\mathrm{LO}} = E_{p,\mathrm{TO}}$ at $\kappa = 0$ (Figure 4.9). The peaks are singularities (van Hove) at L point. The Debye energy (temperature) and Debye model of DOS are also shown. Note that the Debye model does not include the optimal phonons.

Figure 4.12 shows measured individual contributions from the transverse and longitudinal polarizations for $D_p(\omega)$ of Al (FCC) at $T = 80$ K [319]. Similar to Ar

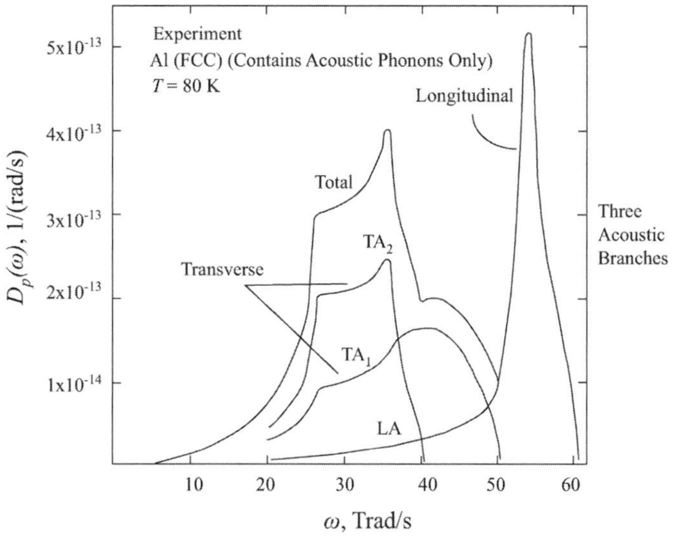

Figure 4.12. Measured phonon angular-frequency DOS for Al (FCC) at T = 80 K [319]. Also shown are the contributions from the three acoustic-phonon modes. The volume is a^3.

FCC, there is no optical phonon in this single-atom primitive cell (basis of one atom, Section 4.3).

The exact phonon dispersion curve varies depending on the existence of polar or nonpolar atoms in the primitive cell. In nonpolar semiconductors (Si and C), the LO and TO branches of phonon dispersion have the same frequency at the zone center Γ (Figure 4.9). For polar semiconductors (like GaAs), the LO and TO frequencies split at the zone center Γ (Figure 4.10), because of long-range dipolar interaction†.

4.7 Debye Average Acoustic Speed and Phonon Specific Heat Capacity

4.7.1 Specific Heat Capacity

The lattice (or phonon) specific heat capacity $c_{v,p}$, is the total value for all phonon modes. The specific heat capacity is given in J/kg-K, J/m^3-K, or J/K (per atom), depending on relevance. Using the Debye model of the phonon DOS, $c_{v,p}$ is derived for monatomic crystalline solids. It has a temperature dependence such that, as expected, it vanishes at 0 K, has a steep dependence as $T = 0$ K is approached, and the temperature dependence disappears at high temperatures (at high temperatures, it reaches a value of $3k_\mathrm{B}$ per atom, associated with 3 degrees of freedom, and equipartition of energy, which gives crystal kinetic and potential energies). Debye,

† The LO modes have an additional electric restoring force, such that [309]

$$\frac{\omega^2_{p,\mathrm{LO}}}{\omega^2_{p,\mathrm{TO}}} = \frac{\epsilon_{e,s}}{\epsilon_{e,\infty}},$$

where $\epsilon_{e,s}$ and $\epsilon_{e,\infty}$ are the static and high-frequency dielectric functions (Section 3.3.5).

noting these trends, suggested a cut-off frequency f_p° beyond which the high-energy phonons do not contribute to the lattice specific heat capacity (Section 4.2.2). His formulation is subsequently given.

For a monatomic, isotropic crystal, the energy per unit volume and atomic mass $M/N_{\rm A} = m$ for N_p phonons, in a volume V, is

$$\begin{aligned} e_p &= \frac{N_{\rm A}}{VM}\sum_i^{N_p} E_{p,i} = \frac{1}{Vm}\sum_i E_{p,i} \\ &= \frac{1}{Vm}\sum_i^{N_p} \hbar\omega_i f_p^{\circ} = \frac{1}{m}\sum_i \hbar\omega_i \frac{1}{e^{x_i}-1} \\ &= \frac{1}{Vm}\sum_i^{N_p} \hbar\omega_i \frac{1}{\exp(\frac{\hbar\omega_i}{k_{\rm B}T})-1} \\ &= \frac{1}{m}\int \hbar\omega \frac{1}{\exp(\frac{\hbar\omega}{k_{\rm B}T})-1} D_p {\rm d}\omega, \end{aligned} \tag{4.52}$$

where the i summation is over the phonon modes, which are N_p in volume V or n_p per unit volume. Note that we have used the integral of the phonon DOS (1/m^3-rad/s) in place of the summation. The lattice specific heat capacity of a solid at constant volume $c_{v,p}$ is found by differentiating e_p (4.52) with respect to temperature T (Table 2.4), i.e.,

$$c_{v,p} = \frac{\partial e_p}{\partial T}|_V = \frac{1}{Vm}\sum_i^{N_p} \hbar\omega_i \frac{\partial}{\partial T}[\exp(\frac{\hbar\omega_i}{k_{\rm B}T})-1]^{-1}. \tag{4.53}$$

We simplify the differentiation on the right-hand side as

$$\frac{\partial}{\partial T}[\exp(\frac{\hbar\omega_i}{k_{\rm B}T})-1]^{-1} = \frac{\hbar\omega_i}{k_{\rm B}T^2}\frac{e^{x_i}}{(e^{x_i}-1)^2}. \tag{4.54}$$

Substituting back into the specific heat expression for an atom of mass m, (4.53) becomes

$$\begin{aligned} c_{v,p} &= \frac{1}{Vm}\sum_i^{N_p} \hbar\omega_i \frac{\hbar\omega_i}{k_{\rm B}T^2}\frac{e^{x_i}}{(e^{x_i}-1)^2} \\ &= \frac{k_{\rm B}}{Vm}\sum_i^{N_p} (\frac{\hbar\omega_i}{k_{\rm B}T})^2 \frac{e^{x_i}}{(e^{x_i}-1)^2} \\ &= \frac{k_{\rm B}}{Vm}\sum_i^{N_p} \frac{x_i^2 e^{x_i}}{(e^{x_i}-1)^2} \\ &\equiv \frac{k_{\rm B}}{m}\int_0^{\omega_{\rm D}} \frac{x^2 e^x}{(e^x-1)^2} D_{p,{\rm D}} {\rm d}\omega \quad x = \frac{\hbar\omega}{k_{\rm B}T}, \end{aligned} \tag{4.55}$$

where we again used the integral over the DOS, along with the Debye cut-off angular frequency (Section 4.2.2). This Debye cut-off frequency $f_\mathrm{D} = \omega_\mathrm{D}/2\pi$ is the highest normal-mode frequency that the crystal can have (Glossary). As described in Section 4.3, there are $3n$ phonon modes (or three polarizations of acoustic phonons, two transversal and one longitudinal), for a monatomic, single-atom lattice crystal. Note that from (4.55), the specific heat capacity per phonon is $k_\mathrm{B}x^2e^x/[e^x - 1]^2$.

In (2.99), we related the number of carriers to the density of states and the probability distribution function. Here we use the Debye approximation, which gives similar results of approximating $f_p^\circ = 1$. Then, using the Debye DOS $D_{p,\mathrm{D}}$ (for three acoustic modes), we use the Debye oscillator conservation constraint, i.e.,

$$\frac{N_p}{V} = n_p = 3n = \int_0^\infty D_p(\omega) f_p^\circ \mathrm{d}\omega = \int_0^{\omega_\mathrm{D}} D_{p,\mathrm{D}}\mathrm{d}\omega = \int_0^{\omega_\mathrm{D}} \frac{3\omega^2}{2\pi^3 u_{p,\mathrm{A}}^3}\mathrm{d}\omega$$

$$= \frac{3}{2\pi^3 u_{p,\mathrm{A}}^3}\int_0^{\omega_\mathrm{D}} \omega^2 \mathrm{d}\omega,$$

$$D_{p,\mathrm{D}} = \frac{3\omega^2}{2\pi^2 u_{p,\mathrm{A}}^3} \quad \text{Debye acoustic-phonon relations.} \tag{4.56}$$

Note that we have used the frequency DOS, $D_p(\omega)$, and integration over frequency space (with cut-off), as compared with energy DOS in (2.99). Here $u_{p,\mathrm{A}}$ is the single, average phonon speed used to represent all polarizations and is generally defined as (4.15)

$$\frac{1}{u_{p,\mathrm{A}}} = \frac{1}{3}(\frac{1}{u_{p,g,\mathrm{L}}} + \frac{2}{u_{p,g,\mathrm{T}}}), \tag{4.57}$$

where L and T are the longitudinal and transverse polarizations.

Evaluating the integral of (4.56) and solving for ω_D gives

$$3 = \frac{3}{2n\pi^2 u_{p,\mathrm{A}}^3}\frac{\omega_\mathrm{D}^3}{3} \tag{4.58}$$

$$\omega_\mathrm{D}^3 = 6n\pi^2 u_{p,\mathrm{A}}^3 \tag{4.59}$$

$$\omega_\mathrm{D} = (6n\pi^2 u_{p,\mathrm{A}}^3)^{1/3}, \tag{4.60}$$

which is the same as (4.28). The approximation in (4.56) has lead to a simple relation between $\omega_\circ$ and $u_{p,\mathrm{A}}$

Noting that, from the definition of the Debye temperature (Glossary), i.e., $\omega_\mathrm{D} = k_\mathrm{B} T_\mathrm{D}/\hbar$, and recalling that $x(T,\omega) = \hbar\omega/k_\mathrm{B}T$, we can write $x_\mathrm{D} = x(T, \omega = \omega_\mathrm{D})$ as

$$x_\mathrm{D} = \frac{\hbar\omega_\mathrm{D}}{k_\mathrm{B}T} = \frac{T_\mathrm{D}}{T} \quad \text{scaled, inverse temperature.} \tag{4.61}$$

Substituting the expressions for f and $D_{p,D}(\omega)$ into the integral form of (4.55), using the Debye cut-off frequency, we have

$$c_{v,p} = \frac{k_B}{m}\int_0^{\omega_D} \frac{x^2 e^x}{(e^x-1)^2} D_p(\omega)\mathrm{d}\omega \tag{4.62}$$

$$= \frac{k_B}{m}\int_0^{x_D} \frac{x^2 e^x}{(e^x-1)^2}\frac{3\omega^2}{2n\pi^2 u_{p,A}^3}\frac{k_B T}{\hbar}\mathrm{d}x$$

$$= \frac{k_B}{m}\int_0^{T_D/T} \frac{12\pi k_B^3 T^3}{n u_{p,A}^3 \hbar^3}\frac{x^4 e^x}{(e^x-1)^2}\mathrm{d}x.$$

Substituting f_D into the expression for T_D gives another form of (4.27), i.e.,

$$T_D = \frac{\hbar\omega_D}{k_B} = \frac{\hbar}{k_B}(6n\pi^2 u_{p,A}^3)^{1/3}, \tag{4.63}$$

which is the same as (4.27), written there for each polarization. We then have

$$c_{v,p} = \frac{k_B}{m}12\pi T^3 \frac{k_B^3}{u_{p,A}^3\hbar^3}\int_0^{T_D/T}\frac{x^4 e^x}{(e^x-1)^2}\mathrm{d}x$$

$$= \frac{k_B}{m}12\pi T^3 \frac{1}{T_D^3}\frac{6\pi^2 n}{(2\pi)^3}\int_0^{T_D/T}\frac{x^4 e^x}{(e^x-1)^2}\mathrm{d}x$$

$$= 9\frac{k_B}{m}(\frac{T}{T_D})^3 n\int_0^{T_D/T}\frac{x^4 e^x}{(e^x-1)^2}\mathrm{d}x$$

Debye model of lattice specific heat capacity. (4.64)

Therefore, to find $c_{v,p}$ per atom we divide the proceeding equation by n. Figure 4.13(a) shows variations of the predicted dimensionless specific heat capacity of a few elements, with respect to temperature. The asymptotic value of $c_{v,p}m/k_B = 3$ is shown and reached for $T \gg T_D$ (end-of-chapter problem). This is called the Dulong–Petit limit for solids. In Figure 2.15, it was shown that this limit is realized in classical MD at temperatures well below the melting temperature. However, in MD, $c_{v,p}$ decrease (can also increase) because of anharmonicity at higher temperatures, whereas the quantum effects cause lower $c_{v,p}$ at lower temperatures (Figure 2.15). Because of anharmonic vibrations, the high-temperature asymptotic value of $c_{v,p}$ may be less than $3k_B$ (per atom), as shown in Figure 2.15, by as much as 10%.

The Debye temperature for elemental crystalline solids is listed in Table A.2. For some compounds, they can be found in [183].

For $T \ll T_D$ there is a $(T/T_D)^3$ trend (end-of-chapter problem), and this is also shown in Figure 4.13(a).

Figure 4.13(b) shows the variations of f_p^o, and scaled $D_{p,D}$, $D_{p,D}f_p^o$, and $\hbar\omega D_{p,D}f_p^o$ [appearing in integral (4.52)], with respect to ω, for Si at $T = 300$ K. Although $\hbar\omega D_{p,D}f_p^o$ does not peak at ω_D, its growth is not as large compared with that

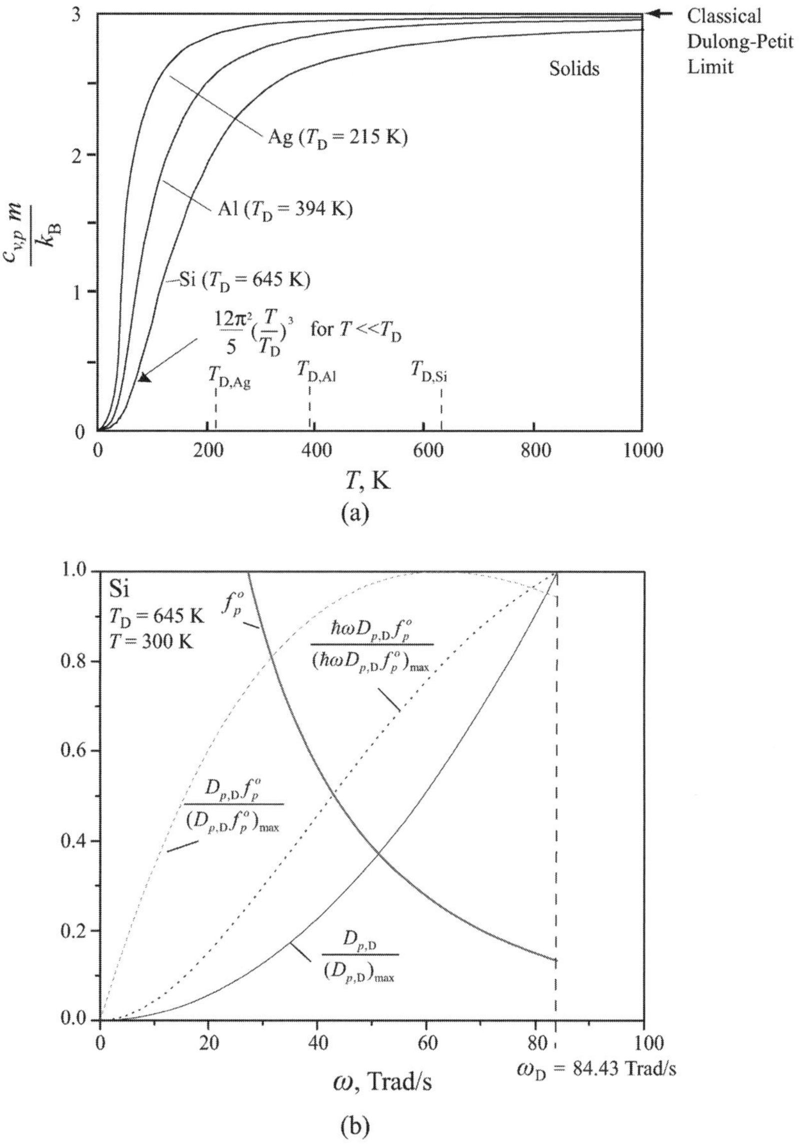

Figure 4.13. (a) Variation of the predicted dimensionless lattice (Debye acoustic phonon) specific heat capacity (per atom) with respect to temperature, for Ag, Al, and Si elements. The classical limit is also shown. (b) Variations of sealed f_p^o, $D_{p,D}$, $D_{p,D} f_p^o$, and $\hbar\omega D_{p,D} f_p^o$ as functions of ω, for Si at $T = 300$ K.

at lower frequencies. As shown in Figure 4.11, the Debye DOS is not an accurate model for Si, especially when the role of optical phonons becomes important (e.g., in scattering of conduction electron by phonons, Section 5.15.3). However, for some applications, such as in total specific heat capacity, this model is easy to use and relatively accurate.

To use the molecular weight, we replace k_B/m with R_g/M ($R_g = k_B N_A$).

For compounds, the average atomic mass $\langle m\rangle$ for N_i atoms of mass m_i per unit volume is

$$\langle m\rangle = \frac{\sum_i N_i m_i}{\sum_i N_i} \quad \text{average atomic mass in polyatomic crystal.} \tag{4.65}$$

Then we can use (4.64) for a compound by replacing m with the average atomic mass $\langle m\rangle$. Note that (4.64) does not include optical phonons present in polyatomic or multiatomic primitive cell crystals.

Expressing the lattice volume by V_c and the number of atoms per lattice as N_c, the sensible heat content $\rho c_{v,p} T$ in the macroscopic energy equation (Table 1.1) for the lattice is

$$\rho c_p T = \frac{3\alpha(T) N_c k_B T}{V_c}, \tag{4.66}$$

where $\alpha(T)$ is dimensionless and expresses the temperature dependence. We note again that (4.64) is based on Debye model of acoustic phonons.

4.7.2 Estimate of Directional Acoustic Velocity

The long wavelength acoustic phonons are the elastic waves in the macroscopic elastic mechanics equation (Table 3.11). Here we discuss the relations among the group velocities of these waves, the elastic moduli, and the interatomic force constants.

(A) *Group Velocity from Directional Spring Constant*

The acoustic modes have all the atoms in the planes perpendicular to the transport direction moving in phase. For a lattice consisting of parallel planes, the transport perpendicular to the these planes is one dimensional, and similar to (4.17), by use of the harmonic theory, we have the sound speed as given by

$$u_{p,g,i} = l_i(\frac{\Gamma_i}{m_c})^{1/2}, \tag{4.67}$$

where l_i is the distance between the planes and m_c is the mass of the primitive cell. The spring constant is defined through (2.54) as

$$\Gamma = \sum_n \frac{\partial^2 \varphi_n}{\partial d_{i,n}^2}, \tag{4.68}$$

where $d_{i,n}$ is the displacement of the nth unit cell along the i direction and φ_n is the potential between the origin unit cell and the nth neighboring unit cell. Here, we consider the interactions only among the nearest neighbors.

(B) *Nearest-Neighbor FCC Structures*

Then, for an FCC structure (each atom has 12 nearest neighbors), along [111], we have $\Gamma_{\rm L} = 2\Gamma$, $\Gamma_{\rm T} = \Gamma/2$, where Γ is the effective spring constant of the pair potential and is defined as

$$\Gamma = (\frac{\partial^2 \varphi_{\rm LJ}}{\partial r^2})_{r_e} \quad r_e \text{ is the equilibrium separation distance.} \tag{4.69}$$

Then using projections, along [100] we have $\Gamma_{\rm L} = 2\Gamma$, $\Gamma_{\rm T} = \Gamma$.

As an example, an Ar solid is an FCC crystal and each primitive cell contains only one atom. Then the distance between the nearest neighbors is $r_e = a/2^{1/2}$. For a L–J potential (2.9), and using the equilibrium condition, i.e., $\partial\varphi_{\rm LJ}/\partial r = 0$, we have $\sigma_{\rm LJ} = 2^{-2/3}a$. Then the effective spring constant becomes $\Gamma = 72\epsilon_{\rm LJ}/2^{1/3}\sigma_{\rm LJ}^2$. Note that $\Gamma_{\rm L} = 2\Gamma = 72 \times 2^{2/3}\epsilon_{\rm LJ}/\sigma_{\rm LJ}^2$, gives a result that is closer to the effective bulk value (2.54). The bulk modulus can also be derived and is (end-of-chapter problem) $E_p = 75\epsilon_{\rm LJ}/\sigma_{\rm LJ}^3$ and $a = 2^{1/2}r_e$, $r_e = 1.09\sigma_{\rm LJ}$. Also, note that the effective bulk (including long-range interactions) Γ given in Table 2.7 has a larger numerical coefficient.

For Ar, from Table 2.2, we have $\sigma_{\rm LJ} = 3.40$ Å, $\epsilon_{\rm LJ} = 0.0104$ eV = 1.67×10^{-21} J, and then $\Gamma = 0.85$ N/m, $m_c = 6.692 \times 10^{-26}$ kg, and $a = 2^{2/3}\sigma_{\rm LJ} = 5.31$ Å.

Along the [100] direction, we have $l = 0.5a$, then $u_{p,g,{\rm L}} = 1362$ m/s and $u_{p,g,{\rm T}} =$ 963 m/s, and the average speed is given by (4.53), so $u_{p,g,{\rm A}} = 1067$ m/s.

Along the [111] direction, we have $l = a/3^{1/2}$, then $u_{p,g,{\rm L}} = 1572$ m/s, $u_{p,g,{\rm T}} =$ 786 m/s and $u_{p,g,{\rm A}} = 943$ m/s.

The Debye temperature of Ar is 89 K [234], and the corresponding average sound speed $u_{p,{\rm A}} = u_{p,g,{\rm A}}$ found from (4.63) is 999 m/s, which is close to the proceeding approximate prediction.

(C) *Nearest-Neighbor SC Structures*

For homogeneous deformation in SC structures, we have the relation for the isothermal bulk modulus $E_p = 1/\kappa_p$, $\kappa_p = -(\partial\rho/\partial p|_T)/\rho$ (Table 3.11), and the spring constant Γ_b through the lattice constant a, as

$$\frac{1}{E_p} = \kappa_p = -\frac{1}{v}\frac{\partial v}{\partial p}|_T = \frac{3a^2 d}{a^3\Gamma_b d/a^2} = \frac{3a}{\Gamma_b} \quad \text{along [111] SC.} \tag{4.70}$$

where we have used the pressure relation $p = \Gamma_b d/a^2$. Then, using (4.67), we have

$$\begin{aligned} u_{p,b} &= l_b(\frac{\Gamma_b}{m_c})^{1/2} = l_b(\frac{3a}{\kappa_p m_c})^{1/2} \\ &= \frac{a}{3^{1/2}}(\frac{3a}{\kappa_p}m_c)^{1/2} = (\frac{a^3}{\kappa_p m_c})^{1/2} \\ &= (\frac{1}{\rho\kappa_p})^{1/2} = (\frac{E_p}{\rho})^{1/2}, \ \rho = \frac{m_c}{a^3}. \end{aligned} \tag{4.71}$$

This is the same as the bulk sound speed given next. Note that the results are slightly different from that for a FCC.

(D) *Bulk and Directional Velocities Using Moduli*

The bulk sound speed is found from

$$u_{p,b} = (\frac{1}{n\langle m\rangle k_p})^{1/2} = (\frac{E_p}{\rho})^{1/2}, \tag{4.72}$$

where $\langle m\rangle$ is average atomic mass (4.65), and n is the atomic number density. The relation among the adiabatic compressibility κ_p $(= 1/E_p)$, the Young modulus E_Y and the shear modulus G is (Table 3.11)

$$\frac{1}{\kappa_p} = \frac{1}{E_p} = \frac{E_\mathrm{Y}}{3(1-2\nu_\mathrm{P})} = \frac{2G(1+\nu_\mathrm{P})}{3(1-2\nu_\mathrm{P})} \quad \nu_\mathrm{P} \text{ is Poisson ratio.} \tag{4.73}$$

The longitudinal and transverse sound speeds are

$$u_{p,\mathrm{L}} = (\frac{E_Y}{n\langle m\rangle})^{1/2} = (\frac{E_Y}{\rho})^{1/2} \quad \text{longitudinal sound speed} \tag{4.74}$$

$$u_{p,\mathrm{T}} = (\frac{G}{n\langle m\rangle})^{1/2} = (\frac{G}{\rho})^{1/2} \quad \text{transverse sound speed.} \tag{4.75}$$

When used in (4.60), the cut-off frequency (for each polarization) and when (4.63) is used, the Debye temperature, can be found from these velocities (thus relating them to the elastic constants).

4.8 Atomic Displacement in Lattice Vibration

In Section 2.5.3(B), the RMS displacement of lattice atoms was given by approximation (leading-order) relation (2.56). Here we derive a general expression, based on the treatment given in [23].

Consider a polyatomic molecule; based on the Newton law (Table 2.5) the force on atom i, $\boldsymbol{F}_i$, is

$$\boldsymbol{F}_i = -\sum_j \frac{\partial \varphi_{ij}}{\partial \boldsymbol{d}_i} = m_i \frac{\mathrm{d}^2 \boldsymbol{d}_i}{\mathrm{d}t^2}. \tag{4.76}$$

With the harmonic approximation for the potentials and for small displacements, we have [similar to (4.37)]

$$\frac{\partial \varphi_{ij}}{\partial \boldsymbol{d}_i} = \frac{\partial^2 \varphi_{ij}}{\partial \boldsymbol{d}_i \partial \boldsymbol{d}_j} \cdot (\boldsymbol{d}_i - \boldsymbol{d}_j). \tag{4.77}$$

Then equation of motion (4.39) is again

$$\boldsymbol{M}\omega^2 \boldsymbol{s}_\alpha = \boldsymbol{D}(\boldsymbol{\kappa})\boldsymbol{s}_\alpha, \tag{4.78}$$

where $\boldsymbol{s}_\alpha$ is a polarization vector, $\boldsymbol{M}$ is the mass matrix, and $\boldsymbol{D}(\boldsymbol{\kappa})$ is the dynamical matrix (4.40)

$$D_{ij}(\kappa) = \sum_i \left(\frac{\partial^2 \varphi_{ij}}{\partial d_i \partial d_j}\right) \exp[i\boldsymbol{\kappa} \cdot (\boldsymbol{x}_i - \boldsymbol{x}_j)]. \quad (4.79)$$

Rearranging (4.78), the eigenvalue equation becomes

$$[\boldsymbol{M}^{-1}\boldsymbol{D}(\boldsymbol{\kappa}) - \omega^2 \boldsymbol{I}] = 0, \quad (4.80)$$

where $\boldsymbol{I}$ is the identity matrix. Here $\boldsymbol{D}(\boldsymbol{\kappa})$ is specified by a complete set of internal displacement coordinates, including bond lengths and angle displacements. This set of coordinates can be represented by a column matrix $\boldsymbol{S}$ [23]. The mean-square relative displacement (MSRD) amplitude matrix $\boldsymbol{\Sigma}$ is defined as

$$\boldsymbol{\Sigma} = \langle \boldsymbol{S}\boldsymbol{S}^\dagger \rangle, \quad (4.81)$$

where $\boldsymbol{S}$ is chosen such that the diagonal elements of $\boldsymbol{\Sigma}$ of the MSRD of the atomic vibrations satisfy $\langle |\boldsymbol{\Delta}_j|^2 \rangle = \Sigma_{jj}$. The superscript † indicates complex conjugate. The coordinate $\boldsymbol{S}$ may be not normal, but we may find a matrix $\boldsymbol{L}$ to transform $\boldsymbol{S}$ to the normal coordinates $\boldsymbol{Q}$ (Glossary), i.e.,

$$\boldsymbol{S} = \boldsymbol{L}\boldsymbol{Q}. \quad (4.82)$$

Then we have

$$\boldsymbol{\Sigma} = \boldsymbol{L}\langle \boldsymbol{Q}\boldsymbol{Q}^\dagger \rangle \boldsymbol{L}^\dagger. \quad (4.83)$$

When $\boldsymbol{Q}$ satisfies

$$\langle \boldsymbol{Q}\boldsymbol{Q}^\dagger \rangle_{\alpha\alpha'} = \left[\frac{\hbar}{2\omega_\alpha} \coth\left(\frac{\hbar\omega_\alpha}{2k_\mathrm{B}T}\right)\right]\delta_{\alpha\alpha'}, \quad (4.84)$$

then we have the relations [81]

$$[\boldsymbol{L}^\dagger \boldsymbol{D}(\boldsymbol{\kappa})\boldsymbol{L}]_{\alpha\alpha'} = \omega_\alpha^2 \delta_{\alpha\alpha'}, \text{ and } \boldsymbol{L}\boldsymbol{L}^\dagger = \boldsymbol{M}^{-1}. \quad (4.85)$$

In the high-temperature regime (within harmonic approximation), we can derive a relationship between the MSRD amplitude matrix and the molecular force fields by using the relations (4.85), and then expanding the coth(x) term in (4.84) as

$$\coth\left(\frac{k_\mathrm{B}T}{2\hbar\omega_\alpha}\right) = \frac{2k_\mathrm{B}T}{\hbar\omega_\alpha} + \frac{\hbar\omega_\alpha}{8k_\mathrm{B}T} + \ldots + . \quad (4.86)$$

Then we have

$$\boldsymbol{\Sigma} = k_\mathrm{B}T\boldsymbol{D}(\boldsymbol{\kappa})^{-1} + \frac{\hbar^2}{16k_\mathrm{B}T}\boldsymbol{M}^{-1}, \quad (4.87)$$

or

$$\Sigma_{jj} = \langle |\boldsymbol{\Delta}_j|^2 \rangle = k_\mathrm{B}TF_{jj}^{-1} + \frac{\hbar^2}{16k_\mathrm{B}T}M_{jj}^{-1}. \quad (4.88)$$

Because $\boldsymbol{\Delta}_j = (\boldsymbol{d}_j - \boldsymbol{d}_\mathrm{o}) \cdot \boldsymbol{s}_j$, where $\boldsymbol{d}_j$ and $\boldsymbol{d}_\mathrm{o}$ are the j atom and the central atom displacements, the MSRD $\langle |\boldsymbol{\Delta}_j|^2 \rangle$ can be rewritten in terms of the MSD

$\langle(\boldsymbol{x}_j \cdot \boldsymbol{b}_j)^2\rangle$, $\langle(\boldsymbol{x}_0 \cdot \boldsymbol{s}_j)^2\rangle$ and the displacement correlation function (DCF) $\langle(\boldsymbol{x}_0 \cdot \boldsymbol{s}_j)(\boldsymbol{x}_j \cdot \boldsymbol{s}_j)\rangle$ such that

$$\langle|\boldsymbol{\Delta}_j|^2\rangle = \langle(\boldsymbol{d}_j \cdot \boldsymbol{s}_j)^2\rangle + \langle(\boldsymbol{d}_{\rm o} \cdot \boldsymbol{s}_j)^2\rangle - 2\langle(\boldsymbol{d}_{\rm o} \cdot \boldsymbol{s}_j)(\boldsymbol{d}_j \cdot \boldsymbol{s}_j)\rangle, \tag{4.89}$$

where $\boldsymbol{s}_j$ is the equilibrium position unit vector of atom j ($\boldsymbol{b}_j$ is the equilibrium position vector of atom j).

In this form, the function can be divided into two parts (one composed of the MSD term and the other of the DCF term) so that the effect of each can be examined separately. For monoatomic crystals, this gives [132, 127]

$$\langle|\boldsymbol{\Delta}_j|^2\rangle = \frac{\hbar}{Nm}\sum_{\kappa,\alpha}(\boldsymbol{s}_{\kappa,\alpha} \cdot \boldsymbol{s}_j)^2\frac{1}{\omega_{\kappa,\alpha}}\coth(\frac{\hbar\omega_{\kappa,\alpha}}{2k_{\rm B}T})[1-\cos(\boldsymbol{\kappa} \cdot \boldsymbol{b}_j)]$$

$$\text{Beni–Platzman–Debye MSRD correlation,} \tag{4.90}$$

where N is the number of atoms of mass m, $\boldsymbol{s}_{\kappa,\alpha}$ is the polarization vector for phonons of momentum $\boldsymbol{\kappa}$, polarization α, and frequency $\omega_{\kappa,\alpha}$. The first term on the right-hand side of (4.90) is two times the MSD (which is independent of the central atom neighbor distance). The second term ensures that only the out-of-phase thermal motion of the atoms along $\boldsymbol{b}_j$ determines the decrease in extended X-ray absorption fine-structure. For a monoatomic cubic crystal, $(\boldsymbol{s}_{\kappa,\alpha} \cdot \boldsymbol{s}_j)^2$ can be replaced with (1/3). Then, by use of (4.56), the Debye approximation for the DOS (for monatomic, cubic crystals, $V\int_0^{\omega_{\rm D}} D_p(\omega)\mathrm{d}\omega = 3nV$ and $\omega = u_{p,{\rm A}}\kappa$), we have

$$\langle|\boldsymbol{\Delta}_j|^2\rangle = \frac{3\hbar}{m\omega_{\rm D}}[\frac{1}{4}+(\frac{T}{T_{\rm D}})^2 g_1] - \frac{3\hbar}{m\omega_{\rm D}}\{\frac{1-\cos(\kappa_{\rm D}b_j)}{2(\kappa_{\rm D}b_j)^2}+$$
$$(\frac{T}{T_{\rm D}})^2[g_1 - \frac{1}{3!}(\kappa_{\rm D}b_j\frac{T}{T_{\rm D}})^2 g_3 + \frac{1}{5!}(\kappa_{\rm D}b_j\frac{T}{T_{\rm D}})^4 g_5 - ...]\}$$

$$\text{Beni–Platzman–Debye MSD model,} \tag{4.91}$$

where

$$g_n = \int_0^{T_{\rm D}/T}\frac{x^n}{e^x-1}\mathrm{d}x, \text{ and } \omega_{\rm D} = \frac{T_{\rm D}k_{\rm B}}{\hbar}. \tag{4.92}$$

In this expression, $\omega_{\rm D}$, $T_{\rm D}$, and $\kappa_{\rm D}$ are the Debye frequency, temperature, and wave vector, respectively. The first term on the right-hand side is the MSD and the second term is the DCF. The MSRD is twice that given by (4.91). The leading term, MSD, was given by (2.57). Calculations show that the DCF has a relatively large contribution to the final value (temperature dependent, ranges from 15% to 40% of MSD for FCC and BCC crystals with respect to the first term).

Examination of (4.91) shows that $\langle|\boldsymbol{\Delta}_j|^2\rangle$ increases as T (e.g., Figure 2.14) and for high melting crystals with small $T_{\rm D}$, this can lead to relatively large displacements at high temperatures. This relation can also be used for an isotropic crystal, for which the corresponding directional Debye temperatures are known. Note that

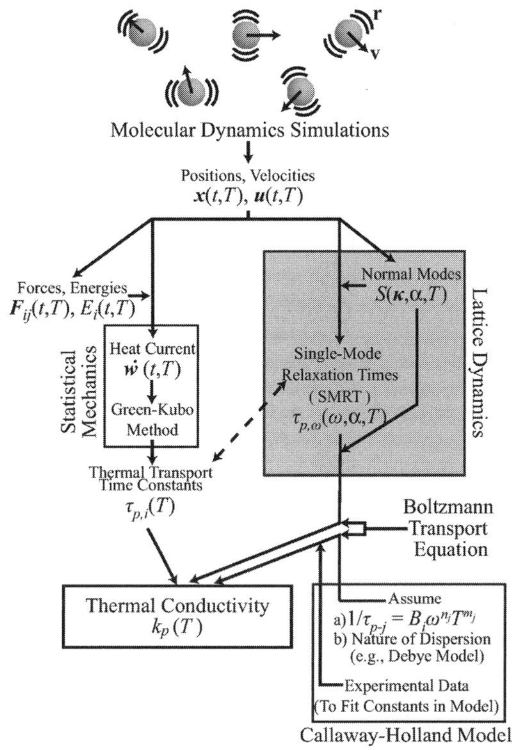

Figure 4.14. The methods of prediction of phonon thermal conductivity, MD and BTE (Callaway–Holland) [233]. The relaxation time τ_{p-j} has a frequency dependence ω^{n_j} and temperature dependence T^{m_j}.

this model is based on harmonic displacement of atoms and is expected to be valid for low temperatures.

4.9 Phonon BTE and Callaway Conductivity Model

4.9.1 Single-Mode Relaxation Time

The single-mode relaxation time (SMRT) for a mode in a phonon system describes the temporal response of the system when that mode is excited and all other modes have their equilibrium populations. This is used in the BTE derivation of the phonon conductivity, as outlined in Figure 4.14. SMRT can be seen as being consistent with an exponential decay of the energy autocorrelations (Table 3.9) using the Green–Kubo theory for thermal conductivity (Section 3.6 and will also be discussed in Sections 4.12-4.13). The limitation of this approach is that the underlying assumption (one mode activated and others at equilibrium) is never realized. This is because the natural decay of the energy in a mode occurs in the presence of other modes that are excited or diminished.

In SMRT approximation, each phonon mode ω has a single, effective phonon relaxation time $\tau_p = \tau_p(\omega_p, \alpha, T)$ representing the various scattering mechanisms.

4.9.2 Callaway Phonon Conductivity Model from BTE

The derivation presented in this section is based on that given by Callaway [53] and Holland [151]. The method presented can be applied to either a quantum or classical system.

The heat flux generated by a change in the population of energy carriers can be expressed as the general form of (3.22), i.e.,

$$\boldsymbol{q} = \sum_{\alpha} \frac{1}{C\pi^3} \int [E_p(\boldsymbol{\kappa}) - \mu]\boldsymbol{u}_p(\boldsymbol{\kappa}) f_p'(\boldsymbol{\kappa}) \mathrm{d}\boldsymbol{\kappa} \quad \text{energy flux vector,} \qquad (4.93)$$

where C is eight for phonons and four for electrons (because of spin degeneracy), the integral is over the first Brillouin zone, E_p is the carrier energy, μ is its chemical potential, $\boldsymbol{u}_p$ is the carrier speed, and f_p' is the deviation of the mode population from the equilibrium distribution. For phonons, the sum is over the mode polarizations (three for a monatomic unit cell), the chemical potential is zero, and the carrier speed is the phonon group velocity. A form of (4.93) is sought that will allow for direct comparison with the Fourier law ($\boldsymbol{q} = -\boldsymbol{K}_p \cdot \nabla T$) and thus generate an expression for the thermal conductivity.

The steady-state BTE (with no external force and source) for a single-phonon mode (Table 3.1) is

$$\boldsymbol{u}_p \cdot \nabla f_p = \frac{\partial f_p}{\partial t}|_s \quad \text{phonon BTE.} \qquad (4.94)$$

If the phonon population is only a function of temperature, then

$$\boldsymbol{u}_p \cdot \nabla f_p = \boldsymbol{u}_p \cdot \frac{\partial f_p}{\partial T} \nabla T = \boldsymbol{u}_p \cdot \nabla T \frac{\partial f_p}{\partial T} \quad \text{introducing temperature gradient.} \qquad (4.95)$$

Let $f_p = f_p^\circ + f_p'$, where f_p° is the equilibrium phonon distribution. Assuming that the deviations from equilibrium f_p' are independent of temperature, we have

$$\frac{\partial f_p}{\partial T} \simeq \frac{\partial f_p^\circ}{\partial T}. \qquad (4.96)$$

The relaxation-time approximation (3.13) is made for the collision term, whereby

$$\frac{\partial f_p}{\partial t}|_s = \frac{f_p^\circ - f_p}{\tau_p} = -\frac{f_p'}{\tau_p} \quad \text{SMRT approximation.} \qquad (4.97)$$

The relaxation time τ_p describes the temporal response of the system when that particular phonon mode is activated (footnote of Section 3.1.4). Under these assumptions, (4.94) becomes

$$f_p' = -\tau_p \frac{\partial f_p^\circ}{\partial T} \boldsymbol{u}_p \cdot \nabla T. \qquad (4.98)$$

Inserting this expression into (4.93) leads to

$$\boldsymbol{q} = -(\frac{1}{8\pi^3}\sum_\alpha \int E_p \tau_p \frac{\partial f_p^\circ}{\partial T}\boldsymbol{u}_p\boldsymbol{u}_p \mathrm{d}\boldsymbol{\kappa}) \cdot \nabla T. \qquad (4.99)$$

Taking E_p (per unit volume) to be independent of temperature, and defined such that

$$\int E_p f_p^\circ \mathrm{d}\boldsymbol{\kappa} = \langle E_p \rangle, \qquad (4.100)$$

where $\langle E_p \rangle$ is the total system energy, the mode specific heat capacity will be given by (4.53) as[†]

$$c_{v,p} = \frac{\partial (E_p f_p^\circ)}{\partial T}|_V = E_p \frac{\partial f_p^\circ}{\partial T}. \qquad (4.101)$$

Thus, for the heat flux

$$\boldsymbol{q} = -(\frac{1}{8\pi^3}\sum_\alpha \int c_{v,p}\tau_p \boldsymbol{u}_p\boldsymbol{u}_p \mathrm{d}\boldsymbol{\kappa}) \cdot \nabla T, \qquad (4.102)$$

where we will use the $c_{v,p}$ given by (4.64). Comparing this result with the Fourier law (Table 1.1, in tensor form) $\boldsymbol{q} = -\boldsymbol{K}_p \cdot \nabla T$ leads to an expression for the phonon thermal conductivity tensor,

$$\boldsymbol{K}_p = \frac{1}{8\pi^3}\sum_\alpha \int c_{v,p}\tau_p \boldsymbol{u}_p\boldsymbol{u}_p \mathrm{d}\boldsymbol{\kappa} \quad \text{phonon conductivity tensor}, \qquad (4.103)$$

where the right hand side is a tensor as a result of the diadic product of $\boldsymbol{u}_p$ with itself. Noting that [same as (4.56), but not summed over modes] the atomic density is given by

$$\int \mathrm{d}\boldsymbol{\kappa} = 8\pi^3 n \quad \text{atomic density}, \qquad (4.104)$$

over the volume of the first Brillouin zone. The expression for the thermal conductivity, if discretized, reduces to an expression similar to that found for gases (Section 6.8.2), $k_p = \sum_i n_i c_{v,p,i} u_{p,i} \lambda_{p,i}/3$, where $c_{v,p,i}$ is the heat capacity per carrier, following from the assumption of a SMRT.

To obtain the thermal conductivity in a direction with unit vector $\boldsymbol{s}$, the component of the group velocity in the desired direction is taken so that

$$k_{p,s} = \frac{1}{8\pi^3}\sum_\alpha \int c_{v,p}\tau_p (\boldsymbol{u}_{p,g}\cdot \boldsymbol{s})^2 \mathrm{d}\boldsymbol{\kappa} = \frac{1}{8\pi^3}\sum_\alpha \int c_{v,p}\tau_p u_{p,g}^2 \cos^2\theta \, \mathrm{d}\boldsymbol{\kappa}, \qquad (4.105)$$

where θ is the angle between $\boldsymbol{u}_{p,g}$ and $\boldsymbol{s}$. It is convenient in some cases to assume that the medium is isotropic, so that the integration can be performed over the wave number as opposed to the wave vector. In this case, $d\boldsymbol{\kappa} = 4\pi\kappa^2 \mathrm{d}\kappa$, and the $\cos^2\theta$

[†] For a quantum-harmonic system, $E_p = \hbar\omega$, and the particles follow the Bose–Einstein distribution. For a classical system, the energy is not discretized, and the forms of E_p and f_p° are not obvious. These specifications are not critical, however, as E_p and f_p° fall out in the derivation, as long as they are defined as given in (4.100).

term is replaced with 1/3, its average value over a unit spherical volume $\langle\cos^2\theta\rangle$ (Chapter 7 problem). This leads to

$$k_p = \frac{1}{6\pi^2}\sum_\alpha \int c_{v,p}\tau_p u_{p,g}^2 \kappa^2 \mathrm{d}\kappa. \tag{4.106}$$

Furthermore, it can be convenient to change the variable of integration from the wave vector to the the angular frequency

$$k_p = \frac{1}{6\pi^2}\sum_\alpha \int c_{v,p}\tau_p u_{p,g}^2 \kappa^2 \frac{\mathrm{d}\kappa}{\mathrm{d}\omega}\mathrm{d}\omega = \frac{1}{6\pi^2}\sum_\alpha \int c_{v,p}\tau_p \frac{u_{p,g}}{u_{p,p}^2}\omega^2 \mathrm{d}\omega$$

$$\text{isotropic phonon conductivity frequency integral,} \tag{4.107}$$

where the definitions of the phonon phase and group velocities given by (4.16) and (4.15) ($u_{p,p} = \omega/\kappa$ and $u_{p,g} = \mathrm{d}\omega/\mathrm{d}\kappa$,) have been used. Note that $\rho c_{v,p} = nc_{v,p}$, where n is the number of atoms per unit volume, and $c_{v,p}$ on the right-hand side is the heat capacity per atom as given by (4.64). One must take care when using this form of the thermal conductivity in cases in which the phonon frequency is not a monotonically increasing function of the wave number e.g., one of the transverse branches in the [110] direction in the FCC crystal, as shown in Figure 4.7(a). In such cases, the integral must be broken into appropriate parts.

Callaway [53] uses a single-phonon speed $u_{p,g}$ and the Debye specific heat capacity (4.64) (per atom), and for the three modes being treated the same, his expression for the phonon thermal conductivity is (end-of-chapter problem)

$$k_p = (48\pi^2)^{1/3}\frac{1}{a}\frac{k_\mathrm{B}^3}{\hbar_\mathrm{P}^2}\frac{T^3}{T_\mathrm{D}}\int_0^{T_\mathrm{D}/T} \tau_p(x)\frac{x^4 e^x}{(e^x-1)^2}\mathrm{d}x \quad \text{Callaway phonon conductivity,} \tag{4.108}$$

where a is the lattice constant $a = n^{-1/3}$ for a cubic lattice, and n is the atomic number density. Note that from (4.55) we have used $k_\mathrm{B}x^2e^x/(e^x-1)^2$ for $c_{v,p}$ in (4.107), since $c_{v,p}$ in (4.107) is per phonon.

Using the concept of the phonon mean free path (1.13) (derivation in Section 6.8.2), the phonon thermal conductivity is written as (note again that $c_{v,p}$ is in J/kg-K)

$$\begin{aligned} k_p &= \frac{1}{3}\rho\sum_\alpha c_{v,p,\alpha}u_{p,\alpha}\lambda_{p,\alpha} \\ &= \frac{1}{3}\rho\sum_\alpha c_{v,p,\alpha}u_{p,\alpha}^2\tau_{p,\alpha}, \end{aligned} \tag{4.109}$$

where the sum is over $3N$ phonon normal modes for an N atom system and $\lambda_{p,\alpha}$ is the modal mean free path.

In following a similar procedure, Callaway [53] developed a more comprehensive form for the thermal conductivity based on a refined expression for the collision term, the use of different relaxation times for the normal and resistive processes was considered and will be discussed in Sections 4.9.3(E) and 4.15 and is shown in Figure 4.37. (This is an end-of-chapter problem.) Equation (4.107) is the general form of these results.

4.9.3 Callaway–Holland Phonon Conductivity Model

Callaway single-mode relaxation-time model (4.109) successfully predicts (when using multiple fitting parameters) the low-temperature thermal conductivity using the Debye approximation (no phonon dispersion) and that there is no phonon dispersion and that the longitudinal and transverse polarizations behave identically. Holland [151] extended the work of Callaway by separating the contributions of LA and TA phonons, including some phonon dispersion, and using different forms of the relaxation times. Better high-temperature agreement is then found. This Callaway–Holland model has been refined to include further detail on the phonon dispersion and relaxation times [316, 328, 8]. The added complexities lead to more fitted parameters. One could argue that the resulting better fits with the experimental data are because of this increase in the number of fitted parameters and not to an improvement of the actual physical model.

The thermal conductivity of an isotropic crystal, with negligible contributions from optical phonons, is given by (4.107) written in Callaway–Holland form as

$$k_p = = k_{p,\mathrm{L}} + k_{p,\mathrm{T}} = \frac{1}{6\pi^2}[\int_0^{\omega_{m\mathrm{L}}} c_{v,p}\frac{u_{p,g,\mathrm{L}}}{u_{p,p,\mathrm{L}}^2}\tau_{p,\mathrm{L}}\omega_\mathrm{L}^2 \mathrm{d}\omega_\mathrm{L}$$

$$+2(\int_0^{\omega_1} c_{v,p}\frac{u_{p,g,\mathrm{T}}}{u_{p,p,\mathrm{T}}^2}\tau_{p,\mathrm{T1}}\omega_\mathrm{T}^2 \mathrm{d}\omega_\mathrm{T} + \int_{\omega_1}^{\omega_{m\mathrm{T}}} c_{v,p}\frac{u_{p,g,\mathrm{T}}}{u_{p,p,\mathrm{T}}^2}\tau_{p,\mathrm{T2}}\omega_\mathrm{T}^2 \mathrm{d}\omega_\mathrm{T})]$$

Callaway–Holland thermal conductivity model. (4.110)

The first term corresponds to longitudinal phonons, and the second and third terms to transverse phonons (two degenerate branches). Here, $c_{v,p}$ is the quantum-harmonic specific heat per normal mode and τ_p is a SMRT. The forms of τ_p are given in [151], and are discussed and summarized in Section 4.9.4. The upper limits of the first and third integrals are the angular frequencies of the phonon branches at the edge of the first Brillouin zone (marked by X in Figure 4.3), signified as $\omega_{m\mathrm{L}}$ and $\omega_{m\mathrm{T}}$. For the transverse phonons, the frequency ω_1 is that at the center of the first Brillouin zone. Also, two different relaxation times are used for the transverse wave in the first and second portions of the Brillouin zone, $\tau_{p,\mathrm{T1}}$ and $\tau_{p,\mathrm{T2}}$.

4.9.4 Relaxation-Time Models

The approximation of adding resistivities to transport in series [Matthiessen rule, (3.53)] is used for combining the contribution from various scattering mechanisms†.

This leads to

$$\frac{1}{\tau_p} = \sum_j \frac{1}{\tau_{p,j}} \quad \text{Matthiessen rule (resistivities in series),} \qquad (4.111)$$

† Matthiessen rule is based on electric resistivity, where contributions of various scatterings are assumed to be additive and the resistivity is proportional to the inverse of the relaxation times, as given by (5.146).

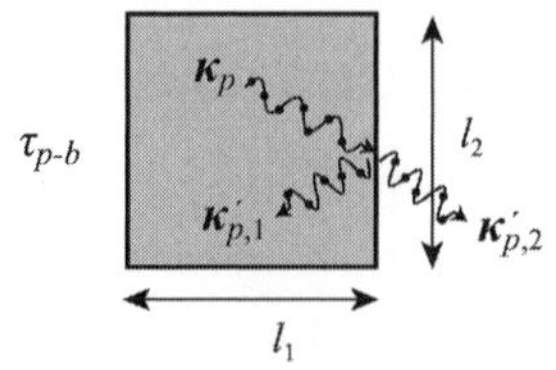

(a) Phonon-Crystalline (i.e., Grain or Casimir) Boundary Scattering

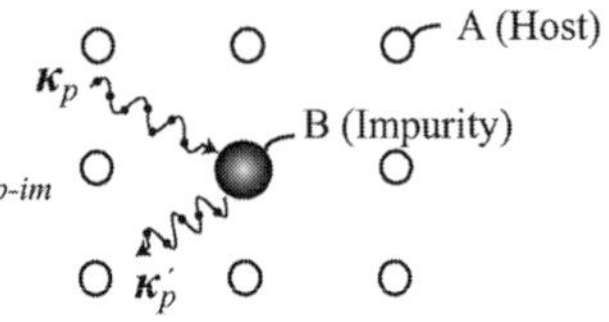

(b) Phonon-Impurity (B) Scattering

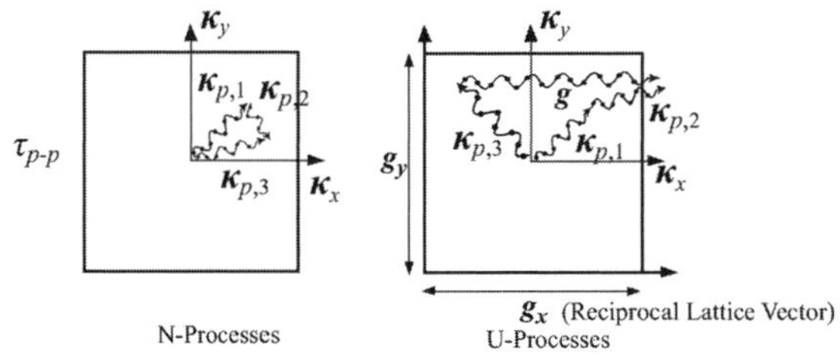

Momentum Conservation In Interphonon Scattering

(c) Three-Phonon Scattering

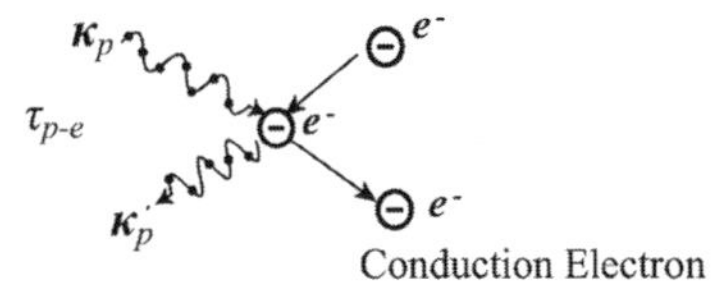

(d) Phonon-Electron Scattering

Figure 4.15. Phonon scattering by (a) crystalline (grain) boundary, (b) impurity, (c) other phonons (N–Processes, in which phonon momentum is conserved, and U–processes, in which the momentum balance allows for flipping into the adjacent Brillouin zone), and (d) electron.

where τ_p is the effective phonon relaxation time and $\tau_{p\text{-}j}$ is phonon scattering that is because of the j scattering mechanism.

Figure 4.15 shows phonon scattering by (a) crystalline (e.g. grain) boundary, (b) impurity, (c) other phonons (normal, or N–Processes, and Umklapp, or U–processes, will be discussed in Section 4.15), and (d) electrons.

Some of the common phonon relaxation-time models are listed in Table 4.2. These involve a frequency-dependent term, a temperature-dependent term, and some empirical constants. Typical constants used in these models are given in Tables 4.2 and 4.3. These constants are typically found by fitting to experimental thermal conductivity results.

Table 4.2. *Some phonon scattering relaxation times [151].*

Mechanism	Relation	Comment
Phonon–crystalline boundaries (e.g., grain)	$\tau_{p\text{-}b}^{-1} = u_{p,\mathrm{A}}/L,\ L = \frac{2}{\pi^{1/2}}(l_1 l_2)^{1/2}$	l_1 and l_2 are crystal linear dimensions
Phonon–displacement	$\tau_{p\text{-}d} = \frac{\pi u_{p,\mathrm{A}}^3}{2nV_c^2}\frac{1}{a_{p\text{-}d}^2\gamma_G^2}(\frac{R}{\Delta R})^2\omega_p^{-4}$	R is radius of impurity atom, ΔR difference is radii of host and impurity, γ_G is Grüneisen constant, n is number density of impurity, V_c is unit-cell volume, and $a_{p\text{-}d}$ is impurity dependent
Phonon–impurity	$\tau_{p\text{-}im} = \frac{4\pi u_{p,\mathrm{A}}^3}{V_c \sum_i x_i(1-\frac{M_i}{M})^2}\omega_p^{-4}$	$V_c = Fa^3$ ($F = 1$ for cubic structures) is unit-cell volume, x_i is mass fraction of impurity atom i
Phonon–isotope impurity	$\tau_{p\text{-}is} = \frac{4\pi u_{p,\mathrm{A}}^3}{a_{p\text{-}is}V_{is}}\omega_p^{-4}$, $3u_{p,\mathrm{A}}^{-1} = (\frac{1}{u_{p,\mathrm{L}}} + \frac{2}{u_{p,\mathrm{T}}})^{-1}$	$a_{p\text{-}is} = 2.1 \times 10^{-4}$ (for Si), V_{is} is isotope atom volume
Three phonons		
N-Process		
Longitudinal	$\tau_{p\text{-}p,\mathrm{L,N}}^{-1} = B_\mathrm{L}\omega_p^2 T^3$	low T
Transverse	$\tau_{p\text{-}p,\mathrm{T,N}}^{-1} = B_\mathrm{T}\omega_p T^4$	low T
Longitudinal	$\tau_{p\text{-}p,\mathrm{L,N}}^{-1} = B'_\mathrm{L}\omega_p^2 T$	high T
Transverse	$\tau_{p\text{-}p,\mathrm{T,N}}^{-1} = B'_\mathrm{T}\omega_p T$	high T
U-Process		
Dispersive	(i) $\tau_{p\text{-}p,\mathrm{U}}^{-1} = B_\mathrm{U}\omega_p^2 T^3 \exp(-\theta/\alpha T)$	Klemens model
Transverse	(ii) $\tau_{p\text{-}p,\mathrm{U}}^{-1} = B_\mathrm{U}\omega_p T^3 \exp(-\theta/\alpha T)$	Klemens model
	(iii) $\tau_{p\text{-}p,\mathrm{TU}}^{-1} = B_\mathrm{TU}\omega_p^2/\sinh x$, $\omega_{p,1} \le \omega_p \le \omega_{p,2}$, $x = \hbar\omega_p/k_\mathrm{B}T$; $= 0,\ \omega_p < \omega_{p,1}$	Callaway model
	(iv) $\tau_{p\text{-}p,\mathrm{U}}^{-1} = B_\mathrm{U}\omega_p^2 T^3$	
	(v) $\tau_{p\text{-}p,\mathrm{U}}^{-1} = B_\mathrm{U}\omega_p^2 T$	high T
Phonon–electron [345]	$\tau_{p\text{-}e}^{-1} = \frac{n_{e,c}^2 e_c^2 u_{p,\mathrm{A}}^2}{n_c k_\mathrm{B} T \sigma_e}\frac{1}{\alpha_1}(\frac{T_\mathrm{D}}{T})^{3/2} = \frac{\sigma_e u_{p,\mathrm{A}}^2}{\mu_e^2 n_c k_\mathrm{B} T\alpha_1}(\frac{T_\mathrm{D}}{T})^{3/2}$, $\alpha_1 = 4^3(\frac{\pi V_a}{6})^{1/2}(\frac{m_{e,e}u_{p,\mathrm{A}}}{h_\mathrm{P}})^{3/2}$	V_a is volume per atom, n_c is number of unit cells per volume, σ_e and μ_e are the electron conductivity and mobility

Table 4.3. *Parameters used in phonon relaxation-time models, for Si and Ge [151].*

Parameters	Si	Ge
$u_{p,\mathrm{T}}$	5.86×10^3, m/s	3.55×10^3, m/s
$u_{p,\mathrm{L}}$	8.48×10^3	4.92×10^3
$u_{p,p}$	6.4×10^3	3.9×10^3
$u_{p,\mathrm{TU}}(\omega > \omega_1)$	2.0×10^3	1.3×10^3
$u_{p,\mathrm{TU}}(\omega > \omega_1)$	4.24×10^3	2.46×10^3
F	~ 0.8	~ 0.8
T_T	180, K	101, K
T_TU	210	118
T_L	570	333
T_4	350	192
T_D	658	376
Γ	2.16×10^{-4}	5.72×10^{-4}
α	2	2
B_T	9.3×10^{-13} 1/K^3	1.0×10^{-11} 1/K^3
B_TU	5.5×10^{-18} s	5.0×10^{-13} s
B_L	2.0×10^{-24} s/K^3	6.9×10^{-24} s/K^3

(A) *Grain Boundary Scattering*

The crystal boundary scattering (crystal size or grain size effect) $\tau_{p\text{-}b}$ is based on diffuse boundary absorption/emission. It is referred to as Casimir boundary scattering and gives a mean free path for a phonon equal to the Casimir length L, which is the length of travel of the phonon before the boundary absorption/reemission. For a rectangular path, l_1 and l_2, it is given as $L = (2/\pi^{1/2})(l_1 l_2)^{1/2}$ [151]. Then with a single phonon speed $u_{p,p} = u_{p,g} = u_{p,\mathrm{A}}$, the relaxation time is

$$\tau_{p\text{-}b}^{-1} = u_{p,\mathrm{A}}/L \quad \text{phonon grain boundary (Casimir) scattering.} \qquad (4.112)$$

In microparticles and nanoparticles (crystalline), this boundary scattering dominates at very low temperatures (below peak in k_p).

(B) *Impurity Scattering*

Impurity (static imperfection) scattering is similar to the Rayleigh scattering of the transverse electromagnetic waves, which is proportional to ω_p^4.

The Klemens [185] derivation of scattering of low-frequency lattice waves by stationary imperfections is based on the quantum treatment of phonon. The general theory of transition probabilities is discussed by Ziman [359] and leads to the FGR. Starting from (4.47), the FGR (3.27) written for this transition rate (for elastic scattering, $\omega = \omega'$) is [185]

$$\frac{1}{\tau_{p\text{-}im}} = a^3 F \int 2|M_{\kappa,\kappa'}|^2 \frac{1}{M^2\omega^2}\frac{\pi}{(2\pi)^3}\frac{\mathrm{d}\kappa'}{\mathrm{d}\omega'}\mathrm{d}\boldsymbol{\kappa}'$$

$$|M_{\kappa',\kappa}|^2 = x_i(1-\frac{M_i}{M})^2\frac{1}{6F}\omega^2\omega'^2, \tag{4.113}$$

where $Fa^3 = V_c$ is the volume of crystal primitive cell and F allows for noncubic structures ($F = 1$ for cubic structures). Here M_i is the molecular weight of impurity species B in host A and x_i is its atomic fraction.

Based on this, Klemens [184] arrived at the phonon mean free path

$$\lambda_{p\text{-}im} = \frac{Fa}{\sum_i x_i(1-\frac{M_i}{M})^2}\frac{1}{(a\kappa)^4}, \tag{4.114}$$

where $a\kappa = 2\pi a/\lambda$ is also called the size parameter, comparing the phonon wavelength with the interatomic spacing.† Then with $\lambda_{p\text{-}im} = u_{p,g}\tau_{p\text{-}im} = u_{p,\mathrm{A}}\tau_{p\text{-}im}$, this is written as

$$\tau_{p\text{-}im} = \frac{4\pi u_{p,\mathrm{A}}^3}{V_c\sum_i x_i(1-\frac{M_i}{M})^2}\omega^{-4} \quad \text{phonon impurity (static imperfection) scattering.} \tag{4.115}$$

The form given in Table 4.2 also makes no distinction between the group and phase velocities, but it can be modified accordingly (Section 4.9.6).

(C) *Phonon–Phonon Scattering*

Interphonon scattering $\tau_{p\text{-}p}^{-1}$ shows a ω^1 and a ω^2, as ω increases. Also, unlike other mechanisms, interphonon scattering is highly temperature dependent. The MD/G-K prediction of $\tau_{p\text{-}p}$ for FCC Ar is reported in [233] and is in general agreement with these trends.

In general, the interphonon-scattering relaxation time is represented as

$$\frac{1}{\tau_{p\text{-}p}} = B\omega_p^n f(T), \tag{4.116}$$

where n is a integer (1, 2, or 3) and $f(T)$ describes the temperature dependence. The temperature dependence is generally of the form T^m, although, as listed in Table 4.2, for three-phonon scattering, more complex relations are used.

To be physically meaningful, the mean free path λ_p of a phonon should be longer than one-half of its wavelength, λ. Noting that $\lambda_p = u_{p,g}\tau_p$, $\lambda = 2\pi/\kappa$, and

† Here the impurity i has a molecular weight M_i and mass fraction x_i. The lattice constant is a, and the host molecular weight is M.

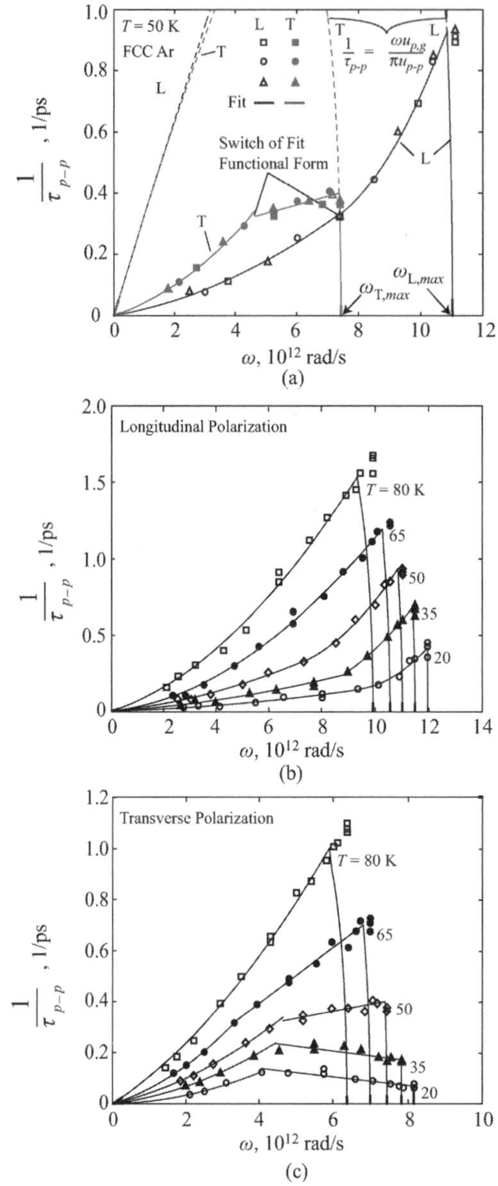

Figure 4.16. (a) Discrete phonon–phonon relaxation times and continuous curve fits $\tau_{p\text{-}p}$ at $T = 50$ K for FCC Ar. Also shown is the minimum physical value of the relaxation time, $\pi u_{p,p}/\omega u_{p,g}$. (b) MD/G-K results and continuous relaxation-time curve fits for the longitudinal polarization at all temperatures considered. (c) MD/G-K results and continuous relaxation-time curve fits for the transverse polarization at all temperatures considered [233].

using the definition of $u_{p,p}$, this limit can alternatively be stated as

$$\frac{1}{\tau_{p\text{-}p}} \leq \frac{\omega u_{p,g}}{\pi u_{p,p}} \quad \text{limit on phonon–phonon-scattering relaxation time,} \qquad (4.117)$$

and is also shown in Figure 4.16(a) for L–J FCC Ar. At a temperature of 50 K, the phonons at the edge of the first Brillouin zone ($\kappa^* = 1$) are outside of the allowed range for both polarizations. As the temperature increases, more of the phonon modes do not satisfy (4.117). At the highest temperature, 80 K, the transition occurs at κ^* values of 0.77 and 0.81 for the longitudinal and transverse polarizations, respectively.

The data for each polarization can be broken down into three distinct regions. The first two are fit with low-order polynomials. For the longitudinal polarization, the first region is fit with a second-order polynomial through the origin, and the second region with a second-order polynomial. For the transverse polarization, the first region is fit with a second-order polynomial through the origin, and the second region with a linear function. The resulting functions are also shown in Figure 4.16(a) and are considered satisfactory fits to the L–J FCC MD/G–K results for k_p. As the temperature increases, the behavior in the two regions becomes similar. For both polarizations at a temperature of 80 K, and for the longitudinal polarization at a temperature of 65 K, a single second order polynomial through the origin is used to fit the data. In the third region, the continuous relaxation-time functions are taken up to the maximum frequency ($\omega_{\mathrm{L},max}$ or $\omega_{\mathrm{T},max}$) by use of (4.117).

The raw data and continuous relaxation-time functions for all temperatures considered are shown in Figures 4.16(b) and (c). The parts of the relaxation-time curves are not forced to be continuous. For both the longitudinal and transverse polarizations, any resulting discontinuities are small and are purely a numerical effect.

The relaxation time functions do not contain the orders-of-magnitude discontinuities found in the Holland relaxation times (Table 4.2), which result from the assumed forms of the relaxation times and how the fitting parameters are determined.

Theoretical calculations predict that in the κ^* range of 0 to 0.2, the longitudinal and transverse curves should follow ω^2 and ω dependencies, respectively [144, 262]. This is not found in the relaxation times predicted by the MD/G–K simulations. The second-order fit found at the high temperatures is consistent with the high-temperature prediction of [318]. Of particular note is the turning over of the low-temperature transverse curves at higher frequencies. In general, it is clear that the extension of the low-frequency behavior to the entire frequency range, as is sometimes done, is not generally suitable. The effect of such an assumption on the thermal conductivity prediction will be considered.

As will be discussed in Section 4.13, the phonon–phonon scattering is divided into N (for normal) and U (for Umklapp) processes and the U–processes are dominant at high temperature. The derivation of the N- and U–Processes are based on quantum mechanics (FGR) [318], and some observations on the interphonon interaction rules based on MD results are presented in [233].

(D) *Electron Scattering*

The phonon–electron relaxation time is found from a momentum balance analysis based on maximum phonon frequency [344] as

$$\frac{1}{\tau_{p\text{-}e}} = \frac{3u_{p,\mathrm{A}}^2}{\mu_e^2}\frac{\sigma_e}{c_{v,p}T} \quad \text{phonon–electron scattering,} \tag{4.118}$$

where σ_e and μ_e are the electrical conductivity and mobility, and $c_{v,p}$ is the phonon specific heat of the phonons allowed to interact with conduction-band electrons. Using the Debye model for $c_{v,p}$ given in Section 4.7, an approximation for (4.118), in terms of the Debye temperature, is given in Table 4.2 [345].

(E) *Displacement Scattering*

The impurities displace atoms and depending on the size mismatch in this replacement ΔR (difference in radius of host and impurity). Then they cause phonon scattering with $\tau_{p,d}$ proportional to ΔR^{-2}, as listed in Table 4.2. This relaxation time is temperature independent, and similar to impurity scattering (4.114), has a ω_p^{-4} dependence.

(F) *Normal and Resistive Relaxation Times*

Callaway [53] divided the various scattering mechanisms into normal $\tau_{p\text{-}p,\mathrm{N}}$ and resistive $\tau_{p,r}$, i.e.,

$$\frac{1}{\tau_p} = \frac{1}{\tau_{p\text{-}p,\mathrm{N}}} + \frac{1}{\tau_{p,r}}, \tag{4.119}$$

where for N–Processes both energy and momentum of scattered phonons are conserved to be within a reciprocal space vector, whereas for U–Processes and other resistive processes the momentum is not conserved. The resistive scattering is further divided into displacement $\tau_{p\text{-}d}$, impurity $\tau_{p\text{-}im}$, Umklapp $\tau_{p\text{-}p,\mathrm{U}}$, crystal boundary $\tau_{p\text{-}b}$, and phonon–electron $\tau_{p\text{-}e}$, i.e.,

$$\frac{1}{\tau_{p,r}} = \frac{1}{\tau_{p\text{-}d}} + \frac{1}{\tau_{p\text{-}im}} + \frac{1}{\tau_{p\text{-}p,\mathrm{U}}} + \frac{1}{\tau_{p\text{-}b}} + \frac{1}{\tau_{p\text{-}e}}. \tag{4.120}$$

(G) *Temperature Dependence of Phonon Conductivity*

Figure 4.17, which shows the k_p-T behavior, summarizes the various phonon-scattering mechanisms and their regimes of importance (or dominance). Starting at $T = 0$, there is no atomic motion and $k_p = 0$. As the temperature increases, the initial increase in k_p is a specific heat effect [Figure 4.13(a)]. Then, first the phonon–grain boundary scattering dominates. As the temperature further increases, the impurity, electron, and finally the interphonon U–Processes dominate. The maximum in k_p occurs at a fraction of Debye temperature, around $0.1T_\mathrm{D}$.

The high-temperature behavior is dominated by interphonon scattering and has a T^{-1} behavior (Slack relation). This will be discussed in Section 4.11. There it will be that the interatomic bond length controls the high-temperature behavior of k_p. The noncrystalline (amorphous) solids generally have a thermal conductivity that increases with temperature and is generally the lowest thermal conductivity for a solid phase. We will discuss these in Section 4.10.

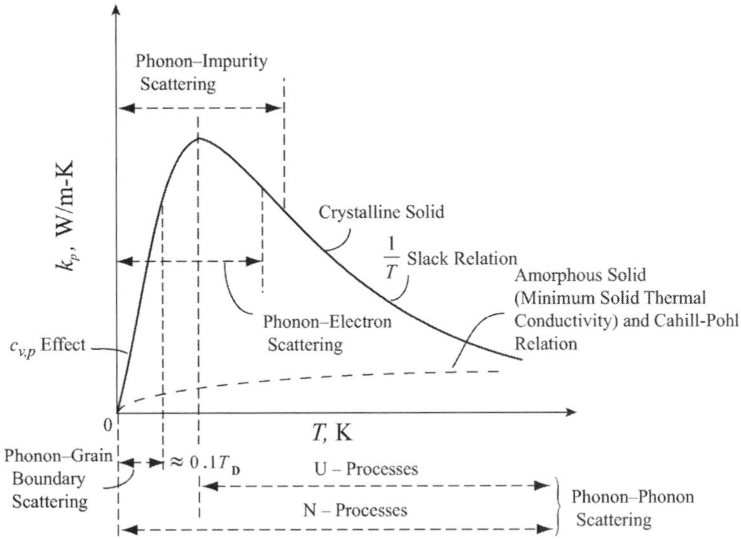

Figure 4.17. Regimes of important (or dominant) phonon-scattering mechanisms in variations of phonon conductivity with respect to temperature [24]. The behavior of an amorphous solid is also shown (the crystalline solid has a higher thermal conductivity).

In the next section, we examine the predicted phonon conductivity of Ge, where phonon-electron scattering is negligible.

4.9.5 Phonon Dispersion Models: Ge As Example

Experimental results for the phonon dispersion of the acoustic modes of Ge in the [100] direction, at a temperature of 80 K, are shown in Figures 4.18(a) and (b) [254]. The horizontal axis is a dimensionless wave number, which is obtained by normalizing the wave number against its value at the edge of the first Brillouin zone, i.e., $\kappa^* = \kappa/\kappa_m = \kappa/(2\pi/a)$. For Ge at a temperature of 80 K, the value of a is 5.651 Å [254]. To allow for the use of (4.110), which assumes isotropic dispersion, this direction is used in the subsequent analysis. The temperature dependence of the dispersion [91] is not considered. The experimental data at a temperature of 80 K are used for all calculations. To assess the importance of accurately modeling the dispersion, five different models are examined.

(A) *Debye Model*

At low frequencies, the phonon frequency is proportional to its wave number, i.e., from (4.17) we have (Figure 4.3)

$$\omega = \kappa u_{p,g,0}, \tag{4.121}$$

where the subscript 0 refers to the low-frequency limit, i.e., $\kappa \to 0$, and $u_{p,g,0}$ is the low-frequency limit of the phonon group velocity. The Debye dispersion model

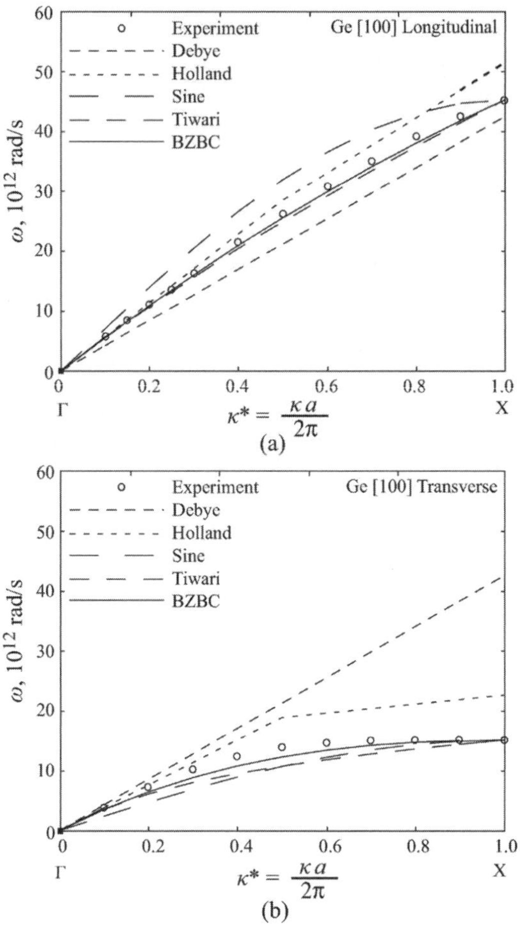

Figure 4.18. Germanium phonon dispersion in the [100] direction. Experimental data [254] and five models used in this study for (a) LA phonons and (b) TA phonons.

uses (4.121) for all frequencies, and does not distinguish between the LA and TA phonons. Note that in this case $u_{p,p} = u_{p,g}$, so that in (4.110), $u_{p,g}/u^2_{p,p}$ becomes $1/u_{p,g}$. We write (4.57) as

$$\frac{1}{u_{p,g,0}} = \frac{1}{3}(\frac{1}{u_{p,g,\mathrm{L},0}} + \frac{2}{u_{p,g,\mathrm{T},0}}). \tag{4.122}$$

With 5142 m/s and 3391 m/s used for $u_{p,g,\mathrm{L},0}$ and $u_{p,g,\mathrm{T},0}$ [254], $u_{p,g,0} = 3825$ m/s for Ge.

(B) *Holland Model*

Holland [151] separated the contributions of LA and TA phonons, and included a partial effect of phonon dispersion by splitting each branch into two linear segments. The change in the slope (and thus the phonon velocities) is assumed to occur at a κ^* value of 0.5. As taken from the experimental data of [254], the phonon velocities

$u_{p,g,L,0}$, $u_{p,g,T,0}$, $u_{p,g,L,0.5}$, and $u_{p,g,T,0.5}$ are 5142, 3391, 4152, and 678 m/s, where 0.5 refers to the segment between κ^* values of 0.5 and 1. Note that the change in slope implies that $u_{p,g}$ is not equal to $u_{p,p}$ in the second region. This effect was neglected by Holland, but is included in the current study.

(C) *Sine Function Model*

In the sine function model, the phonon dispersion relation for each polarization is approximated by that of a linear monatomic chain as in [183] [a special case of (4.7), end-of-chapter problem]

$$\omega_i = \omega_{mi} \sin(\frac{\pi \kappa_i^*}{2}), \tag{4.123}$$

where the label i can be L or T (this notation holds for the rest of the dispersion models). The critical drawback of this model is the nonaccurate behavior at low frequencies ($\kappa \rightarrow 0$) where the asymptotic group velocities (4.15),

$$u_{p,g,i,0} = \frac{\partial \omega_i}{\partial \kappa_i}|_{\kappa_i \rightarrow 0} = \frac{\pi \omega_{mi}}{2\kappa_{mi}}, \tag{4.124}$$

differ from the experimental results. For $u_{p,g,\mathrm{L},0}$ and $u_{p,g,\mathrm{T},0}$, we obtain 6400 and 2130 m/s, compared with the experimental values of 5142 and 3391 m/s.

(D) *Tiwari Model*

In the Tiwari model [328], the dispersion is assumed to be of the form

$$\kappa_\mathrm{L} = \frac{\omega_\mathrm{L}}{u_{p,g,\mathrm{L},0}}(1 + \alpha\omega_\mathrm{L}) \text{ and } \kappa_\mathrm{T} = \frac{\omega_\mathrm{T}}{u_{p,g,\mathrm{T},0}}(1 + \beta\omega_\mathrm{T}^2), \tag{4.125}$$

where α and β are constants, given in [328]. Note that these equations satisfy the Brillouin zone boundary conditions (BZBC) (4.17)

$$\kappa_i(\omega_i = 0) = 0 \text{ and } \frac{\partial \kappa_i}{\partial \omega_i}|_{\omega_i=0} = \frac{1}{u_{p,g,i,0}}. \tag{4.126}$$

This model does not show the observed experimental behavior of

$$\frac{\partial \omega_\mathrm{T}}{\partial \kappa_\mathrm{T}}|_{\kappa_\mathrm{T}^* \rightarrow 1} = 0, \tag{4.127}$$

at the edge of the Brillouin zone X, as shown in Figures 4.3 and 4.18(b).

(E) *BZBC Model*

For the dispersion, a model referred to hereafter as the BZBC model has been introduced in [72]. A quadratic wave-number dependence for LA phonons and a cubic wave-number dependence for TA phonons are used. This model is a modification of the Tiwari model, in that the boundary conditions given by (4.126) and (4.127), and $\omega_i(\kappa_{mi}) = \omega_{mi}$, are applied. The resulting dispersion relations are

$$\omega_{\rm L} = u_{p,g,{\rm L},0}\kappa_m\kappa^* + (\omega_{m{\rm L}} - u_{p,g,{\rm L},0}\kappa_m)\kappa^{*2}$$

$$\omega_{\rm T} = u_{p,g,{\rm T},0}\kappa_m\kappa^* + (3\omega_{m{\rm T}} - 2u_{p,g,{\rm T},0}\kappa_m)\kappa^{*2} + (u_{p,g,{\rm T},0}\kappa_m - 2\omega_{m{\rm T}})\kappa^{*3}. \qquad (4.128)$$

4.9.6 Comparison of Dispersion Models

A comparison of the five dispersion models, along with the experimental data for Ge [254], is shown in Figures 4.18(a) and (b). Here we neglect $p-e$ scattering, since it is expected to be small[151]. For the both the longitudinal and transverse polarizations, both the Tiwari and BZBC models match the experimental data reasonably well over most of the first Brillouin zone. The agreement is not as good for the transverse polarization above κ^* values of 0.7 because of the plateau behavior, which is difficult to fit with a low-order polynomial. The other dispersion curves are unsatisfactory. Of the Tiwari and BZBC models, the BZBC curve gives the best agreement with the experimental data. We note than none of these dispersion relations will be consistent with the experimental phonon density of states because of the isotropic assumption. Additionally, the integral of the volumetric density of states will not go to the expected value of $3n$, where n is the volumetric density of unit cells. For this to occur, one would need to set the dispersion with this result in mind. For these calculations, we are more concerned with matching the experimental dispersion data.

Although the Tiwari and BZBC dispersion curves show some similarities, significant differences become apparent when the quantity they affect in the thermal conductivity expression, $u_{p,g}/u_{p,p}^2$, is considered. This is shown in Figures 4.19(a) and (b). The deviation is most significant for the TA phonons near the edge of the Brillouin zone, and will result in an overprediction of the high frequency contribution to the thermal conductivity by the Tiwari model. This deviation occurs because the appropriate boundary condition at the edge of the Brillouin zone (4.127) has not been enforced in the Tiwari model.

The value of the thermal conductivity is sensitive to impurities, which include the isotopic content of an otherwise pure crystal [122]. The impurity relaxation time[186, 312], (4.115), should be written in terms of phonon phase velocity as,

$$\frac{1}{\tau_{p\text{-}im}} = \frac{V_{\rm c}\sum_i x_i[1-(M_i/M)]^2}{4\pi u_{p,g}u_{p,p}^2}\omega^4. \qquad (4.129)$$

At low frequency, where $u_{p,g} = u_{p,p} = u_{p,g,0}$, the relaxation time displays an ω^{-4} dependence (similar to photon, Rayleigh scattering). Because of phonon dispersion, such an assumption will not be valid at higher frequencies and will lead to an overestimation of the thermal conductivity.

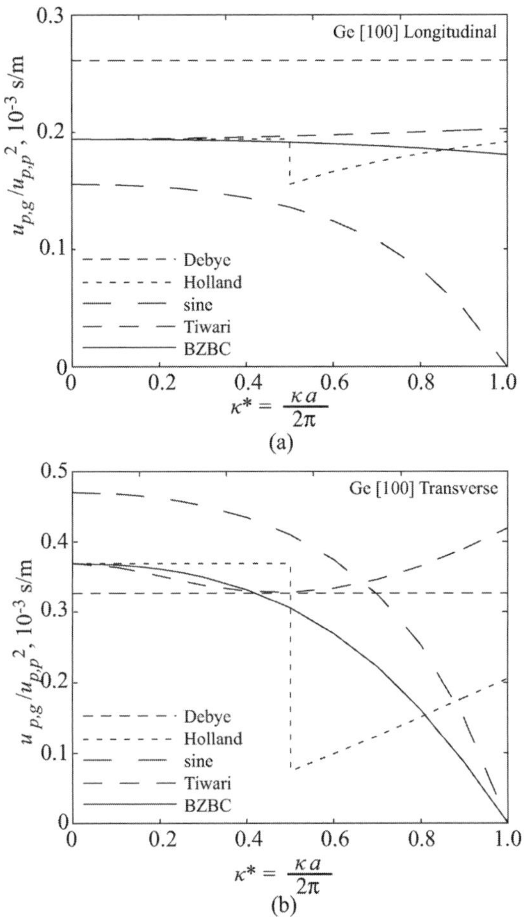

Figure 4.19. $u_{p,g}/u_{p,p}^2$ for the five Ge dispersion models plotted as a function of the normalized wave number for (a) LA phonons and (b) TA phonons [232].

4.9.7 Thermal Conductivity Prediction

(A) *Role of Dispersion Model*

The accuracy of the thermal conductivity model described in Section 4.9.2 depends on the nature of the phonon dispersion and relaxation-time models. To isolate the effects of dispersion, the relaxation-time model used in the subsequent calculations is fixed to that of Holland [151].

In Figures 4.20(a) and (b), the effect of including the difference between the phonon group and phase velocities on the prediction of the thermal conductivity of Ge is shown. The experimental data are taken from Holland [151]. The predicted values in Figure 4.20(a) correspond to (i) the Holland model, that is, no distinction between $u_{p,g}$ and $u_{p,p}$ and a linear two-region treatment, (ii) no distinction between $u_{p,g}$ and $u_{p,p}$, but using the frequency dependence of the group velocity, (iii) different and frequency-dependent $u_{p,g}$ and $u_{p,p}$, and (iv) same as (iii), plus rigorous

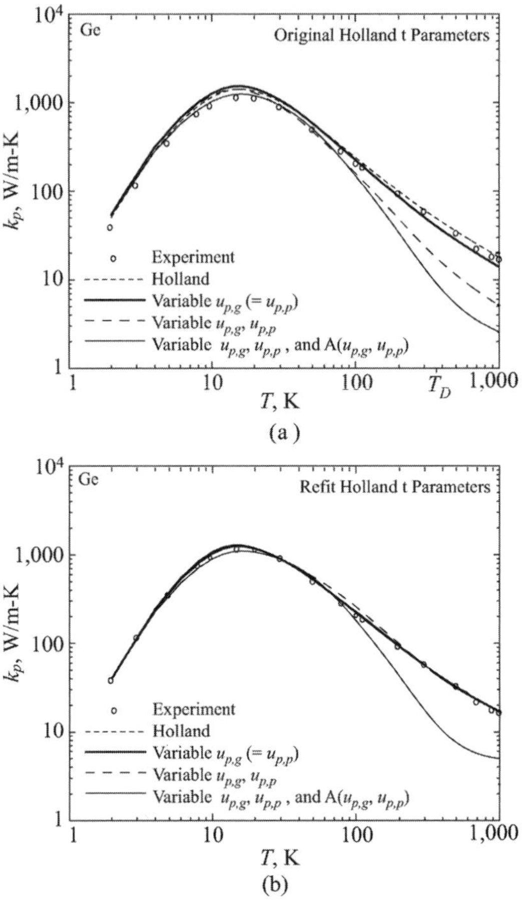

Figure 4.20. Effect of refining the treatment of the dispersion on the prediction of the thermal conductivity of Ge. (a) based on the original Holland fitting parameters, and (b) predictions refit to the experimental data [72].

treatment of the impurity scattering by distinguishing between $u_{p,g}$ and $u_{p,p}$. In (ii)–(iv), the velocities are calculated with the BZBC dispersion model. In (i)–(iii) the impurity scattering is calculated with $u_{p,g} = u_{p,p} = u_{p,g,0}$ for each polarization.

In Figure 4.20(a), the relaxation-time model and fitting parameters originally obtained by Holland are used. In Figure 4.20(b), the curves have been refit to the experimental data. From Figure 4.20(a), it is evident that the fitting parameters are not universal and are strongly dependent on the dispersion model. Thus, to use the values obtained by Holland, one must also use that dispersion model. However, as shown in Figure 4.20(b), by refitting the relaxation time parameters, excellent agreement with the experimental data can be obtained in all cases, except when the impurity scattering is rigorously modeled. In this case, the predicted thermal conductivity becomes lower than the experimental data, especially at high temperatures. Under the Holland dispersion, the plateau in the TA dispersion curve is not properly addressed, and the TA phonons make a significant contribution to the thermal conductivity at high temperatures. Here, with the proper modeling of the TA phonons, this contribution is reduced, and, as such, the role of TA phonons should be reassessed. The effect is not seen at low temperatures, where it is the lower-frequency phonons that dominate the thermal transport.

(B) *Role of Relaxation-Time Model*

The use of a rigorous model for the phonon dispersion has led to an apparent failure of the Holland SMRT approach at high temperatures. The explanation for this must lie in the forms of the relaxation times used. Although exact expressions for the relaxation times can be developed [359], their evaluation is extremely difficult because of the required knowledge of the phonon dispersion and three-phonon interactions. Approximate expressions are generally based on low frequency asymptotes [53, 151] and yet are often applied over the entire temperature and frequency ranges.

Even with the observed disagreement, the importance of modeling the dispersion can still be shown. The Holland relaxation times are plotted in Figure 4.21(a) at temperatures of 80 K for the four cases shown in Figure 4.20(b). The most striking feature of these results is the large discontinuity in the TA relaxation time when κ^* is equal to 0.5, where the functional form is assumed to change [i.e., (4.110)]. As the temperature increases, the size of the discontinuity increases. For example, at a temperature of 900 K, the discontinuity covers four orders of magnitude. The size of the discontinuity decreases as the treatment of the dispersion is refined. This suggests that the observed behavior is more realistic. Such discontinuities are also found in other relaxation-time models [316, 328]. In theory, one would expect the relaxation-time curves to be continuous, and this has been found in MD simulations of the L–J FCC crystal [235].

The effect of the fitting parameters in the relaxation-time model can also be demonstrated by plotting the cumulative frequency dependence of the thermal conductivity. This is shown for the same four cases as Figures 4.21(a) and (b). Note that the thermal conductivity is normalized against the total value for each case. As the treatment of the dispersion is refined, the thermal conductivity curves become smoother.

(C) *Transverse and Longitudinal Phonon Contributions*

The relative contributions of LA and TA phonons to the total heat flow at high temperatures have not been fully resolved. As shown in Figure 4.22(a), the Holland model predicts that the TA phonons are the dominant heat carriers at high temperatures, even though they have lower group velocities than LA phonons. This result was supported in [135] by use of calculations based on a variational method, which does not involve the SMRT approximation. It has been ascertained that, above 100 K, TA phonons are the primary carriers of energy in Si (which has the same crystal structure as Ge) by use of the Monte Carlo simulations [231]. However, their calculations were based on the original Holland relaxation times, which may have

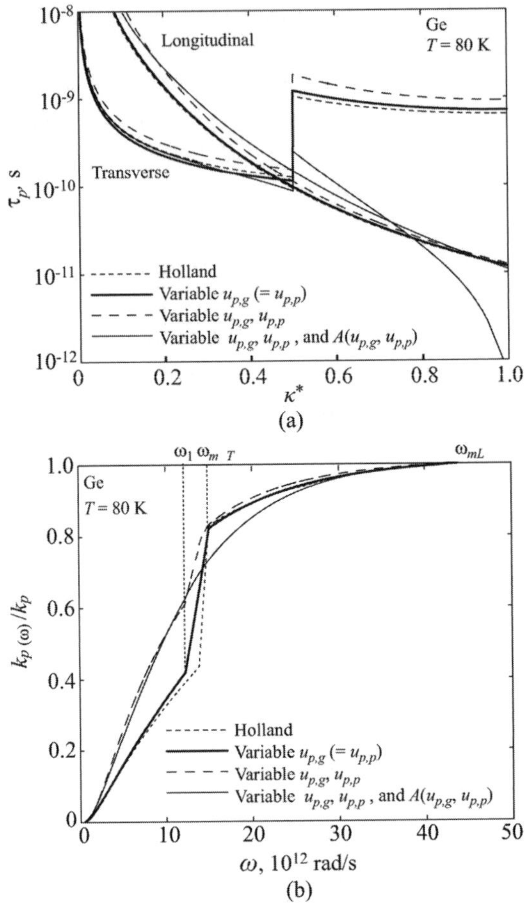

Figure 4.21. (a) Three-phonon relaxation times for refit data from Figure 4.20(b) at $T = 80$ K. (b) Cumulative frequency dependence of the thermal conductivity for refit data from Figure 4.20(b) at $T = 80$ K. The thermal conductivity is plotted as a percentage of the total value for each case. The curves show three distinct regions. The transition between the first and second regions takes place at ω_1, where the form of the TA relaxation time changes. The transition between the second and third regions occurs at $\omega_{m\mathrm{T}}$, after which there is no contribution from TA phonons [i.e., (4.110)] [72].

contributed to this conclusion. In [8] it has been concluded that, although the heat flow in Ge at high temperatures is primarily because of TA phonons, the LA U–processes cannot be ignored (as they are in the Holland formulation). In [165] it has been suggested that LA phonons are the dominant heat carriers in Si near room temperature. In [316] it has been ascertained that LA phonons dominate heat transport in Ge at high temperatures.

In Figure 4.22(b), the relative contributions of LA and TA phonons to the thermal conductivity predicted by the BZBC dispersion model (with refit relaxation-time parameters) are shown. Compared with the results of Figure 4.22(a), the role of TA phonons is quite different when the dispersion and impurity scattering are rigorously modeled. We find a thermal conductivity of 5.2 W/m-K at a temperature

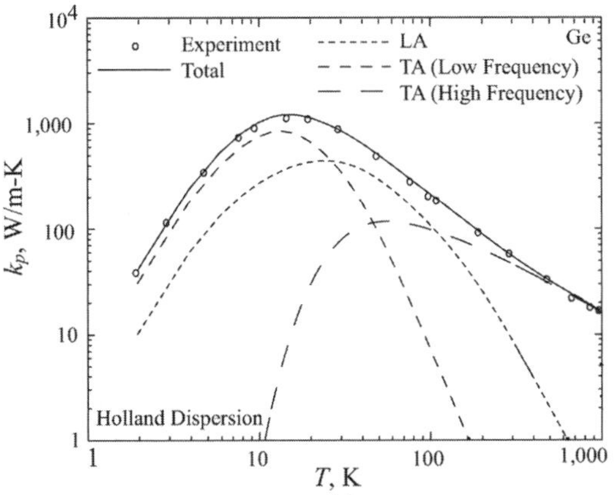

(a)

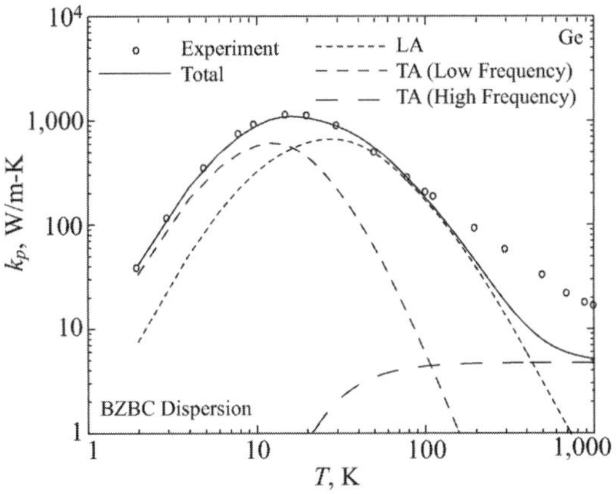

(b)

Figure 4.22. Contributions of LA and TA phonon branches to the thermal conductivity based on (a) Holland dispersion model and (b) BZBC dispersion model [72].

of 1000 K, whereas the experimental value is 17 W/m-K. The high-frequency TA phonons cannot contribute much to the thermal conductivity of Ge because of their low group velocity, which appears directly in the thermal conductivity expression, and also results in a high impurity scattering rate. We note that the expected electronic contribution to the thermal conductivity of Ge at a temperature of 1000 K is 4 W/m-K [314], which is not sufficient to explain the predicted discrepancy.

4.10 Einstein and Cahill–Pohl Minimum Phonon Conductivities

In the Einstein solid thermal conductivity model, the atomic vibrational states are directly used (not related to phonon modes). The complete derivation of the Einstein thermal conductivity is given in Appendix C. He considers the interaction (through a spring constant) of the first, second, and third neighboring atoms with a central atom. He uses a quantum specific heat capacity and a single frequency and arrives at [24]

$$k_{p,\mathrm{E}} = \frac{n^{-1/3}k_\mathrm{B}}{\tau_\mathrm{E}}\frac{x_\mathrm{E}^2 e^{x_\mathrm{E}}}{(e^{x_\mathrm{E}}-1)^2}, \quad x_\mathrm{E} = \frac{\hbar\omega_\mathrm{E}}{k_\mathrm{B}T} = \frac{T_\mathrm{E}}{T}, \quad \tau_\mathrm{E} = \frac{T_\mathrm{E}k_\mathrm{B}}{\hbar\pi}$$

Einstein model from (C.25) in Appendix C, (4.130)

where n is the number density of atoms, and T_E is the Einstein temperature.

Cahill and Pohl [51] extend the proceeding equation to include a range of frequencies (details are given in Appendix C.2), and arrive at

$$k_{p,\text{C-P}} = (\frac{\pi}{6})^{1/3}k_\mathrm{B}n^{2/3}\sum_\alpha u_{p,\alpha}(\frac{T}{T_\alpha})^2\int_0^{T_\alpha/T}\frac{x^3e^x}{(e^x-1)^2}\mathrm{d}x, \quad x = \frac{\hbar\omega}{k_\mathrm{B}T}$$

Cahill–Pohl model from (C.36) in Appendix C, (4.131)

where $u_{p,\alpha}$ and $T_{\mathrm{D},\alpha}$ are the phonon (acoustic) speed and the Debye temperature for polarization α and are related through (4.27), by use of the Debye model.

It is generally accepted that $k_{p,\text{C-P}}$ is the lowest thermal conductivity for a solid, provided the proper values of $u_{p,\alpha}$ or $T_{\mathrm{D},\alpha}$ [they can be related through (4.63)] are available. Based on this $k_{p,\text{C-P}}$ is used for thermal conductivity of amorphous solids and polymers. Figure 4.17 shows the variation of $k_{p,\text{C-P}}$ with respect to temperature. At low temperatures ($T < T_{\mathrm{D},\alpha}$) it increases and for $T > T_{\mathrm{D},\alpha}$ it reaches a plateau. We will have quantitative results for $k_{p,\text{C-P}}(T)$, for amorphous SiO_2, in Figure 4.34.

4.11 Material Metrics of High-T Phonon Conductivity

Starting from the derivation in [167], in which the variational principle is used to obtain an analytical relaxation time for the rare-gas solids, Slack proposed that when heat is mainly carried by acoustic phonons scattered by means of the three-phonon process, the thermal conductivity of crystals with constant volume at high temperatures (normally above 1/4–1/5 of the Debye temperature) can be given by the Slack relation [24, 313]

$$k_p = k_{p,\mathrm{S}} = \frac{3.1\times10^4\langle M\rangle V_a^{1/3}T_{\mathrm{D},\infty}^3}{T\langle\gamma_\mathrm{G}^2\rangle N_\circ^{2/3}}. \qquad (4.132)$$

Here $\langle M\rangle$ is the mean atomic weight of the atoms in the primitive cell, $V_a = 1/n$ is the average volume per atom, $T_{\mathrm{D},\infty}$ is the high-temperature Debye temperature, T is the temperature, $N_\circ$ (Section 4.3) is the number of atoms in the primitive cell, and $\langle\gamma_\mathrm{G}^2\rangle$ is the mode-averaged square of the Grüneisen constant or parameter (Glossary)† at high temperatures. The value of γ_G is listed in Table A.2 for crystalline elements. These values range from 0.5 for strong covalent bonds to 3.0 for weak van

† Grüneisen parameter is related to the solid properties through

$$\gamma_\mathrm{G} = \frac{3\beta E_p}{\rho c_{v,p}},$$

where β is the linear expansion coefficient and E_p is the isothermal bulk modulus (4.70).

der Waals bonds (interactions). In Section 4.11.4, we will discuss the prediction of γ_G. Note that $T_{D,\infty}$ is extracted from the phonon density of states D_p [313], i.e.,

$$T_{D,\infty}^2 = \frac{5h_P^2}{3k_B^2}\frac{\int_0^\infty f^2 D_p(f)\mathrm{d}f}{\int_0^\infty D_p(f)\mathrm{d}f}, \tag{4.133}$$

where h_P is the Planck constant, k_B is the Boltzmann constant, and f is the phonon frequency. However, $T_{D,\infty}$ cannot be conveniently determined for it requires the information of DOS. Because the difference between $T_{D,\infty}$ and the Debye temperature T_D (at 0 K) extracted from the elastic constant or the measurement of heat capacity is normally small, it is customary to use T_D instead of $T_{D,\infty}$ in (4.132). Also, $\langle\gamma_G^2\rangle$ is often replaced with $\langle\gamma_G\rangle^2$ (later, for simplicity we use γ_G to denote $\langle\gamma_G\rangle$), which can be determined from thermal expansion data (Glossary) at high temperatures.

Equation (4.132) is widely tested with pure nonmetallic crystals, and the overall agreement is good, even for complex crystals [24, 313]. The Slack relation illuminates how the atomic structure affects the thermal transport and provides a useful guide to tailoring the thermal transport properties.

4.11.1 Derivation of Slack Relation

To understand the nature of the Slack relation, we can also derive a relation similar to the Slack relation. To be physically meaningful, the mean free path of a phonon mode should be longer than one half of its wavelength [52]. Therefore, a two-segment mean-free-path model can be constructed [154], i.e., when the phonon frequency is below a critical frequency, ω_c, its mean free path will vary according to the relaxation-time model of Roufosse for moderate and high temperatures (above T_D) [287]; when the phonon frequency is above the critical frequency, its mean free path is set to one half of its wavelength. This gives

$$\begin{aligned}\lambda_{p,i}(\omega) &= \frac{u_{p,g,i}}{A_i(1+B_i\omega^2)\omega^2 T}, \quad \omega < \omega_{c,i},\\ &= \frac{\pi u_{p,g,i}}{\omega}, \quad \omega_{c,i} \le \omega \le \omega_{D,i},\\ A_i &= \frac{3\times 10^3 N_A \pi \delta \gamma_G^2 k_B}{2^{1/2}\langle M\rangle u_{p,g,i}^3}, \quad B_i = \left(\frac{4\pi}{3}\right)^{2/3}\frac{5\delta^2}{12\pi^2 u_{p,g,i}^2},\end{aligned} \tag{4.134}$$

where N_A is the Avogadro number. To ensure a continuous $\lambda_{p,i}(\omega)$, ω_c must satisfy

$$\omega_{c,i} A_i(1+B_i\omega_{c,i}^2) = \frac{1}{\pi T}. \tag{4.135}$$

For $B_i \ll \pi^2 A_i^2 T^2$, we have

$$\omega_{c,i} \simeq \frac{1}{A_i \pi T}. \tag{4.136}$$

The critical frequency ω_c decreases with temperature. In other words, as the temperature increases, an increasing number of phonon modes reach the limiting mean free path of one half of their wavelength. Using (4.134), (4.107) can be rewritten as

$$k_{p,\mathrm{A}} = \sum_{i=1}^{3} \frac{k_\mathrm{B}}{2\pi^2 u_{p,g,i} A_i T} \int_0^{\omega_{c,i}} \left[\left(\frac{4\pi}{3} \right)^{2/3} \frac{5\omega^2 \delta^2}{12\pi^2 u_{p,g,i}^2} + 1 \right]^{-1} \mathrm{d}\omega$$

$$+ \left(\frac{\pi}{6} \right)^{1/3} n^{2/3} \sum_{i=1}^{3} u_{p,g,i} \frac{\hbar^4}{T_{\mathrm{D},i}^2 k_\mathrm{B}^3 T^2} \int_{\omega_{c,i}}^{k_\mathrm{B} T_{\mathrm{D},i}/\hbar} \frac{\omega^3 e^{\frac{\hbar\omega}{k_\mathrm{B}T}}}{(e^{\frac{\hbar\omega}{k_\mathrm{B}T}} - 1)^2} \mathrm{d}\omega. \quad (4.137)$$

By setting ω_c equal to ω_D, $u_{p,g,i}$ to the mean phonon speed $u_{p,g}$, and using $\omega_\mathrm{D} = u_{p,g}(6\pi^2 n/N_\circ)^{1/3}$, we have, from (4.137)

$$k_{p,\mathrm{A}} = \frac{4.48 \times 10^3 \langle M \rangle T_\mathrm{D}^3}{T n^{1/3} \gamma_\mathrm{G}^2 N_\circ^{2/3}}, \quad (4.138)$$

which is similar to Slack relation (4.132) except for the constant. The difference in the constants is because of SMRT and a different Hamiltonian used in [287] for three-phonon interactions.

As will be discussed in Section 4.13, the lattice thermal conductivity can be decomposed into three parts [235]:

$$k_p = k_{p,lg,\mathrm{A}} + k_{p,sh,\mathrm{A}} + k_{p,\mathrm{O}}. \quad (4.139)$$

Here $k_{p,lg,\mathrm{A}}$ is the contribution from long-range acoustic phonons, whose mean free path is larger than one-half of their wavelength; $k_{p,sh,\mathrm{A}}$ is the contribution from short-range acoustic phonons, whose mean free path is minimized to one-half of their wavelength; and $k_{p,\mathrm{O}}$ is the contribution from the optical phonons.

According to the proceeding derivation, the Slack relation corresponds to $k_{lg,\mathrm{A}}$ and is valid only when the short-range acoustic or optical phonons are not important. This condition is not always satisfied for crystals with low thermal conductivity, such as zeolites and metal–organic frameworks (MOFs). The Slack relation is valid only when ω_c is comparable with the Debye frequency ω_D and long-range acoustic phonons dominate the thermal transport.

To use the Slack relation, T_D and γ_G must be known, which is the main difficulty in the estimation of the lattice conductivities of new materials. Because these two parameters directly relate the atomic structure to thermal transport, the knowledge of their relations provides more insightful information for the thermal design and allows for the estimation of thermal transport properties of new materials.

In [154] a simple microscopic model is presented for estimating these parameters. This model is based on a phenomenological combinative rule for force constants and a general L–J potential form for a bond.

4.11.2 Force-Constant Combinative Rule for Arbitrary Pair-Bond

The vibration energy is transferred in a crystal through interactions among the atoms, which can be theoretically calculated by quantum-mechanical methods. However, a quantum-mechanical method deals with the electron clouds of the atoms and is very cumbersome for a system involving many particles. Based on the Born-Oppenheimer approximation [200] described in the Glossary, the force-field method uses empirical potentials (fitted to experiments or quantum mechanic calculations), such as L–J and Buckingham potentials, to describe the interactions in the system. In most solids, when the temperature is well below the melting point, the particles only slightly oscillate around their equilibrium positions, and many of their behaviors (including the elastic behavior) can be well described in the framework of the harmonic approximation [9]. In this approximation, the energy of the system can normally be decomposed into four terms corresponding to the bond stretching, bending, torsion, and the nonbonded interactions [200], i.e.,

$$E = \sum_i \frac{\Gamma_i}{2}(\Delta r_i)^2 + \sum_j \frac{\Gamma_{\theta,j}}{2}(\Delta\theta_j)^2 + \sum_l \frac{\Gamma_{\phi,l}}{2}(\Delta\phi_l)^2 + \sum_n \frac{\Gamma_{m,n}}{2}(\Delta r_{m,n})^2, \quad (4.140)$$

where Γ, Γ_θ, Γ_ϕ, and $\Gamma_{m,n}$ are the force constants of the bond length r, bond angle θ, torsion angle ϕ, and the distance between molecules r_m. Normally the stretching interaction is much stronger than the other interactions (by a factor of more than 10), so for a rigid structure, the elastic characteristics are mainly determined by the stretching force constants. The bending and torsion interactions are also important in molecular crystals and for structure stability and deformation.

Because atomic interaction is determined by the electronic structure, potentials and force constants are expected to be transferable if the bond type and surroundings are similar [145]. Here a phenomenological combinative rule is presented for the stretching and the van der Waals force constants.

The general form of two-body potentials can be written as

$$\varphi_{\rm AB}(r) = \varphi_{AB,rep}(r) - \varphi_{AB,att}(r), \quad (4.141)$$

where $\varphi_{\rm AB}$ is the potential energy of the bond A–B, and the subscripts *rep* and *att* represent the repulsive and the attractive terms. The repulsive term is because of the Pauli exclusion principle and/or the electrostatic interactions. It has been shown that the exchange repulsive term for two different atoms can be given as the geometric mean of the corresponding terms for two pairs of equivalent atoms [28], i.e.,

$$\varphi_{\rm AB,\mathit{rep}}(r) = [\varphi_{AA,rep}(r)\varphi_{BB,rep}(r)]^{1/2}. \quad (4.142)$$

The attractive term is because of the interactions of dipoles, electrostatics, or a combination of them. The exchangeability of the dipolar and electrostatic interactions is apparent; thus a similar combinative rule is suggested for the attractive term, i.e.,

$$\varphi_{\mathrm{AB},att}(r) = [\varphi_{AA,att}(r)\varphi_{\mathrm{BB},att}(r)]^{1/2}. \tag{4.143}$$

The potential near the equilibrium position can be described by the general L–J potential model (2.7),

$$\varphi(r) = \frac{A}{r^m} - \frac{B}{r^n}, \tag{4.144}$$

where m and n (Table 2.1) depend on the interaction type; and their values will be discussed in Section 4.11.4. The force constant Γ (4.36) and the bond length r_e at the equilibrium position are given as

$$\Gamma = -\frac{mn\varphi_{\mathrm{o}}}{r_{\mathrm{o}}^2} = \frac{m(m-n)A}{r_{\mathrm{o}}^{m+2}} = n(m-n)B(\frac{Bn}{Am})^{\frac{2+n}{m-n}}$$

$$r_e = (\frac{Am}{Bn})^{\frac{1}{m-n}}, \tag{4.145}$$

where $-\varphi_{\mathrm{o}}$ is the potential energy at the equilibrium position. (4.145) shows that the force constant at the equilibrium position is proportional to φ_{o} when the bond type and the bond length are similar. Note that, at the equilibrium position, the ratio of the magnitudes of the contributions from the repulsive term and the attractive term is $(m+1)/(n+1)$. Therefore, for $m \gg n$ (e.g., for an ionic bond), the force constant is mainly determined by the repulsive term.

From (4.142), (4.143) and (4.145), if Γ_{AA} is defined as the force constant of the potential function $\varphi_{\mathrm{AA}}(r) = \varphi_{\mathrm{AA},rep}(r) - \varphi_{\mathrm{AA},att}(r)$, the force constant of A–B bond Γ_{AB} and its equilibrium bond length $r_{\mathrm{o,AB}}$ can be given as

$$\Gamma_{\mathrm{AB}} = (\Gamma_{\mathrm{AA}}\Gamma_{\mathrm{BB}})^{1/2}, \ r_{e,\mathrm{AB}} = (r_{e,\mathrm{AA}}r_{e,\mathrm{BB}})^{1/2}. \tag{4.146}$$

Note that for ions, the A–A bond may not actually exist. However, because of the similarity of the electronic configuration of the ions in different compounds, we may assign a virtual potential φ_{AA} to the ions, e.g., keeping the interaction that is because of the Pauli exclusion principle as the repulsive term and setting the attractive term as $\varphi_{\mathrm{AA},att} = q^2/r$, where q is the ionic charge. The properties of the virtual potential (e.g., Γ_{AA}) can be extracted from the compounds. In this way, the combinative rule of (4.142) and (4.143) is still valid. Similar relations like (4.146) have been derived in [104] by a 12–6 L–J potential, but they did not consider the effects of bond order and the long-range electrostatic interactions. In addition, it is not appropriate to describe ionic bonds or covalent bonds by 12–6 L–J potential, as will be discussed later.

Note that this combinative rule is only applicable for the bonds with the same bond type (m and n are close) and bond order. In real compounds, a bond with the same atom configuration can have different bond orders. For example, C=O has the bond order of 2, and C–O has the bond order of 1. It is observed that the force constant is approximately proportional to the bond order [349], that is

$$\Gamma_{\mathrm{AB},s} = s\Gamma_{\mathrm{AB},1}, \tag{4.147}$$

where $\Gamma_{AB,s}$ is the force constant of the bond between A and B with the bond order of s. Thus (4.147) can be rewritten as

$$\Gamma_{AB,s} = s(\Gamma_{AA,1}\Gamma_{BB,1})^{1/2}. \quad (4.148)$$

Consequently, we have

$$\Gamma_{AC,s} = s\frac{(\Gamma_{AB,1}\Gamma_{BC,1})^{1/2}}{\Gamma_{BB,1}}. \quad (4.149)$$

According to (4.147), the potential energy φ can be assumed proportional to s, and r_e is expected to be independent of s. For ionic bonds, when this assumption is used, the resulting combinative rule for ionic bond length agrees well with the experiments (the error is less than 3%) [104]. However, this assumption is only moderately accurate for covalent bonds, because the L–J potential does not accurately describe the changes of electron clouds and the energy in the entire range of atomic distance. Generally, for covalent bonds, r_e will decreases slightly when s increases. In [259] an empirical bond order–bond length relationship for covalent bonds is developed:

$$r_{e,s} = r_{e,1} - 0.78(s^{0.33} - 1), \quad (4.150)$$

where $r_{e,s}$ is the equilibrium bond length (in angstroms) with the bond order of s. (4.150) shows good agreements with the experimental results for many bonds [259] and can be used for the estimation of the bond length.

Table A.2 lists the electronegativity of elements. Table 4.4 lists the force constant $\Gamma_{AA,1}$, electronegativity χ, and equilibrium bond length $r_{e,1}$ for most element paris. $\Gamma_{AA,1}$ is extracted from the experimental spectra of diatomic molecules [146] according to (4.148) and (4.149) (the ionic $\Gamma_{AA,1}$ of elements, e.g., O and S, is an average of the values extracted from their compounds). $r_{e,1}$ is extracted from the bond lengths of the diatomic molecules [146]. Table 4.4 shows that $\Gamma_{AA,1}$ of ionic bonds for the elements with high electronegativity χ (e.g., O and Cl) are normally twice that of the corresponding covalent bond. This indicates that the virtual potential of ions is steeper than the covalent potential of the corresponding atoms near the equilibrium position. The electronegativity χ can be used to determine the bond type. Bonds between atoms with a large electronegativity difference (≥ 1.7), are usually considered to be ionic, whereas values between 1.7 and 0.4 are considered polar covalent and values below 0.4 are considered nonpolar covalent bonds [264]. For metallic elements, even though $\Delta\chi$ is small, their electron structures are similar to those in the ionic crystals, for the conduction electrons can move about [183].

Figure 4.23(a) shows that generally the ionic $\Gamma_{AA,1}$ increases as the electronegativity increases. The alkali metals have the lowest $\Gamma_{AA,1}$, whereas the halogen elements have the highest $\Gamma_{AA,1}$. When $1.0 < \chi < 2.5$, most transition-metal elements and semiconducting elements have a $\Gamma_{AA,1}$ around 50 N/m, which is a relatively low value. In general, $\Gamma_{AA,1}$ decreases while the atomic radius increases. However, Figure 1(b) shows that, for covalent bonds, $\Gamma_{AA,1}$ seems to relate to the ratio of χ/z (z is the atomic number) rather than χ. Nitrogen has the highest covalent $\Gamma_{AA,1}$.

Table 4.4. *Electronegativity, equilibrium bond length, and force constants of element pairs (listed in order of atomic number) with the bond order of 1. The data are extracted from [146]. The symbols C and V represent covalent and van der Waals interactions, and the unlabeled are the values for ionic interactions.*

Atom	χ	$r_{e,1}$, Å	Γ, N/m	Atom	χ	$r_{e,1}$, Å	Γ, N/m
H	2.20	0.74[C]	575.67[C]	Br	2.96	2.28[C]	250.83[C], 539.78
He	–	1.04[C]	411.74[C]	Kr	3.00	4.03[V]	1.43[V]
Li	0.98	2.67	25.48	Rb	0.82	3.79	8.25
Be	1.57	1.39	120.62	Sr	0.95	3.05	26.57
B	2.04	1.76[C]	354.90[C]	Y	1.22	–	70.05
C	2.55	1.54[C]	510.5[C]	Zr	1.33	8.41	141.43
N	3.04	1.46[C]	771.20[C]	Nb	1.60	2.36	108.25
O	3.44	1.46[C]	593.57[C] 1305	Ru	1.02	2.17	56.53
F	3.98	1.41[C]	473.82[C] 1960	Ag	1.93	2.59	59.09
Ne	–	3.10[V]	0.12[V]	Cd	1.69	4.28	44.93
Na	0.93	3.08	17.28	In	1.78	2.86	34.41
Mg	1.31	3.89	41.60	Sn	1.96	2.80	58.34
Al	1.61	2.47	49.15	Sb	2.05	2.82[C]	70.64[C]
Si	1.90	2.34[C]	109.04[C]	Te	2.10	2.74[C]	119.46[C]
P	2.19	2.20[C]	201.50[C]	I	2.66	2.66[C]	172.73[C] 343.06
S	2.58	2.08[C]	250.65[C] 536.92	Xe	2.60	4.36[V]	1.74[V]
Cl	3.16	1.98[C]	330.42[C] 705.81	Cs	0.79	4.47	6.97
Ar	–	3.76[V]	0.80[V]	La	1.10	2.83	53.41
K	0.82	3.90	9.84	Ce	1.12	2.74	169.30
Ca	1.00	4.28	34.61	Pr	1.13	–	48.25
Sc	1.36	2.40	77.60	Eu	1.20	–	31.11
Ti	1.54	2.17	107.71	Tb	1.10	–	72.63
V	1.63	2.09	103.03	Ho	1.23	3.11	77.21
Cr	1.66	2.17	87.18	Yb	1.10	2.89	33.74
Mn	1.55	2.59	46.40	Lu	1.27	2.63	78.69
Fe	1.83	2.04	62.53	Hf	1.30	2.44	103.98
Co	1.88	3.20	116.61	Ta	1.50	2.36	179.07
Ni	1.91	2.96	130.00	W	2.36	–	202.54
Cu	1.90	2.22	65.82	Ir	2.20	2.36	110.01
Zn	1.65	3.41	81.15	Au	2.54	2.47	106.80
Ga	1.81	2.43	99.38	Hg	2.00	3.30	32.61
Ge	2.01	2.16	121.79	Tl	1.62	3.07	31.60
As	2.18	2.42	120.48	Pb	2.33	3.03	39.91
Se	2.55	2.34[C]	108.54[C]	Bi	2.02	3.07	49.08

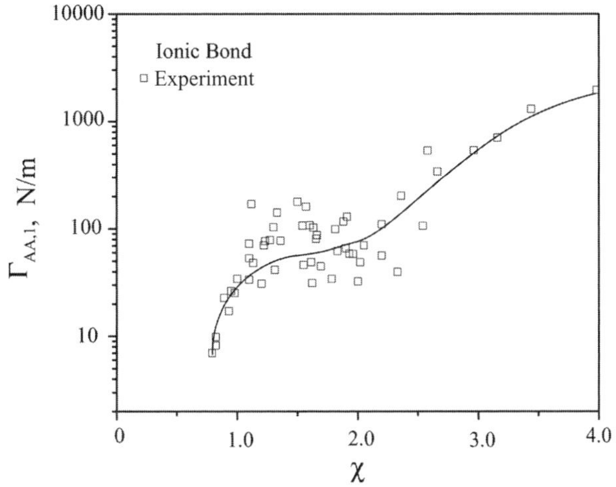

(a)

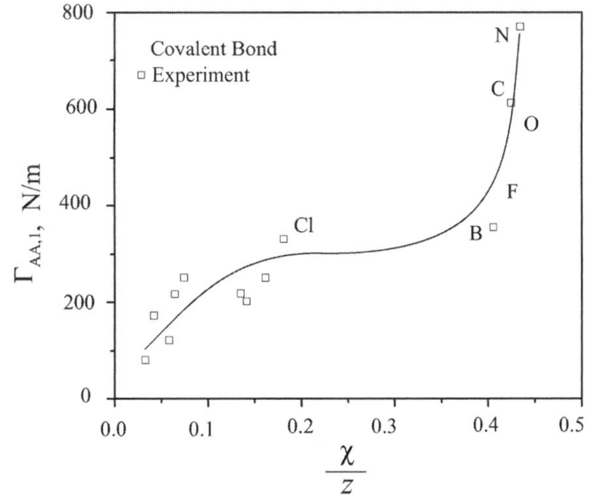

(b)

Figure 4.23. (a) Variation of calculated ionic force constant $\Gamma_{AA,1}$ with respect to the electronegativity. (b) Variation of the calculated covalent force constant $\Gamma_{AA,1}$ with respect to the ratio of the electronegativity and atomic number. The data are extracted from the spectra of diatomic molecules [146]. The curves are used to guide the eyes.

Figure 4.24 compares the experimental results of some bonds in diatomic molecules along with the calculated values. The mean-square error is less than 8%, and the overall agreement is good.

Note that the proceeding force constants and equilibrium bond lengths are derived from the data of gaseous diatomic molecules, for which the intermolecular effects are negligible. For crystal bonds, long-range interactions (mainly electrostatic interactions) from the surroundings may significantly affect the equilibrium bond length and force constant. For example, Na–Cl in a NaCl molecule has a force constant of 110 N/m and a bond length of 2.36 Å [146], while the distance between the

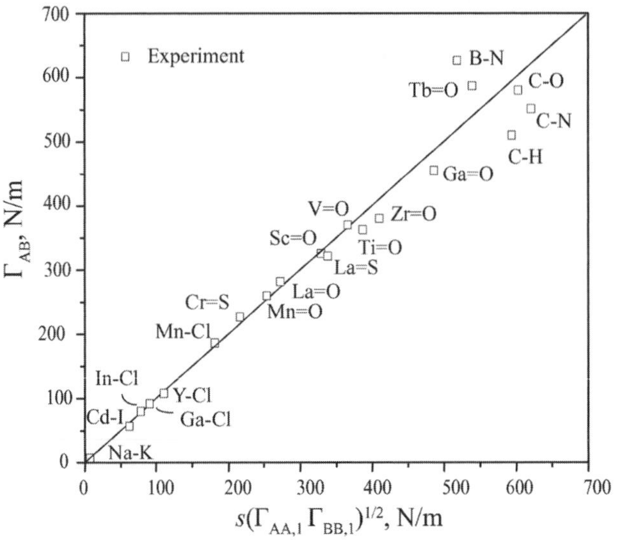

Figure 4.24. Comparison of calculated force constants and the corresponding values from the experimental spectra, for some atomic pairs [146].

nearest Na and Cl ions in a NaCl crystal at $T = 300$ K is 2.83 Å, and the effective force constant of each Na-Cl pair derived from the bulk modulus is only 20 N/m [311]. Thus, a relation between the force constant of a bond in a gaseous diatomic molecule and that in a crystal must be developed to account for the effects of long-range interactions. Here, only the effect of electrostatic interactions is considered. Our approach is to include long-range interactions in an effective bond potential of the nearest-neighbor atoms. A bond (in a diatomic molecule) with a form in (4.144) is considered. Because the repulsive term is a very short-range interaction, we assume that only the long-range attractive term is affected by the surroundings. This effective bond in a crystal can then be represented as

$$\langle\varphi\rangle(r) = \frac{A}{r^m} - \eta\frac{B}{r^n}, \tag{4.151}$$

where η is the correction factor that is because of the long-range interactions (in simple ionic structures, it is related to the Madelung constant). However, η is difficult to determine for complex crystal structures. In practice, according to this assumption and (4.145), the force constant of the bond in the crystal Γ'_{AB} can be simply calculated as

$$\Gamma'_{AB} = \Gamma_{AB}(\frac{r_o}{r'_o})^{m+2}, \tag{4.152}$$

where Γ_{AB} is the force constant of the bond in the diatomic molecule AB, and r'_o is the equilibrium bond length in the crystal. For example, for NaCl, by setting $m = 6.3$ [using approximation method (4.171)], and using the proceeding bond length data, we have $\Gamma'_{NaCl} = 110 \times (2.36/2.83)^{(6.3+2)} = 24$ N/m, which is very close to 20 N/m

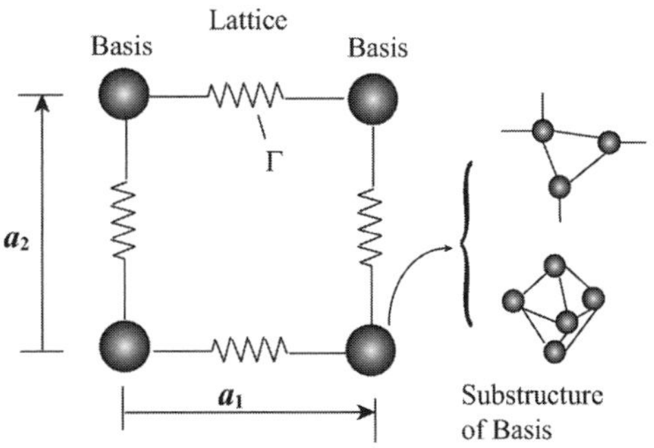

Figure 4.25. Decomposition of a complex crystal into lattice and bases, with equivalent bonds.

derived from the bulk modulus [311]. For ionic bonds, m is large, and (4.152) indicates that Γ'_{AB} is very sensitive to the values of r_o and r'_o, so the experimental values of r_o and r'_o will be preferred. When the experimental value of r_o is unavailable, combinative rule (4.146) can be used.

4.11.3 Evaluation of Sound Velocity and Debye Temperature

If the force constants between atoms are known, the dynamical matrix can be readily constructed to determine the sound velocity. However, for complex polyatomic crystals, the calculation is still very cumbersome, and it is difficult to explicitly relate the numerical results to the complex structure. For the purpose of estimation and design, a simple model that can directly relate the sound velocity and the Debye temperature to the crystal structure is needed.

A real crystal structure can always be considered as an underlying lattice, together with a basis describing the arrangement of the atoms, ions, and molecules within a primitive cell [9]. The acoustic branches of the phonon dispersion correspond to the motion of the mass centers of the primitive cells [9]. Therefore, both the monatomic and polyatomic crystal structures can be modeled as a lattice with rigid bases connected by equivalent bonds with an equivalent force constant, as shown in Figure 4.25.

The elastic response of a solid can be divided into two parts, namely, (i) atomic vibrations at a fixed volume, and (ii) unit-cell volume fluctuations for a fixed atomic configuration (homogeneous deformation). The first part corresponds to the inhomogeneous deformation, in which the bending potentials and the torsion potentials may be important, especially in a flexible structure. The bending potential can also be converted into an equivalent stretching potential between the atoms at the two ends. Because the force constants of the bending potentials and the torsion potentials are normally small, the equivalent force constant will be much reduced by the inhomogeneous deformation. It is difficult to obtain a general simple analytical solution for such an inhomogeneous deformation, and a numerical calculation with a full dynamical matrix (including the bending potentials) is preferred for obtaining the equivalent force constant. However, for many solids, the crystallographic symmetries and the stability of a given phase with respect to small lattice deformations result in the diminishing effects from the first part [10], and the elastic behavior can be described by the equivalent force constants of the stretching potentials. In these cases, the bending potential and the torsion potential may contribute to the stability of the structure, but their contribution to the elastic response is negligible.

When only the bond stretching is considered, it is apparent that only the transport of stretching along the translational unit vector $\boldsymbol{a}$ can contribute to the energy transport in this direction. Thus we define the force constant of a bond along a given unit vector $\boldsymbol{a}$ as [91]

$$\Gamma_{\mu\nu,\boldsymbol{a}} = \frac{\partial^2 \varphi_{\mu\nu}}{\partial x_{\mathbf{a}}^2} = \frac{\partial^2 \varphi_{\mu\nu}}{\partial r_{\mu\nu}^2}\left(\frac{x_{\mathbf{a}}}{r_{\mu\nu}}\right)^2 = (\boldsymbol{a} \cdot \boldsymbol{s}_{\mu\nu})^2 \Gamma_{\mu\nu}, \tag{4.153}$$

where $x_{\mathbf{a}}$ is the projection of the bond length r along $\boldsymbol{a}$, and $\boldsymbol{s}_{\mu\nu}$ is the unit vector pointing from the particle μ to the particle ν. (4.153) shows that the projection of the force constant along $\boldsymbol{a}$ has a factor of $(\boldsymbol{a} \cdot \boldsymbol{s}_{\mu\nu})^2$.

The total deformation of the primitive cell is affected by all the bonds in it. Using (4.153), we may treat the bonds in a primitive cell as springs with the same $\Gamma_{\mu\nu,\boldsymbol{a}}$, and then convert the crystal primitive cell into a network composed of springs. This spring network can be simplified to obtain the equivalent force constant between two bases according to the following rules (we denote the force constants of two bonds as Γ_1 and Γ_2, and that of the equivalent bond of these two bonds as Γ_e)

(i) when the two bonds are in series

$$\Gamma_e^{-1} = \Gamma_1^{-1} + \Gamma_2^{-1}, \tag{4.154}$$

(ii) and when the two bonds are parallel

$$\Gamma_e = \Gamma_1 + \Gamma_2. \tag{4.155}$$

For a monatomic crystal, the primitive cell includes only one atom, and the equivalent force constant is just the force constant of the bond between the atoms.

From the lattice dynamics (Section 4.8), the sound velocities of acoustic branches at the long-wavelength limit are the square roots of the eigenvalues of the matrix [9]

$$-\frac{1}{2M}\sum_{\boldsymbol{R}}(\boldsymbol{s}_\kappa \cdot \boldsymbol{R})^2 \boldsymbol{D}(\boldsymbol{R}), \quad D_{i,j}(\boldsymbol{R}) = \frac{\partial^2 \varphi}{\partial d_i^{\circ} \partial d_j^{\boldsymbol{R}}}, \tag{4.156}$$

where $\boldsymbol{s}_\kappa$ is the unit wave vector, $\boldsymbol{D}(\boldsymbol{R})$ is the dynamical matrix, $\boldsymbol{R}$ is the position vector of the neighbor, d_i is the displacement of the mass center of the primitive cell from the equilibrium position ($\circ$ represents the origin), and M is the mass of the primitive cell.

Using the proceeding simplified model for crystal structures, if only the stretching energy is considered, we can rewrite (4.140) as

$$\varphi = \sum_{\boldsymbol{R}} \frac{\Gamma_{\boldsymbol{R}}}{2}[\boldsymbol{s_R} \cdot (\boldsymbol{d^R} - \boldsymbol{d^\circ})]^2, \quad \boldsymbol{s_R} = \frac{\boldsymbol{R}}{|\boldsymbol{R}|}, \tag{4.157}$$

where $\Gamma_{\boldsymbol{R}}$ is the equivalent force constant between the two bases. Thus $D_{i,j}(\boldsymbol{R}) = \eta_{i,j}\Gamma_{\boldsymbol{R}}$. Note that $\boldsymbol{R}$ is a linear function of the lattice constants, so the acoustic (sound) velocity (Section 4.7.2) will have the form (4.67)

$$u_{p,g,i} = l(s_{k,i}, \{\boldsymbol{a}_i\})(\frac{\Gamma_s}{M})^{1/2}, \quad \Gamma_s = \sum_{\boldsymbol{R}}[\eta(\boldsymbol{R}, s_{k,i})\Gamma_{\boldsymbol{R}}], \tag{4.158}$$

where $\{\boldsymbol{a}_i\}$ is the set of the translational vectors of the lattice. Note that (4.158) has the same form as the formula for the one-dimensional chain [91]. It is instructive to consider a plane wave traveling in a crystal, wherein the lattice consists of parallel planes perpendicular to the wave vector and the atoms in a plane will move in phase. The transportation along the wave vector is essentially one dimensional. From the comparison with the formula of the one-dimensional chain [91], l is indeed the equivalent distance between the planes and normally is the linear function of the lattice constants. The effective force constant Γ_s is the summation of the projections of the equivalent force constant in the polarization $\boldsymbol{s}_{\kappa,i}$, that is, $\eta(\boldsymbol{R}, \boldsymbol{s}_{\kappa,i}) = (\boldsymbol{s_R} \cdot \boldsymbol{s}_{\kappa,i})^2$.

The average sound velocity $u_{p,g,\mathrm{A}}$ can be found from (4.57), i.e.,

$$u_{p,g,\mathrm{A}} = (\sum_{\alpha=1}^{3} \frac{1}{3u_{p,g,\alpha}^3})^{-1/3}. \tag{4.159}$$

For cubic structures, the average sound velocity can be given as

$$u_{p,g,\mathrm{A}} = \frac{1}{3^{1/2}} a (\frac{\Gamma}{M})^{1/2}, \tag{4.160}$$

where a is the lattice constant.

From the longitudinal and transversal sound velocities, we can obtain the polarization-dependent Debye temperature $T_{D,i}$ and the average Debye temperature T_{D} [9]:

$$\begin{aligned} T_{\mathrm{D},\alpha} &= u_{p,g,\alpha}\frac{\hbar}{k_B}(6\pi^2 n)^{1/3} \\ &= \frac{l_\alpha}{V_c^{1/3}}(\frac{\Gamma_\alpha}{M})^{1/2}\frac{\hbar}{k_B}(6\pi^2 N_\circ)^{1/3} \end{aligned}$$

$$T_{\mathrm{D}} = u_{p,g,A}\frac{\hbar}{k_B}(6\pi^2 n)^{1/3} = (\sum_{\alpha=1}^{3}\frac{1}{3T_{\mathrm{D},\alpha}^3})^{-1/3}, \tag{4.161}$$

where n is the number density of atoms, $N_\circ$ is the number of atoms in a primitive cell, and V_c is the volume of a primitive cell. Here $l_\alpha/V_c^{1/3}$ is only a function of the ratio of lattice constants and the polarization, and the Debye temperature relates to

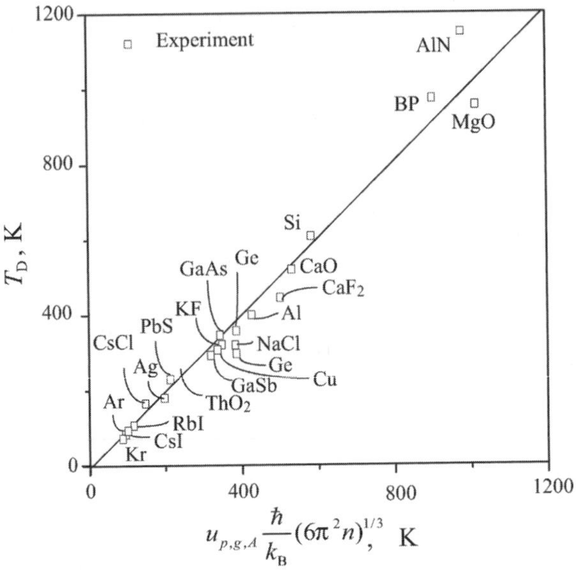

Figure 4.26. Comparison of the predicted and measured Debye temperature for some crystals. The force constants for metallic crystals are calculated according to (4.152) (m is chosen as 8). The experimental values are from [313].

the ratio of the lattice constants rather than their absolute values. It is apparent that if the lattice constant and other parameters are the same except the lattice type, the order of magnitude of T_D is T_D(FCC) > T_D(BCC) > T_D(SC).

Figure 4.26 compares the calculated and experimental Debye temperatures [313] (determined from elastic constants or specific heat capacity measurements) of some crystals. The force constants used in the calculation are from the combinative rule and Table 4.4. The overall agreement is good. The force constants for metals are calculated according to (4.152). It is found that $m = 8$ gives good agreement with experimental values. Again, it is found that, for metallic crystals, the force constants can be reduced significantly by the long-range electrostatic interaction (by a factor of about 5), which results in a low Debye temperature.

4.11.4 Prediction of Grüneisen Parameter

The Grüneisen parameter has been used to represent the volume dependence of the normal-mode frequencies. The overall Grüneisen parameter γ_G (Glossary) is defined as [9]

$$\gamma_G = \frac{\sum_{\kappa,\alpha} \gamma_{G,\kappa,\alpha} c_{v\alpha}(\kappa)}{\sum_{\kappa,\alpha} c_{v\alpha}(\kappa)}, \quad \gamma_{G,\kappa,\alpha} = -\frac{\partial \ln \omega_{\kappa,\alpha}}{\partial \ln V}, \tag{4.162}$$

where the subscript α denotes the branch of a normal mode, c_v is the heat capacity per normal mode, and V is the volume. In the Debye approximation, all the normal-mode frequencies scale linearly with the Debye temperature T_D, and therefore [9]

$$\gamma_G = \gamma_{G,\kappa,\alpha} = -\frac{\partial \ln \omega_D}{\partial \ln V}. \tag{4.163}$$

That is, γ_G represents the relative shift of the Debye angular frequency with respect to the volume.

We consider a crystal containing only one bond type. According to (4.158) and (4.163), because the Debye frequency $\omega_D \propto \Gamma^{1/2}$ [from (4.158) and (4.161)] and the volume $V \sim r_e^3$, we have

$$\gamma_G = -\frac{\mathrm{d}\ln \Gamma}{6\mathrm{d}\ln r_e}. \tag{4.164}$$

Note that γ_G relates only to the bond. In [355] there is defined a "bonding-scaling parameter" γ'_G as

$$\gamma'_{G,i} = -\frac{\mathrm{d}\ln \Gamma'_i}{6\mathrm{d}\ln r'_i}, \tag{4.165}$$

where Γ'_i and r'_i are the force constant and the equilibrium length of the bond i. For the crystals containing only one bond type, the Grüneisen parameter γ_G is equal to the bonding-scaling parameter γ'_G.

We can rewrite (4.165) in terms of a a small relative deviation fraction ϵ

$$\Gamma'_i(\epsilon) \simeq \Gamma'_o(1 - 6\gamma'_G\epsilon), \tag{4.166}$$

where Γ'_o is the initial equivalent force constant. It is apparent that γ'_G represents the intrinsic anharmonicity of a bond, i.e., the relative shift of the force constant with respect to the bond length. It seems reasonable that the γ'_G of each bond is independent of other bonds.

We again consider the crystal containing one bond type to obtain the bond-scaling parameter γ'_G. In [295] a thermodynamic description of Morse oscillators using a statistical treatment is developed. Here a similar approach is applied for the L–J oscillators representing the interatomic potentials. Consider an assembly of independent oscillators with the interatomic potential of (4.144), of which the natural angular frequency (2.54) $\omega_n = \omega = (\Gamma/m_r)^{1/2}$ (m_r is the reduced mass of the oscillator). If (4.144) is expanded in a Taylor series, the vibrational energy $E_{p,l}$ and the mean atomic separation $\langle r_l \rangle$ of the motion with the principle quantum number l of this oscillator can be expressed as [230]

$$E_{p,l} = \hbar\omega(l + \frac{1}{2}) - C_e\hbar^2\omega^2(l + \frac{1}{2})^2, \quad C_e = \frac{5(m+n+3)^2}{48\varphi_o mn}$$

$$\langle r_l \rangle = r_o + C_r r_o \hbar\omega(l + \frac{1}{2}), \quad C_r = \frac{3(m+n+3)}{2\varphi_o mn}. \tag{4.167}$$

Then, we have (a similar derivation is given in [295])

$$\gamma'_G \simeq \gamma'_{G,o}[1 + C_e k_B T \frac{g_1(x_D)}{g(x_D)}], \quad \gamma'_{G,o} = \frac{m+n+3}{6}$$

$$g(x_D) = \int_0^{x_D} \frac{x^3 dx}{e^x - 1}, \quad x_D = T_D/T$$

$$g_1(x_D) = \int_0^{x_D} \frac{x^4(1+e^x)dx}{(e^x-1)^2}. \tag{4.168}$$

Typically, the vibration energy is much smaller than the dissociation energy and $C_e k_B T$ is small, so the temperature dependence of γ'_G is weak. At high temperatures, γ'_G will reach $\gamma'_{G,o}$. It is very interesting to note that $\gamma'_{G,o}$ depends only on m and n, or on the bond type.

(i) *Ideal Ionic Bonds*. The attractive potential is dominated by the electrostatic potential, and the lattice summation of the long-range electrostatic interactions does not change n (the Madelung term); thus $n = 1$. The repulsive term arises from the full-filled shells and the Pauli exclusion principle. The measurements for typical ionic bonds show $m =$ 6–10 [9]. The midpoint $m = 8$ is a reasonable choice for the estimation; therefore $\gamma'_o = 2.0$. In fact, $\gamma'_{G,o} = 2.0$ agrees well with the high-temperature γ_G values of many typical ionic crystals with one bond type [91, 313].

(ii) *van der Waals Interactions*. The attractive term arises from the interaction between dipoles and varies as $1/r^6$, that is, $n = 6$. The widely used 12–6 L–J potential chooses $m = 12$ for the repulsive term. However, it is found that $m = 12$ makes the repulsive term very steep [200]. $m = 12$ gives $\gamma'_{G,o} = 3.5$, a much higher value than the measured results. For example, at high temperatures, γ_G of Ne, Ar, Kr, and Xe are 2.76, 2.73, 2.84, and 2.65 [313], respectively. Considering the repulsive term of van der Waals interaction arises from the same mechanism (i.e., the filled outer shell) as in an ionic bond, it is reasonable to choose the same value 8 for m. This choice gives $\gamma'_{G,o} = 2.83$, which agrees much better with the proceeding experimental results.

(iii) *Nonpolarized Covalent Bonds*. The attractive term is because of the electrostatic interaction, therefore $n = 1$. For m, the case is more complicated, because the distribution of valence electrons differs substantially from that in isolated atoms or ions. The repulsive term includes the electrostatic term because of the Pauli exclusion principle. In fact, the covalent bond is more appropriately described by the Morse potential (Table 2.1) [200]:

$$\varphi = \varphi_o[e^{-2a_o(r-r_o)} - 2e^{-a_o(r-r_o)}]. \tag{4.169}$$

In [295] an empirical relation is suggested, i.e., $a_o r_o = (m+4)/5$. For typical covalent bonds, $a_o r_o \simeq$ 1.0–1.2, thus $m =$ 1–2. Because $m > n$, we choose $m = 2$ and obtain $\gamma'_{G,o} = 1.0$. This value is also in accord with the the relation Γr^6 = constant for covalent bonds, as suggested in [145].

The covalent bond between atoms with different electronegativities is partially polarized (ionic bonds can also be considered highly polarized covalent bonds). Using the relation of the percentage of the ionic character c proposed in [264], γ'_G of a polarized bond can be given as

$$\gamma'_G = \gamma'_{G,AB} = (1-c)\gamma'_{G,cov} + c\gamma'_{G,ion}, \quad c = 1 - e^{-(\chi_A - \chi_B)^2/4}, \tag{4.170}$$

where $\gamma'_{G,cov}$ and $\gamma'_{G,ion}$ represent the bond-scaling parameters of a nonpolarized covalent bond and the ideal ionic bond, respectively. Equation (4.170) together with (4.168) can also be used for the rough estimation of m in an interatomic potential:

$$m \simeq 8 - 6e^{-(\chi_A - \chi_B)^2/4}. \tag{4.171}$$

(iv) *Metallic Bonds.* Although metallic crystals also include ions, they are very different from ionic crystals. The metals can be treated as ions immersed in a sea of free electrons [9]. Thus the interactions between ions can be treated as the summation of the bare interactions between ions and the electron–ion interactions. Both the repulsive term and attractive term include the long-range electrostatic interactions. However, because of the screening effects of free electrons, the interaction between ions decays faster than the pure Coulomb interactions, thus $m > 1$ and $n \geq 1$ (because of the attractions of ions to free electrons, the repulsive term decays faster than the attractive term). The derivation of γ'_G for metallic bonds is complicated. To compare it with experimental results, one also needs to include the contribution from the free electrons (it may be small at high temperatures). However, because the screening effects increase with the increasing electron number density [9], we would expect that, in the metals with high electron number densities, $m \simeq 8$ and $1 \leq n \leq 6$. For simplicity, in this work we set $m = 8$ and $n = 1$ (the same values for ionic bonds). The resulting $\gamma_G = 2.0$ is close to the experimental results of many metals (the alkali metals have a γ_G close to 1.2 because of the poor screening effects).

(v) *Other Interactions.* Some other interactions, e.g., the ion–dipole interaction, may exist in some crystals. These interactions may be considered as the cross terms of the above interactions. Using the combinative rule for potentials [(4.143) and (4.142)], we can have

$$m = \frac{(m_1 + m_2)}{2}, \quad n = \frac{(n_1 + n_2)}{2}, \quad \gamma'_G = \frac{(\gamma'_{G,1} + \gamma'_{G,2})}{2}, \tag{4.172}$$

where the subscripts 1 and 2 denote the individual interactions.

It can be seen that the order of magnitude of γ'_G for bonds is γ'_G(van der Waals bond) > γ'_G(ionic bond) > γ'_G(polarized covalent bond) > γ'_G(nonpolarized covalent bond). Figure 4.27 compares the calculated high-temperature Grüneisen parameters of crystals containing only one bond type with the experimental results (at the Debye temperature) [313], and the overall agreement is good. Note that, for

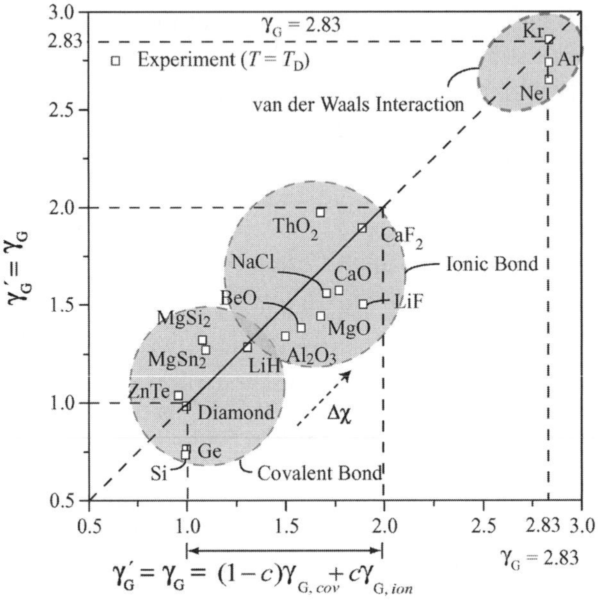

Figure 4.27. Comparison of predicted high-temperature Grüneisen parameters with the experimental results at the Debye temperatures for some crystals [313]. The Grüneisen parameter increases with an increase in the difference of electronegativity [152, 155]. Only for single-bond crystals, $\gamma_G = \gamma'_G$.

ionic crystals, Grüneisen parameters are slightly overestimated. One reason is that the temperature at which the measurements are performed is not high enough. For example, γ_G of NaCl at the Debye temperature is 1.57, but at 800 K, its value is 1.76 [303], compared with 1.71 given by (4.170). Another possible reason is that c in (4.170) determined by Pauling is not very accurate.

Note that (4.165) is valid for the equivalent force constant Γ, so the Grüneisen parameter γ_G of a crystal can be obtained by evaluation of the equivalent γ'_G of the equivalent bond.

For two parallel bonds, (4.155) and (4.166) lead to

$$\gamma'_G = \frac{\Gamma_{10}}{\Gamma_{10}+\Gamma_{20}}\gamma'_{G,1} + \frac{\Gamma_{20}}{\Gamma_{10}+\Gamma_{20}}\gamma'_{G,2}, \tag{4.173}$$

where Γ_{10} and Γ_{20} represent the equilibrium force constants of bond 1 and 2. That is, the equivalent γ'_G of the parallel bonds is the summation of the $\gamma'_{G,i}$ of the bonds weighted by the fraction of force constants.

Similarly, for two bonds in series, the equilibrium requirement gives

$$\epsilon_1 = \frac{\Gamma_{20}(r_1+r_2)}{(\Gamma_{10}+\Gamma_{20})r_1}\epsilon, \quad \epsilon_2 = \frac{\Gamma_{10}(r_1+r_2)}{(\Gamma_{10}+\Gamma_{20})r_2}\epsilon, \tag{4.174}$$

and (4.154) and (4.166) lead to

$$\gamma'_G = (\frac{\Gamma_{20}}{\Gamma_{10}+\Gamma_{20}})^2\frac{r_1+r_2}{r_1}\gamma'_{G,1} + (\frac{\Gamma_{10}}{\Gamma_{20}+\Gamma_{10}})^2\frac{r_1+r_2}{r_2}\gamma'_{G,2}. \tag{4.175}$$

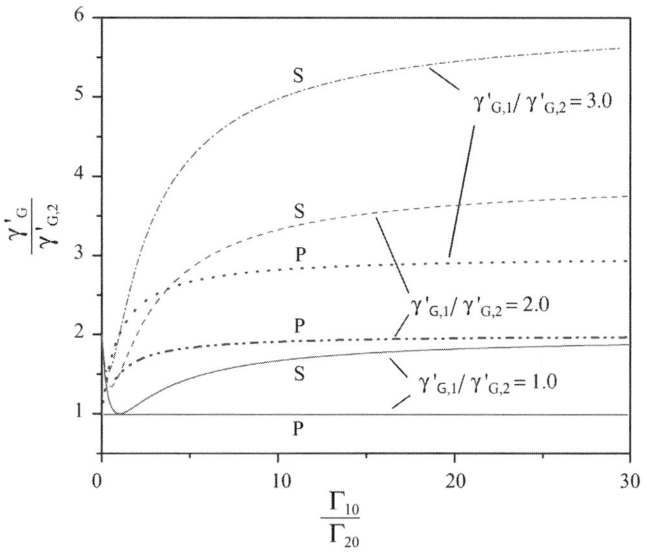

Figure 4.28. Variation of ratio of the equivalent bond-scaling parameter to the smaller bond-scaling parameter of the bonds $\gamma_G'/\gamma_{G,2}'$ with respect to the ratio of the force constants Γ_{10}/Γ_{20}. The symbols P and S denote the parallel and serial arrangements.

Equation (4.175) shows the equivalent γ_G' is related not only to the force constants and $\gamma_{G,i}'$, but also to the bond lengths. Note when $\epsilon_1 = \epsilon_2 = \epsilon$ (homogeneous deformation), (4.175) can be reduced to

$$\gamma_G' = (\frac{\Gamma_{20}}{\Gamma_{10}+\Gamma_{20}})\gamma_{G,1}' + (\frac{\Gamma_{10}}{\Gamma_{20}+\Gamma_{10}})\gamma_{G,2}', \quad (4.176)$$

which does not relate to the bond lengths. Note that $\gamma_{G,i}'$ is found from (4.168).

Assuming $r_1 \simeq r_2$ and $\gamma_{G,2}'$ is the smaller one, the dependence of $\gamma_G'/\gamma_{G,2}'$ on the ratio of force constants Γ_{10}/Γ_{20} is plotted in Figure 4.28.

Figure 4.28 shows that the equivalent γ_G' of both the parallel and serial arrangements is always higher than $\gamma_{G,2}'$. For the parallel arrangement, $\gamma_{G,2}' \le \gamma_G' \le \gamma_{G,1}'$, the stronger bond contributes more to the equivalent γ_G'; when $\gamma_{G,1}'/\gamma_{G,2}' = 1$, the equivalent γ_G' is independent of Γ_{10}/Γ_{20}. For the serial configuration, $\gamma_{G,2}' \le \gamma_G' \le 2\gamma_{G,1}'$, the weaker bond contributes more to γ_G'. For equivalent γ_G', according to (4.175), the lowest value $2\gamma_{G,1}'\gamma_{G,2}'/(\gamma_{G,1}'+\gamma_{G,2}')$ is achieved when $\Gamma_{10}\gamma_{G,1}' = \Gamma_{20}\gamma_{G,2}'$, and the mismatch of $\Gamma_i\gamma_{G,i}'$ of neighboring bonds causes an increase in the anharmonicity. To increase anharmonicity and reduce the sound velocity, the serial arrangement is preferred.

4.11.5 Prediction of Thermal Conductivity

Using the relations for γ_G and T_D, when ω_c is comparable with ω_D, the thermal conductivity can be readily calculated with the Slack relation. The predicted thermal conductivities of some crystals at high temperatures are listed in Table 4.5 and

Table 4.5. *Predicted k_p and Slack model for parameters of some crystals, at listed temperatures. The experimental results [313, 9] are shown in the parentheses, and the calculated results by Slack are shown in the brackets. Slack used $T_{D,\infty}$ from the D_p, which is different from the experimental T_D. γ_G used by him are derived from experiments except for diamond, SiC, Ge, GaAs, and BP (0.7 was used for these crystals). m is found from (4.171).*

Crystals	T, K	T_D, K	$N_\circ$	γ_G	k_p, W/m-K
Ar	84	94 (85)	1	2.83 (2.73)	0.5 (0.4)[3.8]
Kr	66	87 (73)	1	2.83 (2.84)	1.1 (0.5)[0.4]
C(Diamond)	300	2183 (2230)	2	1.0 (0.9)	1292 (1350)
Ge	235	382 (360)	2	1.0 (0.76)	95 (83)[89]
Si	395	584 (625)	2	1.0 (0.56)	76.7 (115)[93]
Cu	300	339 (315)	1	2.0	14.4 (10^a)
Pt	300	194 (230)	1	2.0	9.1 (6^a)
GaAs	220	367 (346)	2	1.01 (0.75)	72 (81)[77]
CaF_2	345	453(510)	3	1.89 (1.89)	7.0 (8.5)[9.1]
MgO	600	1034 (945)	2	1.68 (1.44)	53 (25) [28]
NaCl	230	382(330)	2	1.71(1.57)	11.1 (8.6)[6.3]
c-BN	300	1614	2	1.22	733 (748)
SiC	300	1212 (1079)	2	1.11 (0.76)	463 (490)[461]
BP	670	891 (982)	2	1.0	97.46 (110)[166]

[a] The lattice conductivities are from reference [24]. They are obtained by subtracting the electrical thermal conductivity (derived from Wiedemann–Franz law) from the total thermal conductivity.

shown in Figure 4.29(a), and the measured values and the values calculated in [313] are also given. Note that Slack used $T_{D,\infty}$ calculated from the phonon DOS D_p, which is different from the measured T_D listed in Table 4.5. Table 4.5 shows that the thermal conductivities and the Debye temperatures estimated by our model agree well with the experimental results and the Slack results, but the Grüneisen parameters are normally overestimated in our model, as discussed in Section 4.11.4. The average mean-square error between the estimated values and the experimental results is about 20%. Slack used $T_{D,\infty}$ along with the experimental γ_G (but 0.7 was used for Ge, Si, and SiC for better agreement with the experiments [313]), both of which are normally slightly lower than the values estimated in our model. Note that we also predict the lattice thermal conductivity of Al and Pt by considering only the phonon–phonon scattering. The crystalline metals normally have a low lattice thermal conductivity, not only because of the strong scattering of phonons by free electrons, but also because of their large Grüneisen parameters and small force constants (caused by long-range electrostatic interactions).

Also note that T_D and k_p used in Table 4.5 are different from those in Tables A.1 and A.2. This is because of the range of measured k_p reported for diamond.

When $\omega_c \ll \omega_D$, the thermal transport is dominated by the short-range acoustic phonons and optical phonons. Whereas the acoustic contribution can be calculated with a relation similar to the Cahill–Pohl (C–P) relation [52, 154], the optical part is difficult to determine and it is comparable to the acoustic contribution [154, 235]. However, for some special atomic structures, the phonon mean free path is limited

$$k_p = \frac{4.0 \times 10^{12} N_l \delta \Gamma^{2/3}}{\gamma_G^2 \langle M \rangle N_c^{7/6} T}, \; \text{W/m-K}$$

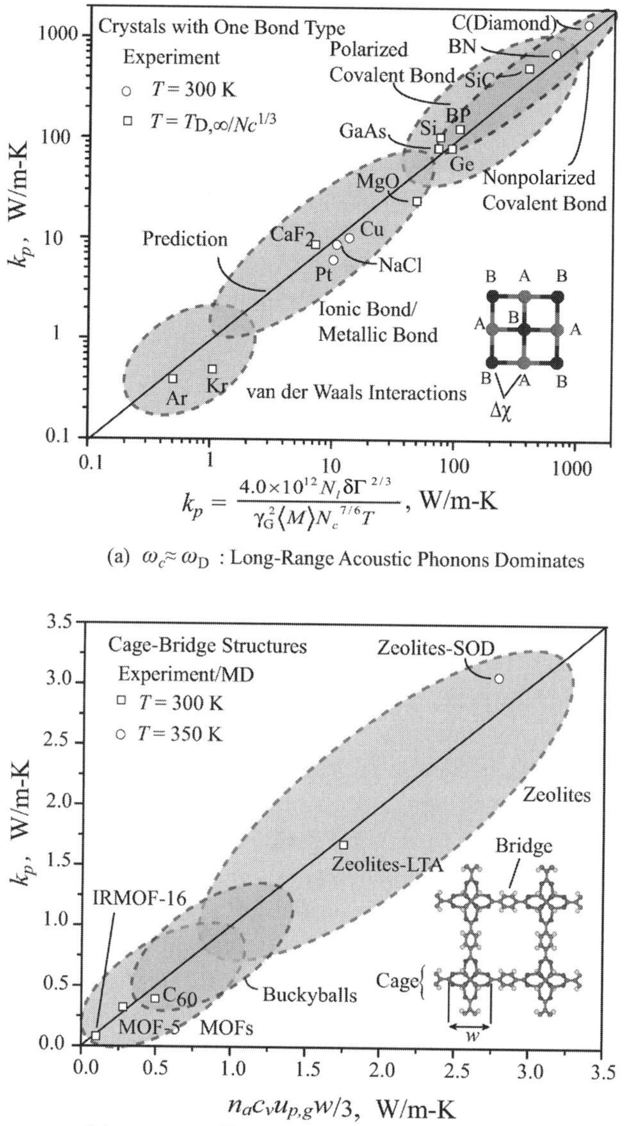

(a) $\omega_c \approx \omega_D$: Long-Range Acoustic Phonons Dominates

$n_a c_v u_{p,g} w/3$, W/m-K

(b) $\omega_c \ll \omega_D$: Short-Range and Optical Phonons Are Important

Figure 4.29. Comparison of the predicted lattice thermal conductivity of some compact crystals with the experimental results. (b) Comparison of the calculated thermal conductivities of some cage-bridge crystals with the experimental or MD results [152, 155].

by the crystal structure, and the thermal conductivities of such crystals often exhibit temperature-independence above the Debye temperature. From (4.109) and $k_p = n c_{v,p} u_{p,A} \lambda_p / 3$ ($c_{v,p}$ is the heat capacity per atom), if the phonon mean free path λ can be determined from the characteristics of the structure, the thermal conductivity can also be easily calculated.

There exist many special structures that can limit the phonon mean free path at high temperatures. For example, in the filled skutterudite structures [333], the fillers act as scatterers and limit the phonon mean free path to be the distance between the scatterers. Here we discuss the cage-bridge structure, which is common for nanoporous crystals, e.g., zeolites, MOFs, and many molecular crystals.

The cage-bridge structure includes complex multiatomic cages connected by relatively simple bridges [see Figure 4.29(b)] (sometimes the cages may also be joined directly without bridges). In such a structure, the atoms in the cage are normally much more than the atoms acting as connectors. Zeolites and MOFs are good examples of such structures. Some siliceous zeolites, e.g., zeolite A (LTA), faujasite (FAU), and sodalite (SOD), contain the complex SOD cage built from SiO_4 tetrahedra [235]. MOF-5, the smallest cubic MOF structure, comprises Zn_4O clusters linked by 1,4-benzenedicarboxylate (BDC) [154].

Many molecular crystals consist of large, complex molecules held together by weak van der Waals interactions or hydrogen bonds. The intramolecular interactions are much stiffer than the intermolecular interactions. They can also be considered a special type of cage-bridge structure, and each complex molecule can be considered a cage.

When $T \geq T_D$, all the vibration modes will have the same contribution to the total vibration energy. Because most atoms lie in the cage, most vibration energy is located in the cage and a fraction of the vibration energy transports to the next cage through connectors. However, the large coordination number difference or bond-stiffness difference makes the connector a bottleneck for the energy transport, and most phonon energy is localized in the cage or reflected at the connectors. In [235] it is shown that the Si–O–Si bonds contribute to the energy localization in FAU- and SOD-zeolites. The work for MOF-5 [154] also showed that the carbon connector limits the transport of phonon energy. An indicator of this phenomenon is the large difference between the phonon partial DOS, weighted by the concentration of atoms) of the cage and that of the connectors (as shown in [154]). Therefore the connectors will act as scatterers in the structure. If the cage is relatively rigid (phonons experience little scattering within the cage), the phonon mean free path will be limited by the distance between the connectors at the boundary of the cage, which is often the same as the cage size w. For molecular crystals, w is essentially the dimension of the molecule. Then we write (4.109) as

$$k_p = \frac{1}{3} n c_{v,p} u_{p,\mathrm{A}} w. \tag{4.177}$$

When the temperature is higher than the Debye temperature, c_v can be simply set as $3k_B$.

This simple mean-free-path model for the cage-bridge structures leads to good agreement with the experimental values or the MD results, as shown in Table 4.6. The values calculated by the Slack relation (k_S) are also shown, and it is apparent that k_S has a lower value.

When the temperature decreases, the high-frequency phonon modes caused by the internal vibrations of the cage will decrease much faster than the low-frequency modes, and the fraction of localized energy will decrease. When the temperature is

Table 4.6. *Comparison of predicted thermal conductivities of some cage-bridge structures calculated by (4.177) and the Slack relation, with the experimental and the MD results.*

Crystals	T, K	$u_{p,\mathrm{A}}$, m/s	w, Å	k_p, W/m-K (4.177)	$k_{p,\mathrm{S}}$	Experiment/MD
MOF-5[a]	300	1184	7.16	0.28	0.025	0.32
IRMOF-16[a]	300	600	7.16	0.10	0.01	0.08
SOD	350	4200[b]	8.88	2.79	1.58	3.09[c]
LTA	300	3200[b]	8.88	1.75	0.47	1.68[c]
C_{60}	260	2000[d]	7.00	0.52	0.007	0.4[e]

[a] values for MOF-5 are taken from [154]. The data for IRMOF-16 are calculated by MD using the same potentials
[b] values are derived from the bulk modulus [10], by setting the Poisson ratio as 0.3
[c] values are taken from MD results from [235]
[d] values are derived from the bulk modulus [93]
[e] values are from reference [354]

much lower than the Debye temperature, the fraction of localized vibration energy will be small and the phonon mean free path will no longer be limited by the cage size.

This simple atomic-structure-based model can be used to quickly estimate the high-temperature thermal conductivity of crystals. On the other hand, some useful insights into the design of materials with desired properties can be extracted.

According to (4.132) and using (4.161), we have

$$k_{p,\mathrm{S}} = \frac{4.0\times 10^{12} N_l \Gamma^{3/2} V_a^{1/3}}{\gamma_{\mathrm{G}}^2 \langle M\rangle^{1/2} N_\circ^{7/6} T}, \qquad (4.178)$$

where N_l is a constant related only to lattice type. Therefore, to increase the thermal conductivity, one may increase the equivalent force constant Γ and lattice constant a, while reducing the mean atomic weight $\langle M\rangle$, $N_\circ$, and the Grüneisen parameter γ_{G}. Here FCC is expected to achieve a high thermal conductivity. The opposite approaches can be used to achieve a low lattice thermal conductivity.

Evidently molecular crystals will normally have a very low thermal conductivity because of the small Γ, large γ_{G} and $N_\circ$.

Table 4.4 and Figure 4.23(a) show that most metals have a low $\Gamma_{\mathrm{AA},1}$, around 50 N/m. Even when they bond with F (which has the highest $\Gamma_{\mathrm{AA},1}$), $\Gamma_{\mathrm{AB},1}$ is expected to be lower than 250 N/m. Also, metal elements normally have a heavy mass and ionic bonds have a relatively high $\Delta\chi$. In comparison, covalent bonds may have a higher $\Gamma_{\mathrm{AB},1}$, lower $\Delta\chi$, and those nonmetallic elements with a high $\Gamma_{\mathrm{AA},1}$ have a relatively light mass. Thus, for high thermal conductivity, covalent crystals are preferred. Among covalent crystals, the compounds of N and C are expected to have a high thermal conductivity, because N and C have the highest $\Gamma_{\mathrm{AA},1}$, moderate χ, light masses, and possibly high bond orders. In general, the sequence of lattice conductivity for crystals is k (nonpolarized covalent crystal) $>$ k (polarized covalent crystal) > k (ionic crystal) > k (molecular crystal), as shown in Figure 4.29(a). Furthermore, the oxidation states of the elements need to match and the mass difference should be small to achieve a small N_o. Materials satisfying these conditions are expected to have a high thermal conductivity, e.g., BN, AlN, BP, and SiC (listed in Table 4.5).

4.12 High-T Phonon Conductivity Decomposition: Acoustic Phonons

Here we consider only the interphonon interactions (phonon–phonon scattering), which are dominant at high temperatures. Three main techniques have been developed to predict the thermal conductivity of a dielectric material by use of MD simulations. These are the G-K approach (an equilibrium method), a direct application of the Fourier law of conduction (a steady-state, nonequilibrium method, sometimes called the direct method), and unsteady methods. The G-K method was discussed in Section 3.6, and here we present some results for the Ar FCC crystal (acoustic phonons only, as there is one atom per primitive cell, Section 4.3) [234]. In Section 4.13, we discuss crystal, with optical phonons.

The net flow of heat in a solid, given by the heat current vector $\dot{\boldsymbol{w}}$ (Table 3.9), fluctuates about zero at equilibrium. In the G–K method discussed in Section 3.5, the thermal conductivity is related to the elapsed time it takes these fluctuations to dissipate, and for an isotropic material is given by (3.78) as

$$k_p = k_{p,\text{G-K}} = k_{p,\text{MD}} \frac{1}{k_\text{B} V T^2} \int_0^\infty \frac{\langle \dot{\boldsymbol{w}}(t) \cdot \dot{\boldsymbol{w}}(0) \rangle}{3} \text{d}t, \qquad (4.179)$$

where $\langle \dot{\boldsymbol{w}}(t) \cdot \dot{\boldsymbol{w}}(0) \rangle$ is the heat current autocorrelation function (HCACF) (Table 3.9). In materials for which the fluctuations are long lived (i.e., the mean phonon relaxation time is large), the HCACF decays slowly. The thermal conductivity is related to the integral of the HCACF and is accordingly large. In materials such as amorphous solids, for which the mean relaxation time of phonons is small, thermal fluctuations are quickly damped, leading to a small integral of the HCACF and a low thermal conductivity. The heat current vector is given by (3.78)

$$\dot{\boldsymbol{w}} = \frac{\text{d}}{\text{d}t} \sum_i \boldsymbol{x}_i E_i, \quad E_i = E_{k,i} + E_{p,i}, \qquad (4.180)$$

where the summation is over the particles in the system, and $\boldsymbol{x}_i$ and E_i are the position vector and energy (kinetic and potential) of a particle. A pair potential, such as the L–J potential, (4.180) can be recast (Table 3.9) as (Chapter 3 problem)

$$\dot{\boldsymbol{w}} = \sum_i E_i \boldsymbol{u}_i + \frac{1}{2} \sum_{i,j} (\boldsymbol{F}_{ij} \cdot \boldsymbol{u}_i) \boldsymbol{x}_{ij}, \qquad (4.181)$$

where $\boldsymbol{u}$ is the velocity vector of a particle, and $\boldsymbol{x}_{ij}$ and $\boldsymbol{F}_{ij}$ are the interparticle separation vector and force vector between particles i and j. This form of the heat current is readily implemented in MD simulation. A derivation of (4.179) and further discussion is given in Appendix B.

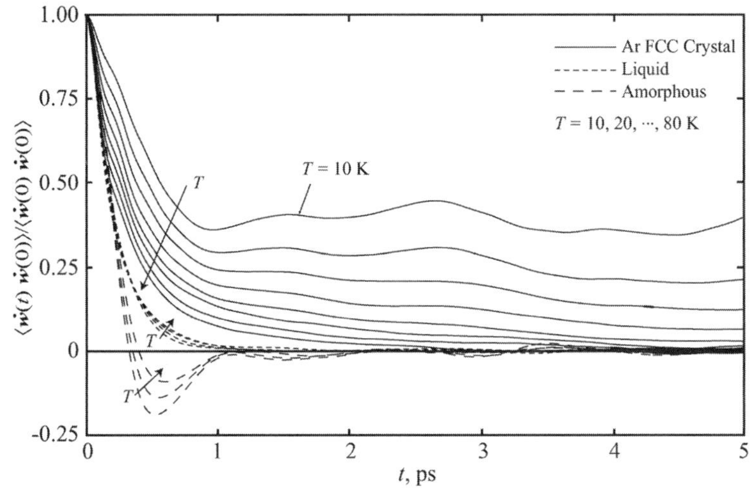

Figure 4.30. Time dependence (decay) of the HCACF for FCC, amorphous, and liquid Ar, for various temperatures [234].

All simulations used in the thermal conductivity calculations consist of 10^6 time steps ($\Delta t = 4.205$ fs) beyond the standard initialization period over which the heat current vector is calculated every five time steps. A correlation length of 5×10^4 time steps starting after 2×10^5 time steps is used in the autocorrelation function. For all cases, five independent simulations (with random initial velocities) are performed and the HCACFs are averaged before finding the thermal conductivity. This ensures a proper sampling of phase space [208]. For the FCC crystal at a temperature of 10 K, where the correlation time is long, 10 independent simulations are performed.

4.12.1 Heat Current Autocorrelation Function

The decay of the HCACF for Ar is shown in Figure 4.30. The HCACF is normalized by its zero time value to allow for comparisons among the different temperatures. Longer time scales are shown for the Ar FCC crystal in Figures 4.31(a), 4.31(b), and 4.31(c) for temperatures of 10, 50, and 80 K, respectively. Note that, as the temperature increases, the HCACF decays of the three phases are approaching each other.

The Ar FCC crystal HCACF shows a two-stage behavior. There is an initial drop, similar for all cases, followed by a longer decay, whose extent decreases as the temperature increases. The oscillations in the secondary decay are believed to be a result of the periodic boundary conditions.

For all cases considered, the integral of the HCACF converges well, and the thermal conductivity can be specified directly by averaging the integral over a suitable range. To remove the subjective judgment, in [208] two methods are proposed by which the thermal conductivity can be specified. In the first-dip (FD) method, the

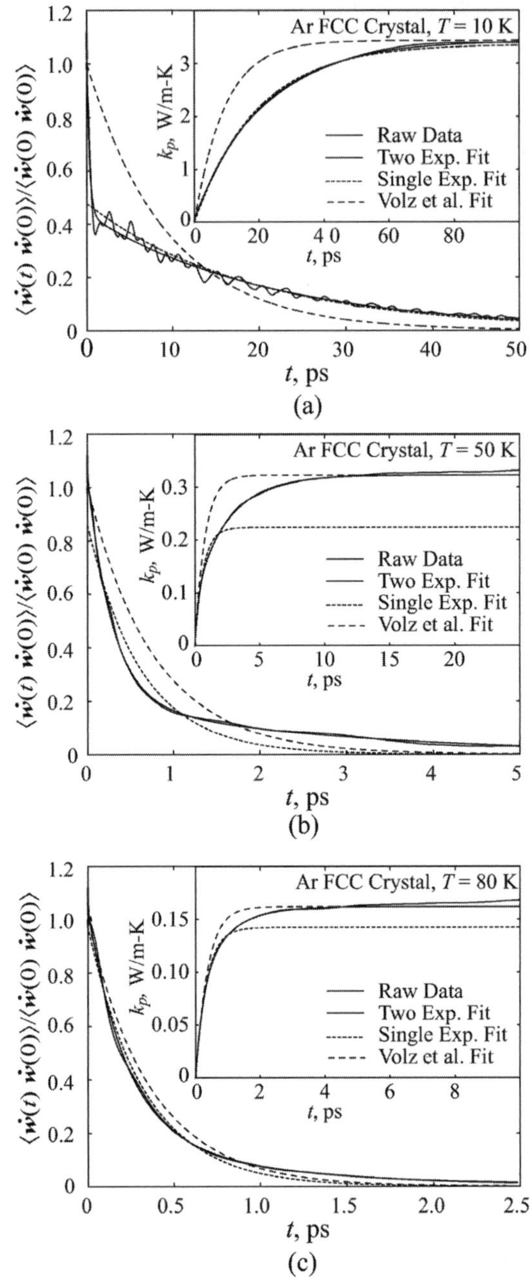

Figure 4.31. Time variation of the raw HCACF and thermal conductivity, and fits of one- and two-term exponential functions and the model of [341], for the FCC crystal at $T =$ (a) 10 K, (b) 50 K, and (c) 80 K. Note the different time scales on the HCACF and thermal conductivity plots for each condition. For a number of cases, the raw data and the two-term exponential fits are indistinguishable [234].

integral is evaluated at the first place where the HCACF goes negative. In the exponential fit (EF) method, an exponential function is fitted to the HCACF beyond a certain point (determined on a case-by-case basis), and this function is then used to calculate the contribution of the tail to the integral. Up to that point the integral is evaluated directly. In the investigation of β-SiC [208], no significant differences were found between the predictions of these two methods.

The liquid HCACF shows a single stage decay, with a time scale comparable with that of the initial drop in the FCC crystal HCACF. Both the FD and EF methods are suitable for specifying the thermal conductivity. The amorphous-phase HCACF shows a very different behavior. It drops below zero in the initial decay and oscillates between positive and negative as it converges to zero. The velocity autocorrelation function for amorphous L–J Ar shows a similar form [218]. This behavior is interpreted as follows. In the FCC crystal, each atom experiences the same local environment. By averaging over time, the same is true for the liquid. This is not the case for the amorphous solid, for which each atom has a distinct local environment. At short time scales, atoms near their equilibrium positions experience the free trajectory of a liquid atom. When the atom eventually feels the effects of the other atoms, the trajectory changes. Because the intended trajectory cannot be completed, the correlation goes negative.

The time scale for this behavior is comparable with that of the liquid HCACF. The FD and EF methods are not appropriate here, and the thermal conductivity must be found from a direct specification of the integral.

The low-temperature oscillations in the autocorrelation decay of Figure 4.30 are because of long-range phonons (the higher thermal conductivity). In this case the finite computational domain and the periodic boundary conditions influence the results (causing the oscillations)

4.12.2 Phonon Conductivity Decomposition

Based on the observed shape of the FCC crystal HCACF, it can be fitted to a sum of two exponential functions as

$$\frac{\langle \dot{\boldsymbol{w}}(t) \cdot \dot{\boldsymbol{w}}(0) \rangle}{3} = A_{sh,\mathrm{A}} \exp(-t/\tau_{p,sh,\mathrm{A}}) + A_{lg,\mathrm{A}} \exp(-t/\tau_{p,lg,\mathrm{A}}), \tag{4.182}$$

as suggested in [62]. The fitting procedure needs special care for large conductivities.

With (4.179), the thermal conductivity is then

$$\begin{aligned} k_p &= \frac{1}{k_\mathrm{B} V T^2}(A_{sh,\mathrm{A}} \tau_{p,sh,\mathrm{A}} + A_{lg,\mathrm{A}} \tau_{p,lg,\mathrm{A}}) \\ &\equiv k_{p,sh,\mathrm{A}} + k_{p,lg,\mathrm{A}}. \end{aligned} \tag{4.183}$$

In (4.182) and (4.183) the subscripts ac, sh, and lg refer to acoustic, short-range, and long-range. The two-stage decay in the HCACF was first observed in [196]. It is in contrast to the Peierls theory of thermal conductivity, which has been found to be consistent with a single-stage decay of the HCACF [196, 207]. In [168] it is suggested that the two stages in the HCACF represent contributions from local dynamics and the dynamics of phonon transport, each having a time constant τ_p and strength A. The use of the term "local" is questionable as, in a crystal, there are no localized vibrational modes. The fit curves for the HCACF and thermal conductivity obtained from (4.182) for temperatures of 10, 50, and 80 K are shown in Figs. 4.31(a), 4.31(b), and 4.31(c), respectively. The fit captures the two-stage decay very well at all temperatures. The fits of a single exponential function with time constant τ_p, according to

$$\frac{\langle \dot{w}(t) \cdot \dot{w}(0) \rangle}{3} = A_1 \exp(-t/\tau_p), \tag{4.184}$$

are also shown in Figs. 4.31(a), 4.31(b), and 4.31(c). The agreement with the raw HCACF is reasonable at low and high temperatures, but poor at the intermediate temperatures.

The two-stage behavior of the FCC crystal HCACF and the resulting decomposition of the thermal conductivity into two distinct components are interpreted in the context of the phonon mean relaxation time. Although the relaxation time is generally taken to be an averaged quantity, it can be applied to an individual phonon. For a given phonon mode, there will thus be some continuous distribution of relaxation times. Physically, the lower bound on the relaxation time corresponds to a phonon with a mean free path equal to one-half of its wavelength. This is the C–P limit, a thermal conductivity model developed for amorphous materials (discussed in Section C.2 of Appendix C) [52, 100]. The first part of the thermal conductivity decomposition $k_{p,sh,\mathrm{A}}$ takes into account those phonons with this limiting value of relaxation time. Phonons with longer relaxation times are accounted for by the second term $k_{p,lg,\mathrm{A}}$, which has a longer decay time.

Self-energy correlations (i.e., an autocorrelation), similar to the nearest-neighbor correlations, are plotted in Figure 4.32 for temperatures of 10, 20, 50 K, and 80 K. Although there is coherent behavior over long time periods at the low temperatures, this effect diminishes as the temperature increases. This is consistent with the temperature trends in $\tau_{p,lg,\mathrm{A}}$ and $k_{p,lg,\mathrm{A}}$. For temperatures of 10 and 20 K (shown separately in the insets of Figure 4.32), exponentials with time constants equal to the appropriate $\tau_{p,lg,\mathrm{A}}$ from the thermal conductivity decomposition are superimposed. The trends in the energy correlation curves are well bounded by the exponentials. The manifestation of both $\tau_{p,sh,\mathrm{A}}$ and $\tau_{p,lg,\mathrm{A}}$ outside of the HCACF supports the use of energy correlation functions for understanding heat transfer in the real-space coordinates. The intermediate time scale in the long-time behavior at lower temperatures (shown as τ_i in Figure 4.32 for the 10 K curve) is thought to be associated with the periodic boundary conditions.

It is interesting to note that the long-time-scale behavior is made up of successive short-time-scale interactions ($\tau_{p,sh,\mathrm{A}}$, also shown in Figure 4.32). At the lower temperatures, the atom-to-atom interactions propagate step by step, leading to behavior with a period of $2\tau_{p,sh,\mathrm{A}}$ over the long time scale. At higher temperatures, as the mean relaxation time gets smaller, the overall behavior approaches that of a

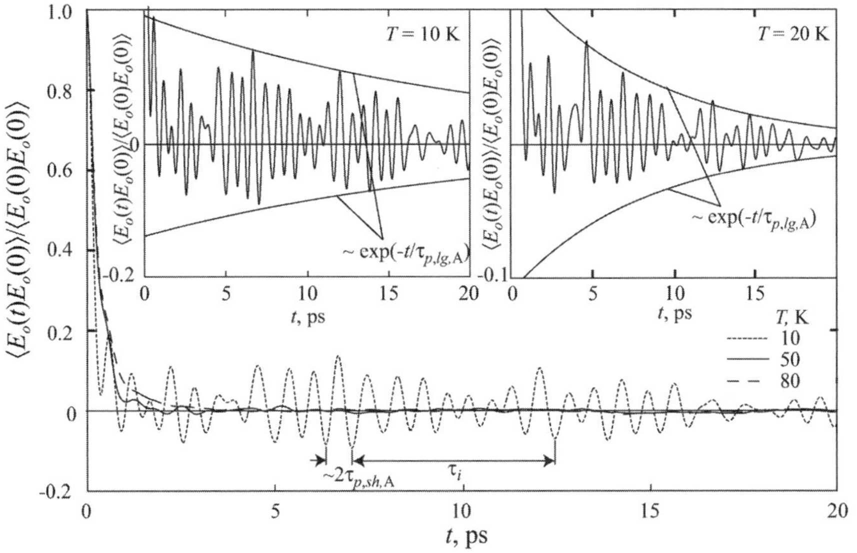

Figure 4.32. Particle energy autocorrelation functions for the Ar FCC crystal at $T = 10$, 50, and 80 K. The energy data correspond to deviations from the mean values. Note the diminishing long-time coherence as the temperature is increased. The inset plots show a smaller scale for the vertical axis for the temperatures of 10 and 20 K, along with curves representing the decay time associated with $k_{p,lg,\mathrm{A}}$ [234].

damped oscillator (i.e., a monotonic decay), as opposed to a set of coupled oscillators (as seen at low temperatures), which can gain and lose energy.

4.12.3 Comparison with Experiment

The predicted thermal conductivities, experimental values [329], and the C–P limit for Ar are shown in Figure 4.33(a) as functions of temperature. Also included are the MD results given in [207], obtained under very similar simulation conditions. In Figure 4.33(b), the decomposition of the FCC crystal thermal conductivity into $k_{p,sh,\mathrm{A}}$ and $k_{p,lg,\mathrm{A}}$ is shown along with the C–P limit (4.131). The FCC crystal MD results are in reasonable agreement with the trend and magnitude of the experimental data (a decrease above the experimental peak value, which is near a temperature of 6 K), justifying the neglection of quantum effects. The data are in good agreement with those in [207].

As given in (4.110), the C-P limit is a quantum, harmonic expression. The MD simulations are classical and anharmonic. As such, for use in Figure 4.33, the classical limit of the C-P limit (4.131) is taken [i.e., the mode specific heat $k_\mathrm{B}x^2e^x/(e^x-1)^2$ equal to k_B], to give

$$k_{p,\text{C-P}} = \frac{1}{2}\left(\frac{\pi}{6}\right)^{1/3}\beta k_\mathrm{B} n^{2/3}\sum_\alpha u_{p,\alpha}. \tag{4.185}$$

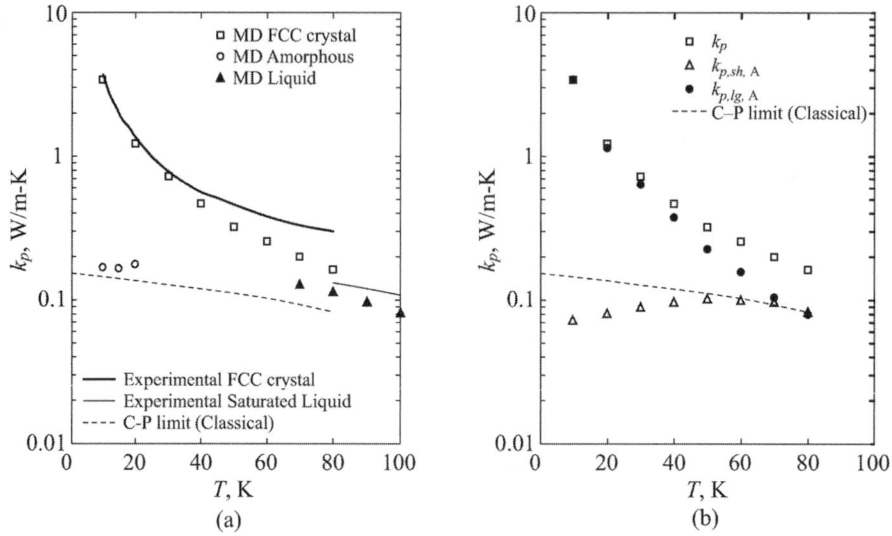

Figure 4.33. Temperature dependence of the experimental and predicted L–J Ar thermal conductivities [234].

The number density, n, is taken from the FCC crystal MD results. Because of anharmonic effects at finite temperatures, the specific heat will deviate from the classical value, which is accounted for by the factor β ($0.5 \leq \beta \leq 1$). The FCC crystal data are used for the calculations, as the amorphous phase is stable only up to a temperature of 20 K. The temperature dependence of the sound speeds, $u_{p,\alpha}$ (of which there are three, one longitudinal and two transverse), is obtained from quasi-harmonic dispersion curves in the [100] direction based on the FCC crystal simulation cell sizes. There will be a difference between the amorphous sound speeds and those for the FCC crystal. As no sound speed data are available for the amorphous L–J phase, the FCC crystal values are used, scaled by a factor of 0.8 (typical for Si and Ge [52, 151]). Under these approximations, the C-P limit is plotted with the understanding that there will be some errors in the calculated values.

4.13 High-T Phonon Conductivity Decomposition: Optical Phonons

Optical phonons contribute to phonon conductivity, but their contributions are weakly temperature dependent and generally small, except for low thermal conductivity crystals. An example is shown for silica structures in [235], where dense (quartz) and cage (zeolites) crystals have been studied by use of MD. The structures are shown in Figure 1.7. The G-K autocorrelation decay decomposition (4.182) for acoustic phonons is now extended to include optical phonons, i.e.,

$$\frac{\langle \dot{\boldsymbol{w}}(t) \cdot \dot{\boldsymbol{w}}(0) \rangle}{3} = A_{sh,\mathrm{A}} \exp(-t/\tau_{p,sh,\mathrm{A}}) + A_{lg,\mathrm{A}} \exp(-t/\tau_{p,lg,\mathrm{A}}) + \sum_i B_{\mathrm{O},i} \exp(-t/\tau_{p,\mathrm{O},i}) \cos(\omega_{p,\mathrm{O},i} t). \quad (4.186)$$

Table 4.7. *Thermal conductivity decomposition for quartz (a direction) at $T = 250$ K. The time constant for the short-range acoustic-phonon component is obtained from the energy correlation analysis [235].*

Component	τ_p, ps	$\omega/2\pi$, THz	$k_{p,i}$, W/m-K
Short-range acoustic $k_{p,sh,\mathrm{A}}$	0.016	–	1.143
Long-range acoustic $k_{p,lg,\mathrm{A}}$	2.37	–	9.494
Optical $k_{p,\mathrm{O}}$			
1	3.19	14.2	0.087
2	1.65	18.3	0.436
3	2.50	21.6	0.052
4	1.24	22.8	0.242
5	2.82	32.8	0.070
			0.887
Total, k_p			11.524

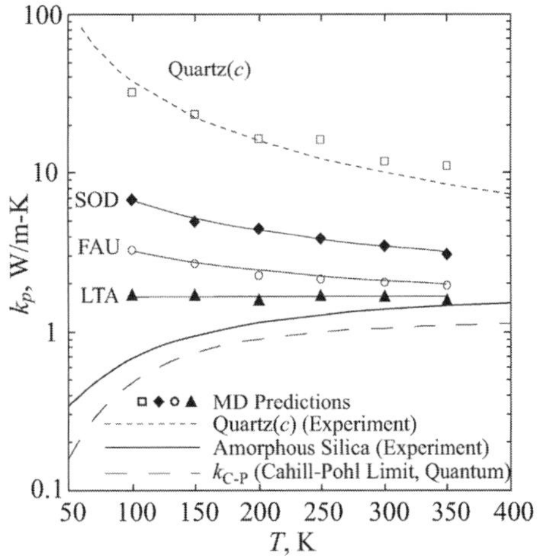

Figure 4.34. MD predicted and experimental thermal conductivities plotted as functions of temperature for silica crystals. All the structures are shown in Figure 1.7. The zeolite MD data are joined by best-fit power-law curves to guide the eye [235].

Then, using the G-K relation listed in Table 3.9, similar to (4.183), we find that the phonon conductivity is

$$k_p = \frac{1}{k_\mathrm{B} V T^2}\left(A_{sh,\mathrm{A}}\tau_{p,sh,\mathrm{A}} + A_{lg,\mathrm{A}}\tau_{p,lg,\mathrm{A}} + \sum_i \frac{B_{\mathrm{O},i}\tau_{p,\mathrm{O},i}}{1+\tau^2_{p,\mathrm{O},i}\omega^2_{p,\mathrm{O},i}}\right)$$

$$\equiv k_{p,sh,\mathrm{A}} + k_{p,lg,\mathrm{A}} + k_{p,\mathrm{O}}. \tag{4.187}$$

The results for quartz at the temperature of 250 K are shown in Table 4.7. The five most distinguished optical-phonon branches are listed (the dispersion results are given in Figure 4.7). For quartz, the optical-phonon contribution is small. Their MD results for the four structures, along with the experimental results for quartz, an amorphous silica, are shown in Figure 4.34.

The contribution of the short- and long-range acoustic phonons, and the optical phonon, are shown separately in Figure 4.35. The optical-phonon contribution

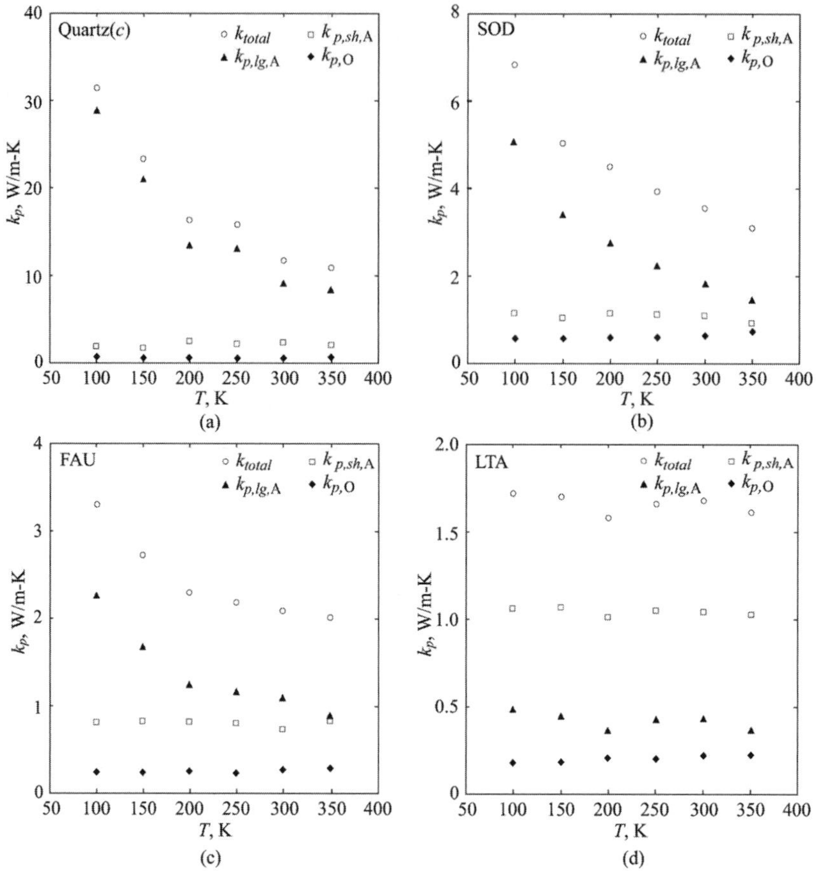

Figure 4.35. Thermal conductivity decomposition for (a) quartz (*c* direction), (b) SOD, (c) FAU, and (d) LTA [235].

is the most significant for zeolite LTA, which has the lowest conductivity. As was mentioned, the optical-phonon contribution is rather temperature independent.

Although not mentioned here, the direct (or nonequilibrium) methods (imposing a temperature gradient) have also been used for the prediction of k_p. Some aspects of these methods are reviewed in [237].

4.14 Quantum Corrections to MD/G-K Predictions

Because of their classical nature, MD simulations cannot explicitly take quantum effects into account. From the standpoint of lattice dynamics, there are two significant points to consider. First, the energy of the phonon modes is quantized in units of $\hbar\omega_k$. This is not true of the classical system, in which the mode energies are continuous. The second point, and the focus of this section, is the temperature dependence of the mode excitations. As predicted by the Bose–Einstein distribution, there are significant temperature effects in the quantum system that are not present in a classical description. The MD approach is thus not suitable near and below the maximum in the crystal-phase thermal conductivity (observed experimentally around one-tenth of the Debye temperature [226], and for Ar at a temperature of 6 K [71]), where quantum effects on the phonon mode populations are important. The thermal conductivity in this region is also strongly affected by impurities and boundary effects, which are not considered here. As such, an MD simulation of a perfect crystal with periodic boundary conditions will lead to an infinite thermal conductivity at zero temperature, as opposed to the experimental value, which goes to zero.

The classical nature of the MD simulations is perhaps most evident when we are considering the predicted specific heat, and how they differ from the quantum-mechanical calculations. The reason for the discrepancy is that in a classical-anharmonic system at a given temperature, all modes are excited approximately equally. The expectation value of the mode energy is about $k_B T$. In a harmonic system, the excitation is exactly the same for all modes, and the expectation value of the energy is exactly $k_B T$. In the quantum system, there is a freezing out of high-frequency modes at low temperatures. Only above the Debye temperature are all modes excited approximately equally. The quantum system also has a zero-point energy not found in the MD system.

There is no simple way to explicitly include quantum effects in the MD simulations. In fact, the whole idea behind the simulations is to save significant computational resources by ignoring quantum effects. That being said, some effort has been made to address the classical-quantum issue by mapping the results of MD simulations onto an equivalent quantum system. Using the results for the L–J FCC crystal, one of these approaches [203, 208, 225, 342], is presented and assessed here. The main idea is to scale the temperature and thermal conductivity (after the simulations have been completed) using simple quantum-mechanical calculations and/or arguments. For the remainder of this section, T_{MD} and k_{MD} are used to represent the temperature and thermal conductivity of the MD system, and T and k are used to represent the values for the real, quantum system.

The temperature in the MD system is calculated from the relation (Table 2.4)

$$\langle \sum_i \frac{1}{2} m_i |\boldsymbol{u}_i|^2 \rangle = \frac{3}{2}(N-1) k_B T_{MD}, \tag{4.188}$$

which equates the average kinetic energy of the particles (summed over the index i) to the expectation value of the kinetic energy of a classical system. For a harmonic system, where equipartition of energy exists between the kinetic and potential energies, and between the modes, the total system energy will be given by $3(N-1)k_B T_{MD}$. The temperature of the real system is found by equating this energy to that of a quantum phonon system (4.50), as suggested in [203, 208] from (4.50), we have

$$3(N-1) k_B T_{MD} = \sum_\alpha \hbar\omega_\alpha \left[\frac{1}{2} + \frac{1}{\exp(\hbar\omega_\alpha / k_B T) - 1}\right], \tag{4.189}$$

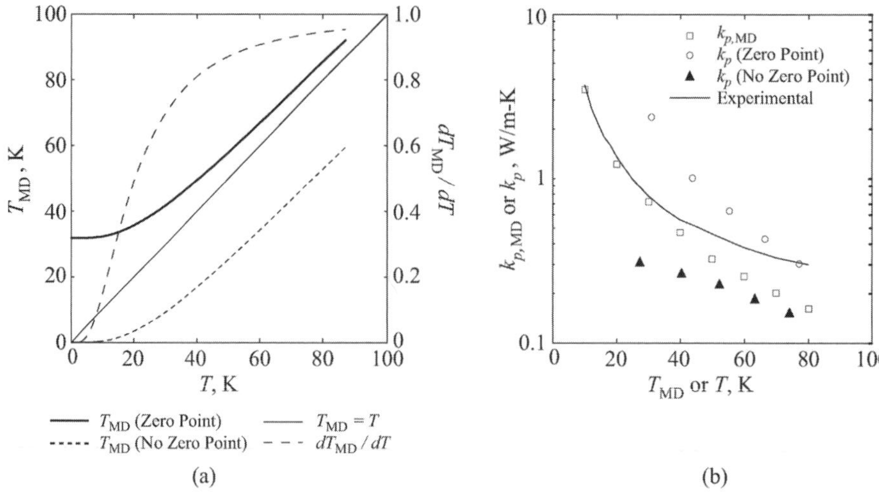

Figure 4.36. (a) Temperature and thermal conductivity scaling relations. (b) Scaled thermal conductivities with raw Ar FCC MD predictions and experimental data [234].

where the summation is over the α normal modes of the system. A similar relation has been proposed without the zero-point energy included [i.e., the factor of $\hbar\omega_\alpha/2$ on the right-hand side of (4.189) is not considered] [225, 342].

For the thermal conductivity, it has been proposed [203, 342, 208] that the conduction heat flux $\boldsymbol{q}$, in the classical and quantum systems, should be the same. Written in one dimension,

$$q_{k,x} = -k_{p,\mathrm{MD}}\frac{\mathrm{d}T_{\mathrm{MD}}}{\mathrm{d}x} = -k_p\frac{\mathrm{d}T}{\mathrm{d}x}, \tag{4.190}$$

such that

$$k_p = k_{p,\mathrm{MD}}\frac{\mathrm{d}T_{\mathrm{MD}}}{\mathrm{d}T}. \tag{4.191}$$

The predicted T_{MD} and $\mathrm{d}T_{\mathrm{MD}}/\mathrm{d}T$ curves for the cases of both including and neglecting the zero-point energy are shown in Figure 4.36(a). The data are plotted up to a temperature of 87 K, the melting point of the MD system.

When the zero-point energy is included, the MD simulations are of interest only at temperatures of 31.8 K and higher. This is an indication of the magnitude of the zero-point energy. The T_{MD} curve approaches T as the temperature is increased, and more modes are excited in the quantum system. The value of T_{MD} will always be higher than T because of the zero-point energy. The thermal conductivity scaling factor starts at zero. This ensures that the thermal conductivity will be zero at zero temperature. As the temperature increases, the scaling factor approaches unity.

When the zero-point energy is not included, the temperature scaling has the same shape as before, but has been shifted downward. In this case, T_{MD} will always be lower than T because of the manner in which the energy is distributed in the modes of the quantum system. The T_{MD} value of zero is relevant in this case, because the associated quantum system can have zero energy. The thermal conductivity

Table 4.8. *Scaled temperatures and thermal conductivities. The first two columns correspond to the raw Ar FCC MD data. The third through fifth columns correspond to the inclusion of the zero-point energy. The last three columns correspond to ignoring the zero-point energy [234].*

			Zero-point			No Zero-point	
T_{MD}, K	$k_{p,MD}$, W/m-K	T, K	dT_{MD}/dT	k_p, W/m-K	T, K	dT_{MD}/dT	k_p, W/m-K
10	3.44	–	–	–	31.0	0.684	2.36
20	1.22	–	–	–	43.9	0.831	1.01
30	0.718	–	–	–	55.4	0.889	0.639
40	0.467	27.3	0.656	0.306	66.4	0.921	0.430
50	0.323	40.6	0.815	0.263	77.2	0.940	0.304
60	0.255	52.3	0.881	0.225	–	–	–
70	0.201	63.3	0.916	0.184	–	–	–
80	0.162	74.2	0.938	0.151	–	–	–

scaling factor is identical to when the zero-point energy is included, as the energies differ only by a scalar.

The scaled thermal conductivities and unscaled Ar FCC MD predictions are given in Table 4.8 and shown in Figure 4.36(b) along with the experimental data. To obtain these results, the T values corresponding to the available T_{MD} values are obtained. Because of the nature of the scaling relation, not all the T_{MD} values have a corresponding T. The appropriate thermal conductivity scaling factor is then determined. Overall, the agreement with the experimental data worsens for either of the scaling possibilities compared with the raw MD data. Others have found an improved agreement (for β-SiC including the zero-point energy [208], and for Si, not including the zero-point energy [342]). This lack of consistency raises a high level of doubt about the validity of this approach and its possible widespread acceptance.

The main idea behind this somewhat *ad hoc* temperature scaling procedure is to map the classical MD results onto an equivalent quantum system. By not including the zero-point energy, a true quantum system is not being considered. For this reason, if such corrections are to be used, the zero-point energy should be included.

As it stands, there are a number ways that this method could be improved. These are related to the harmonic nature of the energy calculations on both sides of (4.189). The classical energy is based on an assumption of equipartition of energy. The average, total energy of the MD system is in fact less than $3(N-1)k_B T_{MD}$ because of anharmonic effects. As shown in Figure 2.15, at a temperature of 20 K, the deviation is 2.6%, and increases to 12.6% at a temperature of 80 K. This correction is straightforward to implement. The phonon-space energy is most easily calculated with the zero temperature, harmonic dispersion relation. However, temperature has a significant effect on the phonon dispersion, and temperature-dependent normal modes would make the the temperature scaling more rigorous.

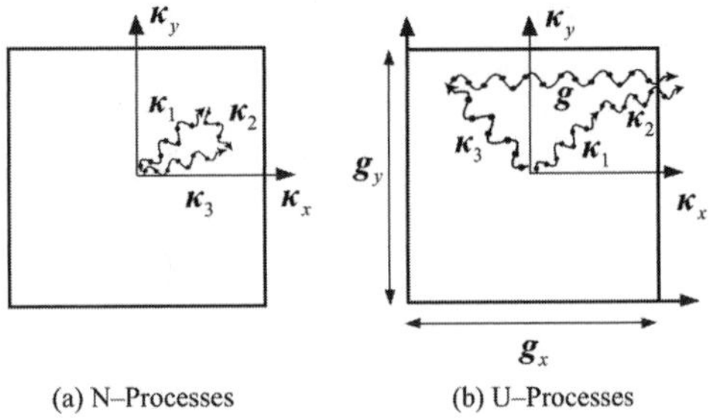

Figure 4.37. The N– and U–interphonon-scattering processes shown in reciprocal lattice space.

That being said, it is unlikely that these modifications would lead to a much improved model. The main drawback of this temperature/thermal conductivity scaling approach, as discussed in [207], is that it is a postprocessing step that maps the entire MD system onto a quantum description. The effects are manifested on a mode-by-mode basis, and on the basis of how the energy is distributed, therefore making corrections on an integrated level simply not suitable. In [207], it is suggested to link the classical MD system to a quantum description through the BTE.

In [62] a more general approach is used to investigate the classical-quantum issue by comparing the general forms of the classical and the quantum HCACFs. They do not find evidence to support the use of quantum corrections with MD thermal conductivity predictions. They argue that this is because long-wavelength phonons are the dominant contributors to the thermal transport, which are active even at low temperatures. This is in contrast to the specific heat, for which it is the high-frequency (short-wavelength) modes that get excited as the temperature of the quantum system is increased and lead to the significant temperature dependence up to the Debye temperature.

As discussed, the MD simulations are classical because it is within this framework that computational costs become reasonable enough to perform simulations of big systems or for long times. When comparing the results of MD simulations with experiments, there are additional factors beyond their classical nature that need to be considered. These include the interatomic potential used, size effects, and the simulation procedures. It is difficult to isolate these effects. Efforts are needed on all fronts to increase the confidence in the results of MD simulations.

4.15 Phonon Conductivity from BTE: Variational Method

Interphonon scattering, which dominates the phonon conductivity at high temperature, has in turn been assumed to be dominated by the three-phonon scattering processes. These are shown in Figure 4.37. There are two types of such interactions: type I ($1 \rightarrow 2+3$) and type II ($1+2 \rightarrow 3$). Because these scatterings change the energy of each phonon involved, it is not expected that they can be accurately presented by a relaxation-time approximation. These three-phonon interactions are divided into the N–Processes, in which the crystal momentum (and energy) is conserved, and the U–Processes, where the crystal momentum is not conserved (however, energy is).

These are shown in Figures 4.15(c) and 4.37, and we write

$$\boldsymbol{\kappa}_1 + \boldsymbol{\kappa}_2 = \boldsymbol{\kappa}_3 \text{ (momentum)}, \ \omega_1 + \omega_2 = \omega_3 \text{ (energy)}$$
$$\text{three-phonon N–Processes} \tag{4.192}$$

$$\boldsymbol{\kappa}_1 + \boldsymbol{\kappa}_2 = \boldsymbol{\kappa}_3 + \boldsymbol{g}, \ \omega_1 + \omega_2 = \omega_3 \ \text{three-phonon U–Processes,} \tag{4.193}$$

where $\boldsymbol{g}$ is the reciprocal lattice vector. The reciprocal lattice vector is not limited to a process that flips the wave-vector summation to the neighboring reciprocal space unit cell only but can involve other cells.

In addition to (4.192) and (4.193), there are other restrictions on 3-phonon interactions, for example, inspection of both energy and momentum conservations (N–Processes) indicates that such interactions cannot involve phonons of the same polarization $(\alpha_1, \alpha_2, \alpha_3)$ [26], whereas the U–Processes allow for the probability of this occurrence.

The Peierls interphonon interactions (scattering) are given by the in-and-out scattering (3.1.3) and the FGR (3.27), as [134]

$$\begin{aligned}\frac{\partial \langle f_{p,1}\rangle}{\partial t}|_s = &\sum_{\boldsymbol{\kappa}_2}\sum_{\alpha_2}[\sum_{\alpha_3} \dot{\gamma}_{p,\alpha_1,\alpha_2,\alpha_3}(\boldsymbol{\kappa}_1, \boldsymbol{\kappa}_2, -\boldsymbol{\kappa}_3)\langle f_{p,3}(f_{p,2}+1)(f_{p,1}+1)\\ &- (f_{p,3}+1) f_{p,2} f_{p,1}\rangle \delta_{\mathrm{D}}(\omega_1 + \omega_2 - \omega_3)\\ &+ \frac{1}{2}\sum_{\alpha_3} \dot{\gamma}_{p,\alpha_1,\alpha_2,\alpha_3}(-\boldsymbol{\kappa}_1, \boldsymbol{\kappa}_2, \boldsymbol{\kappa}_3)\langle (f_{p,1}+1) f_{p,2} f_{p,3}\\ &- f_{p,1}(f_{p,2}+1)(f_{p,3}+1)\rangle \delta_{\mathrm{D}}(\omega_1 - \omega_2 - \omega_3)],\\ &\text{three-phonon scattering rate,}\end{aligned} \tag{4.194}$$

or in terms of the scattering rate the results are

$$\begin{aligned}\dot{\gamma}_{p,\alpha_1,\alpha_2,\alpha_3}(\boldsymbol{\kappa}_1, \boldsymbol{\kappa}_2, \boldsymbol{\kappa}_3) = &\frac{1}{\tau_{p\text{-}p,\alpha1,\alpha2,\alpha3}(\boldsymbol{\kappa}_1, \boldsymbol{\kappa}_2, \boldsymbol{\kappa}_3)} = \frac{\pi\hbar}{4\rho^3 V^2 \omega_1\omega_2\omega_3}\\ &\times [\sum_{i,j,k}\sum_{\alpha,\beta,\gamma} \frac{\partial^3\langle\varphi\rangle}{\partial d_{i,\alpha}\partial d_{j,\beta}\partial d_{k,\gamma}}|_{\mathrm{o}} e_\alpha^{\boldsymbol{\kappa}_1,\alpha_1} e_\beta^{\boldsymbol{\kappa}_2,\alpha_2} e_\gamma^{\boldsymbol{\kappa}_3,\alpha_3}\\ &\times \exp\{i(\boldsymbol{\kappa}_1\cdot \boldsymbol{x}_{o,i} + \boldsymbol{\kappa}_2\cdot \boldsymbol{x}_{o,j} + \boldsymbol{\kappa}_3\cdot \boldsymbol{x}_{o,k})\}]^2\\ &\text{three-phonon scattering relaxation time.}\end{aligned} \tag{4.195}$$

The first triple summation is over the atoms in the system and the second is over the Cartesian coordinates. An example of an MD calculation of this scattering rate, for Ar FCC, is given in [236], where they also discuss the phonon–phonon interaction rules within classical MD simulations.

Note that the scattering matrix element is third-order in the interatomic potentials. The variational method is applied to the BTE [12, 45, 100, 135, 318], in which the phonon distribution function is explicitly involved. The deviation from equilibrium is

$$f_p'(\kappa) = f_p^\circ(\kappa) - \phi_\kappa \frac{\partial f_p^\circ(\kappa)}{\partial E_\kappa} = f_p^\circ(\kappa) + \phi_\kappa \frac{f_p^\circ(1+f_p^\circ)}{k_B T}, \qquad (4.196)$$

where we have used (Table 1.2)

$$f_p^\circ(E_p) = \frac{1}{\exp(\frac{E_p}{k_B T}) - 1}. \qquad (4.197)$$

The left-hand side of the BTE (Table 3.1) can be written as (4.95)

$$-\boldsymbol{u}_p \cdot \nabla T \frac{\partial f_p}{\partial T} = \frac{\partial f_p}{\partial t}|_s. \qquad (4.198)$$

For three-phonon scattering processes, shown by (4.192) and (4.193), we have

$$(\kappa_1, \alpha_1) + (\kappa_2, \alpha_2) \rightarrow (\kappa_3, \alpha_3), \qquad (4.199)$$

where the scattering rate $\partial f_p/\partial T|_s$ is written in terms of the transition rate $\dot{\gamma}_{\kappa_1,\kappa_2}^{\kappa_3}$ [from FGR (3.27)]. Then the BTE becomes

$$\begin{aligned} -\boldsymbol{u}_p(\kappa)\nabla T \frac{\partial f_p(\kappa)}{\partial T} = \int\int (&\{f_p(\kappa_1) f_p(\kappa_2)[1 + f_p(\kappa_3)] \\ &- [1 + f_p(\kappa_1)][1 + f_p(\kappa_2)] f_p(\kappa_3)\}\dot{\gamma}_{\kappa_1,\kappa_2}^{\kappa_3} \\ &+ \frac{1}{2}\{f_p(\kappa_1)[1 + f_p(\kappa_2)][1 + f_p(\kappa_3)] \\ &- [1 + f_p(\kappa) f_p(\kappa_2) f_p(\kappa_3)]\}\dot{\gamma}_{\kappa_1}^{\kappa_2,\kappa_3}) d\kappa_2 d\kappa_3. \end{aligned} \qquad (4.200)$$

With the deviation from the equilibrium given by (4.196), the isotropic phonon thermal conductivity is given by [26]

$$k_p^{-1} = \frac{1}{2k_B T^2} \frac{\int\int\int (\phi_{\kappa_1} + \phi_{\kappa_2} - \phi\kappa_3)^2 \dot{\gamma}_{\kappa_1,\kappa_2}^{\kappa_3} d\kappa_1 d\kappa_2 d\kappa_3}{|\boldsymbol{u}_p \phi_\kappa \frac{\partial f_\kappa^\circ}{\partial T} d\kappa|^2}, \qquad (4.201)$$

which is based on the variational principles and should have the minimum when the trial function ϕ_κ satisfies the BTE. The introduction of trial functions is discussed in [135].

The numerical results of [135] for the predicted phonon resistivity $1/k_p$ of Ge are shown in Figure 4.38. To predict the low-temperature regime, where impurity and crystal boundary effects dominate, these numerical results are included in the model. The U–Processes dominate at high temperature and are influenced by the N–Processes with this influence increasing with decreasing temperature. The resistance is minimum when the interphonon interaction diminishes and the impurity effect begins to dominate. The very low-temperature regime is dominated by crystal boundary scattering.

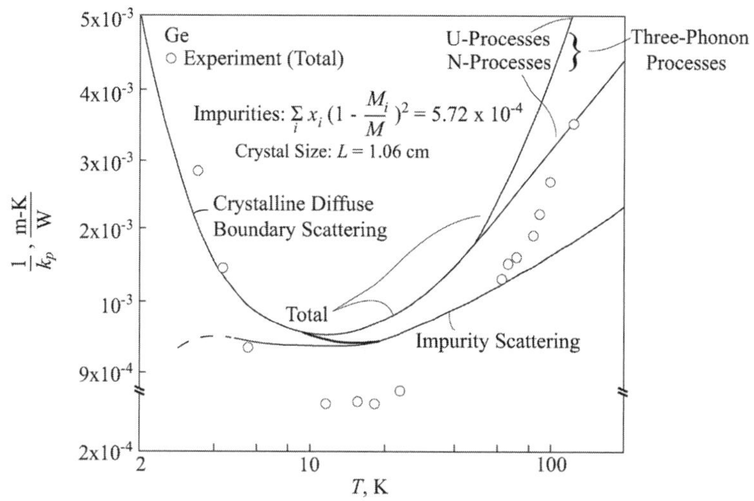

Figure 4.38. Prediction of the variational method, including the three-phonon interactions, for the variation of the phonon resistivity($1/k_p$) of Ge, as a function of temperature. The experimental results are also shown [135].

4.16 Experimental Data on Phonon Conductivity

Figure 4.39 shows typical measured phonon conductivity results as functions of temperature [329]. The theoretical treatments of Sections 4.9 and 4.11 predicts these trends. The results for amorphous phases of Se and Si are also shown. The theoretical treatments of Section 4.10 predict these trends.

For crystalline dielectrics, a maximum in k_p occurs at a fraction of the Debye temperature T_D, whereas for the amorphous phases, the conductivity increases monotonically with temperature and will eventually reaches a plateau (as shown in Figure 4.17). Some intermediate plateaus are also observed.

Figure 4.40 shows how the impurity, pores, and grain size can reduce the phonon conductivity of crystallized ZrO_2. The lattice structure change resulting from impurities are also shown. It is expected that would result in diminishing long-range acoustic-phonon contribution to thermal conductivity (Sections 4.12 and 4.13), i.e., $k_{p,lg,\mathrm{A}} \rightarrow 0$.

4.17 Phonon Boundary Resistance

At interfaces between two materials (discontinuity), the mismatch between phonon transport across the interface results in a resistance to the heat current. This interfacial resistance is generally referred to as the Kapitza resistance, because of phonon scattering.

Currently two theories have been applied to the prediction of the phonon boundary resistance [322]. The first is the acoustic mismatch model (AMM), which assumes scattering because of the difference in acoustic impedance (product of density and speed of sound) and does not distinguish between various phonon

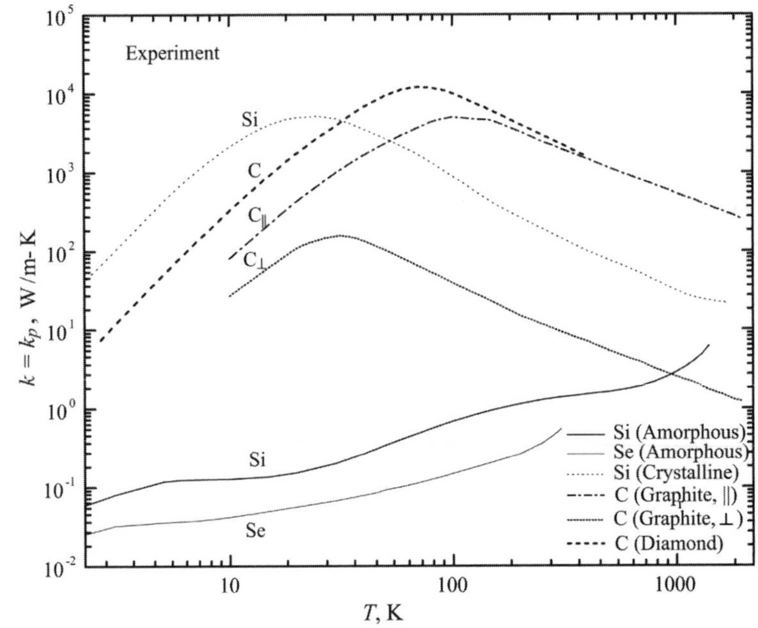

Figure 4.39. Measured variations of measured phonon conductivity, for a few elements in crystalline and amorphous phases, as functions of temperature [329]. For C, the diamond and graphite (anisotropic) crystalline phases are shown. For Si both single crystal and amorphous phases are shown (electronic contribution to k is very small), and for Se only the amorphous phase (low phonon conductivity) is shown.

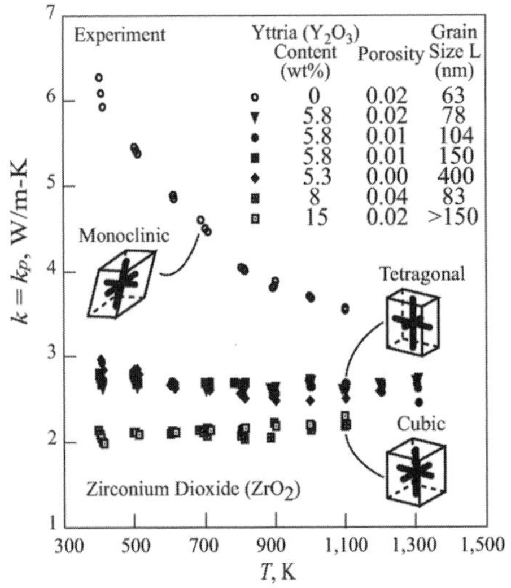

Figure 4.40. Measured variations of high-temperature phonon thermal conductivity of crystallized zirconia (ZrO_2) with respect to temperature. The effects of impurity [yttria, used to make the material a better ion conductor (electrolyte)], grain boundary, and pore scatterings, are shown [301]. These scatterings lower k_p and nearly eliminate the temperature dependence at high temperatures (i.e., the long-range acoustic phonons are suppressed).

wavelengths (for example, if the surface roughness is of the same order of magnitude as the phonon wavelength). The second theory is the diffuse mismatch model (DMM), which assumes that all phonons incident upon the interface will scatter diffusely (i.e., a rough interface with diffuse reflections). For this, the interface roughness $\langle\delta^2\rangle^{1/2}$ is larger than the dominant phonon wavelengths, estimated as $(T_D/T)a$, where a is the lattice spacing. Then for grain boundaries (same materials on both sides) the AMM prediction that no scattering occurs at the interface is reasonable at very low temperatures, where the phonon wavelength is larger ($\kappa \rightarrow 0$) than the interface length scales. As the temperature increases the phonon wavelength decreases, compromising the AMM theory. Both theories have their shortcomings, but at high temperatures, DMM results are in better agreement with the measurements and MD results.

The analysis of [268] shows that, by using the measured phonon DOS (compared with the Debye DOS), the DMM predicts the phonon boundary resistance relating accurately. The details of the DMM derivation are given in Appendix D and here we use the results.

The net conduction heat flux q_k from material 1 at temperature T_1 to material 2 at temperature T_2 is

$$\frac{1}{AR_{p,b}} = \frac{G_{p,b}}{A} = \frac{q_k}{T_1 - T_2} \quad \text{phonon boundary thermal resistance and conductance,} \tag{4.202}$$

where $R_{p,b}$(K/W) is the phonon boundary thermal resistance, $G_{p,b}$ (W/K) is the phonon boundary conductance, A is the area, and q_k is the conduction heat flux across boundary.

Starting with the general expression for the energy transported per unit time from material 1 to material 2, and assuming that the transmission coefficient $\tau_{b,1\text{-}2}$ is independent of the temperature on either side of the interface (only one side of the interface is considered), from (D.14), we have for the dimensional and dimensionless DMM phonon boundary resistance and for $T_1 \rightarrow T_2 \rightarrow T$,

$$\frac{1}{AR_{p,b}} = \frac{G_{p,b}}{A} = \frac{h_P \tau_{b,1\text{-}2}}{8\pi(T_1 - T_2)} \sum_\alpha u_{p,1,\alpha} \int_0^{\omega_D} D_p(\omega_p)\omega_p\Big[\frac{1}{\exp(\frac{h_P\omega_p}{2\pi k_B T_1}) - 1} - \frac{1}{\exp(\frac{h_P\omega_p}{2\pi k_B T_2}) - 1}\Big]d\omega_p,$$

or

$$\frac{1}{(AR_{p,b})^*} \equiv \frac{1}{AR_{p,b}} \frac{8\pi^2\hbar^3}{k_B^4 T_D^3 \tau_{b,1\text{-}2} \sum_\alpha u_{p,1,\alpha}^{-2}} = \frac{1}{T_D^3(T_1 - T_2)} \times \left[T_1^4 \int_0^{T_D/T_1} \frac{x^3}{\exp(x) - 1}dx - T_2^4 \int_0^{T_D/T_2} \frac{x^3}{\exp(x) - 1}dx\right], \tag{4.203}$$

where $\omega_D = \omega_{D,1}$, $T_D = T_{D,1}$, and α is the phonon polarization.

The DMM phonon boundary transmission coefficient is approximated as

$$\tau_{b,1\text{-}2} = \frac{\sum_{\alpha} u_{p,2,\alpha}^{-2}}{\sum_{\alpha} u_{p,1,\alpha}^{-2} + \sum_{j} u_{p,2,\alpha}^{-2}} \quad \text{phonon boundary transmission coefficient.} \tag{4.204}$$

For $T > T_{\rm D}$, (4.203) gives $1/(AR_{p,b})^* = 1/3$ (Figure D.1), for $T_1 \to T_2 \to T$ (independent of T_1, T_2, and T).

For three-phonon wave speeds (two transverse and one longitudinal), we have from (4.57)

$$\sum_{\alpha} u_{p,1,\alpha}^{-2} = \frac{1}{u_{p,1,{\rm L}}^2} + \frac{2}{u_{p,1,{\rm T}}^2}. \tag{4.205}$$

The phonon model speed $u_{p,i,\alpha}$ is related to the Debye temperature $T_{{\rm D},i,\alpha}$ through (4.63) and to the Debye angular frequency $\omega_{{\rm D},i,\alpha}$ through (4.60).

For $\hbar\omega_{\rm D}/k_{\rm B}T \gg 1$, we find an exact solution. Using these, the dimensionless phonon boundary conductance becomes

$$\frac{1}{(AR_{p,b})^*} \equiv \frac{8\pi^2\hbar^3}{(AR_{p,b})k_{\rm B}^4 T_{\rm D}^3 \tau_{b,1\text{-}2} \sum_{\alpha} u_{p,1,\alpha}^{-2}} = 24[1 - C(\frac{T_{\rm D}}{T})](\frac{T}{T_D})^3, \quad \frac{\hbar\omega_{\rm D}}{k_{\rm B}T} \gg 1, \tag{4.206}$$

where $C(T_{\rm D}/T)$ is defined as

$$C(\frac{T_{\rm D}}{T}) = \exp(-\frac{T_{\rm D}}{T})[1 + (\frac{T_{\rm D}}{T}) + \frac{1}{2}(\frac{T_{\rm D}}{T})^2 + \frac{1}{6}(\frac{T_{\rm D}}{T})^3]. \tag{4.207}$$

Figure 4.41(a) shows the measured and the Debye model of D_p for Bi_2Te_3 and Sb_2Te_3 (thermoelectric material discussed in Sections 5.16 and 5.17). The variation of the phonon boundary conductance $AR_{p,b}$ (4.206) as a function of temperature for these thermoelectric materials with a Cu interface is shown in Figure 4.41(b). The minimum is associated with the interplay between D_p and $f_p^{\rm o}$ in (4.203).

For Bi_2Te_3 and Sb_2Te_3, $V_c^{-1} = 5.95 \times 10^{27}$ m^{-3} and 6.40×10^{27} m^{-3}, and these and other properties are given in [83].

The full solution has an asymptotic high-temperature ($T > T_{\rm D}$) limit. As the temperature increases and more and more modes are excited, the phonon boundary resistance reaches its minimum value. The solution for $\hbar\omega/k_{\rm B}T \gg 1$ also has the same asymptotic high-temperature limit.

Figure 4.42 compares the experimental results of [77] with the prediction of (4.203) for TiN deposited (epitaxy) on a MgO substrate. By excluding the optical phonon heat capacity (when the optical phonons are removed from the DOS), a better agreement is found. This is because they are not expected to contribute to transport due to their low velocities.

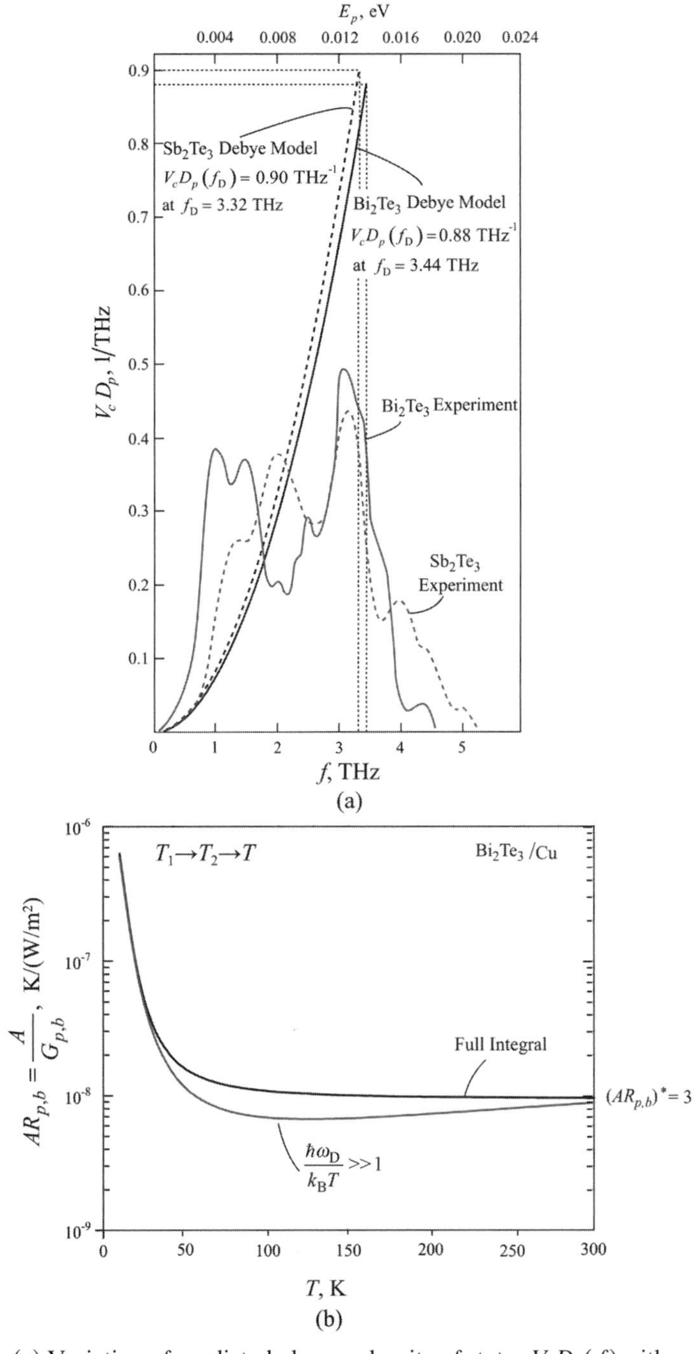

Figure 4.41. (a) Variation of predicted phonon density of states $V_c D_p(f)$ with respect to frequency f, for Bi_2Te_3 and Sb_2Te_3. The experimental results are also shown. The Debye density of states $D_{p,D}$ and the numerical integration results are also shown [83]. (b) The phonon boundary resistance of Bi_2Te_3/Cu as a function of temperature. The low temperature approximation is for $T_1 \rightarrow T_2 \rightarrow T$, and the full integral results are for $T_1 - T_2 = 1$ K.

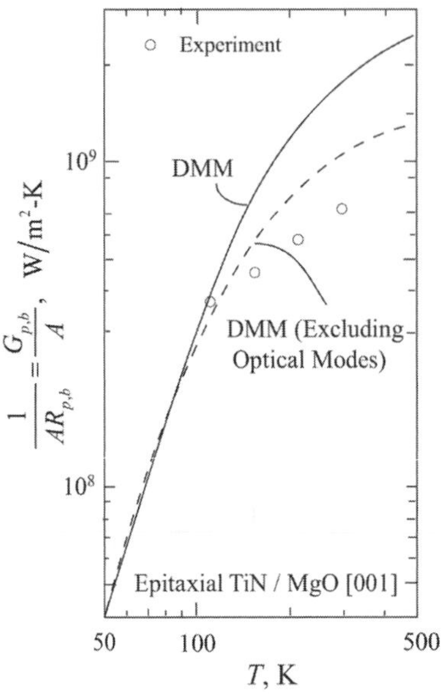

Figure 4.42. Variation of the measured and predicted phonon boundary conductances of the epitaxial TiN/MgO[001] interface as a function of temperature. The solid curve is the DMM prediction for TiN/MgO, and the dashed curve is the DMM with acoustic phonons only. The Debye temperature of TiN is 740 K [77].

4.18 Absorption of Ultrasound Waves in Solids

In previous sections we have discussed phonon heat capacity and thermal conductivity (storage and transport). We now consider an energy conversion (heating) involving phonons. The external sound wave attenuation in a crystalline dielectric can be treated using BTE, allowing for change in the equilibrium distribution function of thermal phonons that is because of the absorption of sound waves. This is referred to as the thermoelectric internal friction [182] and the attenuation is by phonon scattering. The theoretical treatment allows the absorbed energy to transfer to an ideal bath, thus allowing for a steady-state, isothermal treatment [351].

For these dielectric, crystalline solids, in the low-temperature limit, the spectral acoustic absorption coefficient for a sound wave of angular frequency ω is [351]

$$\sigma_{p,\omega} = \gamma_G^2 \frac{c_{v,p} T}{2u_{p,\mathrm{A}}^3} \frac{\omega^2 \tau_p}{1 + 2\omega^2 \tau_p^2}, \quad \gamma_G = \frac{3\beta E_p}{\rho c_v}, \tag{4.208}$$

where γ_G is the Grüneisen constant (or parameter), β is the solid linear thermal expansion coefficient, E_p is the isothermal bulk modulus, and τ_p is the single-phonon relaxation time (4.109), i.e.,

$$\tau_p = \frac{3k_p}{\rho c_{v,p} u_{p,\mathrm{A}}^2}, \tag{4.209}$$

where we have used $u_{p,A}$ as defined in (4.57).

The Grüneisen constant is near unity and is generally assumed independent of temperature. The two frequency limits for (4.198) are

$$\sigma_{p,\omega} = \gamma_G^2 \frac{\omega^2 k_p T}{\rho u_{p,\mathrm{A}}^5} \quad \omega\tau_p \ll 1, \tag{4.210}$$

and

$$\sigma_{p,\omega} = \frac{\pi}{4}\gamma_G^2 \frac{\omega c_{v,p} T}{\rho u_{p,\mathrm{A}}^5} \quad \omega\tau_p \gg 1. \tag{4.211}$$

These show ω^2 dependence at low frequency and a linear dependence at high frequencies.

The acoustic-wave (mechanical) absorption heating rate is given in terms of the external acoustic intensity I_{ac} as [351]

$$\dot{s}_{p\text{-}p,\omega} = \dot{s}_{m,ac} = 2\sigma_{p,\omega} I_{ac} \quad \text{ultrasound heating.} \tag{4.212}$$

Examples of acoustic intensities and spectral acoustic absorption coefficients are given in [174].

4.19 Size Effects

4.19.1 Finite-Size Effect on Phonon Conductivity

When the length (average linear dimension) of the solid L is not significantly larger than the average phonon mean free path λ_p, then the interface scattering occurs [Casimir effect discussed in Section 4.9.4(A)]. A simpler model for this effect is to define the phonon mean free path using $L \rightarrow \infty$, and then including the finite-size effect by using [300]

$$\frac{1}{k_p} = \frac{2}{n k_\mathrm{B} u_{p,\mathrm{A}}}\left(\frac{1}{\lambda_p} + \frac{4}{L}\right), \tag{4.213}$$

where $n = 4/a^3$ (FCC), and $n = 2/a^3$ (BCC), where a is the lattice constant. Here $L/4$ is the average distance a phonon travels after the last anharmonic scattering event. We note that for $L/\lambda_p \rightarrow \infty$, (4.213) becomes $k_p = k_\mathrm{B} n u_{p,\mathrm{A}}/2$, which is 1/2 of k_p in (4.109). This is because the specific heat per atom in (4.213) is taken $3k_\mathrm{B}/2$, i.e., half of total acoustic phonon $c_{v,p}$, to correct for the optical phonons that do not have significant travel speed.

The MD and direct method [232] predictions of [300] for Si and C(diamond) are shown in Figure 4.43(a) and follow the relation just given.

Similar MD predictions (using the direct method, where a temperature gradient is imposed across the sample, also called nonequilibrium method [237]) results are found for the lattice thermal conductivity of metals [141], where k_p is a small fraction of k for these metals. These results are shown in Figure 4.43(b), along with the proportionality of (4.213).

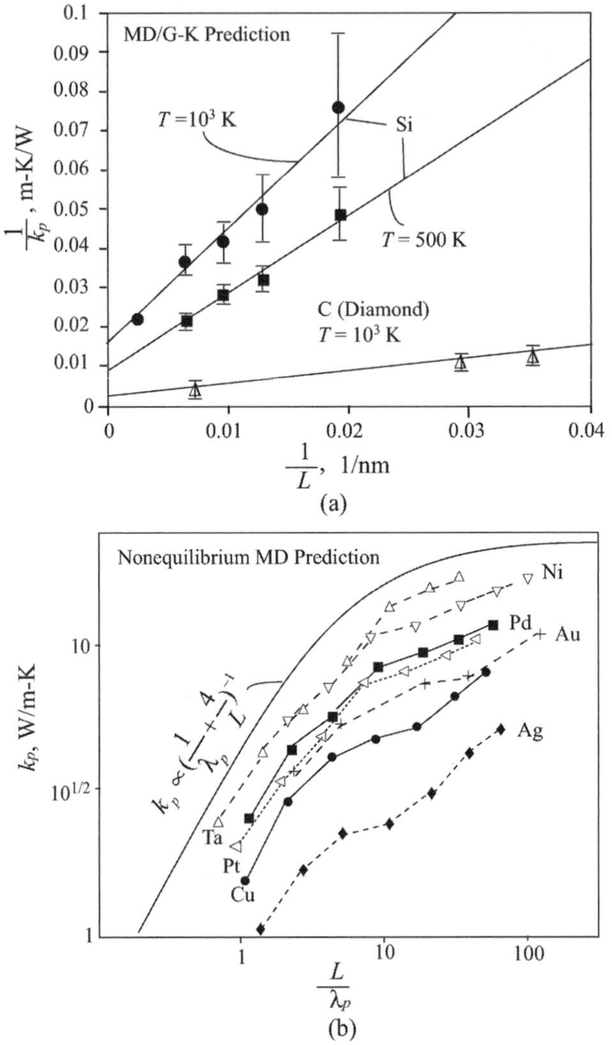

Figure 4.43. (a) Variations of the inverse of predicted lattice conductivity (for nonmetals), with respect to the inverses of computational (solid) linear dimensions [300]. (b) Variations of the inverse of predicted lattice conductivity (for metals), with respect to inverse ratios of computational (solid) linear dimensions to the phonon mean free path [141].

4.19.2 Superlattice Phonon Conductivity

A superlattice is a crystalline solid with periodic, alternating, very thin layers of two or more substances. Their properties (including lattice conductivity) are vastly different in the in-plane (along layers) and in the cross-plane directions. Because of scattering at the layer interfaces, both the in-plane and cross-plane lattice conductivities decrease compared with the thick-layer composites. Phonon boundary scattering is discussed in Section 4.17 and in Appendix D, where the scattering is because of the abrupt charge in the material acoustic impedance $\rho_i u_{p,\mathrm{A},i}$. Figure 4.44(a) shows a superlattice made of two alternating layers (made of substances 1 and 2) having thickness L_1 and L_2. In a superlattice these layer thicknesses are much smaller than

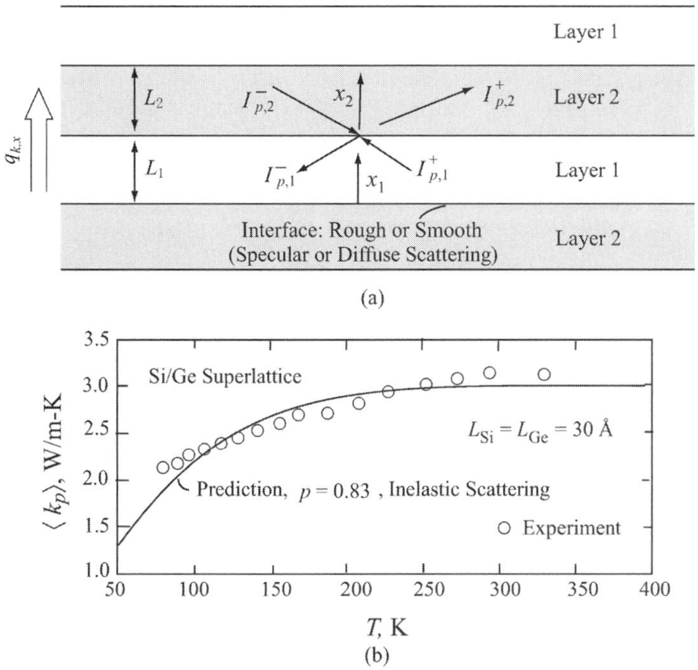

Figure 4.44. (a) Two alternating layers making a superlattice. The heat flow is in the cross-plane direction. The phonon intensity I_p is divided into forward and backward components (so-called two-flux model). (b) Variation of predicted [65] effective lattice conductivity, with respect to temperature, for a Si/Ge superlattice and comparison with experiment [202].

the phonon mean free path of the respective materials, i.e., $L_1 \ll \lambda_{p,1}$ and $L_2 \ll \lambda_{p,2}$ (using a single, average mean free path for each substance). This regime is called the ballistic phonon transport regime, because the boundary scattering dominates the bulk scattering mechanisms.

The one-dimensional BTE has been solved for this two-layer superlattice for the general case of inelastic, diffuse–specular scattering interface [65].

The intensity form of the BTE for a phonon is similar to that derived for a photon in Section 7.6. The resulting effective cross-plane phonon thermal conductivity $\langle k_p \rangle$ of the superlattice is

$$\langle k_p \rangle = \frac{q_{\kappa,x}(L_1+L_2)}{\Delta T|_{L_1+L_2}} = \frac{0.5(L_1+L_2)\rho_2 c_{v,2} u_{p,\mathrm{A},2}\displaystyle\int_0^1 \frac{2p\tau_{s,2\text{-}1}+A}{B}\mu_2 \mathrm{d}\mu_2}{1-2(1-p)\displaystyle\int_0^1 \frac{C}{B}\mu_2 \mathrm{d}\mu_2}$$

cross-plane effective lattice conductivity of superlattice, $\mu = \cos\theta$,

$$p = \begin{cases} 1 & \text{totally specular} \\ 0 & \text{totally diffuse} \end{cases}$$

$$A = (1-p)\tau_{d,2\text{-}1}(1 + p\rho_{s,1\text{-}2} + pD\tau_{s,1\text{-}2})$$

$$B = (1 + p\rho_{s,2\text{-}1})(1 + p\rho_{s,1\text{-}2}) - p^2\tau_{s,1\text{-}2}\tau_{s,2\text{-}1}$$

$$C = (1 + p\rho_{s,1\text{-}2})(-\rho_{d,2\text{-}1} + \tau_{d,1\text{-}2}) + pD\tau_{s,1\text{-}2}(-\rho_{d,1\text{-}2} + \tau_{d,2\text{-}1})$$

$$D = \frac{\mu_2 \mathrm{d}\mu_2}{\mu_1 \mathrm{d}\mu_1}$$

$$\rho_{d,i\text{-}j} = \tau_{d,j\text{-}i} = 1 - \tau_{d,i\text{-}j}$$

$$\tau_{d,i\text{-}j} = \frac{\rho_j c_{v,j} u_{p,\mathrm{A},j}}{\rho_i c_{v,i} u_{p,\mathrm{A},i} + \rho_j c_{v,j} u_{p,\mathrm{A},j}}$$

$$\rho_{s,i\text{-}j} = |\frac{\rho_i u_{p,\mathrm{A},i}\mu_i - \rho_j u_{p,\mathrm{A},j}\mu_j}{\rho_i u_{p,\mathrm{A},i}\mu_i + \rho_j u_{p,\mathrm{A},j}\mu_j}|^2, \quad \mu_i = \cos\theta_i$$

$$\tau_{s,i\text{-}j} = \frac{4\rho_i u_{p,\mathrm{A},i}\rho_j u_{p,\mathrm{A},j}\mu_i\mu_j}{(\rho_i u_{p,\mathrm{A},i}\mu_i + \rho_j u_{p,\mathrm{A},j}\mu_j)^2}$$

$$\frac{\sin\theta_1}{u_{p,\mathrm{A},1}} = \frac{\sin\theta_2}{u_{p,\mathrm{A},2}} \quad \text{elastic scattering}$$

$$\frac{\sin\theta_1}{\sin\theta_2} = [\frac{(\rho c_v u_{p,\mathrm{A}})_2}{(\rho c_v u_{p,\mathrm{A}})_1}]^{1/2} \quad \text{inelastic scattering.} \tag{4.214}$$

The specular parameter p is found to have an intermediate value when the proceeding prediction is matched with the experiment [65].

Figure 4.44(b) shows the predicted and measured [202] results for $\langle k_p \rangle$ as a function of temperature for a Si/Ge superlattice [65]. The specular reflection parameter is set to $p = 0.83$. The bulk k_p for Si and Ge at $T = 300$ K are given in Table A.1 and are 149 and 60.2 W/m-K. The mean free paths for these (based on acoustic-phonon heat capacity only) are 260 and 199 nm [65].

The experimental results [202] show that $\langle k_p \rangle$ does not monotonically decrease with $L_1 = L_2$, suggesting that the phonon tunneling is not completely described by the proceeding treatment.

The cross-plane phonon thermal conductivity has also been calculated for superlattices, by use of the MD-G–K [199], verifying the boundary phonon scattering and reduced effective thermal conductivity.

4.19.3 Phonon Density of States of Nanoparticles

The deviation of lattice vibration from bulk behavior occurs when the solid size decreases. Here we examine MD results for nanoparticles. The Debye DOS has a parabolic distribution resulting from an assumption of an isotropic medium with no dispersion, or any effect of the optical phonons. The measured phonon DOS of nanostructures, however, deviates from that for bulk crystals, because of quantum size effects [352].

The phonon DOS may be calculated with lattice-dynamics calculations or MD simulations [214]. In lattice-dynamics calculations, the dynamical matrix is diagonalized and the vibrational eigenvalues and eigenvectors determined. This approach actually calculates the harmonic modes, e.g., at $T = 0$ K. In MD, the velocity autocorrelation function is calculated for each species and the partial phonon DOS is then obtained by taking the Fourier transforms of this autocorrelation function.

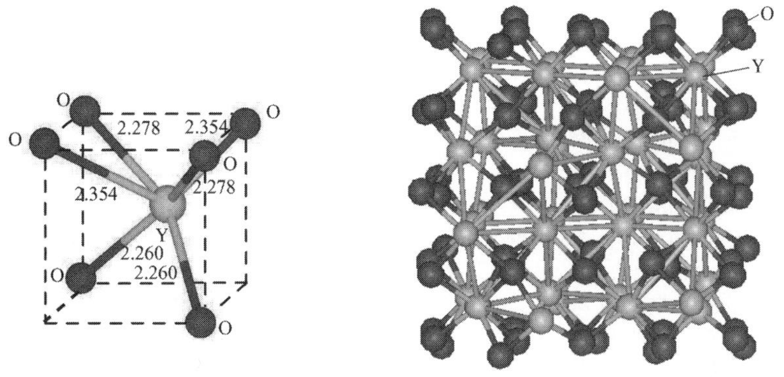

(a) (b)

Figure 4.45. Crystal structure of Y_2O_3. (a) The oxygen ions are arranged in distorted octahedra around the yttrium ions. The listed measured equilibrium bond lengths are in Å. (b) The structure of a cubic unit cell.

The results of these two approaches have been found to agree well with one another at low temperatures [214], as expected. However, at high temperatures the first method is not suitable, because it is harmonic. Also, for systems containing more than a few thousand atoms, it becomes unfeasible to calculate the frequencies of the vibrational modes by diagonalization of the dynamical matrix. For this reason, here $D_p(\omega)$ is calculated with MD simulations [3, 243]. The solid material used is Y_2O_3, which has applications in lasers, including laser cooling [289]. More details are reported in [290].

(A) *Nanoparticle Atomic Structure and MD*

The X-ray diffraction [263] and neutron diffraction [256] experiments have shown that Y_2O_3 has a FCC structure, which is retained in nanocrystals [188]. Eight metal ions are in the positions (1/4, 1/4, 1/4); the remaining 24 occupy the sites $(u, 0, 1/4)$. The 48 oxygen ions are in general positions (x, y, z), arranged in distorted octahedra around the metal ions, the metal–oxygen bonding distances being unequal. The values of u, x, y, z are listed in [263, 256]. The crystal structure is shown in Figure 4.45.

The nanopowder is generated by cutting a sphere out of a much larger bulk crystal, as shown in Figure 4.46. Note that the center of the sphere can be randomly selected; we have many possible configurations given the diameter.

In a MD simulation, we predict the phase-space trajectory of a system of particles by solving the Newton equations. The required inputs are an atomic structure and a suitable interatomic potential, which can be obtained from experiments and/or *ab initio* calculations [70].

The interatomic potential is the modified B-K-S potential (2.10), i.e.,

$$\varphi(r_{ij}) = \frac{q_i q_j}{r_{ij}} + A_{ij} \exp(-\frac{r_{ij}}{r_{o,ij}}) + \frac{C_{ij}}{r_{ij}^6}, \tag{4.215}$$

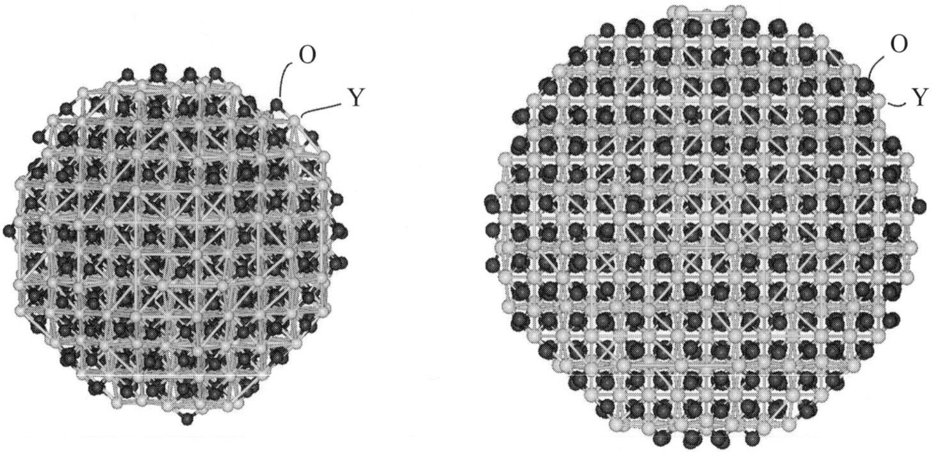

(a) $d_s = 3$ nm (b) $d_s = 4$ nm

Figure 4.46. The structure of the spherical Y_2O_3 nanoparticle (cluster) used in the MD simulations. The boundary is free and the particle structure is relaxed. (a) $d_s = 3$ nm, and (b) $d_s = 4$ nm [289].

Table 4.9. *Parameters used in the Buckingham potential, atomic charges:* $q_Y = 3e_c$, $q_O = -2e_c$.

Atom-Atom	A_{ij}, eV	$r_{o,ij}$, Å	C_{ij}, eV-Å^6	Reference
Y–Y	0	1	0	[43]
Y–O	1345.6	0.3491	0	[43, 206]
O–O	22799	0.149	27.93	[43, 206]

where $\varphi(r_{ij})$ is the interaction energy of atoms i and j, which consists of a Coulomb term and a covalent (short-range) contribution, cast into the usual Buckingham potential [48]. Here q_i is an effective charge of the ith atom, r_{ij} is the interatomic distance between atoms i and j, A_{ij} ($r_{o,ij}$ and C_{ij} are parameters for covalent interactions). For ionic materials, this Buckingham interatomic potential model has been shown to perform well. The parameters in (4.215), obtained from [43, 206], are listed in Table 4.9. This potential set has been verified to reproduce the bulk properties (lattice constant, lattice position, bulk modulus, elastic constant, etc.) well [179].

To determine the DOS of the bulk crystal, MD simulations are carried out in a cubic computation domain that contains $2 \times 2 \times 2 = 8$ unit cells containing $N = 640$ atoms (256 Y and 384 O). The computational method developed in [234] is used. Periodic boundary conditions are applied in all directions. For the nanopowder, the computation domain is a sphere of diameter d_s, which is cut from a much larger bulk crystal, as shown in Figure 4.46. It should be noted that a nanopowder generated in this way may not be neutral in charge; thus some atoms at the surface may need to be removed accordingly to eliminate any net charge of the nanopowder. Also, because the center of the sphere can be randomly selected, we have many possible configurations given the diameter. As such, a number of spherical particles with different configurations are considered in this study, and their behaviors are compared. The free boundary condition is used.

For both bulk crystal and nanocrystals, an initialization period of 5×10^4 time steps is used, with the time step $\Delta t = 1.6$ fs. The system is run in the NVT [Section 2.5.1(A)] ensemble. To set the temperature for the NVT ensemble, the potential energy of the system is monitored every time step. When it reaches a value within 10^{-4}% of the desired value, the ensemble is switched to NVE [Section 2.5.1(A)], and the system is run until the total number of time steps is 1.5×10^5.

(B) *Simulation Results and Analysis*

Typically, the normalized velocity–velocity autocorrelation function is determined for each species in the system. Here we have yttrium and oxygen atoms, and the autocorrelation function for the species α (α = Y, O) is

$$u_\alpha^{2\,*}(t) = \langle \sum_{i_\alpha=1}^{N_\alpha} \boldsymbol{u}_{i_\alpha}(t) \cdot \boldsymbol{u}_{i_\alpha}(0)\rangle / \langle \sum_{i_\alpha=1}^{N_\alpha} \boldsymbol{u}_{i_\alpha}(0) \cdot \boldsymbol{u}_{i_\alpha}(0)\rangle, \quad \alpha = \mathrm{Y}, \mathrm{O}, \tag{4.216}$$

where N_α is the number of atoms of species α, $\boldsymbol{u}_{i_\alpha}$ is the velocity of atom i_α, and $\langle\ \rangle$ is an ensemble average.

For nanocrystals, the internal and surface atoms vibrate differently. The crystal structure of the internal region is similar to that of bulk crystals, which implies that the internal atoms behave as if they were in a bulk crystal. On the other hand, the surface structure deviates much from the bulk structure because surface atoms lose their outer neighboring atoms, leading to different bond lengths, bond angles, etc. It is straightforward to consider the internal and surface regions separately, although the exact boundary between these regions cannot be well defined. In our practice, the spherical shell with a thickness of 3 Å is taken as the surface region, and the more inside region is considered as the internal region. All atoms are hence decomposed into four categories: surface yttrium atoms, internal yttrium atoms, surface oxygen atoms, and internal oxygen atoms. The number and population fraction for these four categories are shown in Table 4.10 for nanoparticles with increasing size. As expected, the surface region takes a smaller portion as the particle becomes larger, and the limit is that the surface effects can be neglected as the system is extremely large to recover the bulk phase.

The autocorrelation function for species α (α = Y, O), region β (β = surface, internal) is given by

$$u_{\alpha\beta}^{2*}(t) = \langle \sum_{i_{\alpha\beta}=1}^{N_{\alpha\beta}} \boldsymbol{u}_{i_{\alpha\beta}}(t) \cdot \boldsymbol{u}_{i_{\alpha\beta}}(0)\rangle / \langle \sum_{i_{\alpha\beta}=1}^{N_{\alpha\beta}} \boldsymbol{u}_{i_{\alpha\beta}}(0) \cdot \boldsymbol{u}_{i_{\alpha\beta}}(0)\rangle,$$

$$\alpha = \text{yttrium, oxygen}, \quad \beta = \text{surface, internal}, \tag{4.217}$$

where the double subscript $\alpha\beta$ denotes atoms of species α and in region β; then $N_{\alpha\beta}$ is the number of atoms of species α and in region β, $\boldsymbol{u}_{i_{\alpha\beta}}$ is the velocity of atom $i_{\alpha\beta}$.

Table 4.10. *Number and fraction of atoms for the nanocrystals, for several diameters d_s.*

Particle	Surface Y		Internal Y		Surface O		Internal O	
diameter d_s, nm	$N_{\alpha\beta}$	%	$N_{\alpha\beta}$	%	$N_{\alpha\beta}$	%	$N_{\alpha\beta}$	%
3	182	19.4	198	21.1	265	28.3	293	31.2
4	349	15.7	512	23.0	543	24.4	825	37.0
5	564	12.9	1203	27.4	839	19.1	1782	40.6
6	825	10.9	2212	30.0	1225	16.1	3340	43.9

The velocity–velocity autocorrelation functions are calculated for these four categories of atoms, and are shown in Figure 4.47. The vibrational frequencies of yttrium atoms are considerably lower than those for oxygen atoms, because the atomic mass for the yttrium atom is much larger. However, no evident difference is observed between the surface and internal regions for the same species, and it will be resolved in the phonon DOS in the next section.

(C) D_p *for Nanoparticles*

The frequency spectrum of the normalized velocity autocorrelation function gives the partial $D_{p,\alpha\beta}(\omega)$ as

$$D_{p,\alpha\beta}(\omega) = \int_0^{\tau} u^{2*}_{\alpha\beta}(t)\cos(\omega t)\mathrm{d}t. \tag{4.218}$$

Generally, the partial phonon DOS calculated in this way can give only the shape of the spectrum, and the absolute values are meaningless. Recognizing that a system with $N_{\alpha\beta}$ atoms has $3N_{\alpha\beta}$ modes, where 3 is the number of degrees of freedom, we can scale $D_{p,\alpha\beta}(\omega)$,

$$D_{p,\alpha\beta,N_{\alpha\beta}} = c_1 D_{p,\alpha\beta}, \tag{4.219}$$

where c_1 is a constant such that

$$\int_0^{\infty} D_{p,\alpha\beta,N_{\alpha\beta}}(\omega)\mathrm{d}\omega = 3N_{\alpha\beta}. \tag{4.220}$$

Then these partial phonon DOSs are addable, and the total phonon DOS of a system is just the summation of the partial phonon DOS,

$$D_p(\omega) = \sum_{\alpha}\sum_{\beta} D_{p,\alpha\beta,N_{\alpha\beta}}. \tag{4.221}$$

It is evident that $D_{p,\alpha\beta,N_{\alpha\beta}}$ is dependent on $N_{\alpha\beta}$, the size of the system. To compare the spectra shapes for systems with different numbers of atoms, it is necessary to define a normalized partial phonon DOS, as

$$D^*_{p,\alpha\beta} = \frac{D_{p,\alpha\beta,N_{\alpha\beta}}}{3N_{\alpha\beta}}. \tag{4.222}$$

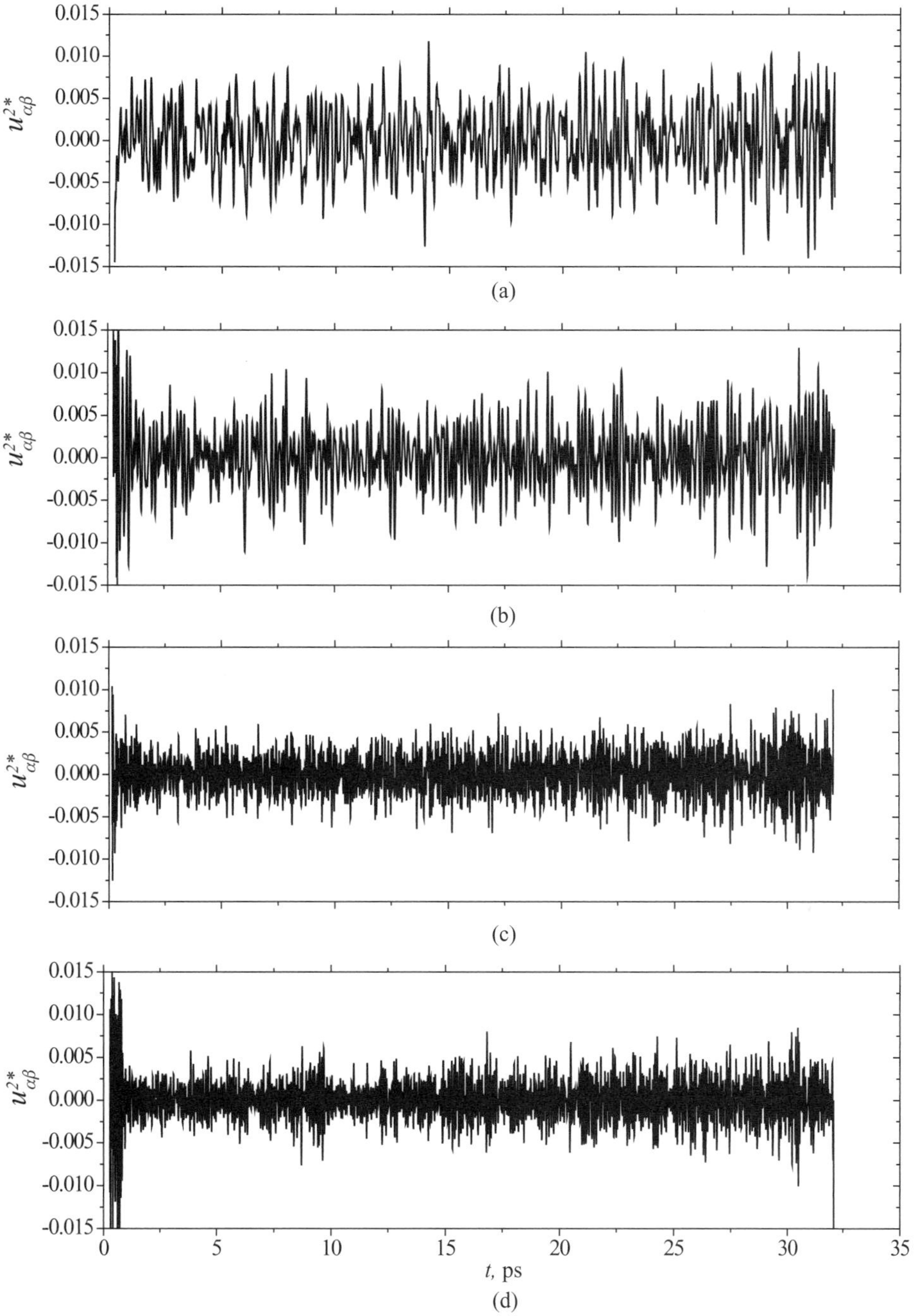

Figure 4.47. Velocity–velocity autocorrelation functions for (a) surface yttrium atoms, (b) internal yttrium atoms, (c) surface oxygen atoms, and (d) internal oxygen atoms [289, 290].

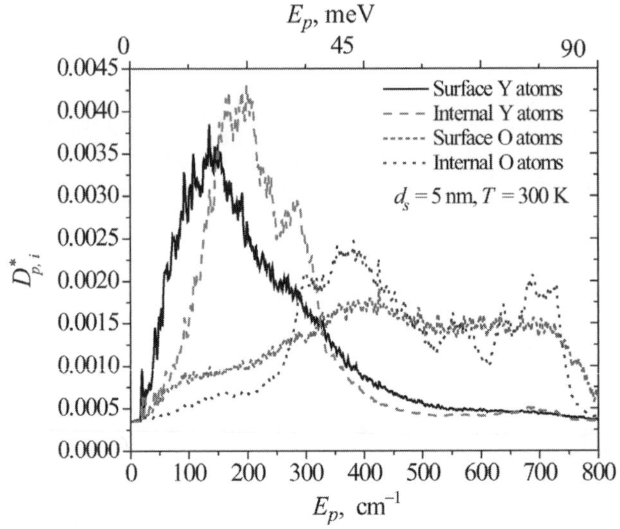

Figure 4.48. Normalized partial phonon DOS of the surface Y atoms, internal Y atoms, surface O atoms, and internal O atoms [290].

Therefore $D^*_{p,\alpha\beta}(\omega)$ satisfies the condition

$$\int_0^\infty D^*_{p,\alpha\beta}(\omega)\mathrm{d}\omega = 1, \tag{4.223}$$

and therefore it is called a normalized partial phonon DOS. The normalized partial $D^*_{p,\alpha\beta}(\omega)$ are calculated for a Y_2O_3 nanoparticle with $d_s = 5$ nm, and the results are shown in Figure 4.48. As indicated by the velocity–velocity autocorrelation function, oxygen atoms have more high-frequency modes than yttrium atoms, because of their lighter mass. For yttrium species, the surface region has more modes in high- and low-frequency tails, whereas the internal region has more modes in the intermediate frequency range. Similar behavior is observed for oxygen species.

These normalized partial phonon DOSs can be used as building blocks for higher-level partial DOSs. For example, the normalized partial DOS for the species α is given by

$$D^*_{p,\alpha} = \sum_\beta \frac{c_{\alpha\beta}}{c_\alpha} D^*_{p,\alpha\beta}, \tag{4.224}$$

where c_α is the population weight for species α, given by

$$c_\alpha = \sum_\beta c_{\alpha\beta}. \tag{4.225}$$

Similarly, the normalized partial DOS for the region β is given by

$$D^*_{p,\beta} = \sum_\alpha \frac{c_{\alpha\beta}}{c_\beta} D^*_{p,\alpha\beta}, \tag{4.226}$$

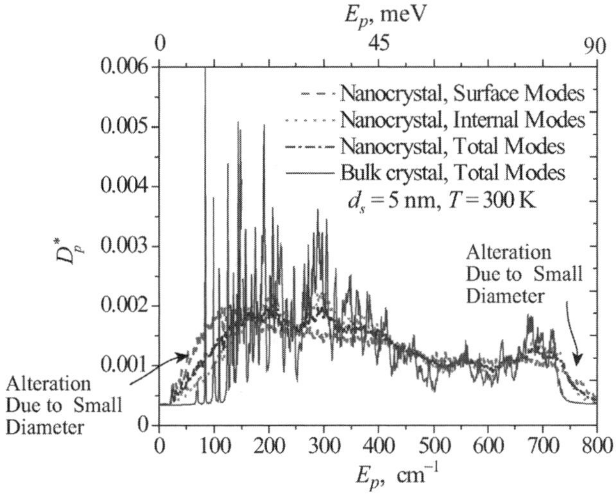

Figure 4.49. Comparison of the phonon DOS of the bulk crystal and nanocrystal for Y_2O_3. The nanocrystal DOS possesses extended low- and high-frequency tails [290].

where c_β is the population weight for region β, given by

$$c_\beta = \sum_\alpha c_{\alpha\beta}. \tag{4.227}$$

The total phonon DOS is obtained by summing over the partial DOS weighted with the population, i.e.,

$$D_p^* = \sum_\alpha \sum_\beta c_{\alpha\beta} D_{p,\alpha\beta}^*. \tag{4.228}$$

The normalized partial $D_{p,\beta}^*(\omega)$ for the surface and internal regions are calculated for the same nanocrystal, and the results are shown in Figure 4.49. Also shown are the total phonon DOS for the nanocrystal and for the bulk material [the bulk D_p is also shown in Figure 4.8(c)]. It can be seen that the phonon DOS of nanocrystals is distinct from that of the bulk crystal, in its broadened peaks, and extended tails at low and high frequencies. The bulk crystal has sharp, well-defined peaks (or modes) related to the rigorous periodic structure, while these peaks are broadened in the nanocrystal, because of the loss to some extent of this periodicity. The high frequency tail in the nanocrystal phonon DOS is believed to result from the surface atoms. Because of the loss of the attraction from their outer neighbors, these atoms have contracted bonds with their inner neighbors, compared with the bulk crystal. This leads to a harder surface and the increased vibrational frequencies.

4.19.4 Phonon Conductivity Rectification in Anisotropic One-Dimensional Systems

Nanotubes have one-dimensional thermal conductivity, and adding heavy atoms to the outside of the nanotube and in a manner to create a gradient in the added-atom

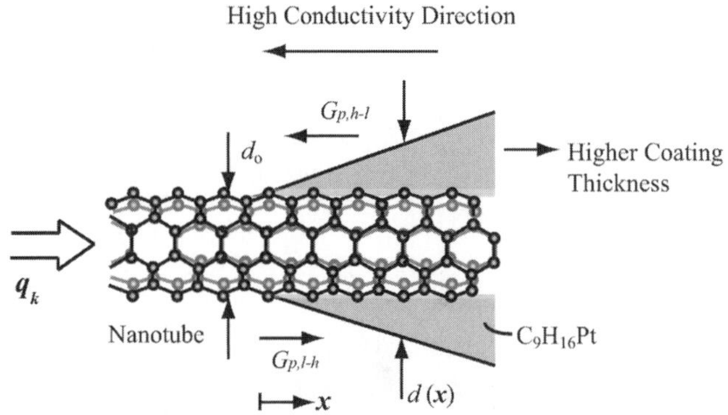

Figure 4.50. Asymmetric phonon conductance for heat flow along decreasing thickness $G_{p,h\text{-}l}$, is larger than the opposite direction $G_{p,l\text{-}h}$ when nanotube of diameter d_0 is nonuniformally coated with amorphous heavy molecules [61].

layer thickness can result in greater conductivity along the decreasing thickness [61]. Figure 4.50 shows the added heavy molecules $C_9H_{16}Pt$ used in experiment reported in [61], on carbon nitride, and boron nitride multiwell nanotubes (MWNT), with outer diameters of $d_\circ = 30$ to 40 nm.

The rectification R is defined in term of asymmetric conductance along decreasng coating thickness $G_{p,h-l}$ and increasing coating thickness $G_{p,l-h}$, i.e.,

$$R = \frac{G_{p,h-l} - G_{p,l-h}}{G_{p,l-h}}, \tag{4.229}$$

and they report up to $R = 0.07$ (7% rectification).

They suggest the transport is by a stable, single phonon soliton (nonperturbative solutions of nonlinear systems) in nanotubes. These localized particle entities collide with each other without changing shape.

The thermal rectifiers have also been suggested in a linear-chain lattice with a nonlinear (anharmonic) interatomic potential [267]. This nonlinear lattice represents soft molecules attached (locally), with temperature dependent vibration, on an atomic chain (such as a nanotube). It is suggested that the variable wall thickness of a nanotube (multilayer nanotubes) will result in similar thermal rectification.

Thermal rectification in nanodevices has been discussed in [302] based on anharmonicity and structural asymmetry (spin-boson nanojunction).

4.20 Problems

Problem 4.1

Describe the Born potential model given by the equation. What is the potential energy of a single bond in the Born potential model? The Born model is similar to the harmonic potential given in Table 2.1, where only the leading terms are taken, and assumes a slight displacement and considers bond stretching Γ_s and bond bending Γ_ϕ, i.e.,

$$\varphi_{i,j} = \frac{1}{2}\Gamma_s[\boldsymbol{s}_{i,j} \cdot (\boldsymbol{d}_j - \boldsymbol{d}_i)]^2 + \frac{1}{2}\Gamma_\phi\{|\boldsymbol{d}_j - \boldsymbol{d}_i|^2 - [\boldsymbol{s}_{i,j} \cdot (\boldsymbol{d}_j - \boldsymbol{d}_i)]^2\},$$

where $\boldsymbol{s}_{i,j}$ is the unit vector in the equilibrium direction from atom i to atom j and $\boldsymbol{d}_i$ and $\boldsymbol{d}_j$ are the displacement vectors for atoms from atom i to atom j, respectively.

Problem 4.2

(a) Using the Born potential model (Table 2.1), also given below, which considers only the nearest-neighbor interactions, write the expression for the potential energy of a one-dimensional (along x) monatomic chain in terms of the atomic displacements from the equilibrium positions. Use an interatomic distance a.

$$\varphi = \frac{1}{2}\Gamma[\boldsymbol{d}(x) - \boldsymbol{d}(x-a)]^2 + \frac{1}{2}\Gamma[\boldsymbol{d}(x) - \boldsymbol{d}(x+a)]^2 \quad \text{Born potential.}$$

(b) Using the expression for the potential energy, determine the force ($F = -\partial\varphi/\partial d$) on a given atom in the crystal in terms of its displacement and the displacements of its neighboring atoms. Write the atomic location using $\exp(\kappa x)$.

(c) Derive the expression for the dynamical matrix, in $\boldsymbol{M}^{-1}\boldsymbol{D}(\kappa) = \omega^2\boldsymbol{I}$, where $\boldsymbol{I}$ is the identity tensor. Note that the single component of the dynamical matrix gives $D = \partial^2\varphi/\partial d^2$.

(d) Derive the dispersion relation using this dynamical matrix and compare the result with (4.6), in the case of $m_1 = m_2 = m$.

Problem 4.3

Consider the case of one-dimensional displacement of a chain of atoms with two different atoms in the primitive basis.

(a) Using the Born potential model (Table 2.1, also Problem 4.2) that considers only the nearest-neighbor interactions, write the expression for the potential energy of atom m_1 in the one-dimensional (along x), two-mass atoms per primitive basis (m_1 and m_2). Use a single spring constant Γ, and the atomic displacements from the equilibrium positions. Start with φ_1 written in terms of relative displacement of atom 1 with respect to its right and left neighbors.

(b) Use the expression for the potential energy to determine the force on a given atom in the crystal in terms of its displacement and the displacements of its neighboring atoms. Use $F_1 = -\partial\varphi_1/\partial d_1$, etc. Note that for harmonic potential used in the dynamical matrix, the second derivative of the potential is used to obtain the components of the matrix. The components of the dynamical matrix have the form given by (4.40).

(c) Derive the expression for the dynamical matrix $\boldsymbol{D}$. Note that $\boldsymbol{M}^{-1}\boldsymbol{D}(\kappa) = \omega^2\boldsymbol{I}$, where $\boldsymbol{I}$ is the identity tensor. Use $D_{ij} = \partial^2\varphi/\partial d_i\partial d_j$.

(d) Derive the dispersion relation using the dynamical matrix (by setting the determinant equal to zero) and compare the result with (4.6). Note that a symbolic solver can be used to determine the determinant of the matrix.

Problem 4.4

Write the primitive lattice vector for the FCC zinc-blende structures, such as SiC, GaAs, and ZnS.

Problem 4.5

The cubic zinc-blende structure may be viewed as two FCC structures displaced from each other by 1/4 of a body diagonal. Write their reciprocal lattice of FCC.

Problem 4.6

Consider SiC crystal; assume that Si is at the origin (see the figure). Around this atom, there are four carbon atoms, all at an equal distance from this central Si. The basis consists of Si at the center and C at $a(0.5, 0.5, 0.5)$. Show that the primitive and reciprocal lattice vectors are those given in the figure.

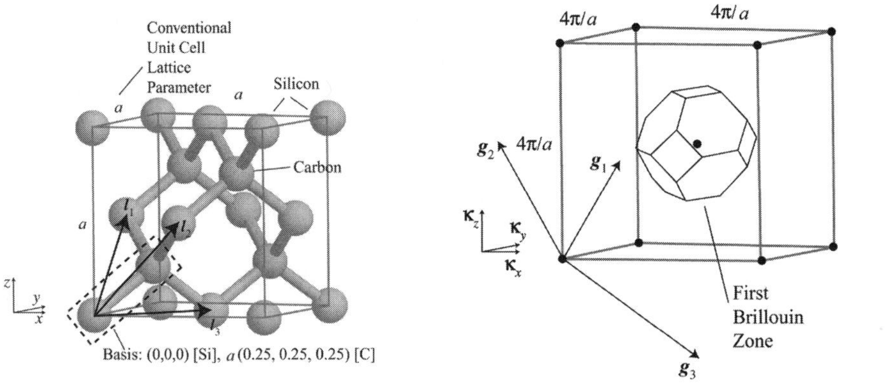

(a) Primitive lattice vectors for SiC: $\boldsymbol{l}_1 = a(0, 0.5, 0.5)$, $\boldsymbol{l}_2 = a(0.5, 0, 0.5)$, $\boldsymbol{l}_3 = a(0.5, 0.5, 0)$

(b) Reciprocal space lattice vectors for SiC: $\boldsymbol{g}_1 = (2\pi/a)(-1, 1, 1)$, $\boldsymbol{g}_2 = (2\pi/a)(1, -1, 1)$, $\boldsymbol{g}_3 = (2\pi/a)(1, 1, -1)$

Problem 4.6. Primitive and reciprocal space lattices for FCC SiC.

Problem 4.7

Show that the following are the unit vectors from C to Si atoms (place Si at the origin) as given as Problem 4.6 for SiC:

$3^{-1/2}a(\boldsymbol{s}_x + \boldsymbol{s}_y + \boldsymbol{s}_z)$

$3^{-1/2}a(-\boldsymbol{s}_x - \boldsymbol{s}_y + \boldsymbol{s}_z)$

$3^{-1/2}a(\boldsymbol{s}_x - \boldsymbol{s}_y - \boldsymbol{s}_z)$

$3^{-1/2}a(-\boldsymbol{s}_x + \boldsymbol{s}_y - \boldsymbol{s}_z)$.

Use the figure in Problem 4.6.

Problem 4.8

Using the Born potential model that considers only the nearest-neighbor interactions, derive the expression for the potential energy of SiC in terms of their displacements from equilibrium.

Use vectors $\boldsymbol{A}$ to $\boldsymbol{H}$ to represent the vectors in the Ath and Hth unit-cell, components of linear superpositions of the FCC unit lattice. Note that

$$\varphi = \sum_{i,j} \varphi_{i,j} = \sum_{i=1}^{4}\sum_{j=1}^{4}\{\frac{1}{2}\Gamma_S[\boldsymbol{s}_{i,j}\cdot(\boldsymbol{d}_j-\boldsymbol{d}_i)]^2 + \frac{1}{2}\Gamma_\phi\{|\boldsymbol{d}_j-\boldsymbol{d}_i|^2 - [\boldsymbol{s}_{i,j}\cdot(\boldsymbol{d}_j-\boldsymbol{d}_i)]^2\}\}$$

$$\begin{aligned}
\varphi =\ & \frac{1}{2}\Gamma_s|\boldsymbol{l}_A\cdot(\boldsymbol{d}_1(\boldsymbol{x})-\boldsymbol{d}_2(\boldsymbol{x})|^2 \\
&+\frac{1}{2}\Gamma_s|\boldsymbol{l}_B\cdot(\boldsymbol{d}_1(\boldsymbol{x})-\boldsymbol{d}_2(\boldsymbol{x}+\boldsymbol{B}))|^2 \\
&+\frac{1}{2}\Gamma_s|\boldsymbol{l}_C\cdot(\boldsymbol{d}_1(\boldsymbol{x})-\boldsymbol{d}_2(\boldsymbol{x}+\boldsymbol{C}))|^2 \\
&+\frac{1}{2}\Gamma_s|\boldsymbol{l}_D\cdot(\boldsymbol{d}_1(\boldsymbol{x})-\boldsymbol{d}_2(\boldsymbol{x}+\boldsymbol{D}))|^2 \\
&+\frac{1}{2}\Gamma_\phi\{|\boldsymbol{d}_1(\boldsymbol{x})-\boldsymbol{d}_2(\boldsymbol{x})|^2-|\boldsymbol{l}_A\cdot(\boldsymbol{d}_1(\boldsymbol{x})-\boldsymbol{d}_2(\boldsymbol{x}))|^2\} \\
&+\frac{1}{2}\Gamma_\phi\{|\boldsymbol{d}_1(\boldsymbol{x})-\boldsymbol{d}_2(\boldsymbol{x}+\boldsymbol{B})|^2-|\boldsymbol{l}_B\cdot(\boldsymbol{d}_1(\boldsymbol{x})-\boldsymbol{d}_2(\boldsymbol{x}+\boldsymbol{B}))|^2\} \\
&+\frac{1}{2}\Gamma_\phi\{|\boldsymbol{d}_1(\boldsymbol{x})-\boldsymbol{d}_2(\boldsymbol{x}+\boldsymbol{C})|^2-|\boldsymbol{l}_C\cdot(\boldsymbol{d}_1(\boldsymbol{x})-\boldsymbol{d}_2(\boldsymbol{x}+\boldsymbol{C}))|^2\} \\
&+\frac{1}{2}\Gamma_\phi\{|\boldsymbol{d}_1(\boldsymbol{x})-\boldsymbol{d}_2(\boldsymbol{x}+\boldsymbol{D})|^2-|\boldsymbol{l}_D\cdot(\boldsymbol{d}_1(\boldsymbol{x})-\boldsymbol{d}_2(\boldsymbol{x}+\boldsymbol{D}))|^2\} \\
&+\frac{1}{2}\Gamma_s|-\boldsymbol{l}_F\cdot(\boldsymbol{d}_2(\boldsymbol{x})-\boldsymbol{d}_1(\boldsymbol{x}+\boldsymbol{F}))|^2 \\
&+\frac{1}{2}\Gamma_s|-\boldsymbol{l}_G\cdot(\boldsymbol{d}_2(\boldsymbol{x})-\boldsymbol{d}_1(\boldsymbol{x}+\boldsymbol{G}))|^2 \\
&+\frac{1}{2}\Gamma_s|-\boldsymbol{l}_H\cdot(\boldsymbol{d}_2(\boldsymbol{x})-\boldsymbol{d}_1(\boldsymbol{x}+\boldsymbol{H}))|^2 \\
&+\frac{1}{2}\Gamma_\phi\{|\boldsymbol{d}_2(\boldsymbol{x})-\boldsymbol{d}_1(\boldsymbol{x}+\boldsymbol{F})|^2-|-\boldsymbol{l}_F\cdot(\boldsymbol{d}_2(\boldsymbol{x})-\boldsymbol{d}_1(\boldsymbol{x}+\boldsymbol{F}))|^2\} \\
&+\frac{1}{2}\Gamma_\phi\{|\boldsymbol{d}_2(\boldsymbol{x})-\boldsymbol{d}_1(\boldsymbol{x}+\boldsymbol{G})|^2-|-\boldsymbol{l}_G\cdot(\boldsymbol{d}_2(\boldsymbol{x})-\boldsymbol{d}_1(\boldsymbol{x}+\boldsymbol{G}))|^2\} \\
&+\frac{1}{2}\Gamma_\phi\{|\boldsymbol{d}_2(\boldsymbol{x})-\boldsymbol{d}_1(\boldsymbol{x}+\boldsymbol{H})|^2-|-\boldsymbol{l}_H\cdot(\boldsymbol{d}_2(\boldsymbol{x})-\boldsymbol{d}_1(\boldsymbol{x}+\boldsymbol{H}))|^2\},
\end{aligned}$$

where $\boldsymbol{l}_A$ to $\boldsymbol{l}_H$ are defined as $\boldsymbol{l}_A = 3^{-1/2}(\boldsymbol{s}_x+\boldsymbol{s}_y+\boldsymbol{s}_z)$, $\boldsymbol{l}_B = 3^{-1/2}(-\boldsymbol{s}_x-\boldsymbol{s}_y+\boldsymbol{s}_z)$, $\boldsymbol{l}_C = 3^{-1/2}(\boldsymbol{s}_x-\boldsymbol{s}_y-\boldsymbol{s}_z)$, $\boldsymbol{l}_D = 3^{-1/2}(-\boldsymbol{s}_x+\boldsymbol{s}_y-\boldsymbol{s}_z)$, $\boldsymbol{l}_F = -\boldsymbol{l}_B$, $\boldsymbol{l}_G = -\boldsymbol{l}_C$, and $\boldsymbol{l}_H = -\boldsymbol{l}_D$.

Considering $\boldsymbol{d}_1(\boldsymbol{x}) = d_{1x}(\boldsymbol{x})\boldsymbol{s}_x + d_{1y}(\boldsymbol{x})\boldsymbol{s}_y + d_{1z}(\boldsymbol{x})\boldsymbol{s}_z$,

$\boldsymbol{d}_2(\boldsymbol{x}) = d_{2x}(\boldsymbol{x})\boldsymbol{s}_x + d_{2y}(\boldsymbol{x})\boldsymbol{s}_y + d_{2z}(\boldsymbol{x})\boldsymbol{s}_z$

$\boldsymbol{d}_1(\boldsymbol{x}+\boldsymbol{B}) = d_{1x}(\boldsymbol{x}+\boldsymbol{B})\boldsymbol{s}_x + d_{1y}(\boldsymbol{x}+\boldsymbol{B})\boldsymbol{s}_y + d_{1z}(\boldsymbol{x}+\boldsymbol{B})\boldsymbol{s}_z$, then,

$$
\begin{aligned}
\varphi = {} & \frac{1}{2}\Gamma_s|3^{-1/2}\{d_{1x}(\boldsymbol{x}) - d_{2x}(\boldsymbol{x}) + d_{1y}(\boldsymbol{x}) - d_{2y}(\boldsymbol{x}) + d_{1z}(\boldsymbol{x}) - d_{2z}(\boldsymbol{x})\}|^2 \\
& + \frac{1}{2}\Gamma_s|3^{-1/2}\{-d_{1x}(\boldsymbol{x}) + d_{2x}(\boldsymbol{x}+\boldsymbol{B}) - d_{1y}(\boldsymbol{x}) + d_{2y}(\boldsymbol{x}+\boldsymbol{B}) + d_{1z}(\boldsymbol{x}) - d_{2z}(\boldsymbol{x}+\boldsymbol{B})\}|^2 \\
& + \frac{1}{2}\Gamma_s|3^{-1/2}\{d_{1x}(\boldsymbol{x}) - d_{2x}(\boldsymbol{x}+\boldsymbol{C}) - d_{1y}(\boldsymbol{x}) + d_{2y}(\boldsymbol{x}+\boldsymbol{C}) - d_{1z}(\boldsymbol{x}) + d_{2z}(\boldsymbol{x}+\boldsymbol{C})\}|^2 \\
& + \frac{1}{2}\Gamma_s|3^{-1/2}\{-d_{1x}(\boldsymbol{x}) + d_{2x}(\boldsymbol{x}+\boldsymbol{D}) + d_{1y}(\boldsymbol{x}) - d_{2y}(\boldsymbol{x}+\boldsymbol{D}) - d_{1z}(\boldsymbol{x}) + d_{2z}(\boldsymbol{x}+\boldsymbol{D})\}|^2 \\
& + \frac{1}{2}\Gamma_\phi\{(d_{1x}(\boldsymbol{x}) - d_{2x}(\boldsymbol{x}))^2 + (d_{1y}(\boldsymbol{x}) - d_{2y}(\boldsymbol{x}))^2 + (d_{1z}(\boldsymbol{x}) - d_{2z}(\boldsymbol{x}))^2\} \\
& - \frac{1}{2}\Gamma_\phi|3^{-1/2}\{d_{1x}(\boldsymbol{x}) - d_{2x}(\boldsymbol{x}) + d_{1y}(\boldsymbol{x}) - d_{2y}(\boldsymbol{x}) + d_{1z}(\boldsymbol{x}) - d_{2z}(\boldsymbol{x})\}|^2 \\
& + \frac{1}{2}\Gamma_\phi\{(d_{1x}(\boldsymbol{x}) - d_{2x}(\boldsymbol{x}+\boldsymbol{B}))^2 + (d_{1y}(\boldsymbol{x}) - d_{2y}(\boldsymbol{x}+\boldsymbol{B}))^2 + (d_{1z}(\boldsymbol{x}) - d_{2z}(\boldsymbol{x}+\boldsymbol{B}))^2\} \\
& - \frac{1}{2}\Gamma_\phi|3^{-1/2}\{-d_{1x}(\boldsymbol{x}) + d_{2x}(\boldsymbol{x}+\boldsymbol{B}) - d_{1y}(\boldsymbol{x}) + d_{2y}(\boldsymbol{x}+\boldsymbol{B}) + d_{1z}(\boldsymbol{x}) - d_{2z}(\boldsymbol{x}+\boldsymbol{B})\}|^2 \\
& + \frac{1}{2}\Gamma_\phi\{(d_{1x}(\boldsymbol{x}) - d_{2x}(\boldsymbol{x}+\boldsymbol{C}))^2 + (d_{1y}(\boldsymbol{x}) - d_{2y}(\boldsymbol{x}+\boldsymbol{C}))^2 + (d_{1z}(\boldsymbol{x}) - d_{2z}(\boldsymbol{x}+\boldsymbol{C}))^2\} \\
& - \frac{1}{2}\Gamma_\phi|3^{-1/2}\{d_{1x}(\boldsymbol{x}) - d_{2x}(\boldsymbol{x}+\boldsymbol{C}) - d_{1y}(\boldsymbol{x}) + d_{2y}(\boldsymbol{x}+\boldsymbol{C}) - d_{1z}(\boldsymbol{x}) + d_{2z}(\boldsymbol{x}+\boldsymbol{C})\}|^2 \\
& + \frac{1}{2}\Gamma_\phi\{(d_{1x}(\boldsymbol{x}) - d_{2x}(\boldsymbol{x}+\boldsymbol{D}))^2 + (d_{1y}(\boldsymbol{x}) - d_{2y}(\boldsymbol{x}+\boldsymbol{D}))^2 + (d_{1z}(\boldsymbol{x}) - d_{2z}(\boldsymbol{x}+\boldsymbol{D}))^2\} \\
& - \frac{1}{2}\Gamma_\phi|3^{-1/2}\{-d_{1x}(\boldsymbol{x}) + d_{2x}(\boldsymbol{x}+\boldsymbol{D}) + d_{1y}(\boldsymbol{x}) - d_{2y}(\boldsymbol{x}+\boldsymbol{D}) - d_{1z}(\boldsymbol{x}) + d_{2z}(\boldsymbol{x}+\boldsymbol{D})\}|^2 \\
& + \frac{1}{2}\Gamma_s| - 3^{-1/2}\{-d_{2x}(\boldsymbol{x}) + d_{1x}(\boldsymbol{x}+\boldsymbol{F}) - d_{2y}(\boldsymbol{x}) + d_{1y}(\boldsymbol{x}+\boldsymbol{F}) + d_{2z}(\boldsymbol{x}) - d_{1z}(\boldsymbol{x}+\boldsymbol{F})\}|^2 \\
& + \frac{1}{2}\Gamma_s| - 3^{-1/2}\{d_{2x}(\boldsymbol{x}) - d_{1x}(\boldsymbol{x}+\boldsymbol{G}) - d_{2y}(\boldsymbol{x}) + d_{1y}(\boldsymbol{x}+\boldsymbol{G}) - d_{2z}(\boldsymbol{x}) + d_{1z}(\boldsymbol{x}+\boldsymbol{G})\}|^2 \\
& + \frac{1}{2}\Gamma_s| - 3^{-1/2}\{-d_{2x}(\boldsymbol{x}) + d_{1x}(\boldsymbol{x}+\boldsymbol{H}) + d_{2y}(\boldsymbol{x}) - d_{1y}(\boldsymbol{x}+\boldsymbol{H}) - d_{2z}(\boldsymbol{x}) + d_{1z}(\boldsymbol{x}+\boldsymbol{H})\}|^2 \\
& + \frac{1}{2}\Gamma_\phi\{(d_{2x}(\boldsymbol{x}) - d_{1x}(\boldsymbol{x}+\boldsymbol{F}))^2 + (d_{2y}(\boldsymbol{x}) - d_{1y}(\boldsymbol{x}+\boldsymbol{F}))^2 + (d_{2z}(\boldsymbol{x}) - d_{1z}(\boldsymbol{x}+\boldsymbol{F}))^2\} \\
& - \frac{1}{2}\Gamma_\phi| - 3^{-1/2}\{-d_{2x}(\boldsymbol{x}) + d_{1x}(\boldsymbol{x}+\boldsymbol{F}) - d_{2y}(\boldsymbol{x}) + d_{1y}(\boldsymbol{x}+\boldsymbol{F}) + d_{2z}(\boldsymbol{x}) - d_{1z}(\boldsymbol{x}+\boldsymbol{F})\}|^2 \\
& + \frac{1}{2}\Gamma_\phi\{(d_{2x}(\boldsymbol{x}) - d_{1x}(\boldsymbol{x}+\boldsymbol{G}))^2 + (d_{2y}(\boldsymbol{x}) - d_{1y}(\boldsymbol{x}+\boldsymbol{G}))^2 + (d_{2z}(\boldsymbol{x}) - d_{1z}(\boldsymbol{x}+\boldsymbol{G}))^2\} \\
& - \frac{1}{2}\Gamma_\phi| - 3^{-1/2}\{d_{2x}(\boldsymbol{x}) - d_{1x}(\boldsymbol{x}+\boldsymbol{G}) - d_{2y}(\boldsymbol{x}) + d_{1y}(\boldsymbol{x}+\boldsymbol{G}) - d_{2z}(\boldsymbol{x}) + d_{1z}(\boldsymbol{x}+\boldsymbol{G})\}|^2 \\
& + \frac{1}{2}\Gamma_\phi\{(d_{2x}(\boldsymbol{x}) - d_{1x}(\boldsymbol{x}+\boldsymbol{H}))^2 + (d_{2y}(\boldsymbol{x}) - d_{1y}(\boldsymbol{x}+\boldsymbol{H}))^2 + (d_{2z}(\boldsymbol{x}) - d_{1z}(\boldsymbol{x}+\boldsymbol{H}))^2\} \\
& - \frac{1}{2}\Gamma_\phi| - 3^{-1/2}\{-d_{2x}(\boldsymbol{x}) + d_{1x}(\boldsymbol{x}+\boldsymbol{H}) + d_{2y}(\boldsymbol{x}) - d_{1y}(\boldsymbol{x}+\boldsymbol{H}) - d_{2z}(\boldsymbol{x}) + d_{1z}(\boldsymbol{x}+\boldsymbol{H})\}|^2
\end{aligned}
$$

Problem 4.9

Use the expression for the potential energy to determine the force on a given atom in the crystal in terms of its displacement and the displacements of its neighboring atoms. Note that force is

$$F_{ij,x} = -\frac{\partial \varphi_{ij}}{\partial x_{ij}},$$

$$F_x = -\frac{\partial}{\partial d_{1,x}} \sum_{i=1}^{4} \sum_{j=1}^{4} \frac{1}{2}\Gamma_S[\boldsymbol{b}_{i,j} \cdot (\boldsymbol{d}_j - \boldsymbol{d}_i)]^2 + \frac{1}{2}\Gamma_\phi\{|\boldsymbol{d}_j - \boldsymbol{d}_i|^2 - [\boldsymbol{b}_{i,j} \cdot (\boldsymbol{d}_j - \boldsymbol{d}_i)]^2\}$$

$$\begin{aligned}
\frac{\partial \varphi}{\partial d_{1x}} = {} & \frac{\Gamma_s}{3}\{d_{1x}(\boldsymbol{x}) - d_{2x}(\boldsymbol{x}) + d_{1y}(\boldsymbol{x}) - d_{2y}(\boldsymbol{x}) + d_{1z}(\boldsymbol{x}) - d_{2z}(\boldsymbol{x})\} \\
& - \frac{\Gamma_s}{3}\{-d_{1x}(\boldsymbol{x}) + d_{2x}(\boldsymbol{x}+\boldsymbol{B}) - d_{1y}(\boldsymbol{x}) + d_{2y}(\boldsymbol{x}+\boldsymbol{B}) + d_{1z}(\boldsymbol{x}) - d_{2z}(\boldsymbol{x}+\boldsymbol{B})\} \\
& + \frac{\Gamma_s}{3}\{d_{1x}(\boldsymbol{x}) - d_{2x}(\boldsymbol{x}+\boldsymbol{C}) - d_{1y}(\boldsymbol{x}) + d_{2y}(\boldsymbol{x}+\boldsymbol{C}) - d_{1z}(\boldsymbol{x}) + d_{2z}(\boldsymbol{x}+\boldsymbol{C})\} \\
& - \frac{\Gamma_s}{3}\{-d_{1x}(\boldsymbol{x}) + d_{2x}(\boldsymbol{x}+\boldsymbol{D}) + d_{1y}(\boldsymbol{x}) - d_{2y}(\boldsymbol{x}+\boldsymbol{D}) - d_{1z}(\boldsymbol{x}) + d_{2z}(\boldsymbol{x}+\boldsymbol{D})\} \\
& + \Gamma_\phi(d_{1x}(\boldsymbol{x}) - d_{2x}(\boldsymbol{x})) \\
& - \frac{\Gamma_\phi}{3}\{d_{1x}(\boldsymbol{x}) - d_{2x}(\boldsymbol{x}) + d_{1y}(\boldsymbol{x}) - d_{2y}(\boldsymbol{x}) + d_{1z}(\boldsymbol{x}) - d_{2z}(\boldsymbol{x})\} \\
& + \Gamma_\phi(d_{1x}(\boldsymbol{x}) - d_{2x}(\boldsymbol{x}+\boldsymbol{B})) \\
& + \frac{\Gamma_\phi}{3}\{-d_{1x}(\boldsymbol{x}) + d_{2x}(\boldsymbol{x}+\boldsymbol{B}) - d_{1y}(\boldsymbol{x}) + d_{2y}(\boldsymbol{x}+\boldsymbol{B}) + d_{1z}(\boldsymbol{x}) - d_{2z}(\boldsymbol{x}+\boldsymbol{B})\} \\
& + \Gamma_\phi(d_{1x}(\boldsymbol{x}) - d_{2x}(\boldsymbol{x}+\boldsymbol{C})) \\
& - \frac{\Gamma_\phi}{3}\{d_{1x}(\boldsymbol{x}) - d_{2x}(\boldsymbol{x}+\boldsymbol{C}) - d_{1y}(\boldsymbol{x}) + d_{2y}(\boldsymbol{x}+\boldsymbol{C}) - d_{1z}(\boldsymbol{x}) + d_{2z}(\boldsymbol{x}+\boldsymbol{C})\} \\
& + \Gamma_\phi(d_{1x}(\boldsymbol{x}) - d_{2x}(\boldsymbol{x}+\boldsymbol{D})) \\
& + \frac{\Gamma_\phi}{3}\{-d_{1x}(\boldsymbol{x}) + d_{2x}(\boldsymbol{x}+\boldsymbol{D}) + d_{1y}(\boldsymbol{x}) - d_{2y}(\boldsymbol{x}+\boldsymbol{D}) - d_{1z}(\boldsymbol{x}) + d_{2z}(\boldsymbol{x}+\boldsymbol{D})\} \\
& + 0 \\
& + 0 \\
& + 0 \\
& + 0 \\
& + 0 \\
& + 0 \\
& + 0 \\
& + 0 \\
& + 0.
\end{aligned}$$

The zero force is because the various contributions exactly cancel each other.

$$
\begin{aligned}
\frac{\partial\varphi}{\partial d_{1x}} = \ & \Gamma_s[\tfrac{4}{3}d_{1x}(\boldsymbol{x}) - \frac{1}{3}d_{2x}(\boldsymbol{x}) - \frac{1}{3}d_{2y}(\boldsymbol{x}) - \frac{1}{3}d_{2z}(\boldsymbol{x}) \\
& - \frac{1}{3}d_{2x}(\boldsymbol{x}+\boldsymbol{B}) - \frac{1}{3}d_{2y}(\boldsymbol{x}+\boldsymbol{B}) + \frac{1}{3}d_{2z}(\boldsymbol{x}+\boldsymbol{B}) \\
& - \frac{1}{3}d_{2x}(\boldsymbol{x}+\boldsymbol{C}) + \frac{1}{3}\Gamma_s d_{2y}(\boldsymbol{x}+\boldsymbol{C}) + \frac{1}{3}d_{2z}(\boldsymbol{x}+\boldsymbol{C}) \\
& - \frac{1}{3}d_{2x}(\boldsymbol{x}+\boldsymbol{D}) + \frac{1}{3}d_{2y}(\boldsymbol{x}+\boldsymbol{D}) - \frac{1}{3}d_{2z}(\boldsymbol{x}+\boldsymbol{D})] \\
& + \Gamma_\phi[4d_{1x}(\boldsymbol{x}) - d_{2x}(\boldsymbol{x}) - d_{2x}(\boldsymbol{x}+\boldsymbol{B}) - d_{2x}(\boldsymbol{x}+\boldsymbol{C}) - d_{2x}(\boldsymbol{x}+\boldsymbol{D})] \\
& - \Gamma_\phi[\tfrac{4}{3}d_{1x}(\boldsymbol{x}) - \frac{1}{3}d_{2x}(\boldsymbol{x}) - \frac{1}{3}d_{2y}(\boldsymbol{x}) - \frac{1}{3}d_{2z}(\boldsymbol{x}) \\
& - \frac{1}{3}d_{2x}(\boldsymbol{x}+\boldsymbol{B}) - \frac{1}{3}d_{2y}(\boldsymbol{x}+\boldsymbol{B}) + \frac{1}{3}d_{2z}(\boldsymbol{x}+\boldsymbol{B}) \\
& - \frac{1}{3}d_{2x}(\boldsymbol{x}+\boldsymbol{C}) + \frac{1}{3}\Gamma_s d_{2y}(\boldsymbol{x}+\boldsymbol{C}) + \frac{1}{3}d_{2z}(\boldsymbol{x}+\boldsymbol{C}) \\
& - \frac{1}{3}d_{2x}(\boldsymbol{x}+\boldsymbol{D}) + \frac{1}{3}d_{2y}(\boldsymbol{x}+\boldsymbol{D}) - \frac{1}{3}d_{2z}(\boldsymbol{x}+\boldsymbol{D})]
\end{aligned}
$$

$$
\begin{aligned}
= & \frac{4}{3}(\Gamma_s + 2\Gamma_\phi)(d_{1x}(\boldsymbol{x})) \\
& - \frac{1}{3}(\Gamma_s + 2\Gamma_\phi)(d_{2x}(\boldsymbol{x}) + d_{2x}(\boldsymbol{x}+\boldsymbol{B}) + d_{2x}(\boldsymbol{x}+\boldsymbol{C}) + d_{2x}(\boldsymbol{x}+\boldsymbol{D})) \\
& - \frac{1}{3}(\Gamma_s - \Gamma_\phi)(d_{2y}(\boldsymbol{x}) + d_{2y}(\boldsymbol{x}+\boldsymbol{B}) - d_{2y}(\boldsymbol{x}+\boldsymbol{C}) - d_{2y}(\boldsymbol{x}+\boldsymbol{D})) \\
& - \frac{1}{3}(\Gamma_s - \Gamma_\phi)(d_{2z}(\boldsymbol{x}) - d_{2z}(\boldsymbol{x}+\boldsymbol{B}) - d_{2z}(\boldsymbol{x}+\boldsymbol{C}) + d_{2z}(\boldsymbol{x}+\boldsymbol{D})) \\
= & \frac{4}{3}(\Gamma_s + 2\Gamma_\phi)d_{1x}(\boldsymbol{x}) \\
& - \frac{1}{3}(\Gamma_s + 2\Gamma_\phi)(1 + e^{i\boldsymbol{\kappa}\cdot\boldsymbol{B}} + e^{i\boldsymbol{\kappa}\cdot\boldsymbol{C}} + e^{i\boldsymbol{\kappa}\cdot\boldsymbol{D}})d_{2x}(\boldsymbol{x}) \\
& - \frac{1}{3}(\Gamma_s - \Gamma_\phi)(1 + e^{i\boldsymbol{\kappa}\cdot\boldsymbol{B}} - e^{i\boldsymbol{\kappa}\cdot\boldsymbol{C}} - e^{i\boldsymbol{\kappa}\cdot\boldsymbol{D}})d_{2y}(\boldsymbol{x}) \\
& - \frac{1}{3}(\Gamma_s - \Gamma_\phi)(1 - e^{i\boldsymbol{\kappa}\cdot\boldsymbol{B}} - e^{i\boldsymbol{\kappa}\cdot\boldsymbol{C}} + e^{i\boldsymbol{\kappa}\cdot\boldsymbol{D}})d_{2z}(\boldsymbol{x}).
\end{aligned}
$$

Other terms $\left[\frac{\partial\varphi}{\partial d_{1y}(\boldsymbol{x})}, \frac{\partial\varphi}{\partial d_{1z}(\boldsymbol{x})}, \frac{\partial\varphi}{\partial d_{2x}(\boldsymbol{x})}, \frac{\partial\varphi}{\partial d_{2y}(\boldsymbol{x})}, \frac{\partial\varphi}{\partial d_{2z}(\boldsymbol{x})}\right]$ can be also obtained by similar mathematical processes.

Problem 4.10

Derive the expression for the dynamical matrix using

$$D_{ij}(\kappa) = \sum_{i=1}^{4}\sum_{j=1}^{4}\sum_{x_p}(\partial^2 \frac{\varphi}{\partial d_i[x_s + x_p]\partial d_j[x_s]})e^{-i(\kappa \cdot x_p)},$$

$$\boldsymbol{D}(\kappa) = \begin{array}{c} \begin{array}{cccccc} d_{1x} & d_{1y} & d_{1z} & d_{2x} & d_{2y} & d_{2z} \end{array} \\ \begin{bmatrix} A & 0 & 0 & B & C & D \\ 0 & A & 0 & C & B & E \\ 0 & 0 & A & D & E & B \\ B^\dagger & C^\dagger & D^\dagger & A & 0 & 0 \\ C^\dagger & B^\dagger & E^\dagger & 0 & A & 0 \\ D^\dagger & E^\dagger & B^\dagger & 0 & 0 & A \end{bmatrix} \end{array} \begin{array}{c} d_{1x} \\ d_{1y} \\ d_{1z} \\ d_{2x} \\ d_{2y} \\ d_{2z} \end{array},$$

where

$$A = 4(\frac{\Gamma_s + 2\Gamma_\phi}{3}),$$

$$B = (\frac{\Gamma_s + 2\Gamma_\phi}{3})(1 + e^{i(\kappa_x+\kappa_y)a/2} + e^{i(\kappa_y+\kappa_z)a/2} + e^{i(\kappa_z+\kappa_x)a/2}),$$

$$C = (\frac{\Gamma_\phi - \Gamma_s}{3})(1 + e^{i(\kappa_x+\kappa_y)a/2} - e^{i(\kappa_y+\kappa_z)a/2} - e^{i(\kappa_z+\kappa_x)a/2}),$$

$$D = (\frac{\Gamma_\phi - \Gamma_s}{3})(1 - e^{i(\kappa_x+\kappa_y)a/2} - e^{i(\kappa_y+\kappa_z)a/2} + e^{i(\kappa_z+\kappa_x)a/2}),$$

$$E = (\frac{\Gamma_\phi - \Gamma_s}{3})(1 - e^{i(\kappa_x+\kappa_y)a/2} + e^{i(\kappa_y+\kappa_z)a/2} - e^{i(\kappa_z+\kappa_x)a/2}).$$

Problem 4.11

Use the Matlab code for the dynamic matrix of SiC. Examine and comment on the structure of the matrix.

Problem 4.12

Use the Matlab code for the dispersion curve (ω-κ relation). It uses the function developed in Problems 4.9 and 4.10. Draw the dispersion curve along Γ–X–L–Γ, Γ–L, X–L, and X. Use the relation $\boldsymbol{M}^{-1}\boldsymbol{D}(\kappa)\boldsymbol{s}_\alpha = \omega^2\boldsymbol{s}_\alpha$ and properties for SiC (force constants $\Gamma_s = 414.14$ kg/s^2, $\Gamma_\phi = 53.405$ kg/s^2, lattice constant $a = 4.35 \times 10^{-10}$ m, and mass of Si and C: $m_1 = 4.66 \times 10^{-26}$ kg, $m_2 = 1.99 \times 10^{-26}$ kg).

Problem 4.13

Use the Matlab code for the DOS of SiC and draw the D_p versus ω curve.

Problem 4.14

Use the Matlab code for the specific heat capacity of SiC (use the DOS developed in Problem 4.13), and plot the $c_{v,p}$ versus T curve.

Problem 4.15

Use the Matlab code for the specific heat of SiC, apply the Debye model and the Einstein model. Draw $c_{v,p}$ as a function of temperature and compare the results with the experimental results. Use the cut-off frequency in the Debye model as $f_D = 1.58 \times 10^{14}$ Hz. Use the frequency in the Einstein model as that of the highest frequency in the DOS found in Problem 4.13.

Problem 4.16

The porous crystal, MOF-5 shown in the figure is made of BDC ($C_8H_6O_4$) links with Zn_4O corners and has a lattice spacing $a = 12.92$ Å.

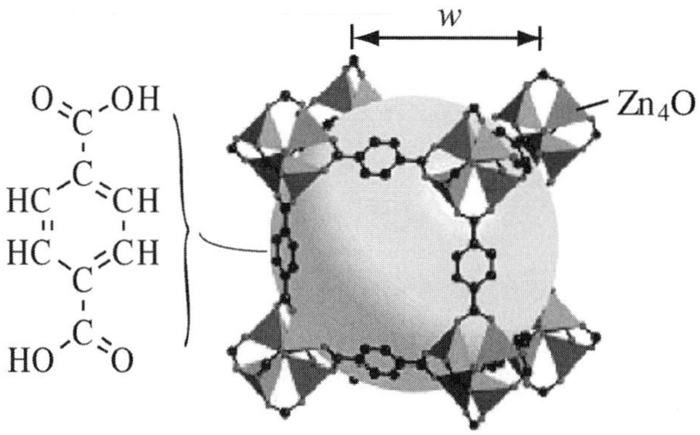

Problem 4.16. Metal–organic porous crystal is made of BDC links and Zn_4O corners.

(a) For a single crystal, determine the number density n. Also, find the molecular weight.

(b) For amorphous solids, we have three acoustic modes, one longitudinal and two transverse. Assuming the longitudinal and transverse velocities are 2500 and 1250 m/s, respectively, use the following relation to predict the Debye temperatures

$$T_{D,i} = \frac{\hbar}{k_B} u_{p,i}(6\pi^2 n)^{1/3}, \qquad i = \mathrm{L}, \mathrm{T},$$

and compare to the simplified average value 102 K.

(c) The C–P model (4.131) for the minimum thermal conductivity of amorphous solids is

$$k_{p,\text{C-P}} = \left(\frac{\pi}{6}\right)^{1/3} k_B n^{2/3} \sum_i \left[u_i \left(\frac{T}{T_{D,i}}\right)^2 \int_0^{T_{D,i}/T} \frac{x^3 e^x}{(e^x - 1)} dx\right],$$

where the summation is over the three acoustic modes. Determine k_{min} at room temperature ($T = 300$ K). What major assumption did we make in using the equation?

(d) The Slack [313] relation for the phonon thermal conductivity of crystals at high temperatures is (4.132), i.e.,

$$k_p = k_{p,\mathrm{S}} = \frac{3.1 \times 10^4 \langle M \rangle T_D^3}{T n^{1/3} \gamma_\mathrm{G}^2 N_\circ^{2/3}},$$

where $\langle M \rangle$ is the average molecular weight of the constituents (4.65), γ_G is the Grüneisen parameter (Section 4.11), and $N_\circ$ is the number of atoms per primitive cell. Assuming $\gamma_\mathrm{G} = 1.5$ (the structure includes ionic and covalent bonds) and using the higher Debye temperature, predict the thermal conductivity at room temperature. Is this higher than $k_{p,\mathrm{C\text{-}P}}$? If not, explain possible reasons.

(e) Compare this with k_p from (4.177), i.e., the mean-free-path model, using $c_v = 3k_\mathrm{B}$ and n from the atomic structure. Use $\lambda_p = w = 7.6$ Å, and the average acoustic phonon velocity $u_{p,\mathrm{A}}$. The MD simulations (Table 4.6) give $k_{p,MD} = 0.32$ W/m-K.

Problem 4.17

The high-temperature thermal conductivity of crystals is represented by (4.132)

$$k_p = k_{p,\mathrm{S}} = \frac{3.1 \times 10^4 \langle M \rangle T_\mathrm{D}^3}{T n^{1/3} \gamma_\mathrm{G}^2 N_\circ^{2/3}},$$

where $\langle M \rangle$ is the average molecular weight of constituents, T_D is the Debye temperature, n is the number density (atom/volume), γ_G is the Grüneisen parameter, and $N_\circ$ is the number of atoms per primitive cell.

(a) For diamond-structure crystals, diamond C, silicon Si, and germanium Ge, determine n.

(b) Predict the room temperature (T = 300 K) thermal conductivity of C, Si, and Ge. Obtain theoretical values of γ_G and T_D from Table 4.5 for these diamond-structure crystals, and use $N_\circ = 2$ (Section 4.3.3).

(c) Determine the percentage error between the predicted and tabulated (Table A.1) values. Note that the measured k_p for diamond may be for crystals containing isotopes, thus this value may be lower than that for pure samples. The Slack model prediction is within about 30% of the experimental results in Table A.1. However, he used his own experimental results for comparison in Table 4.5.

Problem 4.18

(a) Derive (4.5) by substituting (4.3) and (4.4), into (4.1) and (4.2), and setting the determinant of the coefficients to zero.

(b) Plot the dispersion curve (in dimensional form) for a linear chain of Ar atoms, using Table 2.7 for the spring (force) constant. The scales used to convert to dimensional forces are given by (4.8) and (4.24).

(c) Increase the mass of the alternating (every other one) atoms (so the optical phonons are also present) by a factor of 2, and repeat (b). Note that this gives $m_1^* = 1/3$, whereas Figure 4.5 is for $m_1^* = 0.5$ and 0.75.

Problem 4.19

Plot the dimensional phonon DOS (4.24) for Problem 4.18, parts (b) and (c). Comment on the differences. Note that both $D_p(\kappa)$ and $\omega(\kappa)$ relations should be used.

Problem 4.20

(a) Plot the specific heat capacity per atom for a Cu crystal as a function of temperature (up to $2T_{\rm D}$), using the Debye model of lattice specific heat capacity.

(b) Compare with the experimental results, at selected temperatures, given in the following table (in J/kg-K). Note that the high-temperature experimental data do not match the Debye model. The low-temperature data agree better [174]. At higher temperatures, the contribution from conduction electrons (Section 5.8) becomes noticeable.

T, K	$c_{v,p}$, J/kg-K
200	356.1
250	374.1
300	385.0
350	392.6
400	398.6
500	407.7
600	416.7

Problem 4.21

Starting from (4.64) divided by n (so $c_{v,p}$ is per atom), derive the first term of the extended Callaway phonon conductivity model in the extended Callaway lattice conductivity model [53],

$$k_p = (48\pi^2)^{1/3}\frac{1}{a}\frac{k_{\rm B}^3}{\hbar_{\rm P}^2}\frac{T^3}{T_{\rm D}}[g_1(f,T,\tau_p) + \frac{g_2^2(f,T,\tau_p,\tau_{p\text{-}p,n})}{g_3(f,T,\tau_p,\tau_{p\text{-}p,\rm N},\tau_{p,r})}],$$

using (4.119) to divide the phonon scattering into normal and resistive groups.

The integrals g_n, $g_{r,1}$, and $g_{r,2}$ and the relaxation times $\tau_{p,n}$, $\tau_{p,r}$, and τ_p are defined as

$$g_1 = \int_0^{T_{\rm D}/T} \tau_p \frac{x^4 e^x}{(e^x-1)^2}{\rm d}x$$

$$g_2 = \int_0^{T_{\rm D}/T} \frac{\tau_p}{\tau_{p\text{-}p,\rm N}}\frac{x^4 e^x}{(e^x-1)^2}{\rm d}x$$

$$g_3 = \int_0^{T_{\rm D}/T} \frac{\tau_p}{\tau_{p\text{-}p,\rm N}\tau_{p,r}}\frac{x^4 e^x}{(e^x-1)^2}{\rm d}x$$

$$\frac{1}{\tau_p} = \frac{1}{\tau_{p\text{-}p,\rm N}} + \frac{1}{\tau_{p,r}},\quad \frac{1}{\tau_{p,r}} = \sum_i \frac{1}{\tau_{p,r,i}},\quad x = \frac{\hbar\omega}{k_{\rm B}T}.$$

State all the assumptions. The second term involves treatment of the normal and resistive processes, whereas the normal processes conserve momentum.

Start from the thermal conductivity relation based on the mean free path (4.109), i.e., $k_p = \rho c_{v,p} u_{p,\mathrm{A}} \lambda_p/3,\ \lambda_p = u_{p,\mathrm{A}} \tau_p$.

Problem 4.22

Derive the Callaway phonon conductivity model (4.108), starting from (4.106). State all the assumptions. Use (4.55) to write $c_{v,p} = k_\mathrm{B} x^2 e^x/(e^x - 1)^2$. Note that all three acoustic-phonon modes are treated the same in the Callaway model. Use (4.63) for the relation between T_D and $u_{p,\mathrm{A}}$ (in the Debye model only this average phonon speed is used).

Problem 4.23

(a) Plot the variation of the phonon conductivity with respect to temperature, using the Callaway model (4.108) and (4.109) and the following relaxation resistive and the normal relaxation time given in Problem 4.21, and the data for crystalline alumina (sapphire) given below, for $0 \leq T \leq 400$ K.

$$\frac{1}{\tau_{p,r}} = \frac{1}{\tau_{p\text{-}d}} + \frac{1}{\tau_{p\text{-}im}} + \frac{1}{\tau_{p\text{-}p,\mathrm{U}}} + \frac{1}{\tau_{p\text{-}b}}$$

$$= A(\frac{2\pi k_\mathrm{B}}{h_\mathrm{P}})^4 T^4 x^4 + A(\frac{2\pi k_\mathrm{B}}{h_\mathrm{P}})^4 T^4 x^4 + a_\mathrm{U}(\frac{2\pi k_\mathrm{B}}{h_\mathrm{P}})^2 T^3 e^{-T_\mathrm{D}/(\alpha T)} x^2 + \frac{u_{p,\mathrm{A}}}{L}$$

$$\frac{1}{\tau_{p\text{-}p,\mathrm{N}}} = a_\mathrm{N} \frac{k_\mathrm{B}}{\hbar} T^5 x.$$

(b) Plot the various relaxation times, with respect to temperature at $\omega = \omega_\mathrm{D}$ (note that $\hbar\omega_\mathrm{D}/k_\mathrm{B}T = T_\mathrm{D}/T$).

(c) Examine the role of scattering mechanisms on the conductivity, by repeating (a), while suppressing individual scattering contributions (setting relaxation time constant to a very large value):

for crystalline Al_2O_3, we have $a_\mathrm{N} = 2.7\times10^{-13}$ 1/K^4, $A = 4.08\times10^{-46}$ s^3, $a_\mathrm{U} = 1.7\times10^{-18}$ 1/K, $T_\mathrm{D} = 596$ K, $a = 0.35$ nm, $L = 4.12$ mm, $\alpha = 2$, and $u_{p,\mathrm{A}} = 7009$ m/s.

Verify that the peak value is 7097 W/m-K, and occurs at $T = 29$ K. The peak value is sensitive to the accuracy of the integration scheme.

As a check against the experiment, at $T = 300$ K, the measured conductivity is 29.2 W/m-K.

Problem 4.24

Use the Debye–Gaussian DOS model (4.51) to curve fit the FCC Ar phonon MD results given in Figure 4.7. The results should appear similar to those of Figure 4.8(c). Comment on the suitability of this model for representing (a) the peaks, (b) the low-frequency regime, and (c) the area under the curve. (d) What is the magnitude of the dimensionless frequency at which the dimensionless form of the model D_p^* peaks?

Problem 4.25

The cross-plane phonon boundary resistance can become very significant in thin films of thickness L (which can be smaller than the phonon mean free path λ_p). This resistance can also be included as part of the bulk resistance (or conductivity) and one such attempt is given in Liu and Asheghi, *ASME J. Heat Transfer*, 128:75-83, 2006 by modifying (4.109) as

$$k_{p,\perp} = \frac{1}{3}\sum_{\alpha} u_{p,\alpha}^2 \int_0^{T_\alpha/T} c_{v,p,\alpha}\tau_{p\text{-}p,\alpha} z(L^*)\mathrm{d}x, \quad \alpha = \mathrm{L}, \mathrm{T}, \mathrm{TU},$$

where

$$z(L^*) = 1 - \frac{3}{8L^*} + \frac{3}{2L^*}\int_1^\infty \left(\frac{1}{y^3} - \frac{1}{y^5}\right)\exp(-L^* y)\mathrm{d}y$$

$$L^* = \frac{L}{\lambda_p}, \quad \lambda_p = u_{p,\mathrm{A}}\tau_{p\text{-}b}.$$

In this model the length along which $k_{p,\perp}$ is predicted and scaled with respect to the phonon mean free path λ_p and $\tau_{p\text{-}b}$ is the bulk (no boundary effect) relaxation time (4.119).

Using only the three interphonon scattering relaxation times $\tau_{p\text{-}p,\mathrm{L,N}}$, $\tau_{p\text{-}p,\mathrm{T,N}}$, and $\tau_{p\text{-}p,\mathrm{TU}}$ in Table 4.2 (constants given in Table 4.3) for Si, plot k_p as a function of L, for $1 \le L < 1000$ nm, and $T = 300$ K. Assume $c_{v,p,\alpha} = c_{v,p}$ and use (4.55).

Problem 4.26

For the longitudinal phonon in a linear, monatomic lattice, the exponential displacement of the jth atom in the chain given by (4.3) can also be written as

$$d_j = d_o \cos(-j\kappa a + \omega t).$$

For a monatomic chain with the nearest-neighbor interaction (force) constant Γ, show the following about the phonon energy.

(a) Show that the total phonon energy E_p (the sum of kinetic and potential energy) of the wave is

$$E_p = E_{p,k} + E_{p,p} = \frac{1}{2}m\sum_j\left(\frac{\mathrm{d}d_j}{\mathrm{d}t}\right)^2 + \frac{1}{2}\Gamma\sum_j(d_j - d_{j+1})^2,$$

where j is over all atoms.

(b) Use the proceeding expression for d_j and show that the time-averaged, total modal energy per atom is

$$E_p = \frac{1}{4}m\omega^2 a^2 + \frac{1}{2}\Gamma[1 - \cos(\kappa a)]a^2$$
$$= \frac{1}{2}m\omega^2 a^2,$$

using the dispersion relation for monatomic chain $\omega^2 = (4\Gamma/m)\sin^2(\kappa a/2)$. This shows that the modal (and average) kinetic and potential energies of this harmonic lattice motion are equal. When placing a thermostat on this system, for a single degree of freedom, the kinetic energy is equal to $k_B T/2$. Note that $1 - \cos(\kappa a) = 2\sin^2(\kappa a/2)$.

Problem 4.27

Consider a monatomic lattice chain (acoustic phonon only) and show that the equation of motion (4.1) becomes the one-dimensional elastic wave equation,

$$\frac{\partial^2 d}{\partial t^2} = u_{p,\mathrm{L}}^2 \frac{\partial^2 d}{\partial x^2},$$

for $\kappa \rightarrow 0$ (i.e., $u_{p,g} = u_{p,p}$ or linear dispersion).

Note that

$$\frac{\mathrm{d}^2 d}{\mathrm{d}x^2} = \frac{d^{j+1} + d^{j-1} - 2d^j}{a^2},$$

where a is the interatomic spacing.

Note that the elastic, longitudinal, wave phase speed (Section 4.7.2) is found from ω/κ because of (4.9), i.e.,

$$u_{p,\mathrm{L}} = \frac{\omega}{\kappa} = (\frac{\Gamma a^2}{m})^{1/2} = a(\frac{\Gamma}{m})^{1/2} \quad \kappa \rightarrow 0.$$

In Table 3.11 and Section 4.7.2, the elastic wave equation becomes evident. Note that for a constant volume, compression wave speed is $u_{p,\mathrm{L}} = (E_\mathrm{Y}/\rho)^{1/2}$, where E_Y is the Young modulus.

Problem 4.28

The FCC Ar phonon dispersion graph shown in Figure 4.7 can be produced by use of the dynamical matrix.

(a) Using the Matlab code, plot the dispersion curve, showing the three acoustic branches for FCC Ar along Γ–K, from Γ to $\kappa^* = 1$. The $\boldsymbol{\kappa}$ direction results are along Γ to K, and only a portion of the results of Figure 4.7.

(b) Comment about the behavior of the branches in the Γ–K line.

(c) Examine the code and comment about the dimension of the dynamical matrix.

Problem 4.29

(a) Plot the variation of the dimensional phonon boundary conductance $AR_{p,b}$, using (4.203), as a function of temperature ($0 \leq T \leq T_\mathrm{D}$), for Bi_2Te_3 (material 1) bounded by Cu (material 2). The properties to be used are a single Debye temperature $T_\mathrm{D} = T_{\mathrm{D},1} = 165$ K, and $n = 5.95\times10^{27}$ 1/m^3, and use the Debye phonon speed $u_{p,\mathrm{A}}$ from (4.27) to determine $u_{p,1,\alpha}$ (for all modes). For $u_{p,2,\alpha}$ (for Cu), use $u_{p,2,\mathrm{L}} = 4760$, and $u_{p,2,\mathrm{T}} = 2325$ m/s.

(b) Plot $AR_{p,b}$ using the high-temperature approximation (4.206), for $0 \leq T \leq 300$ K, for $T_1 - T_2 = 1$ K, $T_1 = T$, and compare with results of (a). Note that Figure 4.41(b) shows these results.

(c) Comment on the difference between the results of part (a) and part (b).

Problem 4.30

(a) Show that the specific internal energy per unit mass and per unit volume (4.53) can be written as

$$\begin{aligned} e_p &= \frac{1}{m}\int_0^{\omega_D} \hbar\omega f_p^o D_{p,D} d\omega \\ &= \frac{3k_B^4 T^4}{2m\pi^2 u_{p,A}^3 \hbar^3}\int_0^{T_D/T} \frac{x^3}{e^x - 1} dx \\ &= 9\frac{k_B T}{m}(\frac{T}{T_D})^3 n \int_0^{T_D/T} \frac{x^3}{e^x - 1} dx. \end{aligned}$$

(b) For $T \ll T_D$, $T_D/T \rightarrow \infty$, and

$$\int_0^{\infty} \frac{x^3}{e^x - 1} dx = \frac{\pi^4}{15},$$

show that for $T \ll T_D$, (4.64) becomes

$$c_{v,p} = \frac{12\pi^4 k_B}{5m} n(\frac{T}{T_D})^3,$$

which is the so-called Debye T^3-law.

Problem 4.31

(a) By expanding the exponential terms in (4.64), for small x ($T \gg T_D$), show that (per atom)

$$c_{v,p} = 3\frac{k_B}{m},$$

which is the Dulong–Petit limit for harmonic lattice vibration.

(b) Using (4.64), plot $c_{v,p}m/k_B$ (per atom) versus temperature for Cu, and for $T = 300$ K, use the value in Table A.1 to compare. Note that the tabulated value is in J/kg-K.

Problem 4.32

Derive the expression for phonon boundary transmission (D.13) for the Debye phonon DOS model, starting from (D.12).

Problem 4.33

Use the expression for the effective lattice conductivity of a superlattice, (4.214), and determine the effective lattice conductivity $\langle k_p \rangle$ of a Si/Ge superlattice, for the following conditions, and for inelastic scattering:

$(\rho c_v)_{\rm Si} = 0.93 \times 10^6$ J/m^3-K, $(\rho c_v)_{\rm Ge} = 0.87 \times 10^6$ J/m^3-K, $u_{p,{\rm A,Si}} = 1804$ m/s, $u_{p,{\rm A,Ge}} = 1042$ m/s, $L_{\rm Si} = L_{\rm Ge}$, and $p = 0.8$.

(a) Plot $\langle k_p \rangle$ versus L, $L = L_{\rm Si} + L_{\rm Ge}$, for $10 < L < 1000$ Å.

(b) Plot $\langle k_p \rangle$ versus p ($0 \le p \le 0.8$), for $L_{\rm Si} = L_{\rm Ge} = 50$, 200, and 1000 Å.

Note that $(\rho c_v)_i$ is for the acoustic phonons only [65]. Use Table A.1 for densities. Also note that the definite integrals can be evaluated numerically.

5

Electron Energy Storage, Transport and Transformation Kinetics

The solid electric thermal conductivity tensor $\boldsymbol{K}_e$, in addition to the phonon thermal conductivity tensor (i.e., total conductivity $\boldsymbol{K} = \boldsymbol{K}_e + \boldsymbol{K}_p$), determines heat conduction in solids through the Fourier law $\boldsymbol{q}_k = -\boldsymbol{K} \cdot \nabla T$. The average heat capacity of an electron $c_{v,e}$ is small, except at high temperatures. Electrons can also have a net motion under an applied electric field $\boldsymbol{e}_e$, thus creating opportunities for exchange of their gained kinetic energy, e.g., with the lattice through inelastic scattering in Joule heating. The coupling of electronic and thermal transport, known as thermoelectricity, leads to Peltier heating/cooling.

In Section 2.6.5, we examined the electronic energy states of an idealized electron gas by solving the Schrödinger equation for the case of a collection of free electrons. In Section 2.6.6 we also derived the electronic energy states of hydrogen-like atoms, along with the designation of the quantum numbers and atomic orbitals. As atoms gather in a cluster or a bulk phase, their orbiting electrons and their energy states are altered because of various nuclear and electronic interactions (Section 2.2), including representation as interatomic potentials. These interactions may increase or decrease the energy gaps between the electron orbital states of individual atoms. The electrons can gain sufficient energy to be free (conduction) electrons or lack this and be bounded (valence) electrons. If a significant electronic energy gap exists between these two electron states, then this cluster or bulk matter may become an electrical insulator. Once an electron is in the conduction state (or band), its kinetic energy can be increased by imposition of an electrical field or by other excitations (such as absorption of photons).

Figure 5.1(a) gives an idealized presentation of the electronic orbital overlaps and the formation of finer energy states (sublevels). The band formed from the ground-state valence level is called the valence band. The band immediately preceding is called the conduction band. The interval between the top of the valence and the bottom of the conduction bands is called the forbidden energy gap $\Delta E_{e,g}$. For ideal electrical insulators, $\Delta E_{e,g} \rightarrow \infty$.

Figure 5.1(b) gives these energy bands quantitatively for C (Table A.2, $1s^2 2s^2 2p^2$) atoms as they form covalent bonds and diamond equilibrium interatomic spacing (lattice constant in Table A.2) is approached, reached, and passed. The

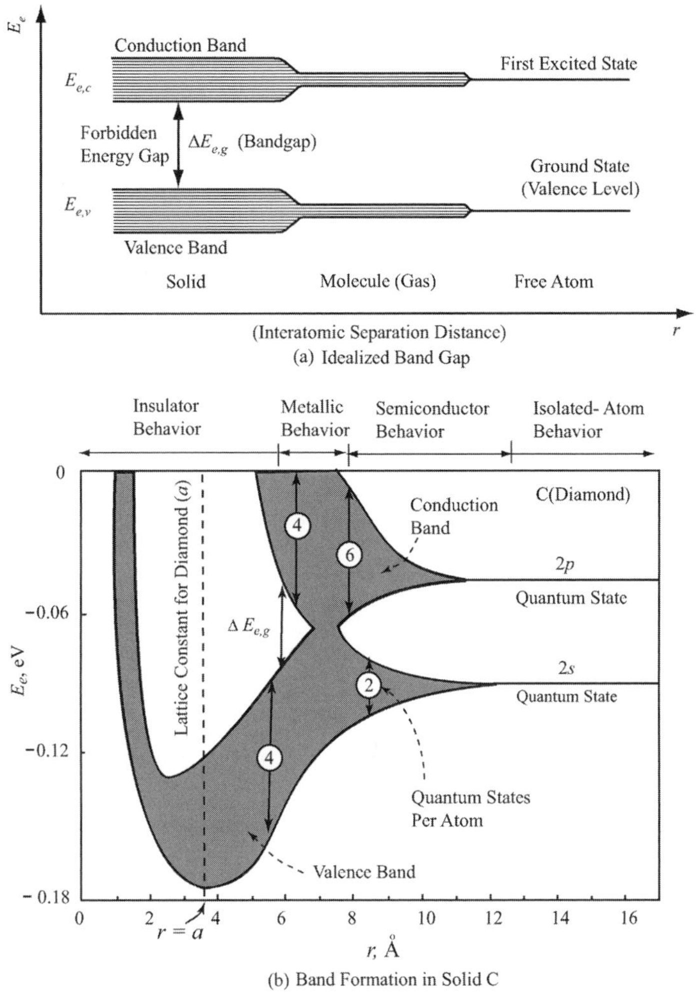

Figure 5.1. (a) Idealization of formation of electronic bands (sublevels) and bandgaps as two atoms approach each other and electronic orbitals (energy levels) overlap. (b) Quantitative energy bands for diamond (C) as a functions of interatomic spacing r [335].

isolated carbon atoms contain six electrons occupying $1s$, $2s$, and $2p$ orbitals in pairs. The $2s$ and $2p$ orbitals are shown in Figure 5.1(b), and as the wave functions overlap by reduced spacing (and because of Pauli exclusion), energy-level splitting occurs, giving $2N$ states (spin splitting) in $2s$ and $6N$ (due to spin splitting of $2p$ orbitals) states in $2p$, where N is the number of neighboring atoms in the crystal. The $2s$ bands will grow to $4N$ due (further splitting and overlap) as the distance is further decreased, but the $2p$ bands decrease to $4N$ (due to overlap). Figure 5.1(b) shows quantum states per atom and suggests that, with an interatomic spacing of about 7 Å, a metallic behavior (no bandgap) can be found in this carbon (C) structure, such as in carbon nanotubes.

Figure 5.2 shows the electronic energy states in bulk solid metals, semiconductors, and insulators. When electrons are allowed to equilibrate with a reservoir (or

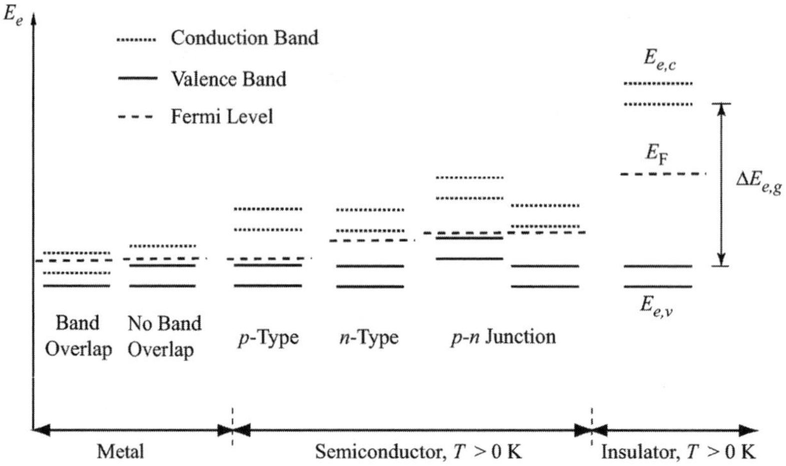

Figure 5.2. The electron energy states in bulk metals, semiconductors ($T > 0$ K), and insulators. Progressively higher Fermi energy (chemical potential) is needed to conduct electrons.

a source), their chemical potential equilibrates with this reservoir (and this is represented by the Fermi energy). Metallic solids have very closely spaced electronic energy states, and there are also possibilities of energy overlap (this will be discussed in Section 5.16). These energy states are readily accessible by the thermal energy $k_B T$. Also shown in Figure 5.2 are the *p*- and *n*-type semiconductors (Glossary), which have moderate gaps (for example $\Delta E_{e,g} = 1.13$ eV for Si). The valence band is fully occupied, and some electrons can be thermally promoted to the conduction band. For insulators these gaps are rather large [$\Delta E_{e,g} = 5.45$ eV for C in the diamond phase, Figure 5.1(b)], compared with the thermal energy, so no promotion of an electron to the conduction band occurs.

In this chapter, we begin with a discussion of electron energy bands in solids. Then we discuss its electronic transport properties. We do not consider electron (or charged particles, in general) beams. When electrons are part of a particle mixture, such as in thermal plasmas (high temperature, charged gases), some of the fluid particle features discussed in Chapter 6 can be used. Finally, we discuss size effects.

5.1 Schrödinger Equation for Periodic-Potential Band Structure

In Section 2.6.5 we found the solution to the Schrödinger equation for free electrons in periodic structure, as it applies to free (conduction) electrons in metals. However, for bound electrons their interaction with the nuclei should be included, as was done in Section 2.6.6 for hydrogenlike, single atoms. Here we extend this to periodic structures (crystal) and potentials.

As two atoms approach, the electronic energy levels of the atom pair split because of the Pauli exclusion principle. The potential energy of the shared electrons is the average of the energy levels of the valence electrons. For an N-atom system, the energy levels split into N levels, with the inner shell overlapping to a lesser extent.

For large N, the spacing can be large, resulting in energy bands. Bands are energy levels for the crystal. Between bands, forbidden energy ranges exist, and each band may be from several atomic energy levels. Electron conductivity requires mobile electrons (i.e., electrons in the conduction bands).

The time-independent Schrödinger equation (Table 2.8) for electrons is

$$[-\frac{\hbar}{2m_e}\nabla^2 + \varphi_c(\boldsymbol{x})]\psi_{e,\boldsymbol{\kappa}}(\boldsymbol{x}) = E_e\psi_{e,\boldsymbol{\kappa}}(\boldsymbol{x}). \tag{5.1}$$

The crystal periodic lattice potential $\varphi_c(\boldsymbol{x})$ is presented (assumed to be the stationary form of the lattice used for the lattice vibration in Chapter 4) as a Fourier series used as the basis of the solution for ψ_e. The symmetry of the reciprocal lattice space (and Brillouin zone, Section 4.3) is most useful, and therefore the crystal potential is given as a series expansion in $\boldsymbol{g}$ as

$$\varphi_c(\boldsymbol{x}) = \sum_{\boldsymbol{g}} \varphi_{\boldsymbol{g}}\exp[i(\boldsymbol{g}\cdot\boldsymbol{x})] \quad \text{periodic potential}, \tag{5.2}$$

where $\boldsymbol{g}$ is the reciprocal lattice vector. The periodic potential is a Fourier series in reciprocal lattice space.

The Bloch wave function, using an orbital centered around a (translational) lattice vector $\boldsymbol{l}$, is the solution for $\psi_{e,\kappa}$ along (2.68), i.e.,

$$\psi_{e,\boldsymbol{\kappa}}(\boldsymbol{x}) = \sum_{\boldsymbol{\kappa}} a_{\boldsymbol{\kappa}}\exp[i(\boldsymbol{\kappa}\cdot\boldsymbol{x})] \quad \text{Bloch plane-wave series}. \tag{5.3}$$

Substituting these $\psi_{e,\kappa}$ and φ_c in (5.1), we have

$$\sum_{\boldsymbol{\kappa}} -\frac{\hbar^2}{2m_e}\kappa^2 a_{\boldsymbol{\kappa}}\exp[i(\boldsymbol{\kappa}\cdot\boldsymbol{x})] + \sum_{\boldsymbol{g}}\sum_{\boldsymbol{\kappa}} \varphi_{\boldsymbol{g}} a_{\boldsymbol{\kappa}}\exp\{i[(\boldsymbol{g}+\boldsymbol{\kappa})\cdot\boldsymbol{x}]\} =$$

$$E_e\sum_{\boldsymbol{\kappa}} a_{\boldsymbol{\kappa}}\exp[i(\boldsymbol{\kappa}\cdot\boldsymbol{x})]. \tag{5.4}$$

Using the orthogonality of $e^{-i(\boldsymbol{\kappa}'\cdot\boldsymbol{x})}$, we have the delta function

$$\frac{1}{V}\int \exp[-i(\boldsymbol{\kappa}'\cdot\boldsymbol{x})]\exp[i(\boldsymbol{\kappa}\cdot\boldsymbol{x})]\mathrm{d}V = \delta_{\boldsymbol{\kappa}'\boldsymbol{\kappa}}, \tag{5.5}$$

where $\delta_{\boldsymbol{\kappa}'\boldsymbol{\kappa}}$ is the Kronecker deltas. Then (5.4) becomes

$$\frac{\hbar^2}{2m_e}\kappa'^2 a_{\boldsymbol{\kappa}'} + \sum_{\boldsymbol{g}} \varphi_{\boldsymbol{g}} a_{\boldsymbol{\kappa}'-\boldsymbol{g}} = E_e a_{\boldsymbol{\kappa}'}. \tag{5.6}$$

The preceding set of algebraic equations is called the central (or secular) equation for quantized energy E_e.

By rearranging, this equation becomes

$$[\frac{\hbar^2(\kappa^2 - g^2)}{2m_e} - E_e]a_{\boldsymbol{\kappa}-\boldsymbol{g}} + \sum_{\boldsymbol{g}'} \varphi_{\boldsymbol{g}'-\boldsymbol{g}} a_{\boldsymbol{\kappa}-\boldsymbol{g}'} = 0 \quad \text{central eigenvalue equation}. \tag{5.7}$$

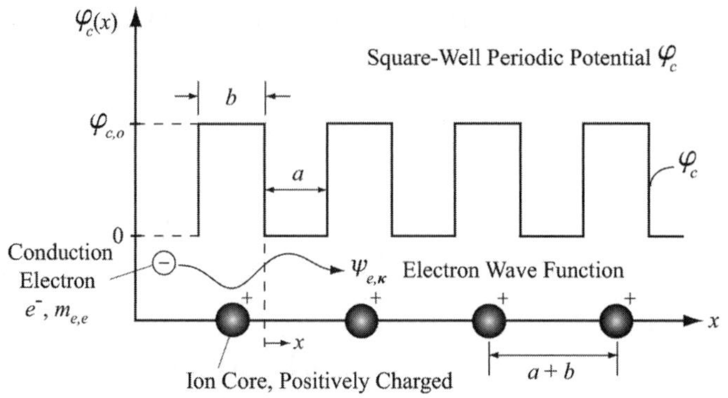

Figure 5.3. A linear, ionic chain lattice and the periodic potential experienced by an electron, causing electronic energy gaps. The electron wave function $\psi_{e,\kappa}(x)$ is expressed as plane waves.

The central role of the periodic potential $\varphi_c(\boldsymbol{x})$ is to create the eigenvalues of electron energy E_e, with energy gaps, and this distinguishes it from the nonperiodic potential of hydrogenlike atom electrons treated in Section 2.6.5.

In (5.7), for a given $\boldsymbol{\kappa}$, only coefficients $a_{\boldsymbol{\kappa}-\boldsymbol{g}}$ are needed, and for $\boldsymbol{\kappa}$ in the first Brillouin zone, $\boldsymbol{g}$ is evaluated in the neighboring zones (multiples of the reciprocal lattice vector). Then (5.7) gives the solution to $E_{e,n}(\boldsymbol{\kappa})$, i.e., the eigenvalues of the central equations, where n is the electron band number.

Note that for the time-independent Schrödinger equation (5.1), we have

$$\mathrm{H}\psi_{e,\boldsymbol{\kappa}}(\boldsymbol{x}) = E_e(\boldsymbol{\kappa})\psi_{e,\boldsymbol{\kappa}}(\boldsymbol{x}), \tag{5.8}$$

or

$$\int \psi_{e,\boldsymbol{\kappa}}^{\dagger}\mathrm{H}\psi_{e,\boldsymbol{\kappa}}\mathrm{d}V = \int \psi_{e,\boldsymbol{\kappa}}^{\dagger}E_e\psi_{e,\boldsymbol{\kappa}}\mathrm{d}V = E_e$$

$$\text{relation between energy and wave function.} \tag{5.9}$$

We now proceed with an example of a linear lattice with a single, periodic crystal potential.

5.2 Electron Band Structure in One-Dimensional Ionic Lattice

To demonstrate the electronic energy gaps occurring for conduction electrons in periodic lattice (ionic) potentials, consider the one- dimensional (linear) atomic chain shown in Figure 5.3 [183].

The one-dimensional form of (5.1) along x, is

$$[-\frac{\hbar}{2m_e}\frac{\mathrm{d}^2}{\mathrm{d}x^2} + \varphi_c(x)]\psi_{e,\kappa}(x) = E_e\psi_{e,\kappa}(x) \quad \text{periodic potential,} \tag{5.10}$$

where $\varphi_c(x)$ is the periodic crystal (lattice) potential, which is idealized as having the square-well potential distribution shown in Figure 5.3. This is called the periodic Kronig–Penny potential model of electron in an ionic lattice.

In region $0 \le x \le a$, $\varphi_c = 0$, and in $-b \le x \le 0$, we have $\varphi_c = \varphi_{c,\mathrm{o}}$ (well potential). Then, the periodic behavior is

$$\varphi_c(x) = \varphi_c(x + a + b). \tag{5.11}$$

In region $0 \le x \le a$, the wave function (5.3) is the solution to (5.10) with $\varphi_c = 0$, and we have

$$\psi_{e,\kappa}(x) = Ae^{i\kappa_1 x} + Be^{-i\kappa_1 x}, \ 0 \le x \le a, \tag{5.12}$$

with $E_e = \hbar^2\kappa_1^2/2m_e$.

For $-b \le x \le 0$, we have $\varphi_c = \varphi_{c,o}$ and the solution

$$\psi_{e,\kappa}(x) = Ce^{\kappa_o x} + De^{-\kappa_o x}, \quad \kappa_o^2 = \frac{2m_e}{\hbar^2}(\varphi_{c,o} - E_e), \ -b \le x \le 0, \tag{5.13}$$

where κ_o is wave number corresponding to energy $\varphi_{c,o} - E_e$. The periodic condition (Brillouin zone boundary) would require that

$$\psi_{e,\kappa}(a \le x \le a + b) = \psi_{e,\kappa}(-b \le x \le 0)e^{i\kappa(a+b)}. \tag{5.14}$$

We determine the four constants are determined by evaluating the two solutions and their first derivatives at $x = 0$ and $x = a$ and setting them equal. These give for $x = 0$ (two equations) and for $x = a$ (two equations)

$$A + B = C + D \tag{5.15}$$

$$i\kappa_1(A - B) = \kappa_o(C - D) \tag{5.16}$$

$$Ae^{i\kappa_1 a} + Be^{-i\kappa_1 a} = (Ce^{-\kappa_o b} + De^{\kappa_o b})e^{i\kappa(a+b)} \tag{5.17}$$

$$i\kappa_1(Ae^{i\kappa_1 a} - Be^{-i\kappa_1 a}) = \kappa_o(Ce^{-\kappa_o b} - De^{\kappa_o b})e^{i\kappa(a+b)}, \tag{5.18}$$

where we have referred to the wave number in $0 \le x \le a$ as κ_1. Note that, from (5.14), $\psi_{e,\kappa}(a) = \psi_{e,\kappa}(-b)e^{i\kappa(a+b)}$.

The nontrivial solution requires that the determinant of the coefficient matrix of the constants to vanish. This gives the characteristic equation (end-of-chapter problem)

$$\frac{\kappa_o^2 - \kappa_1^2}{2r\kappa_1}\sinh(\kappa_o b)\sin(\kappa_1 a) + \cosh(\kappa_o b)\cos(\kappa_1 a) = \cos[\kappa(a + b)]. \tag{5.19}$$

From (5.13) we also have

$$E_e = \frac{\hbar^2\kappa_o^2}{2m_e} + \varphi_{c,o}. \tag{5.20}$$

This is the eigenfunction (or characteristic equation) for the discrete values of $E_e(\kappa)$.

Now consider a simplified case. For $\kappa_o \gg \kappa_1$ and $\kappa_o b \ll 1$, and for the special limiting case of $b \to 0$ and $\varphi_{c,o} \to \infty$, we define

$$\frac{\kappa_o^2 ab}{2} = \frac{m_e ab}{\hbar^2}(\varphi_{c,o} - E_e) \equiv s \quad \text{potential strength } s \text{ is finite.} \tag{5.21}$$

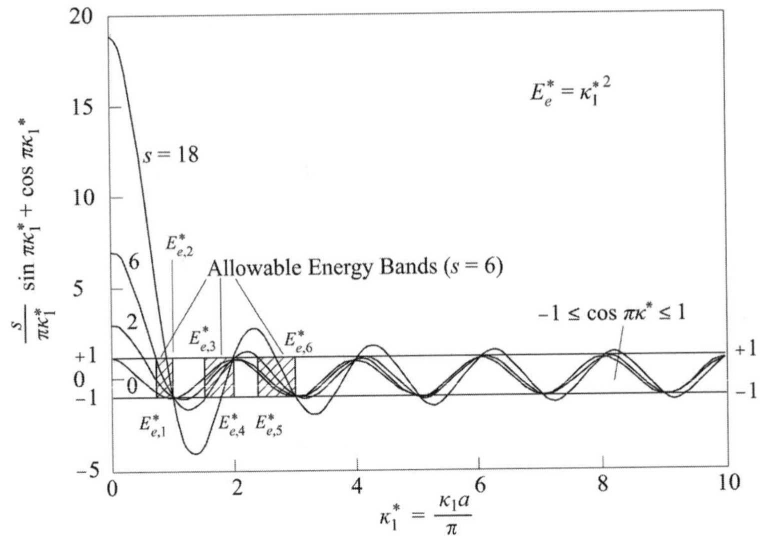

Figure 5.4. The left- and right-hand sides of the characteristic equation, for $s = 2$, 6, and 18. The right-hand side is represented by the bounds (+/−1).

Here s (dimensionless) is the strength of the potential well. For $s = 0$, there is no potential, and we have the free electron (Section 2.6.6).

Then using s and these limits, the characteristic equation (5.19) becomes

$$\frac{s}{\kappa_1 a}\sin(\kappa_1 a) + \cos(\kappa_1 a) = \cos(\kappa a), \ b \to 0, \ \varphi_{c,o} \to \infty. \tag{5.22}$$

Using the electron energy, the dimensionless wave number is defined as

$$\kappa_1^* = \frac{\kappa_1 a}{\pi} = (\frac{2m_e E_e a^2}{\pi^2\hbar^2})^{1/2}, \quad \text{or } E_e = \frac{\hbar^2\kappa_1^2}{2m_e}. \tag{5.23}$$

This makes (5.22) along with the periodicity at the Brillouin zone boundaries give the constrained characteristic equation

$$\frac{s}{\pi\kappa_1^*}\sin(\pi\kappa_1^*) + \cos(\pi\kappa_1^*) = \cos(\pi\kappa^*), \ \ -1 \le \cos(\pi\kappa^*) \le 1, \ \kappa^* = \frac{\kappa a}{\pi}. \tag{5.24}$$

Then only discrete values of κ_1^* are allowed for a given s.

For $s = 2$, 6, and 18, the preceding characteristic equation is plotted in Figure 5.4, and the first three areas where the solution exists are highlighted. For example, for $\kappa_1^* = 0$, no solution exists. Note that $\kappa_1^{*2} = E_e^*$, i.e.,

$$E_e^* = \frac{2m_e E_e a^2}{\pi^2\hbar^2} = \kappa_1^{*2}. \tag{5.25}$$

Figure 5.5 shows the allowed energies in the $E_e^*(\kappa^*)$ space, where $\kappa^* = 0$ is designated as the center and $\kappa^* = 1$ is the edge of the first Brillouin zone. The energy gap at the edge of the Brillouin zone is evident.

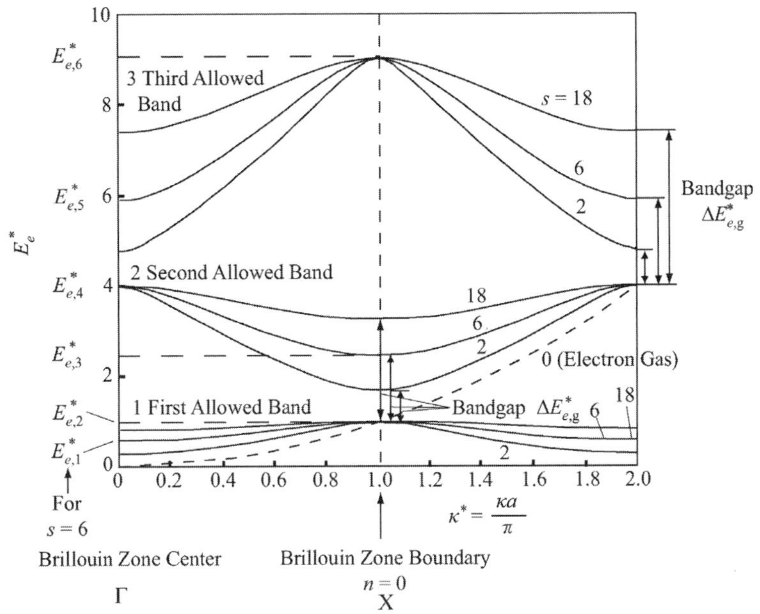

Figure 5.5. The first three electronic bands and the first two bandgaps for the limit ($b \rightarrow 0$) square potential with linear, atomic chain. The results are for $s = 2$, 6, and 18. The electron gas model is also shown.

Numerically, the electron energy equation is found by solving

$$\frac{s}{\pi E_e^{*1/2}} \sin(\pi E_e^{*1/2}) + \cos(\pi E_e^{*1/2}) = \cos(\pi \kappa^*), \qquad (5.26)$$

and allowing for progressively larger eigenvalues E_e^*.

Note that in this simple model the only parameters are s, m_e, and a. Also, note that to obtain the solution for the free electron, we need to use $s = 0$, and also apply the boundary condition $\psi_{e,\kappa}(0) = \psi_{e,\kappa}(L) = 0$. These would lead to the results given in Section 2.6.5.

The solution for $s = 0$ is also shown in Figures 5.4 and 5.5. The parabolic relation between E_e^* and κ^*, (2.94), is evident in Figure 5.5. Also note some similarity between the electron energy of Figure 5.5 and the phonon energy of Figure 4.3, including energy gaps.

5.3 Three-Dimensional Bands Using Tight-Binding Approximation

In semiconductors the valence atomic electrons in s and p orbitals, when brought together in a crystal, become perturbed. The core electrons are not perturbed much and perturbed outer orbitals are presented with the atomic orbitals as their basis in the tight-binding approximation reviewed here.

In the tight-binding method, the energy eigenstates of (5.7) are determined near the band edges, using the outermost (highest principal quantum number) orbitals, thus excluding those electrons tightly bond to their nuclei. In particular, when this assumption is used along with the linear combination of the atomic orbitals (LCAO) to the energy eigenstates, it makes for a simplified treatment for the semiconductors [309].

5.3.1 General LCAO

As we did for the hydrogenlike isolated atoms in Section 2.6.7 for an isolated atom with an electronic structure, we use the Hamiltonian $\mathrm{H_o}$ ($\mathrm{H_o} = -\hbar^2\nabla^2/2m_e + \varphi_e$, where φ_e is the Coulombic potential) and represent the eigenstates of the time-independent Schrödinger equation by

$$\mathrm{H_o}\psi_{e,n} = E_{e,n}\psi_{e,n} \; n = 1, 2, ..., N \text{ for orbitals of isolated atom}, \tag{5.27}$$

where N is the number of atoms. The crystal is now made of these atoms with a known atomic electronic structure represented by $E_{e,n}$.

The LCAO methods allow for the construction of crystal wave function from the preceding relations by use of the Bloch wave form [309]

$$\psi_{e,\kappa}(\boldsymbol{x}) = \sum_{n,\boldsymbol{x}_o} \psi_{e,n}(\boldsymbol{x} - \boldsymbol{x}_o)e^{i(\boldsymbol{\kappa}\cdot\boldsymbol{x})} = \sum_{n,x_o} \phi(\boldsymbol{x} - \boldsymbol{x}_o)e^{i(\boldsymbol{\kappa}\cdot\boldsymbol{x})}, \tag{5.28}$$

where $\boldsymbol{x}_o$ is the center of orbital n, and the basis function ϕ is constructed with the atomic eigenfunctions, i.e.,

$$\phi(\boldsymbol{x}) = \sum_{n=1}^{N} a_n \psi_{e,n}(\boldsymbol{x}). \tag{5.29}$$

The following steps are for the determination of constant a_n.

The perturbation Hamiltonian is due to the neighboring atoms; thus we have

$$\mathrm{H} = \mathrm{H_o} + \mathrm{H}'(\boldsymbol{x}), \tag{5.30}$$

or using (5.1), we have

$$(\mathrm{H_o} + \mathrm{H}')\psi_{e,\kappa} = E_e(\boldsymbol{\kappa})\psi_{e,\kappa}. \tag{5.31}$$

Because $\psi_{e,n}$ functions are orthogonal, by multiplying by $\psi^{\dagger}_{e,m}$ and integrating, we have

$$\int \psi^{\dagger}_{e,m}\{[\mathrm{H_o} + \mathrm{H}'] \sum_{n,\boldsymbol{x}_o} a_n \psi_{e,n}(\boldsymbol{x} - \boldsymbol{x}_o)e^{i(\boldsymbol{\kappa}\cdot\boldsymbol{x})}$$

$$- E_e(\boldsymbol{\kappa}) \sum_{n,\boldsymbol{x}_o} a_n \psi_{e,n}(\boldsymbol{x} - \boldsymbol{x}_o)e^{i(\boldsymbol{\kappa}\cdot\boldsymbol{x})}\}\mathrm{d}\boldsymbol{x} = 0. \tag{5.32}$$

The result is (seperating $\boldsymbol{x}_\text{o} = 0$, and $\boldsymbol{x}_\text{o} \neq 0$ terms)

$$[E_e(\boldsymbol{\kappa}) - E_{e,m}]a_m = -[E_e(\boldsymbol{\kappa}) - E_{e,m}]\sum_{n=1}^{N} a_m$$
$$\times [\sum_{\boldsymbol{x}_\text{o}\neq 0}\int \psi_{e,m}^{\dagger}(\boldsymbol{x})\psi_{e,n}(\boldsymbol{x}-\boldsymbol{x}_\text{o})e^{i(\boldsymbol{\kappa}\cdot\boldsymbol{x}_\text{o})}\text{d}\boldsymbol{x}]$$
$$+\sum_{n=1}^{N} a_n \int \psi_{e,m}^{\dagger}(\boldsymbol{x})\text{H}'(\boldsymbol{x})\psi_{e,n}(\boldsymbol{x})\text{d}\boldsymbol{x}$$
$$+\sum_{n=1}^{N} a_m \sum_{\boldsymbol{x}_\text{o}\neq 0}\int \psi_{e,m}^{\dagger}(\boldsymbol{x})\text{H}'(\boldsymbol{x})\psi_{e,n}(\boldsymbol{x}-\boldsymbol{x}_\text{o})e^{i(\boldsymbol{\kappa}\cdot\boldsymbol{x}_\text{o})}\text{d}\boldsymbol{x}, \qquad (5.33)$$

where

$$E_{e,m}\delta_{mn} = \int \psi_{e,m}^{\dagger}(\boldsymbol{x})\text{H}_\text{o}\psi_{e,n}\text{d}\boldsymbol{x}, \quad \delta_{nm} = \int \psi_{m}^{\dagger}(\boldsymbol{x})\psi_{n}(\boldsymbol{x})\text{d}\boldsymbol{x}, \qquad (5.34)$$

which is of the form given in (5.9). Here δ_{mn} is the Kronecker delta. Note that $\int \psi_{e,m}^{\dagger}(\boldsymbol{x})\psi_{e,n}(\boldsymbol{x}-\boldsymbol{x}_\text{o}) \neq \delta_{mn}$, for $\boldsymbol{x}_\text{o} \neq 0$.

In the tight-binding approximation, the first integral term on the right-hand side of (5.33) is set to zero by assuming negligible overlap between neighboring atomic wave functions, i.e., atomic functions are tightly bound to atoms [309].

For most potentials, the second integral term also vanishes [309].

The preceding equation is solved for the eigenvalues $E_e(\boldsymbol{\kappa})$, and is a set of $N \times N$ coupled equations. Also, for each $\boldsymbol{\kappa}$, there are N solutions, which provide the N-band E_e-$\boldsymbol{\kappa}$ structure.

5.3.2 Example of Tight-Binding Approximation: FCC *s* Orbital

As an example, consider a monoatomic crystal with a single orbital s. Because there is only one energy level, $\{a_m\} = 0$, except for the coefficient, $a_s = 1$. Then (5.33) is only one equation, i.e.,

$$E_e(\boldsymbol{\kappa}) - E_s = \int \psi_{e,s}^{\dagger}(\boldsymbol{x})\text{H}'(\boldsymbol{x})\psi_{e,s}(\boldsymbol{x})\text{d}\boldsymbol{x}$$
$$+\sum_{\boldsymbol{x}_\text{o}\neq 0}\int \psi_{e,s}^{\dagger}(\boldsymbol{x})\text{H}'(\boldsymbol{x})\psi_{e,s}(\boldsymbol{x}-\boldsymbol{x}_\text{o})e^{i(\boldsymbol{\kappa}\cdot\boldsymbol{x}_\text{o})}\text{d}\boldsymbol{x}. \qquad (5.35)$$

This is written as

$$E_e(\boldsymbol{\kappa}) = E_s - \beta_s - \sum_{\boldsymbol{x}_\text{o}} \gamma(\boldsymbol{x})e^{i(\boldsymbol{\kappa}\cdot\boldsymbol{x}_\text{o})}, \qquad (5.36)$$

where

$$\beta_s \equiv -\int \psi_{e,s}^{\dagger}(\boldsymbol{x})\text{H}'(\boldsymbol{x})\psi_{e,s}(\boldsymbol{x})\text{d}\boldsymbol{x} \qquad (5.37)$$

$$\gamma(\boldsymbol{x}_\text{o}) \equiv -\int \psi_{e,s}^{\dagger}(\boldsymbol{x})\varphi_c(\boldsymbol{x})\psi_{e,s}(\boldsymbol{x}-\boldsymbol{x}_\text{o})\text{d}\boldsymbol{x} \text{ overlap energy integral.} \qquad (5.38)$$

Here β_s and γ can be treated as given constants, and then it becomes clear that, because of $\exp[i(\boldsymbol{\kappa}\cdot\boldsymbol{x}_\mathrm{o})]$, the contribution from other than the nearest neighbors may be neglected. Also note that $\gamma(\boldsymbol{x}_\mathrm{o}) = \gamma(-\boldsymbol{x}_\mathrm{o})$.

For example, for the FCC crystals, the 12 nearest neighbors around the atom at $\boldsymbol{x} = 0$ are given by (4.33), i.e.,

$$\boldsymbol{x}_\mathrm{o} = a(\pm 0.5, \pm 0.5, 0);\ a(\pm 0.5, 0, \pm 0.5);\ a(0, \pm 0.5, \pm 0.5), \qquad (5.39)$$

and the first Brillouin zone designations for FCC are shown in Figure 4.6(d).

Then the electron energy equation (or scalar equation) (5.35) becomes

$$\begin{aligned} E_e(\boldsymbol{\kappa}) &= E_s - \beta_s - \gamma \sum e^{i(\boldsymbol{\kappa}\cdot\boldsymbol{x}_o)} \\ &= E_s - \beta_s - \gamma[e^{ia(\kappa_x+\kappa_y)/2} + e^{ia(\kappa_x-\kappa_y)/2} + e^{ia(-\kappa_x+\kappa_y)/2} + e^{ia(-\kappa_x-\kappa_y)/2} + ...] \\ &= E_s - \beta_s - 4\gamma[\cos(\frac{a\kappa_x}{2})\cos(\frac{a\kappa_y}{2}) + \cos(\frac{a\kappa_y}{2})\cos(\frac{a\kappa_z}{2}) \\ &\quad + \cos(\frac{a\kappa_z}{2})\cos(\frac{a\kappa_x}{2})],\ -\frac{2\pi}{a} \le \kappa_i \le \frac{2\pi}{a} \text{ (first Brillouin zone).} \end{aligned} \qquad (5.40)$$

Figure 5.6(a) shows the variation of $E_e(\boldsymbol{\kappa})$ along the $\boldsymbol{\kappa}$ space L–Γ–X–K–Γ–W. This is a typical presentation of the band structure (having only one band s). The results are for $\gamma = 1.0$ eV and $E_s - \beta_s = 0$, and the width of the band is 16γ.

This constant-energy surface in $\{\kappa_x, \kappa_y, \kappa_z\}$ space or the reciprocal lattice is the Fermi surface. The constant-energy surface for (5.40) can be shown in the $\boldsymbol{\kappa}$-space and is an end-of-chapter problem.

Near the zone center Γ, where $\kappa_x = \kappa_y = \kappa_z$ and $\kappa a \ll 1$ (small κ), (5.40) becomes (end-of-chapter problem)

$$E_e(\boldsymbol{\kappa}) = E_s - \beta_s - 12\gamma + \gamma a^2 \kappa^2,$$

parabolic energy relation for $\kappa a \ll 1$, for electron in a crystal, (5.41)

compared with the free-electron energy (Section 2.6.4), i.e.,

$$E_e(\boldsymbol{\kappa}) = E_e(\boldsymbol{p}) = E_{e,0} + \frac{\hbar^2\kappa^2}{2m_e} \text{ parabolic energy relation for electron gas,} \qquad (5.42)$$

which provides the parabolic model for small κ.

Based on this, the effective electron mass $m_{e,e}$ is defined as $\hbar^2/2\gamma a^2$.

Note that the overlap energy integral γ (has units of energy through φ_c) depends on a.

The preceding example can now be generalized. By combining (5.28) and (5.29), the wave function is written as

$$\psi_{e,\kappa}(\boldsymbol{x}) = \sum_{\boldsymbol{x}_i}\sum_{m=1}^{N_\mathrm{o}}\sum_{j=1}^{N_a} a_{m,j}(\boldsymbol{\kappa})\phi_{m,j}(\boldsymbol{x} - \boldsymbol{x}_j - \boldsymbol{x}_{\mathrm{o},i})e^{i(\boldsymbol{\kappa}\cdot\boldsymbol{x}_{\mathrm{o},i})}, \qquad (5.43)$$

where the sum is over the unit cells, m is for the different atomic functions $\phi_{m,j}$ (up to N_o orbitals) used as the basis, and j is the atoms in each unit cell (up to N_a atoms).

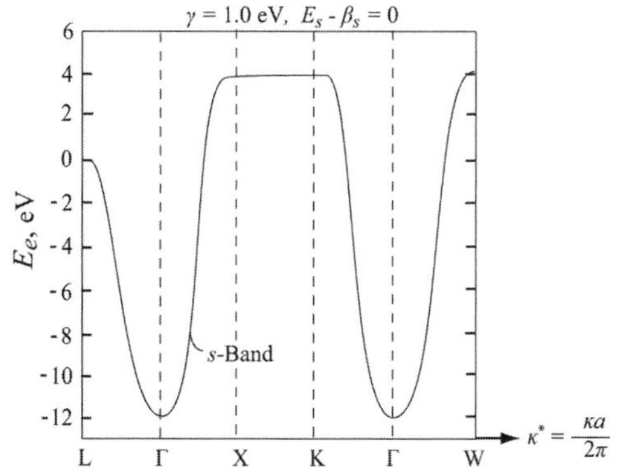

(a) Idealized Prediction for s-Band Using Tight-Binding Approximation

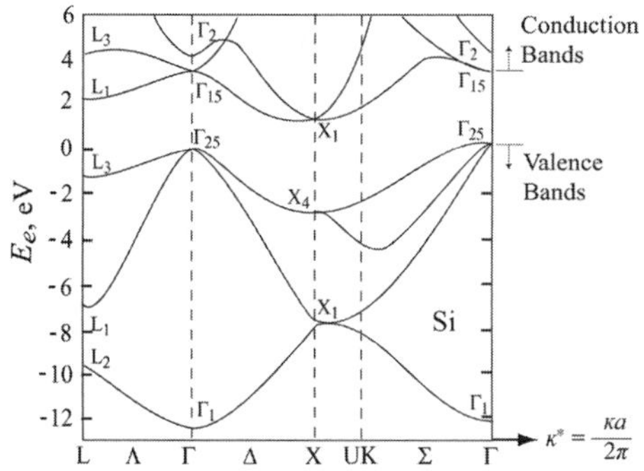

(b) Full-Band Structure of Si (Indirect Bandgap)

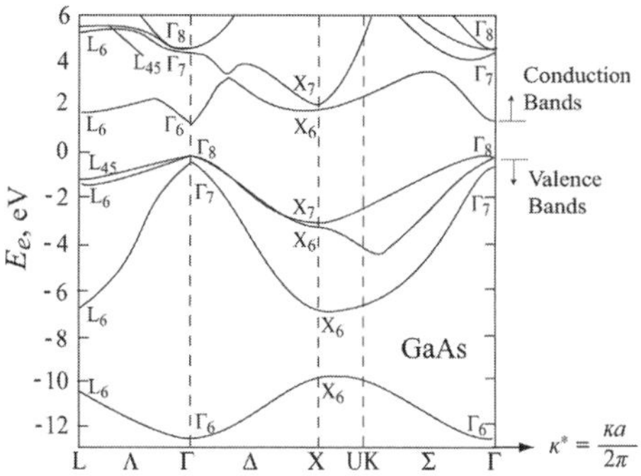

(c) Full-Band Structure of GaAs (Direct Bandgap)

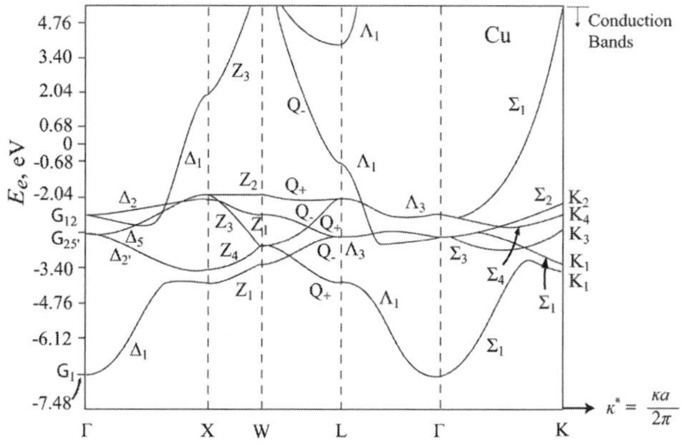

(d) Full-Band Structure of Cu

Figure 5.6. Predicted electron-band structure for (a) an idealized s-band model (tight-binding approximation) and full bands of (b) Si [309], (c) GaAs [309], and (d) Cu [50] at $T = 300$ K.

5.4 Electron Band Structure for Semiconductors and Effective Mass

The electrons in the periodic potential of a crystal are represented by the eigenvalue wave functions in $\boldsymbol{\kappa}$-space. These energy-states that are periodic in $\boldsymbol{\kappa}$-space are separated by certain energy gaps or forbidden energy regions (energies not reached by any real $\boldsymbol{\kappa}$).

This electron band structure $E_{e,n}(\boldsymbol{\kappa})$ found from (5.7) depends on the constituent atoms (their electrical orbits). The semiconductors have the most complex band structures. Table A.2 (perodic table) gives the electron orbits, and for example Si has a $1s^2 2s^2 2p^6 3s^2 3p^2$ electron structure and the last $s-$ and $p-$orbit do tend to influence the bandgaps separating the valence and conduction bands [Figure 5.1(b)].

Each band has $2N$ (because of spin degeneracy) allowed $\boldsymbol{\kappa}$ states and each unit cell (basis) in the crystal contributes electrons to these states.

Figures 5.6(a) and (b) show the predicted electron band structures for Si and GaAs, respectively. There are four valance bands ($E_{e,v} \leq 0$) for Si and three conduction bands. There are many band overlaps. Figure 5.6(c) shows the band structure for Cu. Note the lack of bandgap in Cu.

The three-dimensional eigenfunction $E_e(\boldsymbol{\kappa})$ for semiconductors is rather complex, and therefore simplification is made in their presentation. Figure 5.7(a) shows the model band structure for diamond and zinc-blende crystals (e.g., Si, GaAs). For them, the conduction band has three minima, namely at $\Gamma(\boldsymbol{\kappa} = 0)$, at L (along the $\langle 111 \rangle$ direction at boundary of the first Brillouin zone), and at Δ (along the Δ line). The modal semiconductor also has three valence bands, each having a maximum at $\boldsymbol{\kappa} = 0$, and one is split-off by the spin–orbit interaction [219].

The $\boldsymbol{\kappa}$-space designation for the first Brillouin zone is given in Figure 5.7(b), for FCC lattice. More complete designations are given in Section 4.3.

Figure 5.7(c) shows the six valleys (constant-energy surfaces) in the E_e-$\boldsymbol{\kappa}$-space of Figure 5.6(b), for Si. Note that none of these valleys is located on the Brillouin zone edges. With phonon-induced transitions (scatterings), conduction electrons can transit between these valleys (intervalley scattering). This will be discussed in Section 5.15.3.

In the periodic electron gas treatment of Section 2.6.5, the electron energy was presented by the wave-number parabolic relation (2.94). Using the parabolic band model, an effective electron mass is defined as the electron kinetic energy (as a function of wave number) expressed through the Taylor series expansion as

$$E_{e,k}(\boldsymbol{p}) = E_{e,k}(\boldsymbol{\kappa}) = E_e(0) + \left.\frac{\partial E_e(\kappa)}{\partial \kappa}\right|_{\kappa=0} \kappa + \left.\frac{\partial^2 E_e(\kappa)}{\partial \kappa^2}\right|_{\kappa=0} \frac{\kappa^2}{2} + \cdots +$$

$$\equiv E_e(0) + \frac{\hbar^2 \kappa^2}{2m_{e,e}}, \quad m_{e,e} \equiv \frac{\hbar^2}{\left.\dfrac{\partial^2 E_e(\kappa)}{\partial \kappa^2}\right|_{\kappa=0}}$$

$$\text{effective mass in parabolic energy band,} \tag{5.44}$$

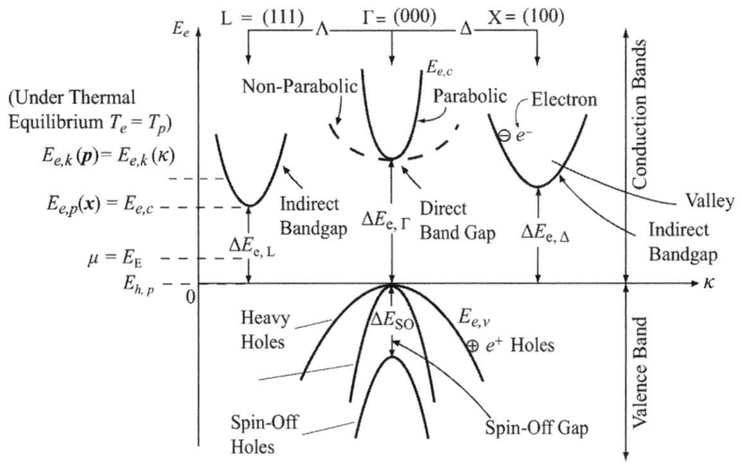

(a) Zinc-Blende Semiconductor Electron-Band Structure Model (Parabolic Bands)

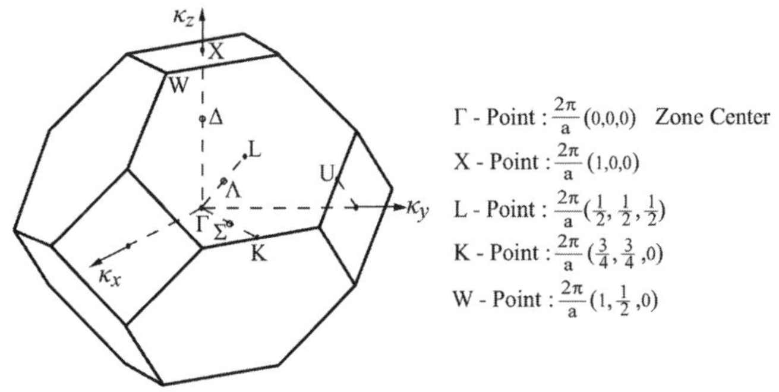

(b) First Brillouin Zone for Face-Centered Cubic (Zinc-Blende) Lattice

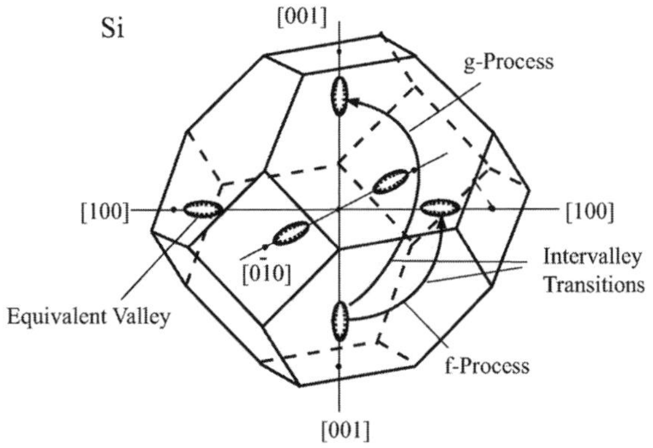

(c) Equivalent Valley in Conduction Bands of Si and Intervalley Transitions

Figure 5.7. (a) Model electronic band structure for diamond and zinc-blende crystal semiconductor. The kinetic energy $E_{e,k}(\boldsymbol{p}) = E_{e,k}(\boldsymbol{\kappa})$ and potential energy $E_{e,p}(\boldsymbol{x})$ are also shown. (b) Standard notation for labeling high-symmetry lines and points in FCC lattice Brillouin zone [219]. (c) Six equivalent valleys (constant-energy surfaces) in the conduction bands of Si (the f and g intervalley scatterings are also shown) [219].

Table 5.1. *Model band parameters of common cubic semiconductors [309] at $T = 300$ K. $m_{e,e}$ is for parabolic and $m_{e,e,\mathrm{L}}$ and $m_{e,e,\mathrm{T}}$ are for nonparabolic band models.*

Species	$\Delta E_{e,\Gamma}$, eV	$\Delta E_{e,L}$, eV	$\Delta E_{e,\Delta}$, eV	ΔE_{SO}, eV	$\frac{m_{e,e}}{m_e}$	$\frac{m_{e,e,\mathrm{L}}}{m_e}$	$\frac{m_{e,e,\mathrm{T}}}{m_e}$
C	11.67	12.67	5.45	0.006	–	1.4	0.36
Si	4.08	1.87	1.13	0.044	–	0.98	0.19
Ge	0.89	0.76	0.96	0.29	–	1.64	0.082
AlSb	2.5	2.39	1.6	0.75	–	1.64	0.23
GaP	2.7	2.7	2.2	0.08	–	1.12	0.22
GaAs	1.42	1.71	1.90	0.34	0.067	–	–
GaSb	0.67	1.07	1.30	0.77	0.045	–	–
InP	1.26	2.0	2.3	0.13	0.080	–	–
InAs	0.35	1.45	2.14	0.38	0.023	–	–
InSb	0.23	0.98	0.73	0.81	0.014	–	–
ZnS	3.8	5.3	5.2	0.07	0.28	–	–
ZnSe	2.9	4.5	4.5	0.43	0.14	–	–
ZnTe	2.56	3.64	4.26	0.92	0.18	–	–
CdTe	1.80	3.40	4.32	0.91	0.098	–	–

where the first derivative $\partial E_e(\kappa)/\partial\kappa|_{\kappa=0} = 0$ because of assumed symmetry and $m_{e,e}$ is the effective electron mass (which can be smaller or larger than the electron mass m_e).

Then for a direct gap (when the minimum of conduction electron and maximum of valence electron occur at $\boldsymbol{\kappa} = 0$), we can use the parabolic approximation

$$E_e(\boldsymbol{\kappa}) = \pm\frac{\hbar^2|\boldsymbol{\kappa}|^2}{2m_{e,e}}, \quad + \text{conduction}, \; - \text{valence}, \; \text{band at } \Gamma, \tag{5.45}$$

which represents a band as a spherical energy surface (but is called parabolic) in $\boldsymbol{\kappa}$-space. This single effective mass indicates an isotropic E_e-$\boldsymbol{\kappa}$.

Then the electron group velocity (4.15) is

$$\boldsymbol{u}_{e,g} = \frac{1}{\hbar}\nabla_\kappa(\boldsymbol{\kappa})E_e(\boldsymbol{\kappa}) = \frac{\hbar\boldsymbol{\kappa}}{m_{e,e}}, \tag{5.46}$$

and the momentum is (de Broglie relation)

$$\boldsymbol{p} = \hbar\boldsymbol{\kappa} = m_{e,e}\boldsymbol{u}_{e,g} \quad \text{de Broglie relation.} \tag{5.47}$$

For bands at L and along Δ, the ellipsoidal (called nonparabolic) surface is given by

$$E_{e,k}(\boldsymbol{p}) = E_e(\boldsymbol{\kappa}) = \frac{\hbar^2}{2}\left(\frac{\kappa_\mathrm{L}^2}{m_{e,e,\mathrm{L}}} + \frac{\kappa_\mathrm{T}^2}{m_{e,e,\mathrm{T}}}\right)$$

$$\text{nonparabolic energy band at L and along } \Delta. \tag{5.48}$$

Table 5.1 gives the model band parameters of common cubic semiconductors. For the nonparabolic bands, the conduction band is flatter (larger effective mass) and thus have a larger DOS (to be shown in Section 5.7). As an approximation a single wave vector is also used with a correction, called the Kane model [219],

$$E_e(\boldsymbol{\kappa})\left[1 + \frac{E_e(\boldsymbol{\kappa})}{\Delta E_{e,g}}\right] = \pm\frac{\hbar^2\kappa^2}{2m_{e,e,\mathrm{o}}} \quad \text{Kane nonparabolic band model,} \tag{5.49}$$

where $m_{e,e,\mathrm{o}}$ is the average effective mass equal to $(m_{e,e,\mathrm{L}}m_{e,e,\mathrm{T}}^2)^{1/3}$.

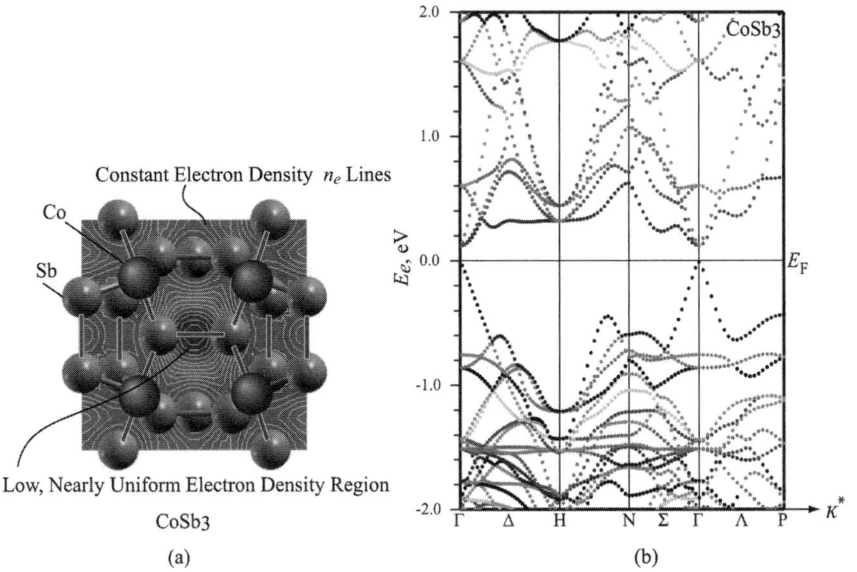

Figure 5.8. (a) Atomic structure of $CoSb_3$. The constant electron density lines in a unit cell are also shown. (b) Calculated band structure of $CoSb_3$, using WIEN2k.

We will further discuss the Kane model (including relaxation times) in Section 5.17.3, in connection with thermoelectric property predictions.

As will be shown, when using the BTE, it is more convenient to use $E_e(\boldsymbol{p})$ in place of $E_e(\boldsymbol{\kappa})$ by using (5.47) for the convection term.

The bandgap energy $\Delta E_{e,g}$ is temperature dependent (for most semiconductors, decreases with increase in temperature; exceptions include PbTe). Empirical relations such as $\Delta E_{e,g}(T) = \Delta E_{e,g} - aT^2/(T+b)$ are used. For Si and the Δ bandgap, $a = 0.473 \times 10^{-3}$ eV/K, and $b = 636$ K. As will be discussed, Table 5.6 includes some data on $\Delta E_{e,g}$ at 0 and 300 K.

5.5 *Ab Initio* Computation of Electron Band Structure

Amongst methods for calculation of the electronic band structures are the nearly free-electron model [222], tight-binding method [9, 309], Green function [9, 190], orthogonalized plane wave method [78, 143], and pseudopotentials [9, 78, 194]. When the unit cell contains a large number of atoms, only the density functional theory (DFT) offers sufficient accuracy and computational efficiency. DFT is based on one-to-one correspondence between the ground-state density of a many-electron system and the external potential. The accuracy of this method depends on the relation between the energy of the electrons and their density. The most accurate approximation is the generalized gradient approximation (GGA). The combination of GGA, full-potential, linearized augmented plane-wave (LAPW), and local orbits (LO) methods, leads to the most accurate scheme for the band structure calculation, which is an all-electron scheme (including relativistic effects) [78]. This method is used in an available computer program WIEN2k [78, 33] and is widely tested. Figure 5.8(a) shows the band structure of $CoSb_3$, a high-temperature thermoelectric material, and its structure is also shown in Figure 5.8 (b), calculated using WIEN2k. The input data is given in the footnotes.†

WIEN2k can also take into account the spin–orbit coupling, which is important for calculations of valence band structure and bandgap, especially for heavy atoms [34, 78].

Another program using GGA is VASP, which uses the pseudopotentials to deal with the core electrons. Since VASP only needs to deal explicitly with the valence electrons (rather than full electrons as in WIEN2k), VASP is much more ef-

† I. Input crystal structure

(1) Click "Execution StructGen™" to start the struct-file generator
(2) Input the configuration of atoms as given in table below

Title	CoSb3
Lattice	204_Im-3
a	9.0385 Å(select the Ang button)
b	9.0385 Å
c	9.0385 Å
α, β, γ	90°, 90°, 90°
Atom	Co, enter position (0.25, 0.25, 0.25)
Atom	Sb, enter position (0.0, 0.3351, 0.1602)

(3) Save structure.
(4) Click "save file and clean up"

II. Initialize the calculation

(1) Click "Execution initialize calc"
(2) Click "x nn" (use "2" for nn-band length factor)
(3) Click "view outputnn" (don't change any parameter)
(4) Click "x sgroup"
(5) Click "View outputsgroup". Do not use the struct-file generated by "sgroup"
(6) Click "x symmetry"
(7) Click "copy struct_st"
(8) Click "x lstart" [use "GGA (Perdew-Burke-Ernzerhof96)" for exchange correlation potential and "- 6.0 Ry" for energy to separate core and valence states]
(9) Click "view outputst"
(10) Click "check CoSb3.in1_st"
(11) Click "check CoSb3.in2_st"
(12) Click "check CoSb3.inm_st"
(13) Click "Prepare input files"
(14) Click "x kgen". Use 3000 for the k-mesh, and choose "Yes" to shift mesh
(15) Click "view CoSb3.klist"
(16) Click "x dstart"
(17) Click "view CoSb3.outputd"
(18) Do not perform spin-polarized calculation
(19) Click "Continue with SCF"

III. Submit the calculation and start SCF cycle (for convergence criteria, choose 0.001 Ry for energy, 1 mRy/au for force, and 0.0001 e_c for charge)

IV. Calculate band structure.

(1) Click "Band Structure" in "Tasks" menu
(2) Click "Create CoSb3.klist_band"
(3) Click "x lapw1 -band" to Calculate eigenvalues
(4) Edit CoSb3.insp. Select an energy range from -2.0 to 2.0 eV, and set Fermi energy to be 0.56372 Ry (this is slightly different from the value calculated by the program)
(5) Click "x spaghetti" to calculate band structure
(6) Plot band structure (needs "ghostscript" installed).

ficient and can handle much larger system. Though VASP makes a pseudopotential approximation, its accuracy is acceptable and similar to that of the LAPW method in many cases [34, 193, 150]. In general, WIEN2k is suitable for the accurate investigation of small systems, while for large systems VASP is suggested because of its high efficiency.

5.6 Periodic Electron Gas Model for Metals

The Fermi surface separates the occupied from unoccupied levels. Then the Fermi momentum $\hbar\kappa_{\rm F} = p_{\rm F}$ and its energy $E_{\rm F}$, is the highest occupied energy level. As we discussed in Section 2.6.5, the Fermi wave vector is given by (2.101) as $\kappa_{\rm F} = (3n_{e,c}\pi^2)^{1/3}$, where $n_{e,c}$ is the number density of conduction electrons. The Fermi velocity is $u_{\rm F} = p_{\rm F}/m_e$.

The Fermi energy is defined through (2.94) as

$$E_{\rm F} \equiv \frac{\hbar^2\kappa_{\rm F}^2}{2m_e} \quad \text{parabolic energy relation for metals} \tag{5.50}$$

$$n_{e,c} = \int_0^\infty D_e(E_e) f_e {\rm d}E_e. \tag{5.51}$$

For metals, $n_{e,c}$ is found directly from the number of conduction electrons, i.e., (2.103). For semiconductors, $n_{e,c}$ depends on temperature and on $E_{\rm F}$, as we will show in Section 5.7, or if $n_{e,c}$ and T are known, $E_{\rm F}$ is determined. Using $D_e(E_e) = 3n_{e,c}/2E_{\rm F} = 2^{1/2}m_e^{3/2}E_{\rm F}^{1/2}/(\pi^2\hbar^3)$ (end of chapter problem), for metals, we have [Figure 2.17(c)]

$$n_{e,c} = \int_0^{E_{\rm F}} D_e(E_e) f_e^{\rm o} {\rm d}E_e, \quad D_e(E_e) = \frac{m_e\kappa_{\rm F}}{\pi^2\hbar^2} = \frac{2^{1/2}m_e^{3/2}E_e^{1/2}}{\pi^2\hbar^3}$$

$$= \int_0^{E_{\rm F}} D_e {\rm d}E_e = \frac{1}{3\pi^2}(\frac{2m_e}{\hbar^2}E_{\rm F})^{3/2} \quad \text{metals,} \tag{5.52}$$

which is the same as (2.101). We have used $f_e^{\rm o} = 1$ for $E_e \le E_{\rm F}$, for $E_{\rm F}/k_{\rm B}T \gg 1$ (Chapter 1 problem).

Also note that, by combining (5.52) and (5.52), we have

$$n_{e,c} = \frac{\kappa_{\rm F}^3}{3\pi^2}. \tag{5.53}$$

Also from (5.52), we have the Fermi energy for an electron gas

$$E_{\rm F} = \frac{\hbar^2}{2m_e}(3\pi^2 n_{e,c})^{2/3} \quad \text{metals.} \tag{5.54}$$

For completeness, we also repeat (2.103), i.e.,

$$n_{e,c} = \frac{N_{\rm A}}{M}\rho z_e \quad \text{metals,} \tag{5.55}$$

where $N_{\rm A}$ is the Avogadro number, M is the molecular weight, ρ is the density, and z_e is the number of free electrons per atom. This is the number of valence electrons

Table 5.2. *The number of free electrons per atom, free-electron number densities, Fermi energies, Fermi temperatures, Fermi wave numbers, and Fermi velocities, for metals [9].*

Element	z_e	$n_{e,c}$, 10^{28} m^{-3}	E_F, eV	T_F, 10^4 K	κ_F, 10^{10} m^{-1}	u_F, 10^6 m/s
Li	1	4.70	4.74	5.51	1.12	1.29
Na	1	2.65	3.24	3.77	0.92	1.07
K	1	1.40	2.12	2.46	0.75	0.86
Rb	1	1.15	1.85	2.15	0.70	0.81
Cs	1	0.91	1.59	1.84	0.65	0.75
Cu	1	8.47	7.00	8.16	1.36	1.57
Ag	1	5.86	5.49	6.38	1.20	1.39
Au	1	5.90	5.53	6.42	1.21	1.40
Be	2	24.7	14.3	16.6	1.94	2.25
Mg	2	8.61	7.08	8.23	1.36	1.58
Ca	2	4.61	4.69	5.44	1.11	1.28
Sr	2	3.55	3.93	4.57	1.02	1.18
Ba	2	3.15	3.64	4.23	0.98	1.13
Nb	1	5.56	5.32	6.18	1.18	1.37
Fe	2	17.0	11.1	13.0	1.71	1.98
Mn	2	16.5	10.9	12.7	1.70	1.96
Zn	2	13.2	9.47	11.0	1.58	1.83
Cd	2	9.27	7.47	8.68	1.40	1.62
Hg	2	8.65	7.13	8.29	1.37	1.58
Al	3	18.1	11.7	13.6	1.75	2.03
Ga	3	15.4	10.4	12.1	1.66	1.92
In	3	11.5	8.63	10.0	1.51	1.74
Tl	3	10.5	8.15	9.46	1.46	1.69
Sn	4	14.8	10.2	11.8	1.64	1.90
Pb	4	13.2	9.47	11.0	1.58	1.83
Bi	5	14.1	9.90	11.5	1.61	1.87
Sb	5	16.5	10.9	12.7	1.70	1.96

for each atom in the metallic bond (Section 2.1.2). Table A.2 shows the outmost electron orbitals, and for example, for Cu we have one 4s ($4s^1$) electron, while we have two 2s ($2s^2$) electrons for Be. Bi and Sb used in thermoelectic compounds, have $z_e = 5$, which is 2 from the s orbital and 3 from the p orbital. This also corresponds to one of the oxidation states listed in Table A.2. Table 5.2 lists z_e, $n_{e,c}$, E_F, T_F, κ_F, and u_F for metals. As we discussed in Section 2.6.4, for $n_{e,c}$ conductive electrons per unit volume, the ground state is formed by occupying all single-particle levels with $\kappa < \kappa_F$ and leaving all levels with $\kappa > \kappa_F$ unoccupied.

Then the Fermi velocity based on E_F and m_e is

$$u_F = (\frac{2E_F}{m_e})^{1/2}. \tag{5.56}$$

The Fermi temperature is

$$T_F = \frac{E_F}{k_B}, \tag{5.57}$$

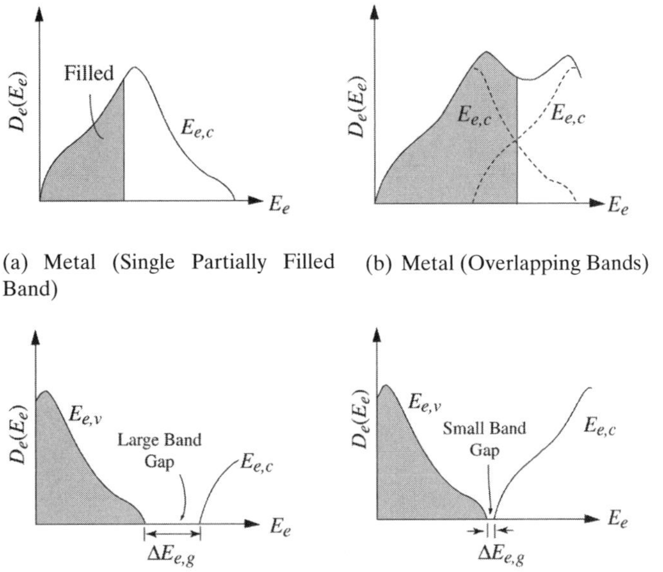

(a) Metal (Single Partially Filled Band) (b) Metal (Overlapping Bands)

(c) Insulator (d) Semiconductor (Direct Band)

Figure 5.9. Electron DOS $D_e(E_e)$ for metals, insulators, and direct-band semiconductors, at $T = 0$ K [46]. The shaded region shows the filled (occupied) states. Thermal excitation (of the order of $k_B T$) moves the electron across the gap for semiconductors.

and the Fermi momentum is

$$p_F = 2(m_e E_F)^{1/2}. \tag{5.58}$$

The electron group velocity $\boldsymbol{u}_{e,g}$ is defined similarly to that for phonons (4.15) as

$$\boldsymbol{u}_{e,g} = \frac{1}{\hbar} \nabla_\kappa E_e(\boldsymbol{\kappa}) \quad \text{electron group velocity.} \tag{5.59}$$

5.7 Electron–Hole Density of States for Semiconductors

The electron DOS $D_e(E_e)$ will allow for determination of the transport properties as well as the electron number density. For metals, insulators, and semiconductors, the general distribution of $D_e(E_e)$ is shown in Figure 5.9 for $T = 0$ K. For metals there is no energy bandgap and for insulators this gap is very large (so all electrons are in the ground state at all temperatures), and for semiconductors at $T = 0$ K all electrons are in the valence band and as the temperature increases more move to the conduction band. The conductive electron density is (5.51)

$$\begin{aligned} n_{e,c} &= \int_0^\infty D_e(E_e) f_e(E_e) \mathrm{d}E_e \\ &= \sum_\kappa f_e[E_e(\boldsymbol{\kappa})]. \end{aligned} \tag{5.60}$$

The electron equilibrium distribution function f_e^{o} is given in Table 1.2.

For a general treatment of electrons, the electron energies $E_e(\boldsymbol{x}, \boldsymbol{p})$ is the sum of the potential and kinetic energy, $E_{e,p}(\boldsymbol{x})$ and $E_{e,k}(\boldsymbol{p})$, and using the Fermi energy in place of the chemical potential μ, we have [219]

$$f_e^{\rm o} = \frac{1}{\exp[\dfrac{E_e(\boldsymbol{x}, \boldsymbol{p}) - \mu}{k_{\rm B} T}] + 1}, \qquad \mu = E_{\rm F}$$

$$= \frac{1}{\exp[\dfrac{E_{e,p}(\boldsymbol{x}) + E_{e,k}(\boldsymbol{p}) - E_{\rm F}}{k_{\rm B} T}] + 1}, \qquad E_e = E_{e,p}(\boldsymbol{x}) + E_{e,k}(\boldsymbol{p}), \qquad (5.61)$$

where $\mu = E_{\rm F}$. †

The potential energy $E_{e,p}(\boldsymbol{x})$ is the same as the lowest energy of the conduction band, as shown in Figure 5.7(a). For the nondegenerative approximation ($E_e - E_{\rm F} \gg k_{\rm B} T$, or for small $n_{e,c}$), and $E_{e,k} = p^2/2m_{e,e}$, we have

$$f_e^{\rm o} = \exp(-\frac{E_e - E_{\rm F}}{k_{\rm B} T}) = \exp(\frac{E_{\rm F} - E_{e,p}}{k_{\rm B} T}) \exp(-\frac{p^2}{2m_{e,e} k_{\rm B} T}). \qquad (5.62)$$

Then the nondegeneracy equilibrium conduction-band electron density is given as based on momentum integration of (3.11); (5.60) becomes

$$n_{e,c} = \frac{1}{\hbar^3} \sum_{\boldsymbol{p}} f_e^{\rm o}(\boldsymbol{p}) = \frac{1}{4\pi^3 \hbar^3} \int_{\boldsymbol{p}} \exp[(-E_{e,p} - \frac{p^2}{2m_{e,e}} + E_{\rm F}) \frac{1}{k_{\rm B} T}] {\rm d}\boldsymbol{p}$$

$$= \frac{1}{4\pi^3 \hbar^3} \int_0^\infty \exp(-\frac{p^2}{2m_{e,e} k_{\rm B} T}) \exp(\frac{-E_{e,p} + E_{\rm F}}{k_{\rm B} T}) 4\pi p^2 {\rm d}p$$

$$= \frac{1}{4} (\frac{2m_{e,e} k_{\rm B} T}{\pi \hbar^2})^{3/2} \exp(\frac{E_{\rm F} - E_{e,p}}{k_{\rm B} T}) \quad \text{for} \quad \frac{E_e - E_{\rm F}}{k_{\rm B} T} \gg 1$$

$$\equiv n_{e,{\rm F}} \exp(\frac{E_{\rm F} - E_{e,p}}{k_{\rm B} T}), \qquad n_{e,{\rm F}} = 2(\frac{m_{e,e} k_{\rm B} T}{2\pi \hbar^2})^{3/2} \qquad (5.63)$$

equilibrium $n_{e,c}$ and effective $n_{e,{\rm F}}$ conduction (nondegenerate) electron densities,

where we have allowed for two possible spin states. The effective density of conduction band $n_{e,{\rm F}}$ has a counter-part for the valence band (or hole density). The intrinsic carrier density n_i is related to $n_{e,{\rm F}}$, $n_{h,{\rm F}}$ and $\Delta E_{e,g}$. ‡

† For the free-electron gas, the chemical potential μ is replaced with the Fermi energy $E_{\rm F}$, but there is a slight temperature dependence given by [9]

$$\mu = E_{\rm F}[1 - \frac{1}{3}(\frac{\pi k_{\rm B} T}{2 E_{\rm F}})^2].$$

‡ The Fermi energy of semiconductors can be found from

$$E_{\rm F} = E_{e,p} + k_{\rm B} T \ln \frac{n_{e,c}}{n_{e,{\rm F}}}.$$

The hole (e^+) counterpart of this is

$$E_{\rm F} = E_{h,p} - k_{\rm B} T \ln \frac{n_{h,c}}{n_{h,{\rm F}}},$$

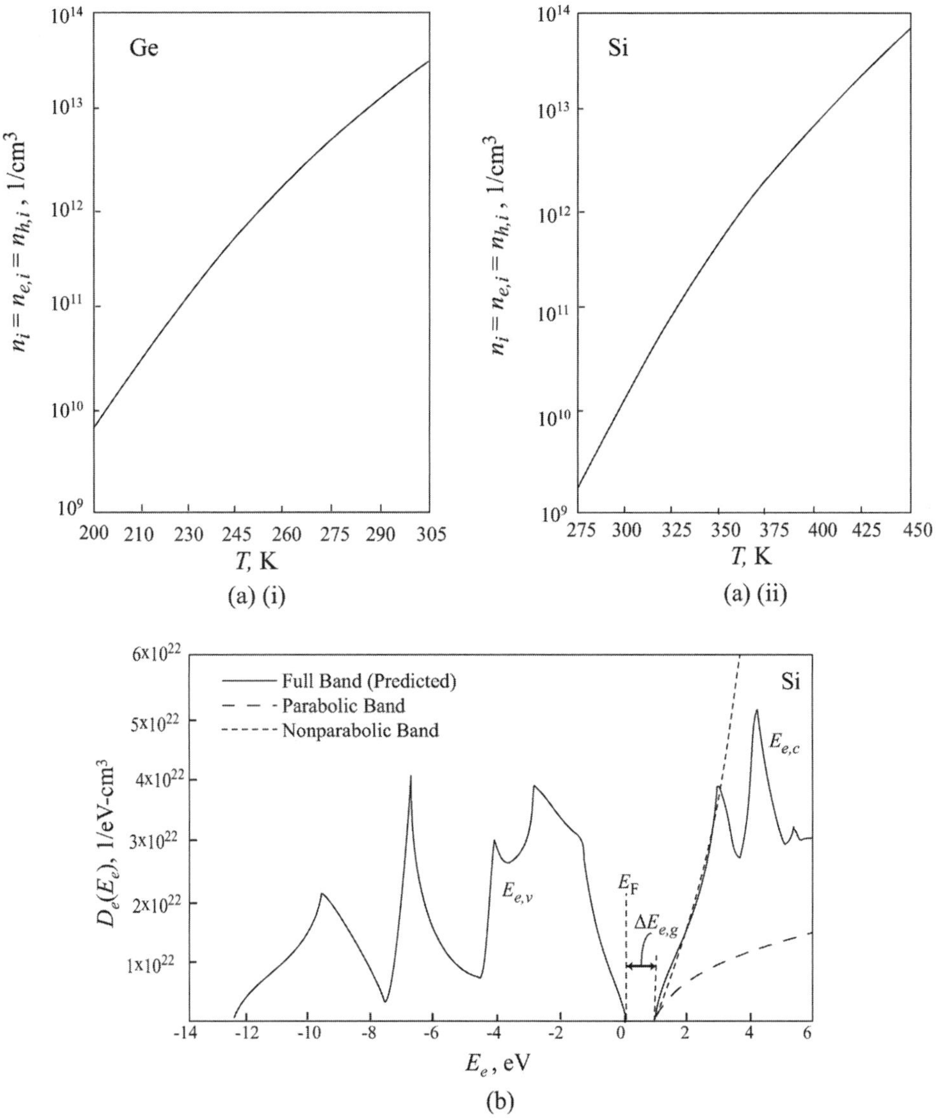

Figure 5.10. (a) Intrinsic electron (equal to hole) density for (i) Ge, and (ii) Si. Because $\Delta E_{e,g}$ is smaller for Ge, its n_e^o is larger than Si (at a given T) [183]. (b) Predicted DOS for the electron bands of Si at $T = 300$ K. Also shown are the parabolic and nonparabolic models [63, 195].

Figures 5.10(a)(i) and (ii) shows the variation of the intrinsic (electron or hole) density $n_i = n_{h,i} = n_{e,i}$ for Ge and Si (end-of-chapter problem).

where $n_{h,c}$ is the valence-band density (conduction hole density) and $n_{h,\mathrm{F}}$ is the effective-valence-band density. By eliminating E_F between the above two relations, the intrinsic carrier density is defined as (the mass action law)

$$n_i = (n_{e,\mathrm{F}} n_{h,\mathrm{F}})^{1/2} \exp(-\frac{\Delta E_{e,g}}{2k_\mathrm{B} T}), \quad \Delta E_{e,g} = E_{e,p} - E_{h,p}.$$

For evaluation of $n_{h,\mathrm{F}}$, the effective hole mass $m_{h,e}/m_e$ is used.
Also, similar to (5.63) the equilibrium conduction hole density is

$$n_{h,c} = n_{h,\mathrm{F}} \exp(\frac{E_{h,p} - E_\mathrm{F}}{k_\mathrm{B} T}), \quad n_{h,\mathrm{F}} = 2(\frac{m_{h,e} k_\mathrm{B} T}{2\pi\hbar^2})^{1/2}.$$

The general (degenerate) form of $f_e^{\rm o}$, (5.61) leads to integrals called the Fermi–Dirac integral of order j, where the integrand is $E_{e,k}^{j}/\{\exp[(E_{e,k}+E_{e,p}-E_{\rm F})/k_{\rm B}T]+1\}$ (Chapter 3 problem). The exponential term is called the Fermi function.

Equation (5.63) relates the equilibrium conduction electron density $n_{e,c}$ to the Fermi energy $E_{\rm F}$, and $n_{e,{\rm F}}$. Here $n_{e,{\rm F}}$ is the effective electron DOS (conduction band) for the case of $E_{e,p}=E_{e,c}=E_{\rm F}$ [Figure 5.7(a)]. We have also used $E_{e,c}$ for the conduction-band-edge energy.

From (5.63), the larger the Fermi energy (chemical potential), the larger are the number of conduction electrons (carriers). Compared with metals that have large $E_{\rm F}$, semiconductors have a small $E_{\rm F}$ (or $n_{e,c}$). Doped semiconductors have larger carrier concentrations than those of shown in Figure 5.10(a)(i) and (ii), however they are yet much smaller than $n_{e,c}$ listed for metals in Table 5.2. Then, $E_{\rm F}$ in semiconductors is generally smaller than $k_{\rm B}T$. We will use this in treatment of the semiconductors in Section 5.9.

The semiconductor Fermi energy can be determined from (5.63), where $n_{e,c}$ is known. The density of electronic states in energy space $D_e(E_e)$ is the number of electric states having energy $E_e(\boldsymbol{\kappa})$ and is given (per volume V) by (4.21), i.e.,

$$D_e[E_e(\boldsymbol{\kappa})]=\sum_{\alpha}\int_{\kappa'}\frac{{\rm d}\boldsymbol{\kappa}'}{(2\pi)^3}\delta_{\rm D}[E_e(\boldsymbol{\kappa})-E_e(\boldsymbol{\kappa}')]$$

$$\equiv\sum_{\boldsymbol{\kappa}',\alpha}\delta_{\rm D}[E_e(\boldsymbol{\kappa})-E_e(\boldsymbol{\kappa}')]. \qquad (5.64)$$

For parabolic energy bands [near the band maximum and minimum given by (5.44)], and for two spins, similar to (5.64)

$$D_e[E_e(\boldsymbol{\kappa})]=\frac{1}{4\pi^3}\int_{\kappa'}\delta_{\rm D}\Big(\frac{\hbar^2\kappa^2}{2m_{e,e}}-\frac{\hbar^2\kappa'^2}{2m_{e,e}}\Big)4\pi\kappa'^2{\rm d}\kappa'$$

$$=\frac{1}{4\pi^3}\int_{\kappa'}\frac{2m_{e,e}}{2\kappa'\hbar^2}\delta_{\rm D}\Big(\frac{\hbar^2\kappa'^2}{2m_{e,e}}-\frac{\hbar^2\kappa^2}{2m_{e,e}}\Big)4\pi\kappa'^2{\rm d}\Big(\frac{\hbar^2\kappa'^2}{2m_{e,e}}-\frac{\hbar^2\kappa^2}{2m_{e,e}}\Big)$$

$$=\frac{2^{1/2}m_{e,e}^{3/2}E_e^{1/2}}{\pi^2\hbar^3}$$

$$\text{DOS parabolic energy band (same as electron gas),} \qquad (5.65)$$

i.e., $D_e(E_e)$ is proportional to $E_e^{1/2}$. This is the same as (5.52) for metals, except here we use the effective mass and bond energy $E_e(\boldsymbol{\kappa})$[†].

The same is obtained using the definition given in the Glossary, $D_e(\kappa)=4\pi\kappa^2{\rm d}\kappa/(2\pi)^3$, i.e., (4.19), and noting that ${\rm d}\kappa/{\rm d}E_e=m_{e,e}/\hbar^2\kappa(m_{e,e}/2\hbar^2E_e)^{1/2}$ and $4\pi\kappa^2{\rm d}\kappa/{\rm d}E_e=4\pi m_{e,e}\kappa/\hbar^2=4\pi 2^{1/2}m_{e,e}^{3/2}E_e^{1/2}/\hbar^3$, allowing for the particle spin of 2, and

[†] The Kane nonparabolic band model gives the $D_e(E)$ as (Section 5.17.3)

$$D_e(E)=\frac{2^{1/2}m_{e,e,{\rm o}}^{3/2}}{\pi^2\hbar^3}E_e^{1/2}\Big(1+\frac{E_e}{\Delta E_{e,g}}\Big)^{1/2}\Big(1+2\frac{E_e}{\Delta E_{e,g}}\Big).$$

divided by $(2\pi)^3$. This shows that the electron dispersion relation is quadratic for the parabolic model.

Figure 5.10(b) shows the DOS $D_e(E_e)$ for model (parabolic and nonparabolic) and full-electron-band structure for Si. The parabolic model bands do not predict the full-band results, leaving out the high energy electrons. The nonparabolic band approximation, (5.48), overpredicts the density of high-energy electrons. However, for the conduction electrons only, the parabolic model is satisfactory. Note that for electron gas (or free electrons) the DOS is that given by (5.65), except m_e is used instead of $m_{e,e}$.

5.8 Specific Heat Capacity of Conduction Electrons

When the conduction electron is in thermal equilibrium with the lattice, (phonon), i.e., $T_e = T_p = T$, below we will show that electron specific heat is not large compared with that of the phonon (lattice). In Sections 5.18 and 5.19, we will discuss the case of local thermal nonequilibrium, $T_e \neq T_p$.

For metals, in the electron-gas approximation the electrons have an equilibrium energy pdf (fermion) that gives their energy with respect to the chemical potential μ and thermal energy $k_\mathrm{B}T$ as (Table 1.2)

$$f_e^\circ = \frac{1}{\exp(\dfrac{E_e - \mu}{k_\mathrm{B}T}) + 1} = \frac{1}{\exp(\dfrac{E_e - E_\mathrm{F}}{k_\mathrm{B}T}) + 1}, \qquad (5.66)$$

where μ is assumed to be equal to the electron Fermi energy E_F. In solids, T is the lattice temperature. As will be shown, because E_F is large, the high energy electrons are not excited until very high temperatures and this tends to vanish the equilibrium electron heat capacity at low and moderate temperatures.

The specific heat of the electron gas is defined similarly to (4.53) and given in terms of the electron DOS D_e as

$$c_{v,e} = \frac{\partial(\langle E_e\rangle - E_\mathrm{F})}{\partial T}|_V, \qquad (5.67)$$

where the energy of a system of electrons is given as

$$\langle E_e\rangle = \int_0^\infty (E_e - E_\mathrm{F}) f_e^\circ D_e(E_e)\mathrm{d}E_e. \qquad (5.68)$$

Then,

$$c_{v,e} = \int_0^\infty (E_e - E_\mathrm{F}) \frac{\partial f_e^\circ}{\partial T} D_e(E_e)\mathrm{d}E_e, \qquad (5.69)$$

or using (5.84), we have

$$c_{v,e} = \int_0^\infty (E_e - E_\mathrm{F}) D_e(E_e) \frac{\partial}{\partial T} \frac{1}{\exp(\dfrac{E_e - E_\mathrm{F}}{k_\mathrm{B}T}) + 1} \mathrm{d}E_e. \qquad (5.70)$$

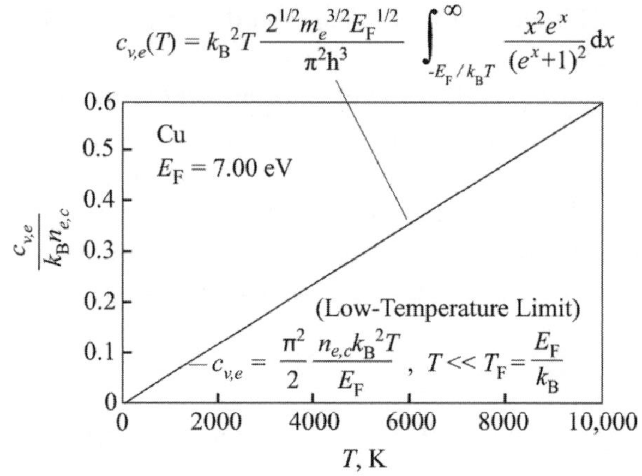

Figure 5.11. Variation of predicted specific heat capacity of Cu conduction electrons, with respect to temperature. The low-temperature limit is also shown. Because the Fermi energy is rather very large (compared with $k_B T$), the electron specific heat capacity is small.

We can approximate this by noting that $D_e(E_e) \simeq D_e(E_F)$ and then moving this outside the integral; then by also making a variable substitution $x = (E_e - E_F)/k_B T$, we have

$$c_{v,e} = k_B^2 T D_e(E_F) \int_{-E_F/k_B T}^{\infty} \frac{x^2 e^x}{(e^x + 1)^2} dx. \tag{5.71}$$

On using (5.52), we have

$$\begin{aligned} c_{v,e} &= k_B^2 T \frac{3n_{e,c}}{2E_F} \int_{-E_F/k_B T}^{\infty} \frac{x^2 e^x}{(e^x + 1)^2} dx \\ &= k_B^2 T \frac{2^{1/2} m_e^{3/2} E_F^{1/2}}{\pi^2 \hbar^3} \int_{-E_F/k_B T}^{\infty} \frac{x^2 e^x}{(e^x + 1)^2} dx \\ &\quad \text{electron-gas specific heat capacity.} \end{aligned} \tag{5.72}$$

We can use (5.52) to introduce $n_{e,c}$ as the coefficient of (5.72).

Figure 5.11 shows the specific heat capacity (divided by $k_B n_{e,c}$) of Cu electrons, where E_F (7.00 eV for Cu) is given in Table 5.2. Because $E_F/k_B T$ is rather large at moderate temperatures, the integral in (5.72) is small, and therefore the specific heat capacity of electron gas is small.

Because at moderate temperatures $k_B T \ll E_F$, and by using (2.101) for $n_{e,c}$ and using that in (5.55), we have (end-of-chapter problem)

$$D_e(E_F) = \frac{3n_{e,c}}{2E_F} \quad \text{low-temperature limit}, \tag{5.73}$$

and now by using this in (5.71) for $E_F \gg k_B T$ we have (end-of-chapter problem)

$$c_{v,e} = (\frac{\pi^2 k_B T}{2E_F}) n_{e,c} k_B \quad \text{for } k_B T \ll E_F \text{ electron gas}, \tag{5.74}$$

which, if the classical behavior for a particle were used, would be $3n_{e,c}k_B/2$, which is independent of the temperature and extremely large for low and moderate temperatures.

Similarly, using (5.63) for $E_F/k_B T \ll 1$, we also have the same result for a nondegenerate semiconductor with parabolic bands, i.e.,

$$c_{v,e} = \frac{\pi^2}{2}\frac{k_B^2 T}{E_F} n_{e,c}(T) \text{ for } T \ll \frac{E_F}{k_B} = T_F \quad \text{nondegenerate semiconductors.} \tag{5.75}$$

where $c_{v,e}$ here is per unit volume. This shows a linear temperature dependence of electron specific heat capacity at moderate temperatures in addition to the $n_{e,c}(T)$ dependence given by (5.63).

This is unlike the lattice specific heat, shown in Figure 4.13(a) and given by (4.64). Note that here $T_e = T_p = T$ is the lattice temperature, because the kinetic energy of the electrons (which leads to their temperature) was not defined and used here.

The free-electron specific heat in metals and semiconductors is much less than $3k_B/2$, which may be assigned based on its temperature. Therefore, in general for solids, only the lattice specific heat is used.

When thermal nonequilibrium is used between the free electrons and the lattice (Sections 5.18 and 5.19) or when gases with thermal nonequilibrium are used between electrons and heavier species (Section 6.13), the temperature of the electron is defined based on its kinetic energy, and then the classical limit of translational kinetic energy is used for the specific heat capacity.

5.9 Electron BTE for Semiconductors: Thermoelectric Force

The theoretical treatment of electron transport is given in [359] and for semiconductors in [219] and is briefly reviewed here using the treatment of [219].

Using the bond structure model of Figure 5.7)(a), the equilibrium distribution (5.61) can be written as

$$f_e^{\text{o}} = \frac{1}{\exp[\dfrac{E_{e,k}(\boldsymbol{p}) + E_{e,p}(\boldsymbol{x}) - E_F(\boldsymbol{x})}{k_B T}] + 1} \equiv \frac{1}{\exp(E_e^* - E_F^*) + 1}, \tag{5.76}$$

where E_F is the quasi (nonequilibrium) Fermi energy. Assuming that under an applied force $\boldsymbol{F}_e$, the distribution is altered to $f_e^{\text{o}} + f_e'$ (f_e^{o} is symmetric and f_e is asymmetric), with $f_e' \ll f_e^{\text{o}}$, the relaxation-time approximation of the BTE (Table 3.4) becomes

$$f_e = f_e^{\text{o}} + f_e' \tag{5.77}$$

$$\left.\frac{\partial f_e}{\partial t}\right|_s = \frac{\partial f_e^{\text{o}}}{\partial t} - \frac{\partial f_e'}{\partial t} = -\frac{\partial f_e'}{\partial t} \simeq -\frac{f_e'}{\tau_e(\boldsymbol{p})} \tag{5.78}$$

$$\boldsymbol{u}_e \cdot \nabla(f_e^{\text{o}} + f_e') + \boldsymbol{F}_e \cdot \nabla_p(f_e^{\text{o}} + f_e') = -\frac{f_e'}{\tau_e(\boldsymbol{p})}. \tag{5.79}$$

Assuming that the applied force $\boldsymbol{F}_e = -\nabla E_{e,p}$, we have

$$\boldsymbol{u}_e \cdot \nabla f_e^{\rm o} - \nabla E_{e,p} \cdot \nabla_{\boldsymbol{p}} f_e^{\rm o} = -\frac{f_e'}{\tau_e(\boldsymbol{p})} \tag{5.80}$$

or

$$\boldsymbol{u}_e \cdot \frac{\partial f_e^{\rm o}}{\partial E_e^*} \nabla E_e^* - \nabla E_{e,p} \cdot \frac{\partial f_e^{\rm o}}{\partial E_e^*} \nabla_{\boldsymbol{p}} E_e^* = -\frac{f_e'}{\tau_e}. \tag{5.81}$$

Using (5.76), we have

$$\nabla E_e^* = \frac{1}{k_{\rm B} T}(\nabla E_{e,p} - \nabla E_{\rm F}) + (E_{e,k} + E_{e,p} - E_{\rm F}) \nabla \frac{1}{k_{\rm B} T} \tag{5.82}$$

$$\nabla_{\boldsymbol{p}} E_e^* = \frac{\boldsymbol{u}_e}{k_{\rm B} T}. \tag{5.83}$$

Then solving for f_e' in (5.81), we have the solution to the asymmetric (deviation) distribution function as

$$\begin{aligned} f_e' &= \frac{\tau_e(\boldsymbol{p})}{k_{\rm B} T}(\frac{\partial f_e^{\rm o}}{\partial E_e^*}) \boldsymbol{u}_e \cdot [\nabla E_{\rm F} + T(E_{e,k} + E_{e,p} - E_{\rm F}) \nabla \frac{1}{T}] \\ &\equiv \frac{\tau_e(\boldsymbol{p})}{k_{\rm B} T}(-\frac{\partial f_e^{\rm o}}{\partial E_e^*}) \boldsymbol{u}_e \cdot \boldsymbol{F}_{te}, \end{aligned} \tag{5.84}$$

where

$$\boldsymbol{F}_{te} = -\nabla E_{\rm F} + T(E_{e,k} + E_{e,p} - E_{\rm F}) \nabla \frac{1}{T}. \tag{5.85}$$

Here $\boldsymbol{F}_{te}$ is called the generalized thermoelectric force. This can be written in indexed form as

$$F_{te,i} = -\frac{\partial E_F}{\partial x_i} + T(E_{e,k} + E_{e,p} - E_{\rm F}) \frac{\partial}{\partial x_i} \frac{1}{T}, \tag{5.86}$$

which has an electrochemical potential $E_{\rm F}$, and a thermal component $1/T$. This shows the effect of gradients in an electrochemical potential, and in the inverse of absolute temperature, on the nonequilibrium distribution function. So, in the Onsager relation (3.56), these should appear as the thermoelectric transport forces, i.e., (3.61) and (3.62). However, as will be discussed in Section 5.12, $\boldsymbol{e}_e$ and ∇T are more directly comparable with experimental measurables.

5.10 Electron Relaxation Time and Fermi Golden Rule

Electrons can be scattered by a scattering potential φ_s, as it appears in the Schrödinger equation given in Table 3.5. Here the transition rate is expressed by FGR (3.27), where the complete derivation is given in Appendix E. The transition rate is derived from the time-dependent Schrödinger equation, using the perturbation of the unperturbed Hamiltonian (no scattering) solution. We briefly discuss the steps below.

Assuming that the solution for $\varphi_s = 0$ ($\mathrm{H} = \mathrm{H_o}$) is given by Ψ_κ^{o} (time dependent), similar to (2.67), we have (for a one-dimensional electron wave)

$$\Psi_\kappa^{\mathrm{o}}(\boldsymbol{x}, t) = \sum_\kappa a_\kappa(t)\psi_{\kappa_{\mathrm{o}}}(\boldsymbol{x})\exp[-i\frac{E_e(\boldsymbol{\kappa}_{\mathrm{o}})t}{\hbar}], \quad \mathrm{H_o}\psi_\kappa = E_e(\boldsymbol{\kappa})\psi_\kappa. \tag{5.87}$$

Consider that this scattering results in a change of the electron wave packet (wave vector) from $\boldsymbol{\kappa}$ to $\boldsymbol{\kappa}'$.

The solution for $\varphi_s = 0$ ($\mathrm{H} = \mathrm{H_o}$) is used to construct the function for $\varphi_s \neq 0$ ($\mathrm{H} = \mathrm{H_o} + \mathrm{H}'$, $\mathrm{H}' = \varphi_s$ is the perturbation Hamiltonion), starting from the Schrödinger equation

$$i\hbar\frac{\partial\Psi}{\partial t} = (\mathrm{H_o} + \mathrm{H}')\Psi = (\mathrm{H_o} + \varphi_s)\Psi, \tag{5.88}$$

where $\mathrm{H_o}$ is the Hamiltonian for $\varphi_s = 0$.

Now using (5.87) in (5.88), we have

$$i\hbar\sum_\kappa\frac{\partial a_\kappa}{\partial t}\psi_\kappa\exp[-\frac{i\,E_e(\boldsymbol{\kappa})t}{\hbar}] = \varphi_s\sum_\kappa a_\kappa(t)\psi_\kappa\exp[-\frac{i\,E_e(\boldsymbol{\kappa})t}{\hbar}]. \tag{5.89}$$

Using the definition (Table 2.8) of the wave function, the probability of finding the electron with wave vector $\boldsymbol{\kappa}_{\mathrm{o}}'$ is

$$P(\boldsymbol{\kappa}') = \lim_{t\to\infty}|a_{\kappa'}(t)|^2. \tag{5.90}$$

For an electron wave packet centered at $\boldsymbol{\kappa}_{\mathrm{o}}$ entering and interacting with φ_s and emerging at $\boldsymbol{\kappa}_{\mathrm{o}}'$, at $t = 0$ we have $a_{\mathrm{o},\kappa}(t=0) = 1$ and $a_{\mathrm{o},\kappa}(t=0) = 0$ for $\boldsymbol{\kappa} \neq \boldsymbol{\kappa}_{\mathrm{o}}$. From the definition of the electron-scattering rate, from $\boldsymbol{\kappa}$ to $\boldsymbol{\kappa}'$, we have [219]

$$\dot{\gamma}_e(\boldsymbol{\kappa}, \boldsymbol{\kappa}') = \lim_{t\to\infty}\frac{|a_{\kappa'}(t)|^2}{|a_{\mathrm{o},\kappa}|^2 t} = \lim_{t\to\infty}\frac{|a_{\kappa'}(t)|^2}{t}. \tag{5.91}$$

Now we determine $a_{\kappa'}$ for $\varphi_s = 0$ by first multiplying both sides of (5.89) by $\psi_{\kappa'}^\dagger\exp[i\,E_e(\boldsymbol{\kappa}')t/\hbar]$, using the orthogonality of ψ_κ, and then integrating over the unit cell (assuming Bloch wave form; details are given in Appendix E). Then (5.89) becomes

$$i\hbar\frac{\partial a_{\kappa'}}{\partial t} = \sum_\kappa M_{\kappa',\kappa}a_\kappa\exp[\frac{i[E_e(\boldsymbol{\kappa}') - E_e(\boldsymbol{\kappa})]t}{\hbar}]. \tag{5.92}$$

Here $M_{\mathrm{o},\kappa',\kappa}$ is called the matrix element [appearing in (3.27)] of the scattering potential (or transition) between electron states $\boldsymbol{\kappa}'$ and $\boldsymbol{\kappa}$, and here becomes (Appendix E) [219]

$$\begin{aligned} M_{\mathrm{o},\kappa',\kappa} &= |\langle\psi_{\kappa'}|\varphi_s|\psi_\kappa\rangle| = |\langle\boldsymbol{\kappa}'|\varphi_s|\boldsymbol{\kappa}\rangle| \\ &\equiv \int_{cell}\psi_{\kappa'}^\dagger(x)\varphi_s\psi_\kappa(\boldsymbol{x})\mathrm{d}\boldsymbol{x}. \end{aligned} \tag{5.93}$$

For example, using the Bloch planar wave in a one-dimensional form of the matrix element results in [219]

$$M_{\text{o},\kappa',\kappa} = \frac{1}{L}\int_{-L/2}^{L/2} e^{-i\kappa' x}\varphi_s(x)e^{i\kappa x}\mathrm{d}x \quad \text{one-dimensional matrix element.} \tag{5.94}$$

For scattering by a periodic, point defect, this leads to a simple expression.†

In general, φ_s is three dimensional and in most cases is given as a series solution involving the Bloch waves.

Using (3.27), the electron-scattering rate $\dot{\gamma}_e$ (5.91) is then represented in terms of the preceding matrix element as

$$\dot{\gamma}_e(\kappa', \kappa) = \sum_{\kappa'}\{\frac{2\pi}{\hbar}|M_{\text{o},\kappa',\kappa}|^2\delta_{\text{D}}[E_e(\kappa') - E_e(\kappa) - \hbar\omega] + \frac{2\pi}{\hbar}|M_{\text{o},\kappa',\kappa}|^2\delta_{\text{D}}[E_e(\kappa') - E_e(\kappa) + \hbar\omega]\}. \tag{5.95}$$

The first term is nonzero when energy $\hbar\omega$ is absorbed and the second when it is emitted. For elastic scattering, $\hbar\omega = 0$. This is similar to the form given in (3.12).

In addition to the energy conservation, there is a requirement of momentum conservation. For example, for electron–phonon interaction, the momentum conservation is written as shown in Figures 4.15(c) and 4.37, i.e.,

$$\kappa - \kappa' \pm \kappa_p = \begin{cases} 0 & \text{N–processes} \\ \boldsymbol{g} & \text{U–processes.} \end{cases} \tag{5.96}$$

As an example, consider elastic scattering by a delta function perturbing potential, such that (5.95) is written as [219]

$$\dot{\gamma}_e(\boldsymbol{p}', \boldsymbol{p}) = \frac{2\pi}{\hbar}\frac{\varphi_\text{o}^2}{L^2}\sum_{\boldsymbol{p}'}\delta_{\text{D}}(E_e' - E_e) = c\sum_{\boldsymbol{p}'}\delta_{\text{D}}L(E_e' - E_e), \tag{5.97}$$

where c is a constant. Then performing the integral similar to (5.63), we have for the momentum relaxation time

$$\begin{aligned}\frac{1}{\tau_e(\text{momentum})} &= \dot{\gamma}_e(\boldsymbol{p}', \boldsymbol{p}) \\ &= \frac{c}{8\pi^2\hbar^3}\int_0^\infty\int_0^\pi\int_0^{2\pi}\delta_D\left(\frac{p'^2}{2m_{e,e}} - \frac{p^2}{2m_{e,e}}\right)p'^2\mathrm{d}p'\sin\theta\mathrm{d}\theta\mathrm{d}\phi \\ &= c\frac{(2m_{e,e})^{2/3}}{4\pi^2\hbar^3}E_e^{1/2}(\boldsymbol{p}) = \frac{c}{2}D_e(\boldsymbol{p}) = \frac{\pi}{\hbar}\frac{\varphi_\text{o}^2}{L^2}D_e(\boldsymbol{p}),\end{aligned} \tag{5.98}$$

where we have used (5.65) for $D_e(\boldsymbol{p})$.

† For $\varphi_s(x) = \varphi_\text{o}\delta(x)$, (5.94) becomes

$$M_{\text{o},\kappa',\kappa} = \frac{1}{L}\int_{-L/2}^{L/2}\varphi_\text{o}(x)e^{-i(\kappa'-\kappa)x}\mathrm{d}x = \frac{\varphi_\text{o}}{L},$$

i.e., the interaction matrix element is constant.

Here we have related τ_e and $\dot{\gamma}_e$ by assuming isotropic scattering and the domination of the out-scattering, as discussed in Section 5.11. Equation (5.98) indicates that the higher the electron DOS $D_e(\boldsymbol{p})$ (the higher the final energy), the higher the scattering rate. Note that here the energy is conserved (elastic scattering), and this is the momentum relaxation time. In this example, $\hbar\omega = 0$.

Because, in general, scattering is anisotropic, i.e., the deflection of $\boldsymbol{p}$ is rather small, the relaxation time is extended to include this. Also, this results in a distinction between the momentum (which is direction dependent) and the energy relaxation. The momentum relaxation time is (3.23), i.e., given by

$$\frac{1}{\tau_e(\text{momentum})} = \sum_{p'} \dot{\gamma}_e(\boldsymbol{p}, \boldsymbol{p}')(1 - \frac{|\boldsymbol{p}'|}{|\boldsymbol{p}|} \cos\theta) \text{ momentum relaxation time,} \quad (5.99)$$

where θ is the polar angle between $\boldsymbol{p}$ and $\boldsymbol{p}'$. This weights each collision by the fractional change in the direction of momentum. For isotropic scattering, $\tau_e^{-1}(\text{momentum}) = \sum_{p'} \dot{\gamma}_e(\boldsymbol{p}, \boldsymbol{p}')$.

The energy relaxation time is given by (3.24), i.e.,

$$\frac{1}{\tau_e(\text{energy})} = \sum_{p'} \dot{\gamma}_e(\boldsymbol{p}, \boldsymbol{p}')[1 - \frac{E_e(\boldsymbol{p}')}{E_e(\boldsymbol{p})}] = \sum_{p'} \dot{\gamma}_e(\boldsymbol{p}, \boldsymbol{p}')[\frac{E_e(\boldsymbol{p}) - E_e(\boldsymbol{p}')}{E_e(\boldsymbol{p})}]$$

$$\text{energy relaxation time.} \quad (5.100)$$

This weights each collision by the fractional change in the energy.

For elastic scattering, $\boldsymbol{p} = \boldsymbol{p}'$, $E_e(\boldsymbol{p}') = E_e(\boldsymbol{p})$, and $\tau_e(\text{energy}) \rightarrow \infty$.

5.11 Average Relaxation Time $\langle\langle \tau_e \rangle\rangle$ for Power-Law $\tau_e(E_e)$

The scattering alters the distribution function f_e (increase of f_e is called an in-scattering and decrease is called an out-scattering process). The rate of scattering is given by (3.8), as described in Section 3.1.3, i.e,

$$\frac{\partial f_e}{\partial t}|_s = \sum_{p'} f_e(\boldsymbol{p}')[1 - f_e(\boldsymbol{p})]\dot{\gamma}_e(\boldsymbol{p}', \boldsymbol{p}) - f_e(\boldsymbol{p})[1 - f_e(\boldsymbol{p}')]\dot{\gamma}_e(\boldsymbol{p}, \boldsymbol{p}'), \quad (5.101)$$

where $\dot{\gamma}_e(\boldsymbol{p}', \boldsymbol{p})$ is the transition rate of scattering from state $\boldsymbol{p}'$ to state $\boldsymbol{p}$. Using the relaxation-time approximation of the scattering term (3.11), we have for $f_e(\boldsymbol{p}) \ll 1$

$$\left.\frac{\partial f_e}{\partial t}\right|_s = \sum_{p'} f_e(\boldsymbol{p}')\dot{\gamma}_e(\boldsymbol{p}', \boldsymbol{p}) - f_e(\boldsymbol{p})\dot{\gamma}_e(\boldsymbol{p}, \boldsymbol{p}')$$

$$\equiv -\frac{f_e'}{\tau_e(\boldsymbol{p})}, \quad (5.102)$$

where $\tau_e(\boldsymbol{p})$ is the relaxation time between collisions.

For isotropic scattering and for nondegenerate semiconductors ($f_e \ll 1$), and considering out-scattering only, we have

$$\frac{1}{\tau_e(\boldsymbol{p})} = \sum_{\boldsymbol{p}'} \dot{\gamma}_e(\boldsymbol{p}, \boldsymbol{p}') \text{ isotropic, out-scattering.} \tag{5.103}$$

This will allow us to use the FGR (5.95) for the determination of the relaxation time (Section 5.15).

As an example for evaluation of the average relaxation time, consider transport in a uniform, one-dimensional electric field $e_{e,x}\boldsymbol{s}_x$, so BTE (5.79) becomes

$$-e_c e_{e,x}\boldsymbol{s}_x \cdot \nabla_p f_e = -\frac{f_e'}{\tau_e(\boldsymbol{p})}, \tag{5.104}$$

where $f_e = f_e^{\rm o} + f_e'$, and $f_e^{\rm o} \gg f_e'$.

Now, using $\nabla_p f_e^{\rm o} \gg \nabla_p f_e'$, we have

$$-e_c e_{e,x}\boldsymbol{s}_x \cdot \nabla_p f_e^{\rm o} = -\frac{f_e'}{\tau_e(\boldsymbol{p})}. \tag{5.105}$$

From the equilibrium distribution (5.76) and using $E_{e,k} = p^2/2m_{e,e}$, we have (5.83) again

$$\nabla_p f_e^{\rm o} = -\frac{\boldsymbol{p}}{m_{e,e}}\frac{1}{k_{\rm B}T} f_e^{\rm o} = -\frac{\boldsymbol{u}_e}{k_{\rm B}T}. \tag{5.106}$$

Then, using this in (5.105), we have

$$f_e' = -\frac{e_c e_{e,x}\tau_e u_{e,x}}{k_{\rm B}T} f_e^{\rm o}. \tag{5.107}$$

This shows that deviation from equilibrium is caused by the imposed field $e_{e,x}$.

The average velocity is called the drift velocity $\bar{u}_{ex} = \langle u_{e,x}\rangle$ and is

$$\langle u_{e,x}\rangle \equiv \frac{\sum_{p} u_{e,x}(f_e^{\rm o} + f_e')}{\sum_{p}(f_e^{\rm o} + f_e')} = \frac{\sum_{p} u_{e,x} f_e'}{\sum_{p} f_e^{\rm o}}, \tag{5.108}$$

where in the denominator we use $f_e^{\rm o} \gg f_e'$, and because $f_e^{\rm o}$ is symmetric in $u_{e,x}$, the product $u_{e,x} f_e^{\rm o}$ is odd and that summation vanishes in the numerator. Then, using (5.63) and (5.107), we have

$$\langle u_{e,x}\rangle = -e_c e_{e,x}\frac{\sum_{p} u_{e,x}^2 \tau_e f_e^{\rm o}}{n_{e,c}k_{\rm B}T}, \tag{5.109}$$

where T is the lattice temperature. Next, we use the average equilibrium kinetic energy $\langle E_{e,k}\rangle = \langle E_e\rangle \sum_p E_{e,k} f_e^{\rm o} = (3/2)k_{\rm B}Tn_{e,c}$ (based on equipartition of

Table 5.3. *Power-law constant s for common electron-scattering mechanisms [219].*

Mechanism i	s
Acoustic phonon	−1/2
Ionized impurity(weakly screened)	+3/2
Ionized impurity(strongly screened)	−1/2
Neutral impurity	0
Piezoelectric (semiconductors)	+1/2

translational energy), and $u_{e,x}^2 = u_e^2/3$ (based on spherical symmetry, will show this in Section 5.12), to write for the drift velocity † along x

$$\langle u_{e,x}\rangle = \frac{-e_c e_{e,x}}{m_{e,e}} \frac{\frac{1}{3}\sum_p \frac{1}{2} m_{e,e} u_e^2 \tau_e(E_e) f_e^{\mathrm{o}}(E_e)}{\frac{1}{3}\langle E_e\rangle}$$

$$\equiv \frac{-e_c}{m_{e,e}} \frac{\langle E_e \tau_e(E_e)\rangle}{\langle E_e\rangle} e_{e,x}$$

$$= \frac{-e_c}{m_{e,e}} \langle\langle\tau_e\rangle\rangle\, e_{e,x}$$

$$\equiv -\mu_{e,x} e_{e,x}, \quad \langle\langle\tau_e\rangle\rangle(\text{momentum}) \equiv \frac{\langle E_e \tau_e(E_e)\rangle}{\langle E_e\rangle}$$

$$\text{average relaxation time,} \qquad (5.110)$$

where $\mu_{e,x}$ (along x) is the electron mobility, generally presented in cm^2/V-s. Therefore the electron mobility and its relaxation time are equivalent. We have also used (5.31) for definition of $n_{e,c}$. We assume a power-law relation for the relaxation time, i.e.,

$$\tau_e(E_e) = \tau_e[E_e(\boldsymbol{p})] = \tau_{e,\mathrm{o}}[\frac{E_e(\boldsymbol{p})}{k_\mathrm{B} T}]^s \quad \text{energy-dependent relaxation time,} \qquad (5.111)$$

where $\tau_{e,\mathrm{o}}$ is a constant and s depends on the type of scattering. Note that this also shows the temperature dependence T^{-s}, in addition to any temperature dependence of $\tau_{e,\mathrm{o}}$.

In Section 5.15 we will examine electron–phonon scattering closely and summarize the relaxation-time relation for the various scattering mechanisms given in Table 5.3.

Table 5.3 gives the value of the power-law constant s for common elastic scattering mechanisms. These are for phonon, impurity, and piezoelectric scatterings, as also shown in Figure 5.12. Note the momentum and energy conservation in the

† Typical drift speed in Cu conductors is 10^{-4} m/s, whereas the random thermal speed of a free electron at room temperature is 1.2×10^5 m/s [$(3k_\mathrm{B} T/m_e)^{1/2}$], and the Fermi velocity (average velocity of an electron in an atom at $K = 0$) is 1.6×10^6 m/s.

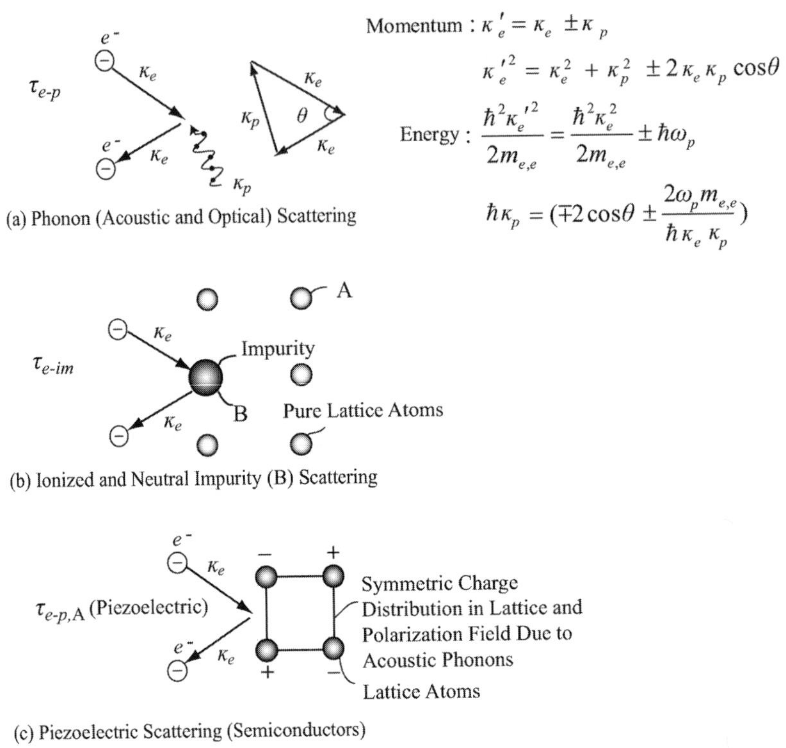

Figure 5.12. Electron scattering by (a) phonon (acoustic and optical), (b) impurity (ionized and neutral), and (c) piezoelectric (semiconductors).

interactions, as shown for the electron–phonon scattering. If the absorbed/emitted phonon energy is small, it is assumed elastic, but this generally is not the case for the optical phonons. As is in Figure 3.3, the scattering of electrons from state $\boldsymbol{\kappa}'_e$ to state $\boldsymbol{\kappa}$ may not be elastic. In Section 5.15, the derivation of the electron–phonon relaxation time (and the relation $s = -1/2$ for acoustic phonons) will be given.

From the power-law relaxation time (3.25), we find that the averaged relaxation time (5.110) becomes†

$$\langle\langle\tau_e\rangle\rangle = \tau_{e,o}\frac{\sum_p \frac{p^2}{2m_{e,e}}\left(\frac{p^2}{2m_{e,e}k_\mathrm{B}T}\right)^s \exp\left(-\frac{p^2}{2m_{e,e}k_\mathrm{B}T}\right)}{\sum_p \frac{p^2}{2m_{e,e}} \exp\left(-\frac{p^2}{2m_{e,e}k_\mathrm{B}T}\right)}$$

† Using $y = p^2/2m_{e,e}k_\mathrm{B}T$, (5.112) is written as

$$\langle\langle\tau_e\rangle\rangle = \tau_{e,o}\frac{\int_0^\infty y^{s+3/2}e^{-y}\mathrm{d}y}{\int_0^\infty y^{3/2}e^{-y}\mathrm{d}y},$$

which leads to the final form of (5.112), using (5.113).

$$= \tau_{e,\text{o}} \frac{\displaystyle\int_0^\infty (\frac{p^2}{2m_{e,e}k_\text{B}T})^s \exp(-\frac{p^2}{2m_{e,e}k_\text{B}T})p^4 \text{d}p}{\displaystyle\int_0^\infty \exp(-\frac{p^2}{2m_{e,e}k_\text{B}T})p^4 \text{d}p}$$

$$\equiv \tau_{e,\text{o}} \frac{\Gamma(s+\frac{5}{2})}{\Gamma(\frac{5}{2})}, \quad \text{for } \tau_e[E_e(\boldsymbol{p})] \equiv \tau_{e,\text{o}}[\frac{E_e(\boldsymbol{p})}{k_\text{B}T}]^s$$

power-law energy dependent average relaxation time. (5.112)

Among the properties of the gamma function $\Gamma(z)$ are

$$\Gamma(z) \equiv \int_0^\infty y^{z-1} e^{-y} \text{d}y$$

$$\Gamma(n) = (n-1)!$$

$$\Gamma(n+1) = n\Gamma(n)$$

$$\Gamma(\frac{1}{2}n) = \frac{(n-2)!!\pi^{1/2}}{2^{(n-1)/2}}$$

$$\Gamma(\frac{1}{2}) = \pi^{1/2}, \quad \Gamma(\frac{3}{2}) = \frac{\pi^{1/2}}{2}, \quad \Gamma(\frac{5}{2}) = \frac{3\pi^{1/2}}{4}$$

$$\Gamma(\frac{7}{2}) = \frac{15\pi^{1/2}}{8}, \quad \Gamma(\frac{9}{2}) = \frac{105\pi^{1/2}}{16}, \quad (5.113)$$

where $n!!$ is the double fractional.† When there is multiple scattering, $\mu_e^{-1} = \sum_i \mu_{e,i}^{-1}$, and because the electrical resistivity ρ_e is proportional to μ_e^{-1}, this states that $\rho_e = \sum_i \rho_{e,i}$ [Matthiessen rule (4.111)]. Thus using $\sigma_e(1/\rho_e)$ proportional to μ_e (5.111), we have

$$\dot{\gamma}_e(\boldsymbol{p}) = \frac{1}{\tau_e} = \sum_i \frac{1}{\tau_{e,i}(\boldsymbol{p})}, \quad (5.114)$$

where some of the mechanisms i are listed in Table 5.3. Because the mobility is defined through the drift velocity and electric field as $\langle \boldsymbol{u}_e \rangle = \boldsymbol{u}_e = -\mu_e \boldsymbol{e}_e$, generally (5.110) is used to determine $\langle\langle \tau_e \rangle\rangle$(momentum). The momentum relaxation time is also determined through (5.113), with $\tau_{e,\text{o}}$ given empirically. For isotropic scattering, the energy relaxation time is related to the momentum relaxation time (examples are given in Table 5.4 and Section 5.19).

5.12 Thermoelectric Transport Tensors for Power-Law $\tau_e(E_e)$

In Section 3.4, from the Onsager transport relations, we discussed presentation of fluxes in terms of gradients. In presence of an electric field $\mathbf{e}_e$ and a temperature gra-

† $n!! = n(n-2)\ldots 5 \times 3 \times 1$ for n odd, and $n!! = n(n-2)\ldots 4 \times 2$, for n even.

dient ∇T, the electrical and heat flux vectors are given in terms of the macroscopic gradients (and including the coupling tensors $\boldsymbol{L}_{et}$ and $\boldsymbol{L}_{te}$) by

$$\boldsymbol{j}_e = \boldsymbol{L}_{ee} \cdot \boldsymbol{e}_e + \boldsymbol{L}_{et} \cdot \nabla T \tag{5.115}$$

$$\boldsymbol{q} = \boldsymbol{L}_{te} \cdot \boldsymbol{e}_e + \boldsymbol{L}_{tt} \cdot \nabla T. \tag{5.116}$$

However, as was shown by (5.86), in an electron BTE, the gradients in the Fermi energy ∇E_{F} and the inverse of temperature $\nabla(1/T)$ are the relevant potential gradients. These fluxes are also related to the microscopic transport, starting with the electrical current density given by

$$\boldsymbol{j}_e \equiv -e_c \boldsymbol{j}_n = -\frac{e_c}{\hbar^3} \sum_{\boldsymbol{p}} \boldsymbol{u}_e f_e'. \tag{5.117}$$

The heat flow rate can be written using the thermodynamic relation of Table 2.4 as (using $\mu = E_{\mathrm{F}}$, as the quasi-Fermi electrochemical potential)

$$\mathrm{d}Q = \mathrm{d}E_e - E_{\mathrm{F}} \mathrm{d}N_e, \tag{5.118}$$

where the E_{F} may vary with position. Now using $E_e = E_{e,k} + E_{e,p}$ and (5.82) and (5.114), we have

$$\boldsymbol{q} = \boldsymbol{q}_E - E_{\mathrm{F}} \boldsymbol{j}_n = \frac{1}{\hbar^3} \sum_{\boldsymbol{p}} (E_{e,k} + E_{e,p} - E_{\mathrm{F}}) \boldsymbol{u}_e f_e'. \tag{5.119}$$

Next, using (5.84) in (5.117) and (5.118), we have

$$\boldsymbol{j}_e = -\frac{e_c}{\hbar^3 k_{\mathrm{B}} T} \sum_{\boldsymbol{p}} \boldsymbol{u}_e \tau_e (-\frac{\partial f_e^{\mathrm{o}}}{\partial E_e^*})(\boldsymbol{u}_e \cdot \boldsymbol{F}_{te}) \tag{5.120}$$

$$\boldsymbol{q} = \frac{1}{\hbar^3 k_{\mathrm{B}} T} \sum_{\boldsymbol{p}} \boldsymbol{u}_e \tau_e (-\frac{\partial f_e^{\mathrm{o}}}{\partial E_e^*})(E_{e,k} + E_{e,p} - E_{\mathrm{F}})(\boldsymbol{u}_e \cdot \boldsymbol{F}_{te}). \tag{5.121}$$

We write $\boldsymbol{F}_{te}$ in terms of $\nabla(E_{\mathrm{F}}/e_c)$ and $\nabla(1/T)$, i.e., (5.86), and also begin to use index notation.

We now define the thermoelectric (TE) transport tensors $\boldsymbol{A}_{ee}$, $\boldsymbol{A}_{et}$, $\boldsymbol{A}_{te}$, $\boldsymbol{A}_{tt}$, and we also introduce their components (elements) $\alpha_{ee,ij}$, etc., by writing (5.120) and (5.121) as

$$\boldsymbol{j}_e = \boldsymbol{A}_{ee} \cdot \nabla \frac{E_{\mathrm{F}}}{e_c} + \boldsymbol{A}_{et} \cdot \nabla \frac{1}{T}, \quad j_{e,i} = \alpha_{ee,ij} \frac{\partial}{\partial x_j} \frac{E_{\mathrm{F}}}{e_c} + \alpha_{et,ij} \frac{\partial}{\partial x_j} \frac{1}{T} \tag{5.122}$$

$$\boldsymbol{q} = \boldsymbol{A}_{te} \cdot \nabla \frac{E_{\mathrm{F}}}{e_c} + \boldsymbol{A}_{tt} \cdot \nabla \frac{1}{T}, \quad q_i = \alpha_{te,ij} \frac{\partial}{\partial x_j} \frac{E_{\mathrm{F}}}{e_c} + \alpha_{tt,ij} \frac{\partial}{\partial x_j} \frac{1}{T}. \tag{5.123}$$

Note that the subscript j here is a repeated summation index (from $\boldsymbol{u}_e \cdot \boldsymbol{F}_{te}$). Then, we have the Onsager transport coefficients for the $\nabla(E_{\rm F}/e_c)$ and $\nabla(1/T)$ forces as

$$\alpha_{ee,ij} = \frac{e_c^2}{\hbar^3 k_{\rm B} T} \sum_{\boldsymbol{p}} u_{e,i} u_{e,j} \tau_e (-\frac{\partial f_e^{\rm o}}{\partial E_e^*}) \tag{5.124}$$

$$\alpha_{et,ij} = \frac{-e_c}{\hbar^3 k_{\rm B} T} \sum_{\boldsymbol{p}} u_{e,i} u_{e,j} \tau_e T(E_{e,k} + E_{e,p} - E_{\rm F})(-\frac{\partial f_e^{\rm o}}{\partial E_e^*}) \tag{5.125}$$

$$\alpha_{te,ij} = \frac{-e_c}{\hbar^3 k_{\rm B} T} \sum_{\boldsymbol{p}} u_{e,i} u_{e,j} \tau_e (E_{e,k} + E_{e,p} - E_{\rm F})(-\frac{\partial f_e^{\rm o}}{\partial E_e^*}) \tag{5.126}$$

$$\alpha_{tt,ij} = \frac{1}{\hbar^3 k_{\rm B} T} \sum_{\boldsymbol{p}} u_{e,i} u_{e,j} \tau_e T(E_{e,k} + E_{e,p} - E_{\rm F})^2(-\frac{\partial f_e^{\rm o}}{\partial E_e^*}). \tag{5.127}$$

Note that the Onsager conditions among the transport coefficients, i.e., (3.59), applying. For nondegenerate conductors (most semiconductors), we have $E_{\rm F}/k_{\rm B}T \ll 1$, and for degenerate conductors, (metals) we have $E_{\rm F}/k_{\rm B}T \gg 1$. To evaluate (5.124) to (5.127), further specifications are needed, and here we consider the simplest of them, i.e., a nondegenerate distribution $[e^{(E_e^* - E_{\rm F})^*} \gg 1]$ in (5.76), and spherical (energy surfaces), parabolic energy (energy variation with momentum or wave vector) semiconductor bands, centered around $\boldsymbol{\kappa}' = \{0,0,0\}$. Then from (5.62) we have

$$f_e^{\rm o} = e^{-(E_e^* - E_{\rm F}^*)}, \quad -\frac{\partial f_e^{\rm o}}{\partial E_e^*} = f_e^{\rm o} \text{ nondegenerate distribution.} \tag{5.128}$$

The integration over $\boldsymbol{p}$ is in both direction and magnitude. The direction integral (for spherical symmetry) gives

$$\frac{1}{\displaystyle\int_0^{\pi} \sin\theta {\rm d}\theta} \int_0^{\pi} u_{e,i} u_{e,j} \sin\theta {\rm d}\theta = \frac{1}{3} u_e^2 \delta_{ij}, \tag{5.129}$$

because $u_{e,i} u_{e,j} = u_{e,i}^2 \delta_{ij}$ and $u_e^2 = 3u_{e,i}^2$; here δ_{ij} is the Kronecker delta.

Then for $\alpha_{ee,ij}$, we have (using the average kinetic energy $\langle E_{e,k} \rangle = 3k_{\rm B}T/2 = 3m_{e,e}\langle u_{e,i}^2 \rangle/2$, where T is the lattice temperature), similar to (5.110),

$$\begin{aligned} \alpha_{ee,ij} &= \frac{e_c^2}{\frac{3}{2} k_{\rm B} T} \frac{1}{\hbar^3} \sum_{\boldsymbol{p}} \frac{u_e^2}{2} \tau_e f_e^{\rm o} \delta_{ij} \\ &= \frac{e_c^2 \frac{1}{\hbar^3} \sum_{\boldsymbol{p}} f_e^{\rm o}}{\frac{1}{\hbar^3} \sum_{\boldsymbol{p}} \frac{3}{2} k_{\rm B} T f_e^{\rm o}} \frac{1}{4\pi^3 \hbar^3} \sum_{\boldsymbol{p}} \frac{E_{e,k}}{m_{e,e}} \tau_e f_e^{\rm o} \delta_{ij} \end{aligned}$$

$$= \frac{n_{e,c}e_c^2}{m_{e,e}}\langle\langle\tau_e\rangle\rangle\delta_{ij}, \quad \langle\langle\tau_e\rangle\rangle(\text{momentum}) \equiv \frac{\langle E_{e,k}\tau_e(E_{e,k})\rangle}{\langle E_{e,k}\rangle}$$

$$= n_{e,c}e_c\mu_e\delta_{ij}, \ \mu_e = \frac{e_c\langle\langle\tau_e\rangle\rangle}{m_{e,e}}, \tag{5.130}$$

where, again

$$\langle E_{e,k}\rangle = \frac{1}{4\pi^3\hbar^3}\int_{\boldsymbol{p}} E_{e,k}(p) f_e^{\text{o}}(p)\text{d}\boldsymbol{p} = \frac{1}{8\pi^3\hbar^3 m_{e,e}}\int_{\boldsymbol{p}} p^2 f_e^{\text{o}}(p)\text{d}\boldsymbol{p}. \tag{5.131}$$

The integrals over $\boldsymbol{p}$ are given by (5.112).

The electron drift velocity vector is presented by the electron mobility (and conductivity) as

$$\boldsymbol{u}_e = -\mu_e \boldsymbol{e}_e = -\frac{e_c \boldsymbol{e}_{e,o}}{m_{e,e}}\langle\langle\tau_e\rangle\rangle \ \text{electron drift velocity}, \tag{5.132}$$

which is the vectorial form of (5.110). We also used the definition of $n_{e,c}$ from (5.65) in (5.130). The average electron drift velocity $\boldsymbol{u}_e$ and the time constant $\langle\langle\tau_e\rangle\rangle$ have been used, where the last is given by (5.110).

Now, for $\alpha_{et,ij}$, we have

$$\alpha_{et,ij} = \frac{-e_c}{\hbar^3 k_{\text{B}}}\frac{2}{3m_{e,e}}[(E_{e,p} - E_{\text{F}})\sum_{\boldsymbol{p}} E_{e,k}\tau_e f_e^{\text{o}} + \sum_{\boldsymbol{p}} E_{e,k}^2\tau_e f_e^{\text{o}}]\delta_{ij}$$

$$= \frac{-n_{e,c}e_c T}{m_{e,e}}[(E_{e,p} - E_{\text{F}})\frac{\langle E_e\tau_e\rangle}{\langle E_e\rangle} + \frac{\langle E_e^2\tau_e\rangle}{\langle E_e\rangle}]\delta_{i,j}. \tag{5.133}$$

Now, using the properties of the Γ function, given by (5.113), we have, similar to (5.112),

$$\frac{\langle E_{e,k}^2 f_e^{\text{o}}\tau_e\rangle}{\langle E_{e,k} f_e^{\text{o}}\rangle} = k_{\text{B}}T\tau_{e,\text{o}}\frac{\Gamma(s+\frac{7}{2})}{\Gamma(\frac{5}{2})} = (s+\frac{5}{2})k_{\text{B}}T\langle\langle\tau_e\rangle\rangle. \tag{5.134}$$

using $\mu_e = e_c\langle\langle\tau_e\rangle\rangle/m_{e,e}$, and from (5.63)

$$\frac{E_{\text{F}} - E_{e,p}}{k_{\text{B}}T} = \ln\frac{n_{e,F}}{n_{e,c}} \tag{5.135}$$

$$\alpha_{et,ij} = -\frac{k_{\text{B}}T^2}{e_c}[\ln\frac{n_{e,\text{F}}}{n_{e,c}} + (s+\frac{5}{2})]\alpha_{ee,ij}. \tag{5.136}$$

By inspection of (5.125) and (5.126), we have

$$\alpha_{te,ij} = \frac{1}{T}\alpha_{et,ij}. \tag{5.137}$$

The last coefficient given by (5.124) becomes

$$\begin{aligned}
\alpha_{tt,ij} &= \frac{1}{\hbar^3 k_B} \sum_p u_{e,i}^2 f_e^o \tau_e (E_{e,p} + E_{e,k} - E_F)^2 \\
&= \frac{2}{3\hbar^3 k_B m_{e,e}} \sum_p E_{e,k} \tau_e f_e^o [(E_{e,p} - E_F)^2 + 2E_{e,k}(E_{e,p} - E_F) + E_{e,k}^2] \delta_{ij} \\
&= \frac{n_{e,c} T}{m_{e,e}} (k_B T)^2 \ln(\frac{n_{e,F}}{n_{e,c}})^2 \frac{\langle E_e \tau_e(E_e) \rangle}{\langle E_e \rangle} + \frac{2 n_{e,c} T}{m_{e,e}} (k_B T) \ln(\frac{n_{e,F}}{n_{e,c}})^2 \frac{\langle E_e^2 \tau_e(E_e) \rangle}{\langle E_e \rangle} \\
&\quad + \frac{n_{e,c} T}{m_{e,e}} \frac{\langle E_e^3 \tau_e(E_e) \rangle}{\langle E_e \rangle},
\end{aligned} \tag{5.138}$$

where the angle brackets indicate integral over $\boldsymbol{p}$ space, as shown in (5.110).

Now, using definition (5.124), the Γ function properties (5.113), and the result for $\alpha_{ee,ij}$ (5.130), we have

$$\alpha_{tt,ij} = \frac{k_B^2 T^3}{e_c^2} \{[\ln \frac{n_{e,F}}{n_{e,c}} + (s + \frac{5}{2})]^2 + (s + \frac{5}{2})\} \alpha_{ee,i,j}. \tag{5.139}$$

5.13 TE Transport Coefficients for Cubic Structures

5.13.1 Seebeck, Peltier, and Thomson Coefficients, and Electrical and Thermal Conductivities

Alternatively, coupled thermoelectric (TE) transport equations (5.120) and (5.121) can be written as the electric-field intensity vector and the heat flux vector, in terms of $\boldsymbol{e}_e$, $\boldsymbol{j}_e$, and ∇T. These are the conventional forms through which electrical and thermal transports are defined. Then for isotropic (cubic) materials we have (end-of-chapter problem)

$$\boldsymbol{j}_e = \alpha_{ee} \boldsymbol{e}_e - \frac{\alpha_{et}}{T^2} \nabla T \tag{5.140}$$

$$\boldsymbol{e}_e = \alpha_{ee}^{-1} \boldsymbol{j}_e + \frac{\alpha_{ee}^{-1} \alpha_{et}}{T^2} \nabla T \tag{5.141}$$

$$\boldsymbol{q} = \alpha_{te} \alpha_{ee}^{-1} \boldsymbol{j}_e - \frac{1}{T^2} (\alpha_{tt} - \alpha_{te} \alpha_{ee}^{-1} \alpha_{et}) \nabla T$$

$$\text{cubic semiconductors}, \tag{5.142}$$

or

$$\boldsymbol{j}_e = \sigma_e \boldsymbol{e}_e - \sigma_e \alpha_S \nabla T \tag{5.143}$$

$$\boldsymbol{e}_e \equiv \rho_e \boldsymbol{j}_e + \alpha_S \nabla T, \quad \rho_e = \frac{1}{\sigma_e}, \quad \alpha_{ee} \equiv \sigma_e, \quad \frac{\alpha_{et} \alpha_{ee}^{-1}}{T^2} \equiv \alpha_S \tag{5.144}$$

$$\boldsymbol{q} \equiv \alpha_{\rm P} \boldsymbol{j}_e - k_e \nabla T, \quad \alpha_{\rm P} = \alpha_{\rm S} T, \quad \alpha_{te}\alpha_{ee}^{-1} \equiv \alpha_{\rm P},$$

$$\frac{\alpha_{tt} - \alpha_{te}\alpha_{ee}^{-1}\alpha_{et}}{T^2} \equiv k_e, \qquad (5.145)$$

where $\rho_e(\Omega\text{-m})$ is the electrical resistivity, $\alpha_{\rm S}$(V/K) is the Seebeck coefficient, $\alpha_{\rm P}$(V) is the Peltier coefficient, and k_e(W/m-K) is the electric thermal conductivity (the total thermal conductivity is $k = k_e + k_p$, where k_p is the phonon thermal conductivity). The Seebeck coefficient (also called the thermopower) is the electrical potential induced by temperature differences, causing the diffusion of electrons or holes toward the cold region.

We note that electrical conductivity is the product of conduction electron density, electron charge, and electron mobility, and that electron mobility and relaxation time are equivalent. For insulators, k_p is the sole contributor and for metals (and to a large extent for semiconductors at and preceding room temperature) k_e dominates. For nondegenerate, isotropic (cubic structure) semiconductors, with spherical and parabolic electrical bands, we have

$$\sigma_e = \frac{1}{\rho_e} = \frac{n_{e,c}e_c^2\langle\langle\tau_e\rangle\rangle}{m_{e,e}} = \frac{n_{e,c}e_c^2}{m_{e,e}}\tau_{e,{\rm o}}\frac{\Gamma(s+\frac{5}{2})}{\Gamma(\frac{5}{2})} \equiv n_{e,c}e_c\mu_e$$

$$\text{electrical conductivity} \qquad (5.146)$$

$$\alpha_{\rm S} = \mp\frac{k_{\rm B}}{e_c}[\ln\frac{n_{e,{\rm F}}}{n_{e,c}} + (s+\frac{5}{2})] \qquad \text{Seebeck coefficient}$$

$$= \mp\frac{k_{\rm B}}{e_c}[\frac{E_{\rm F} - E_{e,p}}{k_{\rm B}T} + (s+\frac{5}{2})] \ \ + \text{for } n\text{-type}, - \text{for } p\text{-type} \qquad (5.147)$$

$$\alpha_{\rm P} = \alpha_{\rm S} T \qquad \text{Kelvin relation and Peltier coefficient} \qquad (5.148)$$

$$k_e = \sigma_e\frac{k_{\rm B}^2 T}{e_c^2}(s+\frac{5}{2}) \qquad \text{electronic thermal conductivity,} \qquad (5.149)$$

where μ_e and $\langle\langle\tau_e\rangle\rangle$ are given by (5.130). The n- and p-type semiconductors are defined in the Glossary. Note that $\alpha_{\rm S}$ and $\alpha_{\rm P}$ do not contain the kinetics $\tau_{e,{\rm o}}$, but σ_e and k_e do.

For multiband structures, $\alpha_{\rm S}$ is summed over the number of bands j (Section 5.17). In a two-band structure (one valence and one conduction with no band degeneracy) the energy difference $E_{e,p} - E_{\rm F}$ is replaced with the bandgap energy $\Delta E_{e,g}$.

For match with experiment, a constant is added after $s + 5/2$ in (5.147) and is referred to as the phonon drag [123].

Note that $k_{\rm B}/e_c$ is 86 μV/K, and this gives the order of magnitude for the Seebeck coefficient.

The relation

$$\frac{k_e}{\sigma_e T} = \frac{k_B^2}{e_c^2}(s+\frac{5}{2}) \equiv N_L \quad \text{Wiedemann–Franz–Lorenz law,} \tag{5.150}$$

is called the Wiedemann–Franz–Lorenz law, and N_L is called the Lorenz number and depends on s.

The Thomson coefficient α_T(V/K), is defined as

$$\alpha_T \equiv \frac{\frac{1}{j_e}\frac{dq}{dx}}{\frac{dT}{dx}} = T\frac{d\alpha_S}{dT} \quad \text{Thomson coefficient.} \tag{5.151}$$

This gives $dq_x/dx = \alpha_T j_e dT/dx$, which shows that in the presence of a temperature gradient, an electric conductor can absorb ($\alpha_T > 0$) or release heat.

Then from the linear relation for α_S in (5.147), we have

$$\alpha_T = \mp\frac{k_B}{e_c}\frac{\Delta E_{e,g}}{k_B T} \quad - \text{ for } n\text{-type and } + \text{ for } p\text{-type,} \tag{5.152}$$

where we have used the relation between the effective carrier density and the intrinsic carrier density (Section 5.7) to remove E_F and use the bandgap energy.

The Thomson coefficient expresses the extent of heat flux induced by an electrical current in a nonisothermal conductor.

5.13.2 Electron Mean Free Path for Metals

For a free-electron gas, the Fermi speed is (5.56), and gives

$$u_F = (\frac{2E_F}{m_e})^{1/2}. \tag{5.153}$$

and $c_{v,e}$ is given by (5.72), for $T_e = T_p$ (electrons in thermal equilibrium with the lattice). For $T_e \neq T_p$, $c_{v,e} = 3k_B T/2$, per electron (Section 5.18).

Then the mobility is given in terms of the relaxation time by (5.146) and can in turn be written in terms of the electron mean free path $\lambda_e = u_F\langle\langle\tau_e\rangle\rangle$, and then the electrical conductivity becomes

$$\sigma_e = n_{e,c}e_c^2\frac{\lambda_e}{m_{e,e}u_F}. \tag{5.154}$$

The electronic thermal conductivity can be defined because of the kinetic form of the thermal conductivity, similar to (4.110) for phonons. Then we have [24]

$$k_e = \frac{1}{3}n_{e,c}c_{v,e}u_F\lambda_e. \tag{5.155}$$

Next, using (5.56), (5.73), (5.154), and (5.155), we have

$$\frac{k_e}{\sigma_e T} = N_{L,o} = \frac{\pi^2}{3}\frac{k_B^2}{e_c^2} = 2.442 \times 10^{-8}\ \frac{\text{W-}\Omega}{\text{K}^2} \quad \text{Lorenz number,} \tag{5.156}$$

i.e., the Lorenz number N_L is expected to be a constant for metals, however, as will be shown in Section 5.16 when we review experimental results, N_L decreases substantially as inelastic phonon scattering becomes important at small T/T_D.

5.14 Magnetic Field and Hall Factor and Coefficient

In the presence of a magnetic field, indicated by the magnetic induction vector $\boldsymbol{b}_e$, (5.79) becomes

$$\boldsymbol{u}_e.\nabla(f_e^{\text{o}} + f_e') - (e_e\boldsymbol{e}_c + e_c\boldsymbol{u}_e \times \boldsymbol{b}_e).\nabla_{\boldsymbol{p}}(f_e^{\text{o}} + f_e') = -\frac{f_e'}{\tau_e(\boldsymbol{p})}. \qquad (5.157)$$

Using a treatment similar to that given for the electric field alone, we write the electrical current as [219]

$$\boldsymbol{j}_e = \sigma_e\boldsymbol{e}_e - \sigma_e\mu_{\text{H}}\boldsymbol{e}_e \times \boldsymbol{b}_e, \qquad (5.158)$$

where μ_{H} is the Hall mobility and is related to the drift mobility μ_e, in (5.130), through

$$\mu_{\text{H}} = \mu_e r_{\text{H}} \quad \text{Hall mobility}$$

$$r_{\text{H}} = \frac{\langle\langle\tau_e^2\rangle\rangle}{\langle\langle\tau_e\rangle\rangle^2} = \frac{\Gamma(2s+\frac{5}{2})\Gamma(\frac{5}{2})}{[\Gamma(s+\frac{5}{2})]^2} \quad \text{Hall factor}, \qquad (5.159)$$

where r_{H} is the Hall factor. The Hall coefficient R_{H} is defined as

$$R_{\text{H}} = \frac{r_{\text{H}}}{\mp n_{i,c}e_c} \quad \text{Hall coefficient, } - \text{ for } i = e \text{ and } + \text{ for } i = h. \qquad (5.160)$$

For negative Hall coefficient we have n-type semiconductor, and vice versa for p-type. The magnitude of r_{H} is near unity. Then using Table 5.3 and (5.113), r_{H} is $3\pi/8$ for acoustic phonon and strongly screened ionized impurity scattering, and is 1.93 for weakly screened ionized impurity scattering. In practice, measured R_{H} is used to determine the electron and hole concentration (density) or Fermi energy in semiconductors, which in general is temperature dependent.†

5.15 Electron–Phonon Relaxation Times in Semiconductors

Electron scattering by a phonon is illustrated in Figure 5.12. The electron-scattering rate is given by (5.95), in terms of the interaction matrix element $M_{\kappa',\kappa}$ and the energy and momentum conservation requirements expressed by the Dirac delta function δ_{D}. The matrix element in turn includes the scattering potential φ_s. It is generally assumed that there is only one phonon involved in the electronic transition.

† Using (5.160) and (5.63), we have

$$n_{e,c}(E_{\text{F}}) = -\frac{r_{\text{H}}}{R_{\text{H}}e_c},$$

and this gives the relation between $n_{e,c}$ and E_{F}.

The electron–phonon interaction is mostly inelastic, and the most important ones are the acoustic phonon, polar optical phonon (e.g., Figure 4.9 for Si), nonpolar optical phonons (e.g., Figure 4.10 for GaAs), equivalent intervalley phonon, and piezoelectric phonon (for very low temperatures).

The scattering of electrons (and holes in semiconductors) that is due to dislocation of atoms in the lattice, is referred to as strain and can be related with deformation scattering theory. This involves specification of the deformation potential $\varphi_d(\boldsymbol{x})$ in the Brillouin zone.

The electron perturbation produced by phonons is by the acoustic phonons (relative displacement of atoms and piezoelectric effect in semiconductors) and by the optical phonons (relative displacement of neighboring atoms and in polar crystal produces polarization fluctuation). For these phonon scatterings the perturbation potential relates to atomic displacement d through

$$
\begin{aligned}
\varphi_{s,e\text{-}p} &\sim \varphi_{d,\mathrm{A}} \partial d/\partial x && \text{acoustic phonon (one-dimensional strain)}\\
\varphi_{s,e\text{-}p} &\sim \frac{e_c e_{pz}}{\epsilon_\mathrm{o}\epsilon_e} d && \text{piezoelectric (acoustic phonons)}\\
\varphi_{s,e\text{-}p} &\sim \varphi'_{d,\mathrm{O}} d && \text{nonpolar optical phonon (displacement)}\\
\varphi_{s,e\text{-}p} &\sim q_e d && \text{polar optical phonon,}
\end{aligned}
\quad (5.161)
$$

where $\boldsymbol{d}$ is the atomic displacement vector, $\varphi_{d,\mathrm{A}}$ and $\varphi'_{d,\mathrm{O}}$ are scattering potential constants, and q_e is the effective charge. Here e_{pz}(C/m^2) is the piezoelectric constant.

Both the momentum conservation and energy conservation are imposed by $\delta_\mathrm{D}(\boldsymbol{\kappa}_e \pm \boldsymbol{\kappa}_p - \boldsymbol{\kappa}'_e)$ and $\delta_\mathrm{D}(E_{e,\kappa'} - E_{e,\kappa} \mp \hbar\omega_p)$. Here we drop the electron subscript in $\boldsymbol{\kappa}$ for simplicity.

The displacement vector $\boldsymbol{d}$ is determined as outlined in Section 4.8 and in a form given by (4.70).

5.15.1 Electron–Phonon Wave Function

The quantum lattice-vibration displacement is given by (4.47), and the normal displacement coordinate $Q_{\kappa_p,\alpha}$ is defined as [309]

$$
Q_{\kappa_p,\alpha} = \left(\frac{\hbar}{2m\omega_\alpha}\right)^{1/2}(b^\dagger_{-\kappa} + a_\kappa). \quad (5.162)
$$

Then (4.47) becomes

$$
\boldsymbol{d} = \frac{1}{N^{1/2}} \sum_{\kappa_p,\alpha} \{[Q_{\kappa_p,\alpha}\boldsymbol{s}_\alpha e^{i(\boldsymbol{\kappa}_p\cdot\boldsymbol{x}_\mathrm{o})}] + [Q_{\kappa_p,\alpha}\boldsymbol{s}_\alpha e^{i(\boldsymbol{\kappa}_p\cdot\boldsymbol{x}_\mathrm{o})}]^\dagger\}, \quad (5.163)
$$

where † indicates the complex conjugate and $\boldsymbol{x}_\mathrm{o}$ is the lattice center location.

The electron is presented by an initial state $|\boldsymbol{\kappa}\rangle$ and the phonon by the product state $\prod |f_{p,\kappa_p,\alpha}(Q)\rangle$, which describes the phonon distribution. In the final state (after scattering), the electron is represented by $|\boldsymbol{\kappa}'\rangle$ and the phonon by a new distribution $\prod |f'_{p,\kappa_p,\alpha}(Q_{\kappa_p,\alpha})\rangle$. Then the initial and final states of the electron–phonon wave function for the system, appearing in the matrix element (5.93) are

$$\psi_i = \psi_\kappa(\boldsymbol{x}) \prod_{\kappa_p,\alpha} | f_{p,\kappa_p}(Q_{\kappa_p,\alpha})\rangle \qquad (5.164)$$

$$\psi_f = \psi_{\kappa'}(\boldsymbol{x}) \prod_{\kappa'_p,\alpha} | f_{p,\kappa'_p}(Q_{\kappa_p,\alpha})\rangle \quad \text{electron–phonon wave function.} \qquad (5.165)$$

In the matrix element, the phonon portion is

$$\prod_{\kappa''_p,\alpha''} \prod_{\kappa'_p,\alpha'} \langle f_{p,\kappa''_p,\alpha''} | Q_{\kappa,\alpha} | f_{p,\kappa',\alpha'} \rangle. \qquad (5.166)$$

Now we use the normal coordinate displacement in terms of the phonon creation and destruction operators and note that the only nonvanishing elements are [309]

$$\langle (f_{p,\kappa_p,\alpha} - 1) | b_{\kappa_p} | f_{p,\kappa_p,\alpha} = f^{1/2}_{p,\kappa_p,\alpha} \qquad (5.167)$$

$$\langle (f_{p,\kappa_p,\alpha} + 1) | b^\dagger_{\kappa_p} | f_{p,\kappa_p,\alpha} = (f_{p,\kappa_p,\alpha} + 1)^{1/2}. \qquad (5.168)$$

The interaction matrix element containing $Q_{\kappa_p,\alpha}$ is [309]

$$\frac{\hbar}{2m\omega_{\kappa_p,\alpha}} \delta_{\rm D}(\kappa_p, \kappa'_p, \kappa''_p)\delta_{\rm D}(\alpha, \alpha', \alpha'')[f^{1/2}_{p,\kappa_p}\delta_{\rm D}(f_{p,\kappa'_p} - 1, f_{p,\kappa_p})$$
$$+(f_{p,\kappa_p} + 1)^{1/2}\delta_{\rm D}(f_{p,\kappa'_p} + 1, f_{p,\kappa_p})], \qquad (5.169)$$

which represents the phonon absorption and emission. At equilibrium, we have $f^{\rm o}_p$.

The electronic portion has an interaction Hamiltonian expressed by (5.163), i.e.,

$$\mathrm{H} = \varphi_{s,e\text{-}p} = \sum_{\kappa_p} \{[\varphi_{s,\kappa_p,\alpha} Q_{\kappa_p,\alpha} e^{i(\kappa_p\cdot \boldsymbol{x}_{\rm o})}] + [\varphi_{s,\kappa_p,\alpha} Q_{\kappa_p,\alpha} e^{i(\kappa_p\cdot \boldsymbol{x}_{\rm o})}]^\dagger\}, \qquad (5.170)$$

For Bloch plane waves, the electronic portion of $M_{\kappa',\kappa}$ is

$$\frac{1}{V}\int d^\dagger(f_{p,\kappa'_p}, \boldsymbol{\kappa}')e^{-i(\kappa'\cdot\boldsymbol{x})}\varphi_{s,\kappa_p,\alpha}(\boldsymbol{x})e^{\pm i(\kappa_p\cdot\boldsymbol{x})}d(f_{p,\kappa_p}, \boldsymbol{\kappa})e^{i(\kappa\cdot\boldsymbol{x})}{\rm d}\boldsymbol{x}. \qquad (5.171)$$

The spatial integral is over a primitive cell, giving

$$\frac{1}{V}\int_{\rm cell} \psi^\dagger(f_{p,\kappa'_p}, \boldsymbol{\kappa}')\varphi_{s,\kappa_p,\alpha}\psi^\dagger(f_{p,\kappa_p}, \boldsymbol{\kappa}) \sum_{\boldsymbol{R}} [i(\boldsymbol{\kappa} + \boldsymbol{\kappa}_p - \boldsymbol{\kappa}')\cdot \boldsymbol{x}_{\rm o}]$$
$$= \frac{1}{V}\int_{\rm cell} \psi^\dagger(f_{p,\kappa'_p}, \boldsymbol{\kappa}')\varphi_{s,\kappa_p,\alpha}\psi^\dagger(f_{p,\kappa_p}, \boldsymbol{\kappa}){\rm d}\boldsymbol{x}\delta_{\rm D}(\boldsymbol{\kappa} \pm \boldsymbol{\kappa}_p - \boldsymbol{\kappa}', \boldsymbol{g}), \qquad (5.172)$$

assuming that, because of the small cell size, $\exp[i(\boldsymbol{\kappa} - \boldsymbol{\kappa}')\cdot \boldsymbol{x}] = 1$. For the normal processes $\boldsymbol{g} = 0$ and for the Umklapp processes $\boldsymbol{g} \neq 0$ (Section 4.15). Then we define the interaction (coupling) coefficient $C_{\kappa_p,\alpha}$ as

$$M_{\kappa,\kappa'}(\text{electronic}) = \frac{1}{V} C_{\kappa_p,\alpha} G(\boldsymbol{\kappa}, \boldsymbol{\kappa}')\delta_{\rm D}(\boldsymbol{\kappa} \pm \boldsymbol{\kappa}_p - \boldsymbol{\kappa}', \boldsymbol{g}),$$
$$C_{\kappa_p,\alpha} G(\boldsymbol{\kappa}, \boldsymbol{\kappa}') \equiv \int_{\rm cell} \psi^\dagger(f_{p,\kappa'_p}, \boldsymbol{\kappa}')\varphi_{s,\kappa_p,\alpha}\psi^\dagger(f_{p,\kappa_p}, \boldsymbol{\kappa}){\rm d}\boldsymbol{x}$$
$$= \varphi_{s,\kappa_p,\alpha} G(\boldsymbol{\kappa}, \boldsymbol{\kappa}'), \qquad (5.173)$$

where for the last equality it is assumed that $\varphi_{s,\kappa_p,\alpha}$ does not vary significantly over the unit cell. The quantity $G(\kappa,\kappa')$ represents the overlap of the cell periodic part of the initial and final electronic states.

The electron–phonon matrix elements for Bloch plane waves, $\exp(i\kappa'\cdot x)$ and $\exp(i\kappa\cdot x)$, become

$$M^2_{\kappa,\kappa'} = \langle\psi_f|\varphi_{s,e\text{-}p}|\psi_i\rangle^2$$
$$= \frac{\hbar}{2Nm}\frac{C^2_{\kappa_p,\alpha}G(\kappa,\kappa')}{\omega_{p,\alpha}}[f_p(\omega_{p,\alpha}) + \frac{1}{2} \mp \frac{1}{2}]\delta_{\rm D}(\kappa \pm \kappa_p - \kappa', 0)$$

(+ absorption, − emission of phonons) for the normal processes. (5.174)

The final form of the electron-scattering rate (by phonons) (5.95), for which the equilibrium phonon distribution instead, is [309]

$$\dot{\gamma}_{e\text{-}p} = \frac{1}{\tau_{e\text{-}p}} = \frac{1}{8\pi^2 Nm}\int \frac{C^2_{\kappa_p,\alpha}G(\kappa,\kappa')}{\omega_{p,\alpha}(\kappa_p,\alpha)}[f_p^\circ(\omega_{p,\alpha}) + \frac{1}{2} \mp \frac{1}{2}]$$
$$\times\delta_{\rm D}(\kappa \pm \kappa_p - \kappa')\,\delta_{\rm D}(E_{e,\kappa'} - E_{e,\kappa} \mp \hbar\omega_{p,\alpha}){\rm d}\kappa'$$

electron–phonon scattering, (5.175)

where

$$G(\kappa,\kappa') = 1 \ \text{ for parabolic bands,} \tag{5.176}$$

and $C_{\kappa_p,\alpha}$ is related to the integral of $\varphi_{s,e\text{-}p}$ taken over the cell (5.173). Again, N is the number of unit cells, and m is the mass of the oscillator. For nonparabolic bands, $G(\kappa,\kappa')$ is given in [309] for both electrons and holes.

5.15.2 Rate of Acoustic-Phonon Scattering of Electrons

Acoustic-phonon scattering of electrons (and holes in semiconductors), i.e., fluctuation in electron energy caused by strain (deformation potential scattering), are nearly elastic, and from using (5.161) and (5.163) we have

$$\varphi_{s,e\text{-}p} = \varphi_{d,{\rm A}}\frac{{\rm d}d}{{\rm d}x}$$
$$= \frac{1}{N^{1/2}}\sum_{\kappa_p}\{[i\,Q_{\kappa_p,\alpha}\varphi_{d,{\rm A}}\boldsymbol{s}_\alpha\cdot\kappa_p e^{i(\kappa_p\cdot x_{\rm o})}]$$
$$+[i\,Q_{\kappa_p,\alpha}\varphi_{d,{\rm A}}\boldsymbol{s}_\alpha\cdot\kappa_p e^{i(\kappa_p\cdot x_{\rm o})}]^\dagger\}. \tag{5.177}$$

Then, using (5.173), we have

$$C^2_{\kappa_p,\alpha} = \varphi^2_{d,{\rm A}}\kappa_p^2. \tag{5.178}$$

Using the $\kappa_p \to 0$ approximation for the acoustic phonons (Section 4.1), we have for the acoustic-phonon dispersion

$$\omega_p = u_{p,{\rm A}}\kappa_p. \tag{5.179}$$

Using Table 1.2 for f_p°, we have the classical limit for bosons (including phonons, Figure 1.1)

$$f_p^\circ = \frac{1}{\exp(\frac{\hbar\omega_p}{k_B T}) - 1} \simeq \frac{k_B T}{\hbar\omega_p} \quad \text{for} \quad \hbar\omega_p \ll k_B T. \tag{5.180}$$

Then, using (5.178) to (5.180) in (5.175), we have the acoustic-phonon-scattering rate of electrons as

$$\dot{\gamma}_{e\text{-}p,\mathrm{A}} = \frac{1}{\tau_{e\text{-}p,\mathrm{A}}} = \frac{4\pi^3}{8\pi^2 Nm}\int \frac{\varphi_{d,\mathrm{A}}^2 \kappa_p^2}{\omega_p}[f_p^\circ + \frac{1}{2} \mp \frac{1}{2}]\delta_{\mathrm{D}}(\boldsymbol{\kappa} \pm \boldsymbol{\kappa}_p - \boldsymbol{\kappa}')$$
$$\times\, \delta_{\mathrm{D}}(E_{e,\boldsymbol{\kappa}'} - E_{e,\boldsymbol{\kappa}} \mp \hbar\omega_p)\mathrm{d}\boldsymbol{\kappa}'$$
$$= \frac{\varphi_{d,\mathrm{A}}^2 k_B T}{8\pi^2 \hbar\rho u_{p,A}^2}\int \delta_{\mathrm{D}}(\boldsymbol{\kappa} \pm \boldsymbol{\kappa}_p - \boldsymbol{\kappa}')\delta_{\mathrm{D}}(E_{e,\boldsymbol{\kappa}'} - E_{e,\boldsymbol{\kappa}} \mp \hbar\omega_p)\mathrm{d}\boldsymbol{\kappa}', \tag{5.181}$$

where $\rho = mN/V$. For low-energy phonons $f_p^\circ \gg 1$, so we have neglected the 1/2 terms. Also for low-energy phonons [classical limit, from (5.180), for bosons], $f_p^\circ \simeq k_B T/\hbar\omega_p$.

We also note that, from (5.64),

$$\frac{1}{8\pi^3}\int \delta_{\mathrm{D}}[E_e(\boldsymbol{\kappa}) - E_e(\boldsymbol{\kappa}')]\mathrm{d}\boldsymbol{\kappa} = D_e[E_e(\boldsymbol{\kappa})]. \tag{5.182}$$

Now, for elastic scattering (dropping $\hbar\omega_p$), and because $\boldsymbol{p}_e = \hbar\boldsymbol{\kappa}_e$, we have [309]

$$\frac{1}{\tau_{e\text{-}p,A}} = \frac{\pi\varphi_{d,\mathrm{A}}^2 k_B T D_e[E_e(\boldsymbol{p})]}{\hbar\rho u_{p,\mathrm{A}}^2}. \tag{5.183}$$

Now, using (5.65) for D_e, we have

$$\frac{1}{\tau_{e\text{-}p,\mathrm{A}}(\text{momentum})} = \frac{2^{1/2} m_{e,e}^{3/2}\varphi_{d,\mathrm{A}}^2 k_B T E_e^{1/2}(\boldsymbol{p})}{\pi\hbar^4 \rho u_{p,\mathrm{A}}^2}$$
$$\text{elastic acoustic-phonon scattering,} \tag{5.184}$$

which is listed in Table 5.4. Note that T here is the lattice temperature. Writing this in terms of power-law relation (5.112), we have $s = -1/2$ (Table 5.3), i.e.,

$$\tau_{e\text{-}p,\mathrm{A}}(\text{momentum}) = (\tau_{e\text{-}p,\mathrm{A}})_{\mathrm{o}}[\frac{E_e(\boldsymbol{p})}{k_B T}]^{-1/2},$$
$$(\tau_{e\text{-}p,\mathrm{A}})_{\mathrm{o}}(\text{momentum}) = \frac{\pi\hbar^4\rho u_{p,n}^2}{2^{1/2} m_{e,e}^{3/2}\varphi_{d,\mathrm{A}}^2 (k_B T)^{3/2}}. \tag{5.185}$$

To use this, the perturbation deformation potential $\varphi_{d,\mathrm{A}}$ should be known (of the order of 1 eV, Table 5.5). The average acoustic-phonon speed $u_{p,\mathrm{A}}$ is found from $u_{p,\mathrm{A}} = u_{p,g,\mathrm{A}}$ given by (4.61).

The total scattering is the sum of emission and absorption, and because they are the same here (neglecting $\hbar\omega_p$), $\tau_{e\text{-}p,\mathrm{A}}$ total is twice (5.183).

The variation of $\tau_{e\text{-}p,\mathrm{A}}^{-1}$, with respect to E_e, for Si, is shown in Figure 5.13(a).

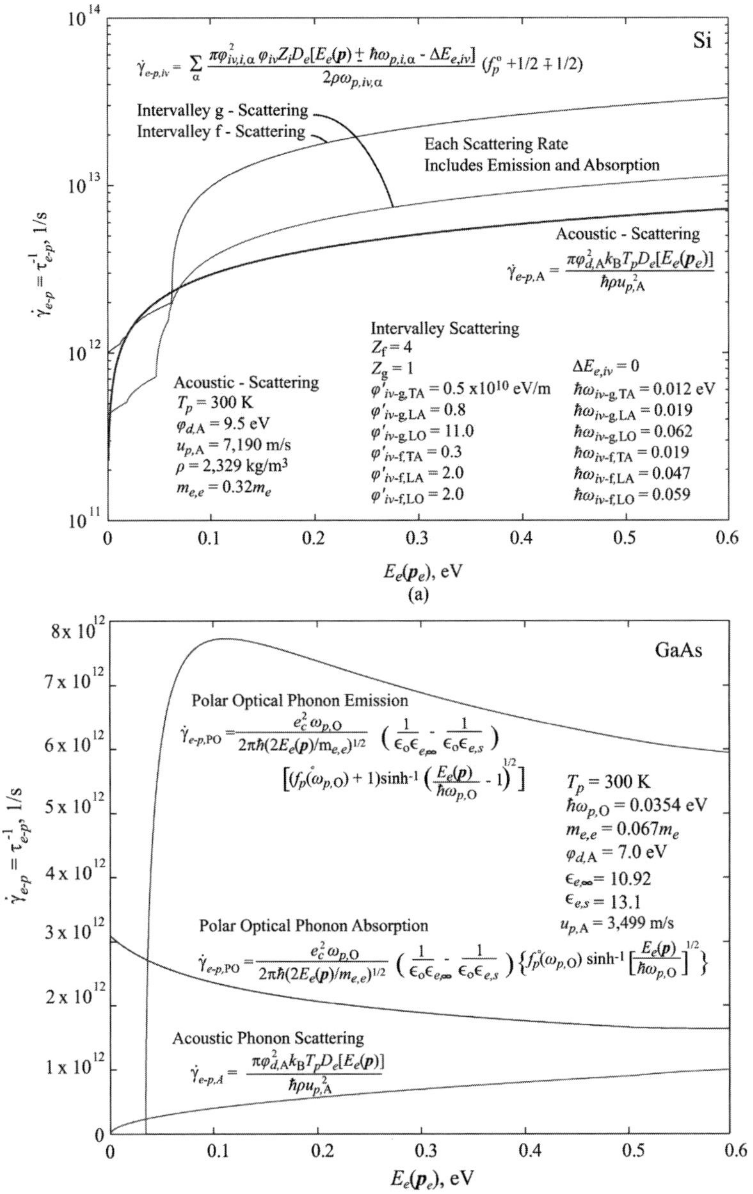

Figure 5.13. (a) Predicted variations of the phonon relaxation times in Si with respect to the electron kinetic energy. For the intervalley scattering the contributions from phonon emission and absorption are added. (b) Variations of the optical-phonon and acoustic-phonon-scattering relaxation times for electrons in GaAs, with respect to the electron kinetic energy. The results are for $T_p = 300$ K.

5.15.3 Rate of Optical-Phonon Scattering of Electrons

(A) *Nonpolar Optical Phonons*

The scattering potential for nonpolar (Section 4.6) optical phonons is given by the perturbation Hamiltonian [309]

$$\varphi_{s,e\text{-}p} = \boldsymbol{\varphi}'_{d,\mathrm{O}} \cdot \boldsymbol{d} \quad \text{optical-phonon perturbation Hamiltonian,} \tag{5.186}$$

where $\boldsymbol{\varphi}'_{d,\mathrm{O}}$ is the nonpolar optical deformation potential derivative vector. Then after evaluation of (5.186) in (5.95), we have [309]

$$C^2_{\kappa_p,\alpha} = \varphi'^2_{d,\mathrm{O}}. \tag{5.187}$$

Using the $\kappa_p \rightarrow 0$ approximation of the optical-phonon dispersion (Section 4.1), we use a constant phonon frequency (dispersionless) $\omega_{p,\mathrm{O}}$ (zone center), and the result for (5.175) is [309]

$$\frac{1}{\tau_{e\text{-}p,\mathrm{O}}(\text{momentum})} = \frac{\pi \varphi'^2_{d,\mathrm{O}}}{2\rho\omega_{p,\mathrm{O}}} f^\circ_p(\omega_{p,\mathrm{O}}) D_e[E_e(\boldsymbol{p}) + \hbar\omega_{p,\mathrm{O}}] \quad \text{nonpolar optical-phonon absorption} \tag{5.188}$$

$$\frac{1}{\tau_{e\text{-}p,\mathrm{O}}(\text{momentum})} = \frac{\pi \varphi'^2_{d,\mathrm{O}}}{2\rho\omega_{p,\mathrm{O}}} [f^o_p(\omega_{p,\mathrm{O}}) + 1] D_e[E_e(\boldsymbol{p}) - \hbar\omega_{p,\mathrm{O}}] \quad \text{nonpolar optical-phonon emission,} \tag{5.189}$$

where we have used (5.64), for $D_e(E)$, with spin degeneracy, and $nm = \rho$, as our analysis is per-unit volume. Typical values of $\varphi'_{d,\mathrm{O}}$ are listed in Table 5.5.

For nonpolar semiconductors, the electron scattering (transition) is dominated by phonon-induced transitions between the valleys in the constant-energy surface and requires optical phonons. Figure 5.7(c) shows the six equivalent conduction-band valleys of Si. The scattering between opposite valleys, e.g., [100] to [$\bar{1}$00] is called the g-process, and that between nonopposite valleys is called the f-process. These are shown in Figure 5.7(c). The momentum is conserved, and the intervalley scattering involves an Umklapp process (5.96). The indirect bands of Si are shown in Figure 5.6(a), and the minimum in the conduction band is not at the zone edge (X-point). Then 0.3 times the dimensionless wave vector away from the zone edge, or $0.3(2\pi/a[100])$ phonon energy, is needed for a g-process. The intervalley (or interband) scattering is similar to the nonpolar optical phonon scattering and the relations for both absorption and emission are given in Table 5.4. For Si, the intervalley scattering is also shown in Figure 5.13(a). The properties are listed in Table 5.5.

Table 5.4. *Some electron scattering relaxation time relations.*

Source of scattering	Relaxation time
Electron-alloy [309]	$\tau_{e\text{-}a}^{-1} = \dfrac{3\pi V_{\rm o}\varphi_s^2 x(1-x) m_{e,e}^{3/2} E_e^{1/2}}{2^{1/2} 8\hbar^4}(\boldsymbol{p})$ $V_{\rm o}$ is $a^3/4$, φ_s is scattering potential, x is alloy atomic fraction
Electron-ionized impurity (generalized form) (momentum) [309]	$\tau_{e\text{-}im}^{-1} = n_{im}\dfrac{1}{1+(2^{1/2})\pi}(\dfrac{ze_c^2}{\epsilon_e\epsilon_{\rm o}})^2 \dfrac{1}{m_{e,e}^{1/2}E_e^{3/2}(\boldsymbol{p})}\times$ $\{\ln[1+\dfrac{8m_{e,e}E_e(\boldsymbol{p})\lambda^2}{\hbar^2}] - \dfrac{1}{1+\dfrac{\hbar^2}{8m_{e,e}E_e(\boldsymbol{p})\lambda^2}}\}$ z is the charge of impurity, λ is screening length
Electron–phonon [219, 309]	momentum (acoustic phonon, elastic): $\tau_{e\text{-}p,{\rm A}}^{-1}(\text{momentum}) = \dfrac{2^{1/2} m_{e,e}^{3/2} k_{\rm B} T \varphi_{d,{\rm A}}^2 E_e^{1/2}(\boldsymbol{p})}{\pi\hbar_{\rm P}^4 \rho u_{p,{\rm A}}^2}$ $\tau_{e\text{-}p,{\rm A}}(\text{energy}) = \dfrac{k_{\rm B}T}{2m_{e,e}u_p^2}\tau_{e\text{-}p,{\rm A}}(\text{momentum})$ momentum (piezoelectric) $\tau_{e\text{-}p,{\rm A}}^{-1} = (\frac{m_{e,e}}{8})^{1/2}(\frac{e_c e_{pz}}{\epsilon_{\rm o}\epsilon_e})^2 \frac{k_{\rm B}T}{\rho u_{p,{\rm A}}^2} E_e^{1/2}(\boldsymbol{p})$ momentum (nonpolar, optical phonon): $\tau_{e\text{-}p,{\rm O}}^{-1} = \dfrac{\varphi_{d,{\rm O}}'^2 f_p^{\circ}(\omega_{p,{\rm O}}) m_{e,e}^{3/2}[E_e(\boldsymbol{p}) + \hbar\omega_{p,{\rm O}}]^{1/2}}{2^{1/2}\pi\hbar^3\rho\omega_{p,{\rm O}}}$, absorption $\tau_{e\text{-}p,{\rm O}}^{-1} = \dfrac{2^{1/2}\varphi_{d,{\rm O}}'^2 [f_p^{\circ}(\omega_{p,{\rm O}})+1] m_{e,e}^{3/2}[E_e(\boldsymbol{p}) - \hbar\omega_{p,{\rm O}}]^{1/2}}{\pi\hbar^3\rho\omega_{p,{\rm O}}}$, emission internally scattering (g- and f-scattering): $\tau_{e\text{-}p,iv}^{-1} = \sum_\alpha \frac{\pi\varphi_{iv,i\alpha} Z_i D_e[E_e(\boldsymbol{p}) \pm \hbar\omega_{p,iv,\alpha} - \Delta E_{e,iv}](f_p^{\rm o} + \frac{1}{2} \mp \frac{1}{2})}{2\rho\omega_{iv,i,\alpha}}$ emission and absorption momentum (polar optical phonon): $\tau_{e\text{-}p,{\rm PO}}^{-1} = \dfrac{e_c^2\omega_{p,{\rm O}} m_{e,e}^{1/2}}{2\pi\hbar[2E_e(\boldsymbol{p})]^{1/2}}(\dfrac{1}{\epsilon_{\rm o}\epsilon_{e,\infty}} - \dfrac{1}{\epsilon_{\rm o}\epsilon_{e,s}})[f_p^o(\omega_{p,{\rm O}})]$ $\sinh^{-1}(\dfrac{E_e(\boldsymbol{p})}{\hbar\omega_{p,{\rm O}}})^{1/2} Z_i$ absorption $\tau_{e\text{-}p,{\rm PO}}^{-1} = \dfrac{e_c^2\omega_{p,{\rm O}} m_{e,e}^{1/2}}{2\pi\hbar[2E_e(\boldsymbol{p})]^{1/2}}(\dfrac{1}{\epsilon_{\rm o}\epsilon_{e,\infty}} - \dfrac{1}{\epsilon_{\rm o}\epsilon_{e,s}})[f_p^o(\omega_{p,{\rm O}})+1]$ $\sinh^{-1}(\dfrac{E_e(\boldsymbol{p})}{\hbar\omega_{p,{\rm O}}} - 1)^{1/2}$ emission energy (polar and nonpolar optical phonon): $\tau_{e\text{-}p,{\rm O}}^{-1}(\text{energy}) = \sum_{\boldsymbol{p}'}\dot{\gamma}_{e\text{-}p}(\boldsymbol{p},\boldsymbol{p}')[1 - \dfrac{E_e(\boldsymbol{p}')}{E_e(\boldsymbol{p})}] =$ $\dfrac{\hbar\omega_{p,{\rm O}}}{E_e(\boldsymbol{p})}\tau_{e\text{-}p,{\rm O}}^{-1}$ (momentum) for high E_e
Electron–photon (momentum) [309]	$\tau_{e\text{-}ph}^{-1} = \dfrac{4\pi^2 e_c^2}{m_e^2\omega_{ph}\epsilon_{\rm o}\epsilon_e} f_{ph}^o D_{ph}(\omega_{ph})\dfrac{2p_{cv}^2}{3}$ absorption $\tau_{e\text{-}ph}^{-1} = \dfrac{4\pi^2 e_c^2}{m_e^2\omega_{ph}\epsilon_{\rm o}\epsilon_e}(f_{ph}^o + 1) D_{ph}(\omega_{ph})\dfrac{2p_{cv}^2}{3}$ emission p_{cv} is the momentum matrix element
Phonon-assisted Electron–photon	$\tau_{e\text{-}ph\text{-}p}^{-1} = \frac{2\pi}{\hbar}(\frac{\omega_{e,g}}{\omega_{ph,i}})^2 \frac{(\boldsymbol{s}_{ph,i}\cdot\boldsymbol{\mu}_e)^2}{2\epsilon_{\rm o}} \frac{a_{i\text{-}p}^2}{2\rho u_{p,{\rm A}}^2} \frac{D_p(E_p) f_p^{\rm o}(E_p)}{E_p} e_{ph,i}$ (Section 7.12)

Table 5.5. *Parameters used in the electron scattering relaxation-time relations, for Si and GaAs [309].*

Parameter	Symbol	Magnitude, Si	Magnitude, GaAs
Mass density (g/cm^3)	ρ	2.329	5.36
Lattice constant	a	5.43	5.642
Low frequency dielectric constant	$\epsilon_{e,s}$	11.7	12.90
High frequency dielectric constant	$\epsilon_{e,\infty}$	–	10.92
Piezolelectric constant (C/m^2)	e_{pz}	–	0.160
LA velocity ($\times 10^5$ cm/s)	$u_{p,\mathrm{LA}}$	9.04	5.24
TA velocity ($\times 10^5$ cm/s)	$u_{p,\mathrm{TA}}$	5.34	3.0
LO phonon energy (eV)	$\hbar\omega_{p,\mathrm{O}}$	0.063	0.03536
Electron effective mass	$m_{e,e}$	–	0.067
(lowest valley)(m_e)	$m_{e,e,l}, m_{e,e,t}$	0.916, 0.19(X)	–
Electron effective mass	$m_{e,e}$	–	0.222(L)
(upper valley) (m_e)	$m_{e,e}$	–	0.58(X)
	$m_{e,e,l}, m_{e,e,t}$	1.59,0.12(L)	–
Nonparabolicity parameter (eV^{-1})	α	0.5(X)	0.610,0.461(L),0.204(X)
Energy separation	$\Delta E_{e,\Gamma\mathrm{L}}$	–	0.29
between valleys (eV)	$\Delta E_{e,\Gamma\mathrm{X}}-$	0.48	
Electron acoustic deformation potential (eV)	$\varphi_{d,\mathrm{A}}$	9.5	7.0(Γ) 9.2(L) 9.0(X)
Electron optical deformation potential ($\times 10^8$ eV/cm)	$\varphi_{e,\mathrm{O}}$	–	3.0(L)
Optical-phonon energy (eV)	$\omega_{p,\mathrm{O}}$	0.0642	0.0343
Hole acoustic deformation potential (eV)	$\varphi_{d,h,\mathrm{A}}$	5.0	3.5
Hole optical deformation potential (eV/cm)	$\varphi'_{d,h,\mathrm{O}}$	6.00	6.48
Intervalley parameters, g-type(X-X) ($\times 10^8$ eV/cm), (eV)	$\varphi'_{iv}, \hbar\omega$	0.5, 0.012(TA) 0.8, 0.019(LA) 11.0,0.062(LO)	
Intervalley parameters, f-type(X-X) ($\times 10^8$ eV/cm), (eV)	$\varphi'_{iv}, \hbar\omega$	0.3, 0.019(TA) 2.0, 9.947(LA) 2.0,0.059(LO)	
Intervalley parameters (X-L) ($\times 10^8$ eV/cm), (eV)	$\varphi'_{iv}, \hbar\omega$	2.0, 0.058 2.0, 0.055 2.0,0.041 2.0,0.017	
Intervalley deformation potential ($\times 10^8$ eV/cm)	$\varphi'_{d,\mathrm{O}}(\Gamma\mathrm{L})$	–	10.0
	$\varphi'_{d,\mathrm{O}}(\Gamma\mathrm{X})$		10.0
	$\varphi'_{d,\mathrm{O}}(\mathrm{LL})$,		10.0
	$\varphi'_{d,\mathrm{O}}(\mathrm{LX})$		5.0
	$\varphi_{d\mathrm{A}}(\mathrm{XX})$		7.0
Intervalley phonon energy (eV)	$(\hbar\omega)_i$	–	0.0278,0.0299 0.0290, 0.0293 0.0299

(B) *Polar Optical Phonons*

In the polar optical phonon–electron interaction, the LO phonon modes create a polarization that is due to an effective charge q_e, which is expressed in terms of low-frequency (static) and high-frequency dielectric constants, $\epsilon_{e,s}$ and $\epsilon_{e,\infty}$ (3.50). The polar optical phonons were discussed in Section 4.6, showing that at the Briollouin zone center $E_{p,\mathrm{LO}} > E_{p,\mathrm{TO}}$ (Figure 4.16).

The polarization and the effective charge density q_e/V_c in (5.161), where V_c is the unit-cell volume, are given by

$$\boldsymbol{p}_e(0) = \frac{1}{4\pi}(\frac{\epsilon_{e,s}-1}{\epsilon_{e,s}})\boldsymbol{d}, \quad \boldsymbol{p}_e(\infty) = \frac{1}{4\pi}(\frac{\epsilon_{e,\infty}-1}{\epsilon_{e,\infty}})\boldsymbol{d}$$

$$\frac{q_e^2}{V_c} = \frac{\epsilon_\mathrm{o}\rho\omega_{p,\mathrm{O}}^2}{\epsilon_{e,\infty}}(\frac{\epsilon_{e,\infty}}{\epsilon_{e,s}} - 1)$$

effective charge in polar optical-phonon scattering, (5.190)

where $(\epsilon_{e,\infty}/\epsilon_{e,s}) - 1$ is a measure of strength of the polar dipole (Section 4.6) [219].

For polar optical-phonon scattering (Section 4.6), the Debye model electric permittivity (dielectric function) of the solid (3.50) is used, and the expressions for $\tau_{e\text{-}p,\mathrm{PO}}$ are listed in Table 5.4.

The variation of $\tau_{e\text{-}p,\mathrm{A}}^{-1}$ and $\tau_{e\text{-}p,\mathrm{PO}}^{-1}$ (emission and absorption) with respect to E_e, for GaAs, is shown in Figure 5.13(b). The phonon dispersion for GaAs is shown in Figure 4.9.

Note that the smallest $\tau_{e\text{-}i}$ controls the electric transport, according to (5.114), and from the results shown in Figure 5.13, this depends on the electron kinetic energy (and also on the lattice temperature T_p).

5.15.4 Summary of Electron-Scattering Mechanisms and Relaxation-Time Relations

A summary of various electron scatterings is given in Table 5.4.† The parameters used in the expressions for two semiconductors, Si and GaAs, are listed in Table 5.5.

These include the electron–alloy, electron-impurity, electron–phonon (acoustic, polar optical, and nonpolar optical), electron–photon and phonon-assisted electron–photon scatterings.

Distinctions are made between the momentum and energy-scattering relaxation times (Section 3.1.6); and we will further discuss these in Section 5.18.

† In addition, for TE materials, such as telluride compounds, more specific relaxation-time models have been developed, including the Kane model further described in Section 5.17.3.

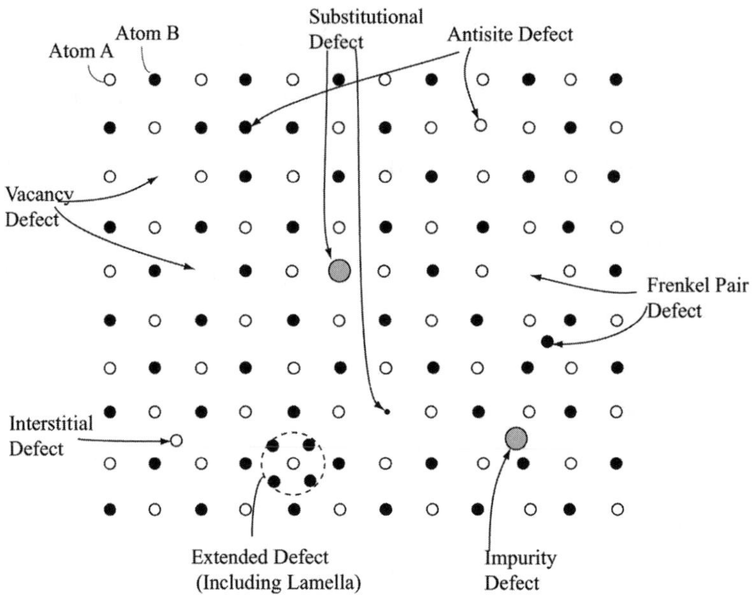

Figure 5.14. Some typical crystalline solid structure defects affecting phonon and electron transport.

5.16 TE Transport Coefficients Data

5.16.1 Structural Defects in Crystalline Solids

Various formation defects cause deviation, from the ideal phonon and electron transport properties, and these should be considered when one is comparing experimental results with the theoretical predictions. A stable crystalline lattice can accept defects up to a threshold maximum concentration (number density) n_d; beyond this point the lattice breaks down, and either a different crystal structure is formed or the structure becomes amorphous. Crystal defects are thermodynamically controlled phenomena and formation of defects, and the most probable defect is simply that which leads to the lowest free energy for the crystal.

An intrinsic defect is a stoichiometric defect, i.e., a structural defect that preserves the overall ideal chemical stoichiometry, e.g., the 1:1 ratio of cation to anion in NaCl. An extrinsic defect is a nonstoichiometric defect, i.e., a defect that leads to a fractional change in stoichiometry. A point defect is a defect involving isolated atom or ion sites in the lattice. The extended (topology) defects include line defects (also called dislocations, including those from lamella) and plane defects. In Frenkel defects, one atom is displaced off its regular lattice site into an interstitial location.

Figure 5.14 shows examples of crystal structural defects that can influence the phonon and electron transport coefficients. In TE materials, both point and extended defects contribute to nonideal properties. The point defects include antisite, impurity, interstitial, substitutional, and vacancy.

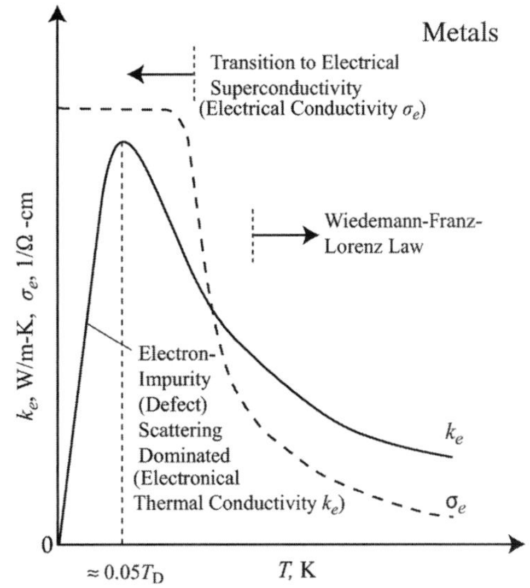

Figure 5.15. Variation of the electrical conductivity and the electronic thermal conductivity with respect to temperature. At low temperatures, the electron-impurity (defect) scattering affects the electronic thermal conductivity, more pronounced resulting in vanishing thermal conductivity, whereas the electrical conductivity reaches a plateau [24].

5.16.2 Metals

For metals (degenerate conductors), the Fermi energy (level) lies well preceding the conduction band edge, and

$$f_e^{\mathrm{o}}(E_e) = \frac{1}{e^{E_e^* - E_{\mathrm{F}}^*} + 1}, \quad E_{\mathrm{F}}^* = \frac{E_{\mathrm{F}}}{k_{\mathrm{B}} T} \gg 1 \quad \text{metals.} \tag{5.191}$$

The Fermi energy for metals is listed in Table 5.2. Then the electrical conductivity (5.124), which when using (5.52) for $n_{e,c}$ becomes [123][†]

$$\sigma_e = \frac{8\pi}{3}\left(\frac{1}{2\pi^2\hbar^2}\right)^{3/2} e_c^2 m_e^{1/2} \tau_{e,\mathrm{o}} \left(\frac{E_{\mathrm{F}}}{k_{\mathrm{B}} T}\right)^{s+3/2}$$
$$\text{electrical conductivity for metals.} \tag{5.192}$$

To obtain (5.192), (5.124) is evaluated under the degenerate conductor treatment of $E_{\mathrm{F}}/k_{\mathrm{B}}T \gg 1$. Wiedemann–Franz–Lorenz law (5.150) holds at higher temperatures, where the electron–phonon scattering dominates. At low temperatures, the electronic thermal conductivity is pronouncedly more affected by impurity scattering (compared to the electrical conductivity). These trends are shown in Figure 5.15. Note that k_e has similar temperature dependence as k_p shown in Figure 4.17. Both k_e and k_p diminish as $T \to 0$. At high temperatures, in metals σ_e is proportional to $1/T$ [will be shown in Figure 5.16(a)], so k_e becomes nearly independent

[†] Note that for a power-law relaxation time we have

$$\int_0^\infty E_e^{j+s+3/2} \frac{\partial f_e^{\mathrm{o}}}{\partial E_e} \mathrm{d}E_e = -(j+s+3/2) \int_0^\infty E_e^{j+s+1/2} f_e^{\mathrm{o}}(E_e) \mathrm{d}E_e,$$

and for metals we have

$$\int_0^\infty E_e^n f_e^{\mathrm{o}}(E_e) \simeq \frac{1}{n+1}\left(\frac{E_{\mathrm{F}}}{k_{\mathrm{B}} T}\right)^{n+1}.$$

of T, while, k_p has the $1/T$ behavior presented by the Slack model (4.132). The peak in k_e and k_p is around 0.05 and 0.1 T_D, respectively. We will discuss this further in the next section, where the experimental results are presented.

Based on acoustic-phonon-dominated scattering ($s = -1/2$ in Table 5.3), (5.192) shows that σ_e is proportional to $1/T$, which is characteristic of metallic electrical conductivity at high temperatures.

Also, for metals the Seebeck coefficient is found from (5.125) and (5.142) as [123]

$$\alpha_S = \mp \frac{\pi^2}{3} \frac{k_B}{e_c} \frac{s + \dfrac{3}{2}}{\dfrac{E_F}{k_B T}} \quad \text{Seebeck coefficient for metals.} \tag{5.193}$$

Note that $k_B/e_c = 86$ μV/K, giving the order of magnitude of the Seebeck coefficient.

The Fermi energies of metals are listed in Table 5.2.

Also, from (5.156), for metals we have

$$k_e = N_{L,o} \sigma_e T = 2.442 \times 10^{-8} \sigma_e T$$

$$\text{Lorenz number for metals at high temperatures.} \tag{5.194}$$

We note that this overpredicts k_e for some metals, especially at moderate temperatures. Then the Lorenz number (or Lorenz ratio) N_L depends on temperature, as shown in Figure 5.16(b) for metals [24]. Note that, for pure metals, the Lorenz number becomes fairly small as $T \to 0$.

Thomson coefficient (5.151) for metals becomes the same as the Seebeck coefficient (5.193).

Figure 5.16(a) shows the measured variation of the electrical conductivity of Ag with respect to temperature. There are three distinct regimes, starting with the constant high conductivity at low temperatures (limited by impurity scattering). Then there is a high-scattering regime (inelastic acoustic-phonon scattering) with a very strong temperature dependence and large drop in conductivity. This is then followed by the elastic-phonon-scattering regime with a T^{-1} dependency. Then (5.192) can predict the high temperature behavior of σ_e, with $s = -1/2$.

Bloch [13] has suggested an approximate analysis for the scaled electrical resistivity, using the Debye temperature. He found that, for all metals, the following relation holds with relatively good accuracy [13]:

$$\frac{\rho_e}{\rho_e(T_D)} = \frac{\sigma_e(T_D)}{\sigma_e} = 3.7(\frac{T}{T_D})^5 \int_0^{T_D/T} \frac{x^5 e^x}{(e^x - 1)^2} dx$$

$$\text{Bloch model for metal electrical conductivity.} \tag{5.195}$$

This relation is shown in Figure 5.17, along with the data for several metals.

Figure 5.18(a) shows the variation of measured thermal conductivity of some metals (including Cu) with respect to temperature.

To examine the validity of the Wiedemann–Franz–Lorenz law, the results from using it along with the experimental results for Cu [149], to predict the thermal

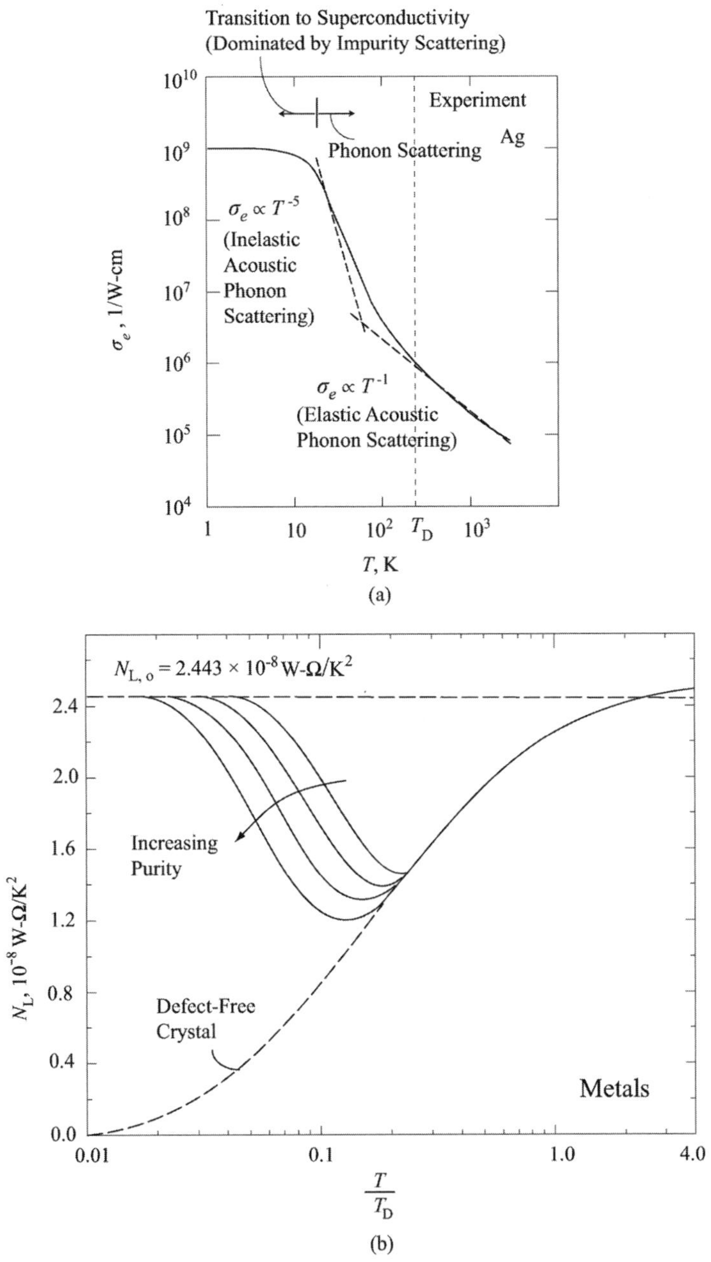

Figure 5.16. (a) Typical temperature dependence of the electrical conductivity for Ag [46]. (b) Variation of the Lorenz number with respect to T/T_D for metals [24].

conductivity, are shown in Figure 5.18(b). The Debye temperature of Cu is 315 K (Table A.2). The peak in the experimental value of k_e (we assume that $k_p \simeq 0$) is located around $T = 10$ K. The Wiedemann–Franz–Lorenz law (with constant Lorenz number) along with the experimental result for σ_e [210] follows the experimental k_e, with better agreement at higher temperatures. Note that the temperature-

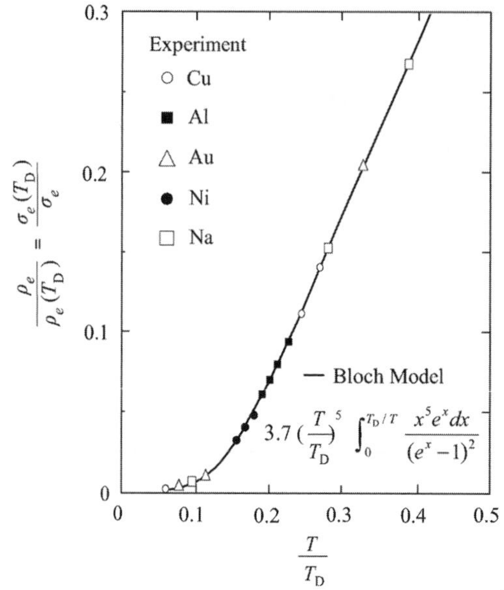

Figure 5.17. Temperature dependence of the electrical resistivity for five different metals as a function of the scaled temperature [13]. The Bloch model is also shown.

dependent Lorenz number of Figure 5.16(b) allows for a match with experimental results.

Also shown in Figure 5.18(b) is that the Bloch model of the electrical conductivity (5.195) overpredicts σ_e at low temperatures, and there is better agreement in the predicted k_e for higher temperatures.

In this example, (5.194) overpredicts k_e and (5.195) should be used at higher temperatures. Also note that the experimental results are for $k = k_p + k_e$. The relation $\sigma_e T$ vanishes as $T \to 0$. Both k_p and k_e are strongly reduced by impurity scattering at low temperatures (k_p is also reduced at low temperatures, due to reduction in $c_{v,p}$).

Because of the lack of data for k_p, we can only suggest the difference between the predicted k_e and the measured k at low temperatures is due to k_p.

The measured Seebeck coefficient of some metals, as a function of temperature, is shown in Figure 5.19. There is a general agreement with the temperature to the power-unity relation given in (5.193) for moderate high temperatures. Platinum has a peak in the positive region and then has a negative α_S at larger temperatures. Normal metals refers to those having $D_e(E_e)$ presented by (5.65), i.e., parabolic relation, and include Na, K, and Al. On the other hand, Pt is an abnormal metal. The diffusion of electrons from a hot to a cold region is influenced by the lattice vibration (phonon) and ion scattering and the electron DOS is therefore altered, resulting in a change of sign in α_S (which is expected to be negative for normal metals).

5.16.3 Semiconductors

The room-temperature ranges of the conduction electron density and the electrical conductivity of various materials are shown in Figure 5.20. Metals are marked with

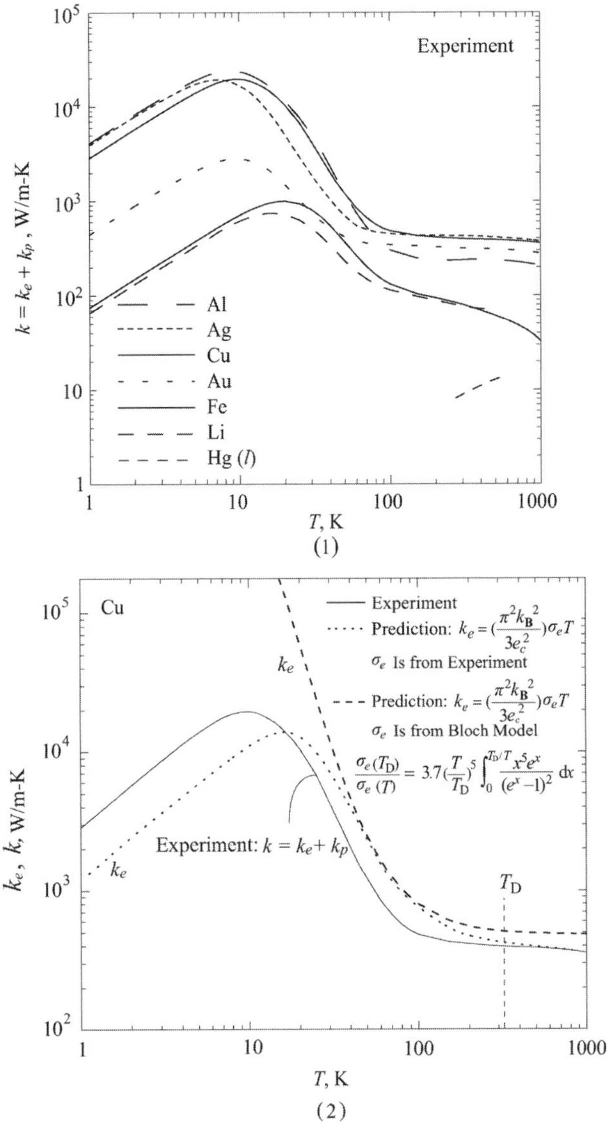

Figure 5.18. (a) Metal (mostly electronic) thermal conductivity for some metals. For Hg, the liquid conductivity is shown [149]. The conductivity for metals is dominated by electronic contribution. (b) Variations of the electronic and total thermal conductivities with respect to temperature, for Cu. In addition to the experimental results, the results obtained with the experimental results for the electrical conductivity along with the use of the Wiedemann–Franz–Lorenz law, are shown. Also shown is the prediction using the Bloch model of the electrical conductivity along with the Wiedemann–Franz–Lorenz law.

having $n_{e,c} > 10^{21}$ 1/cm^3. Doped semiconductors (n- and p-type semiconductors are defined in the Glossary) reach the electron density and conductivity of metals. Metals are also marked with having $\sigma_e > 10^3$ 1/Ω-cm.

The temperature dependences of the electrical conductivity of metals, semiconductors, and insulators are shown in Figure 5.21. Although both metals and

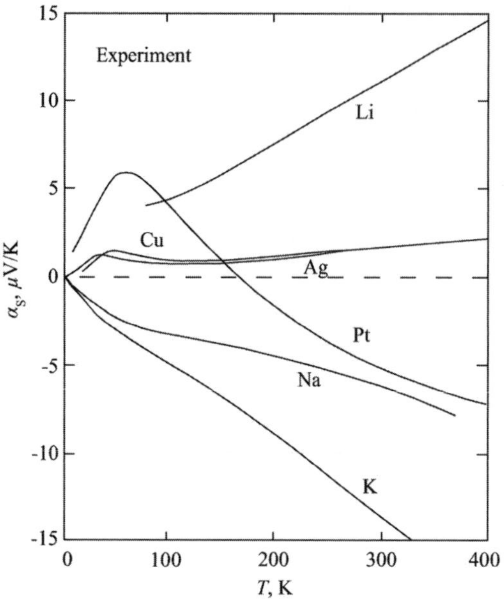

Figure 5.19. Variation of the Seebeck coefficient of some metals with respect to temperature [94].

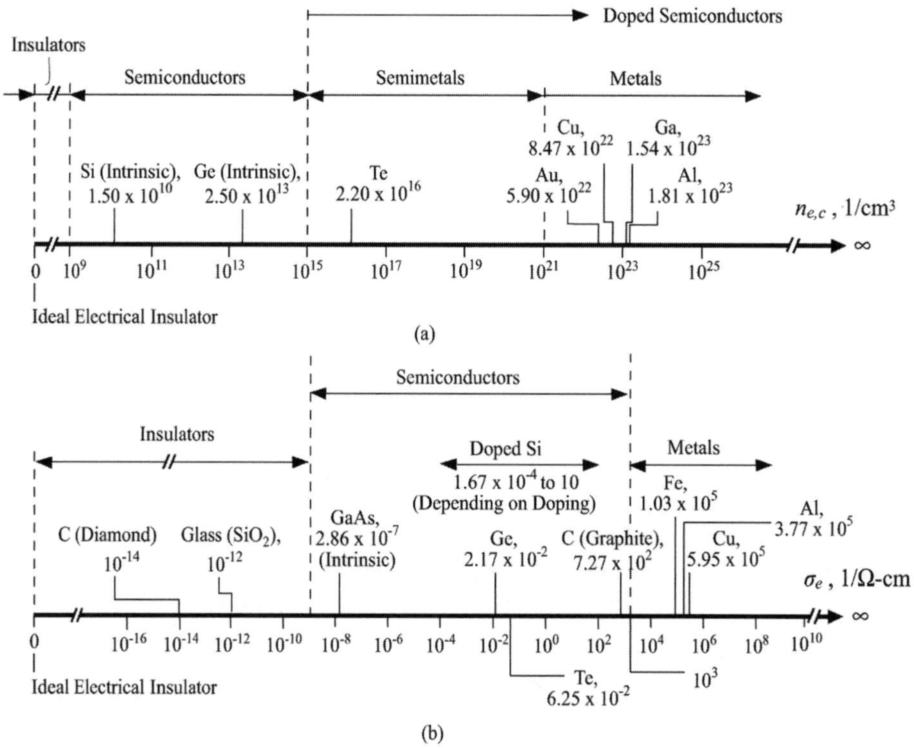

Figure 5.20. (a) Range of the measured conduction electron density $n_{e,c}$ of metals (few electrons per atom), semimetals, semiconductors, and insulators. (b) Range of the electrical resistivity of some metals, semiconductors, and insulators. The results are for $T = 300$ K.

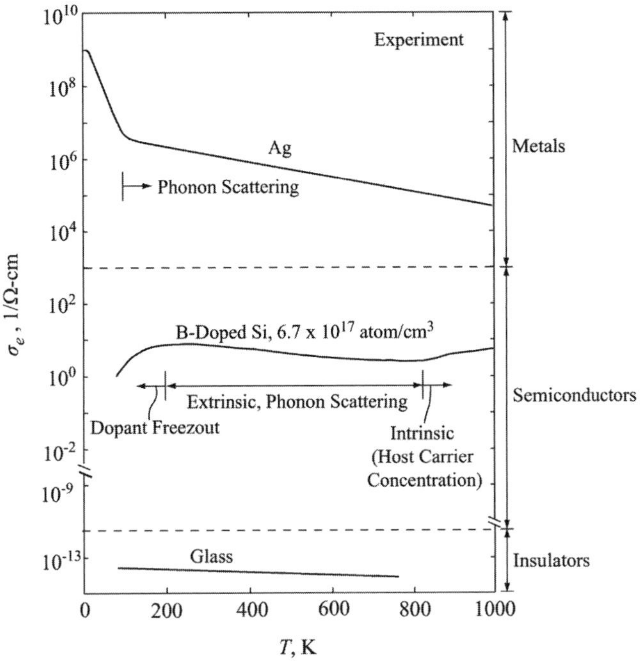

Figure 5.21. Variations of the measured electrical conductivity of a typical metal, semiconductor (intrinsic and doped), and insulator, with respect to temperature.

insulators have their electrical conductivities decreasing with temperature, those of intrinsic semiconductors increase with temperature. Dopants move the behavior of semiconductors toward metals, as expected.

As an example of the electrical and thermal properties of intrinsic and doped semiconductors, 5.22 to 5.25 show the hole (holes h^+ have a change of $+e_c$) concentration $n_{h,c}$, hole mobility μ_h, electrical conductivity σ_e, and thermal conductivity k of doped Si. The total electric conductivity is

$$\sigma_e = e_c(n_{h,c}\mu_h + n_{e,c}\mu_e). \tag{5.196}$$

Table 5.6 lists the effective bandgap at 0 and 300 K, and the electron and hole mobilities at 300 K, for some semiconductors. As mentioned in Section 5.4, for most semiconductors $\Delta E_{e,g}$ decreases with temperature. Exceptions include PbTe.

For low electric fields, the drift velocity–electric field relation is linear, and also, using the energy average relaxation time, the electron mobility is

$$\mu_e = \frac{e_c\langle\langle\tau_e\rangle\rangle}{m_{e,e}}, \tag{5.197}$$

where the contribution from all scatterers are combined using (4.111), i.e., the Matthiessen rule (series).

$$\frac{1}{\tau_e} = \sum_j \frac{1}{\tau_{e,j}} \quad \text{Matthiessen rule (series).} \tag{5.198}$$

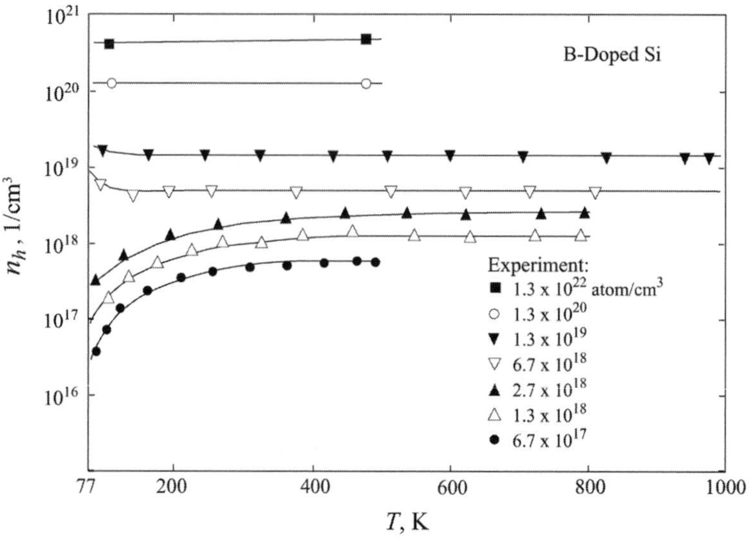

Figure 5.22. Variation of the hole concentration for intrinsic and B-doped Si with respect to temperature [221].

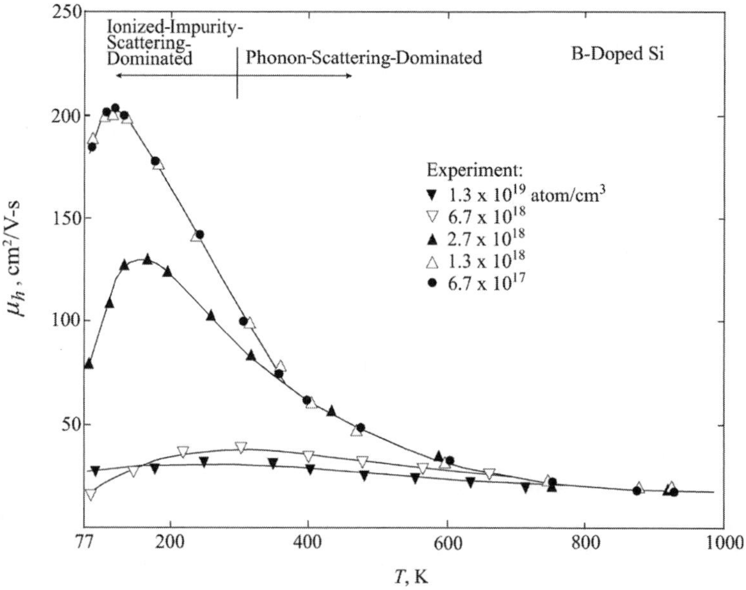

Figure 5.23. Variation of the hole mobility for B-doped Si with respect to temperature [221]. The low-temperature impurity-scattering and high-temperature acoustic-phonon-scattering are evident.

The energy-dependent relaxation time for scattering of electrons by impurity, acoustic phonon, optical phonon, etc., are given in Table 5.4. Then the averaged energy $\langle\langle \tau_e \rangle\rangle$ is found from (5.112).

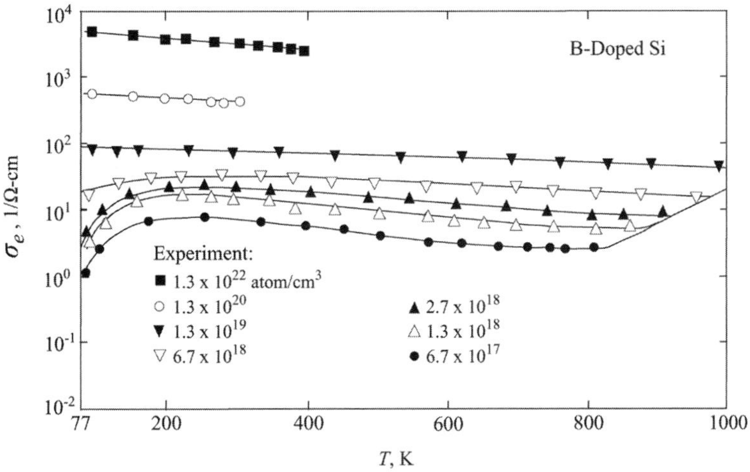

Figure 5.24. Variation of the electrical conductivity for B-doped Si with respect to temperature [221].

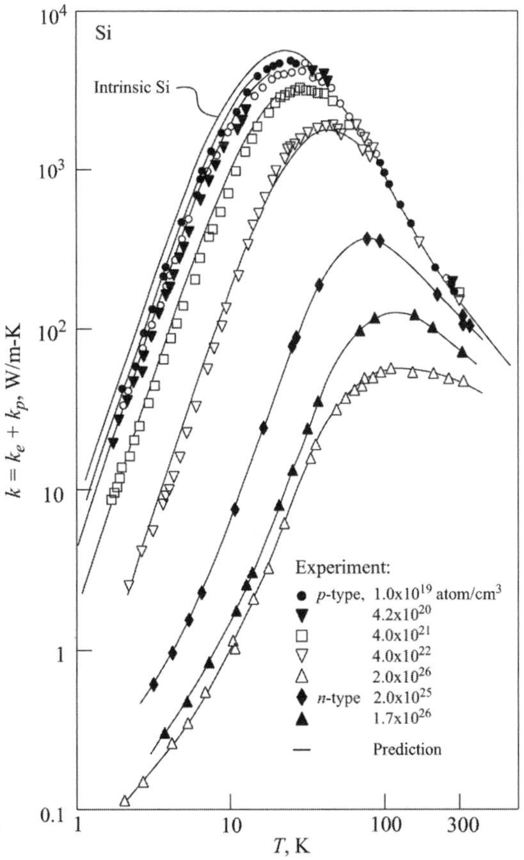

Figure 5.25. Predicted and measured variations of total (mostly phonon) thermal conductivity, for intrinsic and p- and n-type doped Si, with respect to low and intermediate temperatures [26]. The predictions are based on the Callaway SMRT phonon conductivity model. The contribution for the electronic thermal conductivity is very small.

Table 5.6. *The smallest or effective bandgap and the electron and the hole mobilities for some semiconductors. The bandgaps are at $T = 0$ K and $T = 300$ K [309]. For most the bandgap decreases with temperature, for PbTe this is reversed.*

	Bandgap $\Delta E_{e,g}$,eV		Mobility μ_e, μ_h at 300 K, cm^2/V-s	
Material	300 K	0 K	μ_e	μ_h
C	5.47	5.48	1800	1200
Ge	0.66	0.74	3900	1900
Si	1.12	1.17	1500	450
α-SiC	3.00	3.30	400	50
GaSb	0.72	0.81	5000	850
GaAs	1.42	1.52	8500	400
GaP	2.26	2.34	110	75
InSb	0.17	0.23	80000	1250
InAs	0.36	0.42	33000	460
InP	1.35	1.42	4600	150
CdTe	1.48	1.61	1050	100
PbTe	0.31	0.19	6000	4000

For impurity scattering, and using the energy that maximizes τ_{e-im}, the electron mobility becomes (power-law $\tau_{e\text{-}im}$, Table 5.4)

$$\tau_{e\text{-}im} = \tau_{e,\text{o}}(\frac{E_e}{k_\text{B}T})^{3/2}$$

$$\mu_e = \frac{82^{1/2}\epsilon_e(k_\text{B}T)^{3/2}}{\pi^{3/2}e_c^3 m_{e,e}^{1/2} n_{im}[\ln(1+\gamma_m^2) - \gamma_m^2/(1+\gamma_m^2)]}$$

$$\gamma_m \equiv \frac{(24m_{e,e}k_\text{B}T)^{1/2}}{\hbar\lambda} \quad \text{impurity-scattering-dominated mobility,} \qquad (5.199)$$

where λ is the screening length. This shows that the mobility increases with temperature (which dominates the low-temperature behavior) as nearly $T^{3/2}$.

The acoustic-phonon-dominated mobility is (using power-law $\tau_{e\text{-}p,\text{A}}$, Table 5.4)

$$\mu_e = \frac{2^{3/2}e_c\hbar^4\rho u_{p,\text{A}}^2}{3\varphi_{d,\text{A}}^2 m_{e,e}^{5/2}(k_\text{B}T)^{3/2}} \quad \text{acoustic-phonon-scattering-dominated mobility,}$$

$$\tau_{e\text{-}p,\text{A}} = \tau_{e,\text{o}}(\frac{E_e}{k_\text{B}T})^{-1/2}. \qquad (5.200)$$

This shows a $T^{-3/2}$ dependence that represents the high-temperature behavior. Similar results are expected for holes.

The total thermal conductivity is based on parallel transport [174]

$$k = k_e + k_p \quad \text{parallel transport.} \qquad (5.201)$$

The concept of the mean free path used in (4.109) and (5.155) for phonon and electron thermal conductivity by introducting a single velocity and a single mean free path for each carrier gives

$$k = \frac{1}{3}\rho c_{v,p} u_p \lambda_p + \frac{1}{3} n_{e,c} c_{v,e} u_{\mathrm{F}} \lambda_e. \qquad (5.202)$$

As we noted in Section 5.3, $c_{v,e}$ (here per free electron) is rather small at low temperatures (temperature dependent). The $c_{v,e}$ relation is given by (5.71), and the electrons are assumed to be at the lattice temperature, i.e., in thermal equilibrium with phonons.

Among the elements, for metals k_e dominates k, for semiconductors k_p dominates k. For compounds, e.g., in Bi_2Te_3, the contributions of k_e and k_p are nearly equal [83]. Generally, when both k_e and k_p are significant, k is not large.

Figure 5.22 shows the increase in hole concentration with an increase in B concentration in Si, and its dependence on temperature. At high dopant concentrations, the strong temperature dependence, characteristic of intrinsic semiconductors, disappears. In general, mobility of high-electrical-conductivity semiconductors (such as PbTe) is over 1000 cm^2/V-s. Table 5.6 lists the mobility (electron and hole) of some semiconductors at $T = 300$ K [309], where phonon scattering dominates.

Figure 5.23 shows the variation of the hole mobility in the B-doped Si. Note that, at high dopant concentrations, the mobility drops to its value for intrinsic Si. Also, for the moderate dopant concentrations, there is a notable increase in hole mobility at low temperatures, which is due to suppression of phonon scattering.

Figure 5.24 shows that the electrical conductivity of B-doped silicon increases substantially and becomes nearly temperature independent. In contrast, the intrinsic Si has a very strong temperature-dependent electrical conductivity, which is due mostly to the variation of the conduction electron density with temperature.

The thermal conductivity of doped Si is shown in Figure 5.25. As is clear, the phonon-dominated thermal conductivity decrease is due to the presence of dopants. It also shows that the phonon-scattering effect of these dopants surpasses any electronic contributions to thermal conductivity.

Variations of the measured Seebeck coefficient of intrinsic and doped Si, with respect to temperature, are shown in Figure 5.26. From (5.147), because of the $\ln(n_{e,\mathrm{F}}/n_{e,c})$ relation, at low and moderate temperatures, $n_{e,c}$ is low and α_{S} is rather large. The intrinsic Si has an n-type behavior at high temperatures; because of the increase in conduction electron density, the effect of doping disappears and α_{S} tends toward zero at high temperatures.

5.16.4 TE Figure of Merit Z_e

TE materials are used in TE power generation and in TE cooling. In addition to the Seebeck coefficient α_{S}, the electrical resistivity and thermal conductivity play

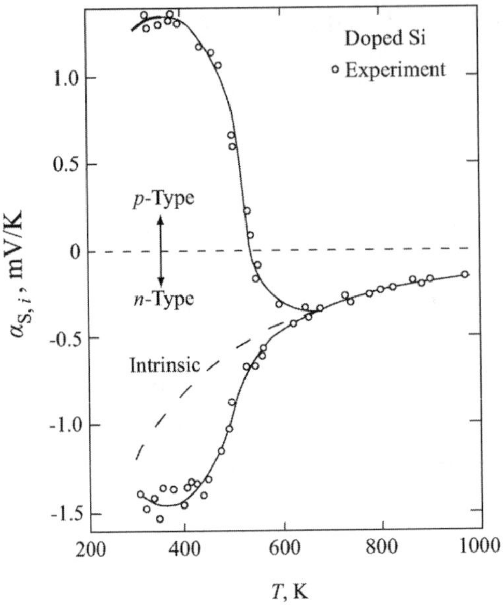

Figure 5.26. Variations of measured Seebeck coefficients, with intrinsic, and p- and n-type behaviors, for doped Si, as functions of temperature [121, 94]. The n-type generally uses As and P, and the p-type doped Si uses B.

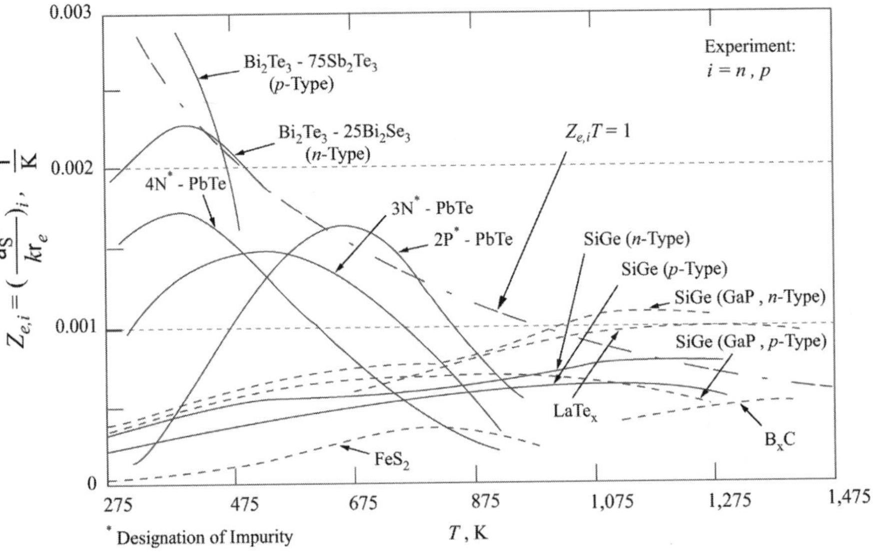

Figure 5.27. Variations of individual (p- or n-type) figures of merit $Z_{e,i}$ for selected TE (semiconductor) material as functions of temperature [288].

significant roles [174]. The TE figure of merit is defined as $Z_e = \alpha_S^2 \sigma_e / k = \alpha_S^2 / \rho_e k$. The product $\alpha_S^2 \sigma_e$ is called the thermoelectric power factor. Because of the high TE figure of merit Z_e, semiconductors have been the choice materials for thermoelectricity. Figure 5.27 shows the measured Z_e with respect to temperature for various high-Z_e materials.

The TE figure of merit can be expressed with (5.147) for α_S (5.146) for σ_e, (5.146) and (5.63) for $n_{e,c}$, (5.201) for k, and (5.150) to relate k_e and σ_e. Then this expression becomes (end-of-chapter problem)

$$Z_e T = \frac{\alpha_S^2 \sigma_e}{k_p + k_e} T = \frac{[\xi - (s + \frac{5}{2})]^2}{[2\frac{k_B^2 T}{e_c} AB \exp(\xi)]^{-1} + s + \frac{5}{2}},$$

$$\xi \equiv \frac{\Delta E_{e,g}}{k_B T}, \quad A \equiv \frac{\mu_e}{k_p}(\frac{m_{e,e}}{m_e})^{3/2}, \quad B \equiv (\frac{m_e k_B T}{2\pi\hbar^2})^{3/2}. \qquad (5.203)$$

The maximum in $Z_e T$ taken with respect to ξ gives (end-of-chapter problem)

$$\xi_o + 4(s + \frac{5}{2})\frac{k_B^2 T}{e_c} AB \exp(\xi_o) \equiv \xi_o + (s + \frac{5}{2})2\beta = s + \frac{1}{2}$$

$$\beta = \frac{k_B^2 \sigma_e T}{e_c^2 k_p} \quad \text{Chasmar-Stratton coefficient.} \qquad (5.204)$$

Here we have used the relation between the intrinsic carrier concentration and the effective band density (Section 5.7) to remove the Fermi energy and replace it with bandgap energy, i.e., $E_{e,c} - E_{e,v} = \Delta E_{e,g}$. The dimensionless coefficient $\beta \equiv 2k_B^2 TAB/e_c = k_B^2 \sigma_e T/e_c^2 k_p$ [note similarity to Wiedemann–Franz–Lorenz law (5.150), except here we have k_p instead of k_e] is referred to as the Chasmar-Stratton coefficient and is < 0.5 for all known materials. Useful TE materials have $\beta \exp(\xi)$ between 0.05 and 0.2, with α_S about $\pm$ 200 μV/K [123].

Figures 5.28(a) and (b) show the crystal structure of Bi_2Te_3. Figure 5.28(c) shows the band structure for Bi_2Te_3 predicted with the tight-binding approximation [201]. The band edges are not located along the high-symmetry lines, but are in the reflection planes. Figures 5.16.4(a) and (b) show the variation of the measured electrical conductivity and Seebeck coefficient with respect to temperature for Pb-doped Bi_2Te_3 [270].

Note that even the crystal with zero Pb concentration has an electrical conductivity that behaves like that of metals (nondegenerate) because of the inherent defects. When comparing theoretical predictions with experimental results (Section 5.16.1), account has to be made of the inherent defects, through specification (prescription) of carrier concentrations. We will discuss this in Section 5.17.3. As Pb concentration increases, σ_e increases and α_S decreases. Similar structures include Sb_2Te_3. As the expected, the phonon conductivity is anisotropic. The atomic spacings are also rather large, resulting in a small total phonon and electron conductivity. The optimal figure of merit is discussed in [288] and in Section 5.17.3.

It is also customary to use the dimensionless TE figure of merit $Z_e T$. The best TE materials (semiconductors) have $Z_e T$ of unity or higher (Figure 5.27).†

† For metals, using $k_e = k$, along with (5.194), and (5.193) with $s = 0$, we have

$$Z_e T = \frac{3\pi^2}{4\xi^2}, \quad \xi = \frac{E_F}{k_B T}.$$

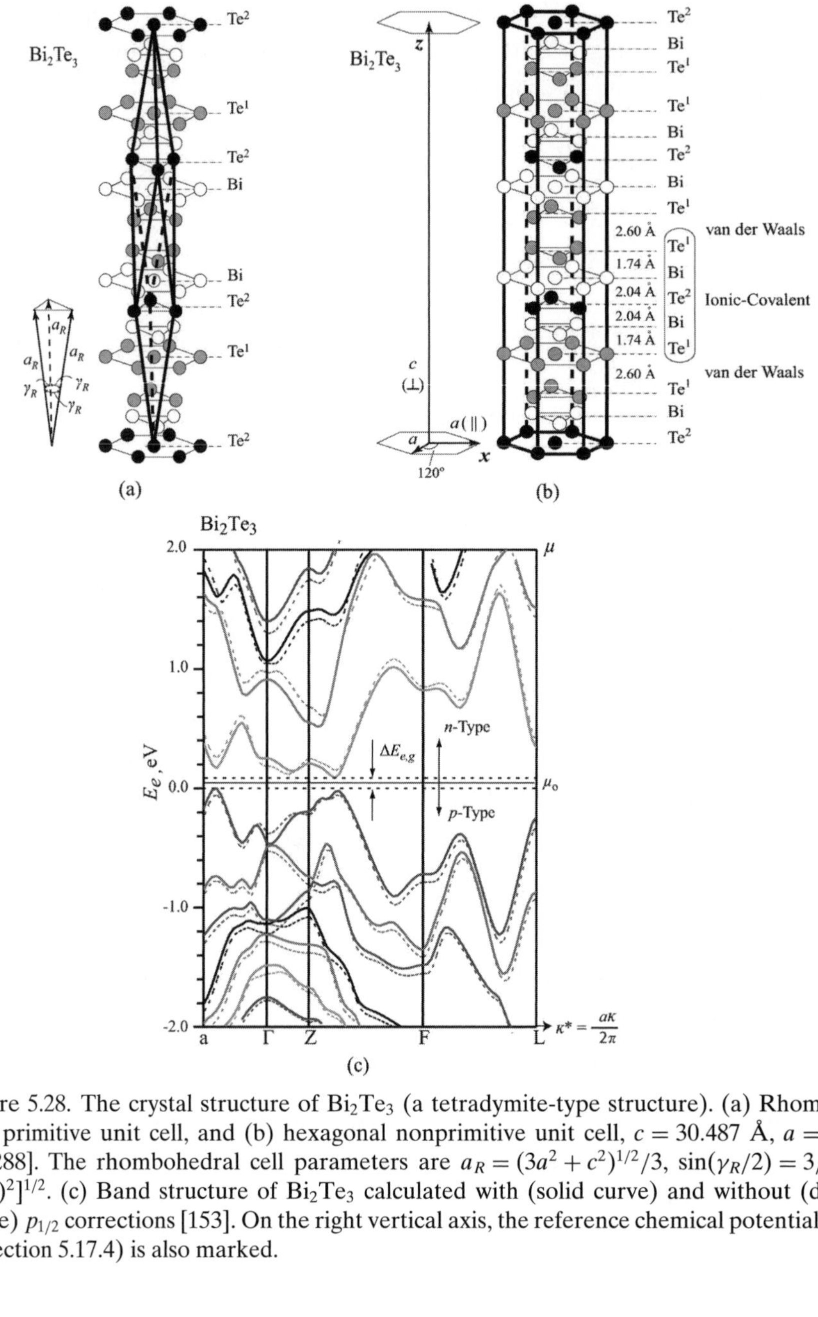

Figure 5.28. The crystal structure of Bi_2Te_3 (a tetradymite-type structure). (a) Rhombohedral primitive unit cell, and (b) hexagonal nonprimitive unit cell, $c = 30.487$ Å, $a = 4.384$ Å [288]. The rhombohedral cell parameters are $a_R = (3a^2 + c^2)^{1/2}/3$, $\sin(\gamma_R/2) = 3/2[3 + (c/a)^2]^{1/2}$. (c) Band structure of Bi_2Te_3 calculated with (solid curve) and without (dashed curve) $p_{1/2}$ corrections [153]. On the right vertical axis, the reference chemical potential (used in Section 5.17.4) is also marked.

The Peltier cooling/heating rate $\dot{S}_{e\text{-}p}$ for a p-type material connected to an n-type material through an electrical conductor (metal) with a current $J_e = j_e A$, is

$$\dot{S}_{e\text{-}p} = \dot{S}_{e,\mathrm{P}} = (\alpha_{\mathrm{S},p} - \alpha_{\mathrm{S},n}) J_e T, \qquad (5.205)$$

where T is the junction temperature.

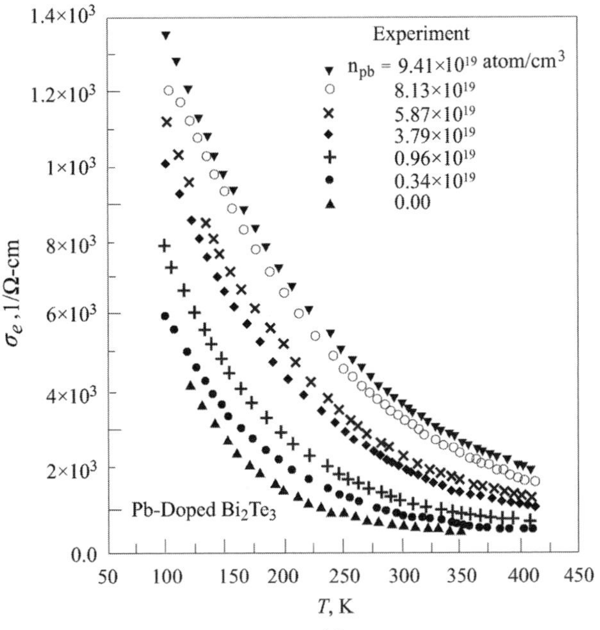

(a)

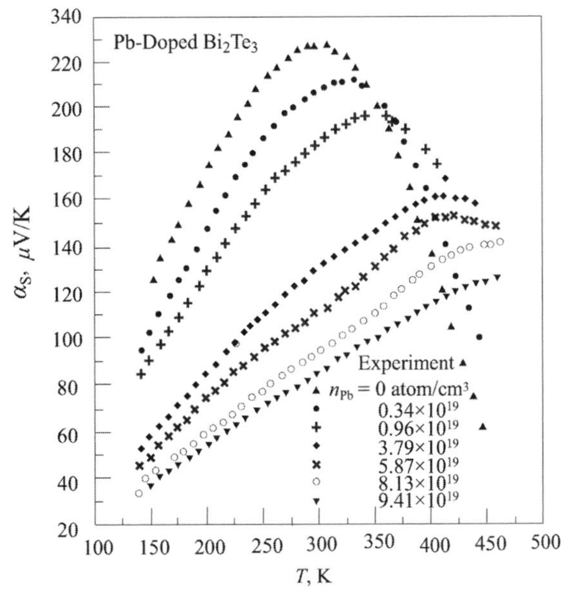

(b)

Figure 5.29. (a) Temperature dependences of the measured electrical conductivity (along c) of the Pb-doped Bi_2Te_3 crystals [270]. Note the metal behavior (monotonic decrease with increase in temperature), even for the pure Bi_2Te_3. This is due to lattice defects. (b) Temperature dependence of the measured Seebeck coefficient (along c) of the Pb-doped Bi_2Te_3 crystals [270].

5.17 *Ab Initio* Computation of TE Transport Property Tensors

Using band structures from *ab initio* calculation, we can also calculate TE transport properties on the basis of appropriate electron-scattering models. Most such calculations are based on the BTE and relaxation-time approximations.

5.17.1 TE Transport Tensors and Variable Chemical Potential

The TE transport properties are derived from the BTE as introduced in Section 5.12, with the relaxation-time approximation. Here we do not assume nondegeneracy and include all electron bands. The most general form of the relations [(5.124) to (5.127)] for TE properties are [153, 223]

$$\boldsymbol{A}_{S,j} = 2\int \frac{\mathrm{d}\boldsymbol{\kappa}}{(2\pi)^3}\boldsymbol{u}_{e,j,\boldsymbol{\kappa}}\delta f_{e,j}(\boldsymbol{\kappa},T)(E_{e,j}-\mu)/(T\boldsymbol{\Sigma}_{e,j})$$

$$\boldsymbol{\Sigma}_{e,j} = 2q_j\int \frac{\mathrm{d}\boldsymbol{\kappa}}{(2\pi)^3}\boldsymbol{u}_{e,j,\boldsymbol{\kappa}}\delta f_{e,j}(\boldsymbol{\kappa},T)$$

$$\boldsymbol{K}^{\circ}_{e,j} = \frac{2}{q_jT}\int \frac{\mathrm{d}\boldsymbol{\kappa}}{(2\pi)^3}\boldsymbol{u}_{e,j,\boldsymbol{\kappa}}\delta f_{e,j}(\boldsymbol{\kappa},T)(E_{e,j}-\mu)^2$$

$$\boldsymbol{K}_{e,j} = \boldsymbol{K}^{\circ}_{e,j} - [2\int \frac{\mathrm{d}\boldsymbol{\kappa}}{(2\pi)^3}\boldsymbol{u}_{e,j,\boldsymbol{\kappa}}\delta f_{e,j}(\boldsymbol{\kappa},T)(E_{e,j}-\mu)]^2/(T\boldsymbol{\Sigma}_{e,j})$$

$$\delta f_{e,j}(\boldsymbol{\kappa},T) \equiv q_j\boldsymbol{u}_{e,j,\boldsymbol{\kappa}}\tau_e[E_{e,j}(\boldsymbol{\kappa})](\frac{\partial f^{\mathrm{o}}_{e,j}}{\partial E_{e,j}})|_{E_{e,j}=\mu}, \qquad (5.206)$$

where $\boldsymbol{A}_S$ is the Seebeck coefficient tensor, $\boldsymbol{\Sigma}_e$ is the electrical conductivity tensor, $\boldsymbol{K}_e$ is the electrical thermal conductivity tensor, j is the band index, q_j is the charge of carrier, $\boldsymbol{\kappa}$ is the wave vector, $\boldsymbol{u}_{e,j,\boldsymbol{\kappa}}$ is the group velocity vector (in the jth band and wave vector $\boldsymbol{\kappa}$), $f^{\mathrm{o}}_e(\boldsymbol{\kappa}) = [e^{E_{e,j}(\boldsymbol{\kappa})-\mu}+1]^{-1}$ is the equilibrium Fermi–Dirac distribution function (note the distribution function is band dependent), μ is the chemical potential, and E_e is the total energy of electrons. Accordingly, the band structure $E_{e,j}(\boldsymbol{\kappa})$, chemical potential μ and relaxation times τ_e are required inputs. (5.206) allows for the investigation of contributions from each band and even each mode of electrons (or holes). The Seebeck coefficient involves $(E_e-\mu)\partial f^{\mathrm{o}}_e/\partial E_e$ and is very sensitive to the magnitude of chemical potential [see Figure 5.30(a)], which in turn is affected by the composition uncertainties (Figure 5.14). The intrinsic chemical potential can be directly obtained based on the intrinsic electron density and DOS of electrons from *ab initio* calculations. However, in real crystals, defects or dopants will also contribute some carriers (hole or electron), resulting in a different chemical potential. The total electron density includes both the extrinsic electron density (can be obtained from the defect density) and the intrinsic electron density. With the assumption that the effects of defects or dopants on the band structure and DOS are

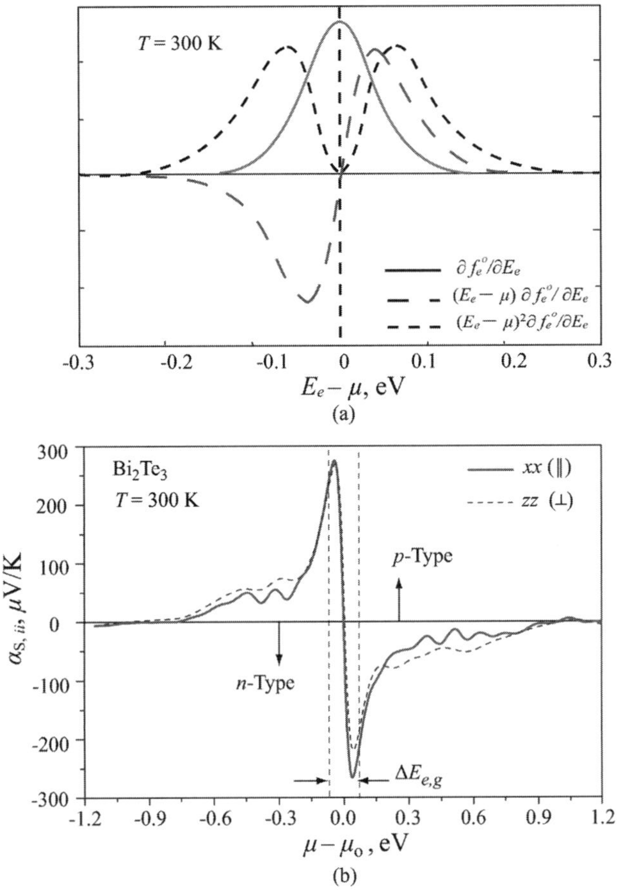

Figure 5.30. (a) Variation of the integrand factor $\partial f_e^\circ/\partial E_e$, $(E_e - \mu)\partial f_e^\circ/\partial E_e$, and $(E_e - \mu)^2\partial f_e^\circ/\partial E_e$, in (5.206), with respect to $E_e - \mu$ [222]. (b) The *ab initio* calculated (BoltzTraP) variation of the directional Seebeck coefficient elements of Bi_2Te_3 with respect to the chemical potential. Only the xx and zz [Figure 5.28(b)] components of the tensor are shown [222].

negligible, the real chemical potential is generally fitted from experiments instead of the intrinsic electron density.

5.17.2 Introduction to BoltzTraP

BoltzTraP [222] is a program for calculating the semiclassical transport coefficients based on the BTE (5.206). To calculate the TE transport coefficients by use of the BTE, the group velocity of electrons should be known. As the interpolated bands pass exactly through the calculated band-energies, the precision of the direct differential approach is limited by possible band crossings where the band derivatives will be calculated incorrectly. BoltzTraP relies on a Fourier expansion of the band-energies where the space group symmetry is maintained by the use of star functions

$$E_{e,j}(\boldsymbol{\kappa}) = \sum_{\boldsymbol{R}} c_{\boldsymbol{R}j} S_{\boldsymbol{R}}(\boldsymbol{\kappa}), \quad S_{\boldsymbol{R}}(\boldsymbol{\kappa}) = \frac{1}{N_p} \sum_{\Lambda} e^{i\boldsymbol{\kappa}\cdot\Lambda \boldsymbol{R}}, \tag{5.207}$$

where $\boldsymbol{R}$ is a lattice vector ($\boldsymbol{x}_o$) and $\{\Lambda\}$ are the N_p point group rotations. The expansion coefficients are given as

$$\begin{aligned} c_{\boldsymbol{x}j} &= E_{e,i}(\boldsymbol{\kappa}_N) - \sum_{\boldsymbol{R}\neq 0} c_{\boldsymbol{R}i} e^{i\boldsymbol{\kappa}\cdot\boldsymbol{R}}, \quad \boldsymbol{R} = 0 \\ &= \rho_R^{-1} \sum_{\boldsymbol{\kappa}\neq\boldsymbol{\kappa}_N} \lambda_{\boldsymbol{R}} (S^*_{\boldsymbol{R}i} - S^*_{0i}), \quad \boldsymbol{R} \neq 0, \end{aligned} \tag{5.208}$$

where $\rho_{\boldsymbol{R}}$ is the roughness function $\rho_{\boldsymbol{R}} = [1 - \frac{3}{4}(|\boldsymbol{R}|/|\boldsymbol{R}_{min}|)^2]^2 + \frac{3}{4}(|\boldsymbol{R}|/|\boldsymbol{R}_{min}|)^6$ ($\boldsymbol{R}_{min}$ is the smallest nonzero lattice vector) and $\lambda_{\boldsymbol{R}}$ are calculated by solving

$$\begin{aligned} \Delta E_{e,j}(\boldsymbol{\kappa}) &= E_{e,j}(\boldsymbol{\kappa}) - E_{e,i}(\boldsymbol{\kappa}_N) \\ &= \sum_{\boldsymbol{\kappa}'\neq\boldsymbol{\kappa}_N} \sum_{\boldsymbol{R}\neq 0} \frac{[S_{\boldsymbol{R}}(\boldsymbol{\kappa}) - S_{\boldsymbol{R}}(\boldsymbol{\kappa}_N)][S^*_{\boldsymbol{R}}(\boldsymbol{\kappa}') - S^*_{\boldsymbol{R}}(\boldsymbol{\kappa}_N)]}{\rho_{\boldsymbol{R}}} \lambda^j_{\boldsymbol{R}}. \end{aligned} \tag{5.209}$$

In this way, the extrapolated energies can be exactly equal to the calculated band-energies and the necessary derivatives for the BTE can be straightforwardly and accurately calculated. This approach has been tested by comparison with the recent calculations in which the group velocity is from the momentum matrix elements, which are in turn calculated directly from the wave functions, and good agreements are found [222].

BoltzTraP uses the BTE [as shown in (5.206)] to calculate the TE transport properties. The relaxation time τ_e, in principle, is dependent on both the band index and the $\boldsymbol{\kappa}$ direction. However, in BoltzTraP, the relaxation time τ_e is assumed to be a constant, so BoltzTraP outputs only σ_e/τ_e. For use of a different relaxation-time model, for example, the Kane model, one needs to modify the code (subroutine FERMIINTEGRALS).

The code of BoltzTraP is interfaced to the band-structure code WIEN2k [33], but can also be easily interfaced to any other band-structure codes. Table 5.7 shows the input parameters in the input file case.intrans, using Bi_2Te_3 as an example.

To run BoltzTrap, a user must enter the directory containing the band-structure file [for WIEN2k, it is case.energy file or case.energyso file (if the spin–orbit coupling is specified)]. The user can run BoltzTrap by the following command:

"path to BoltzTrap"/SRC BoltzTrap1/x trans BoltzTraP

The traces of the tensors calculated in (5.206) are written as functions of μ and T in the output file called case.trace. Users must write their own program to extract the information they need. Figure 5.30(b) shows the variation of the Seebeck coefficient with respect to the chemical potential. It can be seen that the Seebeck coefficient is very sensitive to the change of the chemical potential.

Table 5.7. *Input parameters for BoltzTraP.*

WIEN	bandstyle (format of band structure to be input)
0 0 0 0.0	iskip (not presently used), idebug (output level), setgap (logic switch for bandgap manipulation), shiftgap
0.34517 0.0005 0.3 78.0	eferm (Fermilevel,in Ry), delatae (energygrid, in Ry), ecut (energy span around Fermilevel), number of valence electrons
CALC	CALC (calculate expansion coeff), NOCALC (read from file)
5	lpfac (number of lattice-points per κ-point)
BOLTZ	run mode (only BOLTZ is supported)
0.15	efcut (energy range of chemical potential around Fermilevel)
600. 50.	Tmax (max temperature for integrations), deltat (temperature step)
0.0	(energy range of bands given individual DOS output)

The tensors of $\boldsymbol{\Sigma}_e/\tau_e$ and $\boldsymbol{A}_S$ are written in the file called case.condtens; the tensors of the Hall coefficient are written in the file called case.halltens.

5.17.3 Relaxation Times Based on Kane Band Model

The relaxation-time models simplify the calculation of BTE, but condense all the complexities into the relaxation time τ_e. In principle, the scattering relaxation time can be obtained with the FGR and the perturbation theory. The scattering of electrons is related to the perturbation of the Hamiltonian for an electron, which is

$$\mathrm{H} = -\frac{\hbar^2}{2m_e}\nabla^2 + \frac{e_c^2}{4\pi\epsilon_\circ}\int \frac{n_e g_e(\boldsymbol{r}')}{|\boldsymbol{r}-\boldsymbol{r}'|}\mathrm{d}\boldsymbol{r}' + \varphi_{ec} + \varphi_{ext} + \mathrm{H}' \tag{5.210}$$

$$\begin{aligned}\mathrm{H}' &= \mathrm{H}'_{e\text{-}p,\mathrm{A}} + \mathrm{H}'_{e\text{-}p,\mathrm{O}} + \mathrm{H}'_{e\text{-}p,\mathrm{PO}} + \mathrm{H}'_{e\text{-}v,d} + \mathrm{H}'_{e\text{-}v,\mathrm{C}} + \cdots \\ &= \varphi_{e\text{-}p,\mathrm{A}}\frac{\partial \boldsymbol{d}}{\partial \boldsymbol{r}} + \varphi'_{e\text{-}p,\mathrm{O}}\boldsymbol{d} - \frac{e_c q_e}{V_\circ}\sum_{\kappa_p}\frac{\kappa_p}{\kappa_p^2+\lambda^{-2}}(i\,Q_{\kappa_p}e^{i\boldsymbol{\kappa}_p\cdot\boldsymbol{r}}) + \varphi_{v,c} + \frac{Ze_c^2}{4\pi\epsilon r}e^{-r/\lambda} + \cdots\end{aligned} \tag{5.211}$$

where $g_e(\boldsymbol{r})$ is the electron radial distribution function, φ_{ec} is the exchange-correlation energy, φ_{ext} is the external potential excluding the perturbation, H' is the perturbation Hamiltonian that is due to scatterings, $\mathrm{H}'_{e\text{-}p,\mathrm{A}}$, $\mathrm{H}'_{e\text{-}p,\mathrm{O}}$, $\mathrm{H}'_{e\text{-}p,\mathrm{PO}}$, $\mathrm{H}'_{e\text{-}v,d}$, and $\mathrm{H}'_{e\text{-}v,\mathrm{C}}$ are the perturbation Hamiltonians for the acoustic-phonon scattering, nonpolar optical-phonon scattering, polar optical-phonon scattering, short-range scattering by impurity, and scattering by Coulomb potential, respectively. Here $\varphi_{e\text{-}p,\mathrm{A}}$ and $\varphi'_{e\text{-}p,\mathrm{O}}$ are the deformation potentials for the acoustic and optical phonons, q_e is the effective charge, $\varphi_{v,c}$ is the scattering potential of impurity. $\boldsymbol{d} = N^{-1/2}\sum_{\kappa_p} Q_{\kappa_p}\boldsymbol{s}_{\kappa_p}\exp(i\boldsymbol{\kappa}_p\cdot\boldsymbol{r})$ (N is the number of unit cells and $\boldsymbol{s}_{\kappa_p}$ is the polarization vector) is the normal coordinate form of lattice displacement, $\boldsymbol{\kappa}_p$ is the phonon wave vector, Q_{κ_p} is the normal coordinate, λ is the screening length, and Ze_c is the effective charge of impurity.

The electron relaxation time τ_e for a mode $\boldsymbol{\kappa}$ can be represented as [310]

$$\frac{1}{\tau_e(\boldsymbol{\kappa})} = \sum_i \frac{1}{\tau_{e,i}(\boldsymbol{\kappa})} = \int \frac{d\boldsymbol{\kappa}'}{(2\pi)^3} \dot{\gamma}_{\boldsymbol{\kappa},\boldsymbol{\kappa}',i} (1 - \frac{f'_{\kappa'}}{f'_{\kappa}} \frac{f^\circ_{\kappa}}{f^\circ_{\kappa'}}), \tag{5.212}$$

where f°_κ is the carrier equilibrium distribution , $f'_\kappa = f_\kappa - f^\circ_\kappa$ is the perturbation of the distribution, $\dot{\gamma}_{\boldsymbol{\kappa},\boldsymbol{\kappa}',i}$ is the transition rate from state $\boldsymbol{\kappa}$ to $\boldsymbol{\kappa}'$ by the ith scattering, which can in turn be given by the FGR as [310]

$$\dot{\gamma}_{\boldsymbol{\kappa},\boldsymbol{\kappa}',i} = \frac{2\pi}{\hbar} \delta[E_e - E_e(\boldsymbol{\kappa}')] |M_{\boldsymbol{\kappa},\boldsymbol{\kappa}',i}|^2$$

$$M_{\boldsymbol{\kappa},\boldsymbol{\kappa}',i} = \langle \boldsymbol{\kappa}' | \mathrm{H}'_i | \boldsymbol{\kappa} \rangle = \int \psi^\dagger(\boldsymbol{\kappa}', \boldsymbol{r}) \mathrm{H}'_i \psi(\boldsymbol{\kappa}, \boldsymbol{r}) d\boldsymbol{r}, \tag{5.213}$$

where H'_i is the perturbation Hamiltonian for the scattering mechanism i, and $\psi(\boldsymbol{\kappa}, \boldsymbol{r})$ is the wave function for mode $\boldsymbol{\kappa}$. The Bloch wave function corresponding to the electron wave vector $\boldsymbol{\kappa}$ can be written as

$$\psi(\boldsymbol{\kappa}', \boldsymbol{r}) = \frac{1}{V^{1/2}} \sum_o C'_J \sum_{\boldsymbol{g}} C^{\kappa}_{\boldsymbol{g}} e^{i(\boldsymbol{\kappa}+\boldsymbol{g})\cdot\boldsymbol{r}}, \tag{5.214}$$

where V is the volume, C and C' are coefficients, and the subscripts J and $\boldsymbol{g}$ denote the different orbitals and reciprocal lattice vectors. Therefore,

$$M_{\boldsymbol{\kappa},\boldsymbol{\kappa}'} = \frac{1}{V^{1/2}} \sum_{J'} \sum_{J} C'^{\dagger}_{J'} C'_J \sum_{\boldsymbol{g}'} \sum_{\boldsymbol{g}} e^{i(-\boldsymbol{\kappa}'-\boldsymbol{g}')\cdot\boldsymbol{r}} \mathrm{H}' e^{i(\boldsymbol{\kappa}+\boldsymbol{g})\cdot\boldsymbol{r}}. \tag{5.215}$$

If the wave function and the perturbation potential can be obtained from the first-principle calculation, $\tau_{e,i}$ can be directly determined. This calculation is very challenging, and here we just introduce an analytical relaxation-time model, which is also based on the FGR and incorporates the nonparabolic Kane model for the energy dispersion.

Many semiconductors with narrow bandgaps, e.g., lead chalcogenides, exhibit significant nonparabolicity of their energy bands. The two-band model is one of the models proposed to account for this for theoretical calculations of transport coefficients. The main features of the two-band model are as follows: the conduction- and valence-band extrema are located at the same $\boldsymbol{\kappa}$ point of the $\boldsymbol{\kappa}$-space, for example, Γ point, with the energy separation from other bands at this κ point being essentially larger than the main energy gap; at the same time, the momentum operators have nonzero matrix elements between the states corresponding to the extremal points. The energy dispersion for each valley is of the form (5.29)

$$E_e(1 + \frac{E_e}{\Delta E_{e,g}}) = \frac{\hbar^2 \kappa_{\mathrm{T}}^2}{m_{e,e,\mathrm{T}}} + \frac{\hbar^2 \kappa_{\mathrm{L}}^2}{2m_{e,e,\mathrm{L}}}, \tag{5.216}$$

where $m_{e,e,\mathrm{L}}$ and $m_{e,e,\mathrm{T}}$ are longitudinal and transverse components of the effective mass tensor near the extremum, as given in (5.48), and κ_{L} and κ_{T} are longitudinal and transverse components of the wave vector. The DOS average effective mass is defined as

$$m_{e,e,\circ} = (m_{e,e,\mathrm{L}} m_{e,e,\mathrm{T}}^2)^{1/3} \quad \text{for conduction, direct band.} \tag{5.217}$$

Then the DOS D_e (5.65) becomes [using (5.49) to represent (5.216)]

$$D_e = \frac{2^{1/2} m_{e,e,\circ}^{3/2}}{\pi^2 \hbar^3} E_e^{1/2} \left(1 + \frac{E_e}{\Delta E_{e,g}}\right)^{1/2} \left(1 + 2\frac{E_e}{\Delta E_{e,g}}\right). \tag{5.218}$$

In the framework of the Kane model, five electron-scattering mechanisms are normally considered, namely because of the deformation potential of the acoustic phonons $\tau_{e\text{-}p,\mathrm{A}}$, deformation potential of the optical phonons $\tau_{e\text{-}p,\mathrm{O}}$, polar scattering by the optical phonons $\tau_{e\text{-}p,\mathrm{PO}}$, short-range deformation potential of vacancies $\tau_{e\text{-}v,\mathrm{d}}$, and Coulomb potential of vacancies $\tau_{e\text{-}v,\mathrm{C}}$. According to (5.114), the total scattering relaxation time is expressed as

$$\frac{1}{\tau_e} = \frac{1}{\tau_{e\text{-}p,\mathrm{A}}} + \frac{1}{\tau_{e\text{-}p,\mathrm{O}}} + \frac{1}{\tau_{e\text{-}p,\mathrm{PO}}} + \frac{1}{\tau_{e\text{-}v,d}} + \frac{1}{\tau_{e\text{-}v,\mathrm{C}}}. \tag{5.219}$$

At low temperatures, charge carriers are scattered mainly by the charged vacancy. At low carrier densities, scattering by the Coulomb potential of the vacancies dominates, whereas for high carrier densities, the Coulomb potential is screened by the carriers and scattering by the short-range potential of vacancies dominates. As the temperature increases, scattering by thermal phonons becomes more and more important. When the temperature is above room temperature, scattering by acoustic phonons and optical phonons normally makes the main contribution to the total relaxation time [29].

Within the Kane model, the expressions for different scattering mechanisms are given as follows [29, 153, 278, 356].

(A) *Scattering by Deformation Potential of Acoustic Phonons* $\tau_{e\text{-}p,\mathrm{A}}$

The relaxation time for electrons dispersed on the deformational potential of acoustic phonons, when the Kane model of dispersion is used, can be given as

$$\tau_{e\text{-}p,\mathrm{A}} = \frac{(\tau_{e\text{-}p,\mathrm{A}})_\circ \left(E_e + \frac{E_e^2}{\Delta E_{e,g}}\right)^{-1/2}}{\left(1 + 2\frac{E_e}{\Delta E_{e,g}}\right)[(1-A)^2 - B]}$$

$$A \equiv \frac{\frac{E_e}{\Delta E_{e,g}}(1 - a_\mathrm{A})}{\left(1 + 2\frac{E_e}{\Delta E_{e,g}}\right)}, \quad a_\mathrm{A} = \frac{\varphi_{e\text{-}p,\mathrm{A},v}}{\varphi_{e\text{-}p,\mathrm{A},c}}$$

$$B \equiv \frac{8\frac{E_e}{\Delta E_{e,g}}\left(1 + \frac{E_e}{\Delta E_{e,g}}\right) a_\mathrm{A}}{3\left(1 + 2\frac{E_e}{\Delta E_{e,g}}\right)^2}$$

$$(\tau_{e\text{-}p,\mathrm{A}})_\circ \equiv \frac{2\pi\hbar^4 c_l}{\varphi_{e\text{-}p,\mathrm{A},c}^2 k_\mathrm{B} T (2m_{e,e,\circ})^{3/2}}, \tag{5.220}$$

where $\Delta E_{e,g}$ is the band energy gap, $\varphi_{e\text{-}p,\mathrm{A},c}$ and $\varphi_{e\text{-}p,\mathrm{A},v}$ are the acoustic deformation potential coupling constants for the conduction and valence band, c_l is a combination of elastic constants, a_{A} is the ratio of the acoustic deformation potential coupling constants for the valence and conduction bands, and $m_{e,e,\circ}$ is the DOS effective mass for a single ellipsoid.

(B) *Scattering by Deformation Potential of Optical Phonons* $\tau_{e\text{-}p,\mathrm{O}}$

$$\tau_{e\text{-}p,\mathrm{O}} = \frac{(\tau_{e\text{-}p,\mathrm{O}})_\circ (E_e + \frac{E_e^2}{\Delta E_{e,g}})^{-1/2}}{(1+2\frac{E_e}{\Delta E_{e,g}})[(1-A)^2 - B]}$$

$$A \equiv \frac{\frac{E_e}{\Delta E_{e,g}}(1-a_{\mathrm{O}})}{(1+2\frac{E_e}{\Delta E_{e,g}})}, \quad a_{\mathrm{O}} = \frac{\varphi_{e\text{-}p,\mathrm{O},v}}{\varphi_{e\text{-}p,\mathrm{O},c}}$$

$$B \equiv \frac{8\frac{E_e}{\Delta E_{e,g}}(1+E_e/\Delta E_{e,g})a_{\mathrm{O}}}{3(1+2\frac{E_e}{\Delta E_{e,g}})^2}$$

$$(\tau_{e\text{-}p,\mathrm{O}})_\circ \equiv \frac{2\hbar^2 a_R^2 \rho(\hbar\omega_{p,\mathrm{O}})^2}{\pi\varphi_{e\text{-}p,\mathrm{O},c}^2 k_{\mathrm{B}} T(2m_{e,e,\circ})^{3/2}}, \tag{5.221}$$

where a_R is lattice constant, ρ is the density, $\omega_{p,\mathrm{O}}$ is the frequency of the optical phonons, and $\varphi_{e\text{-}p,\mathrm{O},c}$ and $\varphi_{e\text{-}p,\mathrm{O},v}$ are the optical deformation potential coupling constants for conduction and valence bands. Equation (5.221) is similar to (5.185).

(C) *Scattering by Polar-Optical Phonons* $\tau_{e\text{-}p,\mathrm{PO}}$

In a simple isotropic parabolic model, the relaxation time for polar LO phonons has the form [278]

$$\tau_{e\text{-}p,\mathrm{PO}}^{-1} \sim \frac{1}{u_e\kappa^2}\int_0^{2\kappa} \kappa_p \mathrm{d}\kappa_p, \tag{5.222}$$

where u_e is the velocity of electrons. When the integral takes into account all phonons, we have

$$\tau_{e\text{-}p,\mathrm{PO}} = \frac{\hbar^2 u_e}{2k_{\mathrm{B}} T e_c^2[(\epsilon_\mathrm{o}\epsilon_{e,s})^{-1} - (\epsilon_\mathrm{o}\epsilon_{e,\infty})^{-1}]}. \tag{5.223}$$

Inclusion of nonparabolicity and screening effects will lead to

$$\tau_{e\text{-}p,\mathrm{PO}} = \frac{\hbar^2(E_e + \frac{E_e^2}{\Delta E_{e,g}})^{1/2} F^{-1}}{e_c^2(2m_{e,e,\circ})^{1/2} k_{\mathrm{B}} T[(\epsilon_\mathrm{o}\epsilon_{e,s})^{-1} - (\epsilon_\mathrm{o}\epsilon_{e,\infty})^{-1}](1+2\frac{E_e\epsilon}{\Delta E_{e,g}})}$$

$$F \equiv 1 - \delta\ln(1+\delta^{-1}) - \frac{2\frac{E_e}{\Delta E_{e,g}}(1+\frac{E_e}{\Delta E_{e,g}})}{(1+2\frac{E_e}{\Delta E_{e,g}})^2}[1 - 2\delta + 2\delta^2\ln(1+\delta^{-1})]$$

$$\delta \equiv (2\kappa\lambda_\mathrm{o})^{-2}, \tag{5.224}$$

where $\epsilon_{e,s}$ and $\epsilon_{e,\infty}$ are the static and high-frequency relative permitivities, κ is the carrier wave vector, and λ_o is the screening length of the optical phonons.

(D) *Scattering by Short-Range Deformation Potential of Vacancies* $\tau_{e\text{-}v,d}$

$\tau_{e\text{-}v,d}$ also has a form similar to the relaxation time of electron–acoustic-phonon scattering, because of a similar deformation potential $\varphi_{e\text{-}v,d}$, which is

$$\tau_{e\text{-}v,d} = \frac{(\tau_{e\text{-}v,d})_\circ (E_e + \frac{E_e^2}{\Delta E_{e,g}})^{-1/2}}{(1+2\frac{E_e}{\Delta E_{e,g}})[(1-A)^2 - B]}$$

$$A \equiv \frac{\frac{E_e}{\Delta E_{e,g}}(1-a_v)}{(1+2\frac{E_e}{\Delta E_{e,g}})}, \quad a_v = \frac{\varphi'_{e\text{-}v,d,v}}{\varphi'_{e\text{-}v,d,c}}$$

$$B \equiv \frac{8\frac{E_e}{\Delta E_{e,g}}(1+E_e/\Delta E_{e,g})a_v}{3(1+2\frac{E_e}{\Delta E_{e,g}})^2}$$

$$(\tau_{e\text{-}v,d})_\circ \equiv \frac{\pi\hbar^4}{\varphi^2_{e\text{-}v,d,c} m_{e,e,\circ}(2m_{e,e,\circ})^{1/2} n_v}, \tag{5.225}$$

where n_v is the vacancy density, and a_v is the ratio of the short-range deformation potential coupling constants of vacancies for valence and conduction bands.

(E) *Scattering by Coulomb Potential of Vacancies* $\tau_{e\text{-}v,\mathrm{C}}$

$$\tau_{e\text{-}v,\mathrm{C}} = \frac{\epsilon_s^2 (2m_{e,e,\mathrm{o}})^{1/2}(E_e + \frac{E_e^2}{\Delta E_{e,g}})^{3/2}}{\pi(Ze_c^2)^2 n_v[\ln(1+\xi) - \xi/(1+\xi)](1+2\frac{E_e}{\Delta E_{e,g}})}$$

$$\xi \equiv (2\kappa\lambda_v)^2, \tag{5.226}$$

where Ze_c is the vacancy charge, and λ_v is the screening radius of the vacancy potential, which is given as

$$\lambda_v^{-2} = \frac{4\pi e_c^2}{\epsilon_s} D_e(\mu), \quad \mu = E_\mathrm{F}$$

$$D_e(\mu) \equiv \frac{2^{1/2}(m_{e,e,\circ})^{3/2}}{\pi^2\hbar^3}(\mu + \frac{\mu^2}{\Delta E_{e,g}})^{1/2}(1+2\frac{\mu}{\Delta E_{e,g}}), \tag{5.227}$$

where $D_e(\mu)$ is the DOS at the Fermi level.

5.17.4 Predicted Seebeck Coefficient and Electrical Conductivity

With the knowledge of the relaxation-time models and the band structure, we can predict the key TE properties by using (5.206). Here we use the example of Bi_2Te_3

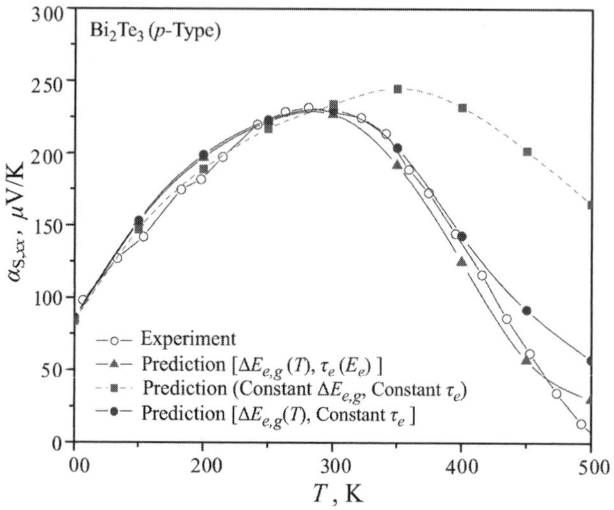

Figure 5.31. Variation of the predicted Seebeck coefficient for p-type Bi_2Te_3 with respect to temperature, compared with the available experimental results [160]. Both a temperature-dependent bandgap $\Delta E_{e,g} = 0.13 - 1.08 \times 10^{-4}T$ eV and a constant $\Delta E_{e,g} = 0.13$ eV are used. Also shown are the predictions made with the energy-dependent relaxation times and the constant relaxation-time model [153].

(anisotropic, tetradymite-type structure shown in Figure 5.28) and modify the BoltzTraP code to incorporate the Kane relaxation-time models. Figure 5.31 shows the variation of $\alpha_{S,xx}$ of Bi_2Te_3 with temperature, predicted with (5.206). The calculation was carried out using BoltzTraP, a software package adopting BTE. Both the energy-dependent relaxation-time models and the constant relaxation-time model were used. The BoltzTraP code was modified and the relaxation-time model was incorporated into the integration, as BoltzTrap assumes a constant relaxation time. To incorporate the temperature dependence of the band structure, the band structure was assumed not to change with temperature, and the conduction band was shifted in the calculation to include the temperature dependence of the bandgap $\Delta E_{e,g}$ (Section 5.4). Figure 5.31 shows that the constant $\Delta E_{e,g}$ leads to much higher values for $T > 300$ K (intrinsic regime), whereas results predicted with the temperature-dependent $\Delta E_{e,g} = 0.13 - 1.08 \times 10^{-4}T$ agree quite well with the experimental results. However, in the extrinsic regime ($T \leq 300$ K), there is only a minor difference between the results with the two different settings for $\Delta E_{e,g}$. As shown in Figure 5.31, with the same temperature-dependent $\Delta E_{e,g} = 0.13 - 1.08 \times 10^{-4}T$ eV, the two relaxation-time models give very similar results in the extrinsic regime, because one kind of carriers dominates the electrical transport. However, some small deviation appears in the intrinsic regime, where the concentrations of the holes and electrons become comparable, and it increases with the increasing temperature. This phenomenon indicates that the temperature dependences of the mobilities of holes and electrons are different.

The band structure calculated with the experimental lattice parameters at 300 K was used in the calculations for α_S in Figure 5.30(b). Temperature changes not only the carrier concentrations but also the lattice parameters. However, the bandstructure calculations adopting the lattice parameters under different temperatures show that the thermal expansion has negligible effects on the band structure. The change of the lattice parameters from 0 to 300 K results in only less than a 2% change in the bandgap. Compared with the actual temperature dependence of the bandgap, it seems that the temperature variation of the bandgap is mainly because of lattice vibration.

Figure 5.30(b) shows the variation of α_S along the xx and zz directions, with respect to the chemical potential μ, at 300 K. Apparently, the two curves are very similar, indicating the isotropy of α_S. Figure 5.30(b) shows that, for p-type Bi_2Te_3, the α_S peaks along the xx and zz directions almost overlap. However, for n-type Bi_2Te_3, the absolute peak value α_S along the xx direction is larger than that along the zz direction, though the peak positions are identical. In Figure 5.30(b), $\mu_\circ$ is the chemical potential value at which $\alpha_S = 0$. It is useful to rewrite the relation for α_S as

$$\alpha_S = -\frac{k_B}{e_c}\frac{\langle E_e - \mu\rangle_{\sigma_e(E_e,\mu)}}{k_B T}, \quad \langle E_e - \mu\rangle_{\sigma_e(E_e,\mu)} = \frac{\int \sigma_e(E_e,\mu)(E_e-\mu)\mathrm{d}E_e}{\int \sigma_e(E_e,\mu)\mathrm{d}E_e}. \tag{5.228}$$

Here $\langle E_e - \mu\rangle_{\sigma_e(E_e,\mu)}$ is the $\sigma_e(E_e,\mu)$-averaged energy deviation from the chemical potential. Then we have

$$\mu_\circ = \langle E_e\rangle_{\sigma_e(E_e,\mu_\circ)}, \tag{5.229}$$

which is the $\sigma_e(E_e)$-averaged energy and close to the middle of the bandgap [marked on the right-hand side of Figure 5.28(c)]. Therefore,

$$\alpha_S = -\frac{k_B}{e_c}\frac{[\langle E_e\rangle_{\sigma_e(E_e,\mu)} - \langle E_e\rangle_{\sigma_e(E_e,\mu_\circ)}] - (\mu - \mu_\circ)}{k_B T}. \tag{5.230}$$

Here $\langle E_e\rangle_{\sigma_e(E_e,\mu)}$ has a simple form, where the nondegenerate approximation is used, i.e.,

$$\begin{aligned}\langle E_e\rangle_{\sigma_e(E_e,\mu)} &\simeq \frac{\sigma_{e,e}\Delta E_{e,g}}{\sigma_{e,e}+\sigma_{e,h}} \\ &= \frac{1}{1+be^{-2\epsilon/(k_B T)}}\Delta E_{e,g},\end{aligned} \tag{5.231}$$

where $\sigma_{e,e}$ and $\sigma_{e,h}$ are the electrical conductivity contributed by electrons and holes, $b = (\mu_h/\mu_e)(m_{e,h}/m_{e,e})^{3/2}$ (μ_h and μ_e are the mobilities of electrons and holes), and $\epsilon = \mu - \Delta E_{e,g}/2$ is the distance of the chemical potential above the middle of the bandgap. For semiconductors with large bandgap ($\Delta E_{e,g} > 10k_B T$), the maximum value of α_S can be estimated as [similar in form to (5.152)]

$$|\alpha_S|_{max} = \frac{k_B}{e_c}\frac{\Delta E_{e,g}}{2k_B T}. \tag{5.232}$$

For small $2\epsilon/k_B T$, we have

$$\langle E_e\rangle_{\sigma_e(E_e,\mu)} \simeq \frac{\Delta E_{e,g}}{1+b}[1+\frac{2b\epsilon/(k_B T)}{1+b}] \tag{5.233}$$

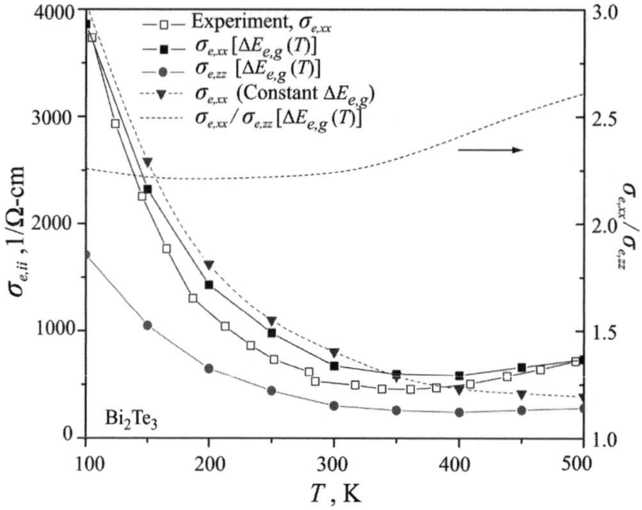

Figure 5.32. Variations of predicted directional electrical conductivities for Bi_2Te_3 with respect to temperature, using the Kane band model and energy-dependent relaxation times, and comparison with the available experimental results [160].

Therefore,

$$\alpha_S \simeq -\frac{k_B}{e_c}\frac{[\frac{2b\Delta E_{e,g}}{(1+b)^2 k_B T} - 1](\mu - \mu_\circ)}{k_B T}. \quad (5.234)$$

For a narrow bandgap semiconductor, as shown in Figure 5.30(b), (5.234) is a good approximation for α_S when $|\mu - \mu_\circ| < \Delta E_{e,g}/2$. When μ is close to the band edge, the effects of the opposite charges become smaller. While μ moves farther into the band edge, $\langle E_e \rangle_{\sigma_e(E_e,\mu)}$ is closer to μ. Therefore, α_S will achieve the maximum near the band edge, and the maximum value can be estimated from (5.234). Assuming $b = 1$ and that the maximum is achieved at the band edge, for Bi_2Te_3 at 300 K, then $|\alpha_S|_{max} \simeq 320$ μV/K, which is close to the results in Figure 5.30(b).

Figure 5.32 shows the predicted electrical conductivity of Bi_2Te_3 along the xx and zz directions, wherein the Kane-band-model-based relaxation times are used. The temperature-dependent bandgap $\Delta E_{e,g} = 0.13 - 1.08 \times 10^{-4} T$ eV are adopted in the calculation. Some parameters, for example, the deformation coupling potentials, are fitted to the experimental results and listed in Table 5.8. Below 300 K the ratio $\sigma_{e,xx}/\sigma_{e,zz}$, is around 2.2 and almost temperature independent. But above 300 K, $\sigma_{e,xx}/\sigma_{e,zz}$ increases with increasing temperature. From the results shown in Figure 5.32, this is because $\sigma_{e,xx}$ changes much faster than $\sigma_{e,zz}$ at high temperatures. Note that the predicted $\sigma_{e,xx}/\sigma_{e,zz}$ is lower than the experimental values (around 2.95). A reason can be the neglect of the direction dependence of the effective masses.

Figure 5.33 shows the variations of average values of various relaxation times, i.e., $\langle\langle \tau_{e\text{-}i} \rangle\rangle$ calculated from the relaxation-time models based on the Kane model (Section 5.17.3), as functions of temperature. The largest is $\langle\langle \tau_{e-v,\mathrm{C}} \rangle\rangle$, and the

Table 5.8. *Parameters used in the Kane relaxation-time models for Bi_2Te_3.*

Parameter	Magnitude	Parameter	Magnitude
$m_{h,e,\circ}/m_e$	0.08	$m_{e,e,\circ}/m_e$	0.06
n_v, 1/m^3	1.04×10^{25}	ρ, kg/m^3	7.86×10^3
$\epsilon_{e,\text{o}}$	400	$\epsilon_{e,\infty}$	69.8
c_l, N/m^2	0.71×10^{11}	$\hbar\omega_\text{O}$, eV	0.0076
Z	0.1	$\varphi_{e\text{-}p,\text{A},c}$, eV	35
$\varphi_{e\text{-}p,\text{O},c}$, eV	40	$\varphi_{e\text{-}v,d,c}$, J-m^3	1.2×10^{-46}
$a_\text{A}, a_\text{O}, a_v$	1.0	n_v, 1/m^3	1.04×10^{25}
$\Delta E_{e,g}$, eV	$0.13 - 1.08 \times 10^{-4} T$	a_R, Å	10.45

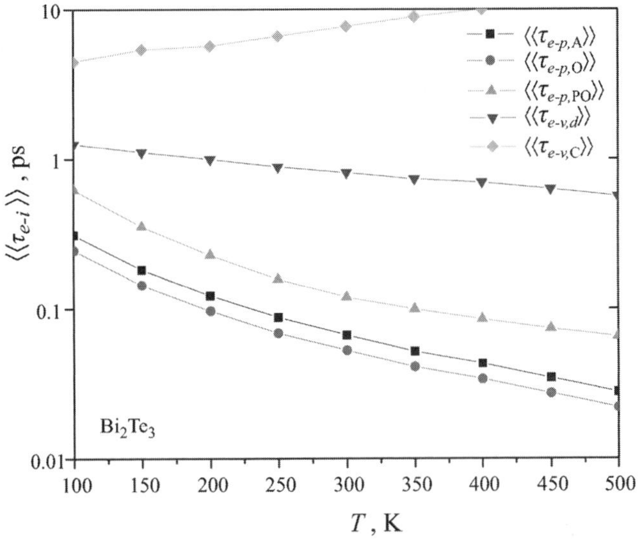

Figure 5.33. Variations of predicted electron relaxation times for Bi_2Te_3 with respect to temperature, using the Kane band model for energy dispersion.

shortest are $\langle\langle \tau_{e\text{-}p,\text{A}} \rangle\rangle$ and $\langle\langle \tau_{e\text{-}p,\text{O}} \rangle\rangle$. The three phonon scatterings, $\langle\langle \tau_{e\text{-}p,\text{A}} \rangle\rangle$, $\langle\langle \tau_{e\text{-}p,\text{O}} \rangle\rangle$, and $\langle\langle \tau_{e\text{-}p,\text{PO}} \rangle\rangle$, make up most of the total $\langle\langle \tau_e \rangle\rangle$.

5.17.5 Electrical and Phonon Thermal Conductivities

For semiconductors, the total thermal conductivity includes two parts, i.e., electrical thermal conductivity k_e and lattice conductivity k_p [(5.201)]. Currently it is not possible to measure k_p and k_e separatively, so it is the total thermal conductivity $k = k_e + k_p$ that is measured, and k_e is generally derived from the Wiedemann–Franz law (5.150). The difference between the measured k and the derived k_e is used for k_p. However, as shown in Figure 5.16, the Lorenz number is not a constant and may be subject to many nonapplicable assumptions. Using (5.206), we can also predict

k_e using the BTE (implemented in BoltzTraP) based on the full-band structure and the relaxation-time models.

(A) *Electric Thermal Conductivity*

Traditionally, the Wiedemann–Franz law (5.150), $k_e = N_L \sigma_e T$, is used to calculate the electric thermal conductivity k_e on the basis of σ_e. However, for semiconductors, N_L may not be the value used for metals [$N_{L,\circ} = (\pi^2/3)k_B^2/e_c^2$, (5.194)], especially when the chemical potential is near the center of the bandgap. When the sample is heavily doped (the chemical potential is deep inside the valence or conduction band), $N_L/N_{L,\circ}$ is close to 1.0. However, for intermediate doping, $N_L/N_{L,\circ}$ can be smaller than 1.0, and the minimum is around 0.7. For small doping concentrations or intrinsic regime, $N_L/N_{L,\circ}$ may be much larger than 1.0. Figure 5.34(a) shows the temperature dependence of the electric thermal conductivity calculated according to Eq. (5.206). The results show that k_e for both directions increase with increasing temperature. Due to the significant changes of N_L in the intrinsic regime, using $N_{L,\circ}\sigma_e T$ underestimates k_e at high temperatures.

(B) *Phonon Thermal Conductivity*

In semiconductors, phonons are scattered by the grain boundary, defects, and other phonons and carriers. According to the Matthiessen rule (4.111) and phonon-scattering mechanisms of (4.120), we have

$$\frac{1}{k_p} = \frac{1}{k_{p\text{-}b}} + \frac{1}{k_{p\text{-}im}} + \frac{1}{k_{p\text{-}p}} + \frac{1}{k_{p\text{-}e}}, \tag{5.235}$$

where $k_{p\text{-}b}$, $k_{p\text{-}im}$, $k_{p\text{-}p}$, and $k_{p\text{-}e}$ are the thermal conductivities limited by the scattering by the grain boundary, defects, phonons (U–processes only), and by charged carriers, respectively.

For semiconductors, the thermal resistivity that is due to the phonon–carrier scattering $1/k_{p\text{-}e}$, is normally very small. We can evaluate $1/k_{p\text{-}e}$ by using the electrical resistivity ρ_e. This relation has been derived by ignoring the difference between the N- and U–processes between carriers and phonons [24, 360], and using the Bloch model for ρ_e (5.195):

$$\frac{1}{k_{p\text{-}e}} = \frac{3.7\rho_e(T_D)}{N_{L,\circ}T}\left(\frac{T_D}{T}\right) I_5 \frac{\pi^2 z_e^2}{27 I_4^2} \tag{5.236}$$

$$\rho_e = 3.7\rho_e(T_D)\left(\frac{T}{T_D}\right)^5 I_5 \quad \text{Bloch model of metal } \rho_e. \tag{5.237}$$

Here

$$I_n = \int_0^{T_D/T} \frac{x^n e^x}{(e^x - 1)^2} \mathrm{d}x. \tag{5.238}$$

In the proceeding relations, z_e is the number of free electrons per atom, N_L is the Lorenz constant, and T_D is the Debye temperature. Then we have

$$\frac{1}{k_{p\text{-}e}} = \frac{\rho_e}{N_{L,\circ}T}\frac{(T_D/T)^6}{27I_4^2}\pi^2 z_e^2. \tag{5.239}$$

For T/T_D increasing from 0.1 to 10, $(T_D/T)^6/I_4(T_D/T)^2$ decreases from 97 to 9. Also, for normal dopant concentrations ($< 10^{19}$ cm^{-3}), z_e is of the order of 10^{-3}. Then for a wide temperature range ($0.1 \leq T/T_D \leq 10$), $1/k_{p,e}$ is only about 10^{-4} of the electrical thermal resistivity $1/k_e$ found from the Wiedemann–Franz law, and is therefore negligible for this semiconductor.

Then, for bulk semiconductors, the lattice thermal conductivity is mainly limited by the phonon–phonon scattering (U–processes). The lattice conductivity can be predicted using the MD simulation and the G–K method, as introduced in Section 4.12. The most important is the appropriate interatomic potentials used to predict the lattice vibrations, including both the anharmonic and harmonic behaviors. Normally those potentials can be developed by fitting the energy surface computed with *ab initio* methods, for example, DFT. This approach, which is rather involved, is used here, and the resulting potentials need to be verified by use of the bulk properties and crystal structures. The potentials used here for Bi_2Te_3 are reported in [153]. Figure 5.34 shows the temperature-dependent, in-plane and cross-plane (Figure 5.28 shows the crystal structure) lattice conductivities of Bi_2Te_3. Note that the predicted in-plane and cross-plane k_p are higher than the experimental results. This is expected, considering the defects (e.g., isotopes, displacements, lamellas, etc.) in a real semiconductor crystal (Figure 5.14), which will reduce the thermal conductivity according to (5.235).

The lattice conductivities in both directions follow the $1/T$ law [153], similar to insulators, i.e., the Slack relation (4.132). The predicted cross-plane thermal conductivity $k_{p,\perp}$ is lower than the in-plane $k_{p,\parallel}$. Because the average cross-plane sound velocity (1631 m/s) is very close to the in-plane sound velocity (1775 m/s), the difference between the two thermal conductivities is mainly due to the anharmonicity along the different directions. This can be verified by the directional Grüneisen parameter $\gamma_{G,i}$ along the direction i. For Bi_2Te_3, the in-plane Grüneisen parameter $\gamma_{G,\parallel}$ is 1.17, and the cross-plane Grüneisen parameter $\gamma_{G,\perp}$ is 1.86 at 300 K [14]. The large difference in the anharmonicity originates from the unique bond characteristics in the layered structure (Figure 5.28), in which the intralayer bonds are covalent but the inter-layer bonds are hybrids of the electrostatic interactions and van der Waals interactions [153]. Furthermore, when the thermal conductivity is decomposed (Sections 4.12 and 4.13), we find that the contribution from the short-range acoustic phonons $k_{p,sh,A}$, and that from the optical phonons $k_{p,O}$, in the two different directions, are almost the same. The difference is due only to the contribution from the long-range acoustic phonons $k_{p,sh,A}$. More details are reported in [153].

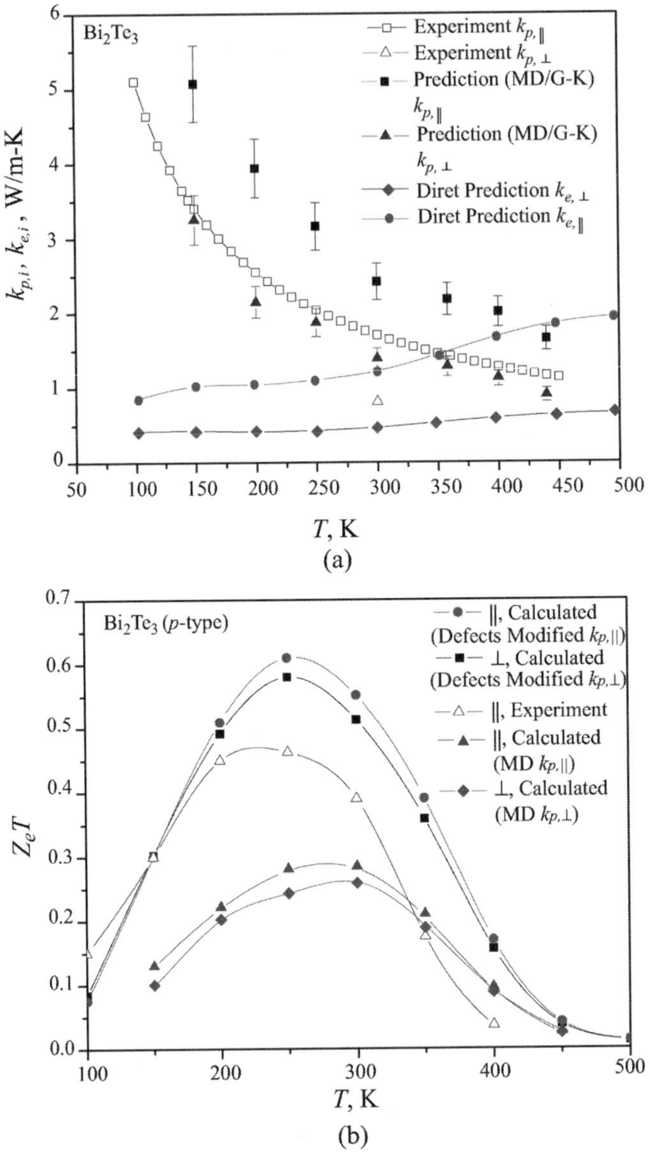

Figure 5.34. (a) Predicted electrical and phonon thermal conductivities of Bi_2Te_3. The electrical thermal conductivity is directly predicted using the BTE and the Kane relaxation-time models. The phonon thermal conductivity is predicted using MD simulation and the G–K method [153]. Available experimental results are also shown [160]. (b) Variation of calculated directional figure of merit for Bi_2Te_3 (along $\|$ and $\perp$), with respect to temperature. Both the results with the directly calculated k_p and that modified by defects are presented. The experimental results [160] are also shown.

(C) *Figure of Merit*

Figure 5.34(b) shows the variation of the figure of merit Z_eT (5.203) for the p-type Bi_2Te_3 specimen of experiment [160] along the $\parallel$ and $\perp$ directions, with respect to temperature. The lower two curves are based on the directly (MD) calculated k_p. Because k_p as well as σ_e is very sensitive to defects, which are always present in fabricated specimens, for comparison between the calculated and measured Z_eT we used a modified k_p and the results are shown with the top two curves. Due to the difficulty in including the various defects, the modified k_p was obtained by fitting the total thermal conductivity to the experimental results [160] at 300 K and then using the Slack $1/T$ law at other temperatures. The experimental results are also shown in Fig. 5.34(b). The Z_eT with a modified k_p is higher than experimental results above 200 K, mainly due to the overestimation of σ_e. The calculated Z_eT reaches its maximum around 250 K. Z_eT along the $\parallel$ direction is higher than that along the $\perp$ direction between 200 K and 400 K, due to the larger ratio $\sigma_{e,\parallel}/\sigma_{e,\perp}$ compared to $k_{p,\parallel}/k_{p,\perp}$. Note that the experimental $\sigma_{e,\parallel}/\sigma_{e,\perp}$ is larger than the calculated results (discussed above), so the difference in the figure of merits along the $\parallel$ and $\perp$ direction is expected to be even larger.

5.18 Electron and Phonon Transport Under Local Thermal Nonequilibrium

Conduction electrons can acquire kinetic energy by an imposed electric field, by absorbing photons, or by other energy transitions. This kinetic energy can be harvested before inelastic scattering (in particular with phonons) results in conversion of this electronic energy to heat. Here we address this electronic carrier transport and interactions, while considering local thermal nonequilibrium between the electrons and phonons (each carrier and subcarrier has a temperature).

In many semiconductor devices, because of spatial variations in carrier (electron and hole) concentrations, the carrier diffusion also becomes important. An average particle (semiclassical) treatment is made starting from the Boltzmann transport equation (with appropriate moments) [36, 219, 226].

5.18.1 Derivations

The BTE (Table 3.1) for carrier $i = e$, is written as

$$\frac{\partial f_i}{\partial t} + \boldsymbol{u}_i \cdot \nabla f_i + \boldsymbol{F}_i \cdot \nabla_{\boldsymbol{p}} f_i = \frac{\partial f_i}{\partial t}|_s + \dot{s}_{f,i\text{-}j}. \tag{5.240}$$

As we discussed in Chapter 3, the BTE describes the statistical particle transport, including electron and phonon. The conservation equations that are encountered in fluid mechanics, heat transfer, and electron transport are found by taking different moments of the BTE.

Taking a function of electron momentum $\phi(\boldsymbol{p}_e)$, its weighed sum (or integral) is (note that here for electrons we have allowed for two spins)

$$\langle\phi(\boldsymbol{p}_e)\rangle \equiv \frac{1}{\hbar^3}\sum_{\boldsymbol{p}_e}\phi(\boldsymbol{p}_e)f_e(\boldsymbol{x},\boldsymbol{p}_e,t). \tag{5.241}$$

Multiplying BTE by $\phi(\boldsymbol{p}_e)$ and summing over the momentum gives the BTE moment equation

$$\frac{1}{\hbar^3}\sum_{\boldsymbol{p}}\phi(\boldsymbol{p})\frac{\partial f_e}{\partial t}+\frac{1}{\hbar^3}\sum_{\boldsymbol{p}_e}\phi(\boldsymbol{p}_e)\nabla\cdot\boldsymbol{u}_e f_e+\frac{1}{\hbar^3}\sum_{\boldsymbol{p}_e}\phi(\boldsymbol{p}_e)\boldsymbol{F}_e\cdot\nabla_{\boldsymbol{p}}f_e$$
$$=\frac{1}{\hbar^3}\sum_{\boldsymbol{p}_e}\phi(\boldsymbol{p}_e)\frac{\partial f_e}{\partial t}|_s+\frac{1}{\hbar^3}\sum_{\boldsymbol{p}}\phi(\boldsymbol{p}_e)\dot{s}_{f,e}. \tag{5.242}$$

The summation implies integrations (3.11) and (3.12), depending on symmetry. Also, the average of relaxation-time approximation (3.13) gives

$$\frac{1}{\hbar^3}\sum_{\boldsymbol{p}}\phi(\boldsymbol{p})\frac{\partial f_e}{\partial t}|_s \equiv -\langle\langle\tau_{e,\phi}^{-1}\rangle\rangle(\langle\phi(\boldsymbol{p}_e)\rangle-\langle\phi(\boldsymbol{p}_e)^\circ\rangle), \tag{5.243}$$

where $\tau_{i,\phi}$ indicates the continuity τ_e(continuity), momentum τ_e(momentum), or energy τ_e(energy) relaxation time, depending on ϕ (Section 5.10).

For $\phi = \boldsymbol{p}_e^\circ$, the BTE gives the continuity equation for a carrier (here written for conduction electrons, but for the holes a similar equation is written)

$$\frac{\partial n_{e,c}}{\partial t}+\nabla\cdot\boldsymbol{u}_e n_{e,c}+\sum_i \dot{n}_{e,c,i}=\frac{\partial n_{e,c}}{\partial t}+\frac{1}{e_c}\nabla\cdot\boldsymbol{j}_e+\sum_i \dot{n}_{e,c,i}=0, \tag{5.244}$$

where we have used $\boldsymbol{j}_e = e_c\boldsymbol{u}_e n_{e,c}$. This is also given in Table 3.6.

For the carrier (electron) momentum, $\phi(\boldsymbol{p}_e)=\boldsymbol{p}^1=m_{e,e}\boldsymbol{u}_e$, we have

$$\langle\phi(\boldsymbol{p})\rangle=\frac{1}{4\pi^3\hbar^3}\sum_{\boldsymbol{p}_e}\boldsymbol{p}_e f_e=n_{e,c}m_{e,e}\boldsymbol{u}_e \tag{5.245}$$

$$\frac{1}{\hbar^3}\sum_{\boldsymbol{p}_e}\boldsymbol{p}_e\nabla\cdot\boldsymbol{u}_e f_e=\nabla_i(n_{e,c}m_{e,e}\boldsymbol{u}_{e,i}\boldsymbol{u}_{e,j}+n_{e,c}k_\mathrm{B}T_e)$$
$$=n_{e,c}m_{e,e}\boldsymbol{u}_e\nabla\boldsymbol{u}_e+\nabla n_{e,c}k_\mathrm{B}T_e. \tag{5.246}$$

These equations give the momentum conservation equation

$$n_{e,c}m_{e,e}[\frac{\partial\boldsymbol{u}_e}{\partial t}+(\boldsymbol{u}_e\cdot\nabla)\boldsymbol{u}_e]=-n_{e,c}e_c\boldsymbol{e}_e-\nabla n_{e,c}k_\mathrm{B}T_e-\frac{n_e m_{e,e}\boldsymbol{u}_e}{\tau_{e\text{-}p}(\text{momentum})}. \tag{5.247}$$

The electron momentum (and energy) relaxation times are given in Table 5.4.

Under a steady-state condition, and by neglecting the gradient of electron drift velocity, (5.247) gives the drift–diffusion equation, i.e.,

$$\boldsymbol{j_e} = n_{e,c}e_c\mu_e\boldsymbol{e_e} + e_c D_e \nabla n_{e,c} + n_{e,c}\mu_e k_B \nabla T_e \quad \text{drift–diffusion equation,} \tag{5.248}$$

where $D_e = k_B T_e \mu_e / e_c$. This is also given in Table 3.6.

For energy, $\phi = E_{e,\kappa}(\boldsymbol{p}_e) = \boldsymbol{p}_e^2/2m_{e,e}$, and we define

$$\langle\phi(\boldsymbol{p}_e)\rangle = \frac{1}{\hbar^3}\sum_{\boldsymbol{p}_e} E_{e,\kappa}(\boldsymbol{p}_e) f_e = \frac{E_e}{V} = \frac{3}{2}n_{e,c}k_B T_e + \frac{1}{2}n_{e,c}m_{e,e}u_e^2. \tag{5.249}$$

Next, we note that

$$\langle\langle\tau_\phi^{-1}\rangle\rangle(\langle\phi(\boldsymbol{p}_e)\rangle - \langle\phi(\boldsymbol{p}_e)^{\mathrm{o}}\rangle) = \langle\langle\tau_{e\text{-}p}^{-1}(\text{energy})\rangle\rangle(\frac{E_e}{V} - \frac{E_e^{\circ}}{V}) \tag{5.250}$$

$$\begin{aligned}\frac{1}{\hbar^3}\sum_{\boldsymbol{p}_e} E_{e,\kappa}\nabla\cdot\boldsymbol{u_e} f_e &= \frac{1}{\hbar^3}\sum_{\boldsymbol{p}_e}\frac{1}{2}m_{e,e}u_e^2\nabla\cdot\boldsymbol{u}_e f_e \\ &= \nabla\frac{1}{2}n_{e,c}m_{e,e}\langle u_e^2\boldsymbol{u}_e\rangle \\ &= \nabla(\frac{E_e}{V}\boldsymbol{u}_e + \boldsymbol{u}_e\cdot n_{e,c}k_B T_e + k_e\nabla T_e).\end{aligned} \tag{5.251}$$

Equation (5.251) is called the heat flux term, as the last term represents heat conduction by an electron.

Using the proceeding equations, we find that the energy conservation equation is [36]

$$\begin{aligned}\frac{\partial(E_e/V)}{\partial t} &= -\nabla\cdot(\frac{E_e}{V}\boldsymbol{u}_e + n_{e,c}k_B T_e\boldsymbol{u}_e + k_e\nabla T_e) + \boldsymbol{j}_e\cdot\boldsymbol{e_e} \\ &\quad - \frac{E_e - E_e^{\circ}}{\langle\langle\tau_e^{-1}(\text{energy})\rangle\rangle} + \sum_i \dot{s}_{e\text{-}i}(\text{energy}) \\ \frac{E_e}{V} &= \frac{3}{2}n_{e,c}k_B T_e + \frac{1}{2}n_{e,c}m_{e,e}u_e^2.\end{aligned} \tag{5.252}$$

Note that the total kinetic energy of the carrier E_e is the sum of translational (3 translational degrees of freedom) and drift motions. Here the carrier temperature T_e is defined by use of the electron collision velocity $\langle u_e'^2\rangle^{1/2}$, such that $3n_{e,c}k_B T_e/2 = m_{e,e}n_{e,c}\langle u_e'^2\rangle^{1/2}/2$. The collision velocity is also called the thermal speed (for the random component of the velocity), where $\langle u_e'\rangle = 0$. This defines the carrier temperature based on the classical equipartition of energy ($k_B T/2$ per degree of freedom). Note that this does not include the quantum effects on the specific heat capacity, as included in (5.72).

By substituting the continuity equation (5.244) and momentum conservation equation (5.247) into the the energy conservation equation (5.252), we have [36]

$$\frac{\partial T_e}{\partial t} + \nabla \cdot \boldsymbol{u}_e T_e = \frac{2}{3n_e k_B}\nabla \cdot k_e \nabla T_e + \frac{1}{3}T_e \nabla \cdot \boldsymbol{u}_e - \frac{T_e - T_{p,O}}{\tau_{e\text{-}p,O}}$$
$$-\frac{T_e - T_{p,A}}{\tau_{e\text{-}p,A}} + \frac{m_{e,e}u_e^2}{3k_B}[\frac{2}{\tau_{e\text{-}p}(\text{momentum})} - \frac{1}{\tau_{e\text{-}p,O}} - \frac{1}{\tau_{e\text{-}p,A}}]$$
$$+\frac{2}{3n_{e,c}k_B}\sum_i \dot{s}_{e\text{-}i}(\text{energy}). \qquad (5.253)$$

This distinguishes between scattering by the acoustic and the optical phonons. Note that this includes the Joule heating (first term in the square brackets).†

The thermal-phonon energy equations are derived from the first law of thermodynamics. The BTE for a phonon is applicable for energy conservation and is written separately for the optical phonon and the acoustic phonon.

5.18.2 Phonon Modal Energy Equations

Starting from (5.240) and using the same momentum power formulation and averages. The optical- and acoustic-phonon systems are treated separatively (having temperatures $T_{p,O}$ and $T_{p,A}$) [115]. The optical-phonon energy conservation equation is based on BTE, with no transport terms (assuming $\boldsymbol{u}_{p,O} = 0$ and $\boldsymbol{F}_{p,O} = 0$ for the optical phonons) [115]:

$$\frac{\partial E_{p,O}}{V\partial t} = -\frac{\partial E_e}{V\partial t}|_s + \frac{\partial E_{p,O}}{V\partial t}|_s. \qquad (5.254)$$

Substituting for scattering terms for (5.253), we have [115]

$$nc_{p,O}\frac{\partial T_{p,O}}{\partial t} = \frac{3n_{e,c}k_B}{2}\frac{T_e - T_{p,O}}{\tau_{e\text{-}p,O}} + \frac{n_{e,c}m_{e,e}u_e^2}{2\tau_{e\text{-}p,O}} - nc_{p,O}\frac{T_{p,O} - T_{p,A}}{\tau_{p,A}}, \qquad (5.255)$$

where n is the number of atoms per unit volume.

The acoustic-phonon energy conservation equation ($\boldsymbol{F}_{p,A} = 0$) is

$$\frac{\partial E_{p,A}}{V\partial t} + \nabla \cdot k_p \nabla T_{p,A} = -\frac{\partial E_{p,O}}{V\partial t}|_s \qquad (5.256)$$

or

$$nc_{p,A}\frac{\partial T_{p,A}}{\partial t} = \nabla \cdot k_p \nabla T_{p,A} + nc_{p,O}\frac{T_{p,O} - T_{p,A}}{\tau_{p,A}} + \frac{3n_{e,c}k_B}{2}\frac{T_e - T_{p,A}}{\tau_{e\text{-}p,A}}. \qquad (5.257)$$

Note that the specific heat is divided into the acoustic and optical components.

† The volumetric Joule heating is found using (5.146) as

$$\rho_e j_e^2 = \frac{m_{e,e}}{e_c^2 n_{e,c}\tau_{e\text{-}p}(\text{momentum})}e_c^2 n_{e,c}^2 u_e^2 = \frac{m_{e,e}n_{e,c}u_e^2}{\tau_{e\text{-}p}(\text{momentum})}.$$

Then, dividing by $3k_B n_{e,c}$, we have the term in (5.253).

In addition to the conservation equations there is the Poisson equation (Table 3.5), which satisfies the Gauss law, i.e.,

$$\nabla^2 \varphi_e = -\frac{e_c}{\epsilon_o \epsilon_e}(n_{e,c} + n_{e,b}), \tag{5.258}$$

where $n_{e,b}$ is the bond electron density, which should be included.

The set of equations derived in this section is used for electron–phonon thermal nonequilibrium analysis of semiconductor devices.

5.18.3 Summary of Conservation (Electrohydrodynamic) Equations

Because of their similarity to the Navier–Stokes equations (Table 3.10), including the mechanical energy equation (6.123), the carrier conservation equations are called the hydrodynamic equations. The summary of carrier balance equations are subsequently given. The steady-state electronic carrier (electron or hole) continuity equation, i.e., (5.244), is

$$\frac{\partial n_{e,c}}{\partial t} + \frac{1}{e_c}\nabla \cdot \boldsymbol{j_e} + \sum_i \dot{n}_{e,c,i} = 0. \tag{5.259}$$

The electronic carrier drift–diffusion equation, i.e., (5.248), is

$$\boldsymbol{j_e} = e_c n_{e,c}\mu_e \boldsymbol{e_e} + e_c D_e \nabla n_{e,c} + n_{e,c}\mu_e k_B \nabla T_e, \quad D_e = \frac{k_B T_e \mu_e}{e_c}. \tag{5.260}$$

The electron energy equation is

$$\begin{aligned}\frac{\partial T_e}{\partial t} + \nabla \cdot \boldsymbol{u}_e T_e = {} & \frac{2}{3n_{e,c}k_B}\nabla \cdot k_e \nabla T_e + \frac{1}{3}T_e \nabla \cdot \boldsymbol{u}_e - \frac{T_e - T_{p,O}}{\tau_{e\text{-}p,O}(\text{energy})} \\ & - \frac{T_e - T_{p,A}}{\tau_{e\text{-}p,A}(\text{energy})} + \frac{m_{e,e}u_e^2}{3k_B}\Big[\frac{2}{\tau_{e\text{-}p}(\text{momentum})} - \frac{1}{\tau_{e\text{-}p,O}(\text{energy})} \\ & - \frac{1}{\tau_{e\text{-}p,A}(\text{energy})}\Big] + \frac{2}{3n_{e,c}k_B}\sum_i \dot{s}_{e\text{-}i}(\text{energy}).\end{aligned} \tag{5.261}$$

The optical-phonon energy equation (including electron kinetic energy transfer) is

$$\begin{aligned}nc_{p,O}\frac{\partial T_{p,O}}{\partial t} = {} & \frac{3n_{e,c}k_B}{2}\frac{T_e - T_{p,O}}{\tau_{e\text{-}p,O}(\text{energy})} + \frac{n_{e,c}m_{e,e}u_e^2}{2\tau_{e\text{-}p,O}(\text{energy})} \\ & - n_{e,c}c_{p,O}\frac{T_{p,O} - T_{p,A}}{\tau_{p,A}(\text{energy})}.\end{aligned} \tag{5.262}$$

The acoustic-phonon energy equation (includes conduction) is

$$\begin{aligned}nc_{p,A}\frac{\partial T_{p,A}}{\partial t} = {} & \nabla \cdot k_p \nabla T_{p,A} + n_{e,c}c_{p,O}\frac{T_{p,O} - T_{p,A}}{\tau_{p,A}(\text{energy})} \\ & + \frac{3n_{e,c}k_B}{2}\frac{T_e - T_{p,A}}{\tau_{e\text{-}p,A}(\text{energy})}.\end{aligned} \tag{5.263}$$

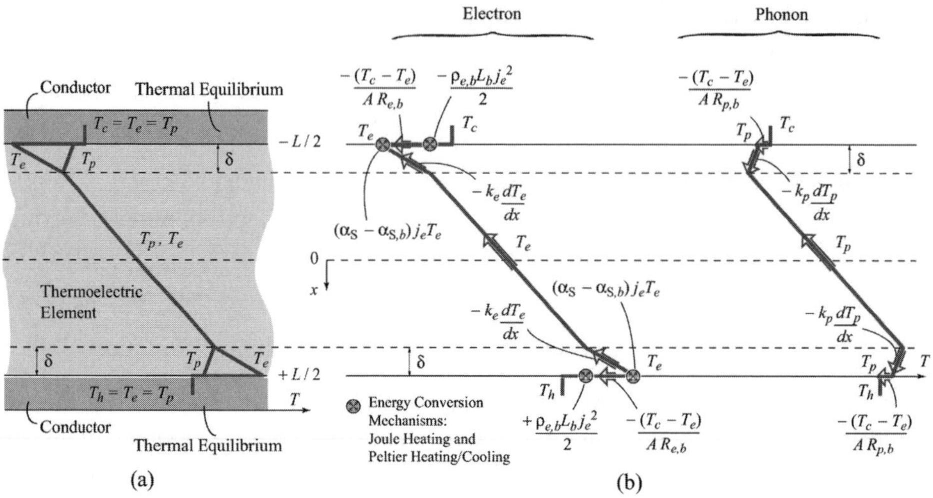

Figure 5.35. (a) Temperature distributions for electron and lattice subsystem, assuming equilibrium in the connectors. (b) Energy conversion mechanisms and heat transfer paths of electron and lattice subsystem [83].

5.19 Cooling Length in Electron–Phonon Local Thermal Nonequilibrium

From the results of Section 5.18, the electron and phonon transport equations, under thermal nonequilibrium, can be written as spectral (energy resolved) or energy-averaged forms. Here the averaged forms are used, those giving electron and lattice temperatures T_e and T_p along with a single, average relaxation time. For simplicity, only the acoustic phonons are considered. To generalize the results, we assume TE transport over a solid conductor, as shown in Figure 5.35. The problem is further described in [83].

The kinetics of electrons and phonons in an electric field or in a temperature field is described by the coupled BTE for electrons and phonons (Section 5.18). In one dimension and for steady state, with constant properties, these equations are the simplified forms of (5.261) [84, 169]

$$-k_e \frac{d^2 T_e}{dx^2} = \rho_e j_e^2 - \frac{n_{e,c} k_B}{\tau_{e\text{-}p}(\text{energy})}(T_e - T_p) \tag{5.264}$$

$$-k_p \frac{d^2 T_p}{dx^2} = \frac{n_{e,c} k_B}{\tau_{e\text{-}p}(\text{energy})}(T_e - T_p), \tag{5.265}$$

where j_e is the electrical current density and $\tau_{e\text{-}p}$(energy) is the average, electron–phonon energy relaxation time. These are the energy equations for the electron and phonon subsystems.

Note that the volumetric Joule heating (3.54), $\boldsymbol{j}_e \cdot \boldsymbol{e}_e = \rho_e j_e^2 = m_{e,e} n_{e,c} u_e^2 / \tau_{e\text{-}p}$ (momentum) (footnote in Section 5.18). In [84, 169], with some justifications, the coefficient 3/2 in (5.262) and (5.263) is replaced with 1, as it appears in (5.264) and (5.265).

Assuming that ρ_e, $n_{e,c}$, $\tau_{e\text{-}p}$(energy), k_e, and k_p are constant, we solve the proceeding equations by defining a center-of-thermal-conductivity temperature $kT_{ctc} = k_e T_e + k_p T_p$ [18]. The solution is

$$T_e = A_1 + A_2\frac{x}{L} + \frac{\rho_e j_e^2[(\frac{L}{2})^2 - x^2]}{2k} + \frac{\rho_e j_e^2 \delta^2}{\gamma k}$$
$$-\frac{1}{\gamma}[A_3\cosh(\frac{x}{\delta}) + A_4\sinh(\frac{x}{\delta})] \qquad (5.266)$$

$$T_p = A_1 + A_2\frac{x}{L} + \frac{\rho_e j_e^2[(\frac{L}{2})^2 - x^2]}{2k} - \frac{\rho_e j_e^2 \delta^2}{k}$$
$$+[A_3\cosh(\frac{x}{\delta}) + A_4\sinh(\frac{x}{\delta})], \qquad (5.267)$$

where

$$\delta = \delta_{e\text{-}p} = (\frac{\tau_{e\text{-}p}}{n_{e,c}k_B}\frac{k_e k_p}{k_e + k_p})^{1/2} \text{ cooling length}$$

$$\tau_{e\text{-}p}(\text{energy}) = \frac{k_B T_p}{2m_{e,e}u_{p,A}^2}\tau_{e\text{-}p}(\text{momentum}),$$

$$\tau_{e\text{-}p}(\text{momentum}) = \langle\langle\tau_e\rangle\rangle(\text{momentum}) = \frac{m_{e,e}\mu_e}{e_c}$$

$$\text{average electron–phonon relaxation time,} \qquad (5.268)$$

where $\gamma = k_e/k_p$ and L is the thermoelectric element thickness.

The relation between $\tau_{e\text{-}p}$(energy) and $\tau_{e\text{-}p}$(momentum) is for the acoustic deformation potential (end-of-chapter problem). The various $\tau_{e\text{-}p}$ are listed in Table 5.4.

In [169], the constant 2 in (5.268) is replaced with similar constants for low and high temperatures (with respect to the Debye temperature).

The four unknown constants A_1, A_2, A_3, and A_4 are determined [82] by the boundary conditions for the flow of heat by electrons and phonons at the boundaries, as presented in Figure 5.35. Also, for simplicity, it is assumed that electrons and phonons are in equilibrium in the connectors, i.e, $T_p = T_e = T_c$ at $x = -L/2$ and $T_p = T_e = T_h$ at $x = L/2$.

The phonon boundary conditions are

$$-\frac{T_c - T_p}{AR_{p,b}} - k_p\frac{dT_p}{dx} = 0, \text{ at } x = -\frac{L}{2} \qquad (5.269)$$

$$-\frac{T_p - T_h}{AR_{p,b}} - k_p\frac{dT_p}{dx} = 0, \text{ at } x = \frac{L}{2}, \qquad (5.270)$$

where $AR_{p,b}$ is the phonon boundary resistance. This resistance is because of a change in acoustic properties at the interface, as discussed in Section 4.13. The electron boundary conditions are

$$-\frac{T_c - T_e}{AR_{e,b}} - k_e \frac{dT_e}{dx} + (\alpha_S - \alpha_{S,b})\, j_e T_e - \rho_{e,b} L_b \frac{j_e^2}{2} = 0, \quad \text{at } x = -\frac{L}{2} \tag{5.271}$$

$$-\frac{T_e - T_h}{AR_{e,b}} - k_e \frac{dT_e}{dx} + (\alpha_S - \alpha_{S,b})\, j_e T_e + \rho_{e,b} L_b \frac{j_e^2}{2} = 0, \quad \text{at } x = \frac{L}{2}, \tag{5.272}$$

where $AR_{e,b}$ is the electron thermal boundary resistance [83] and $\alpha_{S,b}$ and $\rho_{e,b} L_b$ are the boundary Seebeck coefficient and the electrical boundary resistance (multiplied by area).

The first term in the proceeding equations represents the heat flow as defined by the phonon (electron) boundary resistance, and the second term is the heat flow predicted by the Fourier law. The third terms (5.271) and (5.272) represent the Peltier cooling and heating, respectively. The Joule heating at the boundaries is represented by the fourth term.

The bulk Seebeck coefficient α_S is given by (5.273), the n- and p-type thermoelectric materials. The boundary Seebeck coefficient $\alpha_{S,b}$, assuming that tunneling is the dominant electron transport mechanism across the metal/semiconductor interfaces, is given by [18]

$$\alpha_{S,b} = (\frac{k_B}{e_c}) \frac{\pi^2}{3} k_B T (\frac{h_P^2 E_{e,o}}{8\pi^2 m_{e,e}^* d^2})^{-1/2}, \tag{5.273}$$

where $E_{e,o}$ is the height of the boundary electron energy barrier. Equations (5.266) and (5.267) were solved analytically with (5.269) to (5.272). The resulting expressions for the coefficients A_j are in general very complicated. For electrical current j_e equal to zero, the coefficients A_j are

$$A_1 = \frac{T_h + T_c}{2}$$

$$A_2 = \frac{\gamma L^* \coth L^* + \frac{R_e^* R_p^* (1+\gamma)^2}{2R_k^*}}{R_e^* + \frac{R_e^* R_p^* (1+\gamma)^2}{2R_k^*} + \frac{\gamma L^* \coth L^*}{(1+2R_k^*)^{-1}} + R_p^* \gamma^2} (T_h - T_c)$$

$$A_3 = 0$$

$$A_4 = \frac{\gamma^2 R_p^* - \gamma R_e^*}{R_e^* + \frac{R_e^* R_p^* (1+\gamma)^2}{2R_k^*} + \frac{\gamma L^* \coth L_{te}^*}{(1+2R_k^*)^{-1}} + R_p^* \gamma^2} \frac{(T_h - T_c)}{2 \sinh L^*}, \quad \text{for } j_e = 0, \tag{5.274}$$

where the dimensionless parameters (*) are given by

$$R_e^* = \frac{(\frac{1}{AR_{p,b}} + \frac{1}{AR_{e,b}})^{-1}}{AR_{e,b}}, \quad R_p^* = \frac{(\frac{1}{AR_{p,b}} + \frac{1}{AR_{e,b}})^{-1}}{AR_{p,b}},$$

$$R_k^* = \frac{(\frac{1}{AR_{p,b}} + \frac{1}{AR_{e,b}})^{-1}}{\frac{L}{k_e + k_p}}, \quad L^* = \frac{L}{2\delta}. \tag{5.275}$$

More details, as well as a numerical example, are given in [83].

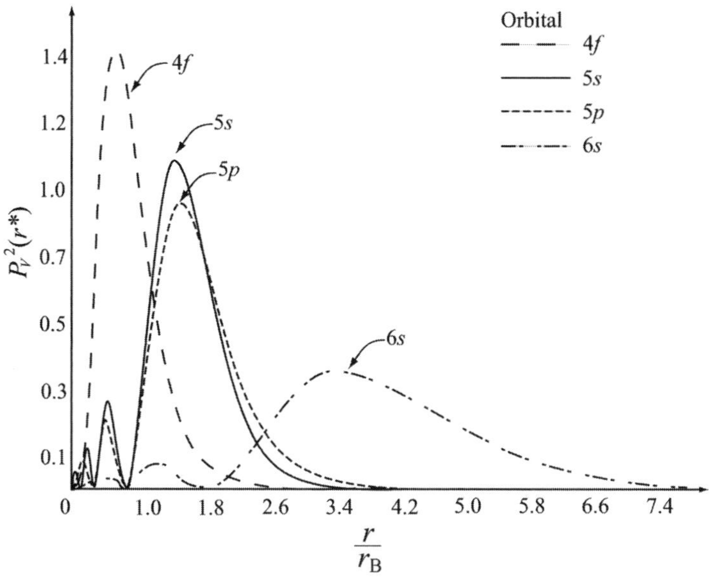

Figure 5.36. Radial electron distribution probabiliy $P_r^2(r) = \psi^\dagger \psi dr$ for $4f$, $5s$, $5p$, and $6s$ electron orbits of Gd^{3+} [111].

5.20 Electronic Energy States of Ions in Crystals

The slightly altered electronic states of doped ions in dielectric crystals allow for precise, spectral electron transitions that can be used, for example, in solid-state lasers, and in laser heating and cooling of solids. The electron state in the crystal host is altered, including the occurrence of energy splitting, and these are briefly reviewed in the following discussion.

We first look at ions in f block that have a weakly interacting $4f$ orbital because the $4f$ shell is deeply embedded inside the $5s$ and $5p$ shells, as seen in Figure 5.36. Because of the relatively weak interaction with the lattice, the energy calculation is much simpler. The Hamiltonian can be written for an individual rare-earth ion and decomposed as [88]

$$\mathrm{H} = \mathrm{H}_{free\ ion} + \mathrm{H}_{ion-lattice}, \tag{5.276}$$

where $\mathrm{H}_{free\ ion}$ is the Hamiltonian of the ion in complete isolation, and $\mathrm{H}_{ion-lattice}$ contains the static interactions of the ion with the host, which is also referred to as the crystal field.

The standard approach for treating $\mathrm{H}_{free\ ion}$ is to employ the central-field approximation, in which each electron is assumed to move independently in a spherically symmetric potential formed by the nucleus and the average potential of all other electrons. The solutions are then included as a product of a radial and an angular function. Whereas the radial function depends on the details of the potential, the spherical symmetry ensures that the angular component is identical to that of a hydrogen atom and can be expressed as spherical harmonics.

Using the radial function dependence on a potential, a semiempirical approach has been employed in which the attempt is made to identify these effective

Table 5.9. *The semiempirical atomic properties of rare-earth ions [253] in* cm^{-1}. *N is the number of electrons in the f shell. (1* cm^{-1} *= 0.12398 meV, Table 1.4.)*

N	Ground state	Ion, z	F_2, cm^{-1}	ξ, cm^{-1}	Ionic radius Å
0	$^1S_0$	La^{3+}, 57			1.061
1	$^2F_{5/2}$	Ce^{3+}, 58		640	1.034
2	$^3H_4$	Pr^{3+}, 59	320	759	1.013
3	$^4I_{9/2}$	Nd^{3+}, 60	327	885	0.995
4	$^5I_4$	Pm^{3+}, 61			0.979
5	$^6H_{5/2}$	Sm^{3+}, 62	370	1200	0.964
6	$^7F_0$	Eu^{3+}, 63	401	1320	0.950
7	$^8S_{7/2}$	Gd^{3+}, 64	408	1470	0.938
8	$^7F_6$	Tb^{3+}, 65	434	1705	0.923
9	$^6H_{15/2}$	Dy^{3+}, 66	420	1900	0.908
10	$^5I_8$	Ho^{3+}, 67	450	2163	0.894
11	$^4I_{15/2}$	Er^{3+}, 68	433	2393	0.881
12	$^3H_6$	Tm^{3+}, 69	447	2617	0.869
13	$^2F_{7/2}$	Yb^{3+}, 70		2883	0.858
14	$^1S_0$	Lu^{3+}, 71			0.848

interactions operating within the f electrons that produce the observed structure [253]. Based on this method of interpretation, $\mathrm{H}_{free\ ion}$ (the free ion Hamiltonian) of the system is written as

$$\mathrm{H}_{free\ ion} = \sum_{k=0,2,4,6} F_k \mathrm{I}^k + \xi S \cdot L + \alpha L(L+1) + \beta G(G_2) + \gamma G(R_7)$$
$$+ \sum_{i=2,3,4,6,7,8} \Upsilon^i t_i + \sum_{k=0,2,4} M^k m_k + \sum_{2,4,6} P^k p_k, \quad (5.277)$$

where F_k are the Slater integral parameters,† which represent the effective Coulomb interaction between $4f$ electrons. Here, I^k are the angular integrals over spherical harmonics, ξ is the spin–orbit parameter, and α, β and γ are coefficients of the Casimir operators for the full rotation group and the nonsymmetry groups G_2 and R_7 [275]. Also, here Υ^i are known as Trees parameters [330], M^k are the spin and orbital magnetic moments of the electrons, and P^k are three-electron electrostatic correlations. The Slater and the spin–orbit parameters for rare-earth ions in $LaCl_3$ are tabulated in Table 5.9.

† The Slater radial integrals are defined as

$$F_k(i.j) = \int_0^\infty \int_0^\infty \frac{r^{<k}}{r^{>k+1}} [P_i(r_1)P_i(r_2)]^2 \mathrm{d}r_1 \mathrm{d}r_2,$$

which is a parameter representing the radial distribution of the electrons.

Table 5.10. *Crystal field parameters of Nd^{3+} in LaF_3 and $LaCl_3$ crystals [87].*

Nd^{3+}: LaF_3 (D_{3h} approximation)	Nd^{3+}: $LaCl_3$ (C_{3h} approximation)
$\alpha = 21.28$	$\alpha = 22.1$
$\beta = -583$	$\beta = -650$
$\gamma = 1443$	$\gamma = 1586$
$\Upsilon^2 = 306$	$\Upsilon^2 = 377$
$\Upsilon^3 = 41$	$\Upsilon^3 = 40$
$\Upsilon^4 = 59$	$\Upsilon^4 = 63$
$\Upsilon^6 = -283$	$\Upsilon^6 = -292$
$\Upsilon^7 = 326$	$\Upsilon^7 = 358$
$\Upsilon^8 = 298$	$\Upsilon^8 = 354$
$M^0 = 2.24$	$M^0 = 2.1$
$M^2 = 1.25$	
$M^4 = 0.84$	
$P^2 = 213$	$P^2 = 255$
$P^4 = 160$	
$P^6 = 106.5$	
$B_0^2 = 216$	$B_0^2 = 163$
$B_0^4 = 1225$	$B_0^4 = -336$
$B_0^6 = 1506$	$B_0^6 = -713$
$B_6^6 = 770$	$B_6^6 = 462$

Then assuming that the $4f$ radial wave functions are for hydrogen, as discussed in Section 2.6.6, the relationship between F_k are obtained as [166]

$$F_4/F_2 = 41/297 = 0.138,$$
$$F_3/F_2 = 7.25/81.143 = 0.0151. \tag{5.278}$$

For the $4f$ shells a good estimate of F_2 is made with

$$F_2 = 12.4(z - 34)\ \text{cm}^{-1}, \tag{5.279}$$

where z is the atomic number of the rare-earth ion. Assuming that the lattice forms approximately hexagonal symmetry (D_{3h}), $H_{ion\text{-}lattice}$ (crystal field Hamiltonian) can be expanded as

$$H_{ion\text{-}lattice} = \sum_{k,q} B_q^k C_q^k = B_0^2 c_0^2 + B_0^4 c_0^4 + B_0^6 c_0^6 + B_0^2 (c_6^6 + c_{-6}^6), \tag{5.280}$$

where B_q^k are the crystal field parameters and c_q^k is a tensor operator acting on the normal coordinate of the electron. The nonvanishing component of c_q^k is dependent on the symmetry of the ion complex and their values should be looked up in references that deal with group theory. Some examples of crystal field parameters are listed in Table 5.10. These parameters are used to map the complete energy spectrum of the rare-earth ions [87]. Detailed examples are shown in [76]. Note

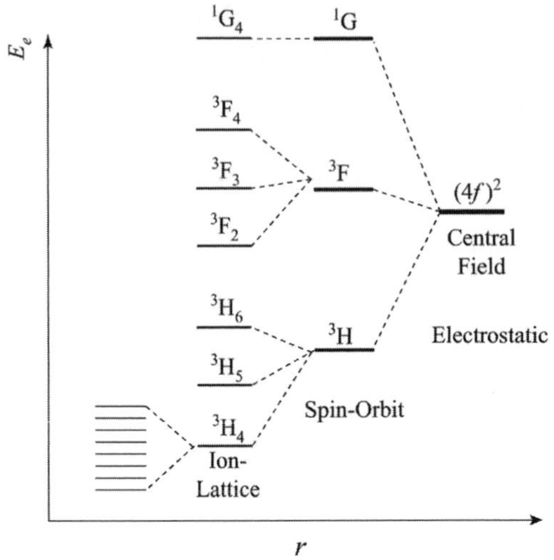

Figure 5.37. Energy diagram illustrating hierarchy of energy splittings, resulting from electron–electron and electron–host (ion) interactions for the f-block atoms (Table A.2) [88].

that, for transition metals, comprehensive examples can be found in [130] using *Dq* parameters.

Because the solutions are constructed from hydrogenic states, the total orbital angular momentum L and the total spin S are the quantum numbers. Here L and S are the vector sums of the orbital and spin quantum numbers for all the $4f$ electrons on the ion. Each f electron contributes an orbital quantum number of 3 and a spin of 1/2. Total orbital angular momenta are specified by the letters S, P, D, F, G, H, I, K,..., to represent L = 0, 1, 2, 3, 4, 5, 6, 7,..., respectively. Russell–Saunders coupling (LS coupling) is most often used. In this scheme, L and S are vectorially added to form the total angular momentum J, and the states are labeled $^{2S+1}L_J$ [131]. The quantum numbers (L, S, J, and another arbitrary one) define the terms of the configuration, all of which are the degeneration of the central-field approximation, as shown in Figure 5.37. The electrostatic interaction lifts the angular degeneracy and produces a spectrum of states with energies depending on L and S, but not on J.

Next in the hierarchy is the spin–orbital, the strongest of the magnetic interactions. Spin–orbit lifts the degeneracy in the total angular momentum and splits the LS terms into J levels. This uses the Hund rules for J for a particular atom [120]. The Hund first rule states that, all other being equal, the state with the highest total spin will have the lowest energy. The Hund second rule states that if a subshell is no more than half filled, then the lowest energy level has $J = |L - S|$; if it is more than half filled, then $J = L + S$ has the lowest energy.

The host has the least influence on the electronic structure and changes the positions of these levels only slightly. The static effects of the host on the rare-earth dopant customarily are treated by replacing the host with an effective crystal field potential at the ion site.

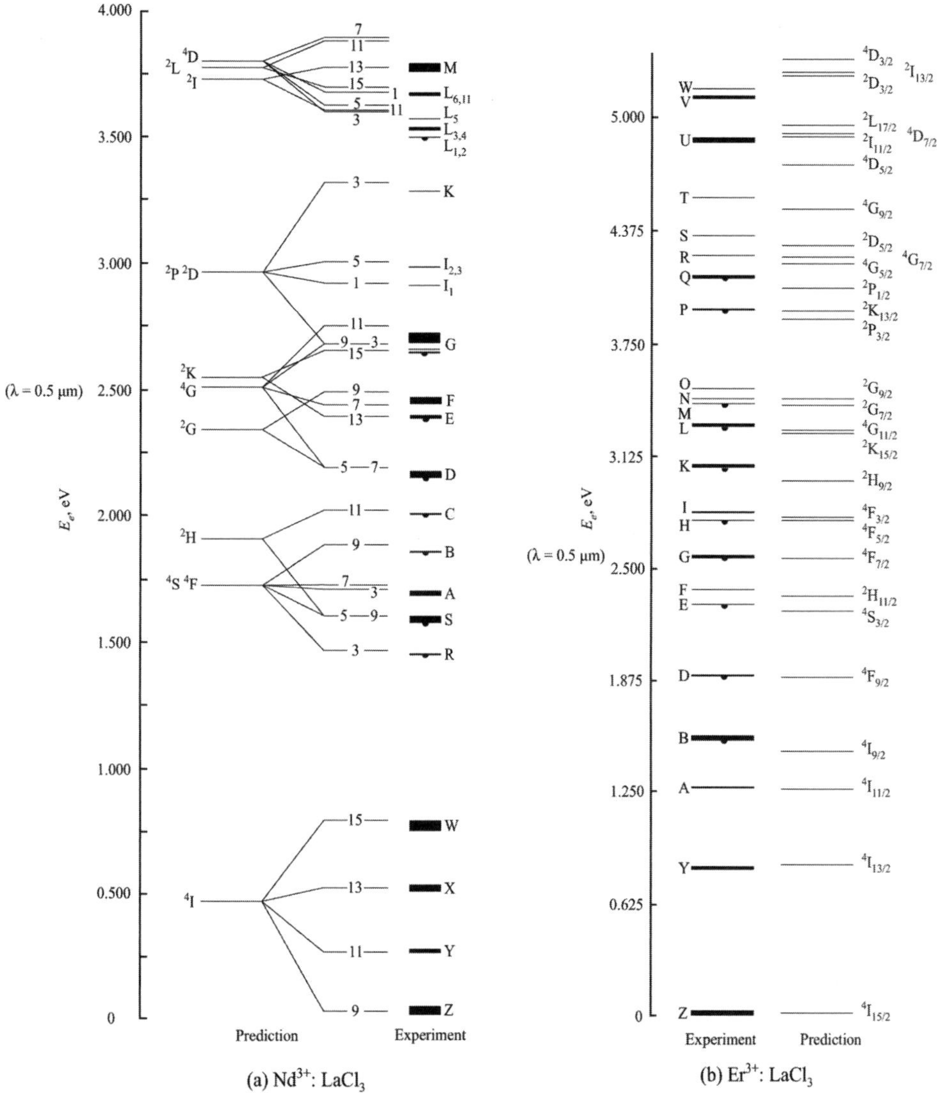

Figure 5.38. Empirical and computed energy levels of (a) $Nd^{3+}(4f^3)$ and (b) $Er^{3+}(4f^{11})$ in ion-doped $LaCl_3$. The semicircles indicate fluorescence (resonance R) in the $LaCl_3$ structure [88, 271].

Two examples of energy levels of rare-earth ion-doped crystals are shown in Figure 5.38.

In Section 7.12, we will discuss application of electron transition of ions embedded in host crystals in the laser cooling of solids.

5.21 Electronic Energy States of Gases

The electronic configurations of gas molecules provide precise, spectral electron transitions similar to the rare-earth ion-doped crystals. This is because gas molecules

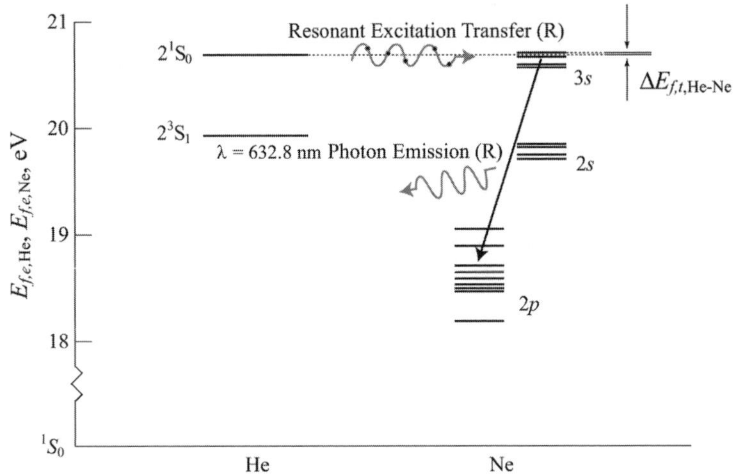

Figure 5.39. A partial energy diagram showing electronic energy states, a translational energy transfer, and a common laser transition found in the He–Ne laser system [348].

and ions doped in crystals are relatively isolated, so they do not form energy bands like atoms in a lattice structure.

The electronic energy states of a gas molecule are defined by the associated wave function of the molecule. The central field approximation and LCAO method can be applied to calculate the energy states of a molecule from the Schrödinger equation, as discussed in Chapter 2.

The notation used to represent the electronic energy states of gas molecules typically includes the LS coupling introduced in Section 5.21. In addition to the coupling, a prefix is included to indicate the excitation level of the atom. For instance, the ground state configuration of Ne is 1S0, and the second excited state with the same quantum spin (opposing spin of the excited electrons) would also be $^1S_0$, which we refer to as 2^1S_0 for clarity. The energy states of He are represented in this notation in Figure 5.39, which shows experimentally determined energy states of He and Ne atoms in a partial energy diagram.

For the excited states of Ne in Figure 5.39, Paschen notation [348] is used to refer to groups of electron configurations more easily. This represents a group of configurations based on the outermost electron shell, where the ground state electron configuration is assumed to be the 1*s* orbital. For example, the ground-state electron configuration of Ne is normally $2s^22p^53s^1$, but becomes 1*s* in the Paschen notation system. The Ne $2s_{2-5}$ electron configurations are equivalent to LS notation $^1P_1$, $^3P_0$, $^3P_1$, and $^3P_2$. Here P indicates that the principal quantum number of the 2*s* subshell is 2, and the other quantum numbers arise from the possible spin configurations of S = 0 or S = 1.

There are at least three basic electronic energy transfers involved in the 632.8 nm He–Ne laser cycle: electrical excitation of He, resonant excitation transfer from He to Ne, and the subsequent laser transition in Ne to a lower energy level.

Electrical discharge is used to electronically excite gas. For discharge to occur, atoms are forced to ionize to conduct electricity. Not only does the ionization energy necessarily excite and free electrons, but traveling electrons also collide with atoms and can stimulate upper electronic energy levels in those atoms. In an electrical discharge through He, the atoms become overexcited and quickly relax to their uppermost stable energy state, He 2^1S_0. A high population of these molecules accumulates.

Resonant excitation transfer is a collisional process that transfers electronic energy between fluid particles. This occurs only when energy states are sufficiently close, so that an electron–electron impact between fluid particles can transfer electronic energy from one fluid particle to the other. The energy coincidence between the He 2^1S_0 and Ne $3s$ energy levels allows for resonant energy transition between the two molecules, electronically exciting the Ne $3s$ energy level. The reversible excitation can be represented as a spin-conserving energy transfer:

$$E_{f,e,\text{He}}(2^1\text{S}_0) + E_{f,e,\text{Ne}}(^1\text{S}_0) = E_{f,e,\text{He}}(^1\text{S}_0) + E_{f,e,Ne}(^3\text{S}_2) - \Delta E_{f,t,\text{He-Ne}}. \tag{5.281}$$

Photon emission of electronic energy states of gases occurs only under special circumstances. Nonradiative decay processes compete with photon emission. If the relaxation times of electronic energy states of a fluid particle are too high, collisional deactivation will take place before radiation occurs. The rate of collisions, particle velocity, and the magnitude of energy transfer all affect the quickness of nonradiative decay processes.

For lasing to occur, more atoms must be in an excited state than in the ground state; this is called population inversion. Ideally, for lasing, atoms should be excited very selectively into a single, upper energy level. It is important to give atoms the same energy states, so they can emit the same photon. If atoms are excited in different energy states they cannot give the same energy photon, and stimulated emission will not occur. When the rate of excitation is high enough to produce a population inversion, stimulated emission occurs.

There are a number of factors in the He–Ne system that collectively produce the population inversion. A high number of He atoms are selectively excited into their stable, upper electronic energy level. The He atoms very selectively excite the Ne $3s^2$ energy level by excitation transfer. Also, because of the large energy gaps between electronic energy states of gases, it is difficult to deactivate the energy levels through collisional processes. The population of excited Ne atoms accumulates, and stimulated emission becomes favorable, producing laser light.

Although the laser produces relatively high-energy, short-wavelength laser light, this transition is not a high power source of laser energy because it produces

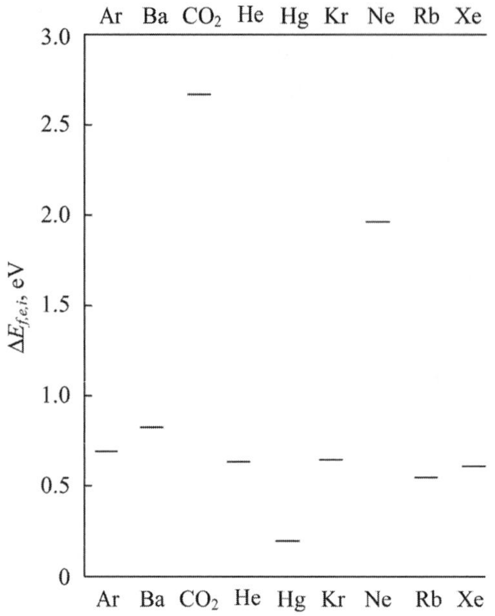

Figure 5.40. Some electronic transition energies for Ar, Ba, CO_2, He, Hg, Kr, Ne, Rb, and Xe gaseous atoms or molecules [248, 348].

a relatively low rate of photon emission. The efficiency of a typical He–Ne laser is around 0.1%. Inefficiencies arise because of the difficulty in forcing an increase in the electronic states of gases, and a significant portion of the energy in the cycle does not produce lasing. The remaining energy is dissipated as heat and must be removed from the laser system. Usually this is accomplished by cycling gases. The major benefits provided by the He–Ne laser system are the high stability that it has compared with that of other lasers, a long operating lifetime, and low manufacturing costs. The energies of some other electronic transitions in gases are shown in Figure 5.40, although there are a number of electronic transition energies for any gas molecule. We will discuss lasers in Section 7.2.

5.22 Size Effects

Thin (comparable with the de Broglie wavelength) layers of semiconductors (intrinsic and doped) can trap (confine) conduction electrons within them, and these electrons behave as quantum wave packets (as compared with particles). This confining structure is called a quantum well (QW). The electron energy bands become discrete and are called the subbands. Confining the electron along a second dimension produces quantum wire and along all three directions gives quantum dot (QD).

The unique characteristic of a QD is that the trapped electron behaves as it belongs to an atom (but subject to the symmetry of the lattice). Because of the discrete energy states, the quantum well, wire, and dot have discrete distributions of the DOS. Figure 5.41 shows the general distribution of the electron DOS $D_e(E_e)$ for these as compared with the bulk semiconductors (idealized), given by (5.65), which show an $E_e^{1/2}$ dependence. Quantum dots have the most discrete $D_e(E_e)$ DOS representing only singularities.

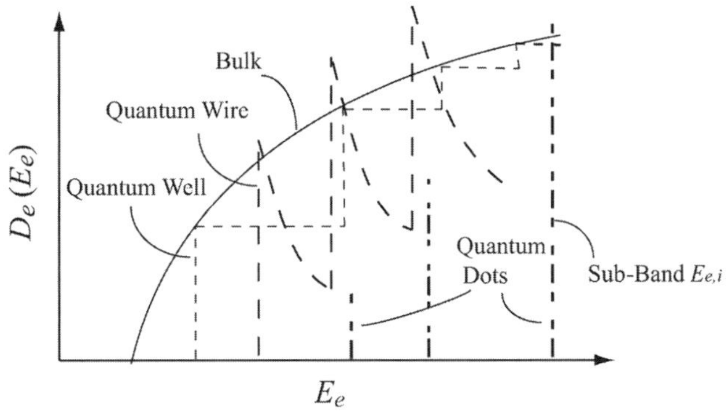

Figure 5.41. Distribution of electron DOS for idealized bulk, QW, quantum wire, and QD semiconductors [67]. The first three have continuous distributions, whereas the quantum dots have singularities (subbands) only.

5.22.1 Quantum Well for Improved TE Z_eT

As indicated above, the quantum wells have $D_e(E_e)$ with discrete distribution, i.e., zero variance around $E_{e,i}$ where $E_{e,i}$ is a sub-band. Here, following [224] it is shown that such discontinuity maximizes the TE figure of merit Z_eT defined by (5.203). We begin by rewriting (5.124) to (5.127), or (5.206), as [224]

$$\sigma_e = e_c^2 \int_{-\infty}^{\infty} (-\frac{\partial f_e^{\text{o}}}{\partial E_e}) G(E_e) \text{d}E_e \tag{5.282}$$

$$T\sigma_e \alpha_{\text{S}} = e_c \int_{-\infty}^{\infty} (-\frac{\partial f_e^{\text{o}}}{\partial E_e}) G(E_e)(E_e - \mu) \text{d}E_e \tag{5.283}$$

$$Tk_{e,\text{o}} = e_c \int_{-\infty}^{\infty} (-\frac{\partial f_e^{\text{o}}}{\partial E_e}) G(E_e)(E_e - \mu)^2 \text{d}E_e \tag{5.284}$$

$$k_e = k_{e,\text{o}} - T\sigma_e \alpha_{\text{S}}^2 \tag{5.285}$$

$$G(E_e) = \sum_{\kappa} u_{e,g,\kappa}^2 \tau_e(\kappa) \delta_{\text{D}}[E_e - E_e(\kappa)] \tag{5.286}$$

$$-\frac{\partial f_e^{\text{o}}}{\partial E_e} = \frac{1}{k_{\text{B}}T} \frac{\exp(\dfrac{E_e - \mu}{k_{\text{B}}T})}{[\exp(\dfrac{E_e - \mu}{k_{\text{B}}T}) + 1]^2}, \tag{5.287}$$

where $G(E_e)$ is the transport distribution function and $k_{e,\text{o}}$ is the electrical thermal conductivity with zero electrochemical potential gradient.

Using $\partial f_e^{\text{o}}/\partial E_e$, the transport coefficients are written in terms of integrals I_n as

$$\sigma_e = \sigma_{e,\text{o}} I_0, \ \sigma_{e,\text{o}} = \frac{e_c^2}{\hbar r_{\text{B}}} \tag{5.288}$$

$$\sigma_e \alpha_{\text{S}} = \frac{k_{\text{B}}}{e_c} \sigma_{e,\text{o}} I_1 \tag{5.289}$$

$$k_{e,o} = (\frac{k_B}{e_c})^2 T\sigma_{e,o} I_2 \tag{5.290}$$

$$I_n = \int_{-\infty}^{\infty} \frac{e^x}{(e^x+1)^2} G^*(x) x^n \mathrm{d}x \tag{5.291}$$

$$G^* = \hbar r_B G(\mu + x k_B T), \tag{5.292}$$

where r_B is the Bohr radius (Table 1.4).

Then we have for (5.203)

$$Z_e T = \frac{\dfrac{T\sigma_e \alpha_S^2}{k_p}}{\dfrac{k_{e,o}}{k_p} - \dfrac{T\sigma_e\alpha_S^2}{k_p} + 1} = \frac{B\dfrac{I_1^2}{I_0}}{BI_2 - B\dfrac{I_1^2}{I_0} + 1}$$

$$\equiv \frac{\xi}{1-\xi+C}, \quad \xi = \frac{I_1^2}{I_0 I_2}, \quad C = \frac{1}{AI_2}, \quad B = (\frac{k_B}{e_c})^2 \frac{T\sigma_{e,o}}{k_p}, \tag{5.293}$$

where B is related to the Chasmar–Stratton coefficient defined in relation to (5.203).

Because $\xi \ll 1$, the maximum $Z_e T$ is at $\xi = 1$, i.e.,

$$Z_e T \le \frac{1}{C} = \frac{k_{e,o}}{k_p}. \tag{5.294}$$

For parabolic bands, we have

$$G(E_e) = D_e(E_e) u_{e,g,x}^2(E_e) \tau_e(E_e), \tag{5.295}$$

and it is shown in [224] that, when $D_e(E_e)$ has a Dirac function behavior, then $\xi \to 1$. Also, when the dimensionless $G^*(x)$ is expressed as

$$G^*(x) = H(x)\delta_D(x - E^*), \tag{5.296}$$

we have for the transport coefficients

$$\sigma_e = \sigma_{e,o} H(E^*) \frac{e^{E^*}}{(e^{E^*}+1)^2}, \quad E^* = \frac{E_{e,m}}{k_B T} \tag{5.297}$$

$$\alpha_S = \frac{k_B}{e_c} E^* \tag{5.298}$$

$$k_{e,o} = (\frac{k_B}{e_c})^2 T\sigma_{e,o} H(E^*) E^{*2} \frac{e^{E^*}}{(e^{E^*}+1)^2}, \tag{5.299}$$

where $E_{e,m}$ is the peak energy with respect to Fermi level.

To maximize $k_{e,o}$ (and $Z_e T$), if $H(E^*) = H$ is assumed constant (from Figure 5.30, we note that α_S has a maximum where $\mu_\circ$ is about a few $k_B T$ from the Fermi level), then the maximum in $E^{*2} e^{E^*}/(e^{E^*}+1)^2$ is at $E^* = 2.4$, i.e., having resonance $E_e = 2.4 k_B T$ above the Fermi level. Using this gives $\alpha_S = 207$ μV/K, as we also found from (5.203). Then the optimal $Z_e T$ is

$$(Z_e T)_{max} = 0.439 (\frac{k_B}{e_c})^2 \frac{\sigma_{e,o} T}{k_p} H. \tag{5.300}$$

Then the QWs having a Dirac delta function $D_e(E_e)$ can maximize the TE figure of merit. Quantum dots offer the possibility of having a single delta function and are the preferred material. These carrier pockets are discussed in [44, 137, 187].

5.22.2 Reduced Electron–Phonon Scattering Rate Quantum Wells

In molecularly constructed (e.g., epitaxy), multilayer (including superlattice) semiconductors, with a layer thickness of the order of 10 Å, the electron confinement also modifies its scattering by phonons [284]. Multiple QW structures consist of a series of QWs made of alternating layers of wells and barriers. When the barrier thickness is large ($\geq$ 40 Å), no significant electronic coupling occurs between neighboring wells and it is called a QW, and for this barrier ($\leq$ 40 Å), electronic coupling occurs and it is called a superlattice.

Confinement of electrons (and holes) influences their phonon-scattering rate, and in the case of harvesting hot electrons (and holes), excess kinetic energy, generated in photon absorption by semiconductors (as in solar cells, Section 7.14), reduction of $\tau_{e\text{-}p}$ is desirable [255]. The electron energy for the case of a QW is the solution to the Schrödinger equation for the extended case of a particle in a box (Section 2.6.6) for a finite barrier height φ_o, and gives [255]

$$E_{e,n} = \frac{\hbar^2 \kappa^2}{2m_{e,e}} - \varphi_o, \ \kappa = \frac{n\pi}{L}, \ n = 1, 2, 3, ..., \tag{5.301}$$

where L is the well width.

As shown in Figure 5.41, the ideal QW D_e is a steplike function with each plateau given by

$$D_e = \frac{n m_{e,e}}{\pi \hbar^2}, \quad n = 1, 2, 3, ... \tag{5.302}$$

The hot-electron cooling time $\tau_{e\text{-}p}$, Figure 5.42(a), is measured with the energy-loss technique, after photon absorption [255], and the average relaxation time (which is because of electron scattering by LO phonons) is found from

$$-\frac{\mathrm{d}E_e}{\mathrm{d}t} = \frac{\hbar\omega_{p,\mathrm{LO}}}{\tau_{e\text{-}p}} \exp(-\frac{\hbar\omega_{p,\mathrm{LO}}}{k_\mathrm{B} T_e}), \tag{5.303}$$

where T_e is the electron temperature. The experimental results for multiple 250-Å GaAs/250-Å $Al_{0.38}Ga_{0.62}As$ QWs are shown in Figure 5.42(b) [255].

The electron temperature was deduced from the time-resolved luminesce spectra [255]. The carrier densities calculated from photon absorption are 10^{19}, 5×10^{18}, and 2×10^{18} 1/cm^3 for the 25-, 12.5-, and 5-mW laser (pumping) powers.

The results for Figure 5.42(b) show that the cooling rate $1/\tau_{e\text{-}p}$ is slower in the metallic QWs (by up to two orders of magnitude). They also show that the higher photogenerated carrier density, the slower the cooling rate, especially in the QWs. This is referred to as the hot-phonon bottleneck, because of the significant nonequilibrium distribution of phonons (in particular, LO) caused by the hot carriers. The

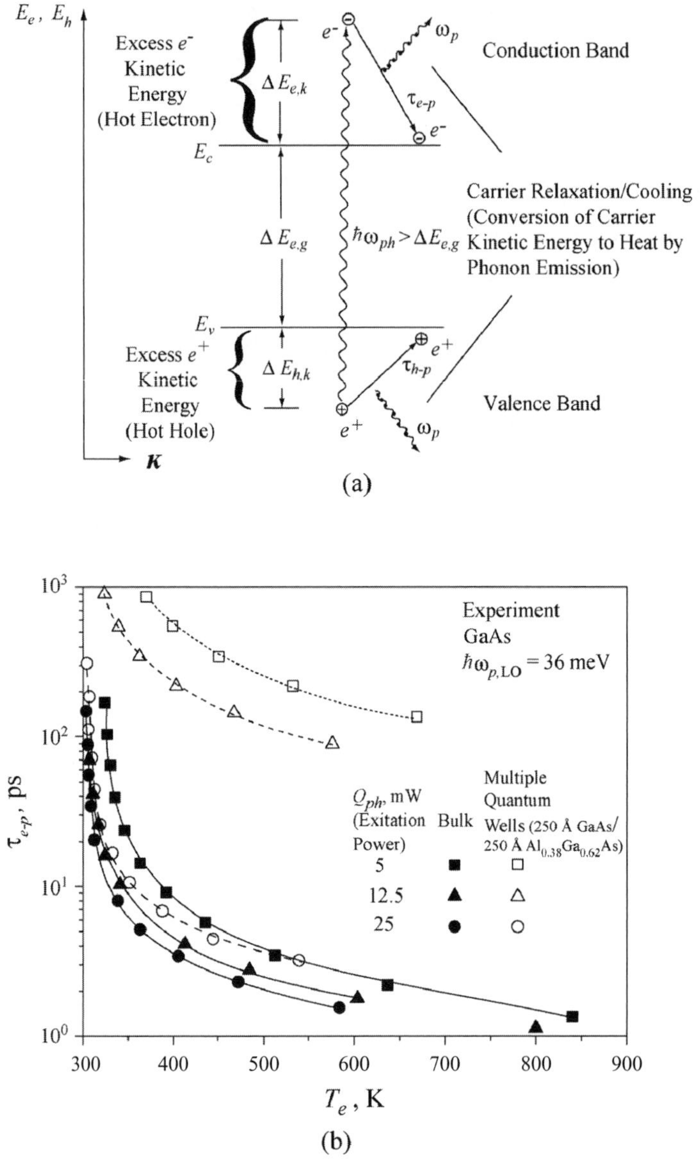

Figure 5.42. (a) Hot electrons (excess kinetic energy that is due to photon absorption) and holes in semiconductors, and phonon emission during their decay. (b) Variation of electron–phonon relaxation times for a GaAs bulk and multiple QWs, with respect to electron temperature. The hot electrons are generated by photon (laser) absorption [255].

relaxation time $\tau_{p\text{-}p}$(LO) is not small enough to return phonons to equilibrium quickly (with the crystal bath) to keep up with the phonon emission rate, especially in the QWs where phonon modes are also modified.

The reduced $\tau_{e\text{-}p}$ shown in Figure 5.42(b) can help in harvesting the hot electrons (through impact ionization) before they cooled [286].

5.23 Problems

Problem 5.1

(a) In the Kronig–Penney model, for $s = 5$, determine the first five allowed electronic energy bands and plot the results similar to Figure 5.4.

(b) Then plot E_e(eV) versus κ(1/nm), using (5.26), similar to Figure 5.5, using the Si lattice constant a from Table A.2. Scan κ^* and obtain roots E_e^*, and then convert to dimensional values.

Problem 5.2

Show that the small κa limit of the s-orbital energy equation for the s-orbital, FCC tight-binding model becomes $E_e(\boldsymbol{\kappa}) = E_s - \beta_s - 12\gamma + \gamma a^2\kappa^2$.

Problem 5.3

Make a three-dimensional $\{\kappa_x, \kappa_y, \kappa_z\}$ contour plot of a constant electron-energy surface, which is based on the s-orbital energy equation for the s-orbital, FCC tight-binding model (5.40). Note that there is no need to specify the constants when using $(E_e - E_s + \beta_s)/|\gamma| = 2$ as the constant energy surface. Also use $-1 \le a\kappa_i/2\pi \le 1$. The contour, is similar to that shown in the figure.

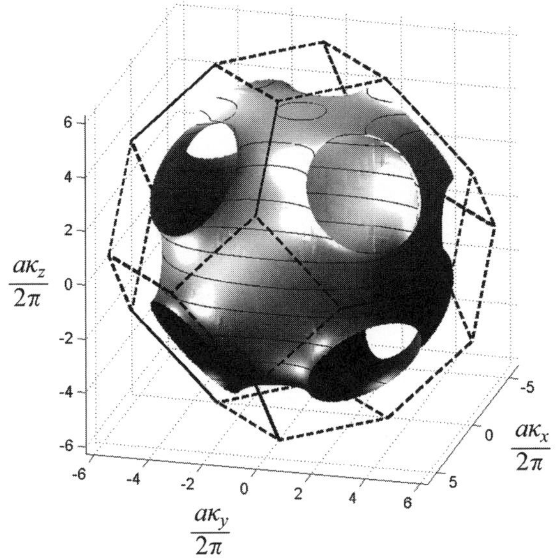

Problem 5.3. Constant energy surface (Fermi Surface), $(E_e - E_s + \beta_s)/|\gamma|$, in reciprocal lattice space, for s-orbital, FCC lattice, with the nearest-neighbor only, tight-binding approximation.

Problem 5.4

(a) Comment on the approximations (tight binding and others) made in the SiC electronic band structure Matlab codes.

(b) Run the SiC Matlab codes and comment on the discrepancies in the predicted absolute values of the band energies.

Problem 5.5

Explain the physical significance of the FGR and the interaction matrix element.

Problem 5.6 Complete the integration to arrive at the energy-averaged relaxation time (5.112), and evaluate the gamma function for the scattering mechanisms given in Table 5.3.

Problem 5.7

Complete the steps and derive (5.130) from (5.124).

Problem 5.8

Complete the steps and derive (5.136) from (5.125).

Problem 5.9

(a) Complete the steps and derive (5.139) from (5.127).

(b) Show how the transport coefficient in (5.140) to (5.142) are related to those appearing in (5.122) and (5.123).

(c) Derive the relations among σ_e, k_e, α_S, and α_P, and α_{ee}, α_{et}, α_{te}, and α_{tt} as given in (5.144) to (5.145).

Problem 5.10

Acoustic lattice vibration also produces electric field $\boldsymbol{e}_e$, and this scatters electrons, most notably at high temperatures. For a one-dimensional motion, the interaction (scattering) potential $\varphi_{s,e\text{-}p}$ is generally

$$\varphi_{s,e\text{-}p}(x,t) = -e_c \int e_{e,x} \mathrm{d}x.$$

Requiring periodic deformation potential gives, using displacement along $\boldsymbol{x}$, for piezoelectric scattering,

$$\epsilon_o \epsilon_e e_{e,x} = -e_{pz} \frac{\partial d_x}{\partial x},$$

or

$$e_{e,x} = -\frac{e_{pz}}{\epsilon_o \epsilon_e} \frac{\partial d_x}{\partial x},$$

where e_{pz} is the piezoelectric constant.

Using these, show that

$$\varphi_{s,e\text{-}p}(x,t) = \frac{e_c e_{pz}}{\epsilon_o \epsilon_e} d_x(x,t).$$

Problem 5.11

Assume the piezoelectric-scattering potential and the displacement distribution

$$\varphi_{s,e\text{-}p} = \frac{e_c e_{pz}}{\epsilon_o \epsilon_e} d, \ d = d_o e^{\pm i(\kappa_p \cdot x)},$$

for the LO phonon scattering of electrons. Here d_o is the magnitude of lattice-vibration displacement.

Then write (5.94) for the matrix element as

$$M_{p',p} = \frac{1}{V_c}\int \frac{e_c e_{pz}}{\epsilon_o \epsilon_e} d_o \exp[\frac{-i(p' - p \mp \hbar\kappa_p)\cdot x}{\hbar}]dx$$
$$= \frac{e_c e_{pz} d_o}{\epsilon_o \epsilon_e}\frac{1}{V_c}\int \exp[\frac{-i(p' - p \mp \hbar\kappa_p)\cdot x}{\hbar}]dx.$$

The displacement is given as [219]

$$d_o^2 = \frac{\hbar}{2\rho V_c \omega_{p,O}}(\frac{k_B T}{f_p^\circ + \frac{1}{2} \mp \frac{1}{2}}), \quad f_p^\circ = \frac{k_B T}{\hbar\omega_{p,O}}.$$

Using $\omega_p/\kappa_p = u_{p,A}$, show that

$$|M_{p',p}|^2 = \frac{e_c^2 e_{pz}^2 k_B T}{2\epsilon_o^2 \epsilon_e^2 E_p \kappa_p^2 V_c}\{\frac{1}{V_c}\int \exp[\frac{-i(p' - p - \hbar\kappa_p)\cdot x}{\hbar}]dx\}^2, \text{ for the } -\frac{1}{2} \text{ case},$$

where, from (4.72), $u_{p,A} = u_{p,b}$ and is

$$u_{p,A} = (E_p/\rho)^{1/2},$$

where E_p is the bulk modulus of elasticity.

Problem 5.12

Show that solutions (5.266) and (5.267) satisfy (5.264) and (5.265).

Problem 5.13

(a) Determine the cooling length δ for Al, Si (intrinsic), and Te, at $T = 300$ K. Use Table A.1 for the total thermal conductivity, and Figure 5.20(b) for electrical conductivity. Then use the Wiedemann–Franz relation (assume phonon scattering is dominant) to find the electronic thermal conductivity. Next use these to determine the phonon conductivity. Also determine the relaxation time from the electrical conductivity (5.146). Figure 5.20(a) gives $n_{e,c}$ for the three elements. The average acoustic phonon speed for Si is given in Figure 5.13(a), for Al it is 6240 m/s, and for $T_e = 2610$ m/s.

(b) Comment on the accuracy of the Wiedemann–Franz relations for semiconductors and metals.

(c) Comment on the ability (spatial constraints) of harvesting hot electrons in these materials. Use data in handbooks, etc., as well as those given in the text.

Problem 5.14

(a) Consider a Sb_2Te_3 thermoelectric element with length L and its two ends at T_h and T_c [83]. Using the one-dimensional solution for the temperature distributions of electrons and phonons in thermal nonequilibrium, plot the electron and phonon temperatures $T_e(x)$, $T_p(x)$, $-2 \le x \le +2$ μm, assuming the coefficients for zero electric current ($j_e = 0$) (A_1 through A_4) given in (5.274) are valid. Use the following conditions and for (i) $J_e = 0$, 15, and (ii) 30 mA.

$T_h = 300$ K, $T_c = 285$ K, $\alpha_S = 171$ μV/K, $\rho_e = 1.04\times10^{-5}$ Ω-m, $k_e = 0.6$ W/m-K, $k_p = 1.5$ W/m-K, $k = k_e + k_p$, $L = 4$ μm, $AR_{k,e,b} = 9.3\times10^{-7}$ K/(W/m^2), $A = 80$ μm^2, $AR_{k,p,b} = 8.0\times10^{-8}$ K/(W/m^2), and $\delta = 156$ nm.

Note that all conductances are per unit area. Also note that the temperatures for $j_e = 0$ and $j_e > 0$ do not match near the two ends. The numerical solution to this problem [i.e., not using (5.274) coefficients] is given in [83].

(b) Comment on the extent of thermal nonequilibrium within the element.

Problem 5.15

Consider conduction boundary resistance at the interface of a metal and a dielectric [227]. The interface is at $x = 0$, and the metal occupies $x \geq 0$.

In the electron–phonon (lattice) thermal nonequilibrium (Section 5.19), the carrier energy equations are given as

$$k_e \frac{d^2 T_e}{dx^2} = \frac{n_e k_B}{\tau_{e\text{-}p}}(T_e - T_p) \quad 0 \leq x < \infty$$

$$k_p \frac{d^2 T_p}{dx^2} = -\frac{n_e k_B}{\tau_{e\text{-}p}}(T_e - T_p) \quad 0 \leq x < \infty.$$

(a) Using $\Delta T = T_e - T_p$, and

$$\frac{1}{\delta^2} = \frac{1}{\delta_e^2} + \frac{1}{\delta_p^2}, \quad \delta_i^2 = \frac{k_i \tau_{e\text{-}p}}{n_e k_B},$$

and using $\Delta T = 0$ for $x \to \infty$, show that $\Delta T = a_1 \exp(-x/\delta)$, where a_1 is a integration constant.

(b) Using this in the electron energy equation, show that

$$T_e = a_1 (\frac{\delta}{\delta_e})^2 \exp(-\frac{x}{\delta}) + a_2 x + a_3.$$

Using $dT_e/dx|_{x=0} = 0$, show that

$$T_e = a_1 (\frac{\delta}{\delta_e})^2 \exp(-\frac{x}{\delta}) + a_1 \frac{\delta}{\delta_e^2} x + a_3$$

$$T_p = a_1 [(\frac{\delta}{\delta_e})^2 - 1] \exp(-\frac{x}{\delta}) + a_1 \frac{\delta}{\delta_e^2} x + a_3.$$

(c) Now, the boundary thermal conditions for T_p and T_e are given in Section 5.19, as

$$k_p \frac{dT_p}{dx} = \frac{T_p - T}{AR_{p,b}} \quad \text{at } x = 0$$

$$k_e \frac{dT_e}{dx} = 0, \ k_p \frac{dT_p}{dx} = k \frac{dT}{dx} \quad \text{at } x = 0,$$

where k and T are for the dielectric, $x \le 0$ (dielectric material), and $1/AR_{p,b}$ is the phonon–phonon boundary conductance. Use this to show that the linear temperature distribution in the dielectric is [starting with $T = a_4(x) + a_5$]

$$T(|x|) = -\frac{k_p}{k}\frac{a_1}{\delta}|x| - a_1[1 + \frac{AR_{p,b}\delta}{k_p} - (\frac{\delta}{\delta_e})^2] + a_3,$$

so the unknown constants are a_1 and a_3.

(d) Now using

$$\frac{1}{AR_b} = \frac{q}{T_e'(0) - T(0)}, \quad q = -k\frac{\mathrm{d}T}{\mathrm{d}x}|_{x=0},$$

where $T_e'(0) = a_3$ is the slip electron temperature, show that the interface (boundary) resistance R_b is

$$AR_b = \frac{\delta(1 + \frac{AR_{p,b,\delta}}{k_p} - (\frac{\delta}{\delta_e})^2]}{k_p}, \quad \frac{1}{\delta^2} = \frac{n_e k_B}{\tau_{e\text{-}p}}(\frac{1}{k_e} + \frac{1}{k_p}) = \frac{n_e k_B}{\tau_{e\text{-}p}}\frac{k_p + k_e}{k_e k_p},$$

where the cooling length $\delta = \delta_{e\text{-}p}$ is the same as that given by (5.268).

Problem 5.16

(a) Using the Taylor series expansion (around solution at $t = t_o$) for the time-dependent wave function

$$\Psi(\boldsymbol{x}, t + t_o) = \sum_{n=0}^{\infty} \frac{t_o^n}{n!}(\frac{\partial}{\partial t})^n \Psi(\boldsymbol{x}, t),$$

satisfying the time-dependent Schrödinger equation (Table 2.8), show that

$$\Psi(\boldsymbol{x}, t + t_o) = \exp(-\frac{i\mathrm{H}t_o}{\hbar})\Psi(\boldsymbol{x}, t).$$

Note that H does not explicitly depend on t. Also, note that the proceeding series represents an exponential function, and that it is related to $\partial/\partial t$.

(b) Show that the expectation value of the dynamical variable $\phi(\boldsymbol{x}, \boldsymbol{p}, t)$ at time $t + t_o$

$$\langle\phi\rangle_{t+t_o} = \langle\Psi(\boldsymbol{x}, t + t_o)|\phi(\boldsymbol{x}, \boldsymbol{p}, t + t_o)|\Psi(\boldsymbol{x}, t + t_o)\rangle$$

is

$$\langle\phi\rangle_{t+t_o} = \langle\Psi^\dagger(\boldsymbol{x}, t)|e^{\frac{i\mathrm{H}t_o}{\hbar}}\phi(\boldsymbol{x}, \boldsymbol{p}, t + t_o)e^{-\frac{i\mathrm{H}t_o}{\hbar}}|\Psi(\boldsymbol{x}, t)\rangle.$$

Note that the Hermiticity of H gives

$$(e^{-\frac{i\mathrm{H}t_o}{\hbar}})^\dagger = e^{\frac{i\mathrm{H}t_o}{\hbar}}.$$

(c) Using the kinetics definition of temperature

$$T = \frac{p^2}{2m},$$

show that in quantum mechanics

$$\langle \boldsymbol{p} \rangle = -i\hbar \int \Psi^\dagger \nabla \Psi d\boldsymbol{x} = \int \Psi^\dagger (\frac{\hbar}{i}) \nabla \Psi d\boldsymbol{x}$$

and

$$\langle T \rangle = -\frac{\hbar^2}{2m} \int \Psi^\dagger \nabla^2 \Psi d\boldsymbol{x}.$$

Problem 5.17

(a) Derive expression (5.203) for the thermoelectric figure of merit.

(b) Show that optimum value ($\partial Z_e T/\partial \xi = 0$, use $\eta = \exp \xi$ and take derivative $\partial Z_e T/\partial \eta$) of ξ_o is found from

$$\xi_o + 4(s + \frac{5}{2})\frac{k_B^2 T}{e_c} AB \exp(\xi_o) = s + \frac{1}{2}$$

or

$$\xi_o \simeq s + \frac{1}{2} \quad \text{for negligible } k_e.$$

(c) Using Table 5.3, show that for acoustic-phonon scattering, $\xi_o = 0$, and for weak ionized impurity scattering, $\xi_o = 2$.

Problem 5.18

(a) Plot $Z_e(T)$ for n-type Bi_2Te_3, using the following parameters [123] and the optimum Fermi energy, $E_F \simeq \xi_o k_B T = (s + 1/2)k_B T$ (for negligible k_e).

$k_p = 0.7$ W/m-K (parallel to c axis), $s = -1/2$, $m_{e,e} = 0.58m_e$, and $\mu_e = 0.120$ m^2/V-s, at 300 K.

(b) Comment on the occurance of a maximum and compare with Figure 5.27.

Problem 5.19

(a) Show that the solution to (5.264) and (5.265) for the case of $T_e = T_p$ at $x \to 0$ (assuming thermal equilibrium at the surface), and finite $T_e - T_p$ for $x \to \infty$ (away from the surface), is

$$T_e - T_p = \frac{\rho_e j_e^2 \delta^2}{k_e}(1 - e^{-x/\delta}),$$

where δ is the cooling length.

(b) Evaluate the $T_e - T_p$ expression for $x \to \infty$ for Si at $T = 300$ K, $k_p = 148$ W/m-K, $k_e = 1$ W/m-K, $\rho_e = 2.33 \times 10^3$ Ω-m, $\delta = 5 \times 10^{-7}$ m, and $j_e = 10^{12}$ A/m^2.

Problem 5.20

(a) Table 5.2 shows that $-E_F/k_B T = -T_F/T$ can be represented by $-\infty$ for moderate temperatures. Using this, show that the integral in (5.71) becomes

$$\int_{-\infty}^{\infty} \frac{x^2 e^x}{(e^x + 1)^2} dx = \frac{\pi^2}{3},$$

by dividing the integral into two equal parts $\int_{-\infty}^{0}$ and $\int_{0}^{\infty}$, and using exp(ax) to keep x positive for both integrals. Then use integration by parts [$u = x^2$, $\mathrm{d}v = e^x(e^x + 1)^{-2}\mathrm{d}x$] to change $(e^x + 1)^{-2}$ to $(e^x + 1)^{-1}$. Then the integral containing $(e^x + 1)^{-1}$ is available in handbooks. Symbolic integral solvers can also be used.

(b) Starting with (2.101), show that

$$\ln n_{e,c} = \frac{3}{2}\ln E_{\mathrm{F}} + c$$

$$\frac{\mathrm{d}n_{e,c}}{n_{e,c}} = \frac{3}{2}\frac{\mathrm{d}E_{\mathrm{F}}}{E_{\mathrm{F}}},$$

and that the free-electron energy DOS is (5.73), i.e.,

$$D_e(E_e) \equiv \frac{\mathrm{d}n_{e,c}}{\mathrm{d}E_{\mathrm{F}}} = \frac{3n_{e,c}}{2E_{\mathrm{F}}}.$$

(c) Use the result of part (b) in (5.71) and the result of part (a), to arrive at (5.75)

$$c_{v,e} = \frac{\pi^2}{2}\frac{n_{e,c}}{E_{\mathrm{F}}}k_{\mathrm{B}}^2 T.$$

Problem 5.21

Using (5.64), show that for $\hbar\omega = 0$ and a constant interaction matrix, the FGR (5.95) becomes

$$\dot{\gamma}_e(\kappa', \kappa) = \frac{2\pi}{\hbar}|M_{\kappa',\kappa}|^2 D_e[E_e(\kappa)],$$

where D_e is the electron DOS. Allow for electron spin.

Problem 5.22

(a) Use $m_{e,e}/m_e = 1.03$ for Si and $T = 300$ K to show that $n_{e,F} = 2.61\times10^{19}$ 1/cm^3.

(b) Use $m_{h,e}/m_e = 0.81$ for Si and $T = 300$ K to show that $n_{h,F} = 1.83\times10^{19}$ 1/cm^3.

(c) Use the relation for intrinsic carrier density (footnote of Section 5.7),

$$n_i = (n_{e,\mathrm{F}} n_{h,\mathrm{F}})^{1/2}\exp\left(-\frac{\Delta E_{e,g}}{2k_{\mathrm{B}}T}\right),\ i = e \text{ or } h,$$

with gap energy $\Delta E_{e,g} = 1.12$ eV, for Si and $T = 300$ K to show that $n_i = 8.736\times10^9$ 1/cm^3. Check this against Figure 5.10(a)(ii).

(d) Using the expression for the intrinsic Fermi energy

$$E_{\mathrm{F}} - E_{\mathrm{F,i}} = k_{\mathrm{B}}T\ln\frac{n_{e,c}}{n_i} \quad \text{for } n\text{-type}$$

$$E_{\mathrm{F}} - E_{\mathrm{F,i}} = -k_{\mathrm{B}}T\ln\frac{n_{h,c}}{n_i} \quad \text{for } p\text{-type},$$

plot $E_{\mathrm{F}} - E_{\mathrm{F},i}$ (x axis) (in eV) versus $n_{e,c}$ (y axis, logarithmic scale), for $n_i < n_{e,c}(n_{h,c}) < 10^{21}$ 1/cm^3 and for $T = 300$ K. This gives the relation between the chemical potential $\mu = E_{\rm F}$ and the extrinsic carrier density $n_{e,c}$ (or $n_{h,c}$), which is separately determined from the measured Hall coefficient, (5.160).

Problem 5.23

The hyperbolic heat conduction equation based on finite wave speed has been suggested for from the electron heat transfer in metals for short-pulse-laser heating [228] in thin-metal films. Start with the conduction energy equation (Table 1.1), i.e.,

$$n_{e,c}c_{p,e}\frac{\partial T_e}{\partial t} + \nabla \cdot \boldsymbol{q}_e = \dot{s},$$

and use the constitutive relation for the thermal wave, i.e.,

$$\boldsymbol{q}_e(\boldsymbol{x}, t + \tau) = -k\nabla T_e$$

$$\boldsymbol{q}_e(\boldsymbol{x}, t) + \frac{\partial \boldsymbol{q}_e}{\partial t}\tau_e + \cdots + = -k\nabla T_e,$$

where τ_e is the electron relaxation time.

Then show that hyperbolic electron heat conduction equation becomes

$$\frac{\partial T_e}{\partial t} - \alpha_e \nabla^2 T_e = \frac{1}{n_{e,c}c_{p,e}}(\dot{s} + \frac{\alpha_e}{u_{e,\omega}^2}\frac{\partial \dot{s}}{\partial t}) - \frac{\alpha_e}{u_{e,\omega}^2}\frac{\partial^2 T_e}{\partial t^2}$$

hyperbolic electron heat conduction equation,

where

$$\alpha_e = \frac{k_e}{n_{e,c}c_{p,e}} \quad \text{thermal diffusivity},$$

and $u_{e,\omega}$ is the electron wave speed (second sound speed), given as

$$u_{e,\omega} = [\frac{p_{\rm F}^2}{3m_e m_{e,e}}(1 + {\rm F_o})]^{1/2}$$

$$\simeq (\frac{1}{3})^{1/2} u_{\rm F}.$$

Here $\rm F_o$ is a dimensionless interaction parameter.

For $\dot{s} = 0$, this becomes

$$\frac{\partial T_e}{\partial t} = \alpha_e \nabla^2 T_e + \frac{\alpha_e}{u_{e,\omega}^2}\frac{\partial^2 T_e}{\partial t^2}, \quad \tau_e = \frac{\alpha_e}{u_{e,\omega}^2}.$$

For $\tau_e \rightarrow 0$, the Fourier law is recovered. The quantum limit of the proceeding equation is discussed in [191].

Problem 5.24

(a) For electron–acoustic-phonon (deformation-potential) scattering, starting from (3.24), replace $E_e(\boldsymbol{p}) - E_e(\boldsymbol{p}')$ with $\hbar\omega_p$ (for emission of one phonon). Then use

$$\dot{\gamma}_{e\text{-}p,{\rm A}} = \frac{\varphi_{d,{\rm A}}^2 \kappa_p^2}{8\pi^2 \rho \omega_p}\delta_{\rm D}(\boldsymbol{p}' - \boldsymbol{p} + \hbar\boldsymbol{\kappa}_p)\delta_{\rm D}(E_{e,\boldsymbol{\kappa}'} - E_{e,\boldsymbol{\kappa}} - \hbar\omega_p),$$

and determine the energy relaxation time $\tau_{e\text{-}p,{\rm A}}$(energy).

Note that, by using the momentum and energy conservations, we have

$$E_e(\kappa') = E_e(\kappa) \pm \hbar\omega_p, \quad \frac{\hbar^2\kappa'^2}{2m_{e,e}} = \frac{\hbar^2\kappa^2}{2m_{e,e}} \pm \frac{\hbar^2\kappa\kappa_p\cos\theta}{2m_{e,e}}$$

$$\kappa' = \kappa \pm \kappa_p, \quad \text{or } \kappa'^2 = \kappa^2 + \kappa_p^2 \pm 2\kappa\kappa_p\cos\theta,$$

and also use the properties of the Dirac delta function, i.e.,

$$\delta_{\rm D}(\kappa' - \kappa \pm \kappa_p)\delta_{\rm D}(E_{e,\kappa'} - E_{e,\kappa} \mp \hbar\omega_p) = \frac{m_{e,e}}{\hbar^2\kappa\kappa_p}\delta_{\rm D}(\pm\cos\theta + \frac{\kappa_p}{2\kappa} \mp \frac{\omega_p m_{e,e}}{\hbar\kappa\kappa_p}).$$

Note that due to lack of spherical symmetry, using (3.12), we have [219]

$$\begin{aligned}\frac{1}{\tau_{e\text{-}p,\rm A}(\text{energy})} &= \frac{\varphi_{d,\rm A}^2}{4\pi E_e\rho u_e}\int_0^\infty\int_{-1}^1 \delta_{\rm D}(-\cos\theta + \frac{\kappa_p}{2\kappa} + \frac{u_{p,\rm A}}{u_e})\mathrm{d}(\cos\theta)\kappa_p^3\mathrm{d}\kappa_p \\ &= \frac{\varphi_{d,\rm A}^2}{4\pi E_e\rho u_e}\int_{\kappa_{p,min}}^{\kappa_{p,max}} \kappa_p^3\mathrm{d}\kappa_p \\ &= \frac{\varphi_{d,\rm A}^2\kappa_{p,max}^4}{16\pi E_e\rho u_e}, \quad \kappa_{p,max} = \frac{2p_e}{\hbar}, \quad E_e = \frac{p_e^2}{2m_{e,e}}.\end{aligned}$$

Note that the ϕ part of the integral yields 2π, since the integral of $\delta_{\rm D}$ over the entire domain (-1 to 1) is equal to 1 (Glossary).

(b) Using $\tau_{e\text{-}p,\rm A}$(energy) just given, and $\tau_{e\text{-}p,\rm A}$(momentum) given by (5.184), show that

$$\tau_{e\text{-}p,\rm A}(\text{energy}) = \frac{k_{\rm B}T}{2m_{e,e}u_{p,\rm A}^2}\tau_{e\text{-}p,\rm A}(\text{momentum}).$$

Problem 5.25

Show that (5.19) is the determinant of the coefficient matrix of the vector of constants $\{A, B, C, D\}$ when set equal to zero.

Problem 5.26

Plot the tight-binding approximated, single s-band structure $E_e(\kappa a/2\pi)$, (5.40), for $\gamma = 1.0$ eV and $E_{s\text{-}\beta} = 0$, and for $\boldsymbol{\kappa}$ along L–Γ–X–K–Γ–W [Figure 5.6(a)].

The coordinates of the Brillouin zone symmetry points for a FCC lattice are given in Figure 5.7(b), and these lead to lines such as

$$\Gamma - \mathrm{X}: 0 \le \kappa_x \le \frac{2\pi}{a}, \quad \kappa_y = \kappa_z = 0$$

$$\Gamma - \mathrm{L}: 0 \le \kappa_x = \kappa_y = \kappa_z \le \frac{\pi}{a}$$

$$\Gamma - \mathrm{K}: 0 \le \kappa_x = \kappa_y \le \frac{3\pi}{2a}, \quad \kappa_z = 0, \text{ etc.}$$

Problem 5.27

(a) Use the software Materials Studio (use DFT method, Section 5.5) for the *ab initio* computation of the electronic band structure in Si. Use the default setting and plot $E_e(\kappa)$, which is along W–L–Γ–X–W–K. This graph is similar to (but not the same as) Figure 5.6(b).

(b) Comment on the direct versus indirect bandgaps, and the lowest energy bandgap in Si.

Problem 5.28

(a) Using (5.146) to (5.149), calculate σ_e, α_S, α_P, and k_e, for the p-type Bi_2Te_3, at $T = 300$ K, assuming that the acoustic-phonon scattering dominates (Table 5.3), and using a single, effective bandgap ($\Delta E_{e,g} = E_{e,p} - E_{h,p}$, at minimum bandgap location) of $\Delta E_{e,g} = 0.13 - 1.08\times10^{-4}T$ in eV (Table 5.8), $n_{e,c} = 6\times10^{24}$ 1/m^3, and $\mu_e = 0.1$ m^2/V-s.

Note that, for the p-type semiconductors, we have

$$\alpha_{S,p} = -\frac{k_B}{e_c}[\frac{E_{v,p} - E_{e,p}}{k_B T} + (s + \frac{5}{2})] = \frac{k_B}{e_c}[\frac{\Delta E_{e,g}}{k_B T} - (s + \frac{5}{2})],$$

where $E_{v,p}$ is the valance-band potential energy.

(b) For $k = k_e + k_p = 1.7$ W/m-K, calculate k_p at $T = 300$ K and then calculate $Z_e T$.

(c) Compare the calculated (predicted) values with those reported in the literature (e.g., *CRC Handbook of Thermoelectrics* [288]). The values vary depending on the doping used, and amongst the reported values are $\alpha_{S,p} = 230$ μV/K and $\rho_e = 10^{-5}$ Ω-m.

Problem 5.29

Show that for an electron in an electromagnetic field [φ_e is the electric field scalar potential, and $\varphi_\circ(\kappa)$ is the other potential] with

$$\mathrm{H} = \frac{(\boldsymbol{p}_e - e_c\boldsymbol{a}_e)^2}{2m_e} + e_c\varphi_e + \varphi_\circ(\boldsymbol{x}),$$

the first-order perturbation $\mathrm{H} = \mathrm{H}_\circ + \mathrm{H}'(t)$ gives

$$\mathrm{H}'(t) = \frac{e_c}{m_e}\boldsymbol{a}_e \cdot \boldsymbol{p}_e,$$

which is a linearly polarized, monochromatic plane wave with $\varphi_e = 0$.

This is used in the derivation of the electron transition rate (transition dipole moment) and the relation between spontaneous and stimulated emissions (Section 7.4.2).

Problem 5.30

Use (5.64) (not allowing for spin) and the plane wave approximation with the periodic boundary condition, $(\kappa_1, \kappa_2, \kappa_3) = (2\pi/L)(n_1, n_2, n_3)$, $n_i = 0, 1, 2, \ldots$, and derive (E.35).

6

Fluid Particle Energy Storage, Transport and Transformation Kinetics

Fluid particle refers to matter in a gas or liquid phase, with the particle being the smallest unit (made of atoms or molecules) in it, for which further breakdown would change the chemical identity of the particle. For a fluid in motion, the convection heat flux vector $\boldsymbol{q}_u = \rho_f c_{p,f} \boldsymbol{u}_f T$ and the surface-convection heat flux vector $\boldsymbol{q}_{ku}$ (which describes the interfacial heat transfer between two phases in relative motion, in which at least one phase is a fluid) are influenced by the specific heat capacity $c_{p,f}$ of the fluid particle (whereas $\boldsymbol{q}_{ku}$ also depends on the isotropic fluid thermal conductivity k_f, viscosity μ_f, and velocity). The fluid (gas or liquid) velocity $\boldsymbol{u}_f$ can be subsonic or supersonic, and for contained gases at low pressures or in small spaces, it is possible for fluid particle–surface collisions to dominate over the interparticle collisions. In this chapter, we examine energy storage and transport in fluids, as well as fluid interactions with surfaces (and the associated fluid flow regimes).

Fluid particles can have five types (forms) of energy: potential, electronic, translational, vibrational, and rotational (Figure 1.1). The electronic energy is part of potential energy, however, here we use the potential energy for interparticle interactions only. Each form of energy can be considered separately, and the total energy is the summation of these five energies. The maximum vibrational and rotational energies a molecule can attain are limited by the dissociation energy φ_e. Above this energy, the atoms of a molecule have an excess of internuclear energy that overcomes the bonding force between them, causing the atoms to dissociate. Similarly, the maximum electronic energy a fluid particle can obtain is the first ionization energy, which is the energy at which an electron has enough energy to escape the pull of the atom. Figure 6.1 gives examples of the magnitudes of these energies for gaseous CO, CO_2, He, Ne, and N_2 used in gas lasers. These are the translational, vibrational, rotational, electronic, dissociation, and ionization energies. As a reference, the average translational energy of a monatomic gas is also shown for $T = 300$ K. For CO_2 and Ne, the energy of photon emission is also shown (vibrational energy). Also shown is the potential energy for nonideal (dense) gases. The rotational energies are the lowest, and the electronic energies are the highest. These energies, and their transformation and transport mechanisms, are discussed throughout this chapter.

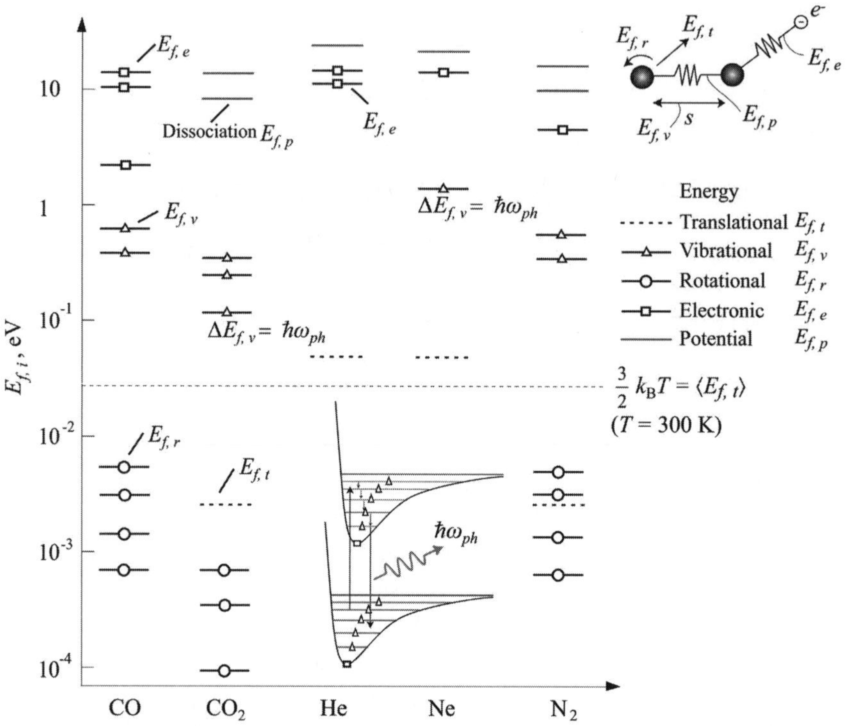

Figure 6.1. Examples of the magnitude of various (translational, vibrational, rotational, electronic, and potential) fluid particle energies [99, 298, 348], for some monatomic (He, Ne), diatomic (CO, N_2), and polyatomic (CO_2) gases. The average monotonic-gas translational kinetic energy $3k_B T/2$, for $T = 300$ K, is also shown. For CO_2 and Ne, their lasing frequencies (and energy) are also shown (related to vibrational energy).

6.1 Fluid Particle Quantum Energy States and Partition Functions

The fluid particle (monatomic, diatomic, or polyatomic) may undergo translational, vibrational and rotational motions (a fluid particle made of N_o atoms has $3N_o$ modes of motion), and dense fluids also have potential energy. The electronic energy of the fluid particle is where energy is stored by displacing its electrons with respect to its nucleus. Then the fluid particle energy (and Hamiltonian) is divided into potential, electronic, translational, vibrational, and rotational energies (Figure 6.1), i.e.,

$$E_f = \sum_i E_{f,i} = E_{f,p} + E_{f,e} + E_{f,t} + E_{f,v} + E_{f,r}$$

$$\mathrm{H}_f = \sum_i \mathrm{H}_{f,i}. \tag{6.1}$$

The partition function Z defined through (2.21), using the fluid particle Hamiltonian, is partitioned into various energies. First it is divided into potential, electronic, and kinetic energy partitions. Then, for each type of motion, i.e., translation, vibration, or rotation, the kinetic energy is further partitioned. Based on the definition of Z, the total partition function is the product of each of these, i.e.,

$$Z_f = \prod_i Z_{f,i} = Z_{f,p} Z_{f,e} Z_{f,t} Z_{f,v} Z_{f,r}. \tag{6.2}$$

The particle in a box model of fluid particles is used to derive the translational quantum energy states. Similarly, other quantum kinetic and electronic energy states are treated (derivations are outlined below). For the case of no potential energy (field free), we have $E_{f,p} = \langle \varphi_f \rangle = 0 \, (Z_{f,p} = 1)$.

6.1.1 Translational Energy and Partition Function

Considering translational motion, the energy is the translational kinetic energy and the Hamiltonian is $\mathrm{H} = -(h_\mathrm{P}^2/2m)\nabla^2$, i.e.,

$$E_{f,t} = \frac{1}{2} m u_f^2 = \frac{1}{2m} p_f^2, \quad u_f = |\boldsymbol{u}_f|, \quad p_f = |\boldsymbol{p}_f|, \tag{6.3}$$

where $\boldsymbol{p}_f$ is the fluid particle momentum vector. Then the time-independent Schrödinger equation (2.62) becomes

$$\mathrm{H}\psi_f = \mathrm{E}_f \psi_f, \quad \frac{h_\mathrm{P}}{8\pi^2 m} \nabla^2 \psi_f + \frac{1}{2m} p_f^2 \psi_f = 0. \tag{6.4}$$

Assuming that the box (containing one gas particle) is impermeable, it would require that $\psi_f = 0$ on the wall (and also outside the box). The solution to ψ_f can be separated into three independent Cartesian-coordinate components, i.e.,

$$\psi_f(\boldsymbol{x}) = \psi_{f,x}(x)\psi_{f,y}(y)\psi_{f,z}(z). \tag{6.5}$$

For a box of volume L^3, the boundary conditions are (similar to electron gas of Section 2.6.5)

$$\psi_{f,i}(0) = \psi_{f,i}(L) = 0, \quad i = x, y, z. \tag{6.6}$$

The solution for each component, e.g., x, is

$$\psi_{f,x}(0) = A_x \sin\left(2\pi \frac{p_{f,x}}{h_\mathrm{P}} x\right) + B_x \cos\left(2\pi \frac{p_{f,x}}{h_\mathrm{P}} x\right). \tag{6.7}$$

While (6.6) the constants are determined and the the eigenvalues are

$$p_{f,x} = \frac{h_\mathrm{P}}{2} \frac{n_x}{L}, \quad n_x = 1, 2, 3, \cdots \tag{6.8}$$

The resulting energy states are

$$E_{f,t,n} = \frac{h_\mathrm{P}^2}{8m}\left(\frac{n_x^2}{L^2} + \frac{n_y^2}{L^2} + \frac{n_z^2}{L^2}\right) \quad \text{translational energy states}, \tag{6.9}$$

where n_x, n_y, n_z are the translational quantum numbers.

Noting that the fluid particle translational energy follows the M–B distribution (Table 1.2), partition function (2.27) for translational energy (6.9) is [55]

$$
\begin{aligned}
Z_{f,t} &= \sum_{i=0}^{\infty} g_{f,t,i} \exp(-\frac{E_{f,t,i}}{k_B T}) \\
&= \prod_{j=1}^{3} \sum_{n_x=1}^{\infty} g_{f,t,i} \exp(-\frac{n_x^2}{L^2}\frac{h_P^2}{8mk_B T}) = \int_0^{\infty} \frac{\pi}{4}(\frac{8mV^{2/3}}{h_P^2})^{3/2} E_{f,t}^{1/2} e^{-E_{f,t}/k_B T} \mathrm{d}E_{f,t} \\
&= V(\frac{2\pi m k_B T}{h_P^2})^{3/2}, \quad V = L^3,
\end{aligned} \tag{6.10}
$$

where the degeneracy is

$$
g_{f,t,i} = \frac{4\pi(8mV^{2/3}h_P^2)^{3/2} E_{f,t,i}^{1/2}}{4}. \tag{6.11}
$$

This can be used to define the criterion for having an ideal gas†.

Using the relation for internal energy shown in Table 2.4, then for the average translational energy, we have (Chapter 2 problem)

$$
\langle E_{f,t} \rangle = k_B T^2 \frac{\partial \ln Z_{f,t}}{\partial T} = \frac{3}{2} k_B T \quad \text{average translational energy.} \tag{6.12}
$$

6.1.2 Vibrational Energy and Partition Function

For polyatomic gases, the vibrational energy is also determined from the Schrödinger equation. For example, for diatomic gases, the quantum vibrational energy is (the treatment of an harmonic oscillator, where $\mathrm{H} = -(h_P^2/2m)\mathrm{d}^2/\mathrm{d}x^2 + \frac{1}{2}\Gamma x$, is given in Section 2.6.4)

$$
E_{f,v,l} = \hbar\omega_f(l + \frac{1}{2}), \quad l = 0, 1, 2, \cdots \quad \text{vibrational energy states,} \tag{6.13}
$$

where l is the vibrational (or vibronic) quantum number. The scale for the vibrational frequency is similar to that used for phonons, (2.51), i.e., for harmonic pair vibration, $\omega_f = (\Gamma/m)^{1/2}$, where Γ is the spring constant and is the second derivative of the interatomic potential evaluated at equilibrium position. Then the partition

† Using (6.10), $Z_{f,t} \equiv \lambda_{dB}^3$, $\lambda_{dB} = h_P^2/(2\pi m k_B T)^{1/2}$ is the thermal de Broglie wavelength (see Glossary), and the dilute- (ideal-) gas criterion is defined as

$$
\frac{\lambda_{dB}^3}{V/N} = \lambda_{dB}^3 n_f = \frac{(h_P^2/2\pi m k_B T)^{3/2}}{k_B T/p} \ll 1 \quad \text{ideal-gas condition,}
$$

where n_f is the number of fluid particles per unit volume (also called quantum concentration).

function becomes

$$Z_{f,v} = \sum_{l=0}^{\infty} \exp[-(l+\frac{1}{2})\frac{\hbar\omega_f}{k_B T}] = \exp(-\frac{\hbar\omega_f}{2k_B T})\sum_{l=0}^{\infty}\exp(-l\frac{\hbar\omega_f}{k_B T})$$

$$\simeq \exp(-\frac{\hbar\omega_f}{2k_B T})\frac{1}{1-\exp(-\frac{\hbar\omega_f}{k_B T})}$$

$$= \frac{\exp(-\frac{T_{f,v}}{2T})}{1-\exp(-\frac{T_{f,v}}{T})}, \quad T_{f,v} \equiv \frac{\hbar\omega_f}{k_B}, \tag{6.14}$$

where $T_{f,v}$ is defined as the characteristic temperature for vibration (this will be given for some diatomic gases in Table 6.2). This is the same as the partition function for the harmonic oscillator given by (2.88).

6.1.3 Rotational Energy and Partition Function

Similarly, the rotational energy for diatomic gas particles, where the rigid rotor Hamiltonian is $\mathrm{H} = -(h_\mathrm{P}^2/2I_f)\nabla^2$, is

$$E_{f,r,j} = \frac{h_\mathrm{P}^2}{8\pi^2 I_f} j(j+1), \quad j = 0, 1, 2, \cdots \quad \text{rotational energy states}, \tag{6.15}$$

where I_f is moment of inertia for the molecule and j is the rotational quantum number. For a homoatomic gas, this is mr_e^2, where r_e is the equilibrium separation distance discussed in Section 2.2.

The partition function for the fluid particle rotational energy is [55]

$$Z_{f,r} = \sum_{j=0}^{\infty} g_{f,r,j}\exp(-\frac{E_{f,r,j}}{k_B T}) = \sum_{j=0}^{\infty}(2j+1)\exp[\frac{-j(j+1)h_\mathrm{P}^2}{8\pi^2 I_f k_B T}]$$

$$\simeq 1 + 3\exp(\frac{-2T_{f,r}}{T}) + 5\exp(\frac{-6T_{f,r}}{T}) + \cdots +$$

$$\simeq \frac{T}{T_{f,r}}(1+\frac{1}{3}\frac{T_{f,r}}{T}+\frac{1}{15}\frac{T_{f,r}^2}{T^2}+\cdots), \tag{6.16}$$

where $T_{f,r}$ is the rotational temperature defined as (will be given for some diatomic and polyatomic gases in Table 6.2)

$$T_{f,r} \equiv \frac{h_\mathrm{P}^2}{8\pi^2 I_f k_B}. \tag{6.17}$$

The practical form of (6.16) uses a symmetry number N_s and is [339],

$$Z_{f,r} = \frac{T}{N_s T_{f,r}} \begin{cases} N_s = 1 & \text{for heteronuclear polyatomic gas} \\ N_s = 2 & \text{for homonuclear polyatomic gas.} \end{cases} \tag{6.18}$$

We will discuss the potential energy $E_{f,p} = \langle\varphi_f\rangle$ and its partition function $Z_{f,p}$ in Section 6.3 when examining dense fluids. It is also customary to combine the

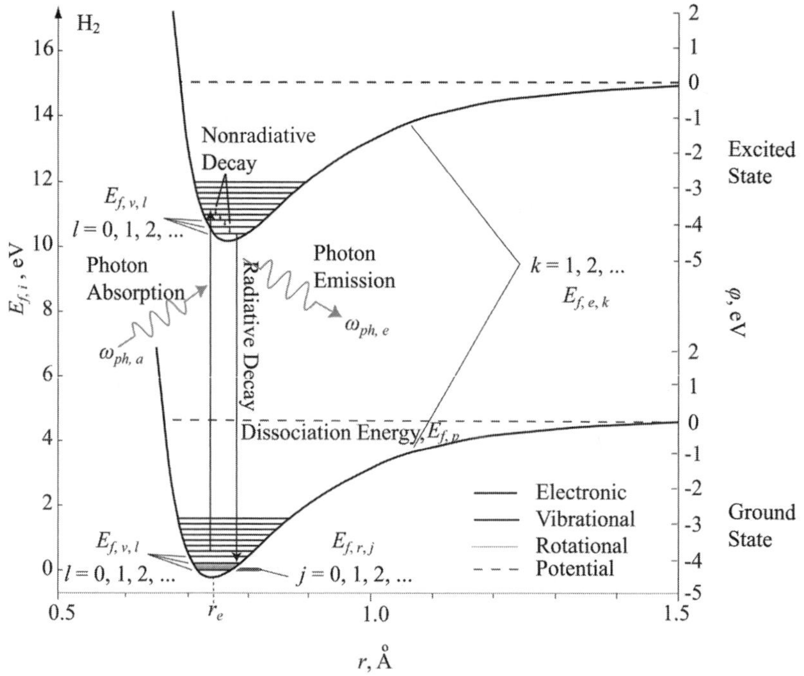

Figure 6.2. Discrete electronic, vibrational, and rotational energies (left) of H_2 for the ground (lowest) and excited state. The variation of the interatomic potential (right) with respect to the separation distance is also shown.

potential and vibrational energies for diatomic and polyatomic gases (Section 6.2). Figure 6.2 shows the discrete electronic, vibrational, and rotational energies of H_2. Their ground states, as well as the excited states are shown. The variation of interatomic potential, with respect to the interatomic spacing, is also shown (L–J potential model). The potential is only slightly different from the excited electronic state (for H_2, but not in general). The large energy transitions are excited by photon (radiation) absorption/emission, and the smaller ones by nonradiative decay/excitement. Note the large energy transitions (around 10 eV). The translational energies are generally small (of the order of millielectronvolts).

6.1.4 Electronic Energy and Partition Function

The partition function for electronic energy states $E_{f,e,k}$ is represented through the electronic excitation temperature $T_{f,e,k}$, where k is the electronic quantum number. Then

$$Z_{f,e} = \sum_{k=0}^{\infty} g_{f,e,k} \exp(-\frac{E_{f,e,k}}{k_B T}) = \sum_{k=0}^{\infty} g_{f,e,k} \exp(-\frac{T_{f,e,k}}{T})$$

$$= g_{f,e,0} + g_{f,e,1} \exp(-\frac{T_{f,e,1}}{T}) \ldots, \quad T_{f,e,k} = \frac{E_{f,e,k}}{k_B}$$

$$E_{f,e,k} \text{ electronic energy states.} \qquad (6.19)$$

Table 6.1. *Degrees of freedom in translational, rotational, and vibrational motions of linear and nonlinear polyatomic molecules made of N_o atoms.*

Motion	Linear polyatomic molecules	Nonlinear polyatomic molecules
Translation, N_t	3	3
Rotation, N_r	2	3
Vibration, N_v	$3N_o - 5$	$3N_o - 6$
Total	$3N_o$	$3N_o$

The electronic energy excitation temperature $T_{f,e,k}$ is generally very large, so generally only the ground state is used (and does not contribute to e_f and $c_{v,f}$).

6.2 Ideal-Gas Specific Heat Capacity

The average specific internal energy e is related to the partition function through (6.12), i.e.,

$$e_f = \frac{R_g}{M} T^2 \frac{\partial \ln Z_f}{\partial T}|_{N,V}, \tag{6.20}$$

where $R_g = k_B N_A$, and product rule (6.2) is used for $Z_{f,i}$, i.e., the contribution from each mechanism of energy storage in the fluid particle.

The ideal gas equation of states (obtained from Z_f, as will be shown in Section 5.3) is

$$p = \frac{R_g}{M}\rho_f T = n_f k_B T,\ n_f = \frac{p}{k_B T} \quad \text{ideal-gas equation of states,} \tag{6.21}$$

where M is the molecular weight, $M = mN_A$.

For ideal, diatomic gases, the electronic energy is assumed negligible, and using (6.20) and the partition functions of Section 6.1 ($E_{f,p} = 0$, no potential energy for ideal gas), we have the specific heat capacity at constant volume as (Table 2.4)(end-of-chapter problem)

$$c_{v,f} = \frac{\partial e_f}{\partial T}|_V \frac{R_g}{M}\{\frac{3}{2} + (\frac{T_{f,v}}{T})^2 \frac{\exp(T_{f,v}/T)}{[\exp(T_v/T) - 1]^2} + 1 + \frac{2}{15}(\frac{T_{f,r}}{T})^2\}$$
$$\text{diatomic ideal gas,} \tag{6.22}$$

where we have used (6.16) for the rotational energy. Here the first term is due to translation, the second is due to vibration, and the last two are due to rotation.

For linear molecules, there are two axes for rotation, and a three for nonlinear molecules. Also, there are $3N_o - 5$ (N_o is the number of atoms in molecule) vibration modes for linear and $3N_o - 6$ modes for nonlinear polyatomic ideal gases. There are $3N_o$ total degrees of freedom per molecule. These are summarized in Table 6.1.

For a linear, diatomic gas there is one vibrational mode. However, it is customary to think of the vibrational mode as having a kinetic and a potential component, so this contribution is R_g/M. At large temperatures, the limit of $7R_g/2M$ is reached.

Table 6.2. *Rotation and vibration temperatures for some diatomic gases and vibrational temperatures for some linear and nonlinear polyatomic molecules [55].*

Molecule	$T_{f,r}$, K	$T_{f,v}$, K	Molecule	$T_{f,v,j}$, K
Cl_2	0.351	808	CCl_4	310(2), 450(3), 660, 1120(3)
H_2	85.3	6215	CH_4	1870(3), 2180(2), 4170, 4320(3)
I_2	0.054	308	CO_2	954(2), 1890, 3360
N_2	2.88	3374	H_2O	2290, 5160, 5360
O_2	2.07	2256	NH_3	1360, 4800, 4880(2), 2330(2)
CO	2.77	3103	NO_2	850(2), 1840, 3200
HCl	15.2	4227	SO_2	750, 1660, 1960
HI	9.06	3266		

At yet higher temperatures, the contribution from the electronic energy storage should be included. Table 6.2 lists the empirical values for $T_{f,v}$ and $T_{f,r}$, for some common diatomic gases. Note that the low-energy rotational modes are excited (occupied) at lower temperatures, compared with the vibrational modes (Figure 6.1), i.e., $T_{f,r} < T_{f,v}$. Also note that the lighter molecules have higher transition temperatures.

For monatomic gases, the only contribution is from the translational motion. Then

$$c_{v,f} = \frac{3R_g}{2M} \quad \text{monatomic ideal gas.} \tag{6.23}$$

For three or more atoms the relative position of the atoms should be included, and for a nonlinear arrangement of atoms, the specific heat capacity for an N_o-atom molecule is assumed to be dominated by translation, rotation, and vibration, and given by [55]

$$c_{v,f} = \frac{R_g}{M}\{3 + \sum_{j=1}^{3N_o-6} (\frac{T_{f,v,j}}{T})^2 \frac{\exp(T_{f,v,j}/T)}{[\exp(T_{f,v,j}/T) - 1]^2}\}$$

$$\text{nonlinear, polyatomic ideal gas.} \tag{6.24}$$

Here the translational and rotational energies are assumed fully occupied compared with the vibrational energy (i.e., the vibrational energy has the highest transition temperature.) Also, note that each vibrational degree of freedom adds one R_g/M to $c_{p,f}$ (combines it with potential energy). The relation for a linear polyatomic molecule is similar, except 3 is replaced with 5/2, and $3N_o - 6$ is replaced with $3N_o - 5$ (Table 6.1). The empirical values for $T_{f,v,j}$ are also listed in Table 6.2 for some common polyatomic gases. Repeated temperatures, in Table 6.2, are indicated in parentheses.

Figure 6.3(a) shows the variations of the specific heat capacity of homonuclear H_2 and N_2 with respect to temperature. From Table 6.2, we note that $T_{f,r}$ for H_2 is only 85.3 K, whereas $T_{f,v}$ is 6215 K. Both of these are lower for N_2. For H_2, the value of 7/2 is reached at about 6000 K. Figure 6.3(b) shows the same for CH_4 and H_2O.

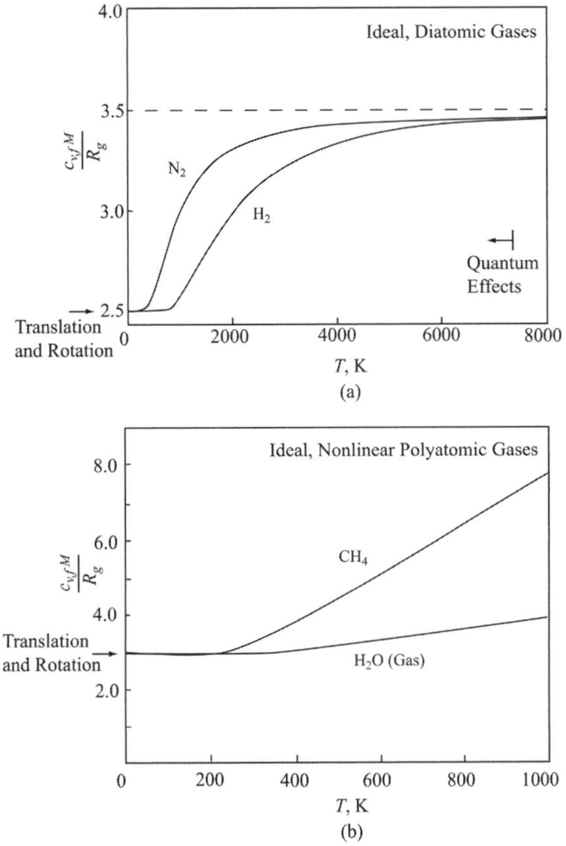

Figure 6.3. Predicted (with empirical transition temperatures, Table 6.2) scaled variation of the specific heat capacity of (a) diatomic, and (b) nonlinear polyatomic gases, with respect to temperature.

Note that for heteronuclear CH_4, the specific heat capacity increases significantly with temperature (the first $T_{f,v,j}$ is 1870 K, Table 6.2).

The two (constant-pressure and constant-volume) specific heat capacities are related through

$$c_{p,f} = c_{v,f} + T(\frac{\partial p}{\partial T}|_v \frac{\partial v}{\partial T}|_p) \equiv c_{v,f} + T\frac{\beta^2}{\rho_f \kappa}, \quad \kappa = \frac{1}{\rho_f}\frac{\partial \rho_f}{\partial p}|_T, \tag{6.25}$$

where β is the volumetric thermal expansion coefficient and E_p is the isothermal compressibility (bulk modulus).

The specific heat capacity per molecule (of mass m) can be written as $\alpha(T)k_B/m$, where $\alpha(T)$ is dimensionless and expresses the degrees of freedom of temperature dependence ($\alpha = 3/2$ for monatomic gases). Then, noting that the product of $\rho_f c_{p,f} T$ appears in the macroscopic energy equation of Table 1.1, for an ideal gas ($p = k_B \rho_f T/m$) (6.21), we have

$$\rho_f c_{p,f} T = p\alpha(T) \quad \text{ideal-gas specific heat capacity,} \tag{6.26}$$

i.e., the sensible heat content per unit volume is proportional to the pressure.

6.3 Dense-Fluid Specific Heat Capacity: van der Waals Model

The partition function for a single-component, dense fluid accounts for the particle interactions and is given as (Table 2.4)

$$Z_f = \frac{1}{N!} Z_{f,p} Z_{f,t} Z_{f,r}, \tag{6.27}$$

where $Z_{f,p}$ is the classical configuration integral accounting for the molecular interactions, $Z_{f,t}$ is the translation component of the molecular partition function, and $Z_{f,r}$ is the rotational component. The factor $1/N!$ accounts for the indistinguishability of the molecules.

The Hamiltonian in the partition function relation (2.21) is written as

$$\mathrm{H} = \langle \varphi_f \rangle_N(\boldsymbol{x}) + \mathrm{H}_{f,t}(\boldsymbol{p}) + \mathrm{H}_{f,r}(\boldsymbol{p}). \tag{6.28}$$

The translation component

$$Z_{f,t} = \frac{1}{h_\mathrm{P}^{3N}} [\int \exp(-\frac{1}{k_\mathrm{B} T} \frac{p^2}{2m}) \mathrm{d}\boldsymbol{p}]^N. \tag{6.29}$$

Then similar to (6.10), (6.27) is written as [55]

$$Z_f = \frac{(2\pi m k_\mathrm{B} T)^{3N/2}}{h_\mathrm{P}^{3N}} Z_{f,r} Z_{f,p}, \tag{6.30}$$

where $Z_{f,p}$ includes the modified volume term, not shown as part of $Z_{f,t}$. The rotational component is given for both linear and nonlinear polyatomic molecules, as [55]

$$Z_{f,r} = \frac{\pi^{(N_{t\text{-}r}-5)N/2}}{N_s^N} (\frac{T}{T_{f,r}})^{(N_{t\text{-}r}-3)N/2}, \tag{6.31}$$

where $N_{t\text{-}r}$ $(= N_t + N_r)$ is the total number of translational and rotational energy storage modes (Table 6.1), N_s is the symmetry number for a polyatomic molecule (the number of ways the molecule can be rotated into itself), and $T_{f,r}$ is the characteristic temperature for rotation. For polyatomic molecules, $N_s = 2$ for H_2O and $N_s = 3$ for NH_3, etc.

The potential component of the molecular partition function (2.21) is

$$Z_{f,p} = \int \exp[-\frac{1}{k_\mathrm{B} T} \langle \varphi_f \rangle_N(\boldsymbol{x}_1, \boldsymbol{x}_2 \cdots \boldsymbol{x}_N)] \mathrm{d}\boldsymbol{x}_1 \boldsymbol{x}_2 \cdots \boldsymbol{x}_N. \tag{6.32}$$

Using the radial distribution function $g(r)$, Section 2.2.5, and for L–J potential (2.47) the total potential energy of the fluid is [similar to (2.49)]

$$E_{f,p} = \langle \varphi_f \rangle_N = \frac{N}{2} \int_d^\infty \frac{N}{V} 4\pi r^2 \varphi_\mathrm{LJ}(r) g(r) \mathrm{d}r = \frac{2\pi N^2}{V} \int_d^\infty r^2 \varphi_\mathrm{LJ}(r) \mathrm{d}r$$

$$\equiv -\frac{a N^2}{V}, \quad g(r) = 0 \text{ for } r < d, \quad g(r) = 1 \text{ for } r \geq d \tag{6.33}$$

$$a = -2\pi \int_d^\infty r^2 \varphi_\mathrm{LJ}(r) \mathrm{d}r = \frac{8\pi \epsilon_\mathrm{LJ} \sigma_\mathrm{LJ}^6}{3d^3} (1 - \frac{\sigma_\mathrm{LJ}^6}{3d^6}), \tag{6.34}$$

where a is the van der Waals coefficient, and d is the diameter of the molecule (this is sometimes taken as the molecular collision diameter, which will be discussed in Section 6.5). Note that the $g(r)$ approximation in (6.33) matches well the distribution in Figure 2.11(b) for liquids.

The integral in (6.32) is over the volume V and is repeated for N molecules leading to V^N. However, for dense fluids, the fluid particles occupy a relatively large percentage of the physical space. These occupied volumes are not accessible and should not be included in the integral over all the coordinates. Therefore, $(V - Nb)^N$ is used instead of V^N, where b is a van der Waals constant representing the mean volume occupied by a fluid particle in the system. Then the potential energy component is [55]

$$Z_{f,p} = \exp(\frac{aN^2}{Vk_\mathrm{B}T})(V - Nb)^N. \tag{6.35}$$

When all three contributions (6.30), (6.31), and (6.35), are substituted into (6.27), the dense-fluid molecular partition function is (end-of-chapter problem)

$$Z_f = \frac{1}{N!}\frac{(2\pi m k_\mathrm{B} T)^{3N/2}}{h_\mathrm{P}^{3N}}\frac{\pi^{(N_{t\text{-}r}-5)N/2}}{N_s^N}(\frac{T}{T_r})^{(N_{t\text{-}r}-3)N/2}\exp(\frac{aN^2}{Vk_\mathrm{B}T})(V - Nb)^N. \tag{6.36}$$

The pressure is given in terms of the partition function from Table 2.4 as

$$p = k_\mathrm{B}T\frac{\partial \ln Z_f}{\partial V}|_{T,N}. \tag{6.37}$$

Substituting partition function (6.36) into the preceding equation and using the Stirling approximation, i.e., $\ln N! = N\ln N - N$, we have (end-of-chapter problem)

$$p = \frac{Nk_\mathrm{B}T}{V - Nb} - \frac{aN^2}{V^2} = n_f k_\mathrm{B}T(\frac{1}{1 - bn_f}) - an_f^2,\ n_f = \frac{N}{V},$$

$$\text{dense-fluid van der Waals equation of states,} \tag{6.38}$$

where n_f is the fluid particle number density.

Note that since V is only in the potential energy partition function, the pressure is only dependent on $Z_{f,p}$. For $bn_f \ll 1$ and $an_f^2 \ll 1$, ideal-gas equation of states (6.21) is obtained. The coefficients of a and b can be related to the L–J coefficients. The definition of a in (6.34) can be simplified as $16\pi\epsilon_\mathrm{LJ}\sigma_\mathrm{LJ}^3/9$ if we assume that $d = \sigma_\mathrm{LJ}$. The parameter b is interpreted as the volume of space occupied by each molecule, which is $b \simeq \pi d^3/6$. Thus assuming that $d = \sigma_\mathrm{LJ}$, we have $b = \pi\sigma_\mathrm{LJ}^3/6$. However, these assumptions underestimate a and b.

Table 6.3 lists the L–J potential parameters and the van der Waals coefficients for a few common compounds. The empirical values for a and b (based on actual d) are also listed, as well as the value of $N_{t\text{-}r}$.

Table 6.3. *L–J parameters and van der Waals coefficients for some common compounds [55].*

Compound	$\epsilon_{\rm LJ}/k_{\rm B}$, K	$\sigma_{\rm LJ}$, Å	a, 10^{-49} Pa-m^6/molecule^2	b, 10^{-29}m^3/molecule	$N_{t\text{-}r} = N_t + N_r$
H_2O	78	3.17	15.20	5.05	6
N_2	95	3.70	3.76	6.40	5
O_2	118	3.58	3.78	5.23	5
CO_2	198	4.33	10.10	7.09	5
CH_4	149	3.78	6.31	7.09	6

Noting that†

$$h = e + pV, \quad e = k_{\rm B} T^2 \frac{\partial \ln Z_f}{\partial T}|_{N,V}, \tag{6.39}$$

and using the van der Waals equation of states (6.38), we find that the specific heat capacity at constant pressure (per unit mass) for a van der Waals fluid is (end-of-chapter problem)

$$c_{p,f} = \frac{\partial h}{\partial T}|_p = \frac{R_g}{M}\{(\frac{N_{t\text{-}r}}{2} + \frac{V/N_{\rm A}}{V/N_{\rm A} - b}) - \frac{[\frac{2a}{(V/N_{\rm A})^2} - \frac{bk_{\rm B}T}{(V/N_{\rm A} - b)^2}](\frac{1}{V/N_{\rm A} - b})}{\frac{2a}{(V/N_{\rm A})^3} - \frac{k_{\rm B}T}{(V/N_{\rm A} - b)^2}}\}$$

van der Waals heat capacity. (6.40)

Note that we first use $N = N_{\rm A}$ to find $c_{v,f}$ per mole and then use M to find $c_{v,f}$ per mass. Note that all three components $Z_{f,i}$ contribute to $c_{p,f}$.

Here we have used the Maxwell relation [55]

$$\frac{\partial h}{\partial T}|_{p,N} = \frac{\partial h}{\partial T}|_{V,N} - \frac{\partial h}{\partial V}|_{T,N} \frac{\partial p/\partial T|_{N,V}}{\partial p/\partial V|_{T,N}}. \tag{6.41}$$

Figure 6.4 shows the predicted specific heat capacities of liquid water obtained with (6.40). The results are about 30% over the experimental results. Unlike gases, the liquid specific heat capacity decreases with temperature. Note that, for triatomic H_2O, the results of Figure 6.4 show that per atom $c_{p,f}/k_{\rm B}$ is larger than 3 at lower (relative) temperatures.

Noting that the kinetic and potential energies are nearly equal, the potential energy of the liquid is released when it is transformed to gas, and this energy is nearly the heat of evaporation. Also note the lower specific heat of gaseous water (per molecule) shown in Figure 6.3(b). We will further discuss this in Section 6.11.5.

† Using (6.39), we have

$$h = \frac{N_{t\text{-}r} N}{2} k_{\rm B} T + \frac{V/N}{V/N - b} N k_{\rm B} T - \frac{2aN^2}{V}.$$

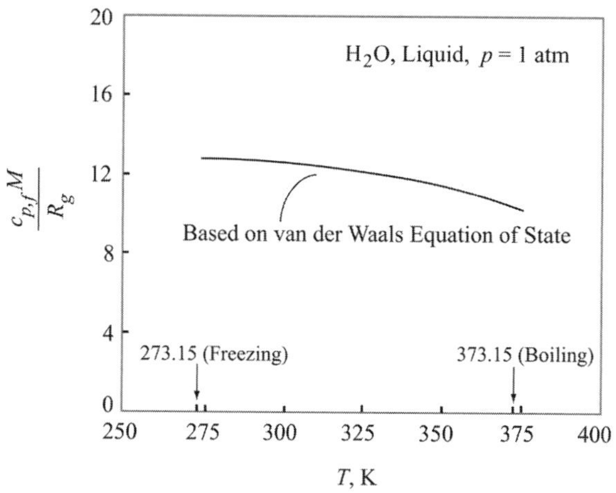

Figure 6.4. Predicted variations of specific heat capacities (at constant pressure) of liquid water (p = 1 atm) with respect to temperature.

The two specific heats ($c_{v,f}$ and $c_{p,f}$) are related through (6.25) and are nearly equal for most liquids. In general, an accurate prediction of the liquid specific heat is more challenging (compared with that of gases) [55].

6.4 Gas BTE, f_f^{o} and Thermal Velocities

The BTE (Table 3.1) requires specification of the collision term on the right-hand side, which here is considered to be the dynamics of the collision of two particles. In the kinetic theory of gases, Boltzmann considered dilute gases and analyzed the elementary collision processes between pairs of molecules. The evolution of the equilibrium distribution function is described by the BTE.

6.4.1 Interparticle Collisions

If there were no collisions, the particles would have positions and momentum distributions such that the probability distribution function at time t and t+dt is that given by (3.5), i.e.,

$$f_f(\boldsymbol{x} + \boldsymbol{u}\mathrm{d}t, \boldsymbol{p} + \boldsymbol{F}_f\mathrm{d}t, t + \mathrm{d}t) = f_f(\boldsymbol{x}, \boldsymbol{p}, t), \qquad (6.42)$$

where $\boldsymbol{F}_f$ denotes an external force acting on fluid particles at point ($\boldsymbol{x}$, $\boldsymbol{p}$).

The collisions (scattering) change this, and they are represented as $\partial f_f/\partial t|_s$ and the BTE (Table 3.1) becomes

$$(\frac{\partial}{\partial t} + \boldsymbol{u}_f \cdot \nabla_x + \boldsymbol{F}_f \cdot \nabla_p) f_f(\boldsymbol{x}, \boldsymbol{p}, t) = \frac{\partial f_f}{\partial t}|_s. \qquad (6.43)$$

Here we consider elastic scattering and $\dot{s}_f = 0$. Boltzmann assumed binary collisions with $\boldsymbol{F}_f = 0$. He also assumed that the velocity and position of a molecule are uncorrelated (assumption of molecular chaos). He constructed a moving-coordinate,

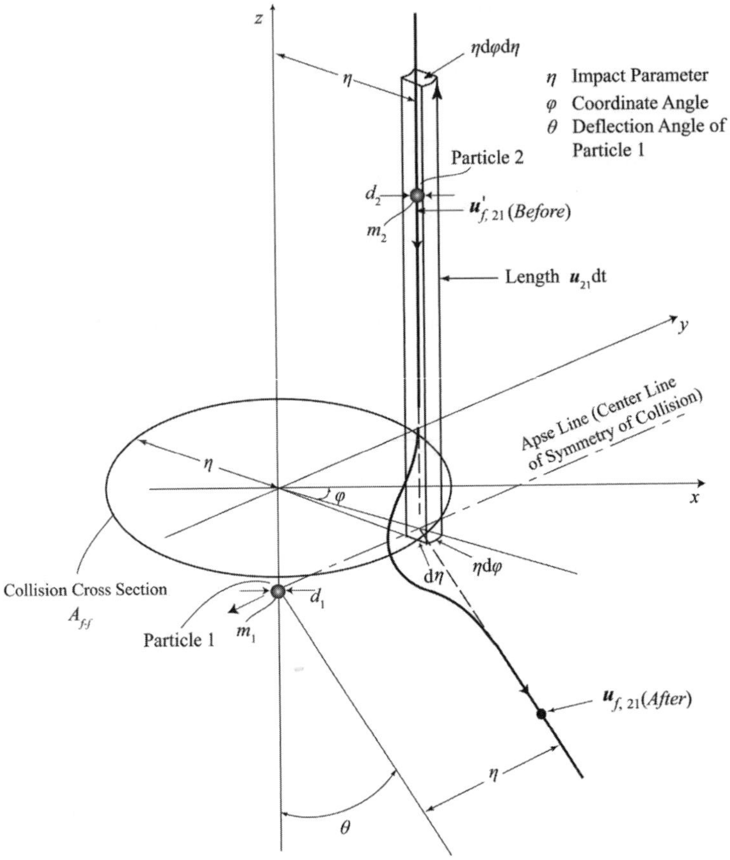

Figure 6.5. A general three-dimensional (moving-coordinate) rendering of the binary encounter of gas particle (hard sphere) 2 with gas particle (hard sphere) 1 with a relative velocity $\boldsymbol{u}_{21}$. The geometrical variables used in the collision integral are shown. The apse line passes through both particle centers [326].

and the relation is subsequently briefly described [326] and is shown in Figure 6.5, where the superscript $'$ indicates before collision.

Each particle is designated by a mass, a diameter (hard sphere), and phase-space coordinates $(\boldsymbol{x}, \boldsymbol{p})$. The effect of the binary collisions is expressed in terms of a differential scattering cross section describing the probability for a change of velocities

$$\{\boldsymbol{u}'_{f,1}, \boldsymbol{u}'_{f,2}\}_{before} \rightarrow \{\boldsymbol{u}_{f,1}, \boldsymbol{u}_{f,2}\}_{after}, \tag{6.44}$$

where ϕ thus denotes the relative orientation of the vectors $(\boldsymbol{u}'_{f,2} - \boldsymbol{u}'_{f,1})$ and $\boldsymbol{u}_{f,21} = (\boldsymbol{u}_{f,2} - \boldsymbol{u}_{f,1})$, as shown in Figure 6.5. In Section 6.5, we will discuss the equilibrium collision (scattering) rate.

Including the nonequilibrium collision term, (6.43) becomes [326]

$$(\frac{\partial}{\partial t} + \boldsymbol{u}_f \cdot \nabla_{\boldsymbol{x}} + \boldsymbol{F}_f \cdot \nabla_{\boldsymbol{p}}) f_{f,1} = \int_\eta \int_\phi \int_{\boldsymbol{u}_2} (f'_{f,1} f'_{f,2} - f_{f,1} f_{f,2}) |\boldsymbol{u}_{f,21}| \eta \, \mathrm{d}\boldsymbol{u}_2 \, \mathrm{d}\psi \, \mathrm{d}\eta$$

BTE with binary collisions (6.45)

where $\boldsymbol{p}_2 = m_2\boldsymbol{u}_2$, and $f_{f,1} \equiv f(\boldsymbol{x}, \boldsymbol{p}, t)$, $f'_{f,1} \equiv f(\boldsymbol{x}, \boldsymbol{p}', t)$, and so on. Note that here f_f is not dimensionless (this will be given in the next section).

Then using

$$\boldsymbol{u}_f \cdot \nabla_{\boldsymbol{x}} f_f \equiv u_x \frac{\partial f_f}{\partial x} + u_y \frac{\partial f_f}{\partial y} + u_z \frac{\partial f_f}{\partial z} \tag{6.46}$$

$$\boldsymbol{F}_f \cdot \nabla_{\boldsymbol{p}} f_f \equiv F_x \frac{\partial f_f}{\partial p_x} + F_y \frac{\partial f_f}{\partial p_y} + F_z \frac{\partial f_f}{\partial p_z}. \tag{6.47}$$

Integrodifferential equation (6.45) describes the spatial–temporal behavior of a dilute gas. Given an initial distribution $f_f(\boldsymbol{x}, \boldsymbol{p}, t = 0)$, the solution $f_f(\boldsymbol{x}, \boldsymbol{p}, t)$ describes the distribution changes over time. Because f_f is difficult to visualize, there are certain moments of f_f that represent averages such as the local particle density, which can be determined as measured. Chapman and Enskog [177] developed a general procedure for the approximate solution of the BTE. For certain simple model systems such as hard spheres, their method produces predictions for $f_f(\boldsymbol{x}, \boldsymbol{p}, t)$ (and its moments). The BTE is generally used to describe nonequilibrium processes. The equilibrium distribution $f_f^o(\boldsymbol{x}, \boldsymbol{p})$ is that solution of BTE that is stationary, i.e., given by (3.6),

$$\frac{\mathrm{D} f_f^o(\boldsymbol{x}, \boldsymbol{p}, t)}{\mathrm{D}t} = 0 \quad \text{equlilibrium distribution.} \tag{6.48}$$

For the case of $\boldsymbol{F}_f = 0$ this gives the M–B equilibrium distribution function f_f^o for a spatially and temporally uniform stationary ideal gas (Section 6.5). We will evaluate the binary collision rate appearing on the right-hand side of ideal gases (6.45) in Section 6.5. Then we use f_f^o and a driving force, along with the collision rate, to evaluate k_f.

6.4.2 Equilibrium Distribution Function for Translational Energy

Consider translational energy only (e.g., monatomic, ideal gas). Gases have a nondegenerate behavior (M–B), except at very low temperatures (Section 7.13.2). Assuming that the molecular velocities are isotropic (in the absence of external fields) and that the distribution of molecular speeds in any one component of velocity is independent of that in any other component, the probability of finding a molecule in a given element of velocity space, $\mathrm{d}\boldsymbol{u}_f$, is then

$$P[(u_x, u_y, u_z), (u_x + \mathrm{d}u_x, u_y + \mathrm{d}u_y, u_z + \mathrm{d}u_z)] = f_f^o(\boldsymbol{u}_f)\mathrm{d}\boldsymbol{u}_f \quad \text{probability,} \tag{6.49}$$

and, using the directional independence, we have variable separation given by

$$P(u_x, u_x + \mathrm{d}u_x) = X(u_x)\mathrm{d}u_x \tag{6.50}$$

$$P(u_y, u_y + \mathrm{d}u_y) = Y(u_y)\mathrm{d}u_y \tag{6.51}$$

$$P(u_z, u_z + \mathrm{d}u_z) = Z(u_z)\mathrm{d}u_z. \tag{6.52}$$

From the second assumption, we have

$$f_f^{\rm o}(\boldsymbol{u}_f)\mathrm{d}\boldsymbol{u} = X(u_x)\mathrm{d}u_x Y(u_y)\mathrm{d}u_y Z(u_z)\mathrm{d}u_z \qquad (6.53)$$

or

$$f_f^{\rm o}(\boldsymbol{u}_f) = X(u_x)Y(u_y)Z(u_z). \qquad (6.54)$$

From the preceding assumption, we know that functions X, Y, and Z should have the same form; therefore

$$f_f^{\rm o}(\boldsymbol{u}_f) = X(u_x)X(u_y)X(u_z). \qquad (6.55)$$

Noting that $u_f^2 = u_x^2 + u_y^2 + u_z^2$, we can choose an appropriate functional form for X (including symmetry around $u_x = 0$, etc.), such as $X(u_x) = A\exp(Bu_x^2)$, then

$$f_f^{\rm o}(\boldsymbol{u}_f) = A^3\exp[B(u_x^2 + u_y^2 + u_z^2)]. \qquad (6.56)$$

The coefficients of A and B are determined based on two constraints, namely the conservation of number density of the fluid particles and the kinetic-theory (translational thermal motion) definition of the temperature (Table 2.4), i.e.,

$$\int_{\boldsymbol{u}_f} f_f^{\rm o}(\boldsymbol{u}_f)\mathrm{d}\boldsymbol{u}_f = [\int_{-\infty}^{\infty} f_f^{\rm o}(u_x)\mathrm{d}u_x]^3 = [2\int_0^{\infty} f_f^{\rm o}(u_x)\mathrm{d}u_x]^3 = 1$$

$$\text{particle number conservation}, \qquad (6.57)$$

$$\frac{m}{3k_{\rm B}}\int_{\boldsymbol{u}_f} u_f^2 f_f^{\rm o}(\boldsymbol{u}_f)\mathrm{d}\boldsymbol{u}_f = \frac{m}{3k_{\rm B}}\langle u_f^2\rangle \;= T \quad \text{definition of temperature}, \qquad (6.58)$$

where

$$\langle u_f^2\rangle^{1/2} \equiv (\frac{3k_{\rm B}T}{m})^{1/2} \quad \text{RMS thermal speed}. \qquad (6.59)$$

Note that alternatively $f_f^{\rm o}$ could be defined such that the integral (6.57) yields the total number of fluid particles N_f or the number of fluid particles per unit volume (number density) n_f [†].

Using the preceding gives

$$A = (\frac{m}{2\pi k_{\rm B}T})^{1/2}, \qquad B = -\frac{m}{2k_{\rm B}T}. \qquad (6.60)$$

[†] To include the number density of fluid particles n_f in f_f^o, we start with (6.49), written as

$$P = \frac{f_f^{\rm o}(\boldsymbol{u}_f)\,\mathrm{d}\boldsymbol{u}_f}{n_f}.$$

Then (6.61), (6.59), and (6.57) become (end-of-chapter problem)

$$f_f^{\rm o}(\boldsymbol{u}_f) = n_f(\frac{m}{2\pi k_{\rm B}T})^{3/2}\exp(\frac{-mu_f^2}{2k_{\rm B}T}), \quad \int_0^{\infty} 4\pi u_f^2 f_f^{\rm o}(\boldsymbol{u}_f)\,\mathrm{d}u_f = n_f.$$

This is the same as the momentum integral ($p_f = mu_f$) given as a Chapter 1 problem.

Then the M–B velocity distribution function becomes

$$f_f^{\rm o}(\boldsymbol{u}_f) = (\frac{m}{2\pi k_{\rm B} T})^{3/2} \exp(\frac{-m\boldsymbol{u}_f \cdot \boldsymbol{u}_f}{2k_B T})$$

M–B velocity vector (momentum) probability distribution function. (6.61)

For example, for the x component of the velocity, we use $\boldsymbol{u}_f = u_{f,x}\boldsymbol{s}_x$.

The distribution of the fluid speed can be obtained by considering the probability of finding the particle in the shell of $u_f < |\boldsymbol{u}_f| < u_f + {\rm d}u_f$, which has the volume of $4\pi u_f^2 {\rm d}u_f$, so $f_f^{\rm o}(u_f){\rm d}u_f = f_f^{\rm o}(\boldsymbol{u}_f)4\pi u_f^2 {\rm d}u_f$. Then

$$f_f^{\rm o}(u_f) = 4\pi u_f^2 f_f^{\rm o}(\boldsymbol{u}_f) = 4\pi(\frac{m}{2\pi k_{\rm B} T})^{3/2} u_f^2 \exp(\frac{-mu_f^2}{2k_{\rm B} T})$$

speed probability distribution (occupancy) function. (6.62)

This is referred to as the Maxwell distribution function.

The average (or mean) speed is defined as †

$$\langle u_f \rangle = \int_0^\infty f_f^{\rm o}(u_f) u_f {\rm d}u_f$$
$$= (\frac{8}{\pi}\frac{k_{\rm B} T}{m})^{1/2} \quad \text{average thermal speed.} \qquad (6.63)$$

For N_2 at $T = 300$ K, $m = 4.652 \times 10^{-26}$ kg, and we have $\langle u_f^2 \rangle^{1/2} = 516.9$ m/s and $\langle u_f \rangle = 476.3$ m/s, i.e., $\langle u_f \rangle$ is slightly smaller. The relation with the speed of sound a_s will be give in Section 6.8.

Noting that $E_f = mu_f^2/2$, we have ${\rm d}E_f = mu_f {\rm d}u_f$; then we have for the energy distribution function

$$f_f^{\rm o}(E_f){\rm d}(E_f) = f_f^{\rm o}(u_f){\rm d}(u_f) = f_f^{\rm o}[(\frac{2E_f}{m})^{1/2}]\frac{{\rm d}E_f}{mu_f}$$
$$= 2\pi(\pi k_{\rm B} T)^{-3/2} E_f^{1/2} \exp(-\frac{E_f}{k_{\rm B} T}){\rm d}E_f, \qquad (6.64)$$

or

$$f_f^{\rm o}(E_f^*) = \frac{2}{\pi^{1/2}} E_f^{*1/2} \exp(-E_f^*), \quad E_f^* = \frac{E_f}{k_{\rm B} T},$$
$$\langle E_f \rangle = \int_0^\infty E_f^* f_f^{\rm o}(E_f){\rm d}E_f = \frac{3}{2} k_{\rm B} T$$

M–B energy probability occupancy distribution function. (6.65)

This is the dimensionless energy M–B energy distribution function listed in Figure 1.1. The mean (average) energy per fluid particle in translational motion is $3k_{\rm B} T/2$ (Chapter 1 problem).

† Note that $\langle u_f \rangle$ is not the net flow, and we use $\overline{u}_f$ for the net flow (Sections 6.11 and 6.12).

6.4.3 Inclusion of Gravitational Potential Energy

When a gravitational potential energy is added to the particle energy, i.e., from (6.1), $E_f = E_{f,t} + E_{f,p} = mu_f^2/2 + mgz$, with $\boldsymbol{g}$ pointing against z, the equilibrium distribution function for $\boldsymbol{F} = m\boldsymbol{g}$, (6.61) becomes

$$f_f^{\rm o}(\boldsymbol{u}_f, \boldsymbol{x}) = (\frac{m}{2\pi k_{\rm B} T})^{3/2} \exp{(-\frac{mgz}{k_{\rm B}T})}\exp(\frac{-mu_f^2}{2k_{\rm B}T})$$

$$E_f = \frac{1}{2}mu_f^2 + mgz, \tag{6.66}$$

satisfying the equilibrium distribution (and under steady state) (6.48), i.e., no collision ideal-gas BTE. Then (6.48) becomes

$$\frac{{\rm D} f_f^{\rm o}}{{\rm D}t} = \boldsymbol{u} \cdot \nabla f_e^{\rm o} + m\boldsymbol{g} \cdot \nabla_{\boldsymbol{p}} f_e^{\rm o} = 0. \tag{6.67}$$

6.5 Ideal-Gas Binary Collision Rate and Relaxation Time

Consider the binary collision of ideal-gas particles (hard spheres) 1 and 2, as shown in Figure 6.5. The mass and diameters are m_1 and m_2, d_1 and d_2, and velocities are $\boldsymbol{u}_1$ and $\boldsymbol{u}_2$, or the relative velocity $\boldsymbol{u}_{f,21} = \boldsymbol{u}_{f,2} - \boldsymbol{u}_{f,1}$. The particle collisions are elastic (the energy remains in the translational energy of the colliding particles). The total momentum and kinetic energy of the two colliding particle are conserved, and if in the kinetic energy conservation it is assumed that the particles have the same mass, it simplifies the analysis (the momentum conservation will be applied for the general cases of m_1 and m_2). Then conservations of momentum and kinetic (translational) energy (elastic collision) are applied and they result in (end-of-chapter problem)

$$(m_1\boldsymbol{u}'_{f,1} + m_2\boldsymbol{u}'_{f,2})_{before} = (m_1\boldsymbol{u}_{f,1} + m_2\boldsymbol{u}_{f,2})_{after}$$

$$\text{momentum conservation (elastic collision)} \tag{6.68}$$

$$|\boldsymbol{u}'_{f,21}|_{before} = |\boldsymbol{u}'_{f,2} - \boldsymbol{u}'_{f,1}|_{before} = |\boldsymbol{u}_{f,21}|_{after} = |\boldsymbol{u}_{f,2} - \boldsymbol{u}_{f,1}|_{after}$$

$$\text{kinetic (translational) energy conservation,} \tag{6.69}$$

where subscripts *before* and *after* indicate before and after the collision.

The number of particles 2 approaching particles 1 during time interval dt and per unit of volume $u_{f,21}\ \eta\ {\rm d}\phi{\rm d}\eta{\rm d}t{\rm d}\boldsymbol{u}_2$ is

$$n_{f,1}n_{f,2}f_{f,2}^{\rm o}u_{f,21}\eta\boldsymbol{u}_2{\rm d}\phi{\rm d}\eta{\rm d}t{\rm d}\boldsymbol{u}_2, \quad u_{f,21} = |\boldsymbol{u}_{f,2} - \boldsymbol{u}_{f,1}|. \tag{6.70}$$

The number of particles 1 colliding with particles 2 is similarly

$$n_{f,1}n_{f,2}f_{f,1}^{\rm o}u_{f,21}\eta{\rm d}\phi{\rm d}\eta{\rm d}t{\rm d}\boldsymbol{u}_1. \tag{6.71}$$

Then the total number of collisions is

$$\dot{n}_{f,12}\ {\rm d}V\ {\rm d}t\ {\rm d}\boldsymbol{u}_1 = \int_\eta \int_\phi \int_{\boldsymbol{u}_2} n_{f,1}n_{f,2}f_{f,1}^{\rm o}\ {\rm d}V{\rm d}\boldsymbol{u}_1\, f_{f,2}^{\rm o}u_{f,21}\eta{\rm d}\phi{\rm d}\eta{\rm d}t{\rm d}\boldsymbol{u}_2 \tag{6.72}$$

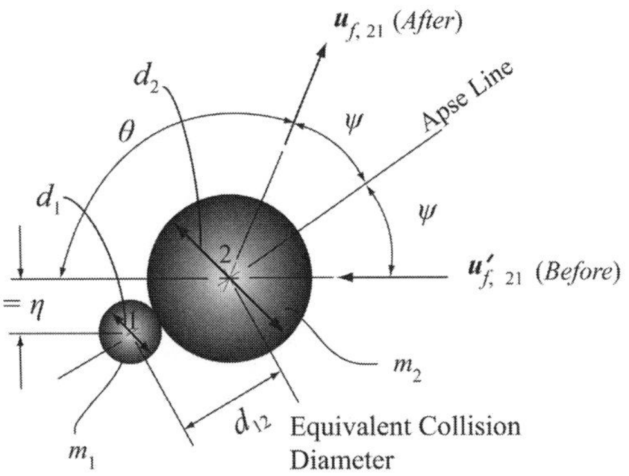

Figure 6.6. Two-dimensional (in moving coordinates) rendering of binary collision of the collision equivalent diameter ideal-gas particles (hard sphere), showing the geometrical variables appearing in collision integral [326]. The equivalent collision diameter d_{12} is also shown.

or

$$\dot{n}_{f,12} = \int_{\boldsymbol{u}_2}\int_{\boldsymbol{u}_1}\int_{\eta}\int_{\phi} n_{f,1} n_{f,2} f^{\rm o}_{f,1} f^{\rm o}_{f,2} u_{f,21} \eta \, {\rm d}\phi {\rm d}\eta {\rm d}\boldsymbol{u}_1 {\rm d}\boldsymbol{u}_2. \qquad (6.73)$$

Now, we define a collision equivalent diameter (the diameter of the sphere of collision) by using the geometrical variables of Figure 6.6, and we have

$$\eta \equiv d_{12} \sin\psi, \quad \eta \, {\rm d}\eta = d_{12}^2 \cos\psi \sin\psi {\rm d}\psi, \quad d_{12} = (d_1 + d_2)/2,$$

$$\text{equivalent collision diameter}, \qquad (6.74)$$

where ψ is related to θ as shown in Figure 6.6. The collision cross-section area $A_{f\text{-}f} = \pi d_{12}^2/4$ is determined with the individual collision diameters. Then we have

$$\dot{n}_{f,12} = \int_{\boldsymbol{u}_1}\int_{\boldsymbol{u}_2} n_{f,1} n_{f,2} f^{\rm o}_{f,1} f^{\rm o}_{f,2} u_{f,21} d_{12}^2 {\rm d}\boldsymbol{u}_1 {\rm d}\boldsymbol{u}_2 \int_0^{\pi/2} \cos\psi \sin\psi {\rm d}\psi \int_0^{2\pi} {\rm d}\phi, \quad (6.75)$$

where

$$u_{f,21} = |\boldsymbol{u}_{f,1} - \boldsymbol{u}_{f,2}| = (u_{f,1}^2 + u_{f,2}^2 - 2\boldsymbol{u}_{f,1} \cdot \boldsymbol{u}_{f,2})^{1/2}. \qquad (6.76)$$

Note that for a general treatment $\boldsymbol{u}_{f,1} \cdot \boldsymbol{u}_{f,2}$ can be set to zero. The integrations over ψ and θ gives π. The equilibrium M–B distribution function for velocity is given by (6.61) and for particle 1 is

$$f^{\rm o}_{f,1}(\boldsymbol{u}_f) = (\frac{m_1}{2\pi k_{\rm B} T})^{3/2} \exp(-\frac{m_1 u_f^2}{2k_{\rm B} T}) \quad \text{M–B distribution (occupancy) function}, \qquad (6.77)$$

where m_1 is $M_1/N_{\rm A}$. Then, in the integration of (6.75), for the approximation we

Table 6.4. *Estimates of collision diameter and mean free path (at 298 K and 1 atm) for some gaseous molecules (or atoms) modeled as hard spheres [177].*

Species	λ_f, nm	d, Å	Species	λ_f, nm	d, Å
Air	64.0	3.72	He	186.2	2.18
Ar	66.6	3.64	Hg	83.2	4.26
CH_4	51.6	4.14	Kr	51.2	4.16
CO_2	41.9	4.59	Ne	132.8	2.59
C_2H_4	36.1	4.95	NH_3	45.1	4.43
C_2H_6	31.5	5.30	N_2	62.8	3.75
HCl	44.4	4.46	O_2	67.9	3.61
H_2	117.7	2.74	Xe	37.6	4.85
H_2O	41.8	4.60			

have

$$\int_{\boldsymbol{u}_1}\int_{\boldsymbol{u}_2} f^{\rm o}_{f,1} f^{\rm o}_{f,2}(u^2_{f,1} + u^2_{f,2} - 2\boldsymbol{u}_{f,1}\cdot\boldsymbol{u}_{f,2})^{1/2}{\rm d}\boldsymbol{u}_1{\rm d}\boldsymbol{u}_2$$

$$\simeq \int_{\boldsymbol{u}_1}\int_{\boldsymbol{u}_2} [(f^{\rm o}_{f,1}u_{f,1}{\rm d}\boldsymbol{u}_1)^2 + (f^{\rm o}_{f,2}u_{f,2}{\rm d}\boldsymbol{u}_2)^2 - 2f^{\rm o}_{f,1}f^{\rm o}_{f,2}(\boldsymbol{u}_{f,1}\cdot\boldsymbol{u}_{f,2}){\rm d}\boldsymbol{u}_1{\rm d}\boldsymbol{u}_2]^{1/2},$$

$$\langle u_{f,i}\rangle = \int_0^\infty f^{\rm o}_{f,i}u_{f,i}{\rm d}u_i = \int_0^\infty f^{\rm o}_{f,i}(u_{f,i})u_{f,i}{\rm d}u_{f,i} \quad \text{average thermal speed,} \tag{6.78}$$

where we have used (6.63).

As was mentioned, the contribution from $\boldsymbol{u}_{f,1}\cdot\boldsymbol{u}_{f,2}$ term is zero. Using M–B distribution (6.77), the other two terms each give the square of the average thermal speed (6.63) for each mass group. Then (6.75) becomes (end-of-chapter problem)

$$\dot{n}_{f,12} = 2n_{f,1}n_{f,2}d^2_{12}[\frac{2\pi k_{\rm B}T(m_1+m_2)}{m_1m_2}]^{1/2}, \tag{6.79}$$

where $n_{f,1}$ and $n_{f,2}$ are the number density of each particle, and the universal gas constant is defined as $R_g = k_{\rm B}N_{\rm A}$, and $M = mN_{\rm A}$. The preceding is the volumetric rate of collision of particles 1 with particles 2. For a monatomic ideal gas, this volumetric collision rate $\dot{n}_{f\text{-}f}$ becomes

$$\dot{n}_{f\text{-}f} = 4n^2_f d^2(\frac{\pi k_{\rm B}T}{m})^{1/2} = \frac{4\pi^{1/2}d^2p^2}{(k_{\rm B}T)^{3/2}m^{1/2}}$$

$$\text{single component gas, binary collision rate,} \tag{6.80}$$

where we have used n_f from (6.21). The estimated (from measurements of properties such as gas viscosity) collision (hard-sphere) diameter, and the mean free path (at 298 K and 10 atm) for some gaseous molecules are listed in Table 6.4 [177]. Note that He has the smallest collision diameter among the atoms and molecules listed (the collision cross-section area is $\pi d^2/4$). The mean free path is also the largest for He (we will discuss the mean free path of ideal gas in the following section). For example, for N_2 at p =1.013×10^5 Pa, and T = 300 K, we have n_f =2.662×10^{25} 1/m^3

and $\dot{n}_{f\text{-}f}$ =2.031×10^{35} 1/m^3-s. We can also define the collision relaxation time as

$$\tau_{f\text{-}f} = \frac{1}{\dot{\gamma}_{f\text{-}f}} = \frac{n_f}{\dot{n}_{f\text{-}f}} = \frac{m^{1/2}}{4\pi^{1/2}d^2(k_\mathrm{B}T)^{1/2}n_f} = \frac{1}{2^{1/2}\pi d^2 n_f\langle u_f\rangle}$$

single component gas binary collision relaxation time, (6.81)

where $\langle u_f\rangle$ is given by (6.63). From (6.81), $\tau_{f\text{-}f}$ increases as m increases, and as d, T, and n_f decrease.

Thus for the preceding N_2 example, we have $\tau_{f\text{-}f}$ =1.311×10^{-10} s.

6.6 Ideal-Gas Mean Free Path and Viscosity

The interparticle collision relaxation time $\tau_{f\text{-}f,i}$ is defined as

$$\tau_{f\text{-}f,i} \equiv \frac{n_{f,i}}{\sum_j \dot{n}_{f,ij}}. \qquad (6.82)$$

Then, using (6.79), we have

$$\tau_{f\text{-}f,i} = \frac{1}{\sum_j 2n_{f,j}d_{ij}^2\,[\frac{2\pi k_B T(m_i+m_j)}{m_i m_j}]^{1/2}}. \qquad (6.83)$$

The mean free path λ_f is defined as

$$\lambda_{f,i} \equiv \tau_{f\text{-}f,i}\langle u_{f,i}\rangle \quad \text{mean free path definition,} \qquad (6.84)$$

where the average molecular (thermal) speed $\langle u_{f,i}\rangle$ (6.63) is

$$\langle u_{f,i}\rangle = (\frac{8}{\pi}\frac{k_\mathrm{B}T}{m_i})^{1/2}. \qquad (6.85)$$

Then

$$\lambda_{f,i} = \frac{1}{\sum_j \pi n_{f,j} d_{ij}^2(\frac{m_j}{m_i}+1)^{1/2}}$$

mean free path for multicomponent, ideal-gas. (6.86)

For a single component ideal-gas, we have

$$\lambda_f = \frac{1}{2^{1/2}\pi n_f d^2},\quad n_f = \frac{p}{k_\mathrm{B}T} \quad \text{single component, ideal-gas mean free path.} \qquad (6.87)$$

Table 6.4 gives the mean free paths for some common gases at $T = 298$ K and p =1.013× 10^5 Pa. These values are of the order of 10 nm, and for smaller-diameter-lighter atoms, H_2 and He, they are of the order of 100 nm. Some other useful relationships can be developed.

Although the net flow of fluid particles across a planar surface is zero, the flux of particles moving along a direction is not. This flux of particles (through a surface with normal vector $\mathbf{s}_n$, where i is along a positive Cartesian coordinate axis) is [55]

$$j_{f,i} = \int_0^\infty n_f u_{f,i} f_f^{\rm o}(u_{f,i}) {\rm d}u_{f,i}$$

$$= \int_0^\infty n_f u_{f,i} (\frac{m}{2\pi k_{\rm B} T})^{1/2} \exp(-\frac{m u_{f,i}^2}{2 k_{\rm B} T})\, {\rm d}u_{f,i} \quad \text{ideal-gas flux,} \qquad (6.88)$$

where we have used (6.61) for only one direction.

Then the gas mass flux moving in a direction is found by multiplying (6.88) by the mass of each particle and can be related to the average thermal velocity according to

$$\dot{m}_f = m j_{f,i} = \frac{1}{4} m n_f (\frac{8 k_{\rm B} T}{\pi m})^{1/2} = \frac{1}{4} m n_f \langle u_f \rangle \quad \text{ideal-gas directional mass flux.} \quad (6.89)$$

The momentum flux (or stress) is related to the dynamic viscosity according to

$$\tau_{ij} = m n_f \int u_{f,i} u_{f,j} f_f^{\rm o}(\boldsymbol{u}_f) {\rm d}\boldsymbol{u}_f \equiv -\mu_f \frac{\partial \overline{u}_{f,i}}{\partial x_j} \quad \text{dynamic viscosity} \qquad (6.90)$$

From this the gas viscosity † is (end-of-chapter problem)

$$\mu_f = \frac{1}{3} m n_f \langle u_f \rangle \lambda_f = \frac{1}{3} \rho_f \langle u_f \rangle \lambda_f = \frac{1}{3} \rho_f \langle u_f \rangle^2 \tau_{f\text{-}f} = \frac{1}{2^{1/2} 3\pi} \frac{m \langle u_f \rangle}{d^2}$$

$$\text{monatomic ideal-gas viscosity.} \qquad (6.91)$$

Note that here we have used the average mean thermal speed $\langle u_f \rangle$ given by (6.63), which is slightly different from the mean thermal speed $\langle u_f^2 \rangle^{1/2}$ given by (6.59).

The pressure is related to the average thermal speed through (6.89)

$$p = n_f k_{\rm B} T = \frac{1}{3} m n_f \langle u_f \rangle^2 = \frac{1}{3} \rho_f \langle u_f \rangle^2. \qquad (6.92)$$

Then, using $\tau_{f\text{-}f} = \lambda_f / \langle u_f \rangle$, we have $\mu_f = p \tau_{f\text{-}f}$, i.e., dynamic viscosity is the product of pressure and relaxation time for ideal gases.

6.7 Theoretical Maximum Evaporation–Condensation Heat Transfer Rate

The mass flux given by (6.89) is used to determine the theoretical maximum heat flux through a liquid–vapor interface A_{lg}, undergoing condensation–evaporation. For a single-component fluid, having a heat of vaporization–condensation Δh_{lg} (or Δh_{gl}) under ideal conditions of no limit to the supply–removal of heat to the liquid–vapor interface, no limit to the supply of make-up liquid–vapor to the interface, and

† Note that the kinematic viscosity is

$$\nu_f = \frac{\mu_f}{\rho_f} = \frac{1}{3} \langle u_f \rangle \lambda_f = \frac{1}{3} \langle u_f \rangle^2 \tau_{f\text{-}f},$$

i.e., product of thermal speed and mean free path.

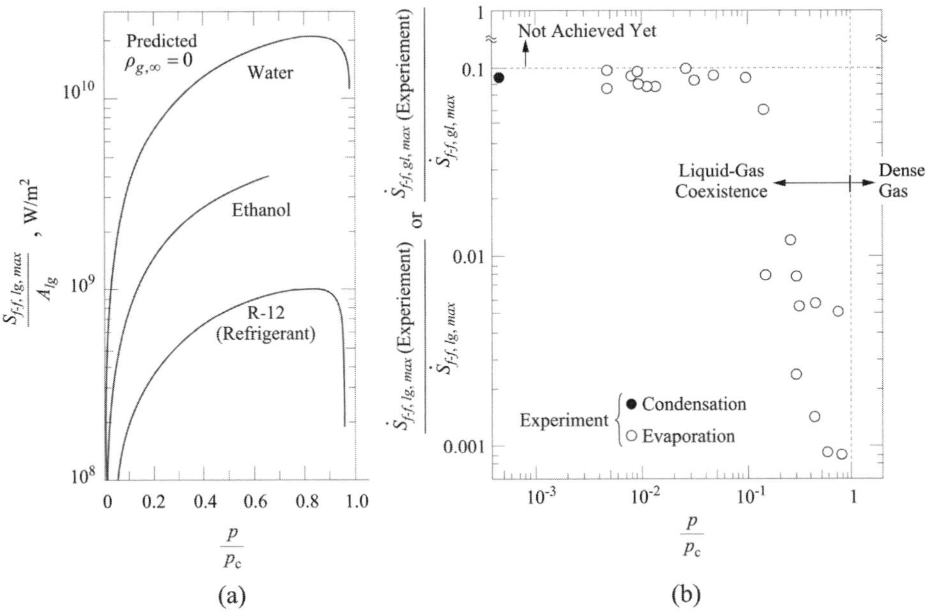

Figure 6.7. (a) Variations of the theoretical maximum heat flux for evaporation for water, ethanol, and Freon-12 with respect to reduced pressure. (b) Variations of the ratio of the measured maximum heat flux for evaporation and condensation to the theoretical limit, with respect to the reduced pressure [118].

complete ideal (no resistance) removal of the produced vapor–liquid, the theoretical maximum heat transport rate to the interface is the product of mass flux (6.89) and the heat of phase change, i.e., [118]

$$\frac{\dot{S}_{f\text{-}f,lg,max}}{A_{lg}} = \dot{m}_{lg,max}\Delta h_{lg} = (\frac{R_g T_{lg}}{2\pi M})^{1/2}\rho_g \Delta h_{lg}, \qquad (6.93)$$

and similarly for $q_{gl,max}$.

Since the saturation temperature $T_{lg}(p_g)$ increases, ρ_g increases, whereas Δh_{lg} decreases (and vanishes at the critical temperature T_c), and $\dot{S}_{f\text{-}f,lg,max}/A_{lg}$ will reach a maximum as the temperature (or pressure) increases. This is plotted in Figure 6.7(a) for water, ethanol, and refrigerant R-12 as $\dot{S}_{f\text{-}f,lg,max}/A_{lg}$ as a function of the reduced pressure p/p_c. Water has the highest heat of evaporation of any natural and synthesized fluid and, as shown in Figure 6.7(a) it also has the highest peak in $q_{lg,max}$ (over 10^{10} W/m²).

The ideal assumptions regarding liquid–vapor supply are not realized in practice because of hydrodynamic resistances, and similarly the heat supply–removal that is due to heat transfer resistances, resulting in achievable values much less than $\dot{S}_{f\text{-}f,lg,max}/A_{lg}$ or $\dot{S}_{f\text{-}f,gl,max}/A_{lg}$. So far, only one-tenth of this limit has been reached in the most exhaustive efforts to reduce these resistances. Figure 6.7(b) shows the experimental results in evaporation and condensation experiments to reach these theoretical maxima.

When the far-field vapor concentration is not zero, i.e, there is no vacuum away from the liquid–vapor interface, the preceding theoretical limit is modified as

$$\frac{\dot{S}_{f\text{-}f,lg,max}}{A_{lg}} = (\frac{R_g}{2\pi M})^{1/2} \Delta h_{lg}(\rho_g T_{lg}^{1/2} - \rho_{g,\infty} T_{lg,\infty}^{1/2}), \tag{6.94}$$

where $\rho_{g,\infty}$ and $T_{lg,\infty}(p_{g,\infty})$ are the far-field vapor density and temperature (assumed to be the saturation condition).

6.8 Ideal-Gas Thermal Conductivity from BTE

6.8.1 Nonequilibrium BTE and Relaxation-Time Approximation

The fluid particle BTE (6.43), under no external force and written for a fluid particle group $f_{f,i}$ is

$$\frac{\partial f_{f,i}}{\partial t} + \boldsymbol{u}_{f,i} \cdot \nabla f_{f,i} = \frac{\partial f_{f,i}}{\partial t}|_s, \tag{6.95}$$

This represents the net rate of change in this fluid particle (molecules) population that is due to collision with particles of different positions and momenta.

Continuing with the collision rate, from Figure 6.6 and also from (6.70), the number of particles of group j approaching particle i, during time interval $\mathrm{d}t$ with $f_{f,j}^{\mathrm{o}}$ is

$$f_{f,j}|\boldsymbol{u}_{f,ji}|\mathrm{d}t\ \eta \mathrm{d}\eta \mathrm{d}\phi \mathrm{d}\boldsymbol{u}_j, \tag{6.96}$$

where $\boldsymbol{u}_{f,ji}$ is the relative velocity, and η and ϕ are defined as in Figure 6.6. Then we rewrite the right-hand side of (6.45) as

$$\frac{\partial f_{f,i}}{\partial t}|_s = \int_\eta \int_\phi \int_{\boldsymbol{u}_j} (f'_{f,i} f'_{f,j} - f_{f,i} f_{f,j})|\boldsymbol{u}_{f,ji}|\mathrm{d}t\ \eta \mathrm{d}\eta \mathrm{d}\phi \mathrm{d}\boldsymbol{u}_j, \tag{6.97}$$

where the prime indicates distribution functions that restore equilibrium. Under equilibrium, we have $f'_{f,i} f'_{f,j} = f_{f,i} f_{f,j}$.

The collision rate signifies the ability of the system to return to the equilibrium distribution function $f_{f,i}^{\mathrm{o}}$; the larger it is, the faster this return.

As discussed in Section 3.1.4, one of the approximations made in simplifying the collision rate is the perturbation approximation; using a perturbation parameter ϵ and the Taylor series expansion, we have

$$f_{f,i} = f_{f,i}^{\mathrm{o}} + (\frac{\partial f_{f,i}}{\partial \epsilon}|_{\epsilon=0})\epsilon + (\frac{\partial f_{f,i}^2}{\partial \epsilon^2}|_{\epsilon=0})\frac{\epsilon^2}{2} + \cdots + \tag{6.98}$$

or

$$\frac{\partial f_{f,i}}{\partial \epsilon}|_{\epsilon=0} = \frac{f_{f,i} - f_{f,i}^{\mathrm{o}}}{\epsilon} \quad \text{for } \epsilon \to 0. \tag{6.99}$$

Using the interparticle collision relaxation time $\tau_{f\text{-}f,i}$ from (6.82), along with noting that the collision restores $f^{\rm o}_{f,i}$, we write

$$\frac{\partial f_{f,i}}{\partial t}|_s = -\frac{f_{f,i} - f^{\rm o}_{f,i}}{\tau_{f\text{-}f,i}}. \tag{6.100}$$

Next we use this to derive the relation for the ideal-gas thermal conductivity.

6.8.2 Thermal Conductivity

Independent of the BTE, the kinetic theory of gas allows for derivation of a relation between the ideal-gas thermal conductivity k_f and the mean free path [326]. Here we derive the relation from the ideal-gas BTE (6.95); by using a treatment similar to those used in Chapters 4 and 5 for phonons and electrons [315]. The relaxation-time approximation of (6.100) gives

$$\frac{\partial f_f}{\partial t}|_s = -\frac{f_f - f^{\rm o}_f}{\tau_{f\text{-}f}}, \tag{6.101}$$

where $f^{\rm o}_f$ is the equilibrium M–B particle distribution function, Table 1.2, and $\tau_{f\text{-}f}$ is the momentum-independent fluid particle (gas) thermal collision relaxation time given by (6.82). Then BTE (6.95) becomes

$$\frac{\partial f_f}{\partial t} + \boldsymbol{u}_f \cdot \nabla f_f + \boldsymbol{F}_f \cdot \nabla_{\boldsymbol{p}} f_f = -\frac{f_f - f^{\rm o}_f}{\tau_{f\text{-}f}}, \tag{6.102}$$

where $\tau_{f\text{-}f}$ is assumed to be independent of $\boldsymbol{p}$.

In the presence of a temperature gradient ∇T, from Onsager relation (3.56), there will be a conduction heat flux (given by the Fourier law) (Table 1.1) and for isotropic behavior we have

$$\boldsymbol{q}_k = -k_f \nabla T, \quad \boldsymbol{q}_{k,j} = -k_f \frac{\partial T}{\partial x_j}, \tag{6.103}$$

where k_f is the gas thermal conductivity. With no net motion, heat flow is due to the thermal motion and deviation of the energy distribution from the thermal equilibrium distribution. Similar to (4.93) and (5.119), the conduction heat flux is also given in terms of the molecular energy flux that is due to a deviation in population $f'_f = f_f - f^{\rm o}_f$ as

$$\boldsymbol{q}_k = \frac{1}{(2\pi\hbar)^3} \int E_f \boldsymbol{u}_f f'_f \mathrm{d}\boldsymbol{p}_f = \frac{1}{(2\pi\hbar)^3} \int E_f \boldsymbol{u}_f (f_f - f^{\rm o}_f) \mathrm{d}\boldsymbol{p}_f, \tag{6.104}$$

where we are using the fluid particle momentum (instead of the wave vector). Under the steady-state condition ($\partial f_f/\partial t = 0$); they (6.102) becomes

$$\boldsymbol{u}_f \cdot \nabla f_f = -\frac{f_f - f^{\rm o}_f}{\tau_{f\text{-}f}}. \tag{6.105}$$

Now we linearize the $\boldsymbol{u}_f \cdot \nabla f_f$ term by approximating f_f with $f^{\rm o}_f$, i.e., using the local equilibrium distribution condition, $\nabla f_f \simeq \nabla f^{\rm o}_f$. Then, we expand $\nabla f^{\rm o}_f$ in terms of ∇T. We use the classical model for $f_f^{\rm o}$, i.e., the equilibrium distribution function for energy given in Table 1.2 as

$$f_f^{\rm o} = \exp(-\frac{E_f - \mu}{k_{\rm B}T}), \quad E_f = \frac{1}{2}mu_f^2, \tag{6.106}$$

where μ is the chemical potential.

Next, expanding $f_f^{\rm o}$, using E_f, T, and μ, we have

$$\begin{aligned}
\boldsymbol{u}_f \cdot \nabla f_f^{\rm o} &= \boldsymbol{u}_f(\frac{\partial f_f^{\rm o}}{\partial E_f}\nabla E_f + \frac{\partial f_f^{\rm o}}{\partial T}\nabla T + \frac{\partial f_f^{\rm o}}{\partial \mu}\nabla \mu)\\
\frac{\partial f_f^{\rm o}}{\partial T} &= -\frac{E_f - \mu}{T}\frac{\partial f_f^{\rm o}}{\partial E_f}\\
\frac{\partial f_f^{\rm o}}{\partial \mu} &= -\frac{\partial f_f^{\rm o}}{\partial E_f}\\
\nabla E_f &= 0.
\end{aligned} \tag{6.107}$$

Then

$$\begin{aligned}
\boldsymbol{u}_f \cdot \nabla f_f^{\rm o} &= -\frac{\partial f_f^{\rm o}}{\partial E_f}\boldsymbol{u}_f \cdot (\frac{E_f}{T}\nabla T - \frac{\mu}{T}\nabla T + \nabla \mu)\\
&= -\frac{\partial f_f^{\rm o}}{\partial E_f}\boldsymbol{u}_f \cdot \nabla T\frac{1}{T}[E_f - (\mu - T\frac{\mu}{T}|_p)],
\end{aligned} \tag{6.108}$$

where p is the pressure.

We now use the thermodynamics relation

$$\mu - T\frac{\partial \mu}{\partial T}|_p = h, \tag{6.109}$$

where h is the specific enthalpy. Then (6.108) becomes

$$\boldsymbol{u}_f \cdot \nabla f_f^{\rm o} = -\frac{\partial f_f^{\rm o}}{\partial E_f}\frac{E_f - h}{T}(\boldsymbol{u}_f \cdot \nabla T). \tag{6.110}$$

Also, using (6.105), we have

$$\frac{\partial f_f^{\rm o}}{\partial E_f} = -\frac{f_h^{\rm o}}{k_{\rm B}T}. \tag{6.111}$$

Using these, the Boltzmann transport equation (6.102) becomes

$$f_f^{\rm o}\frac{E_f - h}{k_{\rm B}T^2}(\boldsymbol{u}_f \cdot \nabla T) = -\frac{f_f - f_f^{\rm o}}{\tau_{f\text{-}f}}, \tag{6.112}$$

or

$$\frac{f_f - f_f^{\rm o}}{f_f^{\rm o}} = -\frac{1}{k_{\rm B}T^2}\tau_{f\text{-}f}(E_f - h)(\boldsymbol{u}_f \cdot \nabla T). \tag{6.113}$$

The thermal conductivity is found by use of the deviation f_f'. We write conduction flux (6.104) for direction j and arrive at

$$q_{k,j} = \frac{1}{(2\pi\hbar)^3}\int u_{f,j}E_f(f_f - f_f^{\rm o})\mathrm{d}\boldsymbol{p}_f$$
$$= -\frac{\tau_{f\text{-}f}}{(2\pi\hbar)^3 k_{\rm B}T^2}\int u_{f,j}E_f(E_f - h)u_{f,j}f_f^{\rm o}\frac{\partial T}{\partial x_j}\mathrm{d}\boldsymbol{p}_f, \qquad (6.114)$$

where we have used the momentum independent τ_{f-f} (6.81). We replace the velocity with E_f and define the integral as

$$q_{k,j} = -\frac{2n_f\tau_{f\text{-}f}}{3mk_{\rm B}T^2}\langle(E_f - h)E_f^2\rangle\frac{\partial T}{\partial x_j}, \quad n_f = \frac{1}{(2\pi\hbar)^3}\int f_f^{\rm o}\mathrm{d}\boldsymbol{p}_f$$
$$\equiv -k_f\frac{\partial T}{\partial x_j}, \qquad (6.115)$$

where from the isotropic assumption (6.57) we have equal translational energy in all three directions, leading to

$$u_{f,j}^2 = \frac{u_f^2}{3} = \frac{2E_f}{3m}. \qquad (6.116)$$

We will subsequently evaluate the integral shown by $\langle\ \rangle$. Note that we have introduced the particle number conservation also through n_f by using (5.63) (footnote of Section 6.4.2). We have used (6.103) in (6.115). We now

$$k_f = \frac{2}{3}\frac{n_f\tau_{f\text{-}f}}{mk_{\rm B}T^2}\langle(E_f - h)E_f^2\rangle. \qquad (6.117)$$

Next the specific enthalpy $h = E_f + pv_f = 5k_{\rm B}T/2$ is used for monatomic gases. Then

$$E_f - h = -k_{\rm B}T. \qquad (6.118)$$

Now, evaluating the integral, we start with

$$\langle(E_f - h)E_f^2\rangle$$
$$= \frac{1}{(2\pi\hbar)^3 n_f}\int(E_f - \frac{5}{2}k_{\rm B}T)E_f^2\exp[-\frac{(E_f-\mu)}{k_{\rm B}T}]\mathrm{d}\boldsymbol{p}_f \quad \text{monatomic gas.} \qquad (6.119)$$

Then, we convert the momentum differential to energy differential using $E_f = p_f^2/2m$, i.e.,

$$\mathrm{d}\boldsymbol{p}_f = 4\pi p_f^2\mathrm{d}p_f = 2\pi m^2E_f\frac{m}{(2E_f)^{1/2}}\mathrm{d}p_f = \frac{8\pi}{2^{1/2}}m^{3/2}E_f^{1/2}\mathrm{d}E_f. \qquad (6.120)$$

Note that in (6.114) similar results would have been found if we had used (6.64) directly and performed the integration over the fluid energy. This shows the consistency of the classical limit for energy distribution function (6.106).

For an ideal-gas, we have $\mu = 0$, and

$$\langle(E_f - h)E_f^2\rangle = \langle E_f^3\rangle - h\langle E_f^2\rangle. \qquad (6.121)$$

Now, returning to the averages similar to (5.112), we have the general form

$$\langle E_f^n\rangle \equiv \frac{\int_{E_f} f_f^{\rm o} E_f^n {\rm d}\boldsymbol{p}_f}{\int_{E_f} f_f^{\rm o} {\rm d}\boldsymbol{p}_f} = \frac{\int_{E_f} f_f^{\rm o} E_f^n \frac{8\pi}{2^{1/2}} m^{3/2} E_f^{1/2} {\rm d}E_f}{\int_{E_f} f_f^{\rm o} \frac{8\pi}{2^{1/2}} m^{3/2} E_f^{1/2} {\rm d}E_f}$$

$$= \frac{\int_{E_f} E_f^{n+1/2} \exp(-E_f/k_{\rm B}T){\rm d}E_f}{\int_{E_f} E_f^{1/2} \exp(-E_f/k_{\rm B}T){\rm d}E_f} = \frac{\Gamma(n+1/2+1)/(k_{\rm B}T)^{-(n+1/2+1)}}{\Gamma(1/2+1)/(k_{\rm B}T)^{-3/2}}$$

$$= (k_{\rm B}T)^n \frac{\Gamma(n+3/2)}{\Gamma(3/2)}, \tag{6.122}$$

where the Γ function is defined in (5.113).

Again, note that we have used classical limit (6.106). Then,

$$\langle E_f^3\rangle - h\langle E_f^2\rangle = (k_{\rm B}T)^3 \frac{\Gamma(9/2)}{\Gamma(3/2)} - \frac{5}{2} k_{\rm B}T (k_{\rm B}T)^2 \frac{\Gamma(7/2)}{\Gamma(3/2)}$$

$$= (k_{\rm B}T)^3 (\frac{105}{8} - \frac{5}{2} \times \frac{15}{4})$$

$$= \frac{15}{4} (k_{\rm B}T)^3, \tag{6.123}$$

where the values of $\Gamma(9/2)$, $\Gamma(7/2)$ and $\Gamma(3/2)$ are also given in (5.113).

Next, we use this in (6.117) to arrive at

$$k_f = \frac{2}{3} \frac{n_f \tau_{f\text{-}f}}{m k_{\rm B} T^2} \frac{15}{4} (k_{\rm B}T)^3 = \frac{5}{2} \frac{n_f \tau_{f\text{-}f}}{m k_{\rm B} T^2} (k_{\rm B}T)^3$$

$$= \frac{5}{2} k_{\rm B} \frac{n_f \tau_{f\text{-}f}}{m} k_{\rm B} T = \frac{1}{3} \rho_f c_{p,f} \langle u_f^2\rangle \tau_{f\text{-}f}, \tag{6.124}$$

where $\rho_f = mN_{\rm A}$ is the density of the gas (kg/m^3), $c_{p,f}$ is the specific heat capacity (J/kg-K), and $\langle u_f^2\rangle^{1/2}$ is the RMS thermal velocity $(3k_{\rm B}T/m)^{1/2}$ given by (6.84), which is also simply found from the equipartition of translational energy, from $m\langle u_f^2\rangle/2 = 3k_{\rm B}T/2$.

Similar to (6.84), the product of $\tau_{f\text{-}f}\langle u_f^2\rangle^{1/2} = \lambda_f$ is the mean free path of the colliding gas molecules (but here we use the RMS thermal velocity). Then by using this relation, we find that the the preceding expression for the gas thermal conductivity becomes

$$k_f = \frac{1}{3} \rho_f c_{p,f} \langle u_f^2\rangle^{1/2} \lambda_f \quad \text{monatomic ideal gas}, \tag{6.125}$$

which is the common form of the result from the so-called elementary kinetic theory† (end-of-chapter problem), except for the appearance of $c_{p,f}$ instead of $c_{v,f}$, i.e.,

$$k_f = \frac{1}{3}\rho_f c_{v,f}\langle u_f^2\rangle^{1/2}\lambda_f = \frac{1}{3}n_f c_{v,f}\langle u_f^2\rangle^{1/2}\lambda_f, \tag{6.126}$$

where in the second expression we have used $c_{v,f}$ per fluid particle (Section 6.2). The kinetic-theory-based treatment of the thermal conductivity (end-of-chapter problem) uses $\langle u_f\rangle$ in place of $\langle u_f^2\rangle^{1/2}$, as was used in the derivation of μ_f relation (6.91).

Note the similarity also with the relation for the gas viscosity, (6.91). Also, note that $\mu_f c_{p,f}/k_f = \mathrm{Pr}$, i.e., the gas Prandtl number, which is equal to unity when (6.126) is used. The Prandtl number of gases is nearly unity [173]. The speed of sound for an ideal gas is

$$a_s \equiv (\frac{\partial p}{\partial \rho_f})_s^{1/2} = (\frac{c_{p,f}}{c_{v,f}}\frac{\partial p}{\partial \rho_f}|_T)^{1/2} = (\frac{c_{p,f}}{c_{v,f}}\frac{E_p}{\rho_f})^{1/2} = (\frac{c_{p,f}}{c_{v,f}}\frac{R_g}{M}T)^{1/2}$$

$$= (\frac{c_{p,f}}{3c_{v,f}})^{1/2}\langle u_f^2\rangle^{1/2} = (\frac{\gamma}{3})^{1/2}\langle u_f^2\rangle^{1/2}, \quad \gamma = \frac{c_{p,f}}{c_{v,f}}. \tag{6.127}$$

Note that $\rho_f^{-1}\partial\rho_f/\partial p|_T$ is the isothermal compressibility $(1/E_p)$ (Table 3.11), and we have used the Maxwell relations among thermodynamic properties [55].

Note that the average thermal speed is defined in (6.63) and is related to $\langle u_f^2\rangle^{1/2}$ through $\langle u_f\rangle = (8/3\pi)^{1/2}\langle u_f^2\rangle^{1/2}$. Also, note that for N_2 at $T = 300$ K, $a_s = 353.1$ m/s. For polyatomic gases a more rigorous derivation gives [205], along with using a_s, i.e.,

$$k_f = \frac{1}{3}\rho_f\frac{15\pi}{32}(c_{v,f} + \frac{9}{4}\frac{R_g}{M})(\frac{8}{\pi\gamma})^{1/2}a_s\lambda_f \quad \text{polyatomic, ideal gas} \tag{6.128}$$

for monatomic gases the coefficient $15\pi/32$ and the term $9R_g/4M$ are dropped. The mean free path is given by (6.87). Figure 6.8(a) shows the variations of measured thermal conductivity of some typical gases with respect to temperature. The gas conductivity increases with temperature, as the speed of sound increases with temperature. Figure 6.8(b) shows the predicted thermal conductivity of gaseous H_2O as a function of temperature using (6.128).

6.9 Liquid Thermal Conductivity from Mean Free Path

Figure 6.9(a) shows variations of measured thermal conductivity of some typical (single component nonmetallic) liquids with respect to saturation temperature. Note that generally for a given pressure, there is a small temperature range over which there is liquid phase. The liquid metals have a thermal conductivity close to their solid-state values (e.g., Hg), as shown in Figure 5.18(a).

† Note that thermal diffusivity of an ideal gas is

$$\alpha_f = \frac{k_f}{\rho_f c_{p,f}} = \frac{1}{3}\langle u_f^2\rangle^{1/2}\lambda_f,$$

i.e., like the kinetic viscosity, it is the product of thermal speed and mean free path.

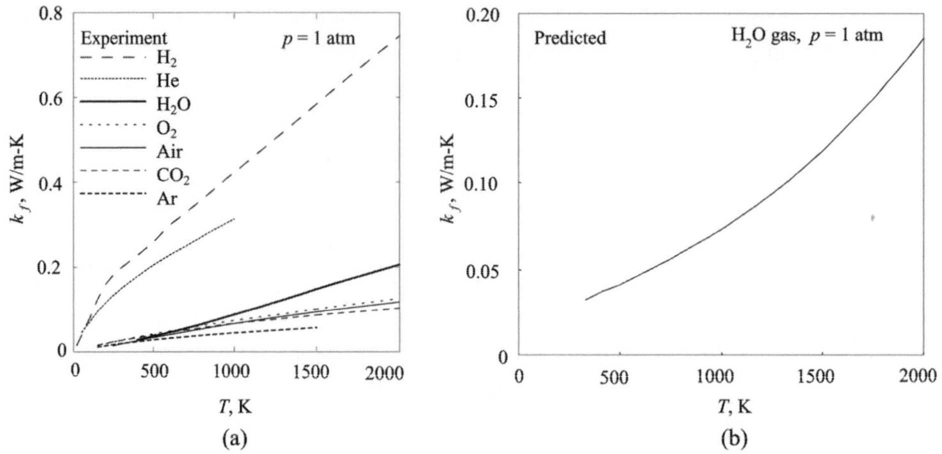

Figure 6.8. (a) Variations of measured thermal conductivity of some typical gases with respect to temperature [329]. (b) Variations of predicted thermal conductivity of gaseous H_2O, using the mean-free-path model (6.128), with respect to temperature.

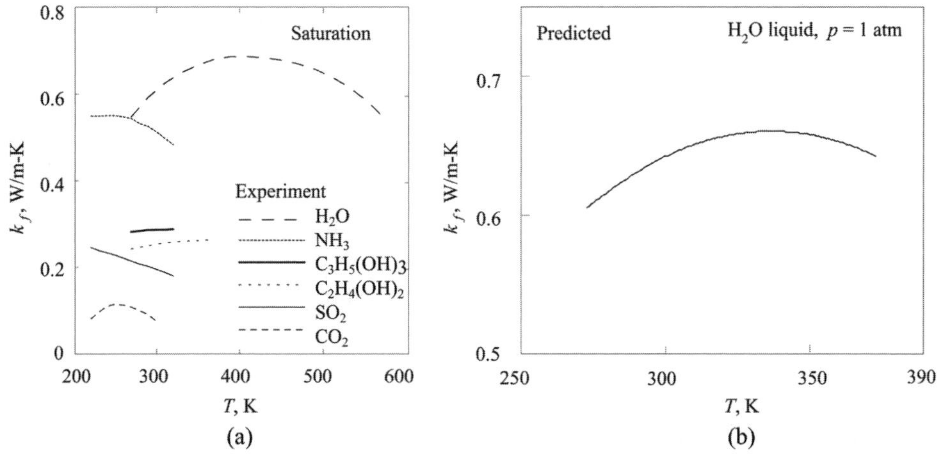

Figure 6.9. (a) Variations of the measured thermal conductivity of some typical (nonmetallic) liquids with respect to saturation temperature [329]. (b) Variations of the predicted thermal conductivity of liquid H_2O with respect to temperature at $p = 1$ atm.

Over the range of equilibrium liquid-phase existence (saturation), the variations in k_f are not very large. The liquid thermal conductivity can be predicted with the result from the Bridgman kinetic theory [42].

As was mentioned in Section 1.1.4, liquid particles collide at a much higher rate than the gases. Therefore, we expect a smaller mean free path for liquids compared with gases. In this theory, Bridgman [31, 42] theory assumes a simple cubic arrangement of atoms, with a spacing of $a = (1/n_f)^{1/3} = (M/\rho_f N_A)^{1/3}$, using (1.10). Bridgman then approximates the fluid thermal conductivity relation (6.126) as

$$k_f = \frac{1}{3}\rho_f c_{v,f} \langle u_f^2 \rangle^{1/2} \lambda_f = \rho_f c_{v,f} a_s n_f^{-1/3} \quad \text{nonmetallic liquid conductivity,} \quad (6.129)$$

by using the mean free path as nearly equal to the fluid particle spacing, i.e., $\lambda_f = 3n_f^{-1/3}$, and the RMS thermal speed equal to the speed of sound a_s [defined in (6.127)]. Assuming rigid spheres, the heat capacity is taken as $Mc_{v,f} = 3R_g$, although for liquid $c_{v,f}$ can be larger. Using this in (6.129), we have

$$k_f = \rho_f \frac{3R_g}{M} a_s (\frac{M}{\rho_f N_A})^{1/3}$$

$$= \rho_f \frac{3R_g}{M} (\frac{c_{p,f}}{c_{v,f}} \frac{\partial \rho_f}{\partial p}|_T)^{1/2} (\frac{M}{\rho_f N_A})^{1/3}. \quad (6.130)$$

Rearranging this, we have

$$k_f = \frac{3R_g}{N_A^{1/3}} \frac{\rho_f^{2/3}}{M^{2/3}} (\frac{c_{p,f}}{c_{v,f}} \frac{1}{\rho_f \kappa_f})^{1/2}$$

$$= \frac{3R_g}{N_A^{1/3}} \frac{\rho_f^{1/6}}{M^{2/3}} (\frac{c_{p,f}}{c_{v,f}} \frac{1}{\kappa_f})^{1/2}, \quad \kappa_f = \frac{1}{\rho_f} \frac{\partial \rho_f}{\partial p}|_T$$

nonmetallic liquid conductivity. (6.131)

For example, for water at $T = 300$ K, this gives $k_f = 0.6437$ W/m-K using $\kappa_f = 4.46 \times 10^{-10}$ Pa^{-1}, which is close to the experimental result of 0.613 W/m-K [174]. Figure 6.9(b) shows the variations of the predicted thermal conductivity of liquid water with respect to temperature for $p = 1$ atm (this limits the temperature range to freezing and boiling temperatures at this pressure). The temperature dependence is through ρ_f, $c_{p,f}/c_{v,f}$, and κ_f, where $\rho_f(T)$ decreases monotonically, $(c_{p,f}/c_{v,f})(T)$ increases monotonically, and $\kappa_f(T)$ has a minimum in the middle of the temperature range.

6.10 Effective Conductivity with Suspended Particles in Brownian Motion

The derivation of the Brownian diffusion by Einstein and that by Langevin are summarized in [73]. These are given here along with a simple effective thermal conductivity model.

6.10.1 Langevin Derivation of Brownian Diffusion

A spherical particle of diameter d_s and mass m_s suspended in a viscous liquid (viscosity μ_f) under isothermal (T_f) conditions and under the thermal motion of the liquid molecules (no other driving force) experiences displacement as governed by the Langevin stochastic particle dynamics equation [73] given in Table 3.7, i.e.,

$$m_s \frac{d^2x}{dt^2} = -3\pi \mu_f d_s \frac{dx}{dt} + F(t) \quad \text{one-dimensional Langevin equation,} \quad (6.132)$$

where the hindering force is the viscous force on the solid, spherical particle (called the Stokes force). This force is $3\pi\mu_f d_s u_s$, where u_s is the particle speed $\mathrm{d}x/\mathrm{d}t$[†]. In (6.132) $F(t)$ is the stochastic (chaotic) driving force that is due to the liquid thermal motion. Multiplying (6.132) by x and using the chain rule, we have

$$\frac{m_s}{2}\frac{\mathrm{d}^2 x}{dt^2} - m_s(\frac{\mathrm{d}x}{dt})^2 = -\frac{3}{2}\pi\mu_f d_s \frac{\mathrm{d}x^2}{\mathrm{d}t} + F(t)x(t). \qquad (6.133)$$

Now taking the average over a long travel time (or many realizations) and assuming that $\langle F(t)x(t)\rangle = 0$, i.e., the thermal fluctuation and the particle positions are not correlated, then

$$\frac{m_s}{2}\frac{\mathrm{d}}{\mathrm{d}t}\langle\frac{\mathrm{d}x^2}{dt}\rangle + \frac{3}{2}\pi\mu_f d_s\langle\frac{\mathrm{d}x^2}{\mathrm{d}t}\rangle = m_s\langle(\frac{\mathrm{d}x}{\mathrm{d}t})^2\rangle. \qquad (6.134)$$

Noting that, from the equipartition of energy, we have

$$\frac{1}{2}m_s\langle(\frac{\mathrm{d}x}{\mathrm{d}t})^2\rangle \equiv \frac{1}{2}m_s\langle u_s^2\rangle^{1/2} = \frac{1}{2}k_\mathrm{B}T_f, \qquad (6.135)$$

then we have

$$\frac{m_s}{3\pi\mu_f d_s}\frac{\mathrm{d}}{\mathrm{d}t}\langle\frac{\mathrm{d}x^2}{\mathrm{d}t}\rangle + \langle\frac{\mathrm{d}x^2}{\mathrm{d}t}\rangle = \frac{2k_\mathrm{B}T_f}{3\pi\mu_f d_s}. \qquad (6.136)$$

The solution to (6.136) is

$$\langle\frac{\mathrm{d}x^2}{\mathrm{d}t}\rangle = \frac{2k_\mathrm{B}T_f}{3\pi\mu_f d_s} + Ae^{-\frac{3\pi\mu_f d_s}{m_s}t}. \qquad (6.137)$$

The viscous relaxation time τ_μ is defined as

$$\tau_\mu = \frac{m_s}{3\pi\mu_f d_s}, \qquad (6.138)$$

which is independent of temperature (except through μ_f), and it is of the order of 10^{-9} s for small particles in liquids.

Then for $t/\tau_\mu \gg 1$, we have

$$\langle\frac{\mathrm{d}x^2}{\mathrm{d}t}\rangle = \frac{2k_\mathrm{B}T_f}{3\pi\mu_f d_s}, \quad \frac{t}{\tau_\mu} \gg 1. \qquad (6.139)$$

Using x_o at $t = 0$, the integration gives

$$\langle x^2\rangle - \langle x_\mathrm{o}^2\rangle = \frac{2k_\mathrm{B}T_f}{3\pi\mu_f d_s}t = \frac{2k_\mathrm{B}T_f}{m_s}t\tau_\mu$$
$$\equiv \langle\delta_x^2\rangle, \qquad (6.140)$$

where $\langle\delta_x^2\rangle^{1/2}$ is the RMS of the displacements of a particle in Brownian motion.

The Einstein (Brownian) diffusion coefficient is defined through

$$\langle\delta_x^2\rangle \equiv 2D_\mathrm{E}t, \qquad (6.141)$$

[†] This gives the Stokes drag coefficient equal to 24/Re_d, where the Reynolds number is $\mathrm{Re}_d=\rho_f u_s d_s/\mu_f$ [174].

or

$$D_{\rm E} = \frac{k_{\rm B} T_f}{3\pi \mu_f d_s} \quad \text{Einstein (Brownian) diffusion coefficient.} \tag{6.142}$$

This shows the solid particle diffuses through liquid more effectively when temperature is raised, or liquid viscosity or particle diameter is decreased.

The Brownian diffusion time (time it takes for a displacement of $d_s/2^{1/2}$) is defined as

$$\tau_{\rm E} = \frac{d_s^2}{D_{\rm E}} = \frac{3\pi \mu_f d_s^3}{k_{\rm B} T} \quad \text{Brownian diffusion time,} \tag{6.143}$$

which becomes very long as the solid particle diameter increases.

6.10.2 Effective Fluid Thermal Conductivity

In possible enhanced thermal conductivity liquids, because of the suspension of nanoparticles, the lack of thermal equilibrium between the particle and the liquid (in a temperature gradient ∇T_f field) also plays a role [158].

Here we take an alternative and simple approach. The effective conductivity of liquids with particle suspensions $\langle k_f \rangle$ may be expressed as a parallel (or additive) contribution of conductivities, i.e.,

$$\langle k_f \rangle = k_f + k_{\rm E} \quad \text{dilute suspension,} \tag{6.144}$$

where $k_{\rm E}$ is the contribution that is due to the Brownian motion of constituents and increases the ability of the particles to diffuse, store heat, and transfer heat through its surface (a continuum or semicontinuum concept). The storage/boundary–resistance /surface–convection is represented by thermal relaxation time (6.82) $\tau_\alpha = \lambda_s/\langle u_s^2 \rangle^{1/2}$, where λ_s is the thermal mean free path of the solid particle. Then, using (6.126), $k_{\rm E}$ can be written as

$$\begin{aligned} k_{\rm E} &= \frac{1}{3} n_s c_{v,s} \langle u_s^2 \rangle^{1/2} \lambda_s = \frac{1}{3} n_s c_{v,s} \langle u_s^2 \rangle \tau_\alpha, \quad \tau_\alpha = R_t c_{v,s} \\ &= \frac{1}{3} n_s c_{v,s} \langle u_s^2 \rangle R_t c_{v,s} = \frac{\pi d_s^3}{18} n_s c_{v,s}^2 \frac{k_{\rm B} T}{m_s} R_t, \end{aligned} \tag{6.145}$$

where n_s is the number of particles per unit volume, $c_{v,s}$ is its heat capacity per particle [(4.66), the classical limit for a monatomic solid $3Nk_{\rm B}T$ for a crystal with $T > T_{\rm D}$], $N = N_{\rm A}\rho_s \pi d_s^3/6M_s$, where ρ_s and M_s are the density and molecular weight of solid particles. N is the number of atoms in the particle, and R_t is the sum of the internal and external resistances to heat flow (heat loss from the particle) [174]. As n_s increases, the mean free path of the solid particle λ_s will be limited by solid interparticle collisions.

Because of the coalescing of the solid particles (as in colloids in general), solid particle clusters are formed. The cluster form favors the dominance of the internal resistance due to relatively small cluster effective conductivity $\langle k \rangle_c$. This low value of $\langle k \rangle_c$ is in part due to the size effect in the thermal conductivity of the solid particles.

Also, note that, where clusters are formed, $\langle d_s \rangle_c$, $\langle n_s \rangle_c$, and $\langle c_{v,s} \rangle_c$ should be used in (6.145), which increases the diameter and reduces the number of particles per unit volume. For a sphere (here the cluster) to lose a substantial amount of its heat content, the Fourier number should be nearly 0.1 [174], i.e.,

$$\mathrm{Fo}_d = \frac{4\langle k \rangle_c \tau_\alpha}{\langle \rho c_p \rangle_c \langle d_s \rangle_c^2} = 0.1, \tag{6.146}$$

or the thermal relaxation (diffusion) time (Table 3.12) given by

$$\tau_\alpha = 0.1\frac{\langle \rho c_p \rangle_c \langle d_s \rangle_c^2}{4\langle k \rangle_c^2} = 0.1\frac{\langle n_c \rangle_s \langle c_{v,s} \rangle_c \langle d_s \rangle_c^2}{4\langle k \rangle_c}. \tag{6.147}$$

Then for clusters, $\langle d_s \rangle_c$ is larger and $\langle k \rangle_c$ is smaller (compared with nonclustered d_s and k_s), resulting in large thermal relaxation time τ_α and larger k_{E}. Note that this thermal relaxation time can be longer than viscous relaxation time (6.138) suggested in some analyses [176].

Note that even with a small volume fraction, the particles can form a colloidal network [percolation threshold, similar to Figure 2.9(b)(v) for soots]. Then the heat conduction through the connected matrix should be considered (instead of the Brownian motion), which assumes well-dispersed particles.

6.11 Interaction of Moving Fluid Particle and Surface

The fluid mean free path presents the probability distance a fluid particle travels before undergoing collision with another fluid particle. In the presence of a condensed phase, i.e., a surface (or interface), there is also the probability that the fluid particle collides with this surface. Representing the geometry for the surface containing the fluid particles by a clearance distance l, then the ratio of the probability of the fluid–fluid collision to the fluid–surface collision is given by the Knudson number Kn_l as

$$\mathrm{Kn}_l = \frac{\lambda_f}{l} \quad \text{Knudson number.} \tag{6.148}$$

For liquids, λ_f is the molecular spacing $(M/\rho_f N_{\mathrm{A}})^{1/3}$, and therefore $\mathrm{Kn}_l \ll 1$.

6.11.1 Fluid Flow Regimes

For large Kn_l, the probability of the fluid particle–surface interaction (collision) is larger than that among fluid particles and this is the free-molecular-flow regime ($\mathrm{Kn}_l > 1$). For small Kn_l, the interfluid particle collisions dominate, and it is called the viscous-flow regime ($\mathrm{Kn}_l \ll 1$). Between these two is the transitional-flow regime ($0.01 < \mathrm{Kn}_l < 1$). Figure 6.10 gives approximate boundaries for these regimes. The viscous-flow regime is described by the Navier–Stokes equations of fluid motion listed in Table 3.10 (momentum and mass conservation equations). These equations can be derived from the BTE, by taking the appropriate moments

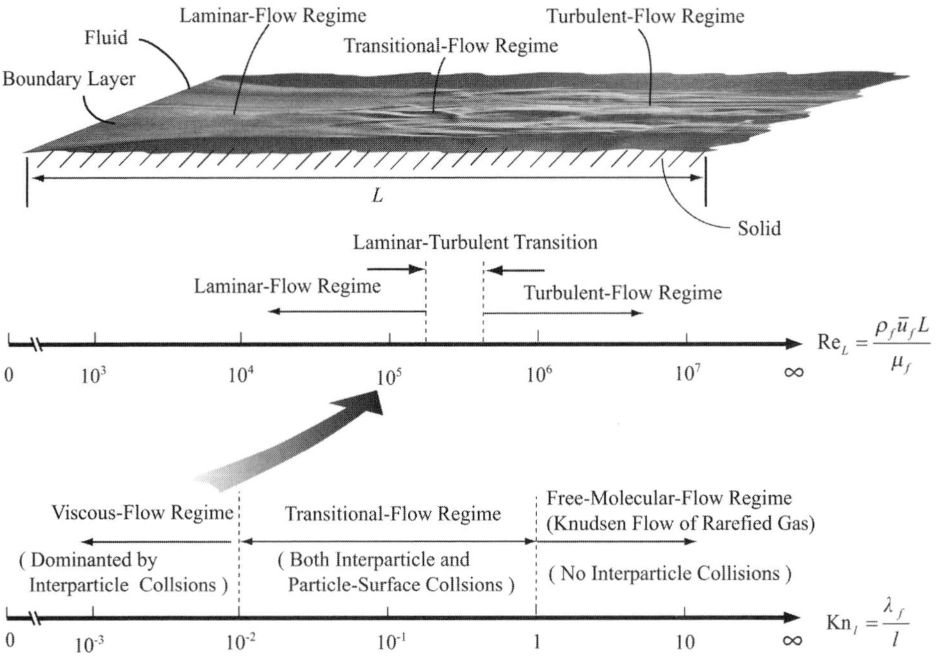

Figure 6.10. The free-molecular, transitional-flow, and viscous-flow regimes, as marked by the magnitude of the Knudson number Kn_l (l is clearance distance) and the expanded viscous-flow regime, as marked by the magnitude of the Reynolds number Re_L (showing laminar, laminar–turbulent transitional, and turbulent-flow regimes).

of this equation [55, 325]. The viscous-flow regime in turn is classified based on the flow Reynolds number Re_L, where L is the characteristic (hydrodynamic) length for the surface (or object). The Reynolds number is the ratio of the inertial to viscous forces, i.e.,

$$\mathrm{Re}_L = \frac{\rho_f \overline{u}_f L}{\mu_f} \quad \text{Reynolds number}, \tag{6.149}$$

where $\overline{u}_f$ is the macroscopic (net) fluid velocity. The low-Reynolds-number flow is characterized by zero-velocity fluctuations, i.e., laminar-flow regime. At a high Reynolds number, the flow becomes unstable and turbulent (chaotic fluctuations), i.e., turbulent-flow regime. One type of turbulent-flow structure is the boundary-layer turbulence, as depicted in Figure 6.10. The regime between the laminar and the turbulent-flow is the transitional, viscous-flow regime. The transition for boundary-layer turbulence is around $\mathrm{Re}_L = 5 \times 10^5$.

Both gas and liquid flows can be sonic or supersonic. The Mach number $\mathrm{Ma} = u_f/a_s$, where a_s is the speed of sound in the fluid, is used to show the effect of reaching and surpassing the speed of sound in fluid flow and its compressibility and shock formation features (in gases, in particular, this results in the formation of various shocks that are discontinuities in pressure, etc.).

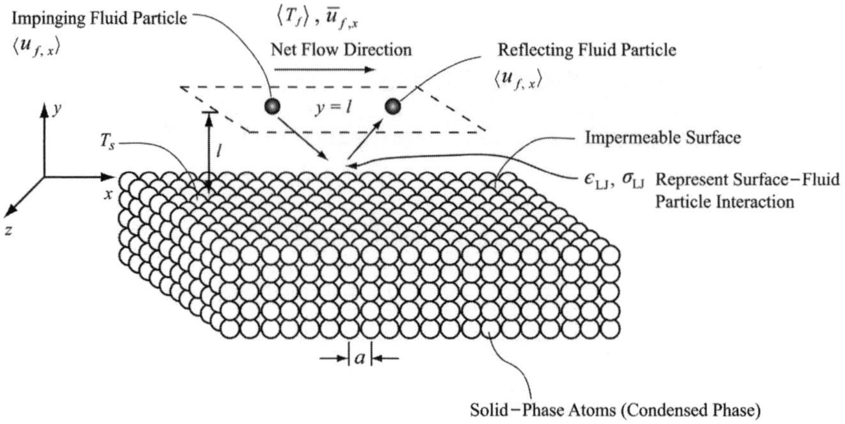

Figure 6.11. Fluid particle–surface interaction showing an impinging fluid particle with a y-component velocity colliding and reflecting from a solid surface.

6.11.2 Knudson-Flow-Regime Surface Accommodation Coefficients

Consider free-molecular flow ($\mathrm{Kn}_l > 1$), as depicted in Figure 6.11. The fluid particles interact only with the surface. The particles approach the surface with velocity (averaged over many realizations) $\langle u_{f,x}\rangle$ and leave with an altered tangential velocity, $\langle u_{f,x}\rangle'$.

Then the Maxwell tangential momentum accommodation coefficient a_u, defined as

$$a_u = \frac{\langle u_{f,x}\rangle - \langle u_{f,x}\rangle'}{\langle u_{f,x}\rangle} \quad \text{momentum accomodation coefficient,} \tag{6.150}$$

represents the tangential velocity slip. Noting that the average fluid velocity at the surface is the average of these two, we have for the surface slip velocity u_i, by using the net flow quantities, i.e., $\overline{u}_{f,x}$,

$$u_i = \frac{\overline{u}_{f,x} + (1 - a_u)\overline{u}_{f,x}}{2} = (\frac{2 - a_u}{2})\overline{u}_{f,x} \quad \text{velocity slip for free-molecular-flow regime.} \tag{6.151}$$

Here we first consider a thermal fluid–solid surface, i.e., $\langle T_f\rangle = T_s = T$. The MD results for a monatomic gas with L–J potential and FCC solid of lattice parameter a, are obtained in [6]. The L–J potential parameters are ϵ_{LJ} and σ_{LJ} for fluid particle-solid interatomic interaction. In the MD simulations, the M–B distribution is used for $\langle u_{f,x}\rangle$. Figure 6.12(a) shows the variation of a_u with respect to σ_{LJ}/a, for $T = 300$ K, $\sigma_{\mathrm{LJ}}/k_{\mathrm{B}} = 100$ K, and three values of a. As the equilibrium interatomic distance represented by σ_{LJ}/a increases, $a_u \rightarrow 0$, i.e., the reflected molecules have the same tangential momentum as those impinging. For a very small interatomic equilibrium distance, all the tangential momentum is lost by the interaction with the surface ($a_u \rightarrow 1$). Figure 6.12(b) shows the variations of a_u with respect to $\epsilon_{\mathrm{LJ}}/k_{\mathrm{B}}T$, for $\sigma_{\mathrm{LJ}} = 2.4$ Å, $a = 4$ Å, and $T = 200,\ 300$, and 400 K. Again, as the interaction energy increases, the reflected tangential momentum decreases ($a_u \rightarrow 1$).

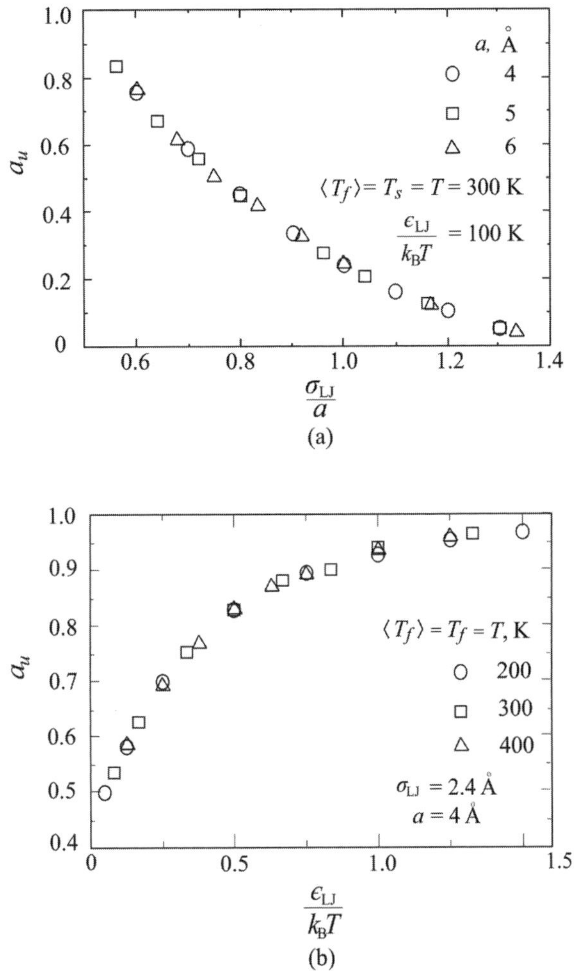

Figure 6.12. (a) Variation of MD predicted momentum accommodation coefficient with respect to the normalized L–J potential separation constant. (b) Same, but with respect to the normalized interaction energy constant [6].

Under thermal nonequilibrium between the gas and the solid surface ($\langle T_f \rangle \neq T_s$), the thermal accommodation coefficient is defined as [35]

$$a_T = \frac{\langle q_{f,y} \rangle - \langle q_{f,y} \rangle'}{\langle q_{f,y} \rangle - \langle q_{f,y}(T_s) \rangle}, \tag{6.152}$$

where $\langle q_{f,y} \rangle$ is the energy flux incident upon the surface, $\langle q_{f,y} \rangle'$ is the energy flux reflected from the surface, and $\langle q_{f,y}(T_s) \rangle$ is the stream of fluid particles leaving the surface and carrying the same mean energy per particle as a stream issuing from a gas in equilibrium at the surface temperature T_s. For $a_T = 1$, there is a complete thermal accommodation (called diffuse reemission).

Tabulated values for a_T are given in [317] and for a very clean surface and weakly interacting fluid particle–surface molecules, a_T can be very small ($a_T \rightarrow 0$). However, for polar gas molecules, we have $a_T \rightarrow 1$.

Table 6.5. *Experimental values for momentum accommodation coefficient a_u for some gas–surface pairs for $T = 300$ K [317].*

Gas	Surface	a_u
Air	machined brass	1.00
Air	old shellac	1.00
Air	fresh shellac	0.79
Air	oil	0.9
Air	glass	0.9
N_2	glass	0.95
CO_2	machined brass	1.00
CO_2	old shellac	1.00
CO_2	oil	0.92
H_2	oil	0.93
He	oil	0.87

For monatomic gases (translational kinetic energy only), $\langle q_{f,y}\rangle$ is proportional to T and the proportionality constant is the same for all quantities in (6.152); then $a_T = (\langle T_f\rangle - \langle T_f\rangle')/(\langle T_f\rangle - T_s)$. However, for polyatomic gases the impact changes the proportions of the rotational and vibrational (and at high temperatures the electronic) energies, and therefore this simplification does not hold.

The MD results for thermal (energy) accommodation of a diatomic gas with both translation and rotational energy are reported in [35]. They simulate the fluid–solid pair N_2–Pt, using the L–J potential, and examine both the directly scattered as well as the adsorbed–desorbed fluid (Section 6.11.5) particles. For $\mathrm{Kn}_l \rightarrow 0$ (viscous-flow regime), we have $a_u = a_T = 1$, and the fluid velocity and temperature at the surface become the solid surface values. These are called the viscous-flow-regime no-slip surface conditions.

Tables 6.5 and 6.6 list experimental values for a_u and a_T for typical gas–surface pairs, stating the surface condition and temperature [317]. For noble gases with very smooth, low-surface-energy solids, α_T is very small (approximately zero). Note that for N_2 on W, there is a small temperature dependence of a_T. The momentum accommodation coefficient is generally close to unity. We will use a_u and a_T in the following Sections 6.11.3 and 6.11.4, and also for size effect on effective conductivity of confined gas in Section 6.14.1.

6.11.3 Slip Coefficients in Transitional-Flow Regime

In the translational-flow regime ($10^{-2} \leq \mathrm{Kn}_l \leq 1$) and in the presence of a gradient in the mean gas velocity $\overline{\boldsymbol{u}}_f$, the tangential surface velocity slip is defined through

$$u_i = \alpha_u \frac{\partial \overline{u}_{f,x}}{\partial y} \quad \text{on a solid surface, velocity slip for translational-flow regime,} \tag{6.153}$$

where α_n is the velocity slip coefficient.

Table 6.6. *Experimental values for thermal accommodation coefficient a_T for some gas–solid-surface pairs [317].*

Gas	Surface	Surface condition (adsorbed gas)	T, K	a_T
Air	bronze	indeterminate		0.88-0.95
	cast iron	indeterminate		0.87-0.96
	aluminum	indeterminate		0.87-0.97
N_2	W	indeterminate	305	0.624
	Pt	indeterminate		0.5
	glass	indeterminate	103	0.38
O_2	Pt, bright	indeterminate		0.81
	Pt, black	indeterminate		0.93
	Pt	saturated	303	0.38
CO_2	W, bright	CO_2	305	0.990
	Pt	saturated	303	0.76
H_2	Pt, bright	indeterminate		0.32
	Pt, black	indeterminate		0.74
	Pt	saturated	303	0.220
	glass	indeterminate	308	0.29
	glass	indeterminate	103	1.0
He	W, flashed	indeterminate	298	0.17
	W, not flashed	indeterminate	298	0.53
	W	clean	303	0.0169
	W	clean	248	0.0153
	W	clean	83	0.0151
	W	K on H_2	298	0.106
	W	H_2 on K	298	0.096
	W	O on K	298	0.22
	W	K on O	298	0.12
	Pt	saturated	303	0.238
	K	clean	298	0.0826
	Na	clean	298	0.0895
	glass	clean	302	0.31
	glass	indeterminate	298	0.35
Ne	W	clean	303	0.0412
	W	clean	243	0.0395
	W	clean	77	0.0495
	W	H_2	295	0.17
	W	N_2	79	0.32
	Pt	saturated	303	0.57
	K	clean	298	0.1987
	Na	clean	298	0.19
	glass	indeterminate	298	0.7
Ar	W	clean	303	0.272
	W	clean	243	0.294
	W	clean	77	0.549
	Pt	saturated	303	0.89
	K	clean	298	0.444
	Na	clean	298	0.459
Kr	W	clean	303	0.462
	W	clean	243	0.498
	W	clean	77	0.926
Xe	W	clean	303	0.773
	W	clean	243	0.804
	W	clean	90	0.942

Also, the flux of molecules impinging on the surface is given by (6.89), and the tangential momentum flux is the product of this molecular flux and the change in the tangential velocity. Then, using (6.90) for this change, we have

$$\tau_{xy} = \frac{1}{4} m n_f \langle u_f \rangle a_u \overline{u}_{f,x}. \tag{6.154}$$

This tangential momentum flux is also the fluid viscous shear stress, and for a Newtonian fluid (6.90) it is

$$\tau_{xy} = \mu_f \frac{\partial \overline{u}_{f,x}}{\partial y}, \tag{6.155}$$

where μ_f is the fluid velocity. Then using (6.153) to (6.155), we have

$$\frac{1}{4} m n_f \langle u_f \rangle a_u \overline{u}_{f,x} = \mu_f \frac{\partial \overline{u}_{f,x}}{\partial y} = \mu_f \frac{u_i}{\alpha_u} = \mu_f \frac{2 - a_u}{2} \frac{1}{\alpha_u}, \tag{6.156}$$

or

$$\alpha_u = \frac{2\mu_f}{m n_f \langle u_f \rangle} \frac{2 - a_u}{a_u} \quad \text{Maxwell velocity slip coefficient.} \tag{6.157}$$

This is the Maxwell velocity slip coefficient relation.

Now noting that $\mu_f = m n_f \langle u_f^2 \rangle^{1/2} \lambda_f / 3$, (6.92) (end-of-chapter problem), it is customary to approximate (6.157) with

$$\alpha_u = \frac{2 - a_u}{a_u} \lambda_f \equiv a_u^* \lambda_f. \tag{6.158}$$

The temperature slip coefficient coefficient α_T is similarly defined through

$$T_i - T_s = \alpha_T \frac{\partial \langle T_f \rangle}{\partial y}, \quad \text{temperature slip coefficient on interface.} \tag{6.159}$$

As an approximation, α_T is related to α_u through the Kennard relation [317]

$$\alpha_T = \alpha_u \frac{2\gamma}{\gamma + 1} \frac{1}{\mathrm{Pr}} \equiv \alpha_T^* \lambda_f \quad \text{Kennard temperature slip coefficient}$$

$$\gamma = \frac{c_{p,f}}{c_{v,f}}, \quad \mathrm{Pr} = \frac{\mu_f c_{p,f}}{k_f}, \tag{6.160}$$

where γ is the specific heat ratio and Pr is the Prandtl number. Note that α_u^* and α_T^* are dimensionless and are of the order of unity. These show that the velocity and temperature slip coefficients are proportional to the gas mean free path (6.87), and, as λ_f decreases, these slip coefficients tend to vanish (viscous-flow regime).

6.11.4 Solid Particle Thermophoresis in Gases

In the presence of a temperature gradient in a gas with solid particle suspensions, the gas kinetic energy difference on different parts of the particle surface can move the particle in the direction of a lower temperature. This is referred to as particle thermophoresis. For the motion to be significant (compared with, for example, the buoyancy force), the particles should be small, compared with the mean free path of gas, i.e., $\mathrm{Kn}_d = 2\lambda_f/d_s$ (d_s is the solid particle diameter) can become large. This problem has been analyzed [323] where for the velocity slip, in addition to α_u^* discussed in Section 6.11.3, is also affected by the temperature variation around the particle. This is presented using the temperature slip using a thermodiffusive slip coefficient $\alpha_{u,T}^*$, such that the tangential (polar) slip velocity is given as

$$u_{\theta,i} = \alpha_u^* \lambda_f r \frac{\partial}{\partial r}\frac{u_\theta}{r} + \alpha_{u,T}^* \frac{2\mu_f}{\rho_f T_f d_s}\frac{\partial T_f}{\partial \theta}, \quad \text{at } r = \frac{d_s}{2}. \tag{6.161}$$

Also, the temperature slip is given similar to (6.159), using α_T^*, as

$$T_f - T_s = \alpha_T^* \lambda_f \frac{\partial T_f}{\partial r}, \quad \text{at } r = \frac{d_s}{2}. \tag{6.162}$$

These allow for fluid temperature variation $T_f(r, \theta)$ around the solid particle.

The result for the thermophoretic force $\boldsymbol{F}_T$ on the solid particle as [323]

$$\boldsymbol{F}_T = -\frac{6\pi \mu_f^2 \alpha_{u,T}^* d_s (\dfrac{k_f}{k_s} + \alpha_T^* \mathrm{Kn}_d)\dfrac{\nabla T_f}{T_f}}{\rho_f (1 + 3\alpha_u^* \mathrm{Kn}_d)(1 + 2\dfrac{k_f}{k_s} + 2\alpha_T^* \mathrm{Kn}_d)}, \tag{6.163}$$

where k_f and k_s are the fluid and solid conductivity, and $\mathrm{Kn}_d = 2\lambda_f/d_s$. In [323], $\alpha_u^* = \alpha_{u,T}^* = 1.14$ and $\alpha_T^* = 2.18$ are suggested.

The thermophoretic velocity is [323]

$$\boldsymbol{u}_T = \boldsymbol{F}_T \frac{1}{3\pi \mu_f d_s}\{1 + \mathrm{Kn}_d[a_1 + a_2 \exp(-\frac{a_3}{\mathrm{Kn}_d})]\}$$

particle thermophoretic velocity. (6.164)

In [323] it is also suggested that $a_1 = 1.20$, $a_2 = 0.41$, and $a_3 = 0.88$. When $\mathrm{Kn}_d \rightarrow 0$, the Stokes velocity $3\pi \mu_f d_s \boldsymbol{F}_T$ is recovered. This is the same as the velocity used in Brownian motion as presented in Section 6.10.

6.11.5 Physical Adsorption and Desorption

When an involatile solid is in contact with a gas, the gas molecules adsorb (stick) to the solid (the solid is called the adsorbent or substrate, and the adsorbed gas is called the adsorbate). Surface adsorption differs from absorption (which requires penetration into the bulk solid phase). In physical adsorption (or physisorption) the gas molecules are held to the surface by the relatively weak van der Waals forces as shown in Figure 6.13(a). In the chemical adsorption (chemisorption), the chemical reaction occurs at the surface and the gas molecules are held to the surface with strong chemical bonds.

The amount of gas adsorbed to a surface at a given temperature depends on the gas pressure and equilibrium condition. This is called the adsorption isotherm and is generally given as the mole of gas per gram of adsorbent as a function of pressure (or normalized by the saturation pressure for a given temperature). Figure 6.13(b) gives the adsorption isotherm for H_2O adsorbed on Pt, for $T = 298$ and

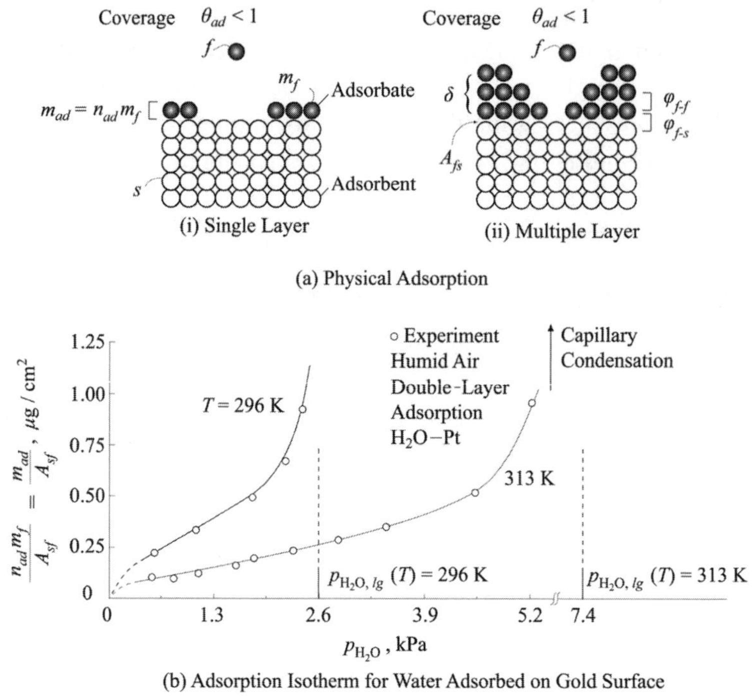

Figure 6.13. (a) Simple and multilayer physical adsorption, and (b) experimental adsorption isotherm for H_2O–Pt at $T = 296$ and 313 K [40].

313 K. The adsorption isotherm is also presented as the amount of fluid particles adsorbed per unit area of solid surface. As the temperature increases, less fluid particles are adsorbed. Also, as the saturation pressure of the fluid is reached the fluid liquifies on the surface and the adsorbed amount increases substantially (capillary condensation) [40].

Desorption is the process of removing the adsorbent gas molecules and is achieved by raising the temperature or lowering the pressure.

The fraction of surface area covered by gas molecules is θ_{ad}, and it is generally allowed that first this fraction reaches unity as the adsorption proceeds and then after completion of a single molecular layer (monolayer), additional layers are formed until the van der Waals forces emanating from the surface are not able to affect the gas molecules sufficiently far from the surface. These are shown in Figure 6.13(a).

The simplest analysis is for simple larger adsorbed fluid particle is called the Langmuir adsorption model. It is for a planar, homogeneous surface, with no reaction among adsorbed species, and a monolayer adsorption only. The rate of adsorption is the integrated flux of gas molecules moving toward the surface with energy sufficiently large to overcome the adsorption barrier E_{ad}. Note that this is proportional to the fraction of surface not covered by adsorbed particle. This is expressed as [30]

$$\frac{\mathrm{d}n_{ad}}{\mathrm{d}t} = n_f(1-\theta_{ad})\int_{u_{f,\mathrm{o}}}^{\infty} f_f^{\mathrm{o}}(\boldsymbol{u}_f)\alpha_{ad}\mathrm{d}u_{f,y} \quad \text{Langmuir adsorption model,} \quad (6.165)$$

where $u_{f,\mathrm{o}}$ is defined from the threshold energy,

$$E_{ad} \equiv \frac{1}{2}mu_{f,\mathrm{o}}^2, \tag{6.166}$$

which defines the threshold gas velocity $u_{f,\mathrm{o}}$ for adsorption and α_{ad} is called the sticking probability (assumed to be independent of $u_{f,y}$ and is generally between 0.8 and 1.0). The equilibrium distribution function $f_f^{\mathrm{o}}(\boldsymbol{u}_f)$ is given by (6.61), and the integration gives

$$\frac{\mathrm{d}n_{ad}}{\mathrm{d}t} = n_f\left(\frac{k_\mathrm{B}T}{2\pi m}\right)^{1/2}\exp\left(-\frac{E_{ad}}{k_\mathrm{B}T}\right)(1-\theta_{ad})\alpha_{ad} \equiv \dot{\gamma}_{ad}n_f(1-\theta_{ad}), \tag{6.167}$$

where $\dot{\gamma}_{ad}$ is the adsorption rate coefficient per fluid particle. The desorption rate is assumed to be proportional to the surface coverage θ_{ad}, i.e.,

$$\frac{\mathrm{d}n_{ad}}{\mathrm{d}t} = \dot{\gamma}_{de}\theta_{ad}\exp\left(-\frac{E_{de}}{k_\mathrm{B}T}\right), \tag{6.168}$$

where $\dot{\gamma}_{de}$ is desorption rate coefficient and E_{de} is the desorption activation barrier. There are other adsorption and desorption kinetic models, in which the dependence on the surface vacancy $1-\theta_{ad}$ and coverage θ_{ad} is not linear. The first-order-models given preceding are among the simpler ones.

At equilibrium, we have $\mathrm{d}n_{ad}/\mathrm{d}t = \mathrm{d}n_{de}/\mathrm{d}t$, and from this we have

$$\frac{\left(\frac{k_\mathrm{B}T}{2\pi m}\right)^{1/2}\alpha_{ad}\exp\left(\frac{\Delta E_a}{k_\mathrm{B}T}\right)}{\dot{\gamma}_{de}}n_f \equiv \frac{\dot{\gamma}_{ad}}{\dot{\gamma}_{de}}n_f = \frac{\theta_{ad}}{1-\theta_{ad}}, \quad \Delta h_{ad} = E_{de} - E_{ad}. \tag{6.169}$$

Rearranging this gives

$$\theta_{ad} = \frac{\frac{\dot{\gamma}_{ad}}{\dot{\gamma}_{de}}n_f}{1+\frac{\dot{\gamma}_{ad}}{\dot{\gamma}_{de}}n_f} = \frac{\frac{\dot{\gamma}_{ad}}{\dot{\gamma}_{de}}p}{1+\frac{\dot{\gamma}_{ad}}{\dot{\gamma}_{de}}p}, \quad p = n_f k_\mathrm{B}T. \tag{6.170}$$

Assuming a monolayer adsorption, where V_{max} is the volume of completely covered ($\theta_{ad}=1$) gas molecules, then the volume of adsorbent is ($\theta_{ad} = V_{ad}/V_{max}$)

$$V_{ad} = \frac{V_{max}\frac{\dot{\gamma}_{ad}}{\dot{\gamma}_{de}}p}{1+\frac{\dot{\gamma}_{ad}}{\dot{\gamma}_{de}}p}, \quad p = n_f k_\mathrm{B}T \quad \text{Langmuir adsorption isotherm.} \tag{6.171}$$

There are other adsorption isotherm models, including the Brunauer–Emett–Teller (BET) isotherm for multilayer adsorption. Examples of isotherms are given in [172].

The net energy Δh_{ad} is the differential (per molecule) isosteric (constant θ_{ad}) enthalpy (or heat) of adsorption and is related to p and T (similarly to the Clausius–Clapeyron relation for liquid–vapor equilibrium) as [205]

$$\frac{\partial \ln p}{\partial T}\Big|_\theta = \frac{\Delta h_{ad}}{k_\mathrm{B}T^2} \quad \text{ideal gas.} \tag{6.172}$$

In general $\Delta h_{ad} = \Delta h_{ad}(T, \theta_{ad})$. This heat (enthalpy) of adsorption is similar to the heat (enthalpy) of evaporation. The specific enthalpy is defined in terms of the specific internal energy (and pressure and specific volume v) as

$$h = e + pv. \tag{6.173}$$

Then we can write for the enthalpy of evaporation, at conditions T and p,

$$\Delta h_{lg} = (c_{g,v} - c_{l,v})T + p(v_g - v_l). \tag{6.174}$$

The enthalpy of evaporation per molecule is the energy required for changing a molecule of a liquid into the gaseous state. This energy breaks down the intermolecular attractive force, and also provides the energy necessary to expand the gas $p(v_g - v_l)$. There is no potential energy (internal energy is the molecular kinetic energy) for an ideal gas. For example, for H_2O the enthalpy of evaporation at $p = 1$ atm and $T = 373.15$ K is 0.39 eV; from Figures 6.3 and 6.4, we note that this is mostly associated with $(c_{g,v} - c_{l,v})T$, compared with $p(v_g - v_l)$.

An example of the prediction of $\theta_{ad}(T)$ and $\Delta h_{ad}(T, \theta_{ad})$ for H_2O adsorption on SiO_2 is given in [274]. They find that Δh_{ad} is larger than the bulk heat of evaporation Δh_{lg} and show that SiO_2 distorts (elongates and changes the bond angle) the hydrogen bond in H_2O compared with the bulk (or free) H_2O.

In high solid–fluid surface area per unit volume A_{sf}/V adsorbers (such as dessicants), up to 25% of the mass of the adsorber occurs. For spherical particles of diameter d packed with porosity ϵ, we have $A_{sf}/V = 6(1-\epsilon)/d$, and as d approaches a few micrometers, A_{sf}/V becomes very large.

6.11.6 Disjoining Pressure in Ultrathin-Liquid Films

The pressure in thin-liquid films is affected by the interatomic forces of the liquid and the solid substrate. The integral force can be presented as the pressure difference between the vapor and the liquid (assuming a single component fluid) across this interface, $p_g - p_l$. For a L–J fluid with a similar L–J (2.9) interaction with the solid atoms, this relation has been derived in [57], i.e.,

$$\varphi_{f\text{-}f} = 4\epsilon_{\mathrm{LJ},f\text{-}f}[(\frac{\sigma_{\mathrm{LJ},f\text{-}f}}{r})^{12} - (\frac{\sigma_{\mathrm{LJ},f\text{-}f}}{r})^{6}] \tag{6.175}$$

$$\varphi_{f\text{-}s} = 4\epsilon_{\mathrm{LJ},f\text{-}s}[(\frac{\sigma_{\mathrm{LJ},f\text{-}s}}{r})^{12} - (\frac{\sigma_{\mathrm{LJ},f\text{-}s}}{r})^{6}]. \tag{6.176}$$

The mean-field potential that a fluid atom experiences at the vapor–liquid interface is $(r^2 = x^2 + z^2)$

$$\langle\varphi_f\rangle = \int_{\delta}^{\infty}\int_{0}^{\infty} n_s\varphi_{f\text{-}s}2\pi x\mathrm{d}x\mathrm{d}z, \tag{6.177}$$

where x is measured for the vapor–liquid (δ is the liquid-film thickness) interface, and n_s is the solid atomic number density. This gives

$$\langle\varphi_f\rangle = -\frac{4\pi n_s \epsilon_{\mathrm{LJ},f\text{-}s}\sigma^6_{\mathrm{LJ},f\text{-}s}}{6\delta^3} + \frac{4\pi n_s \epsilon_{\mathrm{LJ},f\text{-}s}\sigma^{12}_{\mathrm{LJ},f\text{-}s}}{45\delta^9}$$

$$\equiv \frac{A_{\mathrm{H}}}{6\pi n_f \sigma^3_{\mathrm{LJ},f\text{-}s}}[\frac{2}{15}(\frac{\sigma_{\mathrm{LJ},f\text{-}s}}{\delta})^9 - (\frac{\sigma_{\mathrm{LJ},f\text{-}s}}{\delta})^3], \qquad (6.178)$$

where $A_{\mathrm{H}} = 4\pi^2 n_s n_f \epsilon_{\mathrm{LJ},f\text{-}s}\sigma^6_{\mathrm{LJ},f\text{-}s}$ is called the Hamaker constant. Note that this gives a mean surface potential with exponent is 9–3 as compared with 12–6 used for the pairs in (6.175) and (6.176).

The first term is generally very small and negligible; then we have the pressure difference [55]

$$p_l - p_g = p_d = \frac{A_{\mathrm{H}}}{6\pi\delta^3}. \qquad (6.179)$$

This pressure difference is called the disjoining pressure p_d.

6.12 Turbulent-Flow Structure and Boundary-Layer Transport

Turbulent flow is marked by random fluctuations in the fluid velocity $\boldsymbol{u}_f$. For a boundary-layer flow, with the Cartesian coordinate system and with u_f being the velocity in the x direction, Figure 6.14(a) renders this turbulent velocity function. Figures 6.14(c) and (d) show the structures of the viscous and thermal boundary layers for laminar and turbulent flows. The distributions of the velocity and temperature are shown in terms of the dimensionless velocity u_f^* and temperature T_f^*. As with the laminar flow, the near-field nonuniformities are confined to regions adjacent to the surface and are marked by the viscous and thermal boundary-layer thicknesses $\overline{\delta}_\nu$ and $\overline{\delta}_\alpha$. For turbulent boundary layers, the fluctuations vanish far away and very close to the surface. The region adjacent to the surface is called the laminar sublayer. The laminar viscous sublayer has a thickness $\delta_{l,\nu}$, and the laminar thermal sublayer has a thickness $\delta_{l,\alpha}$. These are marked in Figures 6.14(c) and (d). In turbulent boundary layers, although the disturbed region is larger than its laminar counterpart, most of the changes in the velocity and the temperature occur very close to the surface. As such, adjacent to the surface the gradients of the velocity and the temperature are much larger than those in the laminar boundary layers.

It is customary to decompose the fluid velocity into a time-averaged (or mean) $\overline{\boldsymbol{u}}_f$ and a fluctuating component $\boldsymbol{u}'_f$ [148], i.e.,

$$\boldsymbol{u}_f = \overline{\boldsymbol{u}}_f + \boldsymbol{u}'_{f,t}, \quad \overline{\boldsymbol{u}}_f \equiv \frac{1}{\tau}\int_0^\tau \boldsymbol{u}_f \mathrm{d}t \quad \text{mean and fluctuating components of } \boldsymbol{u}_f, \qquad (6.180)$$

where τ(s) is the time period for averaging and is taken to be long enough such that $\overline{\boldsymbol{u}}_f$ no longer changes by any further increase in τ.

(a) Fluid Particle Turbulent Velocity Fluctuations in the x direction, $u_{f,t}$

(b) Turbulent Mixing Length, $\lambda_{f,t}$

(c) Viscous Laminar and Turbulent Boundary Layers, δ_v and $\bar{\delta}_v$

(d) Thermal Laminar and Turbulent Boundary Layers, δ_α and $\bar{\delta}_\alpha$

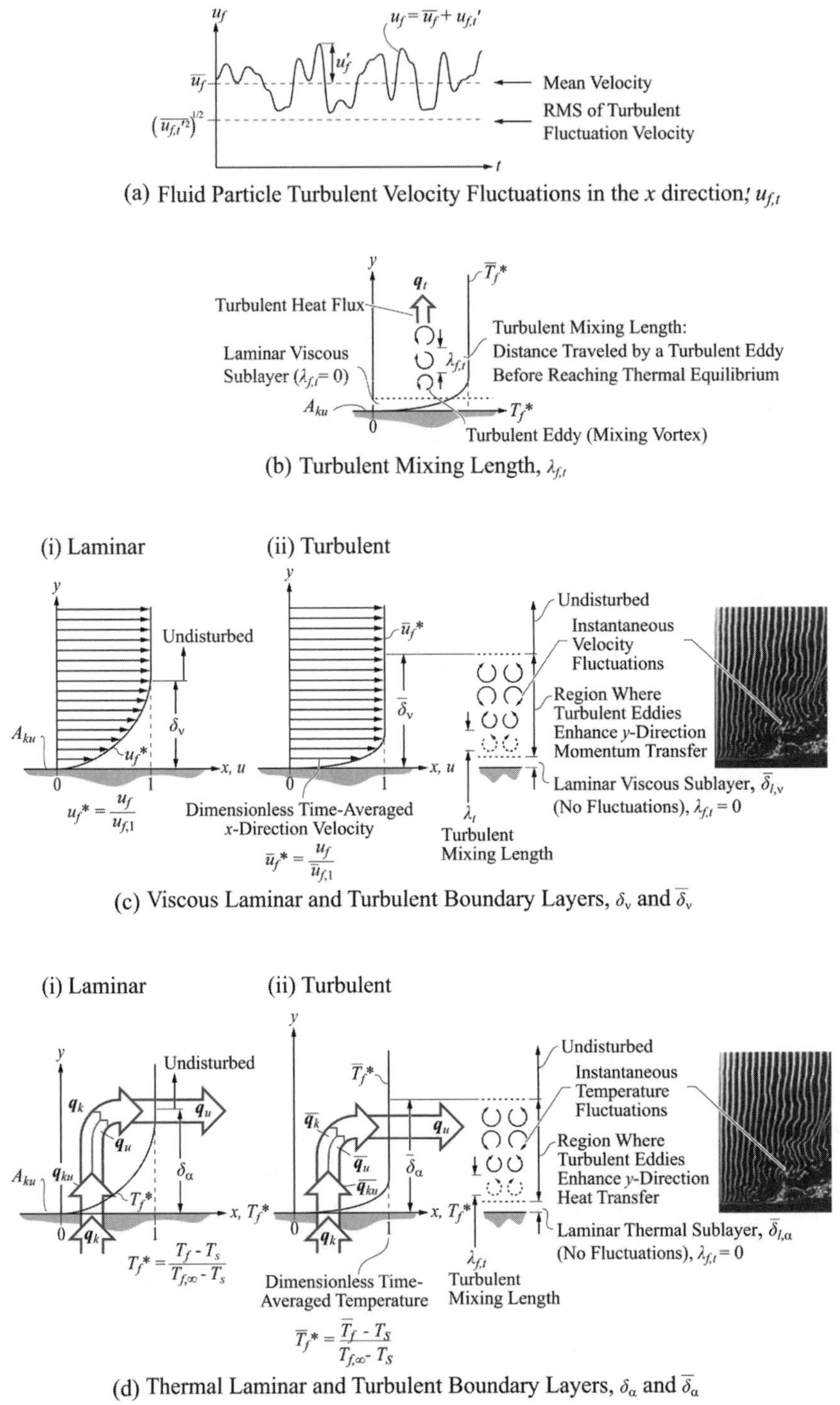

Figure 6.14. (a) Turbulent velocity fluctuations showing the mean and fluctuating components of the x-direction velocity $u'_{f,t}$. (b) A rendering of turbulent mixing length. (c) Structure of the viscous laminar and turbulent boundary layers. (d) Structure of the thermal laminar and turbulent boundary layers, (c) and (d) include renderings of the turbulent eddies (and turbulent mixing lengths) that contribute to the lateral (y-direction) heat transfer.

The temperature is also decomposed as

$$T_f = \overline{T}_f + T_f', \quad \overline{T}_f \equiv \frac{1}{\tau}\int_0^\tau T_f \mathrm{d}t \quad \text{mean and fluctuating components of } T_f. \tag{6.181}$$

Turbulence can be produced within the boundary layer when the flow becomes unstable. This is called the boundary-layer turbulence. The boundary-layer turbulence, in an otherwise parallel laminar flow (far field), occurs at a transition Reynolds number $\mathrm{Re}_{L,t}$. This boundary-layer transition Reynolds number is $\mathrm{Re}_{L,t} = 5 \times 10^5$ [301].

Turbulence can also be present in the far-field flow and this is called the free-stream turbulence. This can be caused by propellers, the presence of grid nets, other interactions with solid surfaces, or by instabilities in the far-field flow. Our discussion here is limited to boundary-layer turbulence.

Although on the solid surface $\boldsymbol{u}_f = 0$, the surface-convection heat transfer is influenced by the velocity and temperature fluctuations near the surface, similar to the laminar-flow case. The turbulent velocity fluctuations are three dimensional, and the component perpendicular to the surface makes the largest contribution to the surface-convection heat transfer.

6.12.1 Turbulent Kinetic Energy Spectrum for Homogeneous Turbulence

The idealized turbulent flow is the homogeneous, isotropic turbulence (spatial homogeneity in mean properties). This is rendered in Figure 6.15(a). In comparison with the boundary-layer turbulence, the homogeneous turbulence is in the free stream (away from any surface). The turbulent kinetic energy equation is obtained from the Navier–Stokes equations given in Table 3.10 and is [148]

$$E_{f,t} = \frac{1}{2}u'_{f,i}u'_{f,i}, \quad \overline{u'^2_{f,i}} = \overline{u'^2_{f,j}} \quad \text{isotropic, turbulent kinetic energy } E_{f,t}. \tag{6.182}$$

$$\frac{\mathrm{D}}{\mathrm{D}t}\overline{E}_{f,t} = -\frac{\partial}{\partial x_i}\overline{u'_{f,i}(\frac{p'}{\rho_f} + E_{f,t})} - \overline{u'_{f,i}u'_{f,j}}\frac{\partial \bar{u}_f}{\partial x_i} +$$
$$\frac{\mu_f}{\rho_f}\frac{\partial}{\partial x_i}\overline{u'_{f,j}(\frac{\partial u'_{f,i}}{\partial x_j} + \frac{\partial u'_{f,j}}{\partial x_i})} - \frac{\mu_f}{\rho_f}\overline{(\frac{\partial u'_{f,i}}{\partial x_j} + \frac{\partial u'_{f,j}}{\partial x_i})\frac{\partial u'_{f,j}}{\partial x_i}}. \tag{6.183}$$

The first term on the right is the convective diffusion of the turbulent mechanical energy, the second is the work of deformation of the mean motion by the turbulent stresses, the third is the viscous shear stress of the turbulent motion, and the last one is the viscous dissipation.

Under homogeneous turbulence only the time-dependent part of the left-hang side and the viscous dissipation on the right are nonzero. For this case, any existing

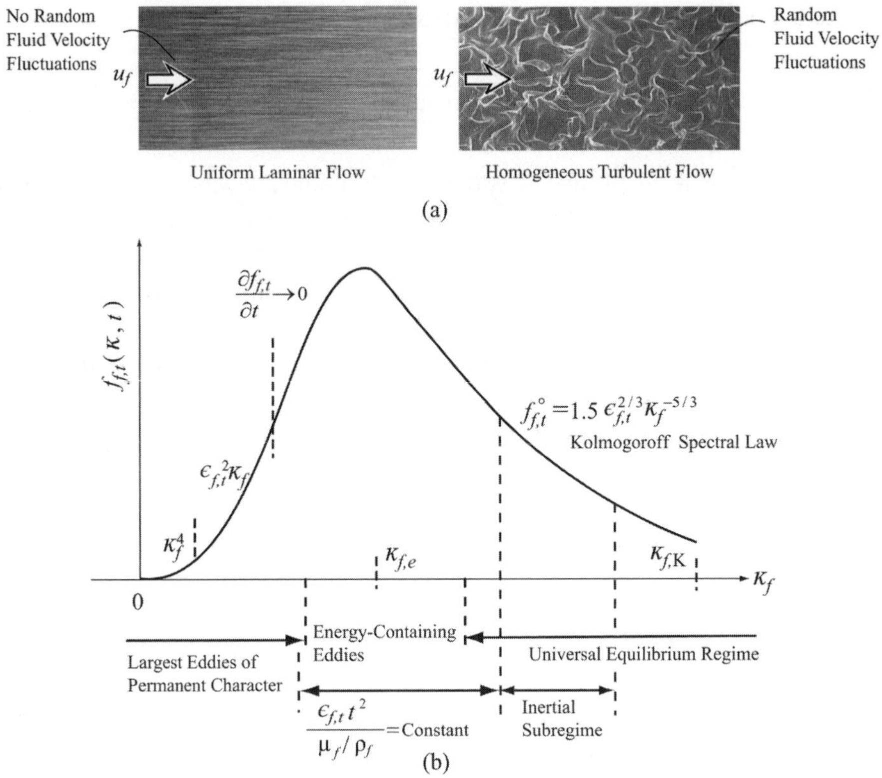

Figure 6.15. (a) Homogeneous turbulence in a unidirectional flow. (b)Turbulent energy spectrum, i.e, turbulent energy distribution function verses turbulent wave number, showing the various regimes [148]. The Kolmogoroff spectrum law applies to the large wave number (inertial) subregime. The results are homogeneous, equilibrium (balance between production and dissipation) turbulence.

turbulent fluctuation decays in time by viscous dissipation. Then

$$\frac{\partial \overline{E}_{f,t}}{\partial t} = \frac{\partial}{\partial t}\frac{3}{2}\overline{u'^2_{f,i}} = \frac{\mu_f}{\rho_f}\overline{\left(\frac{\partial u'_{f,i}}{\partial x_j} + \frac{\partial u'_{f,j}}{\partial x_i}\right)\frac{\partial u'_{f,j}}{\partial x_i}} \equiv \epsilon_{f,t} \quad \text{homogeneous turbulence,} \tag{6.184}$$

where $\epsilon_{f,t}$ is the viscous dissipation rate of the turbulent kinetic energy.

The turbulent velocity fluctuation for an incompressible fluid ($\nabla \cdot \boldsymbol{u}_f = 0$) can be Fourier decomposed by the wave vector $\boldsymbol{\kappa}$ and the angular frequency ω as

$$\boldsymbol{u}'_f(\boldsymbol{x}, t) = \iint \boldsymbol{U}'_f(\kappa_f, \omega)\exp(i\boldsymbol{\kappa}_f \cdot \boldsymbol{x} - \omega t)\,d\boldsymbol{\kappa}_f\,d\omega, \tag{6.185}$$

and the incompressibility requires that

$$\boldsymbol{\kappa}_f \cdot \boldsymbol{U}'_f = 0. \tag{6.186}$$

This states that the only disturbances allowed are transverse (not longitudinal). Such turbulent elements are called turbulent eddies. Then it is said the turbulent velocity field is made of eddies of different sizes (wave numbers). The eddies of large sizes are created by the mean flow, and they in turn exchange their energy with the smaller eddies (this is called turbulent energy cascading).

To describe the turbulent fluctuations (kinetic energy) in terms of distribution over wave number (or frequency), we define the turbulent energy distribution function $f_{t,t}$ through

$$\overline{E}_{f,t} = \frac{3}{2}\overline{u'^2_{f,1}} = \int_0^\infty f_t(\kappa_f, t)\, d\kappa_f \quad \text{turbulent energy distribution function,} \tag{6.187}$$

where the wave number and the angular frequency are assumed to be related through $\kappa_f = \omega/\overline{u}_{f,1}$, where $\overline{u}_{f,1}$ is the mean velocity. Here $f_{f,t}$ has units of m^3/s^2. Figure 6.15(b) gives the variation of the turbulent energy distribution function f_t with respect to the wave number as compiled from measurements and theoretical predictions. This is also called the turbulent energy spectrum. Figure 6.15(b) shows that for the smallest wave number (largest-size) eddies, we have a κ_f^4 relation for the turbulent distribution function and very slow decay (viscosity is not important). The intermediate wave-number eddies contain most of the turbulent kinetic energy and are referred to as the energy-containing eddies. The peak is designated by $\kappa_{f,e}$. The smallest eddy size is the Kolmogoroff length $\kappa_{f,\mathrm{K}} = 2\pi/\lambda_{f,\mathrm{K}}$.

The eddies of length $\lambda_{f,t}$ that have velocity $u_{f,t}$ have Reynolds number $\mathrm{Re}_t = \rho_f u_{f,t}\lambda_{f,t}/u_f$, and for large eddies (larger Re_t) their energy is not dissipated by viscosity, but cascaded to smaller eddies. The eddies with $\mathrm{Re}_t \simeq 1$ are dissipated because of viscosity, and this gives the smallest eddy, which has a length $\lambda_{f,\mathrm{K}}$ (Kolmogoroff length) and velocity $u_{f,\mathrm{K}}$

$$\lambda_{f,\mathrm{K}} = \frac{\mu_f/\rho_f}{u_{f,\mathrm{K}}} \quad \text{Kolmogoroff length, smallest eddies.} \tag{6.188}$$

Using (6.184), Kolmogoroff (Kolmogorov) suggested that the velocity of the smallest eddies must follow (replacing the time derivative with $u_{f,\mathrm{K}}/\lambda_{f,\mathrm{K}}$)

$$\frac{u^3_{f,\mathrm{K}}}{\lambda_{f,\mathrm{K}}} \simeq \epsilon_{f,t} \quad \text{viscous dissipation rate,} \tag{6.189}$$

and using this in (6.188), we have

$$\lambda_{f,\mathrm{K}} = [(\frac{\mu_f}{\rho_f})^3\frac{1}{\epsilon_{f,t}}]^{1/4}, \quad u_{f,\mathrm{K}} = (\frac{\mu_f}{\rho_f}\epsilon_{f,t})^{1/4} \quad \text{Kolmogoroff length and velocity scales.} \tag{6.190}$$

Then, using $\kappa_{f,\mathrm{K}} = 2\pi/\lambda_{f,\mathrm{K}}$, we can show that [246]

$$f^{\mathrm{o}}_{f,t}(\kappa) = 1.5\,\epsilon^{2/3}_{f,t}\kappa_f^{-5/3} \quad \text{turbulent energy distribution function for inertial subregime.} \tag{6.191}$$

This is the Kolmogoroff −5/3 spectral law for equilibrium, homogeneous turbulence.

6.12.2 Boundary-Layer Turbulent Heat Flux

As with (6.180) and (6.181), the convection heat flux vector is similarly averaged, and we have

$$\begin{aligned}\overline{\boldsymbol{q}_u} &= \overline{\boldsymbol{q}_{u,m}} + \overline{\boldsymbol{q}_{u,t}} \\ &= (\rho c_p)_f \overline{\boldsymbol{u}}_f \overline{T}_f + (\rho c_p)_f \overline{\boldsymbol{u}'_{f,t} T'_{f,t}} \\ &\quad \text{mean convection} \qquad \text{turbulent convection} \\ &\quad \text{heat flux vector } \overline{\boldsymbol{q}_{u,m}} \qquad \text{heat flux vector } \overline{\boldsymbol{q}_{u,t}}.\end{aligned} \tag{6.192}$$

The turbulent fluctuations contribute to the heat transfer in a way similar to the molecular fluctuations discussed in Section 6.4. There we used the molecular (thermal) velocity fluctuations and the RMS thermal speed $\langle \overline{u_f^2} \rangle^{1/2}$ and through (3.14) showed how the heat was transfered by these fluctuations. Here, the combined effects of the turbulent velocity fluctuation $(\overline{\boldsymbol{u}'_{f,t} \cdot \boldsymbol{u}'_{f,t}})^{1/2}$, the ability of the fluid to store/release heat $(\rho c_p)_f$, and the ability of the fluctuations to carry the fluid content a short distance before reaching thermal equilibrium also result in heat transfer in the presence of a temperature nonuniformity. This is shown in Figure 6.14(b).

From (6.57), we write the y component (i.e., perpendicular to the surface) of the turbulent convection heat flux as

$$(\overline{q_{u,t}})_y = (\rho c_p)_f \overline{v'_{f,t} T'_{f,t}}. \tag{6.193}$$

We now proceed to relate $\overline{v'_{f,t} T'_{f,t}}$ to mean quantities that are more readily measured or predicted.

6.12.3 Boundary-Layer Turbulent Mixing Length and Thermal Conductivity

Using an analogy with the conduction in gases (which is due to molecular fluctuations), as given by (6.125), the heat transfer that is due to turbulent fluctuations is expressed as [148, 301]

$$(\overline{q_{u,t}})_y = (\rho c_p)_f \overline{v'_{f,t} T'_{f,t}} \equiv -\text{Pr}_t^{-1} (\rho c_p)_f (\overline{v'^2_{f,t}})^{1/2} \lambda_{f,t} \frac{\mathrm{d}\overline{T}_f}{\mathrm{d}y}, \tag{6.194}$$

where $\lambda_{f,t}$ is the turbulent mean free path and is also called the turbulent mixing length [shown in Figure 6.14(b)]. This is a representation of the distance traveled by a fluid particle before reaching equilibrium (either thermal or mechanical). It varies with the distance from the bounding surface. The constant Pr_t (called the turbulent Prandtl number) is determined empirically and is near unity. Here $(\overline{v'^2_{f,t}})^{1/2}$ is the RMS of the turbulent velocity fluctuation in the y direction.

We note that the proportionality of $\overline{v'_{f,t} T'_{f,t}}$ to $\mathrm{d}\overline{T}_f/\mathrm{d}y$, similarly exists for velocity and temperature deviation caused by spatial (instead of temporal) nonuniformities. This is discussed and the proportionality shown in [173].

Because in (6.194) the turbulent heat transfer is related to the gradient of the mean temperature, a turbulent thermal conductivity $k_{f,t}$ is defined by

$$(\overline{q_{u,t}})_y \equiv -k_{f,t}\frac{\mathrm{d}\overline{T}_f}{\mathrm{d}y} \quad \text{turbulent thermal conductivity.} \tag{6.195}$$

Then from (6.195) and (6.194), $k_{f,t}$ is found as

$$k_{f,t} \equiv (k_{f,t})_\perp = \mathrm{Pr}_t^{-1}(\rho c_p)_f(\overline{v_{f,t}'^2})^{1/2}\lambda_{f,t}, \tag{6.196}$$

where subscript $\perp$ indicates perpendicular to the surface.

Next $(\overline{v_{f,t}'^2})^{1/2}$ is related to the gradient of the component of the fluid mean velocity along the surface, i.e., $|\partial\overline{u}_f/\partial y|$. This is because, from the examination of the momentum equation, we find that the magnitude of the fluctuations increases as the velocity gradient becomes larger. When we use the turbulent mixing length $\lambda_{f,t}$ again, and we have $(\overline{v_{f,t}'^2})^{1/2} \equiv \lambda_{f,t}|\partial\overline{u}_f/\partial y|$. Again, this is because as $\mathrm{d}\overline{T}_f/\mathrm{d}y$ was the source for $\overline{v_{f,t}'T_{f,t}'}$, $|\mathrm{d}\overline{u}_f/\mathrm{d}y|$ is the source for $(\overline{v_{f,t}'^2})^{1/2}$. Then (6.61) becomes

$$k_{f,t} = \mathrm{Pr}_t^{-1}(\rho c_p)_f\lambda_{f,t}^2\left|\frac{\partial\overline{u}_f}{\partial y}\right|, \quad \alpha_{f,t} = \frac{k_{f,t}}{(\rho c_p)_f}, \tag{6.197}$$

where $\alpha_{f,t}$ is the turbulent thermal diffusivity (or eddy diffusivity). Note that $\mathrm{Pr}_t = \nu_{f,t}/\alpha_{f,t}$, where ν_t is the turbulent kinematic viscosity.

We can extend the results to the three-dimensional heat flow. The turbulent convection heat flux vector $(\rho c_p)_f\overline{\boldsymbol{u}_{f,t}'T_{f,t}'}$ is proportional to the gradient of $\overline{T}_f$, i.e., $\nabla\overline{T}_f$. Therefore, as $\nabla\overline{T}_f$ becomes larger, $(\rho c_p)_f\overline{\boldsymbol{u}_{f,t}'T_{f,t}'}$ increases. Then, assuming that k_t is isotropic (i.e., the same in all directions), from (6.195) the vectorial form of $\boldsymbol{q}_t$ is

$$\overline{\boldsymbol{q}_{u,t}} = (\rho c_p)_f\overline{\boldsymbol{u}_{f,t}'T_{f,t}'} \equiv -k_{f,t}\nabla\overline{T}_f, \tag{6.198}$$

where we note that the turbulent convection heat flux vector is related to the gradient of mean temperature.

We note that in general $k_{f,t}$ is anisotropic (and should be represented by a tensor).

6.12.4 Spatial Variation of Boundary-Layer Turbulent Mixing Length

For fully-turbulent flows at high Reynolds numbers, $k_{f,t}$ can be larger than k_f (the molecular thermal conductivity) at some locations near the solid surface. Also, $k_{f,t}$ vanishes in the laminar sublayer. Therefore $k_{f,t}$ (or $\boldsymbol{u}_{f,t}'$ and $\lambda_{f,t}$) varies with the distance from the surface and enhances the convection heat transfer. For parallel flows, this enhancement is noticeable in the direction perpendicular to the flow (y direction), where the mean velocity $\overline{v}_f$ is small and the heat transfer is otherwise dominated by molecular conduction.

The smallest turbulent mixing length is the Kolmogoroff length $\lambda_{f,\mathrm{K}}$, (6.190), and the largest is the characteristic length of confining solid L (in the direction of turbulent transport).

Empirical results show that for the flow over a semi-infinite flat plate, the variation of the mixing length in the turbulent boundary layer, adjacent to surface, is approximated as [175]

$$\lambda_{f,t} = \begin{cases} 0 & 0 \le y \le \delta_{l,i} & \text{laminar sublayer (molecular } \lambda_f) \\ \kappa_{\rm vK}(y - \delta_{l,\nu}) & \delta_{l,\nu} \le y \le 0.21\overline{\delta}_{t,\nu} & \text{inner turbulent core} \\ 0.085\overline{\delta}_{t,\nu} & 0.21\overline{\delta}_{\nu} \le y \le \overline{\delta}_{t,\nu} & \text{outer turbulent core} \\ 0 & & \text{outside turbulent boundary layer} \end{cases}$$

$$\text{distribution of turbulent mixing length,} \qquad (6.199)$$

where $i = \alpha$ or ν (thermal or viscous), $\kappa_{\rm vK}$ (von Kármán universal constant) is typically equal to 0.41 and $\overline{\delta}_{t,\nu}$ is the the turbulent viscous boundary-layer thickness that will be subsequently given below [175]. The laminar sublayer thickness is given in terms of the wall shear stress $\overline{\tau}_s$ as $\delta_{l,\nu} = 10.8\mu_f/(\overline{\tau}_s/\rho_f)^{1/2}$, and $\delta_{l,\alpha} = 13.2\,{\rm Pr}\delta_{l,\nu}/10.8$. The mixing length is zero in the laminar sublayer where there are no turbulent fluctuations. Above the sublayer, the mixing length increases linearly with the distance from the bounding surface until $y = 0.21\overline{\delta}_{t,\nu}$, above which $\lambda_{f,t}$ is proportional to the boundary-layer thickness. Outside the boundary-layer (in the free stream) there are no turbulent fluctuations and $\lambda_{f,t} = 0$. In practice it is assumed that $\overline{\delta}_{t,\alpha} \simeq \overline{\delta}_{t,\nu}$, i.e, turbulent thermal and fluctuation boundary layer thickness are equal.

The turbulent wall shear stress related to the wall friction factor and boundary-layer thickness, are

$$\frac{\overline{\tau}_s}{\rho} = \frac{c_f \overline{u}_{f,\infty}^2}{2} \qquad (6.200)$$

$$\frac{c_f}{2} = 0.0296\,{\rm Re}_L^{-0.2} \qquad (6.201)$$

$$\delta_{l,\nu} = 10.8\frac{\mu_f}{(\overline{\tau}_s/\rho_f)^{1/2}}, \quad \frac{\delta_{t,\nu}}{L} = 0.037\,{\rm Re}_L^{-0.2}. \qquad (6.202)$$

Figure 6.16(a) shows the distribution of turbulent mixing length (dimensionless) as a function of the distance (dimensionless) from the edge of the laminar sublayer [4]. The results are deduced from the measured velocity distribution. Figure 6.16(b) shows the variation of the dimensional fluid conductivity $(k_f + k_{f,t})$ with respect to the distance from the solid surface, calculated with the experimental results of [4]. The results are for air at T = 300 K, and ${\rm Re}_L = 3.55 \times 10^5$, which is close to the transition Reynolds number. The laminar sublayer and the turbulent boundary-layer thicknesses are also shown. For this experiment, the total conductivity reaches about 3.0 W/m-K (k_f = 0.026 W/m-K).

The distribution of the mean, x-direction velocity $\overline{u}_f$ is generally determined from empirical relations that include the dependence on y and ${\rm Re}_L$ [175].

Figures 6.14(c) and (d) give a rendering of this enhanced heat transfer. The fluctuations in the velocity and in the temperature are characterized by the turbulent eddies (i.e., small vortices with a large range of sizes and frequencies). The turbulent mixing length $\lambda_{f,t}$ is also shown.

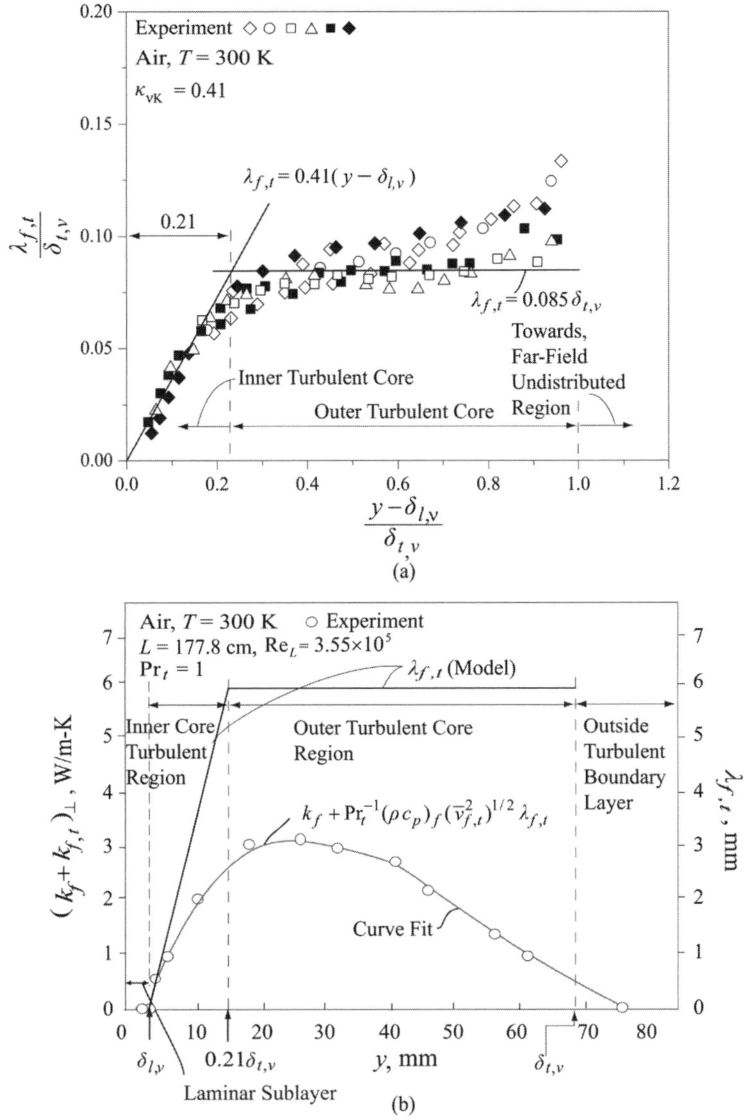

Figure 6.16. (a) Variations of the mixing length deduced from the measured velocity distribution, as fractions of the distance from the edge of the laminar sublayer [4]. (b) Variations of total fluid thermal conductivity (left) and the model mixing length (right) with respect to distance from the surface [4].

6.13 Thermal Plasmas

Electrically charged gases (i.e., plasmas) are heated by Joule heating through imposing an electric current. Gases become charged (i.e., ionized) at high temperatures ($T_f > 2000$ K). When very high temperature gases (i.e., $T_f > 4000$ K) are needed, Joule heating is used and the process is called thermal plasma generation. Figure 6.17 shows the typical temperatures in natural and engineered thermal plasmas and the associated force electron density $n_{e,c}$ (1/cm^3). A thermal plasma is made of electrons, ions, and neutral molecules and atoms. Compared with the conduction

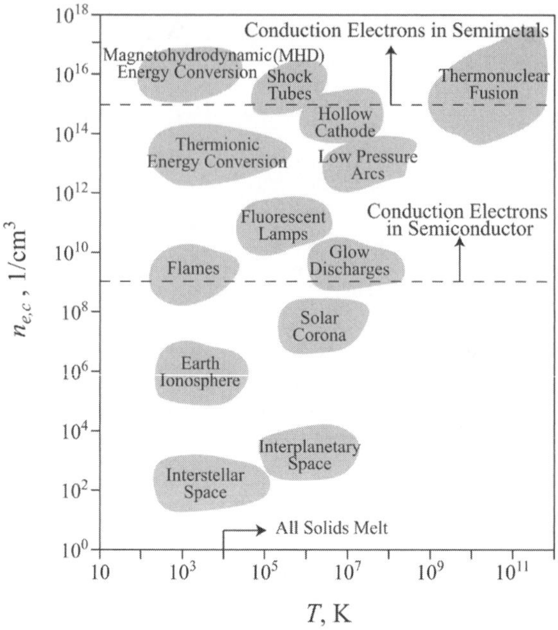

Figure 6.17. Typical temperatures for thermal plasmas and their associated force electron densities [189].

electrons in solids, Figure 5.20 for example, fluorescent lamps have higher conduction electron densities than some semiconductors. Also, the plasma in magnetohydrodynamics (MHD) has higher electron density than semimetals. Because electrons are much lighter than the rest of the species in the plasma, they heat up much faster. Therefore, unless there are many collisions among all the species, there can be a thermal nonequilibrium among them. Once the heat source (e.g., Joule heating) is removed, thermal equilibrium is quickly achieved, except for very short elapsed times and at very low pressures.

Among the common gases used in thermal plasmas are Ar, H_2, O_2, N_2, air, Ar–H_2, Ar–He, and CO_2. Laboratory units with the capability of $\dot{S}_{e,\mathrm{J}} = 30$ to 50 kW and industrial units of up to 1-MW power are used. Because of the high temperatures, there are some desired heat losses to prevent the solids in contact with the plasma from melting. There are also some undesired heat losses. Therefore the temperature in the gas drops over a short distance, but the center of the plasma has temperatures of about 10,000 K.

6.13.1 Free Electron Density and Plasma Thermal Conductivity

Thermal plasmas are used in manufacturing and materials processing, for example in the heating of sprays of paint particles in dry painting, in chemical vapor deposition, in welding, and in surface-material removal. Because of the free electrons, plasmas have a high electrical conductivity σ_e, and, depending on the gas, they can

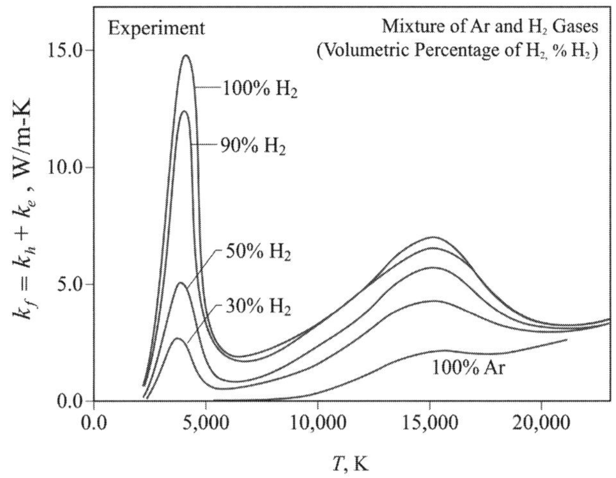

Figure 6.18. Thermal conductivity of Ar–H_2 gas mixture at elevated temperatures [39].

also have a high thermal conductivity. The thermal conductivity can be decomposed as

$$k_f = k_h + k_e = \sum_h \frac{1}{3} n_h c_{v,h} u_h \lambda_h + \sum_e \frac{1}{3} n_e c_{v,e} u_e \lambda_e, \qquad (6.203)$$

where k_h represents the conductivity of heavier species and k_e is for free electrons. Note that u_e used here is based on T_e, which may not be equal to T_h Section (6.13.2), and from (6.63) using the electron mass (Table A.3, $m_e = 5.45 \times 10^{-4}\ m_p$, where m_p is the proton mass), the electron thermal speed is rather large. This results in a large contribution from k_e. The collision diameter for e–h collision is also small, making λ_e large [according to (6.87)]. Figure 6.18 gives the thermal conductivity of Ar–H_2 mixture as a function of temperature and for various hydrogen volume fractions [39]. Note that, around $T = 3500$ K, the conductivity of hydrogen reaches a value of 15 W/m-K (because $k_e > k_h$ in ionized gases, this magnitude is typical of some metallic solids). Because of this high thermal conductivity, such thermal plasmas can very effectively heat the entrained particles (sensible heating rates $\mathrm{d}T/\mathrm{d}t$, of the order of 10^5 to 10^7 K/s are achieved).

The gas is charged by injection of charged particles, by pilot electrodes, by high frequency starting circuits, or by an initial increase in the gas temperature, for example, by an C_2H_2–O_2 torch. For the Joule heating of charged gases, the steady-state energy equation is given in Table 1.1. In thermal plasma generation, the volumetric radiation heat transfer becomes significant and needs to be included. Under local thermal equilibrium for all species in the plasma, the energy equation from Table 1.1 is written as

$$\nabla \cdot (\boldsymbol{q}_k + \boldsymbol{q}_u + \boldsymbol{q}_r) = \dot{s}_{ij} + \dot{s}_{e,\mathrm{J}} = \dot{s}_{ij} + \sigma_e \overline{\boldsymbol{e}_e \cdot \boldsymbol{e}_e} = \dot{s}_{ij} + \sigma_e \overline{e_e^2}$$

$$\text{thermal equilibrium energy equation}, \qquad (6.204)$$

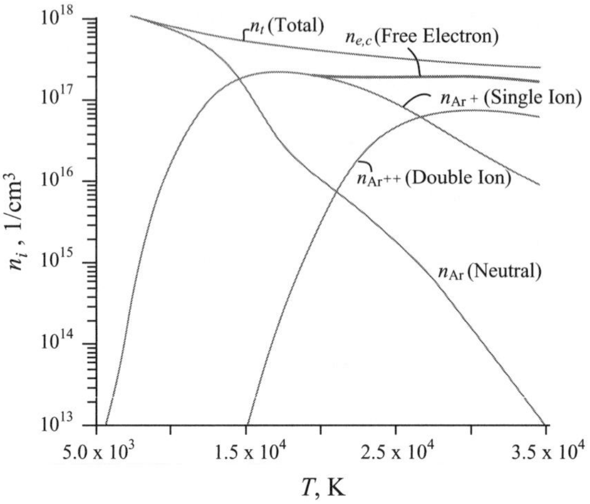

(a) Equilibrium Composition of Argon Plasma (p = 1 atm)

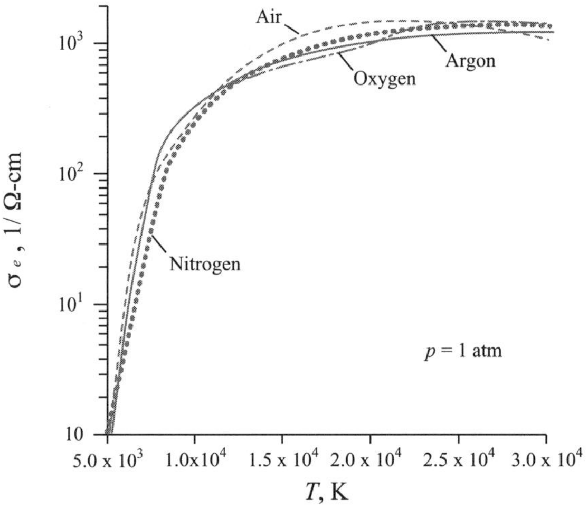

(b) Electrical Conductivity of Air, Argon, Nitrogen, and Oxygen Plasmas

Figure 6.19. (a) Equilibrium composition of argon plasma as a function of temperature at 1–atm pressure. (b) Electrical conductivities of air, argon, nitrogen, and oxygen plasmas as functions of temperature, at 1–atm pressure [197].

where $\dot{s}_{ij}$ is the dissociation-ionization energy conversion rate $\dot{s}_{ij} < 0$, and $\boldsymbol{e}_e$(V/m) is the electric-field intensity vector and can be steady or oscillating (i.e., time periodic). For time periodic fields use (3.52). Then a portion of $\dot{s}_{e,\text{J}}$ is used for the dissociation–ionization. Figure 6.19(a) shows the equilibrium composition of the argon plasma as a function of temperature and at 1-atm pressure. Note that the concentration of free electrons n_e, c rises rapidly as $T = 1.5 \times 10^4$ K is reached and increases at a much lower rate shortly after that. The number of neutral species n_A decreases rapidly for $T > 10^4$ K and becomes relatively rare at $T = 3.5 \times 10^4$ K. In Figure 5.20 it was mentioned that solid semimetals have electron concentrations of

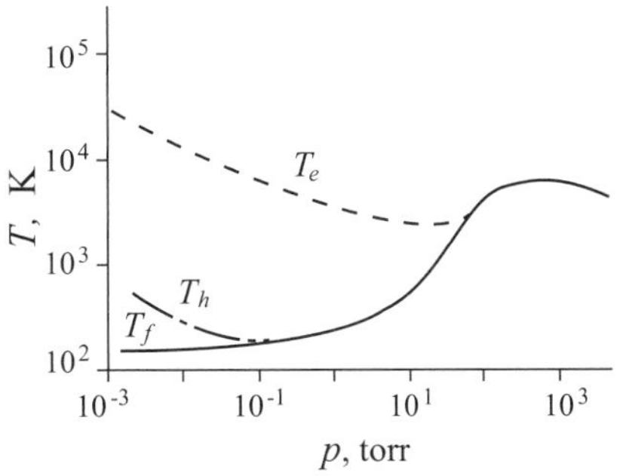

Figure 6.20. The thermal equilibrium, or lack of it, between electrons T_e, ions (heavier charged species) T_h, and the gas temperatures T_f including neutral atoms, as a function of pressure [189]. As the pressure (and collision rate) increases, thermal equilibrium is reached.

10^{15} to 10^{21} electrons/cm^3, and this places these thermal plasmas in the electronic conduction range of semimetals. The electrical conductivities of several gases are given in Figure 6.19(b), as functions of temperature. Again note that for $T > 10^4$ K a plateau is reached. These electrical conductivities are very large for the gas phase, but smaller than those of the metals.

Determination of $\dot{s}_{ij}$ involves determining the most significant dissociation and ionization reactions, which in turn requires knowledge of the associated equilibrium compositions and the energy needed for breaking each of these bonds. This is not addressed here, and details can be found in [197].

Plasmas are multicomponent high-temperature gases containing neutral particles and atoms and molecules, and electrons and ions with continuous and extensive interactions among them. The plasma state of matter and plasma physics are described in [64, 305]. The plasmas are further divided into cold and thermal plasmas. In cold plasmas, the lighter electron temperature is much higher than the other heavier gas constituents (i.e., ions and neutral atoms or molecules), and therefore thermal nonequilibrium exists between the electrons and the remainder gas species referred to as the heavier particles (T_h). In thermal plasmas, there exists a near-thermal equilibrium. In [189], it is shown that the increase in the pressure (or the molecular density) leads to thermal equilibrium. Figure 6.20 shows this trend toward thermal equilibrium with the increase in pressure (1 torr is equal to 1/760 of 1–atm pressure and one atmosphere is 1.013×10^5 Pa), where T_f stands for the gas temperature (electrons, ions, and neutral atoms) and T_e is the electron temperature. These ions are included in the heavier species. For $p > 100$ torr, thermal equilibrium is attained. The neutral atoms have the lowest temperature, thus reducing the average gas temperature. Then the atmospheric pressure plasmas are in equilibrium. The binary collision rate is given by (6.79) and p^2 dependence is shown in (6.80)

Plasmas can be under external applied electric or magnetic fields, and they generate (i.e., induce) a magnetic field as they flow. From the heat transfer characteristics, plasmas are a subset of the MHD, but they are generally treated separately for the following reasons.

Because of the very high temperatures, the radiative heat transfer is generally significant, and the multiple constituents have various constituent size, mass, and initial conditions, and therefore can be in local thermal nonequilibria [96]. The electrical conductivity of the dissociated and ionized gases also becomes significantly large with the increase in temperature, and, in general, the thermoelectrophysical property variations must also be considered. In low-pressure plasmas, the mean free path of the molecules may be of the order of the momentum and thermal boundary-layer thickness, and therefore in a transition-flow regime (Knudsen regime, discussed in Section 6.11), or a molecular-flow regime treatment may be necessary.

The equilibrium population density of the excited states of the various species are determined from the kinetic theory [189] and is in general treated similar to the chemical reactions. Treatments of the species equations and other transport and thermodynamic properties of plasmas are reviewed [96].

An example of plasma flow under an applied field is given in [332], where allowance for a velocity slip is made for electrons flowing at much higher velocities than the rest of the constituents. An example that includes chemical reactions among the constituents is given in [249].

6.13.2 Thermal Nonequilibrium Plasma Energy Equation

The thermal energy conservation equations for the electrons and for the heavier species are written assuming monatomic species. The specific heat capacity at constant pressure for ideally behaving monatomic gases is given by (6.23) using (6.21), i.e, per atom (particle) we have

$$c_p = c_v + k_\mathrm{B} = \frac{5}{2}k_\mathrm{B}. \tag{6.205}$$

The free-electron energy equation, similar to (5.264), but for a free-electron subsystem, becomes

$$\frac{1}{r^2}\frac{\partial}{\partial r}r^2\frac{5}{2}k_\mathrm{B}n_{e,c}T_e u + \frac{1}{r\sin\theta}\frac{\partial}{\partial\theta}\frac{5}{2}k_\mathrm{B}n_{e,c}T_e v\sin\theta = \frac{1}{r}\frac{\partial}{\partial r}k_e r^2\frac{\partial}{\partial r}T_e$$
$$+\frac{1}{r^2\sin\theta}\frac{\partial}{\partial\theta}k_e\sin\theta\frac{\partial}{\partial\theta}T_e - \Delta E_\mathrm{I}\dot{n}_{e,c} + \nabla\cdot\boldsymbol{q}_r + \dot{s}_{e,J} - \dot{s}_{e\text{-}h}, \tag{6.206}$$

where ΔE_I is the ionization energy, $\dot{s}_{e,J}$ is volumetric Joule heating, and $\dot{s}_{e\text{-}h}$ is the rate of energy transfer from the electron to the heavier species. Also, u and v designate r and θ components of the gas velocity. The heavier species energy equation becomes

$$\frac{1}{r^2}\frac{\partial}{\partial r}r^2\frac{5}{2}k_\mathrm{B}n_h T_h u + \frac{1}{r\sin\theta}\frac{\partial}{\partial\theta}\frac{5}{2}k_\mathrm{B}n_h T_h v\sin\theta$$
$$= \frac{1}{r^2}\frac{\partial}{\partial r}k_h r^2\frac{\partial}{\partial r}T_h + \frac{1}{r^2\sin\theta}k_h\sin\theta\frac{\partial}{\partial\theta}T_h + \dot{s}_{e\text{-}h}. \tag{6.207}$$

The $\dot{n}_{e,c}$ is determined from a kinetic model and is a function of the electron temperature and the concentration of the electrons and atoms (as described in Section 6.13.3). The interaction kinetics rate between the electron and heavier species $\dot{s}_{e\text{-}h}$ is divided into the interaction between the electrons and ions and the interaction between the electrons and atoms. These interactions are determined by the proper kinetic models and depend on T_h and T_e as well as on n_a, $n_i (n_h = n_i + n_a)$, and $n_{e,c}$. These kinetic models are discussed in [60] and are reviewed in Section 6.13.4.

The two-medium (two-temperature) descriptions of the energy transport and the fluid dynamics just given for the electrons and the heavier species, with the viscosity (including the second viscosity), density, and the thermal conductivities being strong functions of the temperature, constitute a continuum description that now requires specification of the boundary conditions. The presumed low pressure of the plasma would require the modeling of these boundary conditions through the slip coefficients. These are discussed next.

6.13.3 Species Concentrations For Two-Temperature Plasmas

Consider the single ionization–recombination reaction for a monatomic gas

$$\mathrm{A} \rightleftharpoons \mathrm{A}^+ + e^-. \tag{6.208}$$

The law of mass action having the equilibrium constant in terms of the number densities of atoms, ions, and electrons, n_a, n_i, and n_e (for A, A^+, and e^-), which is derived in [68] for the case of local thermal nonequilibrium between the electrons and the heavier species (T_e and T_h). This is

$$\frac{p_e p_i}{p_a} = 2\frac{Z_i(T_e)}{Z_a(T_e)}\left(\frac{2\pi m_e k_\mathrm{B} T_e}{h_\mathrm{P}^2}\right)^{3/2} k_\mathrm{B} T_e \exp\left(-\frac{\Delta E_I}{k_\mathrm{B} T_e}\right) \quad \text{law of mass action}$$

$$p_e = n_{e,c} k_\mathrm{B} T_e, \quad p_a = n_a k_\mathrm{B} T_h, \quad p_i = n_i k_\mathrm{B} T_h,$$

$$Z_a(T_e) = \sum_k g_{a,k} \exp\left(-\frac{\Delta E_{a,k}}{k_\mathrm{B} T_e}\right), \quad Z_i(T_e) = \sum_k g_{i,k} \exp\left(-\frac{\Delta E_{i,k}}{k_\mathrm{B} T_e}\right), \tag{6.209}$$

or

$$\frac{n_{e,c} n_i}{n_a} = 2\frac{Z_i(T_e)}{Z_a(T_e)}\left(\frac{2\pi m_e k_\mathrm{B} T_e}{h_\mathrm{P}^2}\right)^{3/2} \exp\left(-\frac{\Delta E_I}{k_\mathrm{B} T_e}\right), \tag{6.210}$$

where ΔE_I is the effective ionization energy of the atom and $\Delta E_{a,k}$ and $\Delta E_{i,k}$ are the differences between the kth and ground-state energies. It is assumed that $m_a = m_i = m_h$.

The preceding equation is referred to as the Saha equation.

6.13.4 Kinetics of Energy Exchange Between Electrons and Heavier Species

The energy exchange between electrons and heavier species is based on collision frequency (similar to Section 6.5) extended to multicomponent gas, and is written

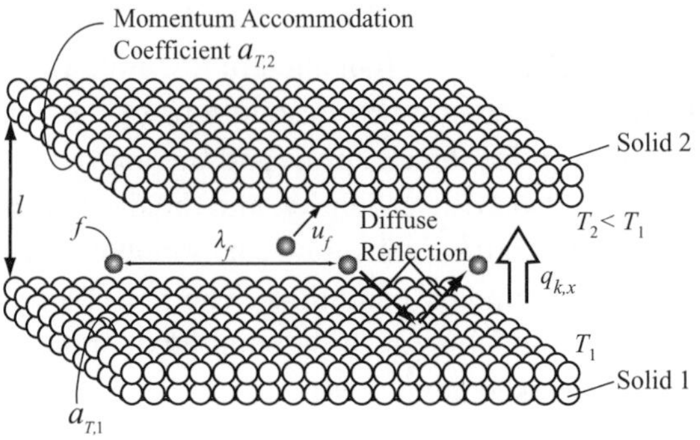

Figure 6.21. Conduction heat transfer across a confined (by two parallel solid surfaces) gas layer where $\mathrm{Kn}_l = \lambda_f/l$ is larger than unity.

as [60, 266]

$$\dot{s}_{e\text{-}h} = \dot{s}_{e\text{-}i} + \dot{s}_{e\text{-}a} \tag{6.211}$$

$$\dot{s}_{e\text{-}i} = \frac{2n_i e_c^4}{3m_n(4\pi\epsilon_o)^2}\left(\frac{8\pi m_e}{k_\mathrm{B}T_e}\right)^{1/2}\frac{1}{T_c}\ln\left[\frac{9(4\pi\epsilon_o k_\mathrm{B}T_e)^3}{4\pi n_{e,c}e_c^6}\right](T_e - T_h) \tag{6.212}$$

$$\dot{s}_{e\text{-}a} = 3k_\mathrm{B}n_{e,c}\frac{m_e}{m_h}\frac{(1-\alpha)\rho_g}{m_h}\left(\frac{8\pi m_e}{k_\mathrm{B}T_e}\right)^{1/2}A_{ea}(T_h - T_e), \tag{6.213}$$

where α is the degree of ionization, $A_{ea} = \pi d_{ea}^2/4$ is the electron-neutral collision cross-section area [157], and d_{ea} is the effective equivalent diameter for this collision (6.74). The collision frequency for electrons (and similarly for ions and neutrals) is expressed using (6.81) and is written as [157]

$$\frac{1}{\tau_e} = \sum_{k=i,a} n'_k A_{ek}(\langle u_e\rangle^2 + \langle u_k\rangle^2)^{1/2}\frac{2m_{ek}}{m_e}, \tag{6.214}$$

from (6.83), $m_{ek} = m_e m_k/(m_e + m_k)$. Empirical relations are used for the cross-section areas, including [157], i.e.,

$$A_{ea}(\mathrm{cm}^2) \simeq (-0.35 + 0.775 \times 10^{-4}T_e) \times 10^{-18}. \tag{6.215}$$

Numerical examples are given in [60].

6.14 Size Effects

6.14.1 Gas Thermal Conductivity in Narrow Gaps

At low gas densities or in small gaps, the probability of collision of fluid particles with the confining solid becomes larger than that of the interparticle collision. This is depicted in Figure 6.21 for two parallel plates, separated by a distance l, confining a gas in thermal motion (no net flow). Using the momentum accommodation coefficient (6.150), the monatomic, free-molecular (fm) regime effective gas thermal

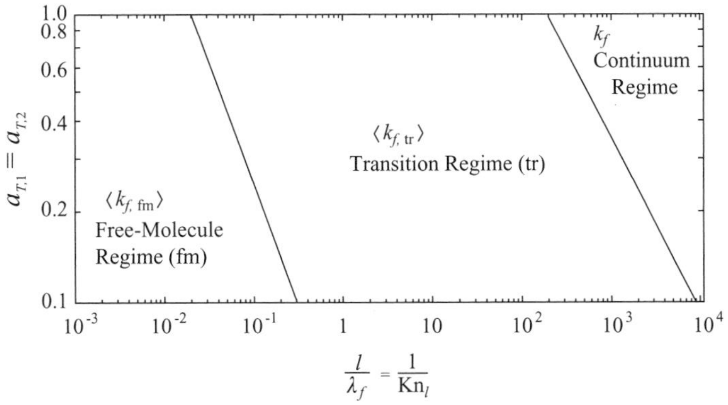

Figure 6.22. Classification of free-molecular, transition, and continuum regimes for heat conduction across a gas confined between parallel plates for $a_{T,1} = a_{T,2}$ [317].

conductivity is [177, 316]

$$\frac{\langle k_{f,fm} \rangle}{l} = \frac{p(c_{v,f} + \frac{R_g}{2})}{\frac{1}{a_{T,1}} + \frac{1}{a_{T,2}} - 1} \frac{1}{(2\pi M R_g T)^{1/2}} \tag{6.216}$$

monatomic gas effective thermal conductivity in fm regime,

where $c_{v,f}$ is the molar heat capacity. This Knudsen number regime is $\mathrm{Kn}_l = \lambda_f/l \rightarrow \infty$, so no Knudsen number appears. For the transition regime (intermediate Knudsen number), the preceding relation is modified as [317]

$$\langle k_{f,t} \rangle = \langle k_{f,fm} \rangle (1 + \frac{4}{15} \frac{1}{\mathrm{Kn}_l} \frac{a_{T,1} a_{T,2}}{a_{T,1} + a_{T,2} - a_{T,1} a_{T,2}})^{-1} \tag{6.217}$$

monatomic gas effective thermal conductivity in transition regime.

This includes the Knudsen number effect and the range of λ_f/l (and momentum accommodation coefficients $a_{T,1}$ and $a_{T,2}$), where (6.216) (free-molecular regime), (6.217) (transition regime), and (6.126) (continuum regime) are valid, and are shown in Figure 6.22. Note that $\mathrm{Kn}_l = 1$ is in the transition regime.

For fluid particle–solid surface pairs with small a_T (Table 6.6), the transition to the continuum regime occurs at higher $1/\mathrm{Kn}_l$, i.e., higher pressure [from (6.87)]. The small thermal accommodation coefficients would generally lead to smaller effective gas thermal conductivity. For example, for Ne–W pair, the transition to continuum regime occurs around 10^7 Pa (at 300 K), due to long mean free path and smaller a_T.

Figure 6.23 shows the variation of the effective thermal conductivity of Ar gas between two W plates separated by 1 μm, at 300 K, as a function of the gas pressure. The thermal accommodation coefficient is from Table 6.6. Note that $\langle k_{f,\mathrm{fm}} \rangle$ increases linearly with p, while $\langle k_{f,\mathrm{tr}} \rangle$ has a middle exponential behavior bounded by a linear and a plateau behavior. There is a slight jump at the continuum regime boundary. For the continuum regime, we have used $k_f = \beta n_f c_{v,f} \langle u_f \rangle \lambda_f/2$, where $\beta = 5/2$ for monatomic gases [326]. Note that for this example, p for the beginning of the continuum regime is over 100 atm.

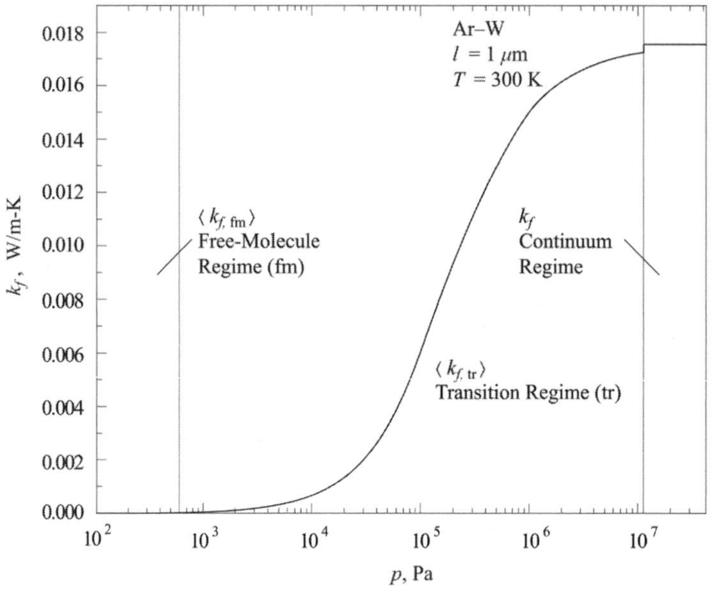

Figure 6.23. Variation of the effective thermal conductivity of Ar gas occupying spacing between two W plates separated by 1 μm, at $T = 300$ K, as a function of pressure.

6.14.2 Thermal Creep (Slip) Flow in Narrow Gaps

Thermal creep (slip) flow of a gas along a surface (nonadsorbing gas, i.e., no gas-surface interactions) with a temperature gradient results in flow from the colder region toward the higher-temperature region. In a tube with a relatively small temperature gradient along its surface, the density does not change appreciably over points a few mean free paths apart. Then the molecules striking it from one direction (i.e., the warmer side) supply it with an impulse proportional to the average thermal speed corresponding to the higher temperature, and those from the other direction give it an impulse proportional to the average thermal speed corresponding to the lower temperature. This difference in impulse (force) acting on the wall results in the differential force on the layer of the gas closer to the wall, thus resulting in the streaming of molecules from the cold to the hot end of the tube. Because the streaming is from the colder side toward the warmer side, the momentum that is due to the streaming gas would also reach the wall, and this would be opposite to the direction of increasing temperature, because it is in this direction that the wall acts on the nearest gas layers with their tangential forces.

Consider a temperature gradient along the x direction on the surface, given by $\mathrm{d}T/\mathrm{d}x$. Using the velocity equilibrium distribution function (6.61), which includes n_f, the probability distribution functions at a reference temperature T_o and at the varying temperature $T(x)$, we have

$$f_f^\mathrm{o}(T) = n_f\left(\frac{m}{2\pi k_\mathrm{B} T}\right)^{3/2} \exp\left(-\frac{mu_f^2}{2k_\mathrm{B} T}\right) \qquad (6.218)$$

$$f_f^\mathrm{o}(T_\mathrm{o}) = n_{f,\mathrm{o}}\left(\frac{m}{2\pi k_\mathrm{B} T_\mathrm{o}}\right)^{3/2} \exp\left(-\frac{mu_f^2}{2k_\mathrm{B} T_\mathrm{o}}\right) \qquad (6.219)$$

or

$$\frac{f_f^{\rm o}(T)}{f_f^{\rm o}(T_{\rm o})} = \frac{n_f}{n_{f,\rm o}}(\frac{T_{\rm o}}{T})^{3/2}\exp[\frac{mu_f^2}{2k_{\rm B}}(\frac{1}{T_{\rm o}} - \frac{1}{T})]. \qquad (6.220)$$

Now using a linear variation in the temperature, we have

$$T = T_{\rm o}(1 + \frac{1}{T}\frac{{\rm d}T}{{\rm d}x}x) \qquad (6.221)$$

$$n_f = n_{f,\rm o}(1 + \frac{1}{T}\frac{{\rm d}T}{{\rm d}x}x)^{-1} \text{ from } p = n_f k_{\rm B}T. \qquad (6.222)$$

Using these, we have

$$\begin{aligned}\frac{f_f^{\rm o}(T)}{f_f^{\rm o}(T_{\rm o})} &= \frac{1}{1+\frac{1}{T}\frac{{\rm d}T}{{\rm d}x}x}\frac{1}{(1+\frac{1}{T}\frac{{\rm d}T}{{\rm d}x}x)^{3/2}}\exp[\frac{mu_f^2}{2k_{\rm B}T_{\rm o}}(1 - \frac{1}{1+\frac{1}{T}\frac{{\rm d}T}{{\rm d}x}x})]\\ &\simeq (1 - \frac{5}{2}\frac{1}{T}\frac{{\rm d}T}{{\rm d}x}x)\exp(\frac{1}{T}\frac{{\rm d}T}{{\rm d}x}x\frac{mu_f^2}{2k_{\rm B}T})\\ &\simeq 1 - \frac{5}{2}\frac{1}{T}\frac{{\rm d}T}{{\rm d}x}x + \frac{1}{T}\frac{{\rm d}T}{{\rm d}x}x\frac{mu_f^2}{2k_{\rm B}T}\\ &= 1 + \frac{1}{T}\frac{{\rm d}T}{{\rm d}x}x(\frac{mu_f^2}{2k_{\rm B}T} - \frac{5}{2}),\end{aligned}$$

where the Taylor series expansion is used twice.

Now the disturbed distribution function is given in terms of the equilibrium distribution function and a perturbation parameter ϕ

$$\begin{aligned}f_f &= f_f^{\rm o}(T)(1+\phi)\\ &= f_f^{\rm o}(T_{\rm o})[1 + \frac{1}{T}\frac{{\rm d}T}{{\rm d}x}x(\frac{mu_f^2}{2k_{\rm B}T} - \frac{5}{2})](1+\phi)\\ &\simeq f_f^{\rm o}(T_{\rm o})[1 + \frac{1}{T}\frac{{\rm d}T}{{\rm d}x}x(\frac{mu_f^2}{2k_{\rm B}T} - \frac{5}{2}) + \phi].\end{aligned} \qquad (6.223)$$

Next the gas BTE (6.102) becomes

$$\begin{aligned}\boldsymbol{u}_f \cdot \nabla f_f &= \frac{\partial f}{\partial t}|_s\\ u_{f,r}\frac{\partial f_f}{\partial r} + u_{f,x}\frac{\partial f_f}{\partial x} &= \frac{\partial f}{\partial t}|_s\\ u_{f,r}\frac{\partial \phi}{\partial r} + u_{f,x}\frac{1}{T}\frac{{\rm d}T}{{\rm d}x}(\frac{mu_f^2}{2k_{\rm B}T} - \frac{5}{2}) &= \frac{\partial \phi}{\partial t}|_s = -\frac{\phi}{\tau_{f\text{-}f}}.\end{aligned} \qquad (6.224)$$

Next, the solution to (6.224) is found and after lengthy steps, the thermal slip (creep) speed at the surface is given by [177, 217]

$$u_{i,T} = \sigma_T\frac{\mu_f}{p}\frac{R_g}{M}\frac{{\rm d}T}{{\rm d}x} \quad \text{thermal slip (creep) speed } u_{i,T} \text{ and constant } \sigma_T, \qquad (6.225)$$

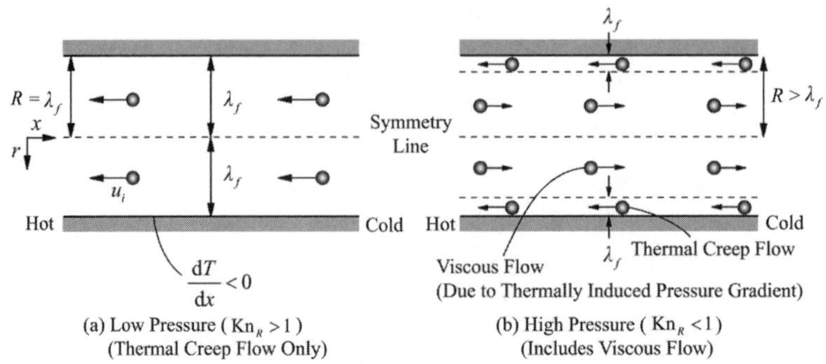

Figure 6.24. (a) Knudsen-, and (b) viscous-flow regimes of the thermal creep flow (temperature gradient along the tube surface)

where σ_T is the thermal slip constant. Then using (6.92) for μ_f, we have

$$u_{i,T} = \frac{\sigma_T}{3}\langle u_f^2\rangle^{1/2}\lambda_f\frac{1}{T}\frac{\mathrm{d}T}{\mathrm{d}x},\quad \sigma_T = \frac{3}{4}\quad \text{Kennard thermal slip constant} \qquad (6.226)$$

Kennard [177] derives a relation that gives $\sigma_T = 3/4$.

As an example for the thermal slip speed of air, at $T = 298$ K and $p = 1.013 \times 10^5$ Pa, we have $\langle u_f^2\rangle^{1/2} = 413.7$ m/s, and $\lambda_f = 64.0 \times 10^{-9}$ m, and then $u_{i,T} = 2.221 \times 10^{-8}\mathrm{d}T/\mathrm{d}x$. On reducing the pressure, for example to 1.013×10^2 Pa, this speed increases by a factor of 10^3. Further refinement of (6.225) is reported in [217] and [304]. In the latter, a value of 1.175 is computed for σ_T, with a relatively small dependence on the mixture composition.

For flow through the tube of radius R, for the case of high $\mathrm{Kn}_R = \lambda_f/R$ (R is the tube radius), we have only the thermal creep flow, and for small Kn_R, we also have the viscous-flow contribution. Figures 6.24(a) and (b) show these two flow regimes. Now the combined the thermal slip (6.226) and the Knudson slip (6.153) for radial geometry give the total velocity slip at the tube wall as

$$u_i = -\alpha_u\frac{\partial \overline{u}_{f,x}}{\partial r} + \sigma_T\frac{\mu_f}{p}\frac{R_g}{M}\frac{\mathrm{d}T}{\mathrm{d}x}\quad \text{at } r = R. \qquad (6.227)$$

The momentum equation is given in Table 3.10, and for $\bar{u}_{f,x}$ and $\mathrm{d}p/\mathrm{d}x$ (using the symmetry at the center line, $r = 0$), we have

$$\bar{u}_{f,x} = -\frac{1}{4\mu_f}(R^2 - r^2 + 2\alpha_u R)\frac{\mathrm{d}p}{\mathrm{d}x} + \sigma_T\frac{\mu_f}{p}\frac{R_g}{M}\frac{\mathrm{d}T}{\mathrm{d}x}. \qquad (6.228)$$

The fluid flow rate is

$$\dot{M}_f = -\frac{\pi}{8}\frac{R^4 pM}{\mu_f R_g T}(1 + 4\frac{\alpha_u}{R})\frac{\mathrm{d}p}{\mathrm{d}x} + \sigma_T\pi\frac{\mu_f R^2}{T}\frac{\mathrm{d}T}{\mathrm{d}x}. \qquad (6.229)$$

Note that for $\dot{M}_f = 0$ we have (for nonadsorbing gas), (6.229) giving the induced pressure gradient as

$$\frac{\mathrm{d}p}{\mathrm{d}x} = \frac{8\sigma_T \mu_f^2 R_g}{MR^2 p(1 + 4\frac{\alpha_u}{R})} \frac{\mathrm{d}T}{\mathrm{d}x} \quad \text{thermally induced pressure gradient for } \dot{M}_f = 0. \tag{6.230}$$

This thermally induced pressure gradient (in the viscous-flow regime) returns the thermal creep flow, resulting in the zero net flow rate. Note that $\mathrm{d}p/\mathrm{d}x$ follows $\mathrm{d}T/\mathrm{d}x$, i.e., pressure is lower where the temperature is lower.

6.15 Problems

Problem 6.1

Derive the specific heat capacity relation (6.22) using the definitions for the ideal gas and the diatomic molecule, translational (6.10), vibrational (6.14), and rotational (6.16) partition functions. For rotational partition function, an approximation $\ln[1 + T_{f,r}/3T + \cdots] \cong T_{f,r}/3T + \cdots$. can be used, but is not necessary.

Problem 6.2

Show that the dense fluid total partition function is (6.36), using the partition function for the individual energies.

Problem 6.3

Derive van der Waals equation of state (6.38), using the dense-fluid total partition function.

Problem 6.4

Derive the relation for the specific heat capacity (at constant pressure) for the van der Waals dense fluid (6.40), using (6.36), (6.38), and (6.39).

Problem 6.5

Using $f_f^{\mathrm{o}}(u_f)$, show that the ideal-gas average thermal speed is $\langle u_f \rangle = (8k_{\mathrm{B}}T/\pi m)^{1/2}$ i.e., (6.63).

Problem 6.6

(a) Using one-dimensional motion, show that the relative speed of colliding particles remains the same before and after elastic collision, i.e., show (6.69) using the momentum and energy conservation during the collision.

(b) Derive (6.79) for the particle collision rate, by completing the integration in (6.75). Take $\boldsymbol{u}_{f,1} \cdot \boldsymbol{u}_{f,2} = 0$ for a general treatment.

(c) Determine the collision rate and the mean free path for ideal gas air at $p = 1$ atm and $T = 300$ K, using the molecular collision diameter from Table 6.4.

Note that the integration over ψ and ϕ are separated and, show that this integral is equal to π.

Problem 6.7

Show that the integral (6.119) will give the same result (6.123) when we use (6.64) for the particle equilibrium distribution function and take the integral directly over the energy (instead of the momentum space).

Problem 6.8

(a) Use the relation for the thermal conductivity of a polyatomic, ideal gas to predict the thermal conductivity of air at $T = 300$ K and $p = 1$ atm. Compare this with the available the measured value. The prediction can be improved using experimental values for $c_{v,f}$, $c_{p,f}$, ρ_f, a_s, and λ_f.

(b) Use the Bridgman liquid thermal conductivity relation to calculate the thermal conductivity of liquid water at $T = 300$ K, $p = 1$ atm.

Problem 6.9

(a) Use the three-dimensional liquid MD code and plot the trajectory of a fluid particle (start near the center of the computational cube).

(b) Plot the velocity distribution function (use x-direction velocity and divide the negative and positive velocities into small bins and enter the number of particles with velocity in a particular bin and then plot this number versus velocity).

(c) Comment on the distribution found in part (b) and compare with the M–B distribution function (6.77).

Problem 6.10

Determine the thermophoretic force on an alumina (Al_2O_3) particle with diameter of 100 nm in air at $T = 800$ K and a temperature gradient of 500 K/mm.

Problem 6.11

Using (6.24) and Table 6.2, plot $Mc_{v,f}/R_g$ for CH_4 (methane) as a function of temperature and comment on its molecular degrees of freedom of motion changing with temperature.

Problem 6.12

(a) Show that (6.34) is the result of integration (6.33).

(b) Assuming $d = \sigma_{LJ}$ and using Table 6.3, evaluate a and b for H_2O and compare with the tabulated values (Table 6.4). Comment on the validity of the assumption in $d = \sigma_{LJ}$.

Problem. 6.13

Consider a one-dimensional ideal gas flow along x, with a velocity $U_{f,x}(y) = u_{f,x} + \overline{u}_{f,x}(y)$, where $\overline{u}_{f,x}(y)$ is the average velocity. Assume that

(i) $(\partial f_f/\partial t)_s = -(f_f - f_f^{\rm o})/\tau_{f\text{-}f}$, where $\tau_{f\text{-}f}$ is the relaxation time to restore equilibrium, which is the time between molecular collisions.

(ii) $\partial f_f/\partial y = \partial f_f^{\rm o}/\partial y$,
(iii) no external forces, and
(iv) steady-state condition.
(a) Show that the gas BTE leads to

$$f_f - f_f^{\rm o} = \tau_{f\text{-}f} U_{f,y} \frac{f_f^{\rm o}}{u_{f,x}} \frac{\partial \overline{u}_{f,x}}{\partial y}.$$

(b) Then [using (6.91) for the definition of the viscous shear stress], show that the gas viscosity is (6.91), i.e., $\mu_f = \rho_f \tau_{f\text{-}f} \langle u_f^2 \rangle /3$, where $\rho_f = n_{fm}$ is the density of the gas, and $\langle u_f^2 \rangle^{1/2}$ is the RMS thermal (fluctuation) speed, which is given by $\langle u_f^2 \rangle = \int u_f^2 f_f^{\rm o}(\boldsymbol{u}_f) \mathrm{d}\boldsymbol{u}_f$.

The definitions of viscous shear stress and viscosity are (6.90)

$$\tau_{xy} \equiv \int_{\boldsymbol{u}} m u_{f,x} u_{f,y} f_f^{\rm o} \mathrm{d}\boldsymbol{u}_f. \quad \tau_{xy} \equiv -\mu_f \frac{\partial \overline{u}_{f,x}(y)}{\partial y}.$$

Note that $\langle u_f^2 \rangle = 3\langle u_{f,y}^2 \rangle$ with the isotropic assumption. Also, note that the integral $\int u_{f,x} u_{f,y} f_f^{\rm o} \mathrm{d}\boldsymbol{u}_f$ is zero, i.e.,

$$\int u_{f,x} u_{f,y} f_f^{\rm o} \mathrm{d}\boldsymbol{u}_f = \int f_{f,z}^{\rm o} \mathrm{d}u_z \int u_x f_{f,x}^{\rm o} \mathrm{d}u_x \int u_y f_{f,y}^{\rm o} \mathrm{d}u_y,$$

where $u_{f,x} f_{f,x}^{\rm o}$ is an odd function of $u_{f,x}$, then $\int u_{f,x} f_{f,x}^{\rm o} \mathrm{d}u_{f,x} = 0$. Similarly, $\int u_{f,y} f_{f,y}^{\rm o} = 0$.

Problem 6.14

Complete the integration (6.88) to show the relation (6.89) for the ideal-gas mass flux.

Problem 6.15

(a) Show that integral (6.177) gives the relation (6.178) for the mean-field potential $\langle \varphi_f \rangle$.

(b) Using $n_s = n_f = 2\times10^{28}$ 1/m^3, $\sigma_{\rm LJ,\it f\text{-}s} = 3.0$ Å, and $\epsilon_{\rm LJ,\it t\text{-}s} = 1.7\times10^{-21}$ J, calculate the Hamaker constant in J and in eV.

(c) Plot the disjoining pressure p_d versus the liquid film thickness, using (6.179), for the fluid-solid pair in (b), and for $5 < \delta < 1000$ Å.

Problem 6.16

Consider the effect of nanoparticles (in Brownian motion) on the effective conductivity of host liquid.

(a) Use effective conductivity $\langle k_f \rangle$ relation (6.144), for water at $T = 300$ K, with alumina (Al_2O_3, five atoms per molecule) nanoparticles, and assume the particles are not in a cluster form (i.e., are well-dispersed). Then determine $\langle k \rangle$ for the following conditions. $d_s = 10$ nm, $n_s = 10^{20}$ 1/cm^3, $c_{v,s} = 15Nk_{\rm B}$, number of molecules per particle $N = N_A \rho_s \pi d_s^3/(6M_s)$, where 5 is the number of atoms in Al_2O_3, $\rho_s = 4000$ kg/m^3, and M_s are the density and molecular weight of the particles, $k_s = 36$ W/m-K, and the water properties at 300 K. Use (6.145) for the diffusion thermal relaxation time, and $\tau_\alpha = 0.1\times3c_{v,s}/2\pi d_s k_s$, with $\langle u_s^2 \rangle = k_{\rm B}\, T_f/m_s$.

(b) Repeat for $d_s = 10^2$ nm.

(c) Comment on the conductivity enhancement that is due to nanoparticles in Brownian motion, is it significant? Note that the larger particles do not remain suspended and segregate (separate).

Problem 6.17

(a) Using (6.128), (6.216), and (6.217), for air–Al (even though this is a diatomic gas mixture) at $T = 300$ K, with a gap size $l = 1$ μm, plot k_f as a function of pressure for 10^2¡p¡2×10^7 Pa. Use Table 6.6 for thermal accommodation coefficient, and Figure 6.22 for marking of the continuum, transition, and free-molecular regimes.

(b) Repeat part (a) for Ne–W using $k_f = \frac{1}{2}\beta n_f c_{v,f}\langle u_f\rangle\lambda_f$ [326], $\beta = 5/2$ for monatomic gases..

(c) Comment on the trends and any jumps in the curves.

Problem 6.18

The high thermal conductivity of thermal plasmas (Figure 6.18) is due to the large number density of charged and neutral particles moving at high thermal speeds.

(a) Consider the electron with the largest thermal speed (smallest mass) to dominate, i.e., use (6.203) in $k_f = k_e = n_{e,c}c_{v,e}u_e\lambda_e/3$. For H_2-only thermal plasma at $T_e = 4000$ K, use k from Figure 6.18, and $n_{e,c} = 10^{16}$ 1/cm^3. Then using $c_{v,e} = 3k_B/2$ for free electrons, and u_e from (6.59), determine λ_e.

(b) Using this λ_e, estimate the electron collision diameter d_e from (6.87).

(c) Comment on the magnitude of d_e and the applicability of this analysis.

Problem 6.19

The mean-free-path concept and fluid particle flux can be directly used for fluid particle thermal conductivity (6.126) for ideal gas. The figure shows the direction of conduction heat flow along direction x, where the temperature distribution is given by $T = T(x)$. The smallest distance we can use as our differential (vanishing length) is the average of the mean free path.

(a) Show that the number of particles that cross $x = 0$ moving in the $+x$ direction with a given speed u, making an angle between θ and $\theta + d\theta$ with the x axis, is proportional to $\sin\theta\cos\theta d\theta$.
Hint: use

$$f_f^o(u_{f,i}) = \left(\frac{m}{2\pi k_B T}\right)^{1/2}\exp\left(\frac{-mu_{f,i}^2}{2k_B T}\right),$$

similar to (6.88).

(b) Show that the average distance traveled along the x direction is $2\lambda_f/3$.

(c) Using this fluid particle (molecular) mass flux (6.89) and the Fourier law of conduction (6.103), show that the conduction heat flux is

$$q_{k,x} = \frac{1}{4}mn_f\langle u_f\rangle(e_{f,1} - e_{f,2})$$

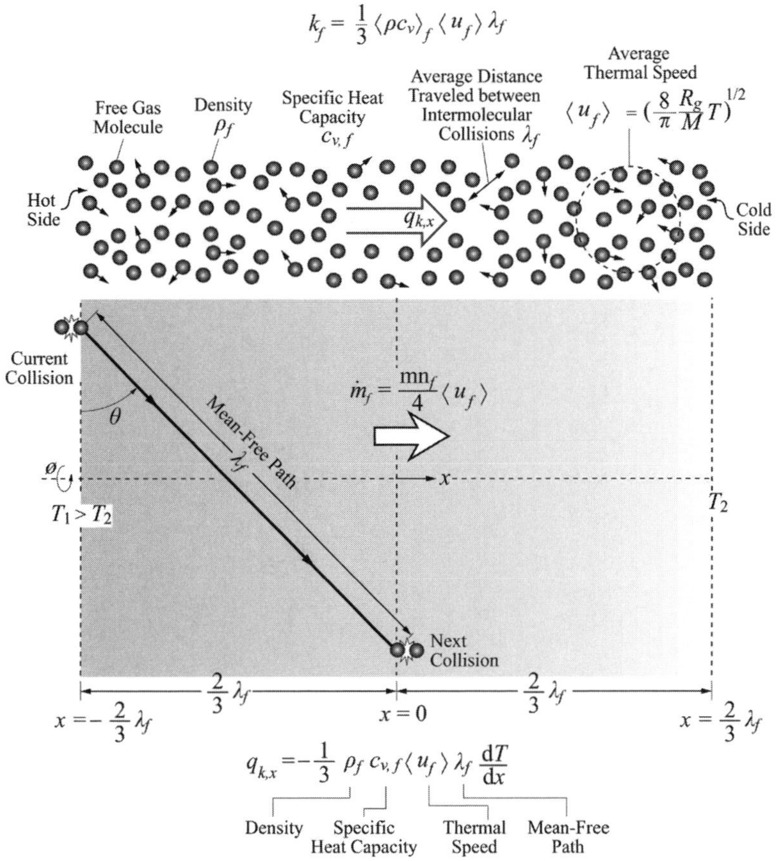

Problem 6.19. Derivation of k_f from kinetic theory of gases.

$$= \frac{1}{4} mn_f \langle u_f \rangle \frac{\mathrm{d}e}{\mathrm{d}T}\frac{\mathrm{d}T}{\mathrm{d}x}(-\Delta x)$$

$$= \frac{1}{4} mn_f \langle u_f \rangle \frac{\mathrm{d}e}{\mathrm{d}T}\frac{\mathrm{d}T}{\mathrm{d}x}(-\frac{4}{3}\lambda_f).$$

(d) Using the definition of $c_{v,f}$ (Table 2.4) show that

$$q_{k,x} = -\frac{1}{3}\rho_f c_{v,f}\langle u_f \rangle \lambda_f \frac{\mathrm{d}T}{\mathrm{d}x}, \quad \rho_f = mn_f. \qquad (6.231)$$

Problem 6.20

(a) Using the available thermodynamics data for saturated water, and (6.93), plot $\dot{S}_{f\text{-}f,lg,max}/A$ as a function reduced pressure p/p_c [as shown in Figure 6.7(a)].

(b) At one atom pressure, the critical heat flux (dryout limit) of a pool of boiling water under normal gravity is about 1.2 MW/m^2. What fraction of (a) is this? Comment on the limitation of pool boiling (buoyancy driven flow) and how this boiling critical heat flux can be increased.

Problem 6.21

Using the fluid particle probability distribution function that satisfies the fluid particle number density n_f for the ideal gas,

$$f_f^{\text{o}} = n_f(\frac{m}{2\pi k_{\text{B}}T})^{2/3}\exp(\frac{-mu_f^2}{2k_{\text{B}}T}),$$

which is also listed in footnote of Section 6.4.2, show that $\frac{1}{m^3}\int f_f^{\text{o}}\text{d}\boldsymbol{u}_f = n_f$.

7

Photon Energy Storage, Transport and Transformation Kinetics

The radiative heat flux vector $\boldsymbol{q}_r$ is given in terms of the photon intensity I_{ph} in Table 1.1. The photon intensity, in turn, is determined by the source of radiation (including emission) and its interactions with matter (including absorption, reemission, and elastic scattering) as it travels at the speed of light in that matter. In addition to thermal emission, which is related to temperature of matter and is generally random in direction over a wide range of wavelengths, there are other stimulated and ordered emissions. Photons are also central in a wide range of energy conversions $\dot{s}_{i\text{-}j}$ (e.g., solar, flames, lasers). In this chapter, we examine various photon emission, absorptions, scatterings, and other interactions. These interactions are strongly dependent on the photon energy $\hbar\omega$, where ω is the angular frequency.

We refer to a propagating, coherent, EM wave and its energy (as described by the Maxwell equations), as well as a quantized wave packet or quasi-particle, both as photons. Historically, radiative heat transfer had been constructed assuming broadband radiation, but with the emergence of lasers, very narrow or discrete photon energy has become common. Here we use discrete, nonequilibrium photon energy distributions, as well as blackbody (Planck law) thermal radiation (as in a photon gas), to treat photon transport and interactions with electronic entities (e.g., electrons in isolated atoms or ions, conduction electrons, valence electrons, molecular dipoles) in matter. We begin by reviewing photon energy and electric energy transitions in matter (atoms or molecules) without first discussing electromagnetic field interactions. These include some interaction (transition) selection rules. However, for a complete quantitative treatment of the interaction, the classical wave nature of a photon is used. This is due to a relative ease of use and the success the classical method had in explaining experimental results.

Blackbody radiation is an important experimental discovery that played a central role in the development of quantum mechanics. Using classical wave theory, the spectral, hemispherical, emissive power is given by the Rayleigh–Jeans law (Sections 1.4 and 7.2). However, it was found that the Rayleigh–Jeans law was applicable only for long wavelengths. The entire spectrum was understood only when Planck suggested that an EM wave with frequency ω exchanges energy with matter in a quantum given by $E_{ph} = \hbar\omega$ (discussed in the quantum–mechanical description of the simple harmonic oscillator in Section 2.6.4). This assumption suggests that light waves have a well-defined energy just as particles do. The derivation of blackbody radiation will be given in the next section.

In the photoelectric experiment light impinges upon a material and electrons are emitted because of the interaction of the light with electrons. Experiments show that, if the frequency of light is below a threshold value, there is no emission of electrons, regardless of intensity. The Planck blackbody radiation was used by Einstein to explain the photoelectric effect (the photoelectric transition rate is reviewed in Section E.3, Appendix E, as an application of the FGR) as light being made of particles with energy $E_{ph} = \hbar\omega$. Electrons are emitted by a single (it can also involve multiple photons, but with much lower probability) photon knocking the electron out; thus the energy per photon must exceed the electron work function.

The quantum treatment of a photon as a harmonic oscillator, along the lines of the treatments given in Sections 2.6.4 and 4.5 (for a phonon), gives the energy of the photon occupation number (probability) $f_{ph}(\boldsymbol{\kappa}, \alpha)$, as

$$E_{ph,\boldsymbol{\kappa},\alpha} = \hbar\omega_\alpha(\boldsymbol{\kappa})[f_{ph}(\boldsymbol{\kappa}, \alpha) + \frac{1}{2}], \tag{7.1}$$

for wave vector $\boldsymbol{\kappa}$ and polarization α. We will further discuss the quasi-particle (quantum) treatment of photon and the radiation field Hamiltonian in Section 7.3.1.

In general, the low-frequency EM waves (for example, microwave) are well treated using the classical wave theory (Maxwell equations), whereas the high-frequency (high-energy) waves are treated using quantum mechanics, when electronic transitions are involved. In this chapter, we start with the quasi-particle treatment, then coherent wave treatment, and make some comparisons.

The photon energy can reach very high values, as shown in Figure 7.1(a), where the various regimes of the EM waves are shown. These are starting from the lowest photon energy, long, medium, and short wave and very-high frequency radio, covering down to wavelength of 1 m, then microwave, infrared, visible, ultraviolet, X-rays, γ-rays, and cosmic rays. This large range for phonons allows for interactions with very low energy entities (such as the rotational energy state in fluid particles) to very high energy entities (e.g., causing ejection of electrons from atoms).

Comparing the energies of different wavelengths (frequencies), we find that $\lambda = 50\,\mu$m (far infrared) has $\omega = 3.770 \times 10^{13}$ rad/s, and $\hbar\omega = 0.02482$ eV (close to $k_B T = 0.02585$ eV for $T = 300$ K, Figure 1.4), $\lambda = 5\,\mu$m (near infrared) has $\omega = 3.770 \times 10^{14}$ rad/s, and $\hbar\omega = 0.2480$ eV (close to the first vibrational transition of N_2, 0.2890 eV, Figure 1.4), $\lambda = 0.5\mu$m (visible green) has $\omega = 3.770 \times 10^{15}$ rad/s, and $\hbar\omega = 2.482$ eV (close to the electronic gap energy of AlSb, 2.50 eV, Table 5.1), $\lambda = 0.05\,\mu$m (ultraviolet) has $\omega = 3.770 \times 10^{16}$ rad/s, and $\hbar\omega = 24.82$ eV (close to first ionization energy of He, 24.58 eV, Table A.2), and $\lambda = 0.005\,\mu$m (X-rays) has $\omega = 3.770 \times 10^{17}$ rad/s, and $\hbar\omega = 248.2$ eV [close to first photoionization energy of Al, Figure 7.1(b)].

Figure 7.1(b) shows examples of spectral (frequency dependent) absorption coefficient $\sigma_{ph,\omega}$ of matter (gas, liquid, and solid, and polymer, oxide, semiconductor,

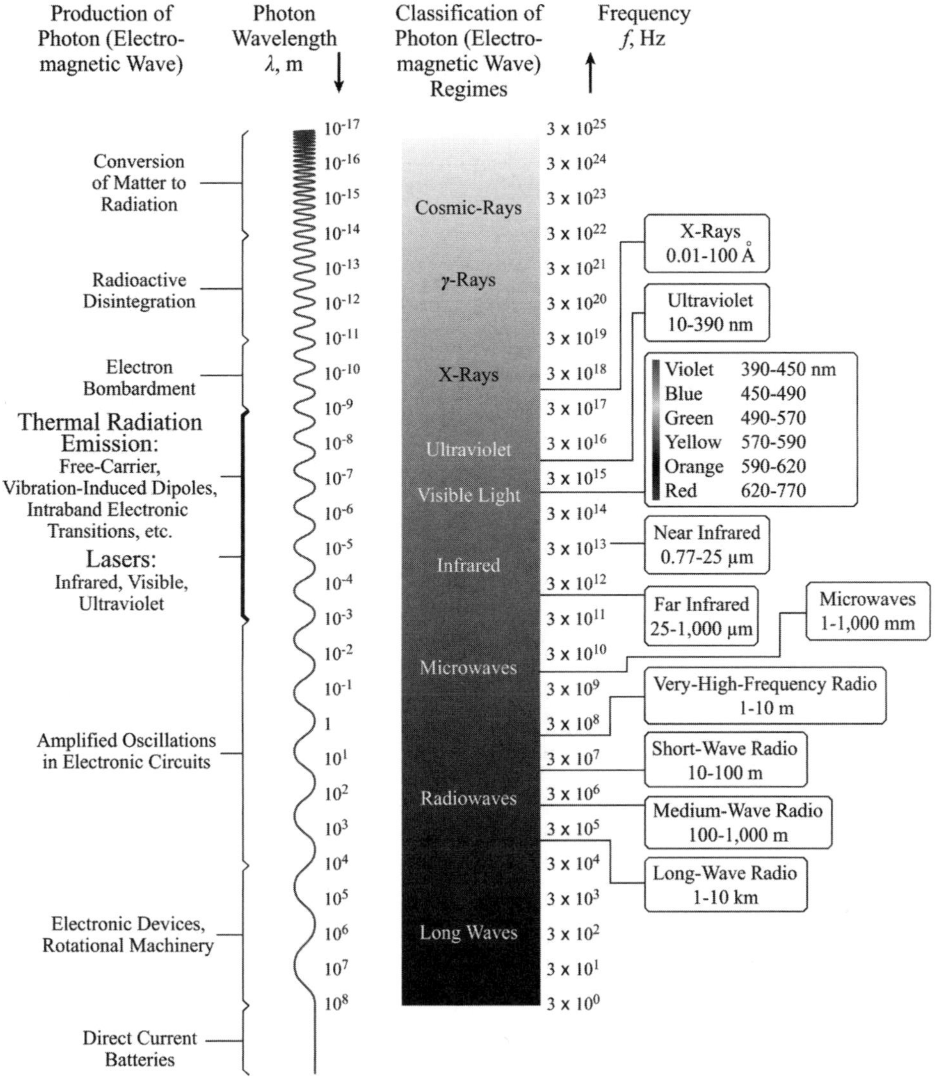

(a) Radiation Spectrum

Figure 7.1. (a) Spectra (wavelength and frequency) of the EM wave (radiation) regimes and their production.

and metal), over a large range of photon energies from the microwave regime to the γ-rays regime. The inverse of $\sigma_{ph,\omega}$ is the spectral photon mean free path $\lambda_{ph,\omega}$. High absorptions are associated with vibrational energy (in gases, liquids, and solids), electronic bandgaps, and free carriers (e.g., electrons).

In this chapter we start with the ideal quasi-particle (photon gas) treatment and the blackbody spectral intensity. Next we examine nonblackbody photon emissions, such as lasers. Then we discuss the quantum and semi-classical treatments of photon–matter interactions. Next we consider two-level electronic transition systems and introduce the Einstein spontaneous and stimulated emission coefficients.

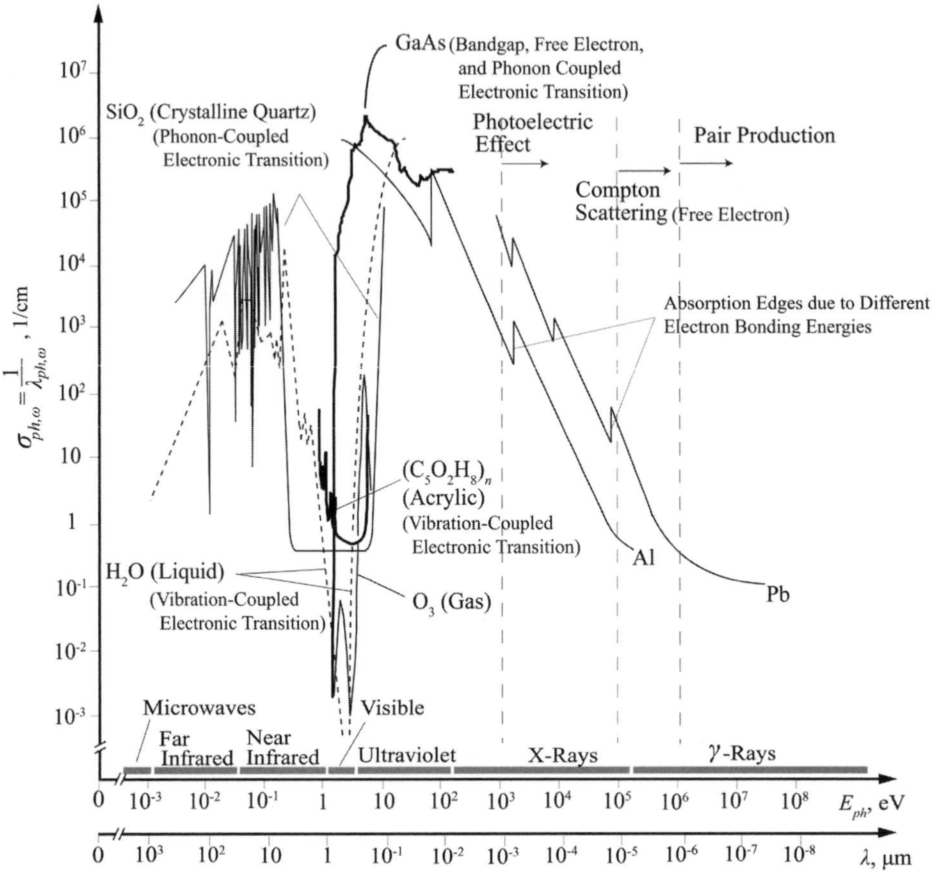

(b) Absorption Spectrum

Figure 7.1(*cont.*) (b). Spectral photon absorption coefficient for typical gas, liquid, and solid phases. For the solid phase, examples of polymer, oxide, semiconductor, and metals are given. The spectral photon mean free path $\lambda_{ph,\omega}$ is the inverse of $\sigma_{ph,\omega}$. As the photon energy progressively increases, it interacts with larger energy entities, starting with low-energy, rotational–vibrational–electronic energy of polyatomic gases, to electron ejection from metals.

Then we use these in the photon BTE to formulate the photon particle transport equation (equation of radiative transport). We will show that, in the optically thick limit, $\lambda_{ph,\omega} \ll L$, where L is the distance traveled by a photon, we can use the Fourier law and define a radiant conductivity. We also examine EM wave treatment of radiation and discuss field localization resulting from coherent interference of the waves as they travel through an ideal, one-dimensional, absorbing multilayer medium. We also contrast this with the particle treatment of this phenomena. Then we examine various photon–matter interactions, including continuous and band absorptions and emissions. Next, examples are given of photon–matter interactions in energy conversion, using laser cooling of solids and gases, and photovoltaics, as examples. Finally, we consider size effects in photon emission and transport.

7.1 Quasi-Particle Treatment: Photon Gas and Blackbody Radiation

Photon particle (or gas) treatment allows us to examine its thermalized (or thermal equilibrium) energy distribution (and also assign pressure and entropy to it as with fluid particle gas). In photon gas the interphoton collisions are absent (however, the wave treatment of the photon allows for interference, etc.), but the photon interacts with electric entities (such as a free electron or oscillating dipole).

We start with the quantum energy of the simple harmonic oscillator, neglecting the zero point ($T = 0$ K, from Figure 1.1, $f_{ph}^{\text{o}} = 0$) energy and expressing the quantum energy as $\hbar\omega$.

For simplicity, we assume only the frequency dependence of the photon energy and write the photon energy (7.1), dropping the 1/2 term, density $e_{ph,\omega}$, similar to (3.38), using the photon distribution function $f_{ph,\omega}$, in frequency range ω to $\omega + \mathrm{d}\omega$ as

$$e_{ph,\omega} = \frac{\mathrm{d}E_{ph,\omega}}{V} = \hbar\omega f_{ph} D_{ph,\omega}\mathrm{d}\omega. \tag{7.2}$$

Similarly, the photon number density is defined.† The blackbody density of state $D_{b,\omega}$ [which is defined similar to (4.19) for phonon and (5.64) for electron] is for dispersionless (linear dispersion relation) phase speed (4.16), which is equal to the group speed (4.15), $u_{ph} = \mathrm{d}\omega/\mathrm{d}\kappa = \omega/\kappa$

$$\begin{aligned} D_b(\kappa) &= \sum_\alpha \int_{\kappa'} \frac{4\pi\kappa^2\mathrm{d}\kappa'}{(2\pi)^3}\,\delta_{\mathrm{D}}(\kappa' - \kappa) \\ D_{b,\omega} &= \frac{1}{\pi^2}\int_{\omega'} \frac{1}{u_{ph}^3}\omega'^2\mathrm{d}(\omega' - u_{ph}\kappa)\delta_{\mathrm{D}}(\omega' - u_{ph}\kappa) \\ &= \frac{\omega^2}{\pi^2 u_{ph}^3}, \qquad \omega = u_{ph}\kappa \quad \text{(dispersion relation)} \\ &\qquad \text{blackbody photon DOS and phase speed,} \end{aligned} \tag{7.3}$$

where the number of polarizations α is 2. Debye phonon DOS (4.26) is based on this, except that Debye introduced the cut-off frequency to constrain the number of oscillators (to $3n$ for monatomic, cubic crystals). Both assume a linear dispersion relation (group and phase velocities are equal and constant). The general dispersion relation for the photon, in terms of the complex index of refraction m_λ ($\kappa = m_\lambda\omega/u_{ph,\text{o}}$) is given in Table 3.5. Then

$$\frac{\mathrm{d}E_{ph,\omega}}{V} = \frac{\hbar\omega^3 f_{ph}\mathrm{d}\omega}{\pi^2 u_{ph}^3}\frac{D_{ph,\omega}}{D_{b,\omega}}. \tag{7.4}$$

† $n_{ph} = \int_0^\infty f_{ph}(\omega)D_{ph}(\omega)\mathrm{d}\omega$. For blackbody emission, this is proportional to T^3 (end-of-chapter problem).

The spectral photon intensity $I_{ph,\omega}$† is related to the photon energy density through (dividing by the solid angle 4π‡ and multiplying by the velocity u_{ph})

$$\begin{aligned} \mathrm{d}I_{ph,\omega}(\omega, T) &= \frac{u_{ph}}{4\pi}\frac{\mathrm{d}E_{ph,\omega}}{V} = \frac{1}{4\pi}u_{ph}\hbar\omega f_{ph}D_{ph}\mathrm{d}\omega \\ &= \frac{u_{ph}}{4\pi}\frac{\hbar\omega^3 f_{ph}\mathrm{d}\omega}{\pi^2 u_{ph}^3}\frac{D_{ph,\omega}}{D_{b,\omega}} \\ &= \frac{\hbar\omega^3 f_{ph}\mathrm{d}\omega}{4\pi^3 u_{ph}^2}\frac{D_{ph,\omega}}{D_{b,\omega}}. \end{aligned} \tag{7.5}$$

The total intensity is found by integrating over angular frequency $I_{ph} = \int \mathrm{d}I_{ph,\omega}$. Similarly, the radiation heat flux is $\mathrm{d}q_{ph} = \mathrm{d}I_{ph}/4$.§ Note that we have not included the angular variations of I_{ph}, although it can be significant and can be included starting from (7.2).

Under equilibrium distribution, $f_{ph} = f_{ph}^{\mathrm{o}}$ (Table 1.2) and for vacuum $u_{ph,\mathrm{o}} = c_{\mathrm{o}}$, the blackbody radiation intensity is given by

$$\begin{aligned} \mathrm{d}I_{b,\omega}(\omega, T) &= \frac{\hbar\omega^3 f_{ph}^{\mathrm{o}}\mathrm{d}\omega}{4\pi^3 u_{ph,\mathrm{o}}^2} \\ &= \frac{\hbar\omega^3}{4\pi^3 u_{ph,\mathrm{o}}^2}\frac{\mathrm{d}\omega}{\exp(\hbar\omega/k_{\mathrm{B}}T) - 1}, \end{aligned} \tag{7.6}$$

† The phonon intensity is defined similarly to (7.5) as

$$\mathrm{d}I_{p,\omega}(\omega) = \frac{u_{p,\mathrm{A}}}{4\pi}\hbar\omega f_p D_p(\omega)\mathrm{d}\omega.$$

‡ The differential solid angle is $\mathrm{d}\Omega = \sin\theta \mathrm{d}\phi\mathrm{d}\theta$, where θ is the polar and ϕ is the azimuthal angle and for azimuthal symmetry $\mathrm{d}\Omega = 2\pi\sin\theta\mathrm{d}\theta = 2\pi\mathrm{d}\cos\theta \equiv -2\pi\mathrm{d}\mu$, where $\mu = \cos\theta$. This is used in Table 1.1. Integration over $-1 \le \mu \le 1$ gives $\Omega = 4\pi$.

§ The radiation heat flux q_{ph} for photon gas is

$$\mathrm{d}q_{ph} = \frac{e_{ph}u_{ph}}{4},$$

which is similar to the mass flux for fluid particles (6.89). Then using (7.2), we have

$$\mathrm{d}q_{ph} = \frac{1}{4}u_{ph}\hbar\omega f_{ph}D_{ph}\mathrm{d}\omega.$$

In general (for directional I_{ph}) the relation between $\mathrm{d}q_{ph}$ and $\mathrm{d}I_{ph}$ is

$$\begin{aligned} \mathrm{d}q_{ph} &= \int_{\Omega=4\pi}\mathrm{d}I_{ph}\mathrm{d}\Omega \\ &= \int_0^{2\pi}\int_0^{\pi/2}\mathrm{d}I_{ph}\cos\theta\sin\theta\mathrm{d}\theta\mathrm{d}\phi, \end{aligned}$$

when the integral is over the solid angle of a unit hemisphere.

or by using the relation $\lambda = 2\pi/\kappa = 2\pi u_{ph,o}/\omega$, we have

$$\mathrm{d}I_{b,\lambda}(\lambda, T) = \frac{\hbar(8\pi^3)u_{ph,o}}{4\pi^3\lambda^3}\frac{2\pi u_{ph,o}\mathrm{d}\lambda/\lambda^2}{\exp(2\pi\hbar u_{ph,o}/\lambda k_\mathrm{B}T) - 1}$$

$$= \frac{4\pi\hbar u_{ph,o}^2}{\lambda^5[\exp(2\pi\hbar u_{ph,o}/\lambda k_\mathrm{B}T) - 1]}\mathrm{d}\lambda \quad \text{Planck law.} \qquad (7.7)$$

The Planck law describes the ideal, quantized (harmonic oscillator) spectral distribution of the EM energy (photon) in thermal equilibrium (called thermal radiation).

The maximum in $I_{b,\lambda}$ ($\partial I_{b,\lambda}/\partial\lambda = 0$) occurs at $\hbar\omega/k_\mathrm{B}T = \hbar\omega/\lambda k_\mathrm{B}T = 4.965$ or $\lambda_{max}T = 2898$ μm-K (called the Wien displacement law), whereas the maximum in $I_{b,\omega}$ ($\partial I_{b,\omega}/\partial\omega = 0$), which gives the most likely frequency, is at $\hbar\omega/k_\mathrm{B}T = 2.821$ (both are end-of-chapter problems).

The blackbody, spectral emissive power $E_{b,\lambda}$ is defined as

$$\mathrm{d}E_{b,\lambda} \equiv \pi\,\mathrm{d}I_{b,\lambda} = \frac{4\pi^2\hbar u_{ph,o}^2}{\lambda^5[\exp(2\pi\hbar u_{ph,o}/\lambda k_\mathrm{B}T) - 1]}\mathrm{d}\lambda$$

$$\text{blackbody spectral emissive power.} \qquad (7.8)$$

This is generally referred to as the the Planck law of blackbody radiation. Almost all (98 %) blackbody thermal emission is between λT =1444 and 22,890 μm-K [174].

This variation of $\mathrm{d}E_{b,\lambda}/\mathrm{d}\lambda$ as a funciton of λ is shown in Figure 7.2(a), for four different temperatures. The Draper point marked there identifies the threshhold of radiation detection by the human eye.

The total, blackbody emissive power E_b is found by integrating the spectral, blackbody emissive power $E_{b,\lambda}$ over the wavelength range ($0 \le \lambda \le \infty$), and this gives

$$E_b(T) = \int_0^\infty E_{b,\lambda}\mathrm{d}\lambda = \sigma_\mathrm{SB}T^4,\ \sigma_\mathrm{SB} = \frac{\pi^2 k_\mathrm{B}{}^4}{60\hbar^3 u_{ph,o}^2} \quad \text{Stefan–Boltzmann law.} \qquad (7.9)$$

This is the Stefan–Boltzmann law, and σ_SB is the Stefan–Boltzmann constant[†], which is $\sigma_\mathrm{SB} = 5.6705\times10^{-8}$ W/m^2-K^4.

For very large wavelengths, the exponential term in (7.8) is approximated by a two-term Taylor series solution (end-of-chapter problem), leading to

$$\mathrm{d}E_{b,\lambda} = \frac{2\pi u_{ph,o}k_\mathrm{B}T}{\lambda^4}\mathrm{d}\lambda \quad \text{Rayleigh–Jeans law for large } \lambda. \qquad (7.10)$$

† Similarly, the phonon Stefan–Boltzmann coefficient (not constant) is arrived at using the Debye model and a single phonon speed, as

$$\sigma_\mathrm{SB}(\text{phonon}) = \frac{\rho c_{v,p}u_{p,\mathrm{A}}}{16T^3}.$$

The same result is obtained using $\hbar \rightarrow 0$ in (7.8), i.e., allowing photons of arbitrary small energy content, as postulated by the classical mechanics. Thus (7.10) is identical to the one derived by Rayleigh and Jeans using the classical mechanics and is called the Rayleigh–Jeans distribution. Because in (7.10), as $\lambda \rightarrow 0$, $\mathrm{d}E_{b,\lambda} \rightarrow \infty$, it is called the ultraviolet catastrophe.

7.2 Lasers (Quantum) and Near-Field (EM Wave) Thermal Emission

7.2.1 Lasers and Narrow-Band Emissions

Laser (light amplification of stimulated emitted radiation) is a device generating or amplifying light (wavelengths ranging from the far-infrared regime to X-rays and γ-rays regimes). The laser beams have high directionality (indicated by their solid angle $\Delta\Omega$, spectral purity (very narrow bands $\Delta\omega$), and intensity. Figure 7.2(b) gives examples of the primary wavelengths emitted by some lasers [307] and covers four orders of magnitude (from ultraviolet to far infrared). Different atoms and atomic combinations (and in different phases and states), creating various quantum energy states and state transition possibilities, are used as the emitting medium. Various excitation mechanisms (energies) are used to create transitions in this medium. Generally mirrors (optical cavity) are used for the amplification (feedback).

Some lasers reach the ideal monochromatic ($\Delta\omega \rightarrow 0$), unidirectional output ($\Delta\Omega \rightarrow 0$), but some, including random lasers, may have significant directional distributions, or might have broadbands. These are illustrated in Figure 7.2(c). As mentioned, various mirrors and cavities are used for creating the directional preference, and various methods are used to create a narrow-band resonance [307].

7.2.2 Classical EM Wave Near-Field Thermal Emission

The fluorescent emission is expected to be random in direction and also has a bandwidth. The radiation entropy aspect of laser cooling has been analyzed in [294]. The radiation that is due to thermally vibrating dipoles and other electric transitions occurring at a distance below the surface and also for an object of linear dimension d or for objects separated by a distance d, when the wavelength of the EM wave λ is nearly the same as the clearance distance $2\pi d/\lambda \leq 1$, the EM field is referred to as the near-field region.† The blackbody radiation discussed in Section 7.1 is the ideal emission in the far-field region, where the distance from the source of emission (local dipoles, etc.) is several times larger than the wavelength. The thermal radiation emission can be presented as the thermal fluctuation of charges in matter. This fluctuating electric field $\boldsymbol{e}_e(\boldsymbol{x}, \omega)$ is given as [251, 280]

$$\boldsymbol{e}_e(\boldsymbol{x}, \omega) = i\omega\mu_o \int_V \boldsymbol{G}_e(\boldsymbol{x}, \boldsymbol{x}', \omega) \cdot \boldsymbol{j}_e(\boldsymbol{x}, \omega)\mathrm{d}\boldsymbol{x}', \qquad (7.11)$$

† $2\pi d/\lambda$ is called the size parameter and is key in the particle scattering theory [172].

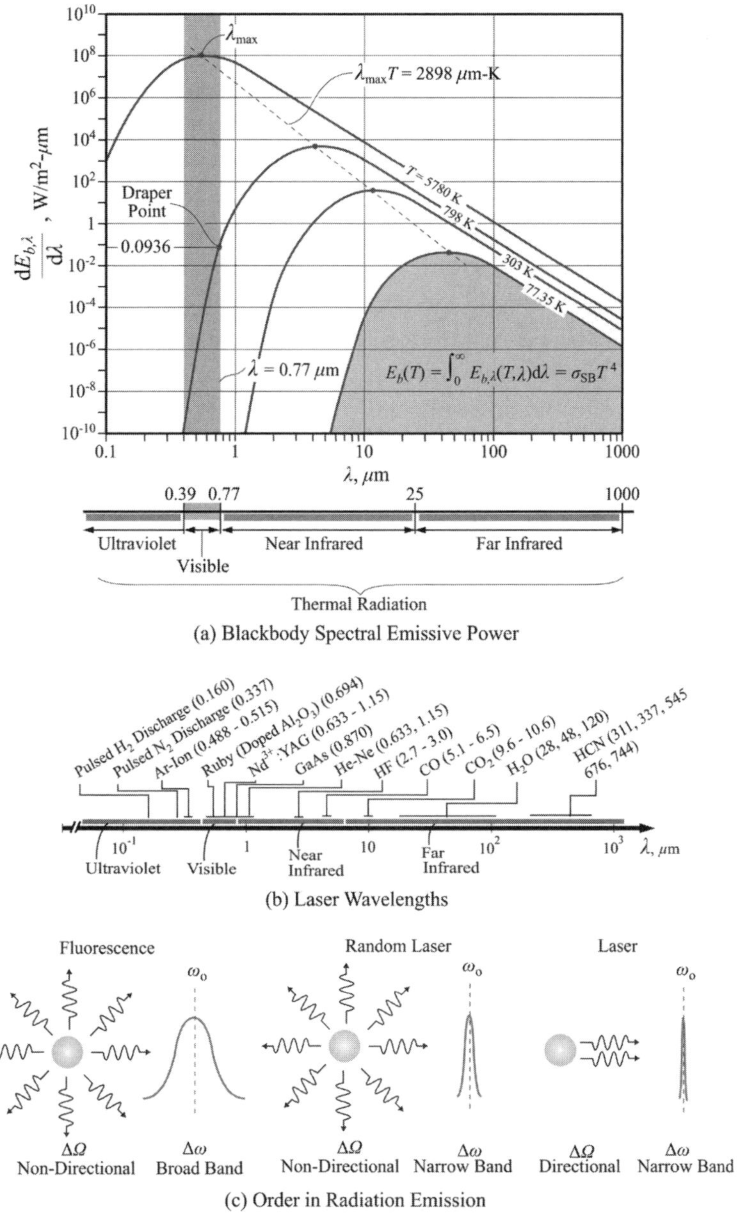

Figure 7.2. (a) Blackbody spectral emission power as a function of wavelength (b) Examples of wavelengths of typical lasers. The numbers in parentheses indicate the wavelength in micrometers (some have multiple wavelengths). (c) Illustration of the solid angle span and bandwidth (around ω_o) of some light sources. The solid angle $\Delta\Omega$ describes the direction of radiation.

where $\boldsymbol{G}_e$ (V/m) is the dyadic Greens function from the source located at $\boldsymbol{x}'$ and $\boldsymbol{j}_e$ is the Fourier transform of the current density vector that is due to thermal fluctuations. The electric field is due to the thermal fluctuations in the dipole moment $\boldsymbol{p}_e(\omega)$. The fluctuation in $\boldsymbol{p}_e$ is represented by a correlation [164], which in turn is related to temperature, frequency, and the imaginary component of the polarizability $\alpha_{e,c}$ (3.41).†

The radiative heat transfer between two spheres is found from the Poynting vector (Table 3.5) normal to the surface and has been numerically calculated in [251]. We subsequently consider the analytical solution in the limit of small clearing distance.

When two dielectric solid surfaces at different temperatures T_1 and T_2 (for example a tip and a surface) are brought very close (submicrometer distance) and in a vacuum, one of the near-field interactions is through fluctuating dipoles in each solid. This is the near-field radiation (Coulomb interaction) dipole–dipole energy transfer (known as the Foster energy transfer [90]). The rate of net radiation energy transfer between a solid 1 and a solid 2 is

$$q_{ph,1\text{-}2}(\omega) = \frac{\omega\epsilon_o}{2}\alpha_{e,c,2}|\boldsymbol{e}_e(\boldsymbol{x})|^2, \quad \alpha_{e,c,j} = \alpha_{e,c,j}(\epsilon_{e,j}) \tag{7.12}$$

where the strongly frequency dependent (at transitions) $\alpha_{e,c,2}$ is the imaginary part of the polarizability (3.41) (ability of an ion to distort the electron of the other ion) of solids 2, and $\boldsymbol{e}_e$ is the incident electric field vector produced by the dipoles excited by lattice vibration. The polarization in turn is related to the electric permittivity ϵ_e, which in turn is a signature of the electric entity of material (Section 3.3.5).‡ It has been shown that (7.12) can be written in terms of temperature T_1 and T_2 as [90]

$$q_{ph,1\text{-}2} = \frac{3}{4\pi^2}\frac{\alpha_{e,c,1}\alpha_{e,c,2}}{d^6}[E_{ph}(\omega, T_1) - E_{ph}(\omega, T_2)], \tag{7.13}$$

where d is the separation distance and photon energy $E_{ph}(\omega, T)$ is

$$E_{ph} = \frac{\hbar\omega}{\exp(\frac{\hbar\omega}{k_B T}) - 1}. \tag{7.14}$$

For d of the order of one nanometer, this energy transfer rate is many orders of magnitude larger than that corresponding to the far-field radiation heat flux for the emission and absorption of photons, which is [90]

$$q_{ph,1\text{-}2} = \frac{\omega^4\alpha_{e,c,1}\alpha_{e,c,2}}{32\pi^3 u_{ph}^2 d^2}[E_{ph}(\omega, T_1) - E_{ph}(\omega, T_2)]. \tag{7.15}$$

Note the vastly different dependence (or lack of) on ω and d for (7.14) and (7.15).

† Writing (3.41) using a polarizability tensor, we have $\boldsymbol{p}_{e,l} = \epsilon_o\alpha_{e,lk}\boldsymbol{e}_{e,k}$, $\boldsymbol{j}_e = -i\omega\boldsymbol{p}_{e,k}$. The correlation function for the dipole moment becomes [164]

$$\langle p_{e,k}(\omega)p_{e,l}^\dagger(\omega')\rangle = \hbar\coth(\frac{\hbar\omega}{2k_B T})\text{Im}[\epsilon_o\alpha_{e,kl}(\omega)]2\pi\delta(\omega - \omega').$$

Also, the mean energy of the harmonic oscillator is [same as (7.1)]

$$\frac{\hbar\omega}{2}\coth\frac{\hbar\omega}{2k_B T} = \hbar\omega[\frac{1}{2} + \frac{1}{\exp(\frac{\hbar\omega}{k_B T}) - 1}] = E_{ph}.$$

‡ For spheres of diameter d_s, the relation is

$$\alpha_{e,c}(\omega) = \text{Im}[\alpha_e(\omega)] = 4\pi d_s^3\frac{3\,\text{Im}[\epsilon_e(\omega)]}{[\epsilon_e(\omega) + 2]^2},$$

where $\epsilon_e(\omega)$ is the frequency dependent electric permittivity (3.44).

The near-field radiation heat transfer is further discussed in Section 7.15.1, under sized effects. The experimental results of [181] for a tip-sample (scanning thermal microscopy) distance below 10 nm show deviation from the preceding theory. These are due to local variations in material-dependent properties, for which the macroscopic semiclassical (Maxwell) description of dielectric properties may fail.

7.3 Quantum and Semiclassical Treatments of Photon–Matter Interaction

The formalism of quantized radiation field, based on quantum operators, is introduced in the next sections. Then we will discuss the photon–matter interaction by using these operators. These are needed for the photon–electron interactions (i.e., Section 7.12).

7.3.1 Hamiltonians of Radiation Field

Quantization of electrodynamics begins by presenting the EM field by the vector potential [296]. Using the EM field described by the vacuum Maxwell equations summarized in Table 3.5, we use the vector potential $\boldsymbol{a}_e$ with the radiation or Coulomb gauge ($\nabla \cdot \boldsymbol{a}_e = 0$, also called transversality condition, as the transverse component of $\boldsymbol{a}_e$ accounts for EM radiation of a moving charged particle) to arrive at the current for Ampere law written as (Chapter 3 problem) [75]

$$\nabla^2 \boldsymbol{a}_e(\boldsymbol{x}, t) - \frac{1}{u_{ph,\mathrm{o}}^2} \frac{\partial^2}{\partial t^2} \boldsymbol{a}_e(\boldsymbol{x}, t) = 0, \tag{7.16}$$

$$\boldsymbol{e}_e = -\frac{\partial \boldsymbol{a}_e}{\partial t}, \qquad \boldsymbol{b}_e = \nabla \times \boldsymbol{a}_e, \tag{7.17}$$

where $\boldsymbol{a}_e$, $\boldsymbol{e}_e$, and $\boldsymbol{b}_e$ are the vector potential and electric field and magnetic field vectors of the EM radiation defined in Section 3.3.1 (Table 3.5).

We assume time–space separations, $\boldsymbol{a}_{e,\alpha}(\boldsymbol{x}, t) = q_\alpha(t)\boldsymbol{a}_\alpha(\boldsymbol{x})$, and use the separation of variables method to find

$$\nabla^2 \boldsymbol{a}_\alpha + \frac{\omega_\alpha^2}{u_{ph,\mathrm{o}}^2} \boldsymbol{a}_\alpha = 0, \qquad \frac{\partial^2 q_\alpha}{\partial t^2} + \omega_\alpha^2 q_\alpha = 0, \tag{7.18}$$

where ω_α is the separation constant (eigenvalues). The solutions are (plane-wave expansion) [296]

$$q_\alpha = q_\alpha e^{-i\omega_\alpha t}, \qquad \boldsymbol{a}_{e,\alpha}(\boldsymbol{x}) = \boldsymbol{s}_{ph,\alpha}\left(\frac{1}{\epsilon_\mathrm{o} V}\right)^{1/2} e^{i\boldsymbol{\kappa}_\alpha \cdot \boldsymbol{x}},$$

$$\boldsymbol{a}_{e,\alpha}(\boldsymbol{x}, t) = |q_\alpha|\left(\frac{1}{\epsilon_\mathrm{o} V}\right)^{1/2} \boldsymbol{s}_{ph,\alpha} e^{i(\boldsymbol{\kappa}_\alpha \cdot \boldsymbol{x} - \omega_\alpha t)}, \tag{7.19}$$

where V is the interaction volume, $\boldsymbol{s}_{ph,\alpha}$ is the unit polarization vector for the α mode and $\kappa_\alpha = \omega_\alpha u_{ph}$ is the wave vector. Note that, unlike EM waves generated by a point charge (i.e., electron having mass and charge), EM wave in vacuum (free space) does not have mass or charge. Therefore the vector potential is written in a quantity per an imaginary particle having the mass and charge of an electron.

Using the preceding, the Coulomb gauge condition gives

$$\nabla \cdot \boldsymbol{a}_{e,\alpha} = 0 = \boldsymbol{s}_{ph,\alpha} \cdot \boldsymbol{\kappa}_\alpha i(\frac{1}{\epsilon_o V})^{1/2} e^{i\boldsymbol{\kappa}_\alpha \cdot \boldsymbol{x}} q_\alpha e^{i\omega_\alpha t}. \quad (7.20)$$

From the preceding, we note that $\boldsymbol{s}_{ph,\alpha} \cdot \boldsymbol{\kappa}_\alpha = 0$; therefore the polarization and propagation vectors are perpendicular (i.e., the field is composed of transverse waves only).

Summing over all modes $\boldsymbol{a}_{e,\alpha}$ (by including negative κ_α and ω_α), (7.19) becomes

$$\boldsymbol{a}_e(\boldsymbol{x}, t) = \sum_\alpha [q_\alpha(t)\boldsymbol{a}_{e,\alpha}(\boldsymbol{x}) + q_\alpha^\dagger(t)\boldsymbol{a}_{e,\alpha}^\dagger(\boldsymbol{x})]. \quad (7.21)$$

Using the preceding, $\boldsymbol{e}_e$ and $\boldsymbol{b}_e$,for a plane wave with polarization along the y direction and motion along the x direction (for example) become [from (7.17)]

$$\begin{aligned}
\boldsymbol{e}_{e,\alpha} &= \frac{i\omega}{u_{ph,o}}(q_\alpha \boldsymbol{a}_{e,\alpha} - q_\alpha^\dagger \boldsymbol{a}_{e,\alpha}^\dagger) \\
&= \frac{i\omega}{u_{ph,o}}|q_\alpha|(\frac{1}{\epsilon_o V})^{1/2}\boldsymbol{s}_y[e^{i(\kappa_{\alpha x}x - \omega_\alpha t)} - e^{-i(\kappa_{\alpha x}x - \omega_\alpha t)}] \\
&= -\kappa_{\alpha x}|q_\alpha|\boldsymbol{s}_y(\frac{1}{\epsilon_o V})^{1/2} 2\sin(k_{\alpha x}x - \omega_\alpha t) \quad (7.22)
\end{aligned}$$

$$\begin{aligned}
\boldsymbol{b}_{e,\alpha} &= \boldsymbol{s}_z \frac{\partial a_{e,y}}{\partial y} \\
&= \boldsymbol{s}_z i\kappa_{\alpha x}|q_\alpha|(\frac{1}{\epsilon_o V})^{1/2}\boldsymbol{s}_z[e^{i(\kappa_{\alpha x}x - \omega_\alpha t)} - e^{-i(\kappa_{\alpha x}x - \omega_\alpha t)}] \\
&= -\kappa_{\alpha x}|q_\alpha|\boldsymbol{s}_z(\frac{1}{\epsilon_o V})^{1/2} 2\sin(\kappa_{\alpha x}x - \omega_\alpha t), \quad (7.23)
\end{aligned}$$

where $\boldsymbol{s}_x$, $\boldsymbol{s}_y$, and $\boldsymbol{s}_z$ are the unit vectors. The Hamiltonian (energy) of the radiation field is given by (assuming a vacuum)

$$\begin{aligned}
\text{H} &= \frac{1}{2}\int (\epsilon_o \boldsymbol{e}_e^2 + \mu_o^{-1}\boldsymbol{b}_e^2)\text{d}V = \int \epsilon_o \boldsymbol{e}_e^2 dV \\
&= \frac{1}{4\pi}\int [\sum_\alpha \frac{i\omega_\alpha}{u_{ph,o}}(q_\alpha \boldsymbol{a}_{e,\alpha} - q_\alpha^\dagger \boldsymbol{a}_{e,\alpha}^\dagger)][\sum_{\alpha'} \frac{i\omega_\alpha'}{u_{ph,o}}(q_{\alpha'} \boldsymbol{a}_{e,\alpha'} - q_{\alpha'}^\dagger \boldsymbol{a}_{e,\alpha'}^\dagger)]\text{d}V.
\end{aligned}$$

(7.24)

Note that magnitude of the magnetic field is the same as the magnitude of the electric field, as evident in (7.22) and (7.23). Then we have for the integrals

$$\int \boldsymbol{a}_{e,\alpha} \cdot \boldsymbol{a}_{e,\alpha'} dV = \frac{1}{\epsilon_o}\delta_{\alpha,\alpha'}$$

$$\int \boldsymbol{a}_{e,\alpha} \cdot \boldsymbol{a}_{e,\alpha} \mathrm{d}V = 0 = \int \boldsymbol{a}^{\dagger}_{e,\alpha} \cdot \boldsymbol{a}^{\dagger}_{e,\alpha} \mathrm{d}V$$

$$\int \boldsymbol{a}^{\dagger}_{e,\alpha} \cdot \boldsymbol{a}_{e,\alpha} dV = \frac{1}{\epsilon_o}. \tag{7.25}$$

Therefore, the classical Hamiltonian of EM field becomes

$$\begin{aligned} \mathrm{H} &= -\frac{1}{4\pi u^2_{ph,o}} \sum_{\alpha,\alpha'} \int [(q_\alpha q'_\alpha \boldsymbol{a}_{e,\alpha} \cdot \boldsymbol{a}_{e,\alpha'} + q^{\dagger}_\alpha q^{\dagger}_{\alpha'} \boldsymbol{a}^{\dagger}_{e,\alpha} \cdot \boldsymbol{a}^{\dagger}_{e,\alpha'}) \\ &\quad - (q_\alpha q'^{\dagger}_\alpha \boldsymbol{a}_{e,\alpha} \cdot \boldsymbol{a}^{\dagger}_{e,\alpha'} + q^{\dagger}_\alpha q_{\alpha'} \boldsymbol{a}^{\dagger}_{e,\alpha} \cdot \boldsymbol{a}_{e,\alpha'})]\mathrm{d}V \\ &= \epsilon_o \sum_\alpha \omega^2_\alpha 2 q_\alpha q^{\dagger}_\alpha \frac{1}{\epsilon_o} \\ &= 2\sum_\alpha \omega^2_\alpha q_\alpha q^{\dagger}_\alpha. \end{aligned} \tag{7.26}$$

Now we replace q_α and $q^{\dagger}_\alpha$ with the following coordinates and momenta

$$Q_\alpha = q_\alpha + q^{\dagger}_\alpha \tag{7.27}$$

$$p_\alpha = \frac{\partial Q_\alpha}{\partial t} = -i\omega_\alpha(q_\alpha - q^{\dagger}_\alpha). \tag{7.28}$$

Then

$$q_\alpha = \frac{1}{2}(Q_\alpha - \frac{1}{i\omega_\alpha}p_\alpha), \qquad q^{\dagger}_\alpha = \frac{1}{2}(Q_\alpha + \frac{1}{i\omega_\alpha}p_\alpha), \tag{7.29}$$

and (7.26) becomes

$$\mathrm{H} = 2\sum_\alpha \omega^2_\alpha q_\alpha q^{\dagger}_\alpha = \frac{1}{2}\sum_\alpha (p^2_\alpha + \omega^2_\alpha Q^2_\alpha). \tag{7.30}$$

This is the Hamiltonian for the harmonic oscillator.

We have just formulated the dynamics of EM radiation as its mechanical analog, a simple harmonic oscillator. Thus, the EM field can be treated as an ensemble of harmonic oscillators with their frequency and polarization.

In the quantum treatment of the EM field, we treat Q, p, q, and $q^{\dagger}$ as operators, with the following commutation relations (2.72) (Chapter 2 problem)

$$[Q_\alpha, p_{\alpha'}] = i\hbar\delta_{\alpha\alpha'}$$

$$[Q_\alpha, Q_{\alpha'}] = [p_\alpha, p_{\alpha'}] = 0. \tag{7.31}$$

Then (7.31) gives

$$[q_\alpha, q_{\alpha'}^\dagger] = \frac{\hbar}{2\omega_\alpha}\delta_{\alpha\alpha'}$$

$$[c_\alpha, c_{\alpha'}^\dagger] = \delta_{\alpha\alpha'}, \qquad (7.32)$$

where $c^\dagger$ and c are creation and annihilation operators, respectively (Section 2.6.4) and $\delta_{\alpha,\alpha'}$ is the Kronecker delta. The creation and annihilation operators are here defined as

$$c_\alpha = (\frac{2\omega_\alpha}{\hbar})^{1/2} q_\alpha, \qquad c_\alpha^\dagger = (\frac{2\omega_\alpha}{\hbar})^{1/2} q_\alpha^\dagger. \qquad (7.33)$$

Then the Hamiltonian of the EM field (7.30) becomes [similar to those given by (2.85) and (4.50)]

$$\mathrm{H}_{ph} = \sum_\alpha \hbar\omega_\alpha (c_\alpha^\dagger c_\alpha + \frac{1}{2}), \qquad (7.34)$$

similar to (4.49) for phonons. Using these operators, vector potential (7.21) becomes

$$\boldsymbol{a}_e(\boldsymbol{x}, t) = \sum_\alpha (\frac{\hbar}{2\epsilon_o \omega_\alpha V})^{1/2} \boldsymbol{s}_{ph,\alpha}(c_\alpha e^{i\kappa_\alpha\cdot\boldsymbol{x}} + c_\alpha^\dagger e^{-i\kappa_\alpha\cdot\boldsymbol{x}}). \qquad (7.35)$$

We have used $\epsilon_o = 1/\mu_o u_{ph,o}^2$, where ϵ_o is the free-space permittivity and μ_o is the free-space permeability.

The preceding photon Hamiltonian will be used in photon–matter interaction, including in Section 7.12.1, where we develop the theory of laser cooling.

7.3.2 Photon-Matter Interactions

In the semiclassical treatment of EM wave (represented by $\boldsymbol{e}_e$ and $\boldsymbol{h}_e$) interaction with matter, the bulk optical properties of the matter are represented by the complex spectral index of refraction $m_\lambda = n_\lambda - i\kappa_\lambda$ (Table 3.5). Here n_λ is the spectral index of refraction and κ_λ is the spectral index of extinction (loss). The Mie scattering theory is an example of this treatment [178] and results in prediction of scattering and absorption characteristics of the particles. Rayleigh elastic scattering is the limit of this for small particles (size parameter $\pi d/\lambda$ is $\ll 1$, where d is the particle diameter) [172].

In the semiclassical treatment of EM wave–matter interactions, the wave is treated as classical, whereas the atom is treated as a quantum system with discrete EM states undergoing transitions because of absorption of photons. These two treatments are shown in Figure 7.3(a). Figure 7.3(b) shows the classical elastic Rayleigh scattering, with no absorption (the radiation intensity only undergoes scattering), as well as some quantum energy absorption and transitions (inelastic). The examples are fluorescence, in which the photon energy matches the electronic transitions, and Raman scattering, in which the atomic kinetic energy is also excited (in addition to electronic transitions). In the next few sections, the focus will be on the semiclassical treatment of the absorption (and emission), when the detailed electronic

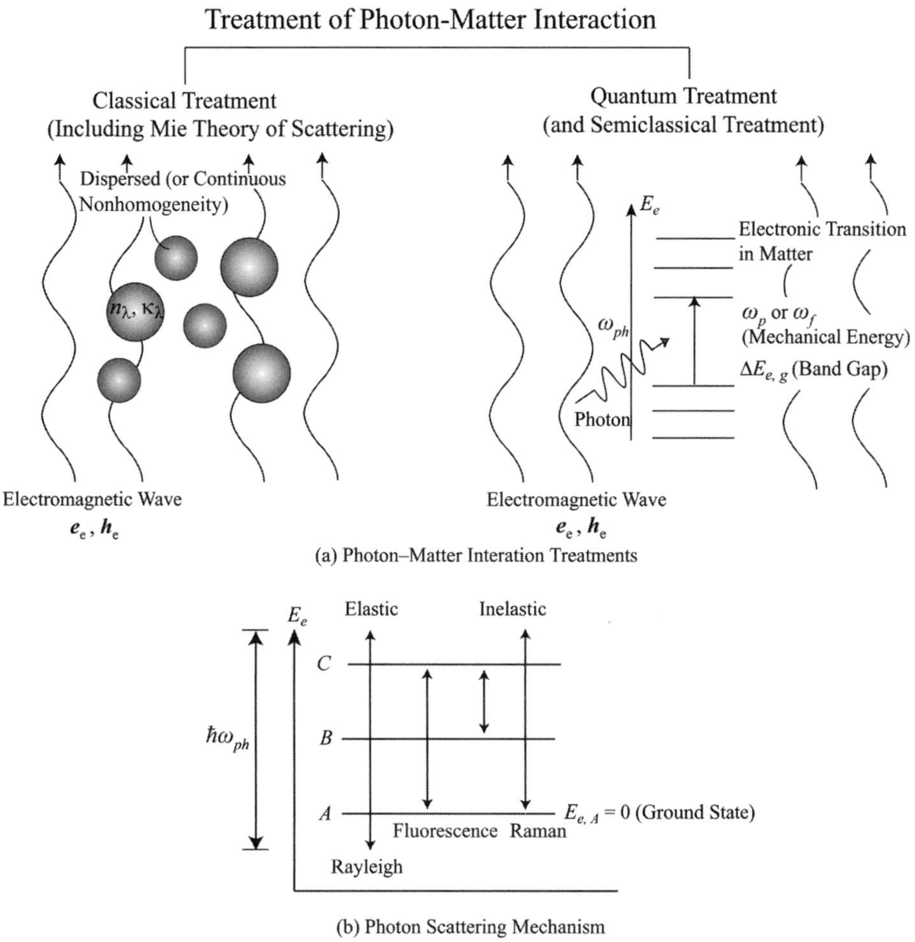

Figure 7.3. (a) Classical (optical constant) treatment and semiclassical (classical EM wave $\boldsymbol{e}_e$ and $\boldsymbol{h}_e$ and discrete electronic energy) treatment of photon–matter interactions. (b) Elastic and inelastic photon–atom interactions showing Rayleigh (no electronic transitions), fluorescence (only electronic transitions) and Raman (electronic and mechanical transitions) scatterings.

transition and the associated atomic kinetic energy changes can be analyzed and used for heating, cooling, or other energy conversions.

In the quantum treatment of EM wave–matter, the photon–electron interaction Hamiltonian becomes [17]

$$\begin{aligned}
\mathrm{H}_{ph\text{-}e} &= -\frac{e_c}{m_e}(a + a^\dagger)\boldsymbol{a}_e \cdot \boldsymbol{p}_e \\
&= -\frac{e_c}{m_e}\left(\frac{\hbar}{2\epsilon_o\omega_\alpha V}\right)^{1/2}(\boldsymbol{s}_{ph,\alpha} \cdot \boldsymbol{p}_e)(a + a^\dagger)(ce^{i\boldsymbol{\kappa}\cdot\boldsymbol{x}} + c^\dagger e^{-i\boldsymbol{\kappa}\cdot\boldsymbol{x}}) \\
&= -\left(\frac{\hbar}{2\epsilon_o\omega_\alpha V}\right)^{1/2}(\boldsymbol{s}_{ph,\alpha} \cdot e_c\frac{\partial \boldsymbol{x}}{\partial t})(a + a^\dagger)(ce^{i\boldsymbol{\kappa}\cdot\boldsymbol{x}} + c^\dagger e^{-i\boldsymbol{\kappa}\cdot\boldsymbol{x}}) \\
&= -\left(\frac{\hbar\omega_\alpha}{2\epsilon_o V}\right)^{1/2}(\boldsymbol{s}_{ph,\alpha} \cdot e_c\boldsymbol{x})(a + a^\dagger)(ce^{i\boldsymbol{\kappa}\cdot\boldsymbol{x}} + c^\dagger e^{-i\boldsymbol{\kappa}\cdot\boldsymbol{x}}),
\end{aligned} \tag{7.36}$$

where $\boldsymbol{s}_{ph,\alpha}$ is the polarization vector of the EM wave, $\boldsymbol{p}_e$ is the dipole moment vector (electric entity here), ω_α is the EM wave frequency, and $a(a^\dagger)$ is the operator describing the internal motion (excitation) of the electron. We have used the commutation relationship† for the transition dipole moment $\boldsymbol{\mu}_e = \langle\psi_f|\boldsymbol{p}_e|\psi_i\rangle = im_e\omega_\alpha\langle\psi_f|\boldsymbol{x}|\psi_i\rangle$.

In a typical atomic transition in the optical regime, the photon wavelength (of the order of several thousand angstrom) is much larger than the linear dimension of the atom (of the order of 1 Å). Then we can use the expansion $e^{-i\boldsymbol{\kappa}\cdot\boldsymbol{x}} = 1 - i\boldsymbol{\kappa}\cdot\boldsymbol{x} - (\boldsymbol{\kappa}\cdot\boldsymbol{x})^2/2 - \ldots$ and take the leading term, which is 1.

This Hamiltonian from the quantum treatment of EM wave–matter interaction will be used in Section 7.12.1, when we develop the interaction rate using the FGR (3.27).

7.4 Photon Absorption and Emission in Two-Level Electronic Systems

As mentioned, the classical description of absorption and emission of photons, based on the complex dielectric constant (and its relationship with the complex index of refraction, is given in [101]). Here we start with a quantum treatment.

We begin the photon absorption by introducing the quantum two-level electronic transitions and the Einstein model for interaction of photon and matter (its electronic structure). Figure 7.4(a) shows the model for absorption and emission in this electron system. When the photon emission is influenced by the absorption of the photon, distinction should be made between the spontaneous photon emission (which is due to fluctuations, including thermal motion) and the stimulated emission (which depends on the absorbed photon). Figure 7.4(b) shows that the stimulated emission follows the direction of incident radiation, whereas the spontaneous emission is random in its angular distribution.

7.4.1 Einstein Population Rate Equation

Consider photon irradiation with frequency ω on an electron system with populations $n_{e,A}$ in the ground state and $n_{e,B}$ in the excited state (because of absorption of photons), as shown in Figure 7.4. The spectral energy density of photon is given by (7.4), and, using this, we find that the population rate equation is

$$\frac{\mathrm{d}n_{e,B}}{\mathrm{d}t} = \underbrace{-\dot{\gamma}_{ph,e,sp}n_{e,B}}_{\text{spontaneous emission}} + \underbrace{\dot{\gamma}_{ph,a}n_{e,A}\frac{\mathrm{d}E_{ph,\omega}}{V\mathrm{d}\omega}}_{\text{stimulated absorption}} - \underbrace{\dot{\gamma}_{ph,e,st}n_{e,B}\frac{\mathrm{d}E_{ph,\omega}}{V\mathrm{d}\omega}}_{\text{stimulated emission}}$$

$$\text{population rate equation.} \quad (7.37)$$

† Note that

$$[\mathrm{H}, \boldsymbol{x}] = \frac{\hbar}{i}\frac{\partial x}{\partial t} = \frac{\hbar}{i}\frac{\boldsymbol{p}}{m_e}$$

$$\frac{\hbar}{im_e}\langle\psi_f|\boldsymbol{p}|\psi_i\rangle = \langle\psi_f|[\mathrm{H}, \boldsymbol{x}]|\psi_i\rangle = \langle\psi_f|\mathrm{H}x - x\mathrm{H}|\psi_i\rangle$$

$$= (E_f - E_i)\langle\psi_f|\boldsymbol{x}|\psi_i\rangle = \hbar\omega_\alpha\langle\psi_f|\boldsymbol{x}|\psi_i\rangle.$$

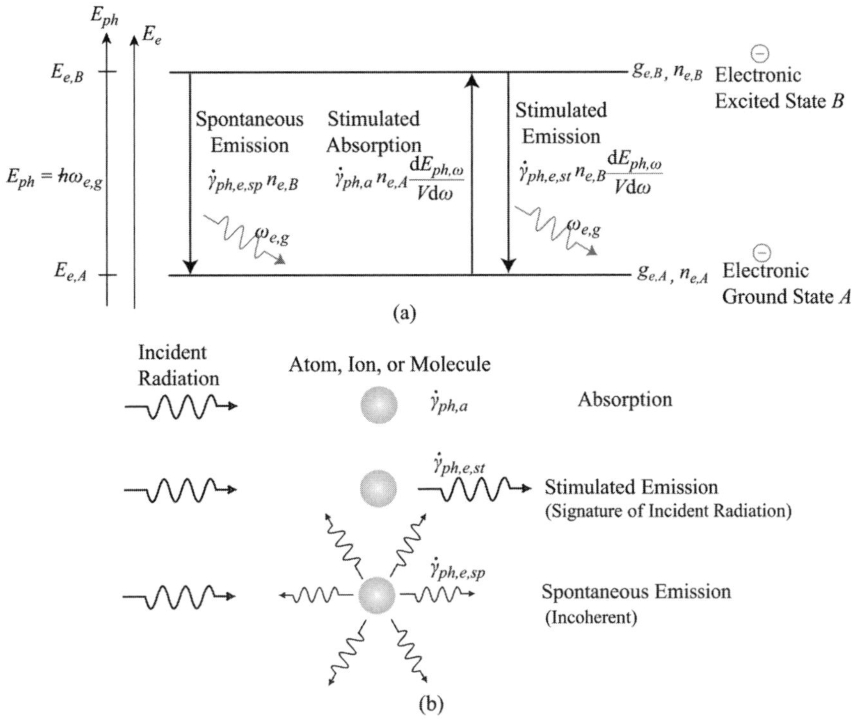

Figure 7.4. (a) Electron transitions and photon emission in an irradiated, two-level electron system. Both spontaneous and stimulated emission are shown, as well as the stimulated absorption. (b) Stimulated emission follows the direction of incident radiation, but spontaneous emission does not.

This is referred to as the Einstein population rate equation [216].[†] Note that $\dot{\gamma}_{ph,e,sp}$ is in units of (1/s) and $\dot{\gamma}_{ph,e,st}$ and $\dot{\gamma}_{ph,a}$ are in (m^3-rad/s)/(J-s). This is needed to make the stimulated absorption and emission proportional to the incident radiation.

The spontaneous emission or relaxation transition is a spontaneous drop from B to A while emitting a photon (EM wave) or phonon at the transition frequency. This is also called fluorescence, and signifies that each electric entity behaves as a random oscillation antenna (e.g., electric dipole antenna). Spontaneous emission is incoherent emission. The stimulated responses or stimulated transitions (absorption and emission) result in each electric entity acting as a miniature, passive resonant antenna and oscillates in response to the applied signal only. This is coherent internal transition.

[†] The radiation (decay) lifetime is

$$\frac{1}{\tau_r} = \frac{1}{\tau_{ph,e,sp}} + \frac{1}{\tau_{ph,e,st}}.$$

Under general condition, spontaneous emission dominates, but in lasers, the stimulated emission dominates.

The steady-state result gives the relation between the coefficients:

$$\dot{\gamma}_{ph,a} n_{e,A} \frac{\mathrm{d}E_{ph,\omega}}{V\mathrm{d}\omega} = \dot{\gamma}_{ph,e,sp} n_{e,B} + \dot{\gamma}_{ph,e,st} n_{e,B} \frac{\mathrm{d}E_{ph,\omega}}{V\mathrm{d}\omega}. \quad (7.38)$$

We note that for $\mathrm{d}E_{ph,\omega}/V\mathrm{d}\omega$ in (7.2), f_{ph} is not required to be the equilibrium distribution and, as in lasers, can have a very narrow band of energy with a population far from equilibrium. One can write $E_{ph,\omega}/V\mathrm{d}\omega$ as a sum of the thermal contribution and external (nonthermal) contribution [216].

7.4.2 Einstein Coefficients for Equilibrium Thermal Source

Under thermal equilibrium the electron distribution functions can be presented by use of the degree of degeneracies $g_{e,A}$ and $g_{e,B}$ (for $g_{e,i} = 1$, there is no degeneracy) defined in the partition function (2.27) as[†]

$$\frac{n_{e,B}}{n_{e,A}} = \frac{g_{e,B}}{g_{e,A}} \exp(-\frac{\hbar\omega_{e,g}}{k_{\mathrm{B}}T}), \ \hbar\omega_{e,g} = E_{e,B} - E_{e,A}. \quad (7.39)$$

Then (7.39) requires that for equilibrium thermal source $\mathrm{d}E_{ph,\omega}/V\mathrm{d}\omega$ be given by (7.4) as $\hbar\omega f_{ph}^{\mathrm{o}} D_{b,\omega}$; we have

$$g_{e,A}\dot{\gamma}_{ph,a} = g_{e,B}\dot{\gamma}_{ph,e,st} \quad \text{equilibrium radiation.} \quad (7.40)$$

This states that, for nondegenerate systems ($g_{e,A} = g_{e,B} = 1$), the stimulated emission and absorption rates are equal.

Also, using (7.39) in (7.38), along with (7.4) for $\hbar\omega \gg k_{\mathrm{B}}T$ requires that for the thermal radiation we have

$$\frac{dE_{ph,\omega}}{Vd\omega} = \frac{\dot{\gamma}_{ph,e,sp}}{\frac{g_{e,A}}{g_{e,B}}\exp(\frac{\hbar\omega}{k_{\mathrm{B}}T})\dot{\gamma}_{ph,a} - \dot{\gamma}_{ph,e,st}}$$

$$= \frac{\hbar\omega^3}{\pi^2 u_{ph}^3}\exp(-\frac{\hbar\omega}{k_{\mathrm{B}}T}) \quad (7.41)$$

$$\frac{\dot{\gamma}_{ph,e,sp}}{\dot{\gamma}_{ph,e,st}} = \frac{\hbar\omega_{e,g}^3}{\pi^2 u_{ph}^3} \quad \text{under thermal source equilibrium condition.} \quad (7.42)$$

Therefore, all three terms are present, i.e., using (7.40) along with (7.42), there is need for the stimulated emission.

[†] There are some special nonequilibrium conditions, namely,

$$\frac{g_{e,1}n_{e,B}}{g_{e,2}n_{e,A}} \ll 1 \quad \text{low excitation condition}$$

$$\frac{g_{e,1}n_{e,B}}{g_{e,2}n_{e,A}} = 1 \quad \text{saturation}$$

$$\frac{g_{e,1}n_{e,B}}{g_{e,2}n_{e,A}} > 1 \quad \text{amplification.}$$

The last one exists in lasers.

The three Einstein coefficients $\dot{\gamma}_{ph,i}$ (also called the A, B_{AB}, and B_{BA} coefficients) are thus related through (7.39) and (7.42) under equilibrium (this also makes for homogeneous and isotropic radiation).

The absorbed power (including stimulated emission), per unit volume, is

$$\dot{s}_{ph\text{-}e} = \hbar\omega_{e,g}\left.\frac{dn_{e,B}}{dt}\right|_{AB} = \hbar\omega_{e,g}(\dot{\gamma}_{ph,a}n_{e,A}\frac{dE_{ph,\omega}}{Vd\omega} - \dot{\gamma}_{ph,e,st}n_{e,B}\frac{dE_{ph,\omega}}{Vd\omega}). \qquad (7.43)$$

The spontaneous radiative lifetime $\tau_{ph,e,sp} = 1/\dot{\gamma}_{ph,e,sp}$ is part of the radiative lifetime and in the absence of stimulated emission becomes the radiative lifetime τ_r (Section 7.12.2).

7.4.3 Spontaneous Versus Stimulated Emissions in Thermal Cavity

Consider the source of photons to be thermal, so f_{ph} in E_{ph} is given by (7.2) as that of an equilibrium boson, i.e.,

$$f^{\text{o}}_{ph} = \frac{1}{\exp(\frac{\hbar\omega}{k_{\text{B}}T}) - 1} \quad \text{thermal equilibrium distribution.} \qquad (7.44)$$

Under steady-state and thermal equilibrium, from (7.38) we have

$$\frac{dE_{ph,\omega}}{Vd\omega} = \frac{\dot{\gamma}_{ph,e,sp}}{\frac{n_{e,A}}{n_{e,B}}\dot{\gamma}_{ph,a} - \dot{\gamma}_{ph,e,st}}, \qquad (7.45)$$

and using (7.39), (7.40), and (7.44) gives

$$\frac{dE_{ph,\omega}}{Vd\omega} = \frac{\dot{\gamma}_{ph,e,sp}}{\dot{\gamma}_{ph,e,st}} f^{\text{o}}_{ph}. \qquad (7.46)$$

Note that the sum of the thermally stimulated and spontaneous emission rates is

$$\dot{\gamma}_{ph,e,st}\frac{dE_{ph,\omega}}{Vd\omega} + \dot{\gamma}_{ph,e,sp} = \dot{\gamma}_{ph,e,sp}(f^{\text{o}}_{ph} + 1). \qquad (7.47)$$

Writing (7.46) for stimulated emission, we have

$$\dot{\gamma}_{ph,e,st}\frac{dE_{ph,\omega}}{Vd\omega} = \dot{\gamma}_{ph,e,sp} f^{\text{o}}_{ph}. \qquad (7.48)$$

We now examine two limits. From (7.44), for $\hbar\omega/k_{\text{B}}T \ll 1$ and $T = 300$ K (radiofrequencies and microwave regimes), we have $f^{\text{o}}_{ph} \rightarrow \infty$, and (7.48) gives

$$\dot{\gamma}_{ph,e,sp} \ll \dot{\gamma}_{ph,e,st}\frac{dE_{ph,\omega}}{Vd\omega} \quad \text{for } \omega \ll \frac{k_{\text{B}}T}{\hbar} = 3.928 \times 10^{13} \text{ rad/s}$$
$$\text{thermally stimulated emission dominates,} \qquad (7.49)$$

i.e., the thermally stimulated emission rate is much larger than the spontaneous emission rate.

Similarly, for $\hbar\omega/k_{\rm B}T \gg 1$ and $T = 300$ K (near-infrared, visible, etc.), we have $f_{ph}^{\rm o} \rightarrow 0$, and from (7.48)

$$\dot{\gamma}_{ph,e,sp} \gg \dot{\gamma}_{ph,e,st}\frac{{\rm d}E_{ph,\omega}}{V{\rm d}\omega} \text{ for } \omega \gg \frac{k_{\rm B}T}{\hbar} = 3.928 \times 10^{13} \text{ rad/s}$$
$$\text{spontaneous emission dominates,} \qquad (7.50)$$

i.e., the spontaneous emission rate is much larger than the thermally stimulated emission (and absorption) of thermal radiation.

As the temperature increases, so does this frequency. Therefore, at high temperatures the thermally excited energy density increases, whereas at low temperatures (including room temperature) it is not that significant when considering transition energy ($\hbar\omega$) in the near-infrared, visible, and ultraviolet regimes.

When the stimulated and spontaneous emission rates are equal, it is called the saturation condition (saturation photon energy density), and using (7.42) we have

$$\frac{{\rm d}E_{ph,\omega}}{V{\rm d}\omega} = \frac{\dot{\gamma}_{ph,e,sp}}{\dot{\gamma}_{ph,e,st}} = \frac{\hbar\omega^3}{\pi^2 u_{ph}^3} \quad \text{saturation energy density.} \qquad (7.51)$$

This corresponds to $f_{ph}^{\rm o} = 1$ in (7.2). From (7.5), using the intensity $I_{ph,\omega}{\rm d}\omega = u_{ph}{\rm d}E_{ph,\omega}/V$ (assuming unidirectional radiation, i.e., not divided by 4π), for $\omega = 3\times10^{15}$ rad/s (red light), and ${\rm d}\omega = 6\times10^{10}$ rad/s (conventional light sources [216]), for the thermal emission light source under saturation (maximum) energy density we have

$$\int I_{ph,\omega}{\rm d}\omega = \frac{\hbar\omega^3{\rm d}\omega}{\pi^2 u_{ph}^3} = 1.924 \times 10^5 \text{ W/m}^2$$
$$\text{maximum narrow-band, directed thermal emission of red light } (f_{ph}^{\rm o} = 1). \qquad (7.52)$$

Laser light sources have intensities much larger than this value (even with narrow ${\rm d}\omega$).

7.4.4 Spectral Absorption Coefficient

We define the spectral absorption coefficient $\sigma_{ph,\omega}$ as

$$\dot{s}_{ph\text{-}e} = \sigma_{ph,\omega}u_{ph}\int_\omega \frac{{\rm d}E_{ph,\omega}}{V} \quad \text{spectral absorption coefficient.} \qquad (7.53)$$

This is related to the Einstein coefficients through (7.58) as

$$\sigma_{ph,\omega} = \frac{\hbar\omega[\dot{\gamma}_{ph,a}n_{e,A} - \dot{\gamma}_{ph,e,st}n_{e,B}]}{u_{ph}\int_\omega {\rm d}\omega}$$
$$= \frac{\hbar\omega\dot{\gamma}_{ph,a}n_{e,A}[1 - \frac{g_{e,A}}{g_{e,B}}\frac{n_{e,B}}{n_{e,A}}]}{u_{ph}\int_\omega {\rm d}\omega}, \qquad (7.54)$$

where $\int_\omega \mathrm{d}\omega$ is over a small bandwidth, centered around ω. The bandwidth (shape) function $g(\omega)$ was introduced in Table 3.5, subject to a normalization condition.

Under thermal equilibrium, we have

$$1 - \frac{g_{e,A}}{g_{e,B}}\frac{n_{e,B}}{n_{e,A}} = 1 - \exp(\frac{-\hbar\omega}{k_\mathrm{B}T}). \tag{7.55}$$

Then

$$\sigma_{ph,\omega} = \frac{\hbar\omega\dot{\gamma}_{ph,a}n_{e,A}[1-\exp(\frac{-\hbar\omega}{k_\mathrm{B}T})]}{u_{ph}\int_\omega \mathrm{d}\omega}. \tag{7.56}$$

Hence we can define the photon absorption cross-section area as

$$A_{ph,a}(\omega) \equiv \frac{\hbar\omega\dot{\gamma}_{ph,a}}{u_{ph}\int_\omega \mathrm{d}\omega} \quad \text{absorption cross-section area.} \tag{7.57}$$

The spontaneous emission power (per unit volume) is

$$\dot{s}_{e\text{-}ph} = \hbar\omega\dot{\gamma}_{ph,e,sp}n_{e,B} \quad \text{spontaneous emission.} \tag{7.58}$$

Then under thermal equilibrium, making the preceding substitutions, we have

$$\begin{aligned}
\dot{s}_{e\text{-}ph} &= \hbar\omega[\frac{\hbar\omega^3}{\pi^2u_{ph}^3}\dot{\gamma}_{ph,a}\frac{g_{e,A}}{g_{e,B}}][n_{e,A}\frac{g_{e,B}}{g_{e,A}}\exp(\frac{-\hbar\omega}{k_\mathrm{B}T})] \\
&= \hbar\omega\dot{\gamma}_{ph,a}n_{e,A}\frac{\hbar\omega^3}{\pi^2u_{ph}^3}\exp(\frac{-\hbar\omega}{k_\mathrm{B}T}) \\
&= \frac{\sigma_{ph,\omega}u_{ph}\int_\omega \mathrm{d}\omega}{1-\exp(\frac{-\hbar\omega}{k_\mathrm{B}T})}\frac{\hbar\omega^3}{\pi^2u_{ph}^3}\exp(\frac{-\hbar\omega}{k_\mathrm{B}T}) \\
&= \sigma_{ph,\omega}\frac{\hbar\omega^3}{\pi^2u_{ph}^3}\frac{1}{\exp(\frac{\hbar\omega}{k_\mathrm{B}T})-1}\int_\omega \mathrm{d}\omega \\
&= \sigma_{ph,\omega}u_{ph}\int_\omega \frac{\mathrm{d}E_{ph,\omega}}{V}
\end{aligned}$$

$$\text{spontaneous emission rate in terms of absorption coefficient.} \tag{7.59}$$

Also, we have from (7.56)

$$\begin{aligned}
\sigma_{ph,\omega} &= \frac{\hbar\omega\dot{\gamma}_{ph,a}n_{e,A}[1-\exp(\frac{-\hbar\omega}{k_\mathrm{B}T})]}{u_{ph}\int_\omega \mathrm{d}\omega} \\
&= \frac{\hbar\omega[\frac{\dot{\gamma}_{ph,e,sp}}{\hbar\omega^3}\pi^2u_{ph}^3\frac{g_{e,A}}{g_{e,B}}][n_{e,B}\frac{g_{e,A}}{g_{e,B}}\exp(\frac{\hbar\omega}{k_\mathrm{B}T})][1-\exp(\frac{-\hbar\omega}{k_\mathrm{B}T})]}{u_{ph}\int_\omega \mathrm{d}\omega} \\
&= \frac{\pi^2u_{ph}^2n_{e,B}\dot{\gamma}_{ph,e,sp}}{\omega^2\int_\omega \mathrm{d}\omega}\frac{1}{f_{ph}^\mathrm{o}}.
\end{aligned} \tag{7.60}$$

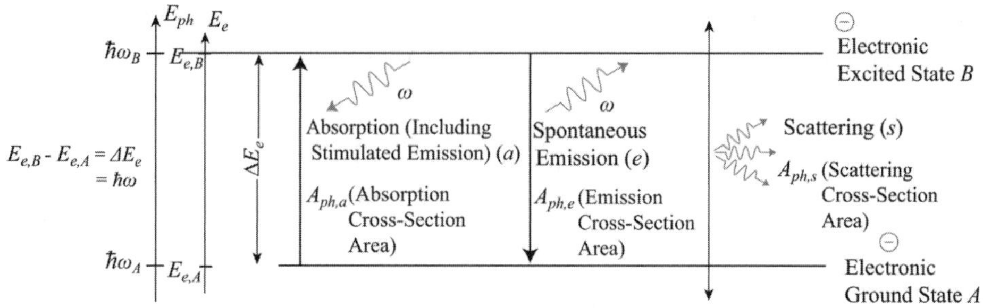

Figure 7.5. A two-level, electron transition photon–electron interaction presentation of the source term in the photon BTE (the absorption and emission are represented by effective cross-sectional areas $A_{ph,a}$ and $A_{ph,e}$). The ground state is designated as A and the excited state as B.

Comparing with (7.56), i.e., $\sigma_{ph,\omega} = n_{e,B} A_{ph,e}/f_{ph}^{\text{o}}$, we have for the photon emission cross-section area in terms of $\dot{\gamma}_{ph,e,sp}$

$$A_{ph,e}(\omega) \equiv \frac{\pi^2 u_{ph}^2 \dot{\gamma}_{ph,e,sp}}{\omega^2 \int_\omega d\omega} \quad \text{emission cross-section area.} \tag{7.61}$$

Note that (7.57) and (7.61) relate the absorption and emission cross-section areas to the Einstein rate coefficients.

In most applications, the spontaneous emission dominates over the stimulated emission.

7.5 Particle Treatment: Photon BTE

The BTE (Table 3.1) written for photon as a particle having momentum $\boldsymbol{p}_{ph} = \hbar\omega \boldsymbol{s}/u_{ph}$, along direction $\boldsymbol{s}$, and for *ph-e* absorption/emission $\dot{s}_{f,ph\text{-}e}$ and any other generation/removal presented by $\dot{s}_{f,ph,i}$, is

$$\frac{\partial f_{ph}}{\partial t} + u_{ph} \boldsymbol{s} \cdot \nabla f_{ph} = \left.\frac{\partial f_{ph}}{\partial t}\right|_s + \dot{s}_{f,ph\text{-}e} + \dot{s}_{f,ph,i}. \tag{7.62}$$

Here we have assumed that the external force $\boldsymbol{F}_{ph}$ is zero.

In Sections 7.8 and 7.9, we will discuss various photon–electron interactions, including those assisted by phonons, in solids (ranging from electrical insulators, to semiconductors, metals and gases). Also in Section 7.12 we will discuss photon–electron interactions in gases, including those assisted by fluid particle kinetic energy.

Here we continue with the two-level electronic transition, as shown in Figure 7.5, to derive the relations for various terms in (7.62). The source term $\dot{s}_{f,ph\text{-}e}$ representing the absorption and emission of photons of angular frequency ω can be written as [297]

$$\dot{s}_{f,ph\text{-}e}(\omega) = u_{ph} \sum_{a,b} \{-n_{e,A}(\omega_A) A_{ph,a}(\omega) f_{ph}(\omega)$$
$$+ n_{e,B}(\omega_B) A_{ph,e}(\omega)[1 + f_{ph}(\omega)]\} \text{ absorption and emission,} \qquad (7.63)$$

where ω_A is the frequency of the lower-energy state A and ω_B is the frequency of the higher-energy state B, and $n_{e,A}$ and $n_{e,B}$ are the number of absorption and emission (spontaneous and stimulated) systems per unit volume. As in Section 7.4, the interaction cross-section areas $A_{ph,a}$ and $A_{ph,e}$ represent the photon absorption (stimulated absorption minus stimulated emission) and emission effectiveness for these electronic transitions. These are generally determined from the FGR (3.27), as will be discussed in Sections 7.8 and 7.9, or empirically.

During the transition from ω_B to ω_A, a photon is emitted, and in the reverse a photon is absorbed.

Now in the scattering term of (7.62), consider a photon of angular frequency ω traveling in the direction $\boldsymbol{s}$ as shown in Figure 7.6. This photon is scattered by a boson or a fermion particle i. The initial photon state is given by ω and $\boldsymbol{s}$, and its final state is given by ω_2 and $\boldsymbol{s}_2$. The particle i has an initial state given by $\boldsymbol{p}_{i,j}$ and final state given by $\boldsymbol{p}_{i,k}$. The photon-scattering cross-section area is $A_{ph,s}$. The general scattering integral (3.8) is written as [297]

$$\left.\frac{\partial f_{ph,\omega}(\boldsymbol{s})}{\partial t}\right|_s = -\sum_{i,j,k} \frac{u_{ph}}{(2\pi\hbar)^3} \int\!\!\int \mathrm{d}\boldsymbol{p}\mathrm{d}A_{ph,s}(\boldsymbol{p}_{i,j}, \omega, \boldsymbol{s} \rightarrow \boldsymbol{p}_{i,k}, \omega_2, \boldsymbol{s}_2)$$
$$\times f_{i,j}(\boldsymbol{p}_{i,j})[1 \pm f_{i,k}(\boldsymbol{p}_{i,k})]\underbrace{\{f_{ph,\omega}(\boldsymbol{s})[1 + f_{ph,\omega_2}(\boldsymbol{s}_2)]}_{\text{out-scattering}}$$
$$\underbrace{- f_{ph,\omega_2}(\boldsymbol{s}_2)[1 + f_{ph,\omega}(\boldsymbol{s})] \times \frac{[1 \pm f_{i,j}(\boldsymbol{p}_{i,j})] f_{i,k}(\boldsymbol{p}_{i,k})}{[1 \pm f_{i,k}(\boldsymbol{p}_{i,k})] f_{i,j}(\boldsymbol{p}_{i,j})}}_{\text{in-scattering}}$$
$$+ \text{ boson, } - \text{ fermion.} \qquad (7.64)$$

The preceding photon absorption/emission and scattering expressions are subsequently further developed and discussed.

7.5.1 Absorption Coefficient and Relation between Photon Absorption and Emission

The absorption and emission cross-section areas in (7.63) are related for the two-level system as [297]

$$A_{ph,a}(\omega) = \frac{g_{e,B}(\omega_B)}{g_{e,A}(\omega_A)} A_{ph,e}(\omega)$$
$$\text{relation between emission and absorption cross-section areas,} \qquad (7.65)$$

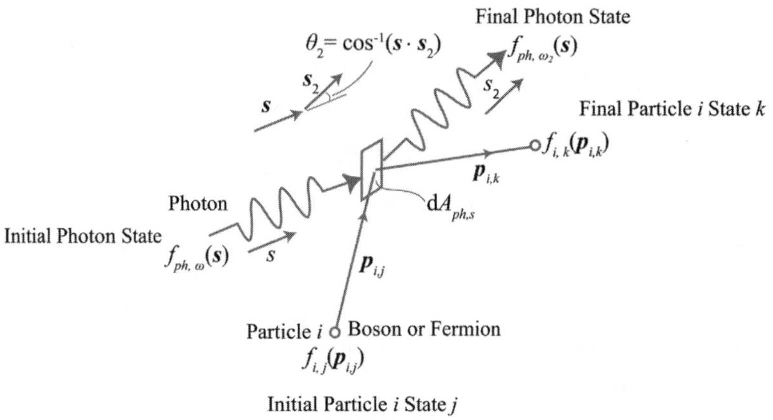

Figure 7.6. General rendering of binary collision and photon scattering by a material particle i (boson or fermion). The scattering is designated by scattering cross-section area $A_{ph,s}$ and scattering direction change $\boldsymbol{s} \cdot \boldsymbol{s}_2$.

where $g_{e,i}$ is the degeneracy for the electronic states A and B with their two corresponding frequencies shown in Figure 7.5.

Using this, (7.63) becomes

$$\dot{s}_{f,ph\text{-}e} = u_{ph} \sum_{A,B} n_{e,A}(\omega_A) A_{ph,a}(\omega)[1 - \frac{g_{e,A}(\omega_A) n_{e,B}(\omega_B)}{g_{e,B}(\omega_B) n_{e,A}(\omega_A)}]$$

$$\times [\frac{1}{\dfrac{g_{e,B}(\omega_B) n_{e,A}(\omega_A)}{g_{e,A}(\omega_A) n_{e,B}(\omega_B)} - 1} - f_{ph,\omega}(s)]$$

$$\equiv u_{ph} \sigma_{ph,\omega} [f_{ph}^n(T) - f_{ph}(\boldsymbol{s})], \tag{7.66}$$

where, similar to (7.60), the radiation (photon) spectral absorption coefficient $\sigma_{ph,\omega}$ is defined as

$$\sigma_{ph,\omega} = \sum_{A,B} n_{e,A}(\omega_A) A_{ph,a} [1 - \frac{g_{e,A}(\omega_A) n_{e,B}(\omega_B)}{g_{e,B}(\omega_B) n_{e,A}(\omega_A)}],$$

$$\text{spectral absorption coefficient.} \tag{7.67}$$

The inverse of $\sigma_{ph,\omega}$ is referred to as the photon mean free path λ_{ph}.

The distribution function (including thermal nonequilibrium) $f_{ph}(\omega, T)$ is

$$f_{ph}(\omega, T) = \frac{1}{\dfrac{g_{e,B}(\omega_B) n_{e,A}(\omega_A)}{g_{e,A}(\omega_A) n_{e,B}(\omega_B)} - 1}. \tag{7.68}$$

Under the special case of local thermal equilibrium (all particles in the differential collision area are at temperature T) for the distribution of absorbing and emitting systems, the population ratio is given by

$$\frac{g_{e,A}(\omega_A) n_{e,B}(\omega_B)}{g_{e,B}(\omega_B) n_{e,A}(\omega_A)} = \exp(-\frac{E_B - E_A}{k_B T}) = \exp(\frac{-\hbar\omega}{k_B T}) \tag{7.69}$$

$$E_B - E_A = \hbar(\omega_B - \omega_A) = \hbar\omega. \tag{7.70}$$

Using this in (7.66), we have

$$\dot{s}_{f,ph\text{-}e}(\omega) = u_{ph}\sigma_{ph,\omega}[f_{ph}^{\text{o}}(T) - f_{ph}(\boldsymbol{s})] \quad \text{local thermal equilibrium.} \tag{7.71}$$

Finally, the absorption coefficient (7.67), using (7.69), becomes

$$\begin{aligned}\sigma_{ph,\omega} &\equiv \sum_{\omega_A} n_{e,A}(\omega_A)A_{ph,a}(\omega)[1 - \exp(-\frac{\hbar\omega}{k_{\rm B}T})] \\ &= \sum_{\omega_B} \frac{n_{e,B}(\omega_B)A_{ph,e}(\omega)}{f_{ph}^{\text{o}}(T)} \quad \text{local thermal equilibrium,}\end{aligned} \tag{7.72}$$

where as in (7.44) we have

$$f_{ph}^{\text{o}}(\omega, T) = \frac{1}{\exp(\frac{\hbar\omega}{k_{\rm B}T}) - 1}. \tag{7.73}$$

In (7.72) the summation is over all two-level systems with energy gap $\hbar\omega$. Then the spectral absorption coefficient given by (7.72) depends on the number density of the available electronic transitions and the absorption/emission cross-section areas. Note that, for $\hbar\omega/k_{\rm B}T \gg 1$, we have $f_{ph}^{\text{o}} = 1$.

Then $f_{ph}(\boldsymbol{s})$ needs to be different than the equilibrium distribution at T, i.e., $f_{ph}^{\text{o}}(T)$, in order to have a net emission–absorption.

7.5.2 Photon–Free Electron Scattering Rate

Photon scattering by free-conduction electrons is called Compton scattering. This allows for a rather simple treatment of scattering rate and this angular distribution (scattering angle θ_2). We sum over the initial and final electron spin states (by doubling the scattering cross section). Then (7.64) written for a fermion becomes

$$\begin{aligned}\left.\frac{\partial f_{ph,\omega}(\omega, \boldsymbol{s})}{\partial t}\right|_s &= -\sum_{i,j,k} \frac{2u_{ph}}{(2\pi\hbar)^3} \int_{\boldsymbol{p}_e} \mathrm{d}\boldsymbol{p}_e \int \mathrm{d}A_{ph,s}(\boldsymbol{p}_e, \omega, \boldsymbol{s} \rightarrow \boldsymbol{p}_{e,2}, \omega_B, \boldsymbol{s}_2) \\ &\quad \times f_e(\boldsymbol{p}_e)[1 - f_e(\boldsymbol{p}_{e,2})]\{f_{ph,\omega}(\boldsymbol{s})[1 + f_{ph,\omega_B}(\boldsymbol{s}_2)] \\ &\quad - f_{ph,\omega_B}(\boldsymbol{s}_2)[1 + f_{ph,\omega}(\boldsymbol{s})] \times \frac{[1 - f_e(\boldsymbol{p}_e)]f_e(\boldsymbol{p}_{e,2})}{[1 - f_e(\boldsymbol{p}_{e,2})]f_e(\boldsymbol{p}_e)}\},\end{aligned} \tag{7.74}$$

where $\boldsymbol{p}_{e,2}$ represents the final momentum of the electron participating in the collision. Assuming equilibrium distribution and using f_e^{o} from Table 1.2, we have

$$\frac{[1 - f_e\boldsymbol{p}_e)]f_e(\boldsymbol{p}_{e,2})}{[1 - f_e(\boldsymbol{p}_{e,2})]f_e(\boldsymbol{p}_e)} = \exp[-\frac{\hbar(\omega - \omega_B)}{k_{\rm B}T}], \tag{7.75}$$

where we have used the conservation of energy between the colliding photon and electron, i.e.,

$$E_e(\boldsymbol{p}_{e,2}) - E_e(\boldsymbol{p}_e) = E_{ph}(\omega_B) - E_{ph}(\omega)$$
$$= \hbar(\omega_B - \omega). \quad (7.76)$$

For the elastic scattering we have $\hbar(\omega - \omega_B) = 0$ and the right-hand side of (7.75) is nearly unity.

The differential scattering cross-section area for nonpolarized radiation is approximated as the Thomson cross-section area,† i.e.,

$$\mathrm{d}A_{ph,s} = (\frac{e_c^2}{4\pi m_e \epsilon_o u_{ph}^2})^2 \frac{1+\cos^2\theta_2}{2}\mathrm{d}(\cos\theta_2)\mathrm{d}\phi_2,$$
$$\mathrm{d}\Omega = \sin\theta_2 \mathrm{d}\theta_2 \mathrm{d}\phi_2 = \mathrm{d}\cos\theta_2 \mathrm{d}\phi_2, \quad (7.77)$$

where $\mathrm{d}\Omega$ is the differential solid angle of the scattering angle θ_2 and

$$\theta_2 = \cos^{-1}(\boldsymbol{s}\cdot\boldsymbol{s}_2), \quad (7.78)$$

and ϕ_2 is the azimuthal angle.

Now substituting this in (7.74), we have for the scattering rate (and its angular distribution)

$$\left.\frac{\partial f_{ph,\omega}(\omega, \boldsymbol{s})}{\partial t}\right|_s = -\pi u_{ph} n_{e,c} (\frac{e_c^2}{4\pi m_e \epsilon_o u_{ph}^2})^2 \int_{-1}^{1} (1+\cos^2\theta_2)$$
$$\times[f_{ph,\omega}(\boldsymbol{s}) - f_{ph,\omega_B}(\boldsymbol{s}_2)]\mathrm{d}(\cos\theta_2), \quad \frac{\hbar\omega}{m_e u_{ph}^2} \ll 1, \quad \frac{p_e}{m_e u_{ph}} \ll 1$$
$$\Phi(\theta_2, \phi) = 1+\cos^2\theta_2 \ \ \text{phase function for free-electron scattering.} \quad (7.79)$$

The angular distribution (inside the integral) of the scattering is represented by the scattering phase function $\Phi(\theta_2, \phi)$.

As given in (5.63), we have

$$n_{e,c} = \frac{2}{(2\pi\hbar)^3}\int_{\boldsymbol{p}_e} f_e(\boldsymbol{p}_e)\mathrm{d}\boldsymbol{p}_e. \quad (7.80)$$

† Derivation is given as an end-of-chapter problem.
The classical free-electron motion (displacement) (Table 2.5) under an oscillating EM field gives [a simplified form of (3.45)]

$$m_e \frac{d^2x}{dt^2} = -e_c e_{e,o} \sin\omega t,$$

or

$$x = \frac{e_c e_{e,o}}{m_e \omega^2}\sin\omega t.$$

The oscillating electric dipole (3.40) is

$$\boldsymbol{p}_e = -e_c x \boldsymbol{s}_x = p_{e,o}\sin(\omega t)\boldsymbol{s}_x,$$

where $p_{e,o} = e_c^2 e_{e,o}/m_e\omega^2$.

Here $n_{e,c}$ is the number of free (conduction) electrons per unit volume and the factor 2 is for the two possible spin states.

In the Compton effect, the photon wavelength (X-rays and γ-rays regimes) increases because of absorption by the free electrons (kinetic energy of electron increases). Note that both energy and momentum are conserved in this collision. By performing the integration over θ_2, the scattering cross-section area (7.77) is

$$A_{ph,s} = \frac{8\pi}{3} r_C^2, \quad r_C = \frac{e_c^2}{4\pi m_e \epsilon_o u_{ph,o}^2}, \quad \sigma_{ph,s} = n_{e,c} A_{ph,s}$$

Compton (or Thomson) scattering by free electron, (7.81)

where r_C is the classical (Compton) electron radius (Table 1.4), $n_{e,c}$ is the number density of scatterers, and $\sigma_{ph,s}$ is the scattering coefficient $A_{ph,s} = 6.649\times10^{-29}$ m^2 = 6.649×10^{-25} cm^2. This shows that, for this free electron, the scattering cross-section area is independent of the frequency of the incident EM wave.

The general treatment of photon (classical EM wave) scattering is described in [178], and this includes the Mie scattering that covers the entire range of the size parameter (defined by $\alpha_\lambda = \pi d/\lambda$, where d is the particle diameter). These are also discussed in [172]. The treatment includes the definition of the phase function, which gives the angular distribution of the scattered wave, similar to that in (7.77). For a very small size parameter, $\alpha_\lambda \ll 1$, we have the Rayleigh scattering (which is for bound electrons) that shows $A_{ph,s}$ is proportional to λ^{-4} (or ω^4) of incident EM wave.†

† Starting from the classical description of bound [in an atomic potential (2.54)] electron motion (Table 2.5) in a time-periodic EM field, we have [extension of (3.45)]

$$m_e \frac{d^2 x}{dt^2} - m_e \omega_{n,e}^2 x = -e_c e_{e,o} \sin \omega t,$$

where $\omega_{n,e}$ is natural frequency of the bound electron (similar to harmonic interatomic bonding), we have for the solution

$$x = \frac{e_c e_{e,o}}{m_e(\omega^2 - \omega_{n,e}^2)} \sin \omega t.$$

The dipole moment vector (3.40) of this electron is

$$\boldsymbol{p}_e = -e_c x \boldsymbol{s}_x = -p_{e,o} \sin \omega t \boldsymbol{s}_x, \quad p_{e,o} = \frac{e_c^2 e_{e,o}}{m_e(\omega^2 - \omega_{n,e}^2)}.$$

Then the Rayleigh scattering cross-section area, found similarly to (7.77) (end-of-chapter problem) is

$$\frac{dA_{ph,s}}{d\Omega} = \frac{\omega^4}{(\omega^2 - \omega_{n,e}^2)^2} r_{e,o}^2 \sin \omega,$$

or

$$A_{ph,s} = \frac{8\pi}{3} r_{e,o} \frac{\omega^4}{(\omega^2 - \omega_{n,e}^2)^2}, \quad \sigma_{ph,s} = n_s A_{ph,s} \quad \text{Rayleigh scattering by bound electron.}$$

For $\omega \ll \omega_{n,e}$ (e.g., visible incident radiation versus the ultraviolet vibration), we have

$$\sigma_{ph,s} = n_s \frac{8\pi}{3} r_{e,o} \frac{\omega^4}{\omega_{n,e}^4} = n_s \frac{8\pi}{3} \frac{r_{e,o}}{\omega_{n,e}^4} \frac{(2\pi u_{ph})^2}{\lambda^4}.$$

7.6 Equation of Radiative Transfer

The equation of radiative transfer (ERT) refers to the BTE for photons written in terms of the radiation intensity I_{ph}. This intensity is the general, nonequilibrium intensity (including laser radiation) given by (7.5). Here we derive this equation and also show the results for the case of absorption–emission dominated (no scattering and source/sink) transport under strong absorption (called the optically thick limit).

7.6.1 General Form of ERT

Using the spectral absorption coefficient in (7.66), the BTE (7.62) can be written as

$$\frac{\partial f_{ph}}{\partial t} + u_{ph}\mathbf{s} \cdot \nabla f_{ph} = \left.\frac{\partial f_{ph}}{\partial t}\right|_s + u_{ph}\sigma_{ph,\omega}[f_{ph}(\omega, T) - f_{ph}(\mathbf{s})] + \dot{s}_{f,ph,i}. \quad (7.82)$$

Multiplying (7.82) by $\hbar\omega^3/4\pi^3 u_{ph}^2$ for blackbody density of states, i.e., $I_{ph,\omega} = u_{ph} f_{ph}\hbar\omega\, D_{b,\omega}/4\pi$, and using the definition of intensity (7.5), we have

$$\frac{\partial I_{ph,\omega}(\omega, \mathbf{s})}{u_{ph}\mathrm{d}t} + \mathbf{s} \cdot \nabla I_{ph,\omega}(\omega, \mathbf{s}) = \left.\frac{\partial I_{ph,\omega}(\omega, \mathbf{s})}{u_{ph}\mathrm{d}t}\right|_s$$
$$+ \sigma_{ph,\omega}[I_{ph,\omega}(\omega, T) - I_{ph,\omega}(\omega, \mathbf{s})] + \dot{s}_{ph,i}. \quad (7.83)$$

This is the spectral (and directional) ERT.

Under local thermal equilibrium, (7.83) becomes

$$\frac{\partial I_{ph,\omega}(\omega, \mathbf{s})}{u_{ph}\mathrm{d}t} + \mathbf{s} \cdot \nabla I_{ph,\omega}(\omega, \mathbf{s}) = \left.\frac{\partial I_{ph,\omega}(\omega, \mathbf{s})}{u_{ph}\partial t}\right|_s$$
$$+ \sigma_{ph,\omega}[I_{b,\omega}(\omega, T) - I_{ph,\omega}(\omega, \mathbf{s})] + \dot{s}_{ph,i} \quad \text{local thermal equilibrium}, \quad (7.84)$$

where we have used the blackbody intensity (7.6) for the photon emission.

It is customary [172] to use a scattering coefficient $\sigma_{ph,s,\omega}$ and then an extinction coefficient as the sum of the absorption and scattering coefficients, $\sigma_{ph,ex,\omega} = \sigma_{ph,\omega} + \sigma_{ph,s,\omega}$. This form of the ERT is presented in Section 7.7.3. The radiative net heat flux vector $\mathbf{q}_{ph}$ is given by

$$\mathbf{q}_r = \mathbf{q}_{ph} = \int_0^\infty \int_{4\pi} \mathbf{s} I_{ph,\omega} \mathrm{d}\Omega \mathrm{d}\omega = \int_0^\infty \int_{-1}^{1} \mathbf{s} I_{ph,\omega} \mathrm{d}\mu \mathrm{d}\omega, \quad (7.85)$$

Table 7.1. *Spectral ERT (photon transport equation for absorbing/emitting/scattering media), under local thermal nonequilibrium.*

$$\frac{\partial I_{ph,\omega}(\omega,\mathbf{s})}{u_{ph}\partial t}+\mathbf{s}\cdot\nabla I_{ph,\omega}(\omega,\mathbf{s})=\frac{\partial I_{ph,\omega}(\omega,\mathbf{s})}{u_{ph}\partial t}|_s+$$

$$\sigma_{ph,\omega}[I_{ph,n,\omega}(\omega,T)-I_{ph,\omega}(\omega,\mathbf{s})]+\dot{s}_{ph\text{-}i}$$

$$\mathbf{q}_r=\mathbf{q}_{ph}=\int_0^\infty\int_{4\pi}\mathbf{s}I_{ph,\omega}\mathrm{d}\Omega\mathrm{d}\omega=\int_0^\infty\int_{-1}^1\mathbf{s}I_{ph,\omega}\mathrm{d}\mu\mathrm{d}\omega$$

$$\nabla\cdot\mathbf{q}_r=\nabla\cdot\mathbf{q}_{ph}=\int_0^\infty\int_{4\pi}\nabla\cdot\mathbf{s}I_{ph,\omega}\mathrm{d}\Omega\mathrm{d}\omega=4\pi\int_0^\infty\sigma_{ph,\omega}I_{ph,n,\omega}(\omega,T)\mathrm{d}\omega-$$

$$2\pi\int_0^\infty\sigma_{ph,\omega}\int_{-1}^1 I_{ph,\omega}(\omega,\mathbf{s})\mathrm{d}\mu\mathrm{d}\omega+2\pi\int_0^\infty\int_{-1}^1\dot{s}_{ph,i}\mathrm{d}\mu\mathrm{d}\omega$$

$I_{ph,\omega}$	spectral, directional photon intensity, W/(m^2-sr-rad/s)
$I_{b,\omega}$	blackbody (equilibrium) spectral photon intensity for, W/(m^2-sr-rad/s), used instead of $I_{ph,n,\omega}$
$I_{ph,n,\omega}$	nonequilibrium, but only temperature-dependent spectral emission intensity, W/(m^2-sr-rad/s)
$\mathbf{q}_r$	net radiative heat flux vector, W/(m^2)
$\mathbf{s}$	unit vector
u_{ph}	speed of light, m/s
$\dot{s}_{ph,i}$	photon generation rate other than the absorption/emission in *ph-e* interactions, W/(m^3-sr-rad/s)
$\sigma_{ph,\omega}$	spectral absorption coefficient, 1/m
$\mathrm{d}\Omega$	$\sin\theta\mathrm{d}\theta\mathrm{d}\phi$, solid angle, sr
μ	$\cos\theta$, θ is polar angle

where $\mu=\cos\theta$, and the divergence of the total radiative heat flux, which is used in the energy equation (Table 1.1), is found from†

$$\nabla\cdot\mathbf{q}_r=\nabla\cdot\mathbf{q}_{ph}=\int_0^\infty\int_{4\pi}\nabla\cdot\mathbf{s}I_{ph,\omega}\mathrm{d}\Omega\mathrm{d}\omega=4\pi\int_0^\infty\sigma_{ph,\omega}I_{ph,n,\omega}(\omega,T)\mathrm{d}\omega$$
$$-2\pi\int_0^\infty\sigma_{ph,\omega}\int_{-1}^1 I_{ph,\omega}(\omega,\mathbf{s})\mathrm{d}\mu\mathrm{d}\omega+2\pi\int_0^\infty\int_{-1}^1\dot{s}_{ph,i}\mathrm{d}\mu\mathrm{d}\omega. \quad (7.87)$$

† Absorption of EM radiation is a common heating method, and the volumetric rate of heating is expressed as

$$\dot{s}_{ph,a}=\dot{s}_{e,\sigma}=-\nabla\cdot\mathbf{q}_r=\sigma_{ph}q_{r,\mathrm{o}}e^{-\sigma_{ph}x}, \quad (7.86)$$

for a columnar radiation with $q_r=q_{r,\mathrm{o}}(x=0)$.

When $\nabla \cdot \boldsymbol{q}_r > 0$, then the net radiation heat transfer leaving a differental control volume (Table 1.1) is larger than that arriving, i.e., thermal emission and other sources dominate over absorption. Then $\nabla \cdot \boldsymbol{q}_r$ is balanced with the rest of the heat transfer mechanism, heat storage, and energy conversion, according to the energy equation given in Table 1.1. The solution to the ERT is discussed in [306], among others. These include close-form, exact, and approximate solutions, and numerical solutions. In the P-N methods, moments of ERT and the expansion of I_{ph} in terms of spherical harmonics are used. In the discrete ordinate method, angular integration over a fraction of the total solid angle (4π) is made, thus dividing this into finite angular regimes. For example, one forward region and one backward regions are used (we will employ this in Section 7.7.3). The numerical techniques applied to BTE (Section 3.1.8) are also used for the ERT.

7.6.2 Optically Thick Limit and Radiant Conductivity

When the optical thickness defined by $\sigma^*_{ph} = \sigma_{ph} L$ for a medium of thickness L and absorption coefficient σ_{ph} is larger than 1, the ERT, for an emitting medium with a strong absorption is approximated as follows. Since for optically thick media, the local emission and absorption dominate (absorption dominates over scattering), then in (7.84), under $\dot{s}_{ph,i} = 0$, for steady-state and local thermal equilibrium conditions, and for a plane-parallel geometry radiation with $I_{ph,\omega}$ changing only along the x direction (with $\boldsymbol{s}$ making an angle θ with the x direction), we have

$$\frac{\mathrm{d}I_{ph,\omega}}{\mathrm{d}(x/\cos\theta)} = \sigma_{ph,\omega}(I_{b,\omega} - I_{ph,\omega}) \text{ plane-parallel radiation.} \tag{7.88}$$

Note that we have used $\boldsymbol{s}I_{ph,\omega} \cdot \boldsymbol{s}_x = I_{ph,\omega}\cos\theta$. Taking the integral over all possible frequencies and also assuming that the spatial variations of $I_{ph,\omega}$ and $I_{ph,b,\omega}$ are identical, we have, using (7.9)

$$\frac{\mathrm{d}I_{ph}}{\mathrm{d}(x/\cos\theta)} = \sigma_{ph}(I_b - I_{ph}),$$

$$\pi I_b = \pi\int_0^\infty I_{b,\omega}\mathrm{d}\omega = E_b = \sigma_{\mathrm{SB}}T^4, \tag{7.89}$$

where $x/\cos\theta$ is the photon path as it travels between surfaces located at x and $x + \Delta x$.

The radiation heat flux $q_{ph,x}$ (Table 7.1) is found by the integration of I_{ph} over a unit sphere, i.e.,

$$q_{ph,x} = \int_0^{2\pi}\int_0^{\pi} I_{ph}\cos\theta\sin\theta\mathrm{d}\theta\mathrm{d}\phi, \tag{7.90}$$

where we have used $\boldsymbol{s} = \boldsymbol{s}_x\cos\theta$. Note that this integral is over a complete sphere (since we are interested in the net radiation flux). Also note that $I_{ph,b}$ is independent of θ and ϕ (diffuse radiation).

The equation of radiative transfer (7.90) can be rearranged as

$$I_{ph} = -\frac{\mathrm{d}I_b \cos\theta}{\sigma_{ph}\mathrm{d}x} + I_b. \tag{7.91}$$

Upon integration, we have for $q_{ph,x}$

$$q_{ph,x} = \int_0^{2\pi}\int_0^{\pi}(-\frac{\mathrm{d}I_{ph}\cos\theta}{\sigma_{r,a}\mathrm{d}x} + I_b)\cos\theta\sin\theta\mathrm{d}\theta\mathrm{d}\phi. \tag{7.92}$$

Now, noting that $I_{ph,b}$ is independent of θ and ϕ, we have

$$\begin{aligned} q_{ph,x} &= -\frac{\mathrm{d}I_b}{\sigma_{ph}\mathrm{d}x}\int_0^{2\pi}\int_0^{\pi}\cos^2\theta\sin\theta\mathrm{d}\theta\mathrm{d}\phi + I_b\int_0^{2\pi}\int_0^{\pi}\cos\theta\sin\theta\mathrm{d}\theta\mathrm{d}\phi \\ &= -\frac{\mathrm{d}I_b}{\sigma_{ph}\mathrm{d}x}\times\frac{4}{3}\pi + I_b\times 0 \\ &= -\frac{4\pi}{3\sigma_{ph}}\frac{\mathrm{d}I_b}{\mathrm{d}x}. \end{aligned} \tag{7.93}$$

Now, using $\pi I_b = E_b = \sigma_{\mathrm{SB}}T^4$, we have

$$q_{ph,x} = -\frac{4}{3\sigma_{ph}}\frac{\mathrm{d}E_b}{\mathrm{d}x} = -\frac{4\sigma_{\mathrm{SB}}}{3\sigma_{ph}}\frac{\mathrm{d}T^4}{\mathrm{d}x} = -\frac{16}{3}\frac{\sigma_{\mathrm{SB}}T^3}{\sigma_{ph}}\frac{\mathrm{d}T}{\mathrm{d}x}. \tag{7.94}$$

It can be rewritten as

$$\boldsymbol{q}_{ph} = -k_{ph}\nabla T, \quad k_{ph} = \frac{16}{3}\frac{\sigma_{\mathrm{SB}}T^3}{\sigma_{ph}}, \quad \text{for } \sigma^*_{ph} \gg 1$$

photon thermal conductivity. (7.95)

Here k_{ph} is called the radiant (photon) conductivity and this optically thick limit $\sigma^*_{ph} = \sigma_{ph}L \gg 1$ is also called the radiation diffusion limit.

Note that in replacing $\mathrm{d}I_{ph}/\mathrm{d}x$ with $\mathrm{d}I_b/\mathrm{d}x$ in the ERT, we are eliminating all the distant radiation heat transfer effects by approximating the local intensity, i.e., rewriting (7.6.2)

$$I_{ph} = I_b - \frac{\cos\theta}{\sigma_{ph}}\frac{\mathrm{d}I_b}{\mathrm{d}x}. \tag{7.96}$$

Also note that E_{ph} is the integral of I_{ph} over a unit hemisphere.

Similar to phonon mean free path (4.109), defining the photon mean free path $\lambda_{ph} = 1/\sigma_{ph}$, (7.95) becomes

$$\begin{aligned} k_{ph} &= \frac{16}{3}\sigma_{\mathrm{SB}}T^3\lambda_{ph} = \frac{4\pi^2}{45}\frac{k_{\mathrm{B}}^4T^3}{\hbar^3u^2_{ph,\mathrm{o}}}\lambda_{ph} = \frac{4\pi^2}{45}\frac{k_{\mathrm{B}}^3T^3}{\hbar^3u^3_{ph,\mathrm{o}}}k_{\mathrm{B}}u_{ph,\mathrm{o}}\lambda_{ph} \\ &= \frac{4\pi^2}{45}\frac{k_{\mathrm{B}}^3T^3}{\hbar^3u^3_{ph,\mathrm{o}}}k_{\mathrm{B}}u_{ph,\mathrm{o}}\lambda_{ph} = an_{ph}k_{\mathrm{B}}u_{ph,\mathrm{o}}\lambda_{ph}, \quad \lambda_{ph} = 1/\sigma_{ph} \end{aligned}$$

photon mean free path, (7.97)

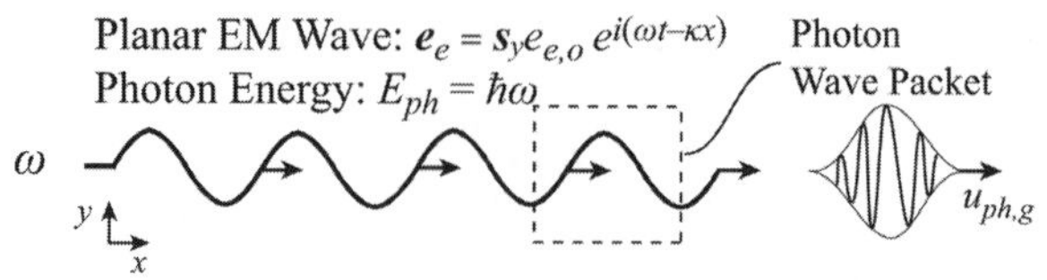

Figure 7.7. Wave–particle duality of photons, illustrated as a planar wave train propagating along the x direction.

where n_{ph} is given by (1.11) (end of chapter problem), and a is a constant.† Note that, similar to (4.109), this presents photon conductivity for optically thick media as a product of the principle carrier heat capacity, speed, and mean free path of the carrier. Also note that, with $k_i = n_i c_{v,i} u_i \lambda_i / 3$, here the photon heat capacity is $n_{ph} c_{v,ph} = 16\sigma_{\rm SB} T^3 / u_{ph,\rm o}$.

7.7 Wave Treatment: Field Enhancement and Photon Localization

Radiative energy is carried by photons, which are in principle both quasi-particles and waves, according to the theory of wave–particle duality. This is depicted as a planar wave train, where a wave is segmented into parts (photons) and presented as a wave packet, in Figure 7.7. Each photon has an energy $\hbar\omega$ and a well-defined phase relation to other photons. The ERT, developed in the last section, can describe the particle behavior, and the EM wave theory (Section 3.3.2) can capture the wave features. For some absorption, emission, and scattering phenomena, the propagation of radiative energy in a homogeneous/heterogeneous media, the particle or wave distinction is not critical, and both ERT and EM treatments can be used. However, some important phenomena are attributed only to the coherent wave nature, and can only be interpreted by the propagating wave theory. These include interference, diffraction, and tunneling. In this section the photon localization and field enhancement, which are due to wave interference effects, will be discussed as examples.

Photon localization is a size effect in photon transport in nanostructures (heterogeneous media). Figures 7.8(a) to (d) show the different regimes of photon (monochromatic irradiation, e.g., laser) scattering in nanopowders [277]. When the photon mean free path λ_{ph} is much larger than the incident wavelength λ_i, photons may experience a single-scattering event and the transport is diffusive [347]. As λ_{ph} decreases, photons begin to undergo multiple-scattering events. When λ_{ph} is

† Comparing (7.95) with (4.109) for phonons, we arrive at the photon counterpart, i.e.,

$$k_{ph} = \frac{16}{3}\sigma_{\rm SB} T^3 \lambda_{ph} = \frac{1}{3} n_p c_{v,p} u_{p,\rm A} \lambda_p,$$

where, as mentioned in Section 7.2, we have

$$\sigma_{\rm SB}(\text{phonon}) = \frac{n_p c_{v,p} u_{p,\rm A}}{16 T^3}.$$

Note that, similar to (4.87), this presents photon conductivity k_{ph} for optically thick media as a product of specific heat capacity, speed, and the mean free path of the carrier.

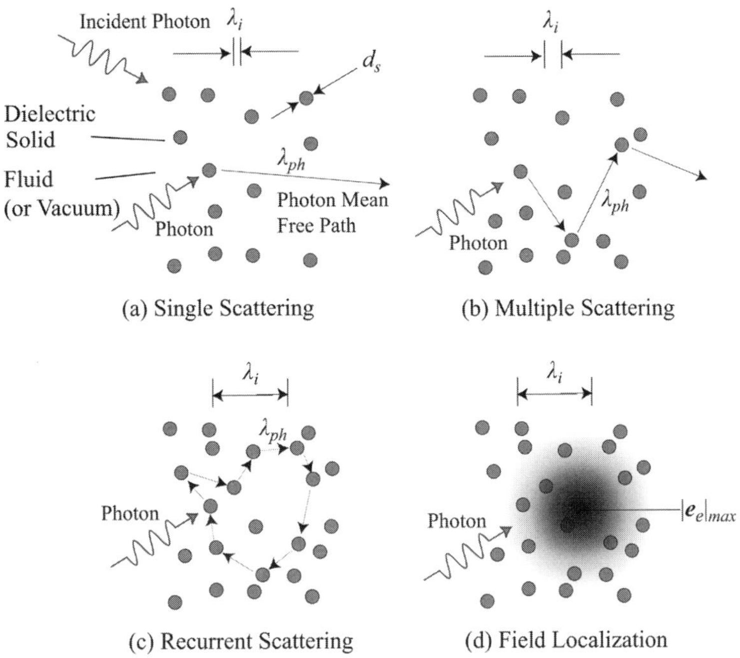

Figure 7.8. (a) Single, (b) multiple, and (c) recurrent scattered photon trajectories in a medium of random scatterers, and (d) a rendering of the electric field intensity distribution for case (c) showing photon localization [277].

comparable with, or smaller than, the incident wavelength, recurrent scattering occurs, i.e., photons may return to the original location after being scattered many times. In this case, photons do not propagate through the medium and are confined in a small spatial region, forming a cavity. This phenomenon, termed photon localization, is the counterpart of the electron localization suggested in [5]. The EM field of the localized photon, as shown in Figure 7.8(d), can be orders of magnitude larger than the incident field [291]. This is demonstrated is subsequent sections.

7.7.1 Photon Localization in One-Dimensional Multilayer

The simplest model of random heterogeneous medium is the one-dimensional multilayer geometry made of parallel solid layers with random thickness, as shown in Figure 7.9. Regions $l = 1$ and $N+1$ are semi-infinite vacuum (air) layers. This multilayer medium has $N/2$ (N is an even number here) solid layers and $N/2-1$ vacuum layers. The coordinates $x_1, x_2, ..., x_N$ are chosen such that the thickness of each layer is random, but obeys a trapezoidal (uniform) distribution (i.e., all thicknesses in a range have equal probability). This multilayer medium has a finite dimension in the direction of the EM wave propagation x, and an infinite length in the plane normal to x. We assume a dielectric solid and use the optical properties

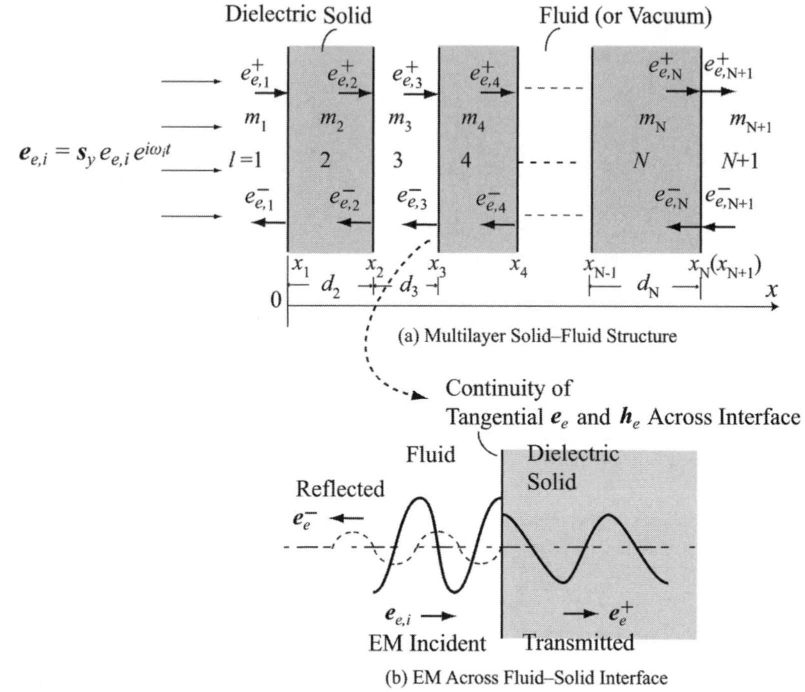

Figure 7.9. (a) Model random heterogenous medium, consisting of parallel solid (dielectric) and air (vacuum) layers with random thicknesses. The porosity (void fraction, ϵ) is prescribed. (b) Incident transmitted and reflected $\boldsymbol{e}_e$ (tangential components of $\boldsymbol{e}$ and $\boldsymbol{h}_e$ are conserved among incident, transmitted, and reflected $\boldsymbol{e}$ and $\boldsymbol{h}_e$).

Figure 3.4, i.e., the complex refractive index $m_{s,\omega}[= n_{s,\omega} - i\kappa_{s,\omega},$ (3.30) and Table 3.5], which depends on the frequency (or wavelength). All the quantities and parameters are scaled with respect to the incident angular frequency ω_i. Vacuum has a refractive index $m_f(m_f = n_f = 1, \kappa_f = 0)$.

For no current flowing and with uniform charge distribution, the EM wave equation is simplified and we have the one-dimensional, time-periodic Helmholtz equation, (3.35) as

$$\frac{\partial^2 e_e(x)}{\partial x^2} + \kappa_o^2 m_i^2 e_e(x) = 0, \qquad \boldsymbol{e}_e = \boldsymbol{s}_y e^{i\omega t} = \boldsymbol{s}_y e_e(x) e^{i\kappa_o c_o t}, \tag{7.98}$$

where κ_o is the vacuum wave number and m_i is the local complex index of refraction at the incident frequency. The wave form is (3.36). Here we have used an arbitrary y-direction polarization. For the medium shown in Figure 7.9(a), the field is divided into two components, the forward (transmitted) component $e_{e,l}^+$ and the backward (reflected) component $e_{e,l}^-$. the solution to (7.98) at a particular location x in the lth layer is given by

$$e_e(x) = e_{e,l}^+ e^{i\kappa_l(x-x_l)} + e_{e,l}^- e^{-i\kappa_l(x-x_l)}, \quad l = 1, 2, ..., N+1, \tag{7.99}$$

where x_{N+1} takes the value of x_N, as there are only N interfaces, and $\kappa_l = m_l\omega_i/c_o$ is the wave vector, where c_o is the speed of light in vacuum. The boundary conditions require that the tangential electric and magnetic field intensities be continuous across each interface as shown in Figure 7.9(b) [215]. The relationship between the amplitudes of the lth and (l+1)th interfaces are given in the matrix form [358][†]

$$\begin{bmatrix} e^+_{e,l} \\ e^-_{e,l} \end{bmatrix} = \boldsymbol{T}_l^{-1}\boldsymbol{T}_{l+1}\boldsymbol{P}_{l+1}\begin{bmatrix} e^+_{e,l+1} \\ e^-_{e,l+1} \end{bmatrix}, \quad l = 1, 2, ..., N, \tag{7.100}$$

where

$$\boldsymbol{T}_l = \begin{bmatrix} 1 & 1 \\ m_l & -m_l \end{bmatrix}, \quad l = 1, 2, ..., N+1, \tag{7.101}$$

and $\boldsymbol{T}_l^{-1}$ is the inverse of $\boldsymbol{T}_l$, and

$$\boldsymbol{P}_l = \begin{bmatrix} e^{-i\kappa_l(x_l - x_{l-1})} & 0 \\ 0 & e^{i\kappa_l(x_l - x_{l-1})} \end{bmatrix}, \quad l = 2, 3, ..., N+1. \tag{7.102}$$

Hence

$$\begin{bmatrix} e^+_{e,j} \\ e^-_{e,j} \end{bmatrix} = \begin{bmatrix} M^{(j)}_{11} & M^{(j)}_{12} \\ M^{(j)}_{21} & M^{(j)}_{22} \end{bmatrix}\begin{bmatrix} e^+_{e,N+1} \\ e^-_{e,N+1} \end{bmatrix}, \quad j = 1, 2, ..., N, \tag{7.103}$$

where

$$\begin{bmatrix} M^{(j)}_{11} & M^{(j)}_{12} \\ M^{(j)}_{21} & M^{(j)}_{22} \end{bmatrix} = \prod_{l=j}^{N} \boldsymbol{T}_l^{-1}\boldsymbol{T}_{l+1}\boldsymbol{P}_{l+1}, \quad j = 1, 2, ..., N. \tag{7.104}$$

For a wave incident from medium 1, we have $e^-_{e,N+1} = 0$. Therefore,

$$\frac{e^+_{e,j}}{e^+_{e,1}} = \frac{M^{(j)}_{11}}{M^{(1)}_{11}} \tag{7.105}$$

and

$$\frac{e^-_{e,j}}{e^+_{e,1}} = \frac{M^{(j)}_{21}}{M^{(1)}_{11}}. \tag{7.106}$$

The use of (7.105) and (7.106) in (7.99) yields the field everywhere. For this time-periodic EM field, the magnetic field is given by the Faraday law for a time-periodic field (Table 3.5) and becomes

$$\boldsymbol{h}_e(x) = \frac{1}{i\omega_i\mu_o\mu_e}\nabla \times \boldsymbol{e}_e(x), \tag{7.107}$$

where μ_e is the relative magnetic permeability.

[†] The $\boldsymbol{T}$-matrix method uses the linearity of the Maxwell equations and is applied to planar and axisymmetric (revolution) geometries. The incident and scattered electric fields are expanded in Taylor series, and the coefficient designated as T_{ij} signifies the transition between them.

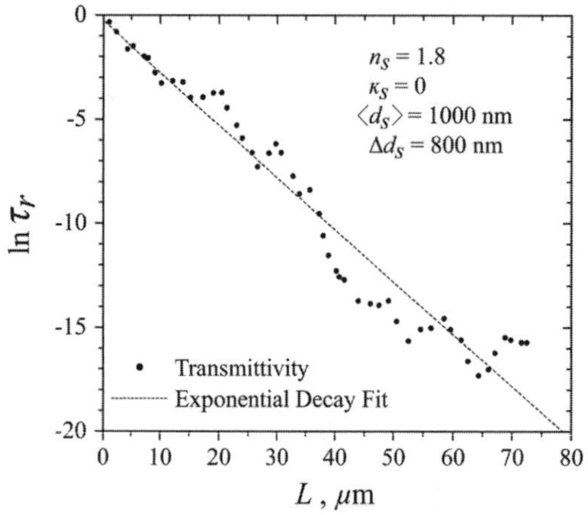

Figure 7.10. Variation of the transmissivity as a function of the medium thickness. The dots are the transmissivities obtained by adding layers to the medium, one at a time, and the dashed line is an exponential decay fit [291].

The EM energy (3.39) transmissivity is defined as

$$\tau_r = \frac{|\boldsymbol{e}^-_{e,N+1}|^2}{|\boldsymbol{e}_{e,i}|^2} \text{ transmissivity.} \tag{7.108}$$

This transmissivity is computed for a model multilayer with 50 layers in which the layer thicknesses d_s follows a uniform distribution between $\langle d_s \rangle \pm d_s = 1000 \pm 800$ nm, and with the porosity $\epsilon = 0.35$, $n_s = 1.8$ and $\kappa_s = 0$ (n_s and κ_s are for doped yttria compacts off resonance [282], $\lambda_i = 1000$ nm). To generate such a multilayer, we have an infinite number of possibilities, each of which is a realization. The nonabsorbing material is used to remove the possibility of causing exponential decay by absorption. To investigate the transmissivity as a function of the sample thickness, we add one layer to the sample at one time and obtain a series of transmissivities. The dots shown in Figure 7.10 represent these transmissivities for one typical realization, where there are 1, 2, ..., 50 layers. These dots are fitted well by an exponential decay line, indicating that localization exists. The localization length of this realization can be obtained from the slope of the fitted line

$$\lambda_{ph} = -\frac{\mathrm{d}L}{\mathrm{d}(\ln \tau_r)}. \tag{7.109}$$

There are infinite possible realizations for this model multilayer, and localization behavior is observed in all realizations that were tried. This supports the statement that random systems in one dimension always exhibit localization [238]. The localization length of this model multilayer is obtained by averaging the localization lengths of numerous realizations. For Figure 7.10 results, λ_{ph} is 3.2 μm, which is 3.2 times $\langle d_s \rangle$.

7.7.2 Coherence and Electric Field Enhancement

The localization is due to the constructive interference among the multiply scattered waves, even though the transmissivity profile in Figure 7.10 does not show any explicit signature of interference. The local field amplitude, rather than the

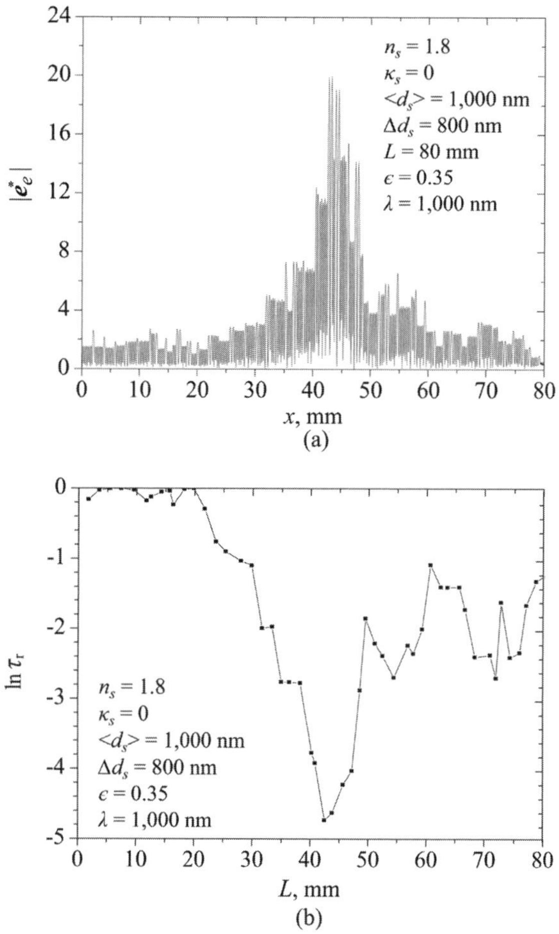

Figure 7.11. (a) Local field distribution in a random multilayer, with field enhancement shown. (b) Variations of the transmissivity as functions of the medium thickness in the same multi-layer [291].

transmissivity, has been examined and strong field enhancement was observed [282], which provides an exact signature of strong, constructive wave interference.

(A) *One-Dimensional Multilayer Geometry*

The local electric field is determined in a model multilayer geometry with 50 layers and subject to the same irradiation as in the previous section. There are also infinite possible realizations for this model multilayer, and the field results for one of them is shown in Figure 7.11(a), where the dimensionless electric field is defined as $|\boldsymbol{e}_e^*| = |\boldsymbol{e}_e|/|\boldsymbol{e}_{e,i}|$.

From Figure 7.11(a), it is evident that there is field enhancement, i.e., there are peaks in the field inside the medium that can be a few orders of magnitude larger than the incident field. Thus the energy density of the electric field can be two or even more orders of magnitude larger than the incident value. In periodic multilayer geometry, the electric field is also spatially periodic, resulting in no isolated peaks inside the medium (even if the field in this case can also be higher than the incident field). The physical basis of the field enhancement is EM wave interference. In the random multilayer system, the waves will multiply transmit and reflect at all interfaces, and interfere with each other. At some sites for some realizations, the interference is so ideally constructive that it results in an extremely large field. Thus the large field enhancement is directly attributed to randomness and cannot be observed in homogeneous or spatially periodic media. Note the coherence requirement that the medium size be smaller than the coherence length must be satisfied in order to observe the field enhancement. The coherence length is $\lambda^2/\Delta\lambda$ for a central wavelength λ and a spectrum width $\Delta\lambda$ [139]. Here we assume a monochromatic wave, thus satisfying the coherence condition ($\Delta\lambda$ is 0 and the coherence length is infinite). The coherence length of many lasers is several kilometers, satisfying the coherence condition. Note how the transmissivity evolves layer by layer in such a realization, leading to field enhancement, as shown in Figure 7.11(b). Different from Figure 7.10, in which the transmissivity reveals an exponential decay behavior, the variation of transmissivity with respect to the sample thickness, as shown in Figure 7.11(b), has a large fluctuation. The localization occurs and diffusion breaks down, because after local field enhancement, the transmissivity increases, instead of decreases, with an increase in the sample thickness. Note that the transmissivity for large L is much larger than that shown in Figure 7.10, as more photons have to travel through and interfere with each other to establish a large local field. Although realizations allowing field enhancement generally lead to a larger transmissivity, the field is trapped in a wavelength size region, forming a random resonator.

(B) *Two-Dimensional Geometry*

A two-dimensional random porous medium is shown in Figure 7.12(a). The solids are an array of infinitely long (in the y direction) dielectric cylinders with random radii and locations. We again examine the EM wave propagation subject to incidence of a planar EM wave upon this structure, traveling in the x direction. A code that is based on the finite-element method, high-frequency structure simulator (HFSS), is used to numerically solve the field distributions. A finite computation domain is chosen, resulting in six surfaces or boundaries. The two x–z boundaries are taken as periodic to simulate the infinite length in the y direction. The two x–y boundaries are also taken as periodic. The y–z planes are set to be the incident boundary at $x = 0$ and the radiation boundary at $x = L$.

A computation domain of $5\times1\times5$ μm^3 is chosen. Again, the local electric field in this region is determined for a normal incident EM wave of wavelength $\lambda = 1000$ nm for a random medium composed of 20 long cylinders with a random diameter $\langle d_s\rangle \pm \Delta d_s = 1000 \pm 400$ nm, and with $n_s = 1.8$ and $\kappa_s = 0$ in air (vacuum). The results are shown in Figure 7.12(b), in which contours of the constant dimensionless electric field are used. Similar to the one-dimensional case, there are sites of field

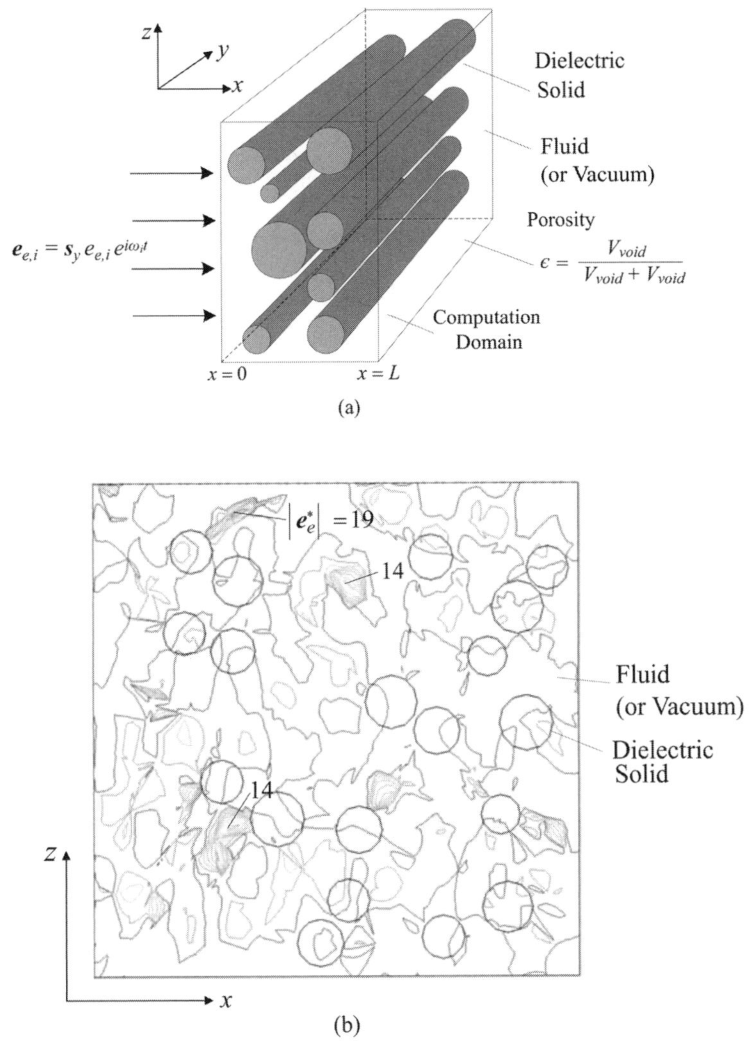

Figure 7.12. (a) The two-dimensional geometry composed of infinitely long cylinders. (b) The two-dimensional distribution of the dimensionless field magnitude $|e_e|^*$ for a realization. The circles are the sections of the cylinders. The porosity is 0.85 [291].

enhancement within the random porous medium that can be a few orders higher in magnitude compared with the incident field.

7.7.3 Comparison with Particle Treatment (ERT)

(A) *Formulation and Two-Flux Approximation*

The radiation transport within the same composite described in previous sections can also be treated by using the ERT. In this treatment, the scattering properties of a single particle are derived first, using small and large particle-size approximations. The effective scattering properties are then determined for a cluster of particles,

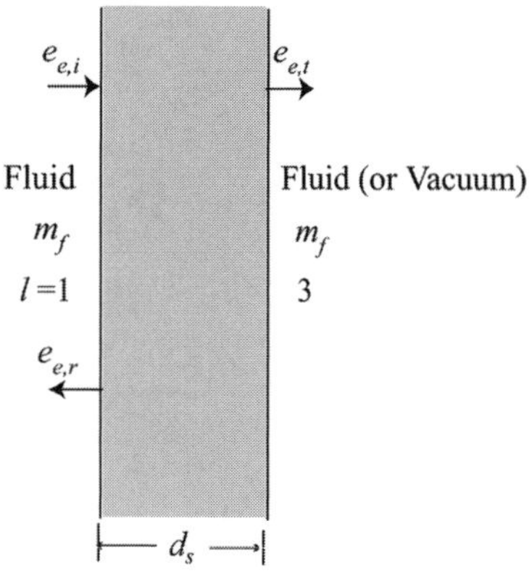

Figure 7.13. One-dimensional transmission and reflection in a single, effective (heterogeneous) layer.

based on the porosity and particle size distribution. The ERT is solved by use of some approximations, such as the two-flux approximation [172].

For a one-dimensional multilayer system, a layer serves as a single scatterer. For the normal incidence of a planar radiation upon a planar particle, as shown in Figure 7.13, the scattering is by reflection. Using the transfer matrix method we find that, the incident, transmitted, and reflected fields are related through

$$\begin{bmatrix} e_{e,i} \\ e_{e,r} \end{bmatrix} = \boldsymbol{T}_1^{-1}\boldsymbol{T}_2\boldsymbol{T}_2\boldsymbol{P}_2^{-1}\boldsymbol{T}_3 \begin{bmatrix} e_{e,t} \\ 0 \end{bmatrix}. \tag{7.110}$$

Thus $e_{e,r}$ and $e_{e,t}$ are derived in terms of $e_{e,i}$, and the transmissivity (7.108) is a function of the layer thickness d_s for a given m_f and m_s, i.e.,

$$\tau_r(d_s) = \frac{|e_{e,t}|^2}{|e_{e,i}|^2} = \frac{I_{ph,t}}{I_{ph,i}}. \tag{7.111}$$

The extinction (sum of absorption and scattering, Section 7.6.1) coefficient of the single particle is defined as

$$\sigma_{ph,ex}(d_s) = \frac{-\ln \tau_r(d_s)}{d_s}. \tag{7.112}$$

We now turn to the particle (ERT) treatment. For a random multilayer geometry with porosity ϵ, where the layer thickness d_s follows a uniform distribution in the range of $\langle d_s \rangle \pm \Delta d_s$, the effective extinction spectral coefficient becomes

$$\langle \sigma_{ph,ex} \rangle = \frac{\int_{\langle d_s \rangle - \Delta d_s}^{\langle d_s \rangle + \Delta d_s} -\ln \tau_r(d_s) \mathrm{d}d_s}{\int_{\langle d_s \rangle - \Delta d_s}^{\langle d_s \rangle + \Delta d_s} d_s \mathrm{d}d_s}(1-\epsilon). \tag{7.113}$$

Then the effective spectral absorption coefficient is given in term of κ_s as (Table 3.5)

$$\langle \sigma_{ph,a,w} \rangle = \frac{4\pi\kappa_s}{\lambda_i}(1-\epsilon). \tag{7.114}$$

The effective scattering coefficient is found from

$$\langle \sigma_{ph,s} \rangle = \langle \sigma_{ph,ex} \rangle - \langle \sigma_{ph,a} \rangle. \qquad (7.115)$$

Because the scattering occurs only in the backward direction, the scattering phase function, $\Phi(\theta)$ defined in (7.79), is a delta function, i.e.,

$$\Phi(\theta) = \delta(\theta - \pi), \quad \int_{-1}^{1} \Phi(\theta) \mathrm{d}\cos\theta = 2. \qquad (7.116)$$

Then ERT (7.83), replacing the scattering term with the expression containing the phase function Φ [172], and under no emission, becomes

$$\begin{aligned}\cos\theta \frac{\mathrm{d}I_{ph}(x,\theta)}{\mathrm{d}x} &= -(\langle \sigma_{ph,s} \rangle + \langle \sigma_{ph,a} \rangle) I_{ph}(x,\theta) \\ &+ \frac{\langle \sigma_{ph,s} \rangle}{2} [\int_{0}^{1} I_{ph}(x,\theta_i) \Phi(\theta_i - \theta) \mathrm{d}\cos\theta_i \\ &+ \int_{-1}^{0} I_{ph}(x,\theta_i) \Phi(\theta_i - \theta) \mathrm{d}\cos\theta_i]. \end{aligned} \qquad (7.117)$$

For normal irradiation on a one-dimensional geometry, the radiative intensity has only two components, namely, forward and backward. Thus the two-flux approximation [172] can be used to solve (7.117). By taking $\theta = 0$ and defining $I_{ph}(x,0) = I_{ph}^{+}$, and $I_{ph}(x,\pi) = I_{ph}^{-}$, (7.117) gives for the forward intensity

$$\frac{\mathrm{d}I_{ph}^{+}}{\mathrm{d}x} = -(\langle \sigma_{ph,s} \rangle + \langle \sigma_{ph,a} \rangle) I_{ph}^{+} + \langle \sigma_{ph,s} \rangle [(1-B) I_{ph}^{+} + B I_{ph}^{-}], \qquad (7.118)$$

where

$$B = \frac{1}{2} \int_{-1}^{0} \Phi(\theta_i) \mathrm{d}\cos\theta_i = 1. \qquad (7.119)$$

Equation (7.118) can then be simplified to

$$\frac{\mathrm{d}I_{ph}^{+}}{\mathrm{d}x} = -(\langle \sigma_{ph,s} \rangle + \langle \sigma_{ph,a} \rangle) I_{ph}^{+} + \langle \sigma_{ph,s} \rangle I_{ph}^{-}. \qquad (7.120)$$

Similarly, by taking $\theta = \pi$, (7.117) gives for the backward intensity

$$-\frac{\mathrm{d}I_{ph}^{-}}{\mathrm{d}x} = -(\langle \sigma_{ph,s} \rangle + \langle \sigma_{ph,a} \rangle) I_{ph}^{-} + \langle \sigma_{ph,s} \rangle I_{ph}^{+}. \qquad (7.121)$$

The boundary conditions for (7.120) and (7.121) are

$$I_{ph}^{+}(0) = 1, \quad I_{ph}^{-}(L) = 0. \qquad (7.122)$$

(B) *Photon Diffusion in Nonabsorbing, Random Media*

Here we show that the particle-based ERT is not capable of predicting photon localization. Again, a nonabsorbing medium ($\sigma_{ph,a} = 0$) is used to remove the possibility of causing exponential decay by absorption. Equations (7.120) and (7.121) are simplified to

$$\frac{\mathrm{d}I^+_{ph}}{\mathrm{d}x} = -\langle\sigma_{ph,s}\rangle I^+_{ph} + \langle\sigma_{ph,s}\rangle I^-_{ph}$$

$$-\frac{\mathrm{d}I^+_{ph}}{\mathrm{d}x} = -\langle\sigma_{ph,s}\rangle I^-_{ph} + \langle\sigma_{ph,s}\rangle I^+_{ph}. \qquad (7.123)$$

The solutions subject to the boundary conditions given by (7.122) are

$$I^+_{ph}(x) = \frac{1 + \langle\sigma_{ph,s}\rangle(L - x)}{1 + \langle\sigma_{ph,s}\rangle L}$$

$$I^-_{ph}(x) = \frac{\langle\sigma_{ph,s}\rangle(L - x)}{1 + \langle\sigma_{ph,s}\rangle L}. \qquad (7.124)$$

Then the transmissivity is found from

$$\tau_r(L) = I^+_{ph}(L) = \frac{1}{1 + \langle\sigma_{ph,s}\rangle L}. \qquad (7.125)$$

In random porous media, $\langle\sigma_{ph,s}\rangle L$ is generally much larger than 1 (optically thick), and $\tau_r(L)$ has an inverse dependence, i.e., $\tau_r(L) = 1/[\langle\sigma_{ph,s}\rangle L]$, which is characteristic of the classical diffusion [340] with a mean free path $\lambda_{ph} = 1/\langle\sigma_{ph,s}\rangle$. Thus photon localization cannot be predicted by the ERT, because the random medium is represented as an effective medium in order to define the effective scattering and absorption properties, where EM wave interference (coherence) effects are not now filtered (i.e, removed).

7.8 Continuous and Band Photon Absorption

7.8.1 Photon Absorption in Solids

Photon absorption by matter is described by interaction of EM waves with electric entities (e.g., free electron and electric dipoles) as discussed in Section 7.3. The spectral photon absorption coefficient $\sigma_{ph,\omega}$ of a solid is experimentally found through the simplified form of the ERT (Table 7.1) under unidirectional (along x) radiation, steady-state, no $\dot{s}_{ph,i}$, and the dominance of absorption, i.e.,

$$\frac{\mathrm{d}I_{ph,\omega}}{\mathrm{d}x} = -\sigma_{ph,\omega} I_{ph,\omega}, \quad \frac{I_{ph,\omega}}{I_{ph,\omega}(x = 0)} = \exp(-\sigma_{ph,\omega} x)$$

ERT (Beer–Lambert law) for columnar, absorption dominated radiation, (7.126)

where $\sigma_{ph,\omega}$ is assumed constant. Note that (7.126) is also obtained for a columnar radiation ($\cos\theta = 1$), under no emission, from (7.88). The theoretical prediction of

$\sigma_{ph,\omega}$ is through (7.54), where the interaction (transition) probability rate (FGR) (5.95) gives $\tau_{ph,a} = 1/\dot{\gamma}_{ph,a}$, and using the absorption time constat $\tau_{ph,a}$

$$\sigma_{ph,\omega} = \frac{\hbar\omega\dot{\gamma}_{ph,a}n_e}{u_{ph}\int \mathrm{d}\omega} = \frac{\hbar\omega n_e}{\tau_{ph,a}u_{ph}e_{ph}} \quad \text{absorption time constant,} \qquad (7.127)$$

where e_{ph} is given by (7.2).

Using the spectral optical property κ_λ (spectral index of extinction, Table 3.5), $\sigma_{ph,\omega}$ is also given as

$$\sigma_{ph,\omega} \equiv \frac{2\omega\kappa_\lambda}{u_{ph}} = \frac{4\pi\kappa_\lambda}{\lambda} \quad \text{relation with index of extinction.} \qquad (7.128)$$

In turn, κ_λ is related to σ_e, μ_e, and ϵ_e through the relation given in Table 3.5, using the classical treatment. In doped ions of otherwise transparent solids, and in dilute gases and solutions, the absorption coefficient is given in terms of the absorption cross-section area $A_{ph,a}$ per absorber, and the absorber concentration n_a, as

$$\sigma_{ph,\omega} = A_{ph,a}n_a \quad \text{absorption cross-section area relation,} \qquad (7.129)$$

and (7.126) is used to measure $A_{ph,a}$. In general, $A_{ph,a}$ depends on the temperature and pressure (e.g., residual stress in solids). The photon absorption cross-section area is much smaller than the molecular collision cross-section area (Table 6.4).

In addition, the photon spectral absorptance $\alpha_{ph,\omega}$ is defined as

$$\alpha_{ph,\omega} = 1 - \exp(-\sigma_{ph,\omega}L) \quad \text{spectral absorptance,} \qquad (7.130)$$

where L is the path length of photons. For small $\sigma_{ph,\omega}L$ (optically thin limit), we have $\alpha_{ph,\omega} = \sigma_{ph,\omega}L$, i.e., absorptance is equal to the optical thickness.[†]

Figure 7.14 shows the hypothetical absorption coefficient of a semiconductor-like crystal, caused by various absorption mechanisms. A rendering of these various photon absorption mechanisms is shown in Figure 7.15. These electronic (including semiconductors), phonon-mediated (exiton), phonon (optical), and other photon absorptions are subsequently described.

In Figure 7.14, starting at high frequencies (ultraviolet regime), the absorption of a photon is due to electronic transitions (intraband transition for electrons and holes), resulting in photoconductivity. For high photon energy (greater than 10 eV, as also marked in Figure 1.4) the fall in the absorption coefficient is rather small, but in the visible–near-infrared regime the fall is rather rapid and is called the semiconductor low-energy broadening of the fundamental absorption edge (energy gap $\Delta E_{e,g}$). In this regime, for ionic crystals, exciton absorption becomes significant.

As the far infrared is approached, the intraband conduction or valence electronic transition occurs (free-carrier absorption). This extends to far-infrared and microwave regimes, and the magnitude of $\sigma_{ph,\omega}$ depends on the mobile electron orbital density. In metals this is the dominant absorption.

[†] $\alpha_{ph,\omega} = 1 - e^{-\sigma_{ph,\omega}L} \simeq 1 - 1 + \sigma_{ph,\omega}L + \cdots = \sigma_{ph,\omega}L$ optically thin limit.

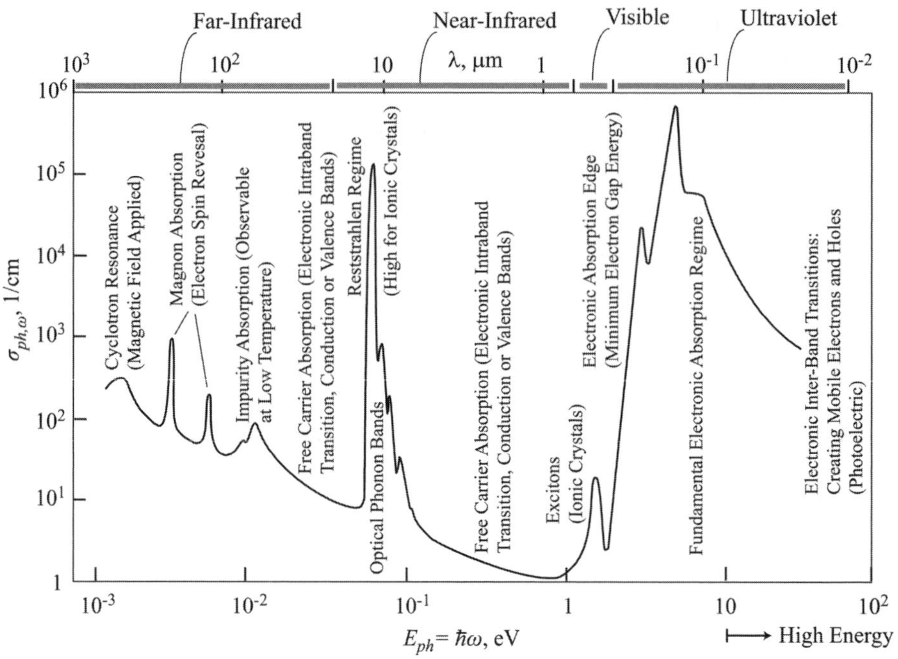

Figure 7.14. Generalized photon absorption spectrum of crystalline semiconductor (typical groups 13–15 elements in Table A.2) [101]. The far-infrared to ultraviolet photon energy regimes are also shown.

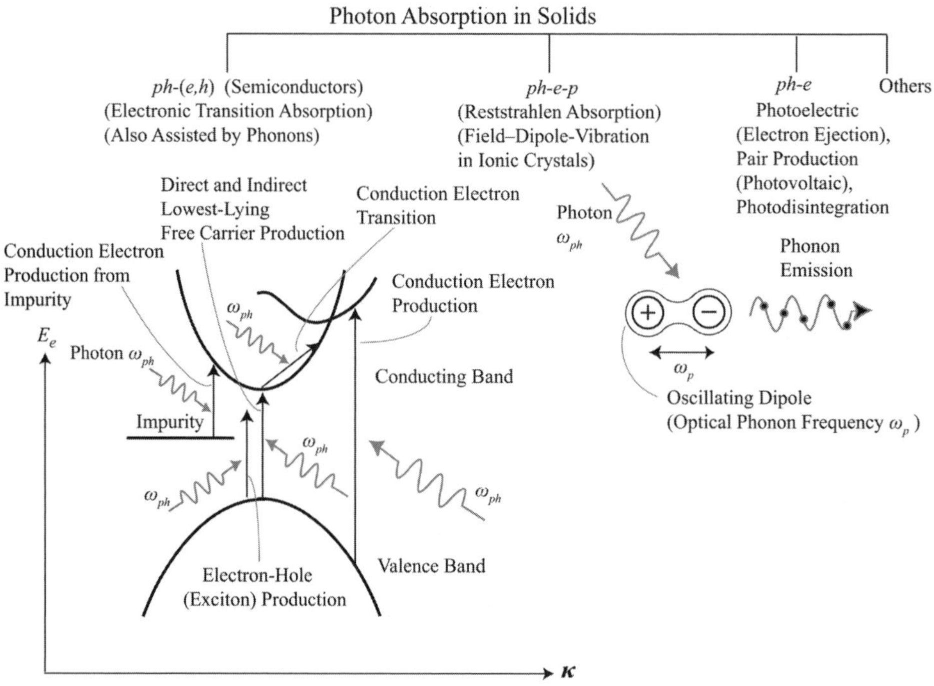

Figure 7.15. Various mechanisms of photon absorption in solids. Examples of photon absorption in metal, semiconductor, and dielectric (electric insulator) are given in Figure 7.1(b).

At the end of the near-infrared regime and beginning of the far-infrared regime (photon energy of 0.02 to 0.05 eV, of the order of $k_B T$, for moderate T), the photon–phonon interactions dominate and very high absorption coefficients are found (highest for ionic crystals) along with high reflection.

Impurities also result in the absorption of low-energy photons, and this is detectable only at low temperatures (otherwise the impurities ionize); at yet lower photon energies, absorption that is due to electron spin reversal occurs because of paramagnetic imparities. Some of the low energy absorptions may require imparities of a magnetic field.

The classical optical relation or the FGR is used to arrive at the absorption coefficient, as previously outlined.

The free-electron spectral absorption coefficient for metals is given as (Chapter 3 problem; note that $I_{ph,\omega}$ is proportional to $|\boldsymbol{e}_e|^2$, so $\sigma_{ph,\omega} = 2/\delta_{ph\text{-}e}$, where δ is the skin depth) [92, 109]

$$\sigma_{ph,\omega} = \frac{2\omega\kappa_\omega}{u_{ph}} = (\frac{2n_{e,c}e_c^2\langle\langle\tau_e\rangle\rangle\omega}{\epsilon_o\epsilon_e m_e u_{ph}^2})^{1/2} \quad \text{metals}, \tag{7.131}$$

where $\langle\langle\tau_e\rangle\rangle$ is the average momentum electron relaxation time of (5.132) and (5.146) determined from the electrical conductivity.

For semiconductors, the absorption coefficient is defined for the interband (between valence and conduction electrons) transitions [92], similar to (7.127), as

$$\sigma_{ph,\omega} = \frac{\hbar\omega\dot{\gamma}_{ph\text{-}e}}{I_{ph,\omega}} \quad \text{direct interband transition}, \tag{7.132}$$

where $\dot{\gamma}_{ph\text{-}e}$ is the rate of interband transitions per unit volume and per unit time, and $I_{ph,\omega}$ is the incident electromagnetic flux. Then we use the FGR (3.27) along with the interaction matrix element (5.93). The perturbation Hamiltonian is related to the local dipole (end of chapter problem), and leads to the matrix element (transition between valence and conduction electrons)

$$|M_{v,c}|^2 = (\frac{e_c}{m_{e,e}u_{ph}})^2|\boldsymbol{a}_e|^2|\langle\psi_v|e_c\boldsymbol{x}|\psi_c\rangle|^2 = (\frac{e_c}{m_{e,e}u_{ph}})^2|\boldsymbol{a}_e|^2|\boldsymbol{\mu}_e|^2, \tag{7.133}$$

where $\boldsymbol{a}_e$ is the vector potential (Table 3.5) and $e_e\boldsymbol{x} = \boldsymbol{p}_e$ is the local dipole of electronic transition, the expression for $\sigma_{ph,\omega}$ is [92]

$$\sigma_{ph,\omega} = \frac{e_c^2}{\epsilon_o^2 m_{e,e}^2 u_{ph} n_\omega \omega}|\langle\psi_v|e_c\boldsymbol{x}|\psi_c\rangle|^2 D_{ph\text{-}e}[f_e^o(E_{e,v}) - f_e^o(E_{e,c})]$$

$$\text{direct interband transition}, \tag{7.134}$$

where n_ω is the spectral index of refraction, $D_{ph\text{-}e}$ is the joint DOS, and $|\langle v|\boldsymbol{p}_e|c\rangle|$ couples the states with the same $\boldsymbol{\kappa}$ in the valence and conduction bands.

An analytical relation for the resonant photon absorption–emission for the transition rate of the conduction band to the heavy-hole band and the conduction band to the light-hole band (Figure 5.7) is given in [20]. A similar relation is derived for the indirect band, in which phonon energy is absorbed or emitted to match the different $\boldsymbol{\kappa}$ of the valence and conduction bands. There are two coupling coefficients, namely, *ph-e-p* coupling involving phonon absorption ($a_{ph\text{-}e\text{-}p,a}$), and *ph-e-p* coupling involving phonon emission ($a_{ph\text{-}e\text{-}p,e}$).

Table 7.2 lists absorption of photons by atoms, ions, or molecules capable of electronic transition through the transition dipole moment $\boldsymbol{\mu}_e$. The relations to the absorption cross-section area and the Einstein coefficients are also used. Table 7.2 also summarizes the spectral absorption coefficient $\sigma_{ph,\omega}$ for photon absorption that is due to electrons and for conduction (free) electrons (free-carrier absorption) in metals and semiconductors; also listed are $\sigma_{ph,\omega}$ for semiconductor interband (direct- and indirect- band) absorption transitions.

From Table 7.2, the frequency dependence for free carrier absorption in metals is $\omega^{1/2}$, whereas for semiconductors it is ω^{-2}, i.e., much stronger frequency dependence. The frequency dependence of the direct band spectral absorption is $(\hbar\omega - \Delta E_{e,g})^{1/2}/\omega$, after proper substitutions. The phonon absorption–emission assisting in indirect-band absorption gives a $(\hbar\omega - \Delta E_{e,g} \pm \hbar\omega_p)^2$ dependence, which again is much stronger frequency dependent than that for the direct bands.

The inverse of $\sigma_{ph,\omega}$ is referred to as the photon mean free path $\lambda_{ph,\omega} = 1/\sigma_{ph,\omega}$, (7.97). The photon mean free path in metals is very short and is called the skin depth (the metals are also highly reflective). The derivation of $\sigma_{ph,\omega}$ is a Chapter 3 problem.

Absorption (attenuation) of EM waves by dielectrics was discussed in Section 3.3.5. Figure 7.1(b) gives example of $\sigma_{ph,\omega}$ for a metal (Al), a semiconductor (GaAs), a dielectric (SiO_2), and an organic solid (acrylic). These are predictable with the relations previously discussed.

7.8.2 Photon Absorption in Gases

The interaction of a photon (EM field) and fluid particle (or atom and molecule in general) is considered as the interaction of an electric field $\boldsymbol{e}_e$ (magnetic field interaction is considered for magnetic materials) with an electric transition dipole moment $\boldsymbol{\mu}_e$, i.e.,

$$\boldsymbol{\mu}_e = \langle\psi_f|e_c\boldsymbol{x}|\psi_i\rangle, \tag{7.135}$$

where ψ_i and ψ_f are the initial- and final- state wave functions of the electron.

The perturbation Hamiltonian is

$$\text{H}' = -|\boldsymbol{\mu}_e|e_{e,o}\cos(\omega t), \quad \boldsymbol{e}_e = e_{e,o}\cos(\omega t)\boldsymbol{s}_\alpha, \tag{7.136}$$

where $\boldsymbol{s}_\alpha$ is the polarization vector.

Table 7.2. *Some relations for spectral photon absorption and scattering coefficients in isolated atoms, ions, or molecules, and in solids (semiconductors and metals).*

Mechanism	Relation
Electronic transition (absorption) in atom, ion, or molecule	$\sigma_{ph,\omega} = n_{e,A} A_{ph,a}(\omega) = n_{e,A} \frac{\pi^2 u_{ph}^2}{\omega_{e,g}^2 \int_\omega d\omega} \dot{\gamma}_{ph,e,sp}$ $n_{e,A}$ is the number density of ground-state atoms (or ions or molecules), $\omega_{e,g}$ is the transition angular frequency, $\dot{\gamma}_{ph,e,sp} = 1/\tau_{ph,e,sp} = 1/\tau_r = \omega_{e,g}^3 \lvert\boldsymbol{\mu}_e\rvert^2 / 3\pi\epsilon_o \hbar u_{ph}^3$, $\boldsymbol{\mu}_e$ is the transition dipole moment, $\int_\omega d\omega$ is the bandwidth
Photoelectric (hydrogen) (Appendix E)	$\sigma_{ph\text{-}e} = n_{\rm H} A_{ph\text{-}e},\ A_{ph\text{-}e} = \frac{512\pi^2}{3} \frac{r_{\rm B}^2 e_c^2}{4\pi\epsilon_o \hbar c} (\frac{E_{e,1}}{\hbar\omega})^4 \frac{e^{-4\gamma \cot^{-1}\gamma}}{1-e^{2\pi\gamma}},$ $\gamma = (\frac{\hbar\omega}{E_{e,1}} - 1)^{-1/2}$ $n_{\rm H}$ is number density of H atoms, $E_{e,1}$ is ground-state energy
Molecular vibrational absorption	$\sigma_{ph,\omega} = \frac{4\pi^2 n_{f,v}\omega}{3\hbar u_{ph} Z_f} (e^{-E_{f,i}/k_{\rm B}T} - e^{-E_{f,f}/k_{\rm B}T}) \lvert\langle\psi_f\lvert\boldsymbol{p}_e\rvert\psi_i\rangle\rvert^2 f(\omega, \omega_{i\text{-}f})$ quantities defined under (7.138)
Compton (free-electron) scattering	$\sigma_{ph,s} = n_{e,c} \frac{8\pi}{3} r_{e,o}^2 = n_{e,c} \times (6.649 \times 10^{-29} {\rm m}^2),\quad r_{e,o}^2 = \frac{e_c^2}{4\pi m_e \epsilon_o u_{ph,o}^2}$ $n_{e,c}$ is the number density of free electrons
Rayleigh (bound-electron) scattering	$\sigma_{ph,s,\omega} = n_{e,b} \frac{8\pi}{3} r_{e,o}^2 \frac{\omega^4}{(\omega^2 - \omega_{n,e}^2)^2}$ $n_{e,b}$ is the number density of bound electrons
Semiconductor interband electronic transitions [92]	indirect-band absorption: with phonon absorption $\sigma_{ph,\omega} = a_{ph\text{-}e\text{-}p,a} \frac{(\hbar\omega - \Delta E_{e,g} + \hbar\omega_p)^2}{\exp(\hbar\omega_p/k_{\rm B}T) - 1},$ where $a_{ph\text{-}e\text{-}p,a}$ is the phonon absorption coupling constant indirect-band absorption: with phonon emission $\sigma_{ph,\omega} = a_{ph\text{-}e\text{-}p,e} \frac{(\hbar\omega - \Delta E_{e,g} \hbar\omega_p)^2}{1 - \exp(-\hbar\omega_p/k_{\rm B}T)},$ where $a_{ph\text{-}e\text{-}p,e}$ is the phonon emission coupling constant direct-band abosorption: $\sigma_{ph,\omega} = \frac{e_c^2}{\epsilon_o^2 m_{e,e}^2 u_{ph} n_\omega \omega} \lvert\langle\psi_v\lvert e_c \boldsymbol{x}\rvert\psi_c\rangle\rvert^2 D_{ph\text{-}e}[f_e^o(E_{e,v}) - f_e^o(E_{e,c})],$ where n_ω is the index of refraction, $D_{ph\text{-}e}$ is the joint density of states, and $\lvert\langle\psi_v\lvert e_c\boldsymbol{x}\rvert\psi_c\rangle\rvert = \lvert\langle v\lvert\boldsymbol{p}_e\rvert c\rangle\rvert$ couples the states with the same $\boldsymbol{\kappa}$ in the valence and conduction bands for GaAs, $\sigma_{ph,\omega} = 1.9 \times 10^{-3} \frac{(\hbar\omega - \Delta E_{e,g})^{1/2}}{\hbar\omega}$
Free carrier absorption [92, 38]	absorption in metals: $\sigma_{ph,\omega} = (2\sigma_e \mu_o \mu_e \omega)^{1/2} = (\frac{2 n_{e,c} e_c^2 \langle\langle\tau_e\rangle\rangle\omega}{\epsilon_o \epsilon_e m_e u_{ph}^2})^{1/2}$ absorption in semiconductors: $\sigma_{ph,\omega} = \frac{\omega_{pl}^2}{(4\pi)^{1/2} u_{ph} \omega^2 \langle\langle\tau_e\rangle\rangle},$ ω_{pl}(unscreened plasma frequency) $= (\frac{n_{e,c} e_c^2}{\epsilon_o m_{e,e}})^{1/2}$

The discrete, internal, translational, vibrational, rotational, and electronic energy states of the fluid particle (including the gas phase) were shown in Figure 6.1. Therefore we expect resonant absorption at the discrete energies, and because these internal energies coexist, there are band broadenings when, for example, a vibrational transition is accompanied by one or more rotational transitions.

The spectral absorptance $\alpha_{ph,\omega}$ is defined for a gas having a length L along a collimated beam of frequency ω and is given by (7.130), where $\sigma_{ph,\omega}$ is the absorption coefficient. Figure 7.16(a) shows the near-infrared spectra of absorptance of CO_2, H_2O, and CH_4, for the given temperature, total pressure p, and partial pressure p_i (for CH_4). Using the ideal gas relation (6.21), $p_i = n_{f,i}k_BT$, and as will be shown $\sigma_{ph,\omega}$ is proportional to $n_{f,i}$ which are excited. There are a few absorption bands for each molecule, and strength (absorption) of these bands (and their widths) depend on the temperature, total pressure, and partial pressure of these species. For a given temperature, the product of the partial pressure and the beam path length is used to specify the gas conditions. Details on the band models and their calculation are given in [245].

The vibrational energy transitions are larger than the rotational energy transitions (examples will be given in Section 7.13), and also these transitions occur simultaneously. This is the reason for the band-type behavior of the gas photon absorption and emission. The combined rigid-rotor and harmonic-oscillator models gives an energy transition (higher-order Hamiltonians are also used when accuracy requires), which combines (6.14) and (6.16), i.e.,

$$E_{f,v\text{-}r} = (l + \frac{1}{2})\hbar\omega_f + j(j+1)\frac{\hbar^2}{2I_f}, \quad l, j = 1, 2, ... \qquad (7.137)$$

Note that allowed transition $\Delta l = \pm 1$ combined with $\Delta j = 0, \pm 1, ...,$ creates distinct branches in each band. Some of the significant bands in common combustion/environmental gases are listed in Table 7.3 (H_2O, CO_2, CO, NO, SO_2, and CH_4) [98].

The vibrational degrees of freedom expresses the modes of vibration, and for example, for CO_2, which is a linear molecule, there is a symmetric mode, a bending mode, and an asymmetric (two oxygen atoms oscillate against the carbon atom). The vibrational quantum numbers, l_1, l_2, and l_3 in parentheses in Table 7.3 are for the symmetric, bending, and asymmetric modes.

Figure 7.16(b) shows variations of the measured spectral photon absorption cross-section area (per molecule) of CO_2 as functions of the photon wavelength for $T = 298$ K and a total pressure of 1 atm (dilute CO_2 concentration in N_2). CO_2 has absorption peaks at 2.0, 2.7, 4.3, 4.8, 9.4, 10.4, and 15 μm, and Figure 7.16(b)(i) shows the strongest bands at 4.3 and 15 μm (with another but less significant band at 2.7μm). Note the band broadening that is due to rotational energy transitions. There is also band broadening that is due to gas particle collisions and due to the Doppler effect (relative velocity of the absorbed photon and the fluid particle thermal translational motion). The Doppler broadening is a signature of the gas temperature.

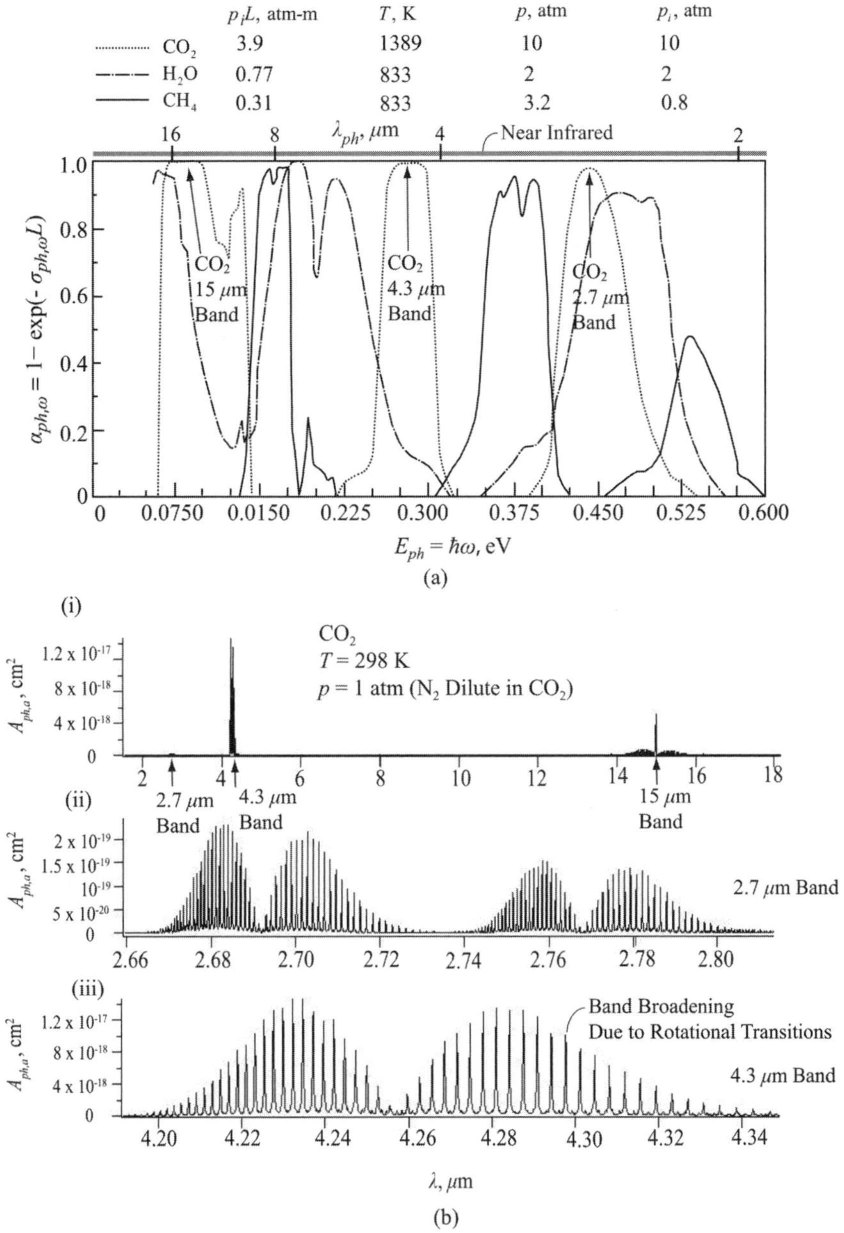

Figure 7.16. (a) Near-infrared variations of the spectral absorptance for CO_2, H_2O, and CH_4 gases, with respect to photon energy, for the listed temperature and total and partial pressures [306]. Each absorption band has a bandwidth that depends on the pressure and temperature. (b) Variations of the measured photon absorption cross-section area of CO_2 (at $T = 298$ K and $p = 1$ atm total pressure) as functions of wavelength. The partial pressure of CO_2 is low (N_2 is the background gas). Note the strong absorption around 4.2–4.3 μm which is shown in more detail in (b)(iii) [89].

Table 7.3. *Major gas absorption bands for some combustion/environmental gases [98]. The CO_2 bending vibrational modes are degenerate (two modes), but here are shown as only one.*

	Band center, μm	Vibrational degrees of freedom	$\pm l_1, \pm l_2 \pm l_3$ $\pm l_1, \pm l_2 \pm l_3$
H_2O	71	3	(0,0,0), Δj only
	6.3	3	(0,1,0)
	2.7	3	(0,2,0)
	1.87	3	(0,1,1)
	1.38	3	(1,0,1)
CO_2	15	3	(0,1,0)
	10.4	3	(−1,0,1)
	9.4	3	(0,−2,1)
	4.3	3	(0,0,1)
	2.7	3	(1,0,1)
	2.0	3	(2,0,1)
CO	4.7	1	(1)
	2.35	1	(2)
NO	5.34	1	(1)
SO_2	19.27	3	(0,1,0)
	8.68	3	(1,0,0)
	7.35	3	(0,0,1)
	4.34	3	(2,0,0)
	4.0	3	(1,0,1)
CH_4	7.66	4	(1,0,0,1)
	3.31	4	(0,0,1,0)
	2.37	4	(1,0,0,1)
	1.71	4	(1,1,0,1)

The natural (fundamental) frequency $\omega_n = \omega_f$ can be estimated with the interatomic spring (force) constant using (2.55).

For quantum calculation, for atmospheric gases, the spectral absorption coefficient of the transition electric dipole resonance is used [260]. This is based on the FGR (3.27) and (7.54), and is given by

$$\sigma_{ph,\omega} = \frac{4\pi^2 n_{f,v}\omega}{3\hbar u_{ph} Z_f}(e^{-E_{f,i}/k_B T} - e^{-E_{f,f}/k_B T})|\langle\psi_f|\boldsymbol{p}_e|\psi_i\rangle|^2 g(\omega,\omega_{i\text{-}f}), \quad (7.138)$$

where $n_{f,v}$ is the number density of the particular vibrational state of the gas molecule, $E_{f,i}$ and $E_{f,f}$ are the lower and upper energies for the transition, Z_f is the partition function, $\boldsymbol{\mu}_e = \langle\psi_f|\boldsymbol{p}_e|\psi_i\rangle$ is the transition dipole moment, and includes $|\psi_i\rangle$ and $|\psi_f\rangle$, which are the wave functions of the upper and lower states (7.135, and $g(\omega,\omega_{i\text{-}f})$ is the line-shape function (Table 3.5) to allow for line broadening that is due to molecular collision, etc. Note that the matrix element $|\boldsymbol{\mu}_e|$ is that defined by (3.29) and similarly in (5.93). Equation (7.138) is also listed in Table 7.2. In Section 7.12.6 we discuss the *ab initio* computation of $|\boldsymbol{\mu}_e|$ for a molecular complex.

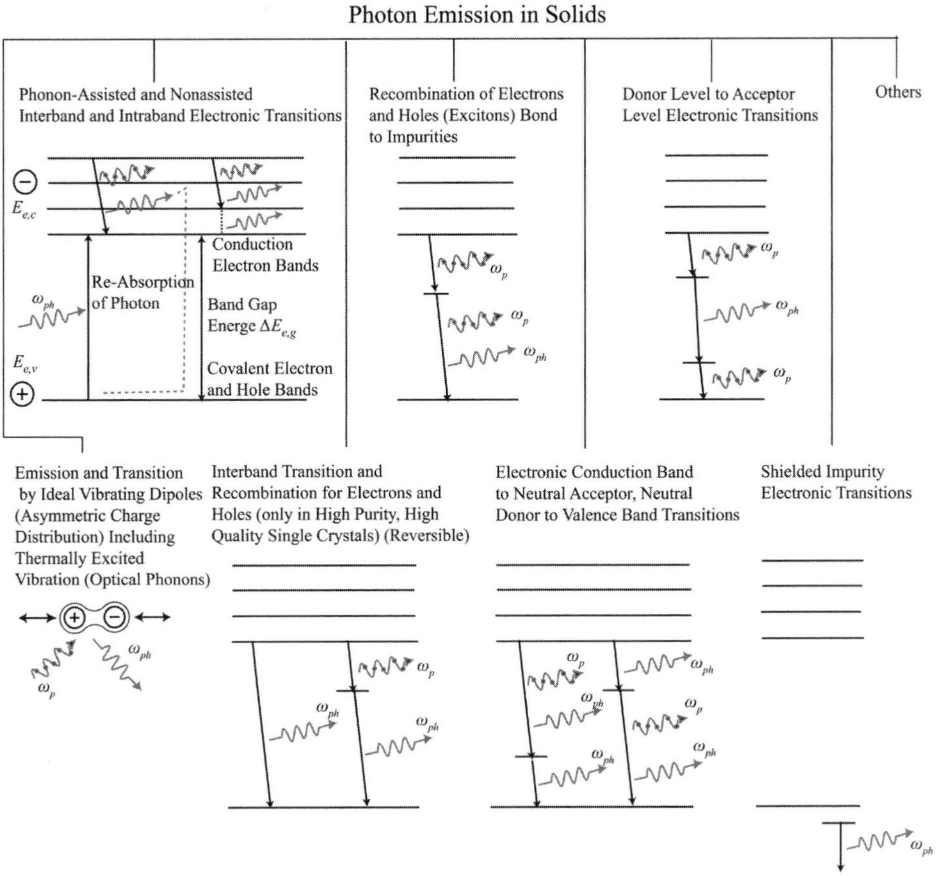

Figure 7.17. Various mechanisms of photon emission in solids [101]. These in turn determine the electric permittivity (dielectric function) of the material.

Figure 7.1(b) gives example of $\sigma_{ph,\omega}$ for O_3 (ozone) at 300 K at 1 atm. It also gives $\sigma_{ph,\omega}$ for H_2O at the same thermodynamic conditions.

7.9 Continuous and Band Photon Emission

7.9.1 Emission Mechanisms

Although in some processes the photon emission is the reverse of a photon absorption process, in many cases it is not. The most commonly encountered photon emission mechanisms, which are electronic, but may be phonon assisted, are shown in Figure 7.17.

Photons can be emitted by vibrating dipoles of a multiatom crystal. This is limited in frequency to the highest optical phonons, although multiple phonon contributions are possible (but the probability decreases with the number involved in the interaction).

In the phonon-assisted interband and intraband electron transitions, the electrons in the upper conduction band come to thermal equilibrium with a lattice (i.e., thermalize) rather quickly (Sections 5.19 and 5.22.2). The emitted photon for this mechanism has a range of energy (broad spectral range). In this process, if the photon energy exceeds the gap energy $\Delta E_{e,g}$, it maybe reabsorbed, causing further electronic transitions. The intraband emissions would require special excitation contributions (electron avalanches).

The interband recombination of electrons and holes with photon emission is the reverse of strong photon absorption, but relative to other processes is less likely (except for high-purity crystals). The interband transitions can occur in direct and indirect gap transitions. The emission band width should be of the order of $k_B T$, because of thermalization of electrons and holes.

The impurity-related emissions, including acceptors and donors, can be due to annihilation by the exciton band to impurities. Again these are phonon emission or absorption assisted. Transitions between band edges and acceptor or donor impurities are common, and can be phonon assisted. The electrons captured at donor impurity levels can recombine with holes captured at acceptor levels and emit photons.

The optical transition from excited states to ground states of impurities can be influenced by the lattice (strong coupling) or slightly perturbed (well-shielded). The well-shielded impurities result in transitions and photon emission spectra with sharp lines (related to equivalent undisturbed atomic lines). The energy levels of triply ionized rare-earth ion impurities are deduced from fluorescence experimental results. Solid-state lasers use such ion-embedded crystals (such as neodymium ion Nd^{3+} in yttria–aluminum–garnet, YAG).

In Section 7.10, the surface emission of Si will be examined, with particular attention to the lattice-vibration regime. In Section 7.15.1 we consider the near-field emission of Si. The lattice vibration in turn influences the frequency dependence of the dielectric function ϵ_e (Section 3.3.5) and the signature of phonon-resonance-induced peaks in ϵ_e are evident in the near-field emission.

7.9.2 Reciprocity between Spectral Photon Absorption and Emission for Multiple-Level Electronic Systems

We now consider the cross-section area for the photon absorption and emission processes arising for the generalized two electronic energy levels, the upper and lower states, as shown in Figure 7.18(a). The two levels are characterized by the energies $E_{e,i}$ and $E_{e,j}$ and the degeneracies $g_{e,i}$ and $g_{e,j}$. The emission and absorption cross-section areas can be determined as a summation of the individual cross-section areas $(A_{ph,e})_{ij}$ and $(A_{ph,a})_{ji}$, between the levels. Assume the multiplets A and B are each under their own thermal equilibrium; we have

$$A_{ph,e}(\omega) = \sum_{i,j} g_{e,j}\{\frac{\exp[-(E_{e,j} - \Delta E_{e,AB})/k_B T]}{Z_B}\}(A_{ph,e})_{ji} g_{e,i} \quad (7.139)$$

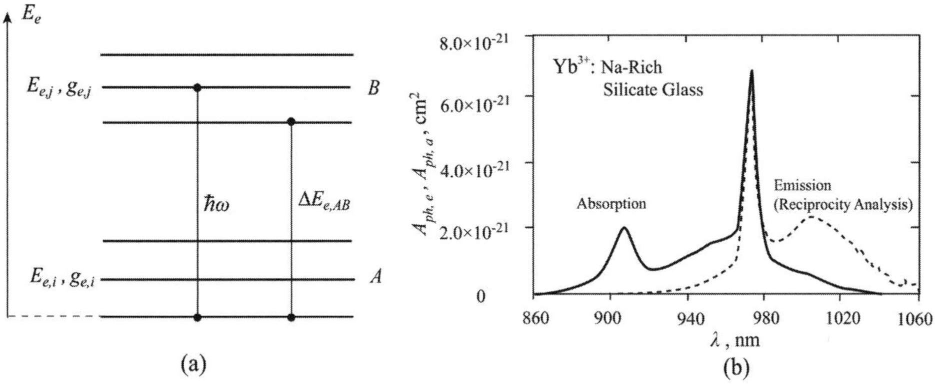

Figure 7.18. (a) Electronic transitions in a two multiplets (A and B) with sublevels (i and j). The absorbed photon energy is $\hbar\omega$, and the electronic transition is $\Delta E_{e,AB}$. (b) Same showing the absorption and emission cross-section areas for Yb^{3+} in a silicate-blend glass [107].

$$A_{ph,a}(\omega) = \sum_{i,j} g_{e,i}[\frac{\exp(-E_{e,i}/k_B T)}{Z_A}](A_{ph,a})_{ij} g_{e,j}, \tag{7.140}$$

where $\Delta E_{e,AB}$ is the energy gap between the two lowest levels of the upper and lower multiplets. Here Z_A and Z_B are the partition functions of the lower and upper multiplets, which are given by

$$Z_B = \sum_j g_{e,j} \exp[-(E_{e,j} - \Delta E_{e,AB})/k_B T] \tag{7.141}$$

$$Z_A = \sum_i g_{e,i} \exp(-E_{e,i}/k_B T), \tag{7.142}$$

and the energy difference is

$$E_{e,j} - E_{e,i} = \hbar\omega. \tag{7.143}$$

We now divide (7.139) by (7.140), and using $(A_{ph,e})_{ji} = (A_{e,a})_{ij}$, we have

$$\begin{aligned} \frac{A_{ph,e}(\omega)}{A_{ph,a}(\omega)} &= \frac{Z_A}{Z_B} \exp[-(E_{e,j} - \Delta E_{e,AB} - E_{e,i})/k_B T] \\ &= \frac{Z_A}{Z_B} \exp[-(\hbar\omega - \Delta E_{e,AB})/k_B T]. \end{aligned} \tag{7.144}$$

For a simple two-level system (without multiplets), we have $Z_A = Z_B$, and $\hbar\omega = \Delta E_{e,AB}$. Then (7.144) reduces to the simple relation $A_{ph,e}(\omega) = A_{ph,a}(\omega)$. Figure 7.18(b) shows the measured absorption cross-section area $A_{ph,a}(\lambda)$ and the emission cross-section area determined from the reciprocity relation [107].

For the general case, we have

$$\dot{s}_{e\text{-}ph,\omega} = \sigma_{ph,\omega} u_{ph} \int_\omega \frac{\mathrm{d}E_{ph,\omega}}{V} = n_{e,B} A_{ph,e} u_{ph} \int_\omega \frac{\mathrm{d}E_{ph,\omega}}{V}. \tag{7.145}$$

7.10 Spectral Surface Emissivity

A general treatment (near and far field) of thermally excited surface EM emission is given in [164]. Here we loosely build on the conventional far-field definition of the surface emissivity (this is rather a volumetric description and falls short; the surface treatment outlined in Section 7.15.1 leads to the proper definition of emissivity). The spectral (and directional) emissivity $\epsilon_{ph,\omega}(\theta, \phi, T)$ is defined using the blackbody spectral emissive power $E_{b,\omega}$ (7.8), i.e.,

$$\epsilon_{ph,\omega}(\theta, \phi, T) = \frac{\mathrm{d}E_{ph,\omega}(\theta, \phi, T)}{\mathrm{d}E_{b,\omega}(T)}, \quad (7.146)$$

where $E_{ph,\omega}$ is the spectral emissive power. Assuming that $f_{ph} = f_{ph}^{\mathrm{o}}$, from (7.1) we have

$$\epsilon_{ph,\omega,T}(\theta, \phi) = \frac{D_{ph,\omega}(T)}{D_{b,\omega}(T)} \quad \text{for } f_{ph}^{\mathrm{o}} \text{ photon gas}, \quad (7.147)$$

where the blackbody photon DOS is defined by (7.3).

We briefly discuss semiconductors, insulators, and metals.

Figures 7.19(a) and (b) show the variations of measured normal ($\theta = 0$), spectral emissivity of crystalline Si. Figure 7.19(a) is for $T = 323$ K and the low-energy results are from [329]. The blackbody DOS is also shown, as well as the DOS $D_{ph}(\omega) = \epsilon_{ph,\omega} D_b(\omega)$. The visible regime is dominated by direct and indirect interband transitions.

Figure 7.19(b) shows the variations of normal ($\theta = 0$) spectral emissivity $\epsilon_{ph,\omega}(\omega, T)$ for Si as functions of energy for four different temperatures [299]. This is the low-energy regime, where the lattice vibration dominates (dipole emission). The emissivity first increases with an increase in temperature and then decreases. The peak around $\lambda = 9$ μm is associated with the Si–O bond from oxygen impurity.

Comparing Figure 7.19(a) with the spectral absorption characteristic of semiconductors (Figures 7.14 and 7.15), we see that the high absorption and emission regimes are very similar, as the dominant mechanisms for absorption and emission in the respective regimes are the same.

The near-field and far-field emissions of Si have been predicted in [114] and will be discussed in relation to size effects in Section 7.15.1. The near-infrared regime behavior is dominated by nonpolar, optical-phonon-induced transition dipoles, and by intraband transitions for electrons and holes. Also shown in Figure 7.19(a) is the normal spectral emissivity of Si predicted in [114]. The general agreement around the bandgap and the high-energy phonon regimes is rather good, validating the dielectric function model used (Section 7.15.1). As the temperature increases, the higher-energy phonons are excited; however, the Si phonon DOS shown in Figure 4.11 nearly cuts off at $E_{ph} = 0.064$ eV (optical phonon). Then the increase in temperature results in smaller occupancy of the low energy states, and therefore their reduced efficiencies.

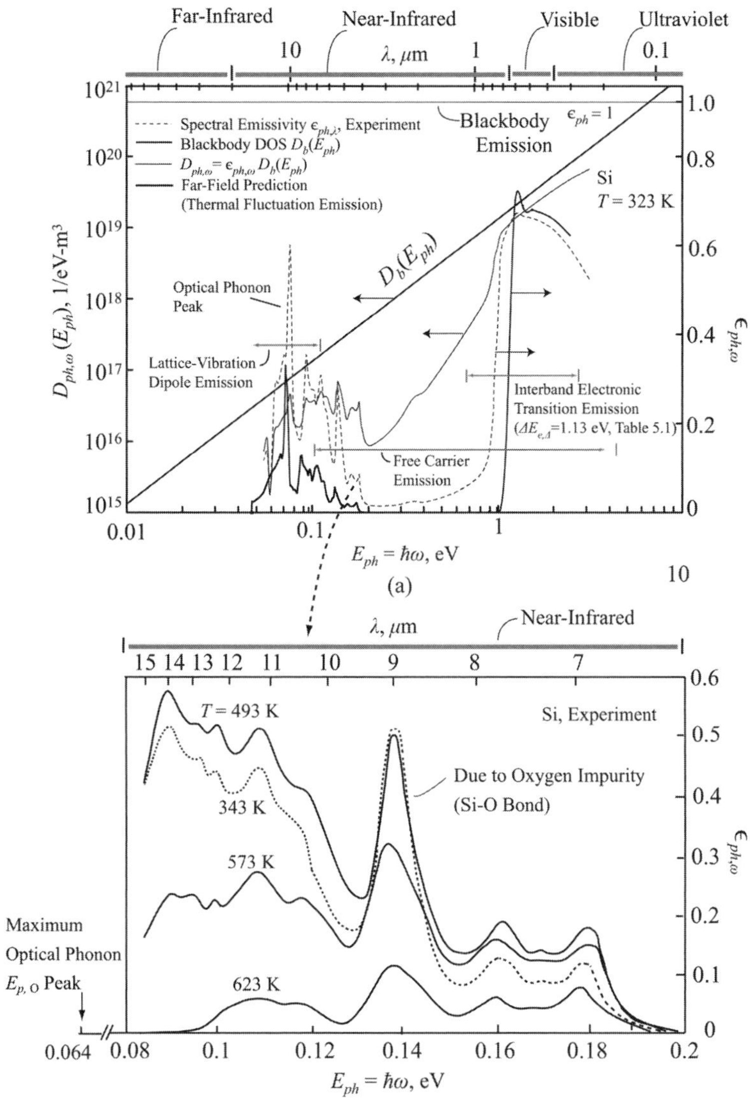

(b)

Figure 7.19. Spectral, normal emissivity $\epsilon_{ph,\omega}(\omega, T)$ of Si. (a) DOS $D_{ph}(E_{ph})$ and $D_b(E_{ph})$, and $\epsilon_{ph,\omega}(\omega, T)$ for low- and high-energy regimes [329]. Also shown are the predicted normal spectral emissivity [114]. (b) The lattice-vibration (near-infrared) dipole radiation (emission) regime and its temperature dependence [299]. The phonon DOS for Si is shown in Figure 4.11, and the peak in the optical-phonon modes is also shown there.

The interband electronic transition regime of Figure 7.19(a) begins near the $\Delta E_{e,\Delta}$ bandgap of 1.13 eV for Si (Table 5.1). The spectral reflectivity and emissivity of SiC in the near-infrared regime has been predicted in [117].

The role of optical phonons is through dependence of the electric permittivity on frequency (polar, optical phonons, Section 4.6), and in turn the dependence of the surface (tangent to surface) EM waves (or its wave vector $\kappa_{||}$) on the electric permittivity. For SiC, the following photon dispersion and dielectric function relations are used [117]

$$\kappa_{\|} = \frac{\omega^2}{u_{ph}^2}\frac{\epsilon_e}{\epsilon_e + 1} \qquad \text{photon dispersion relation,} \tag{7.148}$$

$$\epsilon_e(\omega) = \epsilon_{e,\infty}(1 + \frac{\omega_{p,\mathrm{LO}}^2 - \omega_{p,\mathrm{TO}}^2}{\omega_{p,\mathrm{TO}}^2 - \omega^2 - i\tau_e^{-1}\omega})$$

$$\text{oscillator model for dielectric function.} \tag{7.149}$$

For SiC, they use $\hbar\omega_{p,\mathrm{LO}} = 0.127$ eV (10.32 μm), $\hbar\omega_{p,\mathrm{TO}} = 0.0983$ eV (12.61 μm), $\hbar\tau_e^{-1}$ (damping constant) $= 0.000590$ eV, and $\epsilon_{e,\infty} = 6.7$. Then, between 10.32 and 12.61 μm, this dispersion relation has a complex entity and behavior [117], and in turn determines the high near-field emission (coherent and near-resonant emission), and far-field emissivity (high emissivity).

Using the classical EM treatment for the normal, spectral emissivity [306], we have

$$\epsilon_{ph,\omega,n} = \frac{4n_\lambda}{(n_\lambda + 1)^2 + \kappa_\lambda^2}. \tag{7.150}$$

Also, for metals at long wavelengths (infrared), the two components of the complex index of refraction n_λ and κ_λ (Table 3.5) are equal (Chapter 3 problem) and given by

$$n_\lambda = \kappa_\lambda = [\frac{\lambda\mu_o c_o}{4\pi\rho_e(T)}]^{1/2} \quad \text{metals at long wavelengths.} \tag{7.151}$$

Using this, we note that $\epsilon_{ph,\omega,n}$ decreases with an increase in λ, and also increases with T (as ρ_e increases with T for metals, Figure 5.16). The $\epsilon_{ph,\omega,n}(\lambda, \rho_e)$ is called the Hagen–Rubens relation.

For a surface, the sum of the spectral reflectivity, absorptivity, and transmissivity is equal to unity [174]. Here our treatments of photon transport are volumetric and this reference to surface property is for the sake of completeness and is used in the treatment of energy-conversion devices. Also, as a surface property, the spectral emissivity strongly depends on the surface topology (e.g., roughness), and surface composition (e.g., oxidation adsorption).

7.11 Radiative and Nonradiative Decays

Using the classical description of a bound, oscillating electron, a damping rate $\dot{\gamma}_e$ can be added for any energy loss.† In general, this energy decay in electronic transitions

† The one-dimensional classical equation of motion for conduction electron (Table 2.5) becomes [i.e., (3.45)]

$$\frac{\mathrm{d}^2x}{\mathrm{d}t^2} + \dot{\gamma}_e\frac{\mathrm{d}x}{\mathrm{d}t} + \omega_{n,e}^2 x = -\frac{e_c}{m_e}e_{e,x},$$

where $\omega_{n,e} = (\Gamma_e/m_e)^{1/2}$ is the natural or resonance angular frequency, $\dot{\gamma}_e = 1/\tau_e$, and Γ_e is the force constant, similar to interatomic force constant (2.54).

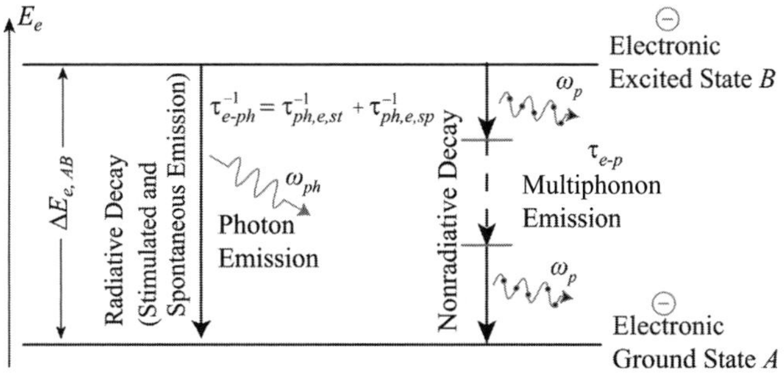

Figure 7.20. Radiative and nonradiative decay processes in a two-level system. The radiative decay is generally dominated by spontaneous photon emission, $\tau_{ph,e,sp}$.

includes both radiative (electromagnetic radiation) and nonradiative contributions. The radiative decay is the total (spontaneous and stimulated emission) and is in general dominated by the spontaneous emission [307]. The nonradiative decay includes inelastic collision of atoms with each other, or with the molecular and lattice vibration (phonon emission or absorption in solids).

The multiphonon relaxation process (purely nonradiative) is more clearly observed in rare-earth-element doped crystals (Section 5.20). When an electron of the ion is excited from the ground state to the excited state, it will not remain in the excited state long and returns to the ground state, with energy released equal to the energy gap $\Delta E_{e,AB}$ between the two states. As shown in Figure 7.20, this energy can be released through a radiative decay process by emitting a photon with an energy of $\Delta E_{e,AB}$, or through a purely nonradiative decay process by emitting several phonons (vibrational quanta of the host crystal), and the number of phonons emitted is given by $N_p = \Delta E_{e,AB}/\hbar\omega_p$, for energy per phonon $\hbar\omega_p$.

To calculate the total decay rate, the interaction of the isolated rare-earth ion with its crystalline host is interpreted as being due to the modulation of the crystalline electric field by the lattice vibrations. Based on the FGR (3.27) and weak coupling, the total decay rate, which is defined as the fraction of excited electrons decaying per second, is modeled as [32, 271]

$$\dot{\gamma}_{e\text{-}ph\text{-}p}(T) = \frac{1}{\tau_{e\text{-}ph\text{-}p}} = \dot{\gamma}_{e\text{-}ph}(T=0)(1+f_p^{\circ})^{N_p}$$

N_p-phonon nonradiative decay model (effective phonon-mode model), (7.152)

where $\tau_{e\text{-}ph\text{-}p}$ is the lifetime of excited electrons, N_p is the number of phonons emitted to fill the energy gap between the two states, and f_p° is the phonon equilibrium distribution (occupation) function (Table 1.2), i.e.,

$$f_p^{\circ} = \frac{1}{\exp(\hbar\omega_p/k_{\rm B}T)-1}. \qquad (7.153)$$

Relation (7.152) shows strong dependence of $\gamma_{e\text{-}ph\text{-}p}(T)$ on the number of phonons involved. Finally the temperature-dependent decay rate is given by

$$\dot{\gamma}_{e\text{-}ph\text{-}p}(T) = \dot{\gamma}_{e\text{-}ph}(T=0)[1 - \exp(-\hbar\omega_p/k_{\rm B}T)]^{-N_p}$$

$$N_p\text{-phonon nonradiative decay model.} \qquad (7.154)$$

Here $\dot{\gamma}_{e\text{-}ph}(T=0)$ is the radiative decay rate, because $\dot{\gamma}_{e\text{-}ph\text{-}p}(T) = \dot{\gamma}_{e\text{-}ph\text{-}p}(T=0)$ at $T=0$ when $f_p^{\rm o}=0$ and the phonon modes are all initially in their ground state. As the temperature is raised, the phonon modes become thermally populated and the multiphonon relaxation rate increases. Thus the radiative decay rate is constant, whereas the nonradiative decay rate increases with temperature. The radiative lifetime is

$$\dot{\gamma}_{e\text{-}ph} = \frac{1}{\tau_{e\text{-}ph}} = \dot{\gamma}_{e\text{-}ph}(T=0). \qquad (7.155)$$

Now, using Matthiessen rule (4.111), we have

$$\dot{\gamma}_{e\text{-}ph\text{-}p} = \dot{\gamma}_{e\text{-}ph} + \dot{\gamma}_{e\text{-}p}, \quad \dot{\gamma}_{e\text{-}p}(T) = \frac{1}{\tau_{e\text{-}p}} = \dot{\gamma}_{e\text{-}ph\text{-}p}(T) - \dot{\gamma}_{e\text{-}ph}$$

$$\text{Matthiessen rule (series),} \qquad (7.156)$$

or

$$\frac{1}{\tau_{e\text{-}p}} = \frac{1}{\tau_{e\text{-}ph\text{-}p}} - \frac{1}{\tau_{e\text{-}ph}}, \qquad (7.157)$$

where $\tau_{e\text{-}ph}$ is called the radiative lifetime ($\tau_{e\text{-}ph}^{-1} = \tau_{ph,e,st}^{-1} + \tau_{ph,e,sp}^{-1}$, stimulated emission is neglected other than in optical cavities) and $\tau_{e\text{-}p}$ is the nonradiative lifetime. In Section 7.12.6, we discuss *ab initio* calculations of the nonradiative decay time, using the Huang–Rhys theory of coupling.

An oscillating atom, like an oscillating classical dipole, emits EM energy at its discrete oscillation frequency, and this energy decays in time (emission). The classical electron oscillator (Section 3.3.5) has the radiative decay time of (3.42)

$$\tau_{e\text{-}ph,{\rm o}} = \tau_{ph,e,sp,{\rm o}} = \dot{\gamma}_{ph,e,sp,{\rm o}}^{-1} = \frac{6\pi\epsilon_{\rm o}m_e u_{ph}^3}{e_c^2\omega^2} = \frac{3\epsilon_{\rm o}m_e u_{ph}}{2\pi e_c^2}\lambda^2 = 4.504\times10^4\lambda^2$$

$$\text{classical oscillating electron radiative decay time,} \qquad (7.158)$$

for spontaneous emission in vacuum.† This gives $\tau_{ph,e,sp,{\rm o}} = 1.1\times10^{-8}$ s for $\lambda = 0.5$ μm (visible regime), and the larger wavelengths have noticeably larger decay times. The radiation broadening refers to the emission spectra line that has this decay signature and the linewidth is defined as $\Delta\lambda = \lambda^2\dot{\gamma}_{ph,e,sp}/2\pi u_{ph}$. For this example, $\Delta\lambda = 1.2\times10^{-4}$ Å.

† The average radiation power is [215] (Chapter 3 problem)

$$P_e = \frac{\omega^4 p_{e,o}^2}{12\pi\epsilon_{\rm o}u_{ph,{\rm o}}^3},$$

for an oscillating dipole moment.

Table 7.4. *Examples of radiative decay (emission) times* $\tau_{ph,e,sp} = \tau_{ph,e,sp} = \dot{\gamma}_{e\text{-}ph}^{-1}$ *of some laser and phosphorous materials [307, 353].*

Material and Transition	λ, nm	$\tau_{e\text{-}ph}$, s
Na		
$3s \rightarrow 3p$	589	15.9×10^{-9}
$3s \rightarrow 4p$	330	3.5×10^{-7}
He–Ne		
$3s_2 \rightarrow 4p_4$	633	7.0×10^{-7}
$2s_2 \rightarrow 2p_4$	1153	2.3×10^{-7}
$2s_2 \rightarrow 3p_4$	3392	1.04×10^{-6}
Nd^{3+}: YAG		
$^4F_{3/2} \rightarrow {}^4I_{3/2}$	1064	1.22×10^{-3}
Ruby		
$^2E \rightarrow {}^4A_2$	694	4.3×10^{-3}
Mn^{2+}: γ-$Zn_3(PO_4)_2$	635	100×10^{-3}
Ce^{3+}: $Ca_2MgSi_2O_7$	600	100×10^{-3} to 1
Mn^{2+}: MgF_2	585	> 1
Ce^{3+}: CaS	507	7.2×10^2
Eu^{2+},Al^{3+}: CaS	650	3.6×10^3
Ti^{4+}: Y_2O_2S	565	1.8×10^4

Some ions have radiative decay times close to those of preceding classical electron oscillator. Table 7.4 gives examples of the radiative decay times for some laser phosphorous (Glossary) materials . These are dominated by spontaneous emission (unless the excited-state population reaches saturation, i.e., becomes the same as the ground state). For some of these solids, the decay time is orders of magnitude larger than $\tau_{ph,e,sp,o}$. For the alkali metal Na, the radiation lifetime is close to that predicted by (7.158), but for rare-earth metals it is not as small as that predicted by (3.43) The decay time $\tau_{e\text{-}ph}$ for Nd^{3+}: $LaCl_3$ of Figure 5.38, for various fluorescent transitions, are reported in [7], and are between 20 to 170 μs.

Phosphorous materials have decay times significantly longer than that of the laser materials. The excited electrons in these materials form a singlet that is surrounded by levels that are not allowed by quantum mechanic exclusions. Although the electrons eventually decay to the ground levels, the time it takes for them to reach the ground state is long, thus the luminescence is prolonged (used in displays).

The temperature dependence of (7.152) has been confirmed by experiments. For example, in a Y_2O_3 crystal doped with Yb^{3+}, the lifetime of the excited electrons ($^2F_{5/2}$ to $^2F_{7/2}$) is measured as a function of temperature [281]. The results are

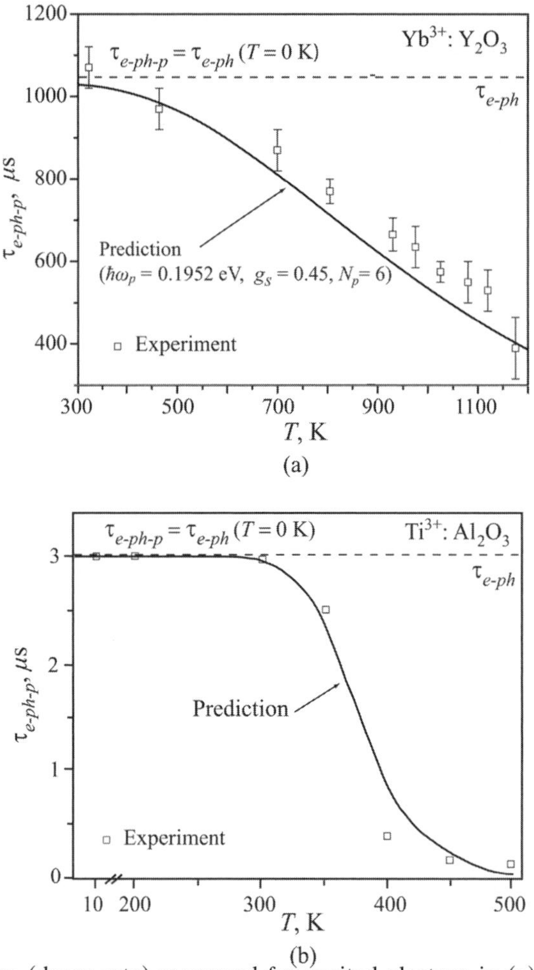

Figure 7.21. Lifetime (decay rate) measured for excited electron in (a) Yb^{3+} doped yttria [281], and (b) Ti^{3+}-doped alumina (sapphire) [271]. For (a) the prediction based on O–O cutoff frequency is also shown [180]. Here g_s is a coefficient that depends on the structure of the crystals.

shown in Figure 7.21(a). The results are predicted for the multiphonon decay rate using (7.154) with $\hbar\omega_p = 0.1952$ eV and $N_p = 6$ (predicted by the metrics introduced in [180]). This suggests that the decay is a six phonon relaxation process, with the energy of 0.1952 eV per phonon, which corresponds to the highest available phonon mode of the O–O pair. Note that for very large temperatures (reaching the melting temperature of the host crystal), $\tau_{e\text{-}ph\text{-}p}$ tends to zero [from (7.154)]. The prediction of (7.154) is also shown in Figure 7.21(b) for Ti^{3+}: Al_2O_3, and is is good agreement is found with experiments.

Other multiphonon relaxation-time models have been suggested [283], and are more accurate representations of experimental results at low temperatures. Because the population of the excited state deviates from that given for the thermal equilibrium, it needs to be determined using the rate equation. Using (7.37), at steady state, the excitation events from the ground state must be balanced by the decay events from the excited state, i.e.,

$$n_{e,A}(\omega_1)A_{ph,a}(\omega)f_{ph}(\omega) - n_{e,B}(\omega_2)A_{ph,d}(\omega)[1 + f_{ph}(\omega)] = 0, \tag{7.159}$$

where $A_{ph,d}$ is the decay cross section and is related to $A_{ph,e}$ through

$$\eta_{e\text{-}ph} = \frac{A_{ph,e}}{A_{ph,d}} = \frac{\dot{\gamma}_{e\text{-}ph}}{\dot{\gamma}_{e\text{-}ph\text{-}p}(T)} = \frac{\dot{\gamma}_{e\text{-}ph}}{\dot{\gamma}_{e\text{-}ph} + \dot{\gamma}_{e\text{-}p}} = \frac{\tau_{e\text{-}ph\text{-}p}}{\tau_{e\text{-}p}} \quad \text{quantum efficiency}, \tag{7.160}$$

where $\eta_{e\text{-}ph}$ is the quantum efficiency. Then we have

$$\frac{n_{e,B}}{n_{e,A}} = \frac{A_{ph,a}(\omega)f_{ph}(\omega)}{A_{ph,d}(\omega)[1 + f_{ph}(\omega)]} = \frac{A_{ph,a}(\omega)f_{ph}(\omega)}{\frac{A_{ph,e}(\omega)}{\eta_{e\text{-}ph}}[1 + f_{ph}(\omega)]}. \tag{7.161}$$

Using $\dot{s}_{f,ph}$ given by (7.66), we have the heating rate as

$$\dot{s}_{f,ph} = u_{ph}\sigma_{ph,\omega}[f^n_{ph}(T) - f_{ph}(\mathbf{s})], \quad \sigma_{ph,\omega} = A_{ph,a}n_a. \tag{7.162}$$

Here n_a is the number density of absorption sites, and the nonequilibrium distribution function is

$$f^n_{ph}(T) = \frac{1}{\dfrac{g_{e,B}(\omega_2)n_{e,A}(\omega_1)}{g_{e,A}(\omega_1)n_{e,B}(\omega_2)} - 1} = \frac{1}{\dfrac{g_{e,B}(\omega_2)}{g_{e,A}(\omega_1)}\dfrac{A_{ph,a}(\omega)f_{ph}(\omega)}{\frac{A_{ph,e}(\omega)}{\eta_{e\text{-}ph}}[1 + f_{ph}(\omega)]}}$$

$$= \frac{\eta_{e\text{-}ph}f_{ph}(\omega)}{1 + f_{ph}(\omega) - \eta_{e\text{-}ph}f_{ph}(\omega)}, \tag{7.163}$$

where we have used (7.65).

When the spontaneous emission dominates over the stimulated emission $[f_{ph}(\omega) \ll 1]$, then f^n_{ph} is simplified to

$$f^n_{ph}(T) = \eta_{e\text{-}ph}f_{ph}(\mathbf{s}). \tag{7.164}$$

Note that $f^n_{ph}(T)$ is not the equilibrium distribution, and here it is proportional to the local distribution function $f_{ph}(\mathbf{s})$.

In Section 7.12.6 we will discuss *ab initio* calculations of $\dot{\gamma}_{e\text{-}ph}$ and $\dot{\gamma}_{e\text{-}p}$, for a simple ionic complex.

7.12 Anti-Stokes Fluorescence: Photon-Electron-Phonon Couplings

7.12.1 Laser Cooling of Ion-Doped Solids

Phonon assisted electronic transition by photon absorption, i.e., phonon-assisted, dipole-moment transition, is the cooperative process (Glossary) governing laser cooling of solids. The concept of laser cooling (optical refrigeration) by the anti-Stokes luminescence in solids was introduced in [273], where the thermal vibrational energy can be removed by the anti-Stokes fluorescence in which a material is excited with photons having an energy below the mean fluorescence energy. Initially, it was believed that optical cooling by the anti-Stokes fluorescence contradicted the second law of thermodynamics [336, 337], suggesting that the cycle of excitation and fluorescence is reversible, and hence an energy yield greater than one would be equivalent to the complete transformation of heat to work. This was cleared up when entropy was assigned to radiation [198]. The entropy of a radiation field is proportional to its frequency bandwidth and also to the solid angle through which it propagates. Because the incident laser light has a small bandwidth and propagates in a well-defined direction, it has almost zero entropy. On the other hand, the fluorescence is broadband and is emitted in all directions, so if the power of the emission is equal to or greater than the incident beam, the emission has a comparatively large entropy.

Figures 7.22(a) to (e) show the energy diagram, as well as the steps identifying the mechanisms, in laser cooling of an ion-doped dielectric (otherwise transparent to the laser irradiation) solid. There the host crystal lattice is idealized as transparent to the pumping laser. Typical host crystals include $Y_3Al_5O_{12}$(YAG), $LiYF_4$, $LiSrAlF_6$, $Ca_5(PO_4)_3F$, $LaMgAl_{11}O_{19}$, $Gd_3Sc_2Ga_3O_{12}$, $La_2Be_2O_5$, Al_2O_3, $BeAl_2O_3$, and MgF_2 [271]. Some of its atoms are replaced with optically active, doped ions (e.g., Yb^{3+}), and thus become slightly absorbing. In quantum-mechanics the ion is represented by an effective transition dipole moment(or dipole operator), defined as a quantum mechanical spatial integral of the classical dipole moment $e_c\boldsymbol{x}$ (7.135), i.e.,

$$\boldsymbol{\mu}_e = \int \psi_f^* q \boldsymbol{x} \psi_i \mathrm{d}\boldsymbol{x}, \tag{7.165}$$

where q (here equal to e_c) is the electric charge, $\boldsymbol{x}$ is the position vector, and ψ_i and ψ_f are the initial- and final-state wave functions of the two level system. The electromagnetic field, which has a polarization vector $\boldsymbol{s}_{ph,\alpha}$, may interact with the ion if the coupling factor $\boldsymbol{s}_{ph,\alpha} \cdot \boldsymbol{\mu}_e$ is nonzero (i.e., they are not orthogonal). The principal mechanisms of the photon–electron–phonon interactions resulting in the cooling effect in the solid, are shown in Figures 7.22(a)–(e). In step (a), the electron in the ground state is coupled to a phonon and forms a combined state.

In (b), when the medium is irradiated by laser light with a frequency $\omega_{ph,i}$ that is below the resonance frequency $\omega_{e,g}$ for the energy gap ($F_{5/2}$ to $F_{7/2}$ transition of the $4f$–$4f$ orbital transition, 10,250 cm^{-1} for Yb^{3+} ion in Y_2O_3, similar to results given in Figure 5.38), the electron has some probability of being excited by absorbing a photon from the pumping field and the coupled phonon, such that $\hbar\omega_{ph,i} + \hbar\omega_p = \hbar\bar{\omega}_{ph,e}$, where $\bar{\omega}_{ph,e}$= mean emission angular frequency. According to the Born–Oppenheimer approximation, this electronic transition is instantaneous and the ion core does not move. In (c), after the electron is in the excited state, the charge distribution around the ion has been alternated, and the lattice will relax to new equilibrium positions. In (d), the electron then undergoes a radiative decay and returns to the ground state, emitting a photon with frequency $\omega_{ph,e}$. It is followed in (d) by another lattice-relaxation process that restores the lattice to its original positions.

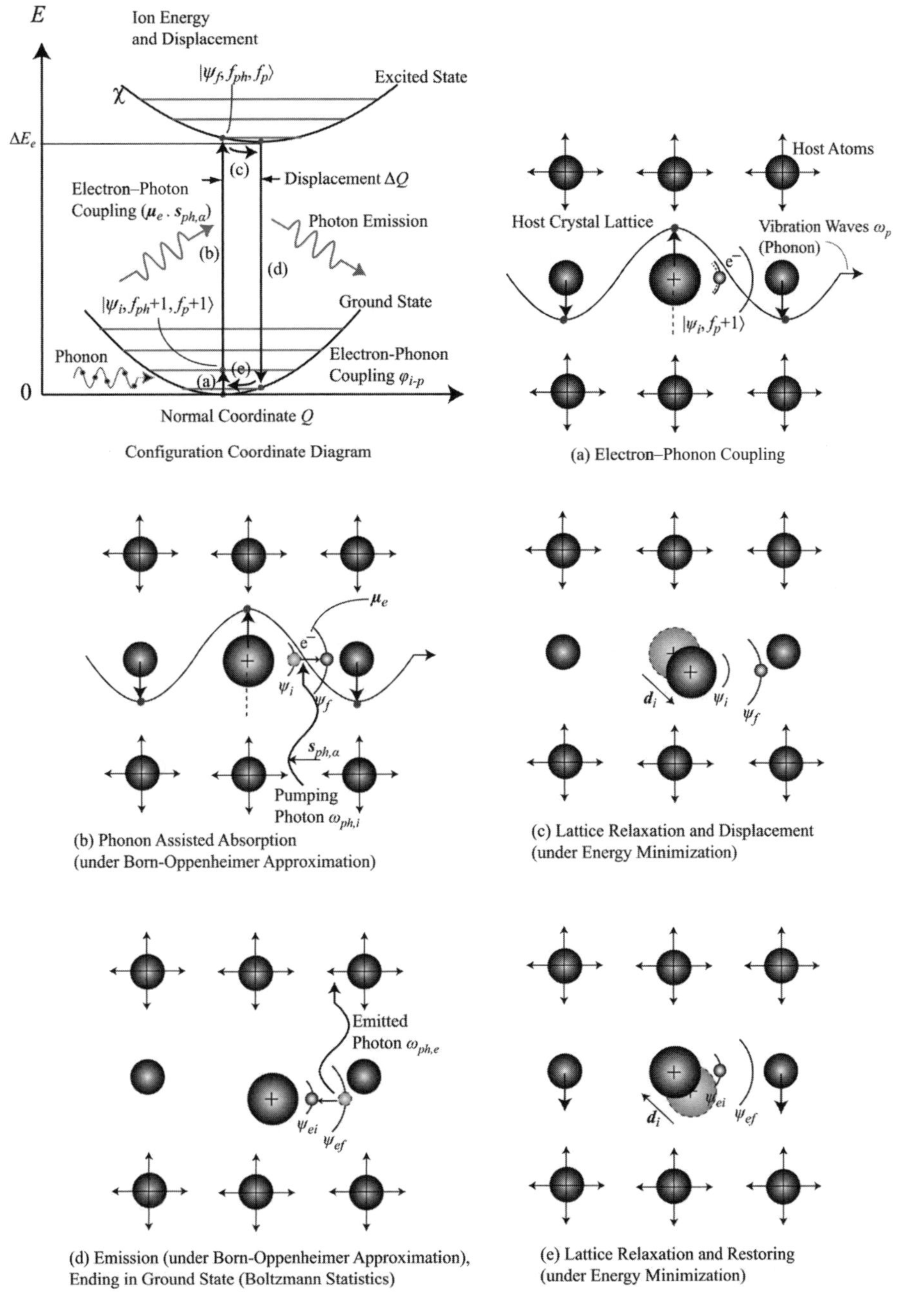

Figure 7.22. (a) The energy diagram, (b)–(e) five steps in the photon–electron–phonon couplings in the energy transfer cycle for the laser cooling of ion-doped dielectric solids.

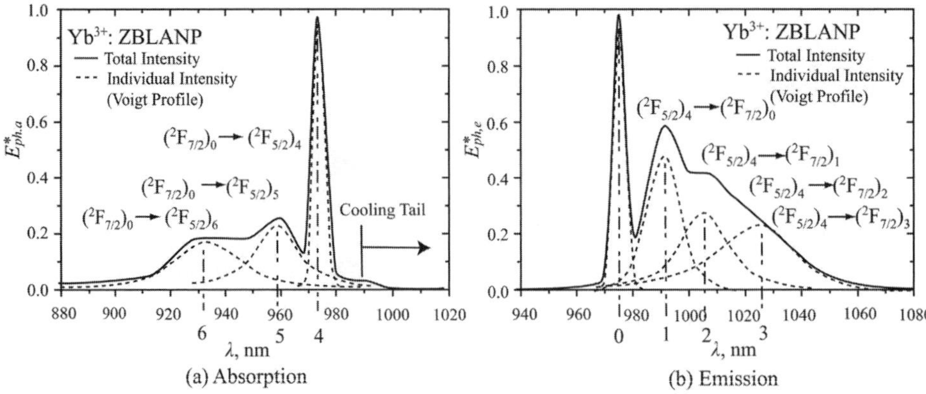

Figure 7.23. (a) Dimensionless absorption spectrum of Yb^{3+}: ZBLANP at $T = 10$ K. The transitions $(^2F_{5/2})_4 \rightarrow (^2F_{7/2})_{0,1,2,3}$, from the first excited manifold to four ground level manifolds are extrapolated (Voigt profile) and are also shown. (b) Dimensionless emission spectrum for the same [204].

If the average emitted photon frequency $\bar{\omega}_{ph,e}$ is larger than $\omega_{ph,i}$, the medium loses thermal energy and is cooled. A local temperature decrease was detected by a photothermal deflection technique, and the cooling efficiencies achieved in ytterbium-doped glass (up to 2%) were more than 10^4 times those observed in Doppler cooling of gases (Section 7.13.2). Since then, various ytterbium-doped glasses and crystals have been cooled [97, 102, 105, 124, 140, 220, 250, 279]. For the f-orbital transitions (rare-earth ions), an approximate relation for $\boldsymbol{\mu}_e$ is developed based on the transition charge distribution overlap [180].

7.12.2 Laser Cooling Efficiency

As shown in Figure 7.23(a), because of the ion–lattice coupling, the resonant, zero-phonon line in the absorption spectrum is broadened by the phonon sidebands, in which the electronic excitation may be assisted by absorbing one or more lattice phonons. The average emission wavelength $\bar{\lambda}_{ph,e}$ is independent of the excitation wavelength $\lambda_{ph,i}$ [86, 336]. This is in accordance with Vavilov's empirical rule [336], which states that the shape of the emission spectrum of a transition is essentially independent of the wavelength at which it is excited. Although this is not true for all materials, in the case of ytterbium in crystals, the time scale of thermal interactions ($\sim 10^{-12}$ s to 10^{-9} s) is orders of magnitude shorter compared with the time scale for optical transitions (of the order of 10^{-3} s) [103]. This implies that no matter what photon wavelength $\lambda_{ph,i}$ is used to excite the ytterbium transition, the population distribution in the upper manifold reaches thermal equilibrium before relaxing to the ground-state manifold, so, the emission spectra is independent of the pumping wavelength. The absorption spectrum (dimensionless $E^*_{ph,e}$) of Yb^{3+}: ZBLANP is shown in Figure 7.23(a) [204]. Apart from the strongest resonance transition $(^2F_{5/2})_4 \rightarrow {}^2(F_{7/2})_0$, there exist 3 phonon sideband transitions designated as (1,2,3). When the mean absorbed phonon energy is larger than the average emitted phonon (the 3 vibronic transitions), cooling occurs. The conditions for cooling and heating are

$$\hbar\omega_{ph,i} - \hbar\bar{\omega}_{ph,e} = \hbar\omega_p < 0 \text{ cooling (anti-Stokes process)}$$

$$\hbar\omega_{ph,i} - \hbar\bar{\omega}_{ph,e} = \hbar\omega_p > 0 \text{ heating (Stokes process)}, \qquad (7.166)$$

where $\omega_{ph,i}$ is the incoming photon frequency, $\bar{\omega}_{ph,e}$ is the average emission photon frequency, and ω_p is the single phonon frequency.

The emission spectrum (dimensionless $E^*_{ph,a}$) of Yb^{3+}: ZBLANP in Figure 7.23(b) shows that there are also three sub-levels in the excited state manifold. Note that the shift of emission to the long wavelengths in Figure 7.23(b) is that also predicted by the reciprocity discussed in Section 7.9.2, and also shown in Figure 7.21(b). In Figure 7.18(b), the mean emission wavelength $\bar{\lambda}_{ph,e}$, shown as the vertical line, is defined as

$$\bar{\lambda}_{ph,e} = \frac{\int_0^\infty E_{ph,\lambda}\lambda \mathrm{d}\lambda}{\int_0^\infty E_{ph,\lambda}\mathrm{d}\lambda}. \qquad (7.167)$$

The absorption at wavelengths longer than $\bar{\lambda}_{ph,e}$ is defined as the cooling tail. This is a phonon-assisted absorption process, as was shown in Figure 7.22. Note that $\sigma_{ph,\omega}$ and $E_{ph,\omega}$ are related through the absorption and emission cross-section area, given by reciprocity relation (7.144). Then the laser cooling efficiency is defined as the ratio between the net cooling power and the absorbed laser power, i.e.,

$$\eta_{p\text{-}ph} = \frac{\bar{\omega}_{ph,e} - \omega_{ph,i}}{\omega_{ph,i}} = \frac{\bar{\omega}_{ph,e} - (\omega_{e,g} - N_p\omega_p)}{\omega_{e,g} - N_p\omega_p} = \frac{\lambda_{ph,i} - \bar{\lambda}_{ph,e}}{\bar{\lambda}_{ph,e}}$$

$$\text{laser cooling efficiency}, \qquad (7.168)$$

where N_p is the number of phonons involved in one transition. It is evident that $\eta_{p\text{-}ph}$ increases as $N_p\omega_p$ increases, indicating that higher-energy phonons are desirable, to couple to the anti-Stokes electron excitation.

The cooling efficiency $\eta_{p\text{-}ph}$ increases as the pumping wavelength $\lambda_{ph,i}$ increases. However, the excitation probability $\dot{\gamma}_{ph\text{-}e}$ (which determines the absorption coefficient $\sigma_{ph,\lambda,i}$, and also the extinction index $\kappa_{\lambda,i}$) becomes too small if the pumping wavelength is too long. Thus the cooling efficiency $\eta_{p\text{-}ph}$ of different materials (or structures) are compared at the same excitation probability. We consider the excitation probability in the cooling tail, where the energy $\hbar(\omega_{e,g} - \omega_{ph,i})$ is provided by annihilation of the lattice phonons. This multiphonon-assisted anti-Stokes excitation process is one of the multiphonon phenomena, which also include multiphonon Stokes excitation, multiphonon relaxation, and multiphonon energy transfer. For the multiphonon anti-Stokes excitation, the process involving the smallest number of phonons is the most probable. Thus, as a first approximation, only the cut-off phonons have been assumed to couple effectively to the anti-Stokes excitation of the electron, although phonons of various energies are available in the lattice [11]. However, this approach generally does not give a detailed fit to experimental spectra. To resolve the experimental spectra, different phonon modes should be used according to different pumping wavelengths, so that the energy requirement $\hbar\omega_{ph,i} + \hbar\omega_p = \hbar\omega_{e,g}$ is satisfied without assuming any arbitrary broadening. To do this, the phonon DOS obtained by lattice dynamics calculations should be included. A thorough model (for weak couplings) has been developed for this purpose [105, 106] by considering one phonon absorption, and is confirmed in modeling the laser cooling in bulk materials. Here, the solution procedure is briefly introduced in the next section. Later, we discuss strong couplings and the derivations leading to *ab initio* computations.

7.12.3 Photon–Electron–Phonon Transition Rate Using Weak Coupling Approximation

Because a much longer pumping wavelength than the resonance is used in laser cooling, the probability of a purely electronic transition (the first order process) between electronic sublevels becomes very small. While the phonon-assisted transition, a second-order process, begins to contribute significantly to absorption. As a result, the absorption turns out to be a combination of the first-order and second-order transitions. Because the first-order process does not involve phonons, only the second-order process is analyzed here for the purpose of understanding the role of phonons in laser cooling.

In the second-order process, the ion in its ground state absorbs an incident photon and a phonon, and moves up to the excited state. The probability per unit time of such a process can be evaluated using the perturbation theory. The Hamiltonian for the physical system considered is given by [105]

$$\mathrm{H} = \mathrm{H}_e + \mathrm{H}_p + \mathrm{H}_{ph} + \mathrm{H}_{ph\text{-}e} + \mathrm{H}_{e\text{-}p}. \tag{7.169}$$

The first term is† (neglecting the 1/2 energy term)

$$\mathrm{H}_e = \hbar\omega_{e,g} a^\dagger a, \tag{7.170}$$

i.e., the Hamiltonian of the ion electronic levels, where $\hbar\omega_{e,g}$ is the energy difference between the optically active energy levels of the dopant ion (considered as a two-level ion) and $a^\dagger$. (a) is the creation (annihilation) operator of an electronic excitation. Again, we have neglected the 1/2 term appearing in the Hamiltonian by using the ladder operator (Section 2.6.5 footnote)

The second term is (Section 4.5)

$$\mathrm{H}_p = \sum_p \hbar\omega_p b^\dagger b, \tag{7.171}$$

† Following the footnote of Section 2.6.4, we use $(a, a^\dagger)$ for electron, $(b, b^\dagger)$ for phonon , and $(c, c^\dagger)$ for photon.

i.e., the phonon field Hamiltonian, where ω_p is the phonon frequency and $b^\dagger(b)$ is the creation (annihilation) operator of a phonon in mode p. Again, we have neglected the 1/2 term.

The third term is (Section 7.3.1)

$$\mathrm{H}_{ph} = \hbar\omega_{ph,i} c^\dagger c, \tag{7.172}$$

i.e., the electromagnetic laser field Hamiltonian, where $\omega_{ph,i}$ is the pumping frequency and $c^\dagger(c)$ is the creation (annihilation) operator of a photon.

The fourth term is that developed in Section 7.3.2, i.e., (7.36), which we write as

$$\begin{aligned}\mathrm{H}_{ph\text{-}e} &= -\boldsymbol{s}_{ph,i} \cdot \boldsymbol{\mu}_e (\frac{\hbar\omega_{ph,i}}{2\epsilon_o V_s})^{1/2} (c^\dagger + c)(a^\dagger + a) \\ &= C_{ph\text{-}e}(c^\dagger + c)(a^\dagger + a), \quad C_{ph\text{-}e} \equiv -\boldsymbol{s}_{ph,i} \cdot \boldsymbol{\mu}_e (\frac{\hbar\omega_{ph,i}}{2\epsilon_o V_s})^{1/2}, \end{aligned} \tag{7.173}$$

i.e., the photon–electron interaction Hamiltonian, where $\boldsymbol{s}_{ph,i}$ is the polarization vector of the photon, $\boldsymbol{\mu}_e$ is the transition dipole moment of the electronic transition, ϵ_o is the vacuum permittivity, and V_s is the interacting volume. Note that from (3.36) the term $[\hbar\omega_{ph,i}/(2\epsilon_o V_s)]^{1/2}$ is the vacuum electric field per photon.

(A) *Acoustic Phonons*

The fifth term is the electron–phonon interaction Hamiltonian that can be described as distortion of the ligand ions affecting the crystal field. Such a distortion is a function of the local strain; therefore we can expand the crystal field potential in powers of such a strain. The Local strain is defined by the strain term $\epsilon_{i,j}$ as

$$\epsilon_{i,j} = \frac{1}{2}(\frac{\partial \boldsymbol{d}_i}{\partial x_j} + \frac{\partial \boldsymbol{d}_j}{\partial x_i}) \quad (i, j = 1, 2, 3), \tag{7.174}$$

which is also given in Table 3.11. For simplicity we shall assume that

$$\epsilon \approx \frac{\partial \boldsymbol{d}}{\partial x} |_{x=0}, \tag{7.175}$$

thus not taking into any account the anisotropy of the elastic waves. We take as origin for the coordinate axes to point of the lattice in which the nucleus of the ion is located.

The displacement $\boldsymbol{d}$ is given by (4.47), and its derivative is expressed as [271]

$$\frac{\partial \boldsymbol{d}}{\partial x} |_{x=0} = (\frac{\hbar\omega_p}{2\rho u_p^2})^{1/2} (b - b^\dagger). \tag{7.176}$$

Then the crystal field Hamiltonian, H_c in terms of strain is

$$\mathrm{H}_c = \mathrm{H}_o + \varphi_{e\text{-}p,\mathrm{A}}\epsilon, \tag{7.177}$$

where H_o is the static crystal field Hamiltonian term and higher-order terms are neglected. The the interaction Hamiltonian becomes

$$\begin{aligned}
\mathrm{H}_{e\text{-}p} &= \varphi_{e\text{-}p,\mathrm{A}}\epsilon \\
&= \varphi_{e\text{-}p,\mathrm{A}}(\frac{\hbar\omega_p}{2\rho u_p^2})^{1/2}(b - b^\dagger) \\
&= C_{e\text{-}p,\mathrm{A}}(b - b^\dagger), \quad C_{e\text{-}p} \equiv \varphi_{e\text{-}p,\mathrm{A}}(\frac{\hbar\omega_p}{2\rho u_p^2})^{1/2},
\end{aligned} \tag{7.178}$$

i.e., the electron–phonon interaction Hamiltonian, where $\varphi_{e\text{-}p,\mathrm{A}}$ is the acoustic interaction potential that is analogous to (5.177), ω_p is the phonon frequency, u_p is the average speed of sound, and ρ is the mass density.

This process appears in the second-order term of the perturbation expansion, and the transition (photon-induced, phonon-assisted absorption) rate $\dot{\gamma}_{ph\text{-}e\text{-}p}$, which (is the dominant rate controlling mechanism; thus we have a weak phonon coupling) is given by the FGR (3.27) [105], i.e,

$$\dot{\gamma}_{ph\text{-}e\text{-}p} = \sum_f \dot{\gamma}_{e,i\text{-}f} = \frac{2\pi}{\hbar}\sum_f |M_{f,i}|^2 \delta_{\mathrm{D}}(E_f - E_i), \tag{7.179}$$

where E_i and E_f are, respectively, the initial and final energies of the electron system. The interaction matrix element $M_{f,i}$ admits a perturbative expansion given by [105]

$$\begin{aligned}
M_{fi} &= \langle f|\mathrm{H}_{int}|i\rangle + \sum_m \frac{\langle f|\mathrm{H}_{int}|m\rangle\langle m|\mathrm{H}_{int}|i\rangle}{E_{e,i} - E_{e,m}} \\
&+ \sum_{m,n} \frac{\langle f|\mathrm{H}_{int}|m\rangle\langle m|\mathrm{H}_{int}|n\rangle\langle n|\mathrm{H}_{int}|i\rangle}{(E_{e,i} - E_{e,m})(E_{e,i} - E_{e,n})} + \cdots,
\end{aligned} \tag{7.180}$$

with $\mathrm{H}_{int} = \mathrm{H}_{e\text{-}ph} + \mathrm{H}_{e\text{-}p}$. The summations on m and n include all the intermediate phonon and photon states.

We calculate the transition probability $\dot{\gamma}_{ph,a}$ between initial $|i\rangle = |\psi_i, f_{ph} + 1, f_p + 1\rangle$ and final $|f\rangle = |\psi_f, f_{ph}, f_p\rangle$ states of the system, where the first ket element ψ_i refers to the ion (electronic) state, the second one, f_{ph}, to the photon number in the interacting volume V_s, and the third one, f_p, to the phonon distribution function. These types of processes appear only in the second-order perturbation expansion of the $M_{f,i}$ interaction matrix, which gives [17, 271]

$$\begin{aligned}
M_{fi,2nd} &= \sum_m \frac{\langle f|\mathrm{H}_{int}|m\rangle\langle m|\mathrm{H}_{int}|i\rangle}{E_{e,i} - E_{e,m}} \\
&= \sum_m \Bigg[\frac{\langle \psi_f, f_{ph}, f_p|\mathrm{H}_{e\text{-}ph}|\psi_m, f_{ph}+1, f_p\rangle}{E_i - (E_m - \hbar\omega_p)} \\
&\quad \times \langle \psi_m, f_{ph}+1, f_p|\mathrm{H}_{e\text{-}p}|\psi_i, f_{ph}+1, f_p+1\rangle \\
&\quad + \frac{\langle \psi_f, f_{ph}, f_p|\mathrm{H}_{e\text{-}p}|\psi_m, f_{ph}, f_p+1\rangle}{E_i - (E_m - \hbar\omega_{ph})} \\
&\quad \times \langle \psi_m, f_{ph}, f_p+1|\mathrm{H}_{e\text{-}ph}|\psi_i, f_{ph}+1, f_p+1\rangle \Bigg] \\
&\simeq \sum_m \frac{\langle \psi_f, f_{ph}, f_p|C_{ph\text{-}e}(a^\dagger + a)(c^\dagger + c)|\psi_m, f_{ph}+1, f_p\rangle}{E_i - (E_m - \hbar\omega_p)} \\
&\quad \times \langle \psi_m, f_{ph}+1, f_p|a_{e\text{-}p}C_{e\text{-}p}a^\dagger a(b - b^\dagger)|\psi_i, f_{ph}+1, f_p+1\rangle \\
&= \sum_m \frac{\langle \psi_f, f_{ph}, f_p|C_{ph\text{-}e}(a^\dagger + a)(c^\dagger + c)|\psi_m, f_{ph}+1, f_p\rangle}{E_i - (E_m - \hbar\omega_p)} \\
&\quad \times \langle \psi_m, f_{ph}+1, f_p|C_{e\text{-}p}a^\dagger a(b - b^\dagger)|\psi_i, f_{ph}+1, f_p+1\rangle \\
&= \sum_m \frac{\langle \psi_f, f_{ph}, f_p|C_{ph\text{-}e}(a^\dagger + a) f_{ph}^{1/2}|\psi_m, f_{ph}, f_p\rangle}{E_i - (E_m - \hbar\omega_p)} \\
&\quad \times \langle \psi_m, f_{ph}+1, f_p|C_{e\text{-}p}a^\dagger a f_p^{1/2}|\psi_i, f_{ph}+1, f_p\rangle \\
&= \sum_m C_{ph\text{-}e} f_{ph}^{1/2} C_{e\text{-}p} f_p^{1/2} \frac{\langle \psi_f|(a^\dagger + a)|\psi_m\rangle\langle \psi_m|a^\dagger a|\psi_i\rangle}{E_i - (E_m - \hbar\omega_p)}. \qquad (7.181)
\end{aligned}$$

The energy symbol with no subscript indicates the total energy (whereas subscripts e indicate electron system only). The approximation is based on not allowing transition to states lower than the lowest excited manifold, so we have phonon absorption followed by photon absorption [Figure 7.22(a)]. Because the phonon energy $\hbar\omega_p$ is much smaller than the energy gap, it cannot by itself induce an electronic transition. In the perturbation theory the intermediate wave function ψ_m is then approximated as unperturbed, i.e., $\psi_m = \psi_i$, and $E_m = E_i$. Hence, using the definition of matrix element (5.93), (7.181) becomes

$$\begin{aligned}
M_{f,i,2nd} &= \sum_p C_{ph\text{-}e} f_{ph}^{1/2} C_{e\text{-}p} f_p^{1/2} \frac{\langle \psi_f|(a^\dagger + a)|\psi_i\rangle\langle \psi_i|a^\dagger a|\psi_i\rangle}{E_i - (E_i - \hbar\omega_p)} \\
&= \sum_m C_{ph\text{-}e} f_{ph}^{1/2} C_{e\text{-}p} f_p^{1/2} \frac{\langle \psi_f|(a^\dagger + a)|\psi_i\rangle\langle \psi_i|a^\dagger a|\psi_i\rangle}{E_i - (E_i - \hbar\omega_p)} \\
&= \sum_m C_{ph\text{-}e} f_{ph}^{1/2} C_{e\text{-}p} f_p^{1/2} \frac{1}{\hbar\omega_p}, \qquad (7.182)
\end{aligned}$$

where we have used $\langle\psi_i|a^\dagger a|\psi_i\rangle = 1$, $\langle\psi_f|a|\psi_i\rangle = 0$, and $\langle\psi_f|a^\dagger|\psi_i\rangle = 1$. These are due to the use of the number operator, applying the lowering operator to the ground state, and the definition of the raising ladder operator (Section 2.6.4), respectively.

Substituting (7.182) into (7.179), we have [105]

$$\dot{\gamma}_{ph\text{-}e\text{-}p} = \sum_f \dot{\gamma}_{e,i\text{-}f} = \frac{2\pi}{\hbar}\sum_p \left(\frac{C_{ph\text{-}e}C_{e\text{-}p}}{\hbar\omega_p}\right)^2 f_{ph} f_p \delta_{\rm D}(\hbar\omega_{ph,i} + \hbar\omega_p - \hbar\omega_{e,g}). \tag{7.183}$$

To perform the summation on the phonon modes in (7.183), we must introduce the phonon DOS $D_p(E_p)$. In terms of this distribution function, transition rate (7.183), becomes an integral given as

$$\begin{aligned}\dot{\gamma}_{ph\text{-}e\text{-}p} &= \frac{2\pi}{\hbar}C^2_{ph\text{-}e} f_{ph}\frac{\varphi^2_{e\text{-}p,\mathrm{A}}}{2\rho u_p^2}\int_{E_{min}}^{E_{max}} \mathrm{d}E_p D_p(E_p)\frac{f_p^{\mathrm{o}}(E_p)}{E_p} \\ &\quad\times\delta_{\rm D}(\hbar\omega_{ph,i} + E_p - \hbar\omega_{e,g}) \\ &= \frac{2\pi}{\hbar}C^2_{ph\text{-}e} f_{ph}\frac{\varphi^2_{e\text{-}p,\mathrm{A}}}{2\rho u_p^2}\frac{D_p(E_p)f_p^{\mathrm{o}}(E_p)}{E_p},\end{aligned} \tag{7.184}$$

where E_p is the phonon energy given by $E_p = \hbar\omega_{e,g} - \hbar\omega_{ph,i}$ and $\varphi_{e\text{-}p,\mathrm{A}}$ was defined in (7.178). It implies that the excitation spectra can be well associated with the phonon spectra, as observed in [247]. Here we have used the equilibrium distribution functions for phonons. Then (7.184) is rewritten as

$$\begin{aligned}\dot{\gamma}_{ph\text{-}e\text{-}p} &= \frac{2\pi}{\hbar}\frac{(\mathbf{s}_{ph,i}\cdot\boldsymbol{\mu}_e)^2}{2\epsilon_{\mathrm{o}}}\frac{\varphi^2_{e\text{-}p,\mathrm{A}}}{2\rho u_p^2}\frac{D_p(E_p)f_p^{\mathrm{o}}(E_p)}{E_p}\hbar\omega_{ph,i}\frac{f_{ph}}{V_s} \\ &= \frac{2\pi}{\hbar}\frac{(\mathbf{s}_{ph,i}\cdot\boldsymbol{\mu}_e)^2}{2\epsilon_{\mathrm{o}}}\frac{\varphi^2_{e\text{-}p,\mathrm{A}}}{2\rho u_p^2}\frac{D_p(E_p)f_p^{\mathrm{o}}(E_p)}{E_p}e_{ph,i},\end{aligned} \tag{7.185}$$

where $e_{ph,i}$ is the incident photon energy density (per unit solid volume) also used in (7.2).

(B) *Optical Phonons*

When specific normal modes of vibration dominate the electron–phonon interaction, it is useful to express the electron–phonon interaction Hamiltonian in terms of normal coordinates, which is presented as (5.162) and (5.163). Because the optical deformation potential that is analogous to (5.186), the crystal field Hamiltonian term (7.177) can be replaced with [180]

$$\mathrm{H}_c = \mathrm{H}_{\mathrm{o}} + \varphi'_{e\text{-}p,\mathrm{O}}Q_q. \tag{7.186}$$

Again, the higher-order terms have been neglected. Then the interaction Hamiltonian term (7.178) becomes

$$\begin{aligned}\mathrm{H}_{e\text{-}p} &= \varphi'_{e\text{-}p,\mathrm{O}} Q_q \\ &= \varphi'_{e\text{-}p,\mathrm{O}} (\frac{\hbar}{2m_{\mathrm{AC}}\omega_p})^{1/2} (b_q + b_q^\dagger) \\ &= C_{e\text{-}p,\mathrm{O}} (b_q + b_q^\dagger), \quad C_{e\text{-}p} \equiv \varphi'_{e\text{-}p,\mathrm{O}} (\frac{\hbar}{2m_{\mathrm{AC}}\omega_p})^{1/2},\end{aligned} \tag{7.187}$$

where $\varphi'_{e\text{-}p,\mathrm{O}}$ is the electron–phonon coupling similar to deformation potential (5.186), ω_p is the phonon frequency, m_{AC} is the mass of the reduced oscillating atom, and $b_q(b_q^\dagger)$ is the creation (annihilation) operator in normal coordinates. Note that $(\hbar/2m_{\mathrm{AC}}\omega_p)^{1/2}$ represents the atomic displacement (2.59). Then going through FGR 3.27) similar to acoustic phonons by using the new interaction matrix, we rewrite (7.185) as

$$\dot{\gamma}_{ph\text{-}e\text{-}p} = \frac{1}{\tau_{ph\text{-}e\text{-}p}} = \frac{\pi\hbar}{2m_{\mathrm{AC}}\epsilon_{\mathrm{o}}} (\boldsymbol{s}_{ph,i} \cdot \boldsymbol{\mu}_e)^2 \varphi'^2_{e\text{-}p,\mathrm{O}} \frac{D_p(E_p) f_p^{\mathrm{o}}(E_p)}{E_p^3} \hbar\omega_{ph,i} \frac{f_{ph}}{V_s}. \tag{7.188}$$

Note that now $D_p(E_p)$ is the phonon DOS for normal (local) modes that may be different from $D_p(E_p)$ for the bulk host material. Also, note that D_p maybe represented by the Debye–Gaussian model (4.29) [e.g., Figure 4.8(c)]. The theoretical treatment of $\varphi'_{e\text{-}p,\mathrm{O}}$ is given in [180], and Figures 7.24(a) to (c) show aspects of the material metrics of laser cooling of solids [180]. Figure 7.24(a) shows the phonon assisted electronic excitation for Yb^{3+}. Figures 7.24(b) and (c) show that although strong phonon coupling is desirable in excitation, it is not desired in the decay (since it reduces the quantum efficiency).

7.12.4 Time Scales for Laser Cooling of Solids (Weak Couplings)

In the weak coupling analysis the kinetics of laser cooling cycle has been presented with various characteristic times, namely, $\tau_{ph\text{-}e}$, $\tau_{e\text{-}p}$ and $\tau_{ph\text{-}e\text{-}p}$.

A variety of parameters can be used to describe the photon–electron interaction, e.g., the Einstein $\dot{\gamma}_{ph,e,sp}$ given in (7.37), $\dot{\gamma}_{ph,a}$ given by (7.184), the radiative lifetime ($\tau_r = \tau_{ph\text{-}e}$), the oscillator strength, or the absorption cross-section area. These parameters are all governed by the transition dipole moment, which is a more fundamental quantity. For example, the Einstein Acoefficient $\dot{\gamma}_{ph,e,sp}$ in (7.37) (equivalent to the radiative decay rate $\dot{\gamma}_r$) and the radiative lifetime $\tau_r = \tau_{ph,e,sp} = \tau_{ph\text{-}e}$ (applied to anti-Stokes emission dominated by spontaneous emission, as demonstrated in Section 7.4.3) are related to the transition dipole moment by (3.43) [147]

$$\dot{\gamma}_{ph,e,sp} = \dot{\gamma}_{ph\text{-}e} = \frac{\omega_{e,g}^3}{3\pi\epsilon_{\mathrm{o}}\hbar u_{ph}^3} \langle\mu_{ph\text{-}e}\rangle^2 \tag{7.189}$$

$$\tau_{ph,e,sp} = \tau_{ph\text{-}e} = \frac{1}{\dot{\gamma}_{ph,e,sp}} = \frac{3\pi\epsilon_{\mathrm{o}}\hbar u_{ph}^3}{\omega_{e,g}^3} \frac{1}{\langle\mu_{ph\text{-}e}\rangle^2}, \tag{7.190}$$

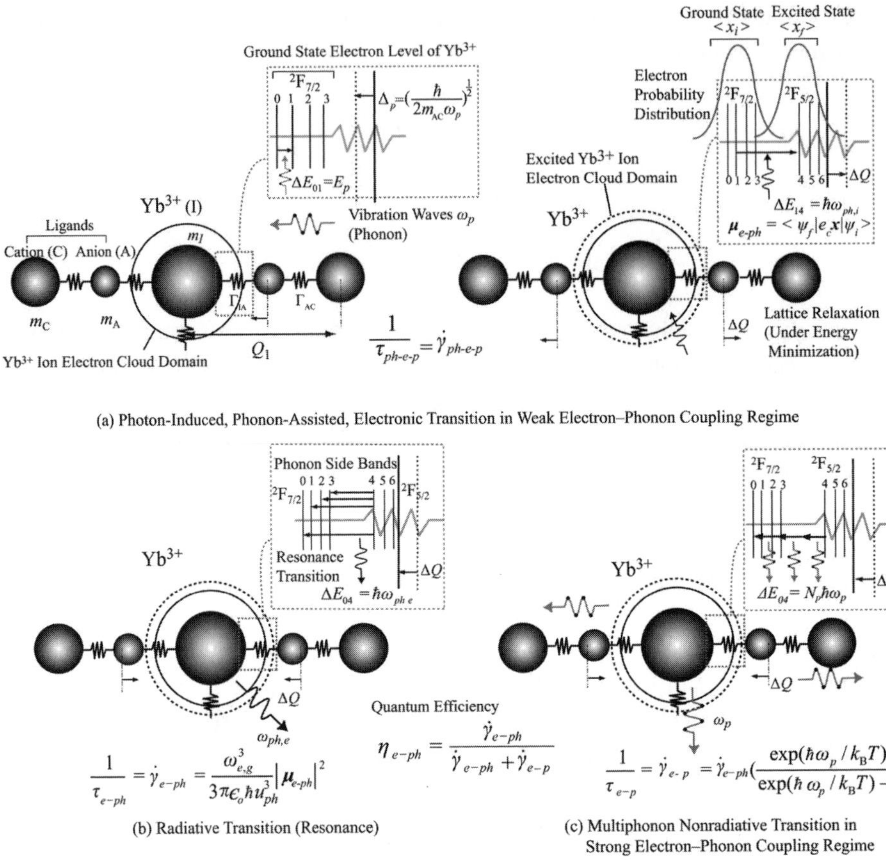

Figure 7.24. Material metrics of the photon–electron–phonon interactions in laser cooling of Yb^{+3}-doped solids. (a) Model for the optical-phonon coupling with a bound electron followed by photon absorption. (b) Purely radiative emission process. The phonon side band transitions are also shown. (c) Purely nonradiative emission process. The quantum efficiency η_{e-ph} is also defined by the ratio of radiative transition rate to total transition rate. The various variables will be discussed in the texts of the related sections [180].

where ϵ_o is the free-space electric permittivity and $\langle \mu_{ph\text{-}e} \rangle$ is the spatial average of the photon polarization and transition dipole moment vectors and is defined as $\langle \mu_{ph\text{-}e} \rangle^2 = \langle (\boldsymbol{s}_{ph,i} \cdot \boldsymbol{\mu}_{ph\text{-}e})^2 \rangle = (\mu_{ph\text{-}e}/3^{\frac{1}{2}})^2$.

The multiphonon decay process is given by [285] [also given by (7.154)

$$\dot{\gamma}_{e\text{-}p} = \dot{\gamma}_{ph\text{-}e}[1 - \frac{\exp(\hbar\omega_{p,c}/k_B T)}{\exp(\hbar\omega_{p,c}/k_B T) - 1}]^{N_p}, \quad N_p = \frac{E_{e,g}}{\hbar\omega_{p,c}}. \qquad (7.191)$$

The photon-induced phonon-assisted transition time $\tau_{ph\text{-}e\text{-}p}$ is given by the inverse of (7.188) assuming that the photon density of state f_{ph} integrates to one for an incoming laser (monochromatic and single-photon interaction). The predicted $\tau_{ph\text{-}e\text{-}p}$ can be directly compared with other transition times by defining the

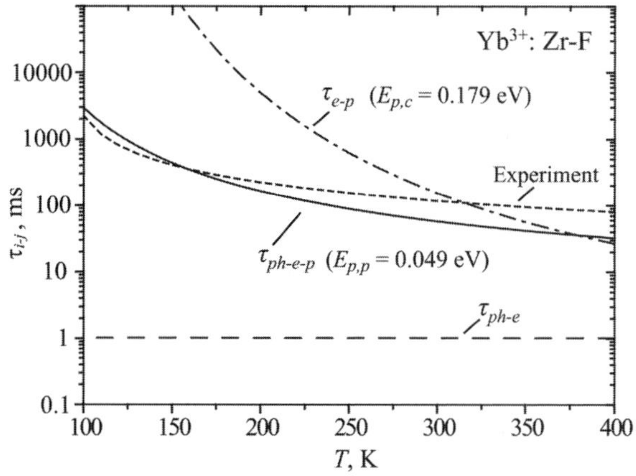

Figure 7.25. Variations of time constants as functions of temperature. The radiative relaxation time is the shortest followed by the phonon-assisted photon absorption at low temperatures. Multiphonon relaxation exhibits strong temperature dependence and is dominant at high temperatures [180]. Comparison with experimental results [124] is also shown.

photon-induced, photon-assisted transition dipole moment $\mu_{ph\text{-}e\text{-}p}$ i.e.,

$$\dot{\gamma}_{ph\text{-}e\text{-}p} = \frac{1}{\tau_{ph\text{-}e\text{-}p}} = \frac{\omega_{e,g}^3}{3\pi\epsilon_o\hbar u_{ph}^3}\frac{3\pi^2\hbar^2 u_{ph}^3\hbar\omega_{ph,i}}{2m_{\mathrm{AC}}\omega_{e,g}^3 V_s}\frac{\varphi_{e\text{-}p,\mathrm{O}}'^2 D_p(E_p) f_p^{\mathrm{o}}(E_p)}{E_p^3}\langle\mu_{ph\text{-}e}\rangle^2$$

$$\equiv \frac{\omega_{e,g}^3}{3\pi\epsilon_o\hbar u_{ph}^3}\langle\mu_{ph\text{-}e\text{-}p}\rangle^2. \tag{7.192}$$

In Figure 7.25, variations of the predicted photon-induced, phonon-assisted transition lifetime with respect to temperature, is compared with the experimental results of [124], for Yb^{3+}: ZrF, and good agreement is found. Also shown is the temperature dependence of $\tau_{e\text{-}p} = \dot{\gamma}_{e\text{-}p}^{-1}$.

The sample temperature depends on the balance between the thermal load (radiative heat transfer from surroundings) and the laser cooling rate. The simplified analysis of the laser cooling of a sample used in [124] is given here. The sample is a 7 mm long, 170 μm diameter optical fibre, subject to laser irradiation from one end, as shown in Figure 7.26. The only external load is the thermal radiation from the surroundings. Because the external radiative resistance is much larger than the internal conductive resistance, the sample can be assumed to be at uniform temperature T, and the macroscopic, steady-state, integral-volume energy equation (Table 1.1) is

$$Q_{ph,s} = \dot{S}_{ph\text{-}e\text{-}p}, \tag{7.193}$$

where the thermal radiation load $Q_{ph,s}$ is given by

$$Q_{ph,s} = (\frac{1}{2}\pi D^2 + \pi DL)\epsilon_{ph}\sigma_{\mathrm{SB}}(T_s^4 - T_\infty^4), \tag{7.194}$$

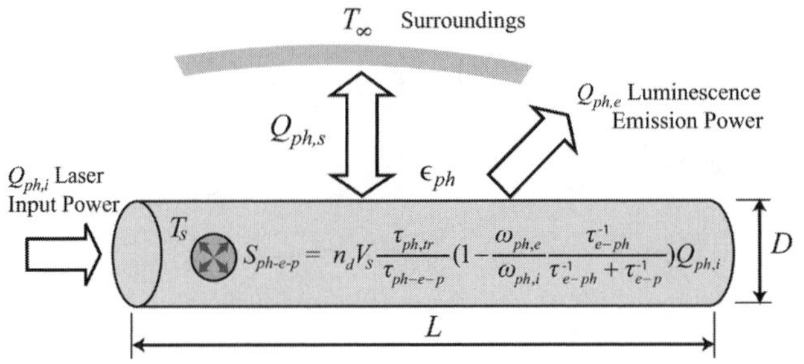

Figure 7.26. Heat transfer mechanism of laser cooling of solids. The solid sample exchanges radiation heat with the surroundings through vacuum. The sample temperature will be determined by the balance between the external radiation (thermal load) and the cooling rate.

where ϵ_{ph} is the total surface emissivity. The energy conversion $\dot{S}_{ph\text{-}e\text{-}p}$ is given by

$$\begin{aligned}
\dot{S}_{ph\text{-}e\text{-}p} &= Q_{ph,a} - Q_{ph,e} \\
&= \hbar\omega_{ph,i}\dot{\gamma}_{ph\text{-}e\text{-}p}\int_{V_s} n_d \mathrm{d}V_s - \hbar\omega_{ph,e}\dot{\gamma}_{ph\text{-}e}\int_{V_s} n_d \mathrm{d}V_S \\
&= (1 - \frac{\omega_{ph,e}}{\omega_{ph,i}}\frac{\dot{\gamma}_{e\text{-}ph}}{\dot{\gamma}_{ph\text{-}e\text{-}p}})\hbar\omega_{ph,i}\dot{\gamma}_{ph\text{-}e\text{-}p}\int_{V_s} n_d \mathrm{d}V_s \\
&= (1 - \frac{\omega_{ph,e}}{\omega_{ph,i}}\eta_{e\text{-}ph})A_{ph,a}n_d u_{ph}\int_{V_s} e_{ph,i}\mathrm{d}V_s \quad A_{ph,a} = \frac{\hbar\omega_{ph,i}\dot{\gamma}_{ph\text{-}e\text{-}p}}{u_{ph}\hbar\omega_{ph,i}n_{ph,i}} \\
&= (1 - \frac{\omega_{ph,e}}{\omega_{ph,i}}\eta_{e\text{-}ph})\sigma_{ph,i}u_{ph}\hbar\omega_{ph,i}n_{ph,i}V_s \\
&= (1 - \frac{\omega_{ph,e}}{\omega_{ph,i}}\eta_{e\text{-}ph})\alpha_{ph,i}Q_{ph,i},
\end{aligned} \tag{7.195}$$

where n_d is the dopant concentration, $\sigma_{ph,i}$ is the absorption coefficient (which is the product of $A_{ph,a}$ and n_d), and $n_{ph,i}$ is the number of photons per unit volume. The spectral absorptance $\alpha_{ph,i}$ is related to the absorption coefficient (for optically thin solids, see footnote of Section 7.8) as,

$$\alpha_{ph,i} = 1 - \exp(-\sigma_{ph,i}L) \simeq \sigma_{ph,i}L, \quad \sigma_{ph,i}L \ll 1, \tag{7.196}$$

and $\dot{Q}_{ph,a} = \alpha_{ph,i}Q_{ph,i}$, where $Q_{ph,i}$ is the incident laser power.

Thus (7.193) becomes

$$(\frac{1}{2}\pi D^2 + \pi DL)\epsilon_{ph}\sigma_{\rm SB}(T_s^4 - T_\infty^4) = (1 - \frac{\omega_{ph,e}}{\omega_{ph,i}}\eta_{e\text{-}ph})\alpha_{ph,i}Q_{ph,i}. \tag{7.197}$$

from which the sample temperature T_s is determined.

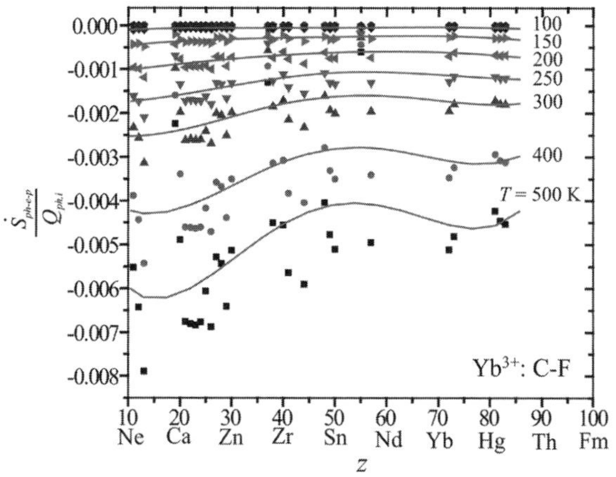

Figure 7.27. Dimensionless cooling rate as a function of atomic number, for discrete values of temperature for Yb^{3+} : C-F where C stands for cation. Some elements are added for reference. Also, a fourth-order polynomial fit is shown to guide the eye. Note that semiconductors and rare-earth materials have been omitted [180].

Using the quantum quantities developed in Section 7.12.1, (7.195) becomes

$$\dot{S}_{ph\text{-}e\text{-}p} = \frac{\pi\hbar}{2\epsilon_{\rm o} m_{\rm AC}} \langle \mu_{ph\text{-}e}\rangle^2 \varphi'^2_{e\text{-}p,{\rm O}} \frac{D_p(E_p) f_p^{\rm o}(E_p)}{E_p^3} \frac{\hbar\omega_{ph,i} n_d L}{u_{ph}} (1 - \frac{\omega_{ph,e}}{\omega_{ph,i}} \eta_{e\text{-}ph}) Q_{ph,i}$$

$$= n_d V_s \frac{\tau_{ph,tr}}{\tau_{ph\text{-}e\text{-}p}} (1 - \frac{\bar{\omega}_{ph,e}}{\omega_{ph,i}} \frac{\tau^{-1}_{ph\text{-}e}}{\tau^{-1}_{ph\text{-}e} + \tau^{-1}_{e\text{-}p}}) Q_{ph,i}, \qquad (7.198)$$

for the optical-phonon absorption, where $\tau_{ph\text{-}e}$, $\tau_{e\text{-}p}$, and $\tau_{ph\text{-}e\text{-}p}$ are the radiative, nonradiative, and phonon-assisted transition time constants respectively, and $\tau_{ph,tr}$ is the photon transit time that is defined by $\tau_{ph,tr} = L/u_{ph}$, where L is the sample length (along the beam). This indicates that, unlike energy transport, the energy conversion process is a product of the time constants of the processes. The cooling rate is directly limited by the the phonon-assisted absorption process, which has a transition rate two orders of magnitude smaller than that of the purely radiative transition. The reabsorption of emitted photons is neglected because it is estimated to be only 0.005% of the total emission rate [292]. Further details are reported in [292] and [180]. The *ab initio* computation of the three carrier interaction rates in the strong coupling regime is introduced in the next section.

7.12.5 Optimal Host Material

Figure 7.27 shows distribution of the maximum cooling rate as a function of the atomic number. The trend is fitted to a 4th-order polynomial to guide the eye. The results show that there are two peaks, between atomic numbers 20 and 30 and 75 and 85. This trend supports the recent successes of blending of light and heavy cations as host materials for laser cooling of solids. However, as the temperature decreases

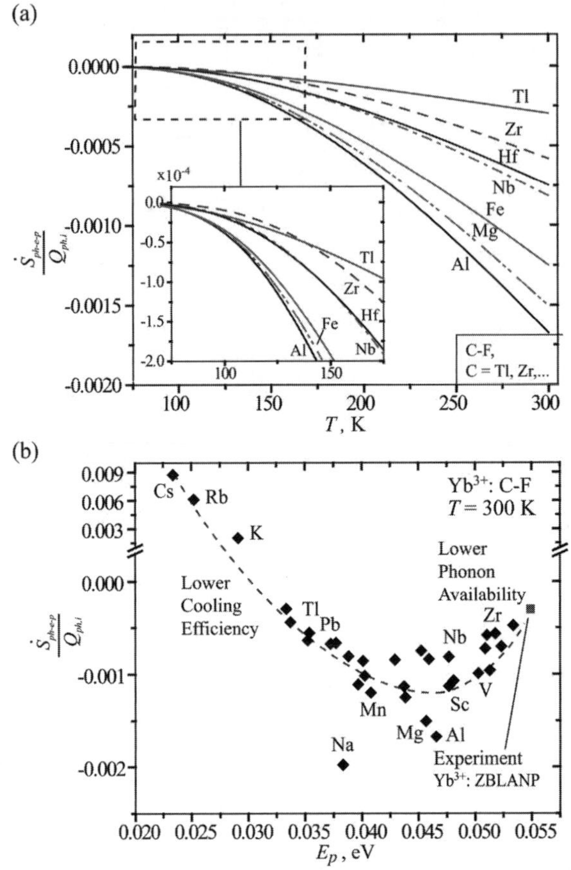

Figure 7.28. (a) Dimensionless cooling rate as a function of temperature, for Yb^{3+}: Tl-F, Zr-F, Hf-F, Nb-F Fe-F, Mg-F, and Al-F. The results are for ideal conditions, i.e., quantum efficiency of one and identical FCC structures. (b) The maximum, normalized cooling rate for various cation-fluoride pairs, as function of phonon energy for Yb^{3+}: C-F. Note that semiconductors and rare-earth materials have been omitted. The dashed line is only intended to guide the eye [180].

and the available high energy phonons diminish rapidly (Bose–Einstein distribution $f_p^{\rm o}$), the cooling rate is quickly suppressed.

The above analysis provides a guide to the selection of ion-host materials for optimal performance. Here we compare various host materials, based on performance by atomic pairs, and we choose F as one of the atoms. Figure 7.28(a) shows variation of dimensionless cooling rate with respect to temperature, for some C-F pairs with Yb^{3+} ion. The crystal structure assumed is FCC, which has C-F pairs as the ion immediate ligands. Here C is Tl, Zr, Hf, Nb Fe, Mg and Al. These structures may not be realized, for example, AlF_3 is an stable, existing compound. However, if a blend of different C-F pairs are made, the contributions of these pair ligands exist at the ion site. Figure 13(a) shows that Al, which has a relatively low phonon peak energy, exhibits high cooling rate over a wide range of temperatures, however, one can expect that the energy removed per transition is low (low capacity). On the

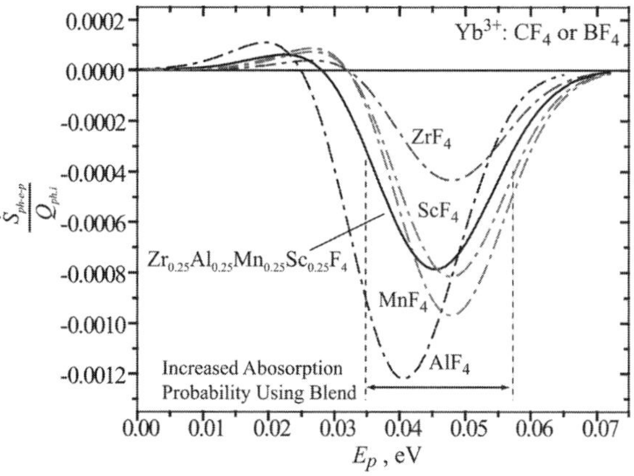

Figure 7.29. Dimensionless cooling rate as a function of phonon energy for diatomic host Yb^{3+}: CF_4 or blend host, BF_4. The cooling rates have been calculated using ideal conditions, i.e., quantum efficiency of one and FCC structure, and are linearly superimposed. Although the magnitude of the cooling rate is moderated, the absorption probability increases as the phonon spectrum broadens. The cooling rates for diatomic hosts are shown in dashed lines and exhibit less broadening [180].

contrary, one can expect the energy removed per transition is high (high capacity) for Zr (due to relatively high phonon peak energy), yet the performance decreases rapidly as the temperature decreases. The inset in Figure 7.28(a) shows that at temperatures near 150 K, it is possible to reach even lower temperature with Tl compared to Zr.

The maximum cooling rate for some cation-fluoride pairs are given in Figure 7.28(b). The results show that the first column alkali-metals from the periodic table are not good candidates for laser cooling at T = 300 K. However, due to the relatively lower phonon energy, these elements are expected to be more suitable at lower temperatures with the exception of Cs and Rb. The results are expected from (7.198), which shows that there are several competing processes in laser cooling of solids. These are, (a) higher phonon peak energy results in more energy removed per transition, (b) lower phonon peak energy results in higher phonon distribution values (c) low cut-off frequency results in higher phonon density of states, and (d) higher cut-off frequency results in higher a nonradiative decay.

Using the above discussions, it is possible to quantitatively predict the cooling performance of a blended material. Figure 7.29 shows the cooling performance of an example blend of host materials. In practice, the composition discussed here may not be realized, however, the blend here provides an example providing a valuable general guide. The figure shows that blending materials that have different phonon peak energies increases the half width of the transition. This, in turn increases the transition probability. One can expect that as the half width broadens, it increases the probability of various phonon modes available in the lattice coupling with the electron in oscillation. This blending strategy is expected to increase the absorption rate as much as factor of 2.

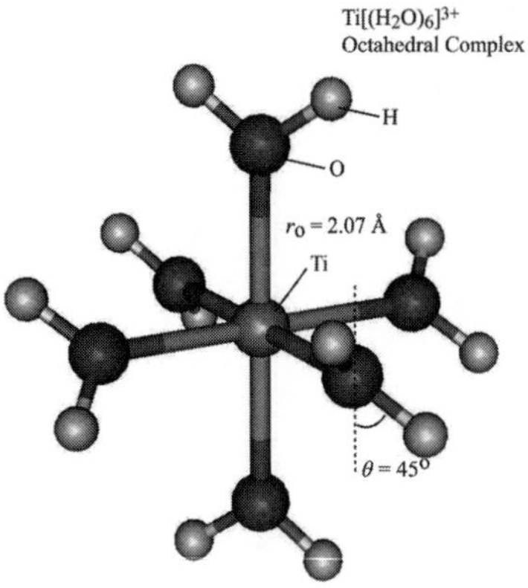

Figure 7.30. The optimized geometry of the ground state of $Ti[(H_2O)_6]^{3+}$ (hydrated titanium) complex.

Results of Figures 7.28(a) and 7.29 suggest using elements Al, Mn, Na and Mg in the host blend. Nevertheless, for optimized cooling performance, a wide range of elements should be present in the blend for increased absorption probability (with the exception of Rb and Cs).

7.12.6 Photon–Electron and Electron–Phonon Transition Rates Using Strong Couplings (Ab Initio Computation)

It was found in the previous section that the absorption rate $\dot{\gamma}_{ph,a}$ is limited by the photon–electron coupling and electron–phonon coupling. It becomes feasible to calculate these coupling parameters by using *ab initio* methods, which may allow for an understanding to link these coupling parameters to the atomic structure, especially in the quantitative level.

(A) *Photon–Electron Coupling (Transition Dipole Moment)*

Here we use the $Ti[(H_2O)_6]^{3+}$ complex as the model system. The calculation is performed using Gaussian 03 Package [113] introduced in Chapter 2, with the B3LYP method and the 6-311+G* basis set. The Ti^{3+} ion has a single unpaired electron, which gives a spin multiplicity of 2 for the complex. To avoid convergence problems of the self-consistent function (SCF), a quadratic convergent (QC) procedure has been applied. No symmetry restriction was prescribed at the start of the calculation, and the optimized geometry converges to the D_{3d} all-vertical symmetry, as shown in Figure 7.30. This symmetry is lower than the O_h point group and higher than C_i. The calculated Ti–O bond length is 2.07 Å, which agrees well with the experimental value 2.03 Å [331] and previously reported ground-state calculations on this complex [170, 324].

The energy levels for the complex at the ground state equilibrium geometry can be calculated using the time-dependent DFT (TDDFT), which is believed to be the most accurate method for excited states. For a free-standing Ti^{3+} ion, it has five degenerate *d* levels, with the orbitals shown in Figure 7.31. As the ion is put into an

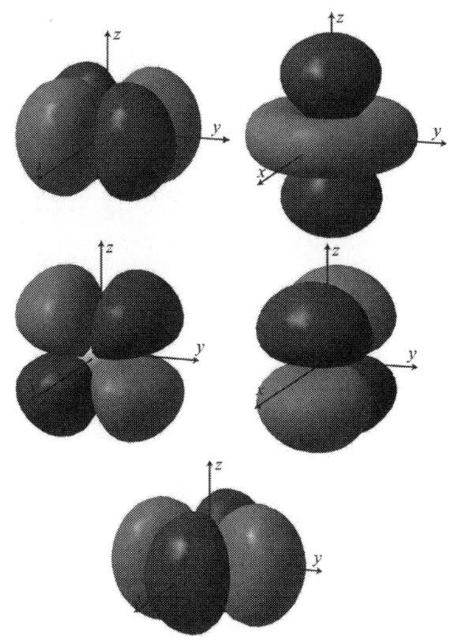

Figure 7.31. The fine degenerate $3d$ orbitals of a single Ti^{3+} ion.

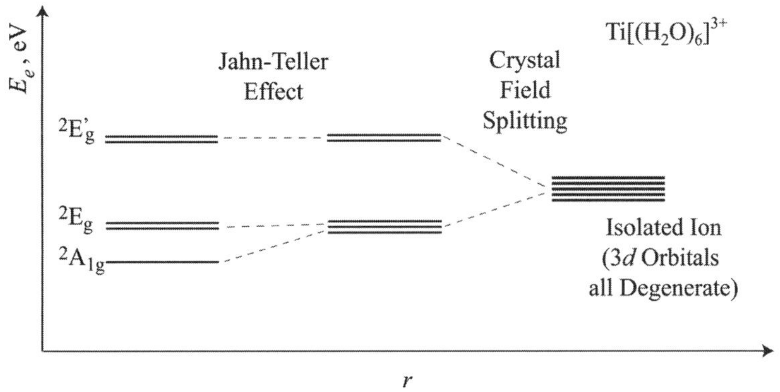

Figure 7.32. Calculated energy levels of the $Ti[(H_2O)_6]^{3+}$ complex at the ground-state equilibrium geometry. The Ti-O bond length is $r_e = 2.07$ Å. The energy multiplets that are due to the crystal field and the Jahn–Teller effect are shown.

octahedral crystal field (Section 5.20) of the six surrounding oxygen atoms, the levels are split into two groups, three levels for the ground state and two levels for the excited state. The Jahn–Teller effect further splits the ground state into two multiplets, with an energy gap of 6046 cm^{-1}. The Jahn–Teller theorem states that any complex occupying an energy level with an electronic degeneracy is unstable against a distortion that removes that degeneracy in first order. The vibronic coupling of ions in solids can cause a local distortion of the lattice in which the atoms move in the direction of normal-mode displacement to lift the electronic degeneracy. A new equilibrium position is achieved in which the local symmetry is lower than the point-group symmetry of the crystal. Here for the $Ti[(H_2O)_6]^{3+}$ complex the symmetry is lowered from O_h to D_{3d}. The evolution of the energy levels is shown in Figure 7.32.

The ground- and excited-state wave functions are calculated with Gaussian (Section 2.2.2, as shown in Figures 7.33 (a) and (b). The *ab initio* calculations showed

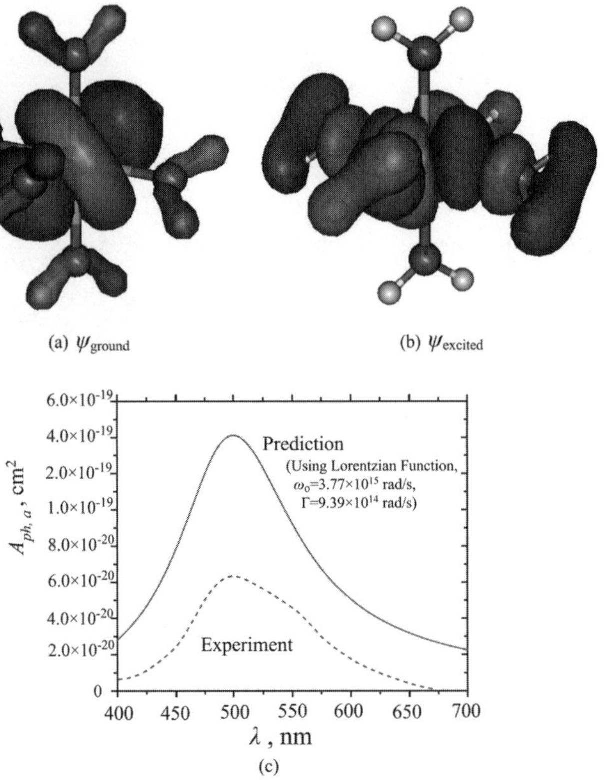

Figure 7.33. Calculated wave functions for (a) ground- and (b) excited-state $T_i[(H_2O)_6]^{3+}$ [293], and (c) predicted and measured absorption bands. The Lorentzian band-broadening function is defined in Table 3.5.

that the energies of the orbitals below the HOMO (highest occupied molecular orbital) are approximately the same for the ground and the excited electronic states, so that the differences of the state energies can be discussed with these singly occupied molecular orbital (SOMO) energies. Based on the theory of LCAO [126], the ground state is mainly composed of the $3d_{z^2}$ orbital, and the excited state is dominated by the $3d_{x^2-y^2}$ orbital.

A photon can be absorbed by an ion if the coupling factor $\boldsymbol{s}_{ph,i} \cdot \boldsymbol{\mu}_e$ is nonzero (i.e., they are not orthogonal), where $\boldsymbol{s}_{ph,i}$ is the polarization vector of the EM field, and $\boldsymbol{\mu}_e$ is the effective transition dipole moment of the ion. The transition dipole moment is defined as a quantum mechanical spatial integral of the classical dipole moment $e_c\boldsymbol{x}$, i.e.,

$$\boldsymbol{\mu}_{e,21} = \sum_{1} |\langle \psi_{f,m_1} | e_c \boldsymbol{x} | \psi_{i,m_2} \rangle|, \qquad (7.199)$$

where e_c is the electron charge, $\boldsymbol{x}$ is the position vector, m_1 is the sublevels of the ground state ψ_{f,m_1}, and m_2 is the sublevels of excited state ψ_{i,m_2}. Note that $\boldsymbol{\mu}_{e,21}$ must be independent of m_2; otherwise, the different m_2 levels would have different transition dipole moments and radiative lifetimes (which are not possible in an isotropic environment). The transition dipole moment between a sublevel m_1 of the ground state and a sublevel m_2 of the excited state is then calculated by

$$\boldsymbol{\mu}_{e,sub} = |\langle \psi_{i,m_1} | e_c \boldsymbol{x} | \psi_{i,m_2} \rangle|, \tag{7.200}$$

which gives 0.005259 D (Table A.3) (1.859×10^{-32} C-m). This value is rather small, indicating that the ground state and excited state have very similar symmetry properties.

Because of a strong electron–phonon coupling in Ti^{3+} systems, the ground state has many vibrational sublevels. If we assume there are N sublevels and each has a similar transition dipole moment, then the total transition dipole moment becomes

$$\boldsymbol{\mu}_{e,21} = N\boldsymbol{\mu}_{e,sub}, \tag{7.201}$$

where N will be determined by fitting to experimental value. The lifetime of Ti^{3+}: Al_2O_3 was measured by spectroscopy experiments to be 3 μs [272]. By substituting (7.201) into (7.190), we have $N = 4000$, indicating that the vibronic effect of Ti doped systems is significant.

(B) *Strong Electron–Vibration Coupling (Electron–Phonon Wave Function)*

When the coupling between the electrons of the optically active ion and the lattice vibration is strong (involving many phonons), transitions become broadband with strong temperature dependence. The atoms in a solid are never completely at rest. The thermal vibrations of the atoms modulate the local crystal field at the site of an optically active ion. This modulation can have several effects on the optical properties of the doped ion. For example, it can modulate the position of the electronic energy levels, leading to a broadening and shifting in peak position of the spectral transition. Also it can cause transitions to occur between electronic energy levels accompanied by the absorption or emission of vibrational energy, but with or without the emission or absorption of photons.

The coupling of an electron to a specific vibrational mode is essentially the change of the electronic property in response to the lattice displacement along that vibrational mode. The normal vibrational modes of the $Ti[(H_2O)6]^{3+}$ complex can be conveniently calculated with Gaussian, after the geometry is optimized. The total degrees of freedom for the complex, (neglecting the H atoms we have seven atoms), is found from Table 6.2 to be 21, 3 translational, 3 rotational, and 15 vibrational. Among all calculated vibrational modes, we consider only those that are also observed for the octahedral TiO_6 core. The modes that are due to the hydrogen atoms only do not contribute significantly to the electron–phonon coupling, because these modes are screened by the more inner oxygen atoms. The five highest vibrational energy normal modes and frequencies are shown in Figure 7.34.

A configuration coordinate diagram is often used to describe transitions between electronic transitions coupled to vibrations. It depicts the variation in the electronic energy levels with respect to the displacement of the normal vibrational coordinate away from its equilibrium position. Here, because the vibration modes

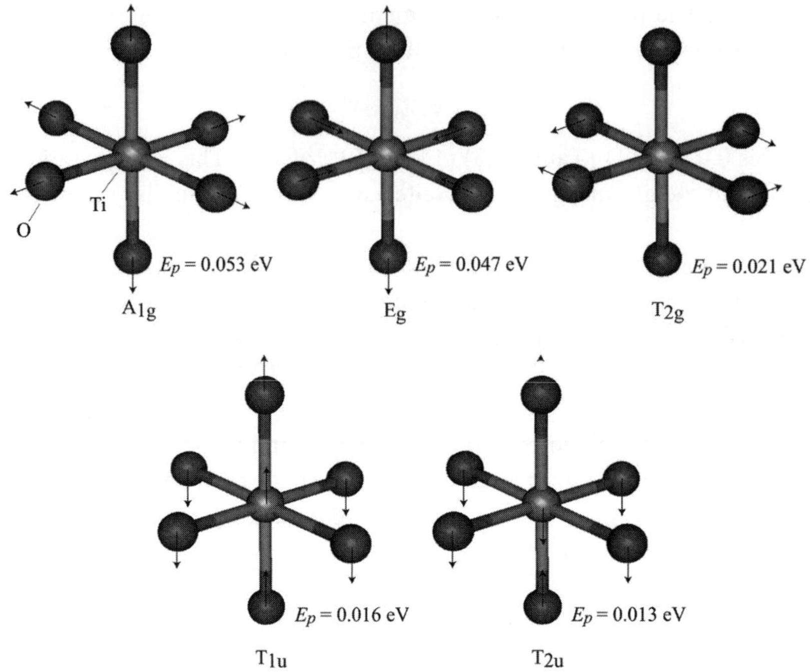

Figure 7.34. The five highest-energy vibrational normal modes along with the frequencies, for $Ti[(H_2O)_6]^{3+}$ complex.

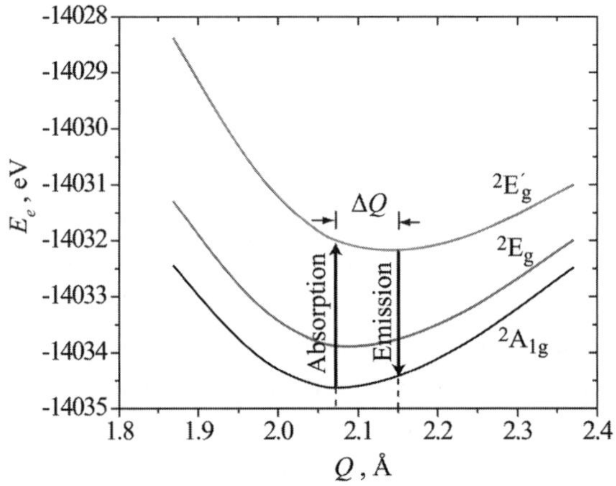

Figure 7.35. Configuration coordinate diagram (energy verses normal coordinate) corresponding to the A_{1g} normal mode, for $Ti[(H_2O)_6]^{3+}$ complex [293].

that are due to hydrogen atoms have very little effect on the energy level of the Ti^{3+} ion, we are concerned only with the vibration of the octahedral TiO_6 core. For these modes the hydrogen atoms move rigidly with the oxygen atoms.

The configuration coordinate diagram is obtained by calculating the energy levels with respect to the normal coordinate of a specific vibrational mode. Shown in Figure 7.35 is the configuration coordinate diagram corresponding to the

A_{1g} normal mode. Five levels resulting from the 3*d* orbitals are shown. As discussed in Figure 7.32, we have the ground state $^2A_{1g}$, the first excited state ^2E_g, which is composed of two nearly degenerate levels, and the second excited state $^2E'_g$, which is also composed of two nearly degenerate levels. The transition between the $^2A_{1g}$ and $^2E'_g$ levels is very important in lasers and luminescent applications and is therefore of concern here. As shown in the figure, the potential energy minimum for the excited state is shifted to the right to that of the ground state, as expected. This shift leads to the well-known Stokes shift in the emission wavelength. For the excited state, the electron is normally farther away from its nuclei than the ground state, repelling the surrounding oxygen atoms. As the result, the equilibrium Ti–O bond length for the excited state becomes longer than that of the ground state, which in turn leads to a smaller force constant and vibrational frequency – the curvature for the excited state is flatter than that for ground state. To take into account the modifications of both normal coordinates (5.163) and frequencies between the electronic states, we can express them in general as follows

$$Q' = Q + \Delta Q \quad \text{shift in normal coordinate} \tag{7.202}$$

$$\omega'_s = \omega_s(1 - F) \quad \text{shift in energy,} \tag{7.203}$$

where ΔQ is the normal coordinate shift of the energy minimum and F is the fraction of the frequency shift. Only those normal modes that have modifications in either normal coordinates or frequencies between the electronic states concerned can contribute to the radiationless transition probability. The Huang–Rhys coupling factor S_s for mode ω_s is

$$S_s = \frac{m\omega_s \Delta Q^2}{2\hbar} \quad \text{Huang–Rhys coupling factor,} \tag{7.204}$$

where M is the mass of the vibrating atom, the O atom here.

In the Born–Oppenheimer approximation, the system wave function ψ is presented as [271]

$$\psi_{i,\nu}(\boldsymbol{x}, Q) = \phi_i(\boldsymbol{x}, Q)\theta_{i,\nu}(Q) \quad \text{electron–phonon wave function.} \tag{7.205}$$

where $\theta_{i,\nu}(Q)$ is the vibrational wave function at a nuclei normal coordinate Q, and $\phi_i(\boldsymbol{x}, Q)$ is the electronic wave function for a fixed position of the nuclei. This implies that the motion of the electron is very rapid compared with the nuclei motion. The Hamiltonian H for the entire system can be chosen as

$$\text{H} = \text{H}_k(Q) + \text{H}_e(\boldsymbol{x}) + \text{H}_{e\text{-}p}(Q), \tag{7.206}$$

where $\text{H}_k(Q)$ is the kinetic energy operator of all nuclear motions, $\text{H}_e(\boldsymbol{x})$ is the electronicenergy operator for electronic states, and $\text{H}_{e\text{-}p}(Q)$ is the electron–lattice interaction potential. In the adiabatic approximation, $\phi_e(\boldsymbol{x}, Q)$ and $\theta_{i,\nu}(Q)$ are solutions of the following Schrödinger equations, i.e., similar to (2.75)

$$[\text{H}_e(r) + \text{H}_{e\text{-}p}(Q)]\phi_i(\boldsymbol{x}, Q) = \varphi_i(Q)\phi_i(\boldsymbol{x}, Q) \quad \text{electronic energy} \tag{7.207}$$

$$[\text{H}_k(Q) + \varphi_i(Q)]\theta_{i,\nu}(Q) = \text{H}_{i,\nu}(Q)\theta_{i,\nu}(Q) \quad \text{phonon energy,} \tag{7.208}$$

where $\varphi_i(Q)$ is the adiabatic potential of the electronic state at the instantaneous position Q and ν signifies the overall vibrational state of the nuclei. Although $\psi_{i,\nu}$ is a good approximation for stationary states, the system oscillates to and from various quantum states of nearly the same energy. This should be interpreted as the transition from one electronic state to another, accompanied by a transition in the quantum states of the nuclear motion. The perturbation Hamiltonian H′ for the nonradiative transition process is given by [211, 271]

$$\mathrm{H}'\psi_{i,\nu}(\boldsymbol{x}, Q) = -\frac{\hbar^2}{2m}\sum_s \frac{\partial\phi_i(\boldsymbol{x}, Q)}{\partial Q_s}\frac{\partial\theta_{i,\nu}}{\partial Q_s} - \frac{\hbar^2}{2m}\sum_s \frac{\partial^2\phi_i(\boldsymbol{x}, Q)}{\partial Q_s^2}\theta_{i,\nu}. \tag{7.209}$$

The nonradiative transition rate $\dot{\gamma}_{e\text{-}p}$ is given by the FGR (3.27):

$$\dot{\gamma}_{e\text{-}p} = \frac{2\pi}{\hbar}\sum_{\nu,\nu'} p_{i\nu}|\langle f\nu'|\mathrm{H}'|i\nu\rangle|^2\delta_{\mathrm{D}}(E_{f,\nu'} - E_{i\nu}), \tag{7.210}$$

where $p_{i\nu}$ is the distribution function for the Boltzmann population of the initial vibrational levels and a δ_{D} function is used for the density of final states to ensure conservation of energy. Using the electronic and vibrational wave functions in (7.210), we have

$$\dot{\gamma}_{e\text{-}p} = \frac{2\pi}{\hbar}\sum_{\nu,\nu'} p_{i\nu}|(-\frac{\hbar^2}{m})\sum_s\langle\phi_f|\frac{\partial\phi_i}{\partial Q_s}\rangle\langle\theta_{f\nu'}|\frac{\partial\theta_{i\nu}}{\partial Q_s}\rangle|^2\delta_{\mathrm{D}}(E_{f\nu'} - E_{i\nu}), \tag{7.211}$$

where Q_s is for mode ω_s.

Here the electronic part of the matrix element can be defined as

$$R_s(fi) = -\frac{\hbar^2}{m}\langle\phi_f|\frac{\partial\phi_i}{\partial Q_s}\rangle. \tag{7.212}$$

The derivative represents how sensitive the electronic wave function is with respect to the displacement along a particular vibrational mode. This is shown in Figure 7.36.

Equation (7.211) has been evaluated in [211] by replacing the δ_{D} function with an integral, and the final result of the transition probability of the nonradiative decay is

$$\dot{\gamma}_{e\text{-}p} = \frac{\pi\omega_s}{2\omega_s\hbar^3}|R_s(fi)|^2\exp[-S_s\coth\frac{\hbar\omega_s}{2k_{\mathrm{B}}T}][(\coth\frac{\hbar\omega}{2k_{\mathrm{B}}T} + 1)\exp(-i\phi' P_i^+)J_{P_i^+}$$
$$\times(S\,\mathrm{csch}\frac{\hbar\omega}{2kT}) + (\coth\frac{\hbar\omega}{2k_{\mathrm{B}}T} - 1)\exp(-i\phi' P_i^-)J_{P_i^-}(S\,\mathrm{csch}\frac{\hbar\omega}{2k_{\mathrm{B}}T})], \tag{7.213}$$

where S is the effective Huang–Rhys coupling factor and includes all modes, $J_{P_i^+}$ is the P_i^+th order Bessel function, and P_i^+ and P_i^- are defined as

$$P_i^+ = \frac{1}{\omega}[-\omega_{AB} - \omega_s + \frac{\rho\omega}{2}\coth\frac{\hbar\omega}{2k_{\mathrm{B}}T}] \tag{7.214}$$

$$P_i^- = \frac{1}{\omega}[-\omega_{AB} + \omega_s + \frac{\rho\omega}{2}\coth\frac{\hbar\omega}{2k_{\mathrm{B}}T}], \tag{7.215}$$

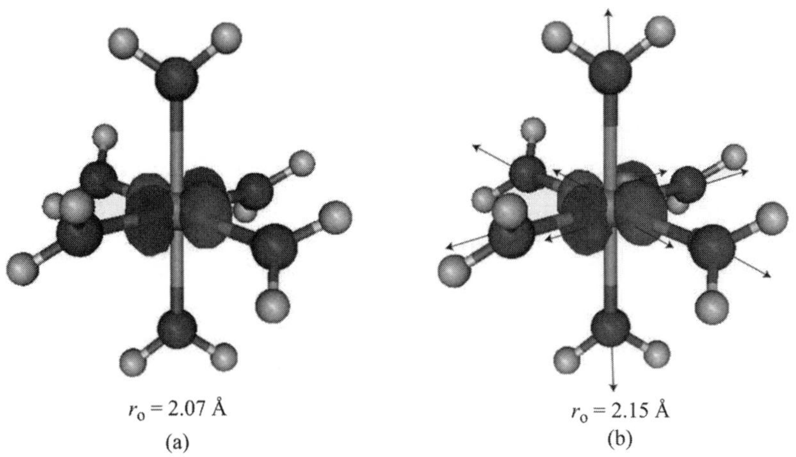

Figure 7.36. The modulations (a) and (b) of the excited state wavefunction with respect to the A_{1g} vibrational mode, for $Ti[(H_2O)_6]^{3+}$.

where ω_{AB} is the gap energy between the ground and the excited state.

Using (7.213), the nonradiative decay rate is calculated as a function of temperature, and is shown in Figure 7.37(a). The total decay rate, $\dot{\gamma}_{ph\text{-}e\text{-}p}$ is the summation of the radiative and nonradiative rates as (7.156), i.e.,

$$\dot{\gamma}_{ph\text{-}e\text{-}p}(T) = \dot{\gamma}_{ph,e,sp} + \dot{\gamma}_{e\text{-}p}(T), \tag{7.216}$$

and the lifetime is given by

$$\tau_{ph\text{-}e\text{-}p} = \frac{1}{\dot{\gamma}_{ph\text{-}e\text{-}p}}. \tag{7.217}$$

The calculated lifetime in this way is shown in Figure 7.37(b). Similar to the results given in Figures 7.21(a) and (b), Figure 7.37(b), at low temperatures, the nonradiative decay rate is negligibly small compared with the radiative decay rate, so that the lifetime remains almost a constant. As the temperature increases, more phonons are activated and involved in the decay process, and the nonradiative decay rate is increasing rapidly. At around 250 K, the nonradiative decay rate becomes comparable or even larger than the radiative decay rate, and the lifetime drops significantly with temperature. At high temperatures, the decay process is dominated by the nonradiative decay. The predicted lifetime is compared with that of the Ti:sapphire laser material, which also has an octahedral crystal field around the Ti^{3+} ion. The measured Ti^{+3}: Al_2O_3 lifetime [Figure 7.21(b)] at low temperatures is much shorter than that predicted for $Ti[(H_2O)_6]^{3+}$, because it was not measured at the zero-phonon line, and in real materials defects and impurities may modify the symmetry properties around the Ti ion significantly. At high temperatures, the nonradiative decay rates follow a similar trend. Exact agreement is not expected because the phonon modes for our cluster are different from those in the Ti^{3+}: Al_2O_3 solids.

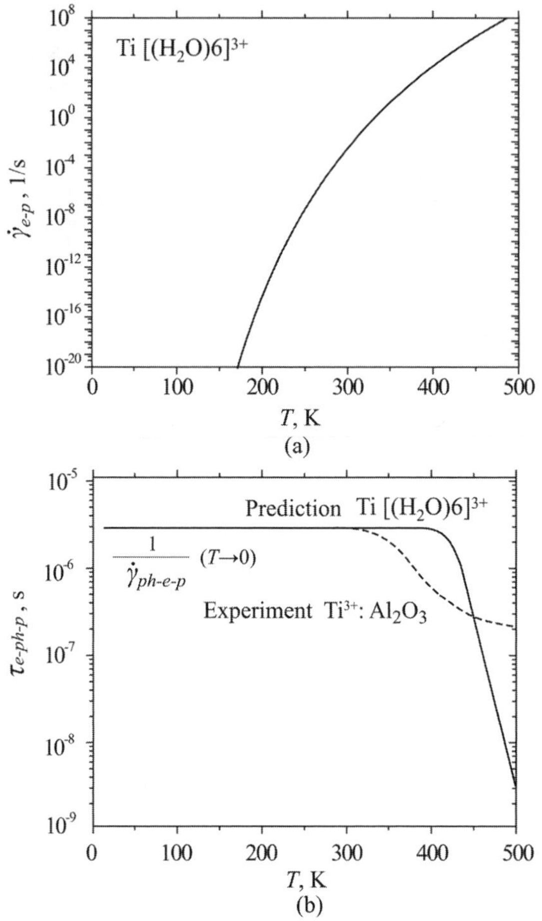

Figure 7.37. (a) Variations of total transition rate with respect to temperature. (b) Variations of lifetime ($1/\tau_d = 1/\tau_r + 1/\tau_{nr}$) with respect to temperature. The experimental results are from Figure 7.37(b) [293].

7.13 Gas Lasers and Laser Cooling of Gases

7.13.1 Molecular-Gas Lasers

In molecular-gas laser systems [348], photons can be emitted from transitions in the vibrational–rotational spectrum of a molecule. The energy of the photon emission is directly related to the fluid particle energies associated with such a laser system.

Unlike solids and liquids, the energy of gaseous fluid particles is completely contained within an individual molecule and not stored in phonons or intermolecular forces. As discussed in Chapter 6, the fluid particle energy is composed of five energies: potential, electrical, translational, vibrational, and rotational energy; however, here we assume that the potential energy is negligible.

Energy states are quantized and have populations as functions of temperature, as given by the M–B distribution. Near room temperature, vibrational and electrical

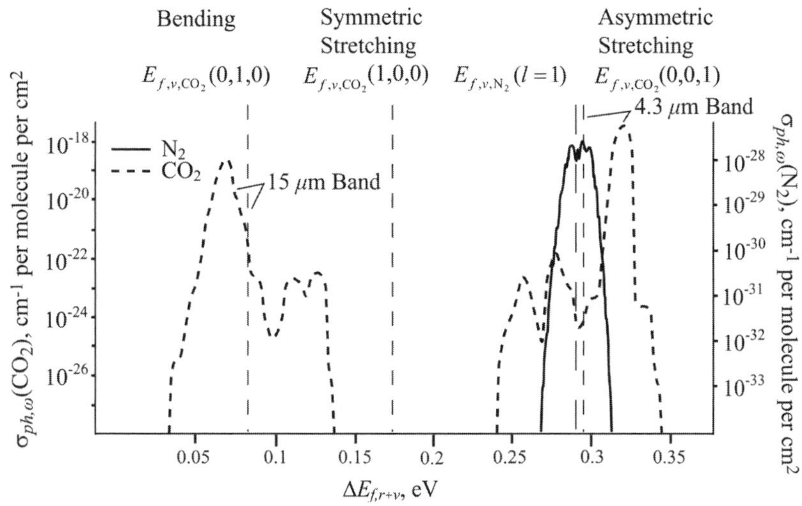

Figure 7.38. Variations of the measured spectral absorption coefficient with respect to energy (infrared regime) for N_2 and CO_2. The rotational–vibrational spectra of N_2 and CO_2 show the first vibrational transition of N_2 and its relation to CO_2 absorption lines [49]. The CO_2 band designations are from Table 7.3.

energy states are not significantly populated. However, transitions to excited vibrational and electronic states can occur through special excitation processes.

Transitions between these energy states are governed by selection rules [16, 80]. The most basic selection rule states that the rotational and vibrational quantum numbers (Section 6.1), j and l, can increase or decrease only by a value of one for any transition:

$$\Delta j = \pm 1, \ \ \Delta l = \pm 1. \tag{7.218}$$

Polyatomic molecules have multiple modes of vibration, and have multiple corresponding vibrational quantum numbers. CO_2 [95] is a triatomic molecule with three modes of vibration, (l_1, l_2, l_3), Table 7.3. These vibrational quantum numbers correspond to modes of vibration referred to as the symmetric stretch, bending, and antisymmetric stretch modes, respectively. Selection rules for this molecule are

$$|\Delta l_2| = \text{even}, \ \ |\Delta l_3| = \text{odd} \tag{7.219}$$

or

$$|\Delta l_2| = \text{odd}, \ \ |\Delta l_3| = \text{even}. \tag{7.220}$$

Based on the distribution of rotational energy states present in a molecule, there are a number of transitions that can occur in a single vibrational transition. This gives many closely spaced absorption lines (in $\sigma_{ph,\omega}$) that make up a rotational–vibrational spectrum, as shown in Figure 7.38. Each absorption line (in $\sigma_{ph,\omega}$) is attributed to a specific rotational–vibrational energy transition from the ground state. The change in vibrational energy between $l = 0$ to $l = 1$ for N_2 occurs at 0.2890 eV. This energy state is, like all energy states in the rotational–vibrational spectrum, quantized and fixed. The spectrum arises because rotational-energy states are not linearly spaced, and as such a change of $j = \pm 1$ is generally not constant.

Not all energy states of a molecule appear in the absorption spectra. According to the selection rules for a triatomic atom, a transition from $l_1 = 0$ to $l_1 = 1$ cannot occur without an accompanying transition in another vibration mode. From the ground state, the (1,0,0) energy level of the CO_2 molecule cannot be accessed and does not show up in the absorption spectra of Figure 7.38. This arises physically because a single collision cannot exert a force that will excite a symmetric stretch mode of vibration from the ground state. A complete set of energy states can be described quantum mechanically.

The energy states of fluid particles can be predicted classically when combined with quantum restrictions from the Schrödinger equation. Quantum restrictions give acceptable rotational energy levels of a fluid particle as $\hbar^2$ times multiples of the rotational quantum number j. The combination of these results gives the rotational energy states defined by (6.15), i.e.,

$$E_{f,r} = j(j+1)\frac{\hbar^2}{2I_f}, \quad j = 0, 1, 2, 3... \text{ rotational energy}, \qquad (7.221)$$

where I_f is the moment of inertia. Values for the rotational energy states of a gas molecule can be predicted from the interatomic spacing of atoms and their masses. From masses and interatomic separations, the moment of inertia is determined by assuming the molecule rotates as a rigid body about its center of mass. This is called the rigid-rotor assumption. The moment of inertia, I_f, is represented as a function of the two masses and their separation distance, i.e.,

$$I_f = \frac{m_1 m_2}{m_1 + m_2} r_e^2 \text{ moment of inertia}. \qquad (7.222)$$

Vibrational energy states of gas molecules are also determined quantum mechanically from the Schrödinger equation. The quantum restrictions yield the acceptable energy levels in terms of a vibrational quantum number l, i.e., from (6.14), we have

$$E_{f,v} = (l + \frac{1}{2})\hbar\omega_f \text{ vibrational energy}, \ l = 0, 1, 2, 3... \qquad (7.223)$$

Values of the vibrational natural frequency of a gas molecule can be predicted from the potential energy function. To simplify the problem, we assume harmonic oscillation about the equilibrium separation, so that the potential is of the form $\varphi_{ij} = \Gamma_{ij}^2/2$. We create this harmonic approximation of the potential from the Morse potential (Table 2.1). By taking the second derivative of the potential evaluated at the equilibrium separation, we determine the value of the bond force constant Γ. The form of the Morse potential is

$$\varphi = \varphi_o\{[1 - e^{-a(r-r_e)}]^2 - 1\} \text{ Morse potential}, \qquad (7.224)$$

Table 7.5. *Morse potential parameters for several molecules [80, 298].*

Molecule	Atom pair	φ_o, eV	r_e, Å	a, 1/Å
Au_2	Au–Au	2.30	2.47	1.69
BH	B–H	3.16	1.11	2.47
Ca_2	Ca–Ca	0.129	4.28	1.10
CO_2	C=O	8.34	1.16	2.24
Cu_2	Cu–Cu	2.03	2.22	1.41
N_2	N–N	9.80	1.10	2.71
Ne_2	Ne–Ne	0.0020	1.75	1.17
NH	N–H	3.24	0.99	2.72
OH	O–H	7.05	0.95	3.18
SiH	Si–H	1.69	1.45	2.34
Xe_2	Xe–Xe	0.0230	4.36	1.53

where φ_o is the potential-well depth r_e is the equilibrium separation, and a is the harmonicity parameter. The harmonicity term varies significantly, depending strongly on the type of bonding present between atoms. The second derivative of this function evaluated at $r = r_e$ is the force constant: (2.54)

$$\Gamma = 2a^2\varphi_o \text{ force constant.} \tag{7.225}$$

The natural frequency of a diatomic bond can then be easily determined from the harmonic oscillation between two molecules about their center of mass. The natural frequency (2.54) becomes

$$\omega_n = (\Gamma\frac{m_1 + m_2}{m_1 m_2})^{1/2},$$

$$\text{natural (fundamental vibrational) frequency for diatomic molecule.} \tag{7.226}$$

Table 7.5 lists the parameters φ_o, r_e and a in the Morse potential (7.224) for several molecules. The same principles can be applied to a polyatomic molecule. For a polyatomic molecule, the multiple modes of vibration make the analysis more complex. The harmonic analysis of these structures can be carried out in the same manner as for mechanical systems: Create equations of motion based on the bond force constants and masses involved, and determine natural frequencies for the system through modal analysis to approximate system behavior.

Substituting the values for the natural frequencies of vibration into (7.223) gives a good approximation to the lower energy states of a molecule. As the quantum number increases, the oscillation becomes increasingly anharmonic, and the harmonic oscillator assumption generally overestimates the energy level.

Translational energies have an impact on the distribution of vibrational and rotational energy levels, as well as the probability of energy transfer between vibrational and rotational energy levels. When a molecule attains an excited vibrational

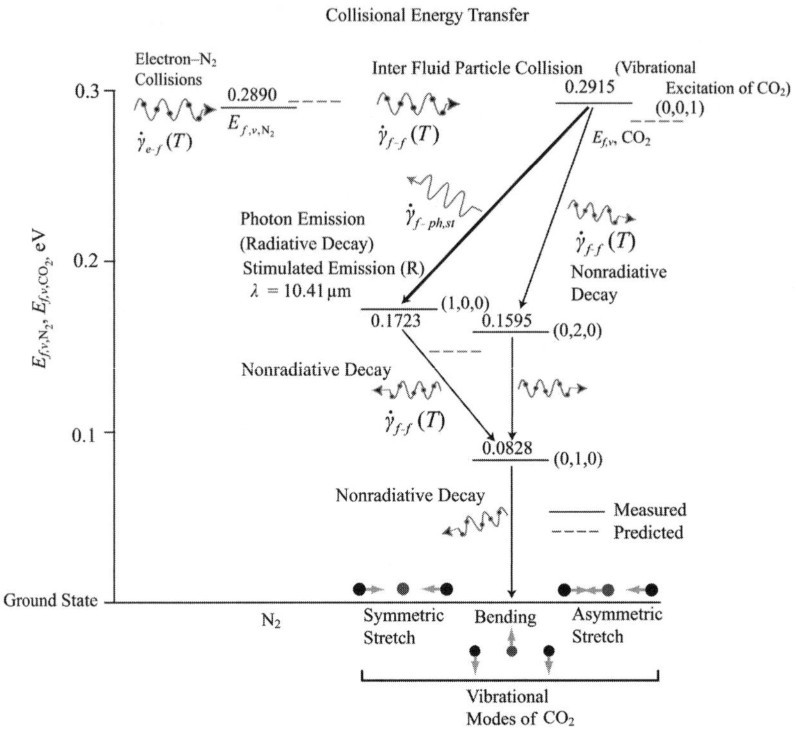

Figure 7.39. The energy-level diagram of a CO_2 laser showing $\lambda_{ph} = 10.6\ \mu$m laser transition and coincident energy levels between the N_2 (l =1) and CO_2 (0,0,1) energy states [348].

or rotational state, with energy above the ground-state distribution, the primary nonradiative transfer mechanism is through collisional deactivation.

In a laser system, the process of radiative emission competes with nonradiative transfer mechanisms (Section 7.11). To decrease the probability of deactivation, it is important to try to reduce the effect of collisional deactivation. Maintaining a high number of molecules in an excited state allows stimulated emission to occur. In the molecular laser system, the stimulated emission $\dot{\gamma}_{ph,f,st}$ that occurs is a transition between the vibrational energy states of the fluid particle (molecule).

An example of a typical molecular laser is the molecular N_2–CO_2 system. Figure 7.39 shows the partial energy diagram representing this molecular laser system. A common CO_2 laser cycle consists of electrically exciting N_2, transferring resonant energy to a CO_2 molecule, and laser emission.

CO_2 is a linear, polyatomic gas [Figure 2.10(b)], and its 9 degrees of freedom (Table 6.2) are 3 translational, 2 rotational, and 4 vibrational. The vibrational modes of CO_2 are symmetric stretching, bending (degenerate, 2 modes), and asymmetric stretching.

Nitrogen molecules are electrically excited into their first vibrational state. Applying an electrical current ionizes the gas and creates a flow of electron motion, which produces electron–fluid particle collisions. Electrical excitation increases the vibrational–rotational energy states of a fluid particle through electron impact and collisional energy transfer mechanisms. Electron impacts vibrationally excite nitrogen. Also, because transitions between vibrational energy levels in the same electronic state for a homonuclear diatomic molecule are forbidden according to the parity selection rule, the population of nitrogen molecules in the first vibrational energy state is allowed to accumulate.

Energy is then transferred through a resonant collision between nitrogen and carbon dioxide. An inelastic collision can transfer vibrational, rotational, and translational energy between fluid particles, if the energy states are sufficiently close. The near coincidence of the $l = 1$ energy state of nitrogen molecule and the (0,0,1) state of carbon dioxide allows this transition to occur easily near room temperature. The major absorption bands of CO_2 are listed in Table 7.3, and the quantum numbers in parentheses are explained in Section 7.8.2.

A transition of vibrational energy can occur radiatively. Radiative decay occurs most naturally into the next lowest energy state. Nonradiative decay may occur first, and is more likely to occur if the particle velocity is high, the collision rate is great, and the change in energy states is small. For carbon dioxide, a radiative decay from the (0,0,1) energy state to the (1,0,0) state will occur if conditions limit the effectiveness of nonradiative decay.

From the harmonic-oscillator approximation of vibrational energy states, the energy of the $l = 1$ state of nitrogen is estimated [from (7.223)] as 0.2920 eV. For the (0,0,1) and (1,0,0) energy states indicated in Figure 7.39, using the C=O bond vibrational frequency as the natural frequencies of carbon dioxide, gives for the first vibrational energy levels 0.1478 eV and 0.2831 eV. Note that the actual frequencies of (0,0,1) and (1,0,0) are different than these. Therefore, in general, the vibrational frequencies are measured.

The observed energy of the photon emitted in the carbon dioxide molecule (Figure 7.39) is $0.2915–0.1723 = 0.1192$ eV. This corresponds to a photon wavelength of 10.41 μm (Table 1.4) in the infrared regime. There are many other transitions that have been observed in the range between 10.4 and 10.6 μm in the CO_2 molecule, which are due to the rotational sublevel and isotope variations in the same vibrational transition.

A number of factors can be controlled in a molecular laser system. Of primary importance in applications are controlling the wavelength, power and efficiency, and stability of a laser system. The energy transition present in the rotational–vibrational spectrum of the lasing fluid particle is what gives the molecular laser its wavelength, and in this way the wavelength of a laser can be chosen.

The output power and stability of a system are also dependent on the fluid particle energies. Output powers and stability are limited by inefficiencies of energy transfer and inconsistent transition energies. Creating stable operating conditions generates a stable laser system, which is done by reducing the variation in the initial distribution of energy states present by cooling. The addition of low pressures can reduce the collision rate, which in turn reduces the effect of collisional deactivation. Both low temperature and pressure can be obtained through gas cycling and control. Inherent energy losses and ineffective energy transfers can be limited by closely matching energy states between fluid particles and using transition energies that are closer to the ground state, so less energy is dissipated as heat.

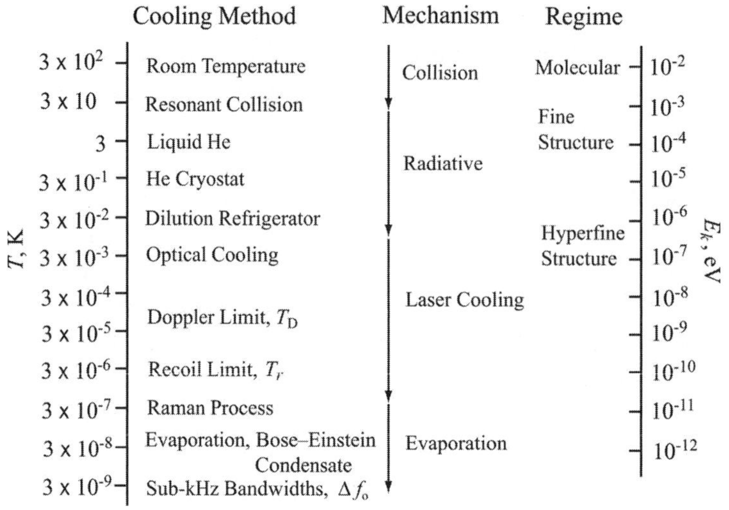

Figure 7.40. Various techniques used for cooling gases to near 0 K. Laser cooling provides major subcryogenic cooling, whereas evaporation allows for reaching the nanokelvin regime [246]. The detunning (Doppler) frequency $\Delta f_o = \Delta\omega_o/2\pi$, determines the Doppler temperature.

7.13.2 Laser Cooling of Atomic Gases

Laser cooling is optical control of atomic motion, and, in regard to gases, the speed of atoms can be substantially reduced, thus potentially reaching the absolute zero temperature. As shown in Figure 7.40, the ordinary cryogenic method extends to 0.1 K, whereas the laser cooling of gases can extend to the sub-μK (10^{-6} K) range [242]. The Bose–Einstein condensate is defined in the Glossary.

The kinetic energy of atomic motion is also shown on the right-hand axis of Figure 7.40, along with the regime of energy structure. The molecular regime (MD scale, Table 2.7) designates atomic/molecular translational, rotational, and vibrational energy. The fine structure relates to atomic scales of Table 1.6. The hyperfine structure defined in Section 1.5.2 is a small perturbation in the fine-structure energy.

Laser cooling of atomic gases involves absorption of photons and the consequential emission of photons, with the net effect of reducing the translational kinetic energy of the constituent atoms [136, 242, 269]. This is also called light pressure cooling. This is shown in Figures 7.41(a) to (c). The frequency of irradiation should be in the lower half of the Doppler-broadened absorption (slightly below the atomic resonance frequency) of the atom. This will ensure that only the atoms moving toward the laser source will absorb, thus losing net energy and momentum by scattering the photons. If the photons come from all directions, the atoms lose energy and the Doppler-shift detunes any photon traveling in the same direction as the atoms.

The laser cooling process results in gas temperatures with a lower limit of the order of $\hbar\dot{\gamma}_{e\text{-}ph}/k_B$, where $\dot{\gamma}_{e\text{-}ph}$ is the spontaneous emission rate of the excited state. The random addition to the average momentum transfer produces a random walk

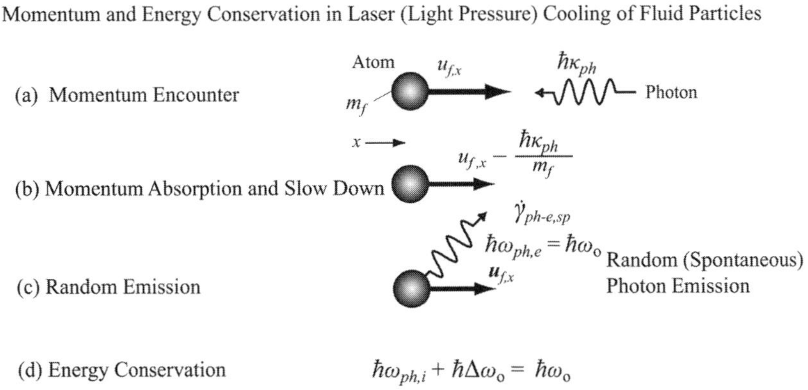

Figure 7.41. (a) A fluid particle with velocity $\boldsymbol{u}_f$ encounters a photon with momentum $\hbar\kappa_{ph}$. (b) After absorbing the photon, the atom is slowed. (c) After emission in a random direction, on average the atom is slower than (a) [269]. (d) Energy conservation for the process.

of the atomic momentum and an increase in the mean-square atomic momentum. This heating is balanced by cooling coming from opposing the atomic motion.

The atoms also emit (dominated by spontaneous) photons (in random directions) and their reabsorption results in heating. The translation-assisted electron excitation (absorption) and emission give $\hbar\omega_{ph,i} + \hbar\Delta\omega_o = \hbar\omega_o$, where $\omega_{ph,i}$ is the incident and $\omega_o = \omega_{ph,e}$ is the emitted photon frequency. Here $\Delta\omega_o$ is the contribution from motion. The Doppler temperature T_D is found from the balance between cooling and heating. The mechanical energy, using the statistical Fokker–Planck equation (3.8) [159] and for a one-dimensional transition along the x direction, is given as [269]

$$\frac{1}{2}m_f\langle u_{f,x}\rangle^2 = \frac{1}{2}k_B T_D = \frac{\hbar\dot{\gamma}_{ph,e,sp}}{4}\left(\frac{\dot{\gamma}_{ph,e,sp}}{2\Delta\omega_o} + \frac{2\Delta\omega_o}{\dot{\gamma}_{ph,e,sp}}\right), \qquad (7.227)$$

where $\dot{\gamma}_{ph,e,sp} = 1/\tau_{e\text{-}p}$ (listed for Na in Table 7.4) is the rate of spontaneous emission of the excited state given by (7.189), and $\Delta\omega_o$ is the angular frequency of the detuning (Doppler frequency) of the laser from the atomic resonance ω_o. The minimum temperature occurs for $\Delta\omega_o = \dot{\gamma}_{ph,e,sp}/2$ which gives

$$T_D = \hbar\dot{\gamma}_{ph,e,sp}/k_B \quad \text{Doppler temperature}, \qquad (7.228)$$

i.e., the Doppler temperature is dependent on the emission rate. It is necessary that the atoms have relatively large electronic transition and because thousands of photons are trapped per atom, the spontaneous emission should be entirely to the ground state. Among atoms used are metastable noble alkali-metal and alkaline-earth atoms. Table 7.6 gives some of strong transitions for some laser materials, including Na. Note that $\omega_o = 2\pi u_{ph}/\lambda_o$. Figure 7.42 shows the electronic transition from $3^2S_{1/2}$ to $3^2P_{3/2}$ in the Na atom. It is necessary to use a repumper in order to avoid off-resonance excitation [269].

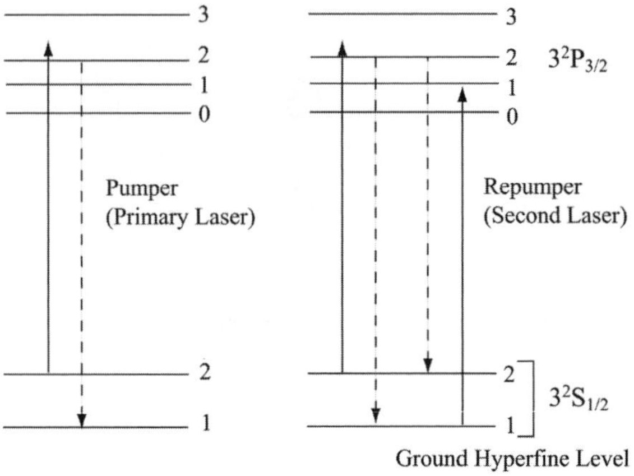

Figure 7.42. Use of a repumping laser to allow many absorption–emission cycles, for gaseous Na atom laser cooling [269].

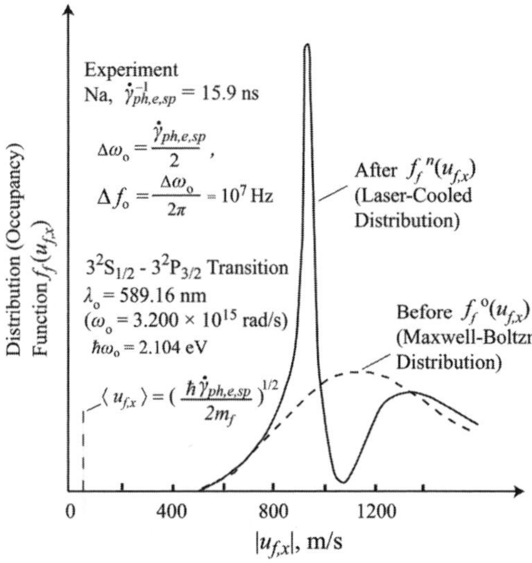

Figure 7.43. Cooling an atomic beam with a fixed frequency laser. The dotted curve is the velocity distribution before cooling (equilibrium distribution f_f^o), and the solid curve is after cooling (equilibrium distribution f_f^n). Atoms from a narrow velocity range are transferred to a slightly narrow range centered on a lower velocity [269]. The transition energy of Na, $3^2S_{1/2} - 3^2P_{3/2}$ is 2.104 eV and $\tau_{e\text{-}ph} = \dot{\gamma}^{-1}_{ph,e,sp} = 15.90$ ns [242].

Then the fluid particle velocity distribution changes from the equilibrium M–B distribution $f_f^o(u_{f,x})$ (6.62) to nonequilibrium distribution $f_f^n(u_{f,x})$, as shown in Figure 7.43. This for $\dot{\gamma}_{ph,e,sp} = 6.25 \times 10^7$ [269] ($\Delta f_o = 10^7$ Hz) gives $T_D = 238.7$ μK and corresponds to $\langle u_{f,x} \rangle^2$ [given by (7.227), for one-dimensional transition] of 29.38 cm/s (also marked in Figure 7.43). The results given in Figure 7.43 indicate that this low temperature is not reached. The experimental results [269] are for Na vapor with resonance wavelength $\lambda_o = 0.5892$ μm (yellow light), or 2.104 eV.

There is a lower-temperature limit, called the recoil temperature T_r, which is defined through

$$\frac{1}{2}m_f\langle u_{f,x}\rangle^2 = \frac{1}{2}\frac{\hbar^2\kappa_{ph}^2}{2m_f} \equiv \frac{1}{2}k_B T_r, \tag{7.229}$$

or

$$T_r = \frac{\hbar^2\kappa_{ph}^2}{2k_B m_f} = \frac{\hbar^2\omega_o^2}{2k_B u_{ph}^2 m_f}. \tag{7.230}$$

This gives for the above Na example, $T_r = 1.20\times10^{-6}$ K $= 1.20$ μK. Both T_D and T_r are marked in Figure 7.40.

The laser cooling of the atomic vapor to very low temperatures ($\simeq 10^{-8}$ K) results in nearly pure condensate (Bose–Einstein condensate, where all gas atoms will be in the ground state, i.e., quantum degenerate gas). From the definition of the Bose–Einstein condensate phase transition given in the Glossary we have

$$\begin{aligned} n_f\lambda_{dB}^3 &= 2.612 \\ &= n_f\left(\frac{h_P^2}{2\pi m_f k_B T_c}\right)^{1/2}, \end{aligned} \tag{7.231}$$

or

$$T_c = \frac{2\pi\hbar^2}{m_f k_B}\left(\frac{n_f}{2.612}\right)^{1/2}, \tag{7.232}$$

where T_c is critical temperature. This shows a dependence on the atomic number density. Below T_c, the ground-state occupation fraction is

$$\frac{n_{f,o}}{n_f} = 1 - \left(\frac{T}{T_c}\right)^3. \tag{7.233}$$

7.14 Photovoltaic Solar Cell: Reducing Phonon Emission

In semiconductors the electronic gap energy $\Delta E_{e,g}$ can be overcome by absorption of a photon of equal energy $\hbar\omega_{e,g}$. When considering a solar irradiation with a multiple photon frequency, there are other interactions and transitions involving holes, electrons, phonons, and photons. Figure 7.44 shows a photon with energy larger than $\Delta E_{e,g}$ absorbed in the semiconductor. The excess energy gives the electron a kinetic energy that is then exchanged in electron–phonon interaction (resulting in phonon emission). There is also the possibility of electron–hole recombination. The electron kinetic energy loss kinetics $\tau_{e\text{-}p}$ is quicker than the recombination time $\tau_{e,rec}$ (unless it is slowed down by confinement, Section 5.22.2).

A schematic of the different interactions involved in a *p-n* single-junction photovoltaic (PV) solar cell are shown in Figure 7.45. The cell is made of layers of *n*-type (e.g., phosphorous-doped), and *p*-type (e.g., boron-doped) semiconductors (e.g., Si), with a depletion region, having lengths l_n (about 0.5 μm), l_p (about 300 μm)

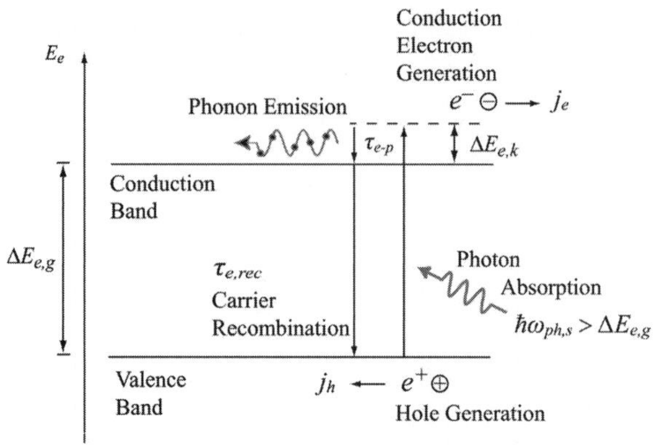

Figure 7.44. A photon of energy larger than the energy gap is absorbed and the excess electron energy (hot electron) results in phonon emission, unless it is slowed down by electron confinement.

and l_d.† These describe the conversion of the photon flux $E_{b,\omega}(T)/\hbar\omega$ (7.8) to the electron current j_e through an external resistance.

All the processes, which are listed in Table 7.6, are involved in the conversion of solar energy to electric power, the efficiency determined by the effectiveness of each process. The solar irradiation flux is the Planck (blackbody) spectral emissive power $E_{b,\omega}(T_s)$, where T_x is the sun temperature (7.8) and the photon flux is $E_{b,\omega}(T_s)/\hbar\omega$. The total irradiation flux from the sun $q_{ph,s}$ is found using (7.6) for solar emission, along with relation (7.90) and expressing the solid angle as a_s, i.e.,

$$q_{r,s} = q_{ph,s} = \frac{a_s\hbar}{4\pi^2 u_{ph}^2}\int_0^\infty \frac{\omega^3}{\exp(\hbar\omega/k_\mathrm{B}T_s) - 1}\mathrm{d}\omega, \tag{7.234}$$

The earth–sun solid angle factor $a_s = 2.18\times10^{-5}$ and for $T_s = 5760$ K gives $q_{ph,s} = 1,350$ W/m^2. For the ambient radiation, T_s is replaced with $T_a = 300$ K and the solid only factor a_s is replaced with $a_s = 1$ (because the solar solid angle is very small). Note that for blackbody emission $E_b = \pi I_b$ (7.8). A faction of $\boldsymbol{q}_{ph,s}$ results in electric power generation $\dot{S}_{ph\text{-}e}/A$, and a fraction results in phonon emission $\dot{S}_{ph\text{-}e\text{-}p}/A$. There is also reflection. The phonon emission leads to a rise in temperature and is transported by conduction $\boldsymbol{q}_k$ (thus decreasing the photovoltac conversion $\dot{S}_{ph\text{-}e}$). These are also shown in Figure 7.45.

In the next section, the PV efficiency is analyzed first for a single-bandgap semiconductor (e.g., Si), then for an ideal multiple-bandgap semiconductor. Finally, the semiempirical efficiency for the actual performances of current PV cells is given.

† The depletion layer combines the p- and n-depletion layers, and for each carrier i is given by

$$l_{d,i} = (\frac{2\epsilon_o\epsilon_e\Delta\varphi_e}{e_c n_i})^{1/2},$$

where $\Delta\varphi_e$ is the built-in potential, and n_i is the carrier i density.

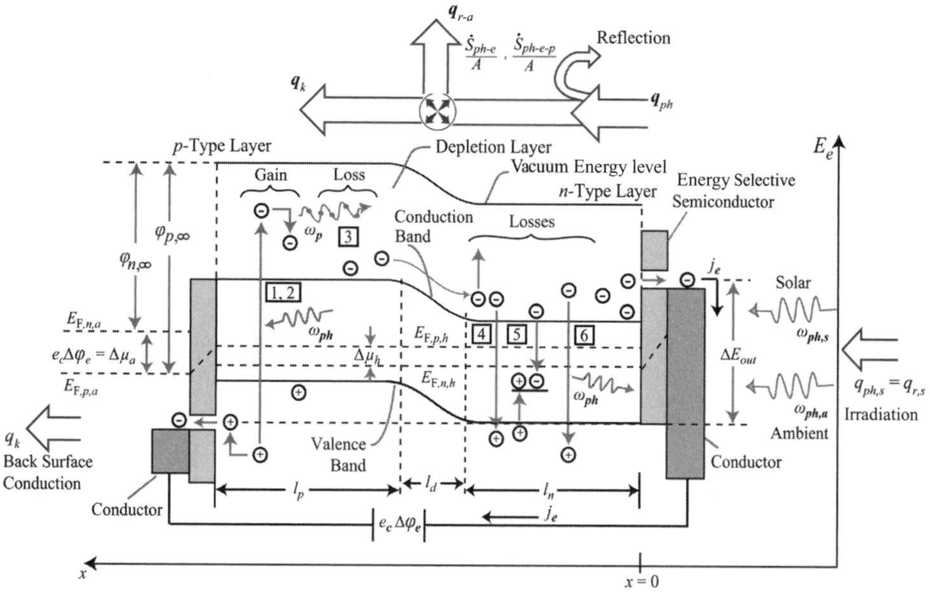

Figure 7.45. Hole, electron, phonon, and photon interactions in the irradiation of a *p–n* junction PV solar cell. The process numbers are defined in Table 7.6.

Table 7.6. *Hole, electron, phonon, and photon interactions (processes) in PV energy conversion.*

	Notation	Process
1	$\dot{n}_{ph\text{-}e,s}$	electronic excitation by a solar photon
2	$\dot{n}_{ph\text{-}e,a}$	electronic excitation by an ambient photon
3	$\dot{s}_{ph\text{-}e\text{-}p}/A$	electron relaxation to conduction band edge through phonon emission
4	$\dot{n}_{e,h\text{-}ph}$	electron–hole recombination resulting in photon emission
5	$\dot{n}_{e\text{-}t}$	impurity trap and defect trap assisted recombination
6	$\dot{n}_{e,h\text{-}k}$	recombination by electron–electron or hole–hole kinetic energy exchange

7.14.1 Single-Bandgap Ideal Solar PV Efficiency

To determine the ideal efficiency of a single direct bandgap PV solar cell we first define some aspects of the power-converting device. All photons with energy above the bandgap of the semiconductor are absorbed and excite one electron–hole pair. The excited carriers are assumed to immediately come to thermal equilibrium with each other at a temperature T_h, but not with the lattice at T_a. The carriers have infinite mobility, which results in constant quasi-Fermi energies across the cell, as illustrated in Figure 7.45. Also, there is no minority carrier flow across the conductors (i.e., there is no surface recombination).

The efficiency of a PV system is defined as

$$\eta_{ph\text{-}e} = \frac{(\Delta\varphi_e j_e)_{max}}{q_{ph,s}} \quad \text{PV efficiency,} \tag{7.235}$$

where $\Delta\varphi_e$ is the voltage potential, which is equal to $\Delta\mu_a/e_c$, and μ_a is the chemical potential of the carriers at ambient temperature. The distribution of energy for fermions in quasi-equilibrium is given by the Fermi–Dirac distribution (5.61). The difference in quasi-Fermi energies for the hot carrier distribution is $\Delta\mu_h = E_{F,n,h} - E_{F,p,h}$. However, we start with $\Delta\mu_a$; as the conductor is at ambient temperature the carriers need to be extracted at ambient temperature (unless we introduce a device for extraction of hot carriers or take other measures to use the hot-carrier energy), ideally with no loss in energy. A scenario in which this is possible is by isentropic cooling of the carriers at constant energy through an energy selective semiconductor, as illustrated in Figure 7.45. The relationship between $\Delta\mu_h$ and $\Delta\mu_a$ is determined by considering the energy conservation equation, (Table 2.4), i.e.,

$$\mathrm{d}E = T_h \mathrm{d}S + \mu_{h,i}\mathrm{d}N_i. \tag{7.236}$$

By the electron traveling through the external resistance, there is the removal of one particle from both the electron gas and the hole gas, with a change in chemical potential of $\Delta\mu_h$. Per carrier, the conservation equation becomes,

$$\Delta E = T_h \Delta S + \Delta\mu_h. \tag{7.237}$$

Similarly for electrons at ambient temperature, we have

$$\Delta E = T_a \Delta S + \Delta\mu_a. \tag{7.238}$$

With isentropic cooling, the change in entropy by removing a particle from the system is constant and with constant energy, equating the two ΔS gives the relationship between the hot-carrier chemical potential and the ambient chemical potential

$$\Delta\mu_a = \Delta\mu_h \frac{T_a}{T_h} + \Delta E_{out}(1 - \frac{T_a}{T_h}). \tag{7.239}$$

In this relationship ΔE_{out} is defined as the utilization pathway for the electrons as illustrated in Figure 7.45.

With infinite carrier mobility, the current can be determined from the steady-state, one-dimensional carrier continuity equation (5.259),

$$\frac{1}{e_c}\frac{\mathrm{d}j_e}{\mathrm{d}x} - \dot{n}_{ph\text{-}e,s} - \dot{n}_{ph\text{-}e,a} + \dot{n}_{e,h\text{-}ph} = 0. \tag{7.240}$$

The carrier continuity equation includes only three of the generation processes listed and named in Table 7.6. The electron relaxation, $\dot{s}_{e-p}$, does not affect the number of carriers and is therefore not included in the equation. For electron generation, the radiation from both the sun $\dot{n}_{ph-e,s}$ and ambient $\dot{n}_{ph-e,a}$ are considered. The generation terms, for an optically thick material ($\sigma_{ph} l > 4$, where $l = l_n + l_d + l_p$ in Figure 7.45), become

$$\dot{n}_{ph\text{-}e,s} = \int_{\omega_{e,g}}^{\infty} \frac{a_s E_{b,s,\omega}}{\hbar\omega l}\mathrm{d}\omega, \quad \dot{n}_{ph\text{-}e,a} = \int_{\omega_{e,g}}^{\infty} \frac{a_a E_{b,a,\omega}}{\hbar\omega l}\mathrm{d}\omega, \tag{7.241}$$

where a_s and a_a are the solid angles mentioned in relation (7.234).

Here we assume that among the recombination processes, the radiative recombination $\dot{n}_{e,h\text{-}ph}$ is dominant (this is not true in a typical cell, as shown in Figure 7.49). Then net radiative recombination rate can be determined by considering the rate of photon emission given by (7.8) with the correction that the system is not in chemical equilibrium and assuming all photons emitted with energy above the bandgap energy are the result of carrier recombination, i.e.,

$$\dot{n}_{e,h\text{-}ph} = \dot{n}_{ph\text{-}e,a}\exp(\frac{\Delta\mu_h}{k_B T_h}). \tag{7.242}$$

After the continuity equation is integrated along x between $x = 0$ and $x = l$, the current is

$$\frac{j_e}{e_c} = \int_{\omega_e,g}^{\infty} \frac{a_s E_{b,s,\omega}}{\hbar\omega} \mathrm{d}\omega + \int_{\omega_e,g}^{\infty} \frac{a_s E_{b,a,\omega}}{\hbar\omega} \mathrm{d}\omega - \int_{\omega_e,g}^{\infty} \frac{a_s E_{b,a,\omega}}{\hbar\omega} \exp(\frac{\Delta\mu_h}{k_B T_h})\mathrm{d}\omega. \tag{7.243}$$

To fully characterize the system, we add the energy balance equation, assuming that there is no loss of energy through phonon emission

$$\frac{j_e \Delta E_{out}}{e_c} = \int_{\omega_e,g}^{\infty} a_s E_{b,s,\omega} \mathrm{d}\omega + \int_{\omega_e,g}^{\infty} a_s E_{b,a,\omega} \mathrm{d}\omega - \int_{\omega_e,g}^{\infty} a_s E_{b,a,\omega} \exp(\frac{\Delta\mu_h}{k_B T_h})\mathrm{d}\omega. \tag{7.244}$$

By defining the bandgap of the semiconductor $\hbar\omega_{e,g}$, the utilization pathway ΔE_{out}, and the voltage $\Delta\varphi_e$, the maximum power out of the cell is found. In the process the carrier temperature is also determined. The maximum efficiency is 66.8% for $\Delta E_{e,g} = \hbar\omega_{e,g} = 0.36$ eV at $T_h = 3684$ K. It is important to note that the carriers cool through phonon emission very quickly, and it is often assumed the carriers are at the ambient temperature. The maximum efficiency assuming complete electron relaxation through phonon emission ($T_h = T_a = 300$ K) is 30.4% at $\Delta E_{e,g} = 1.26$ eV. The efficiencies are plotted for both scenarios as functions of bandgap energy in Figure 7.46. For the analysis of carriers in thermal equilibrium with the lattice, the power losses are from several different mechanisms as shown in Figure 7.47.

7.14.2 Multiple-Bandgap Ideal Solar PV Efficiency

The carrier relaxation through phonon emission occurs very quickly (~fs) as compared with carrier collection (~ns) and carrier–lattice thermal equilibrium is often assumed. One approach to negate energy loss through phonon emission but allowing the carrier–lattice thermal equilibrium assumption is arranging a system of photovoltaic cells in tandem, each with a reducing bandgap energy, as illustrated in Figure 7.48. The incoming photon is then better matched to a given cell, passing through cells with larger bandgap energy, ideally unabsorbed, until reaching a cell with equal or lower bandgap energy, where it will excite an electron–hole pair. Beyond absorbing a larger portion of the solar spectrum, there will also be recycling of the emitted photons from each cell by neighboring cells. To simplify the efficiency analysis, when there is an infinite number of tandem solar cells, photon recycling

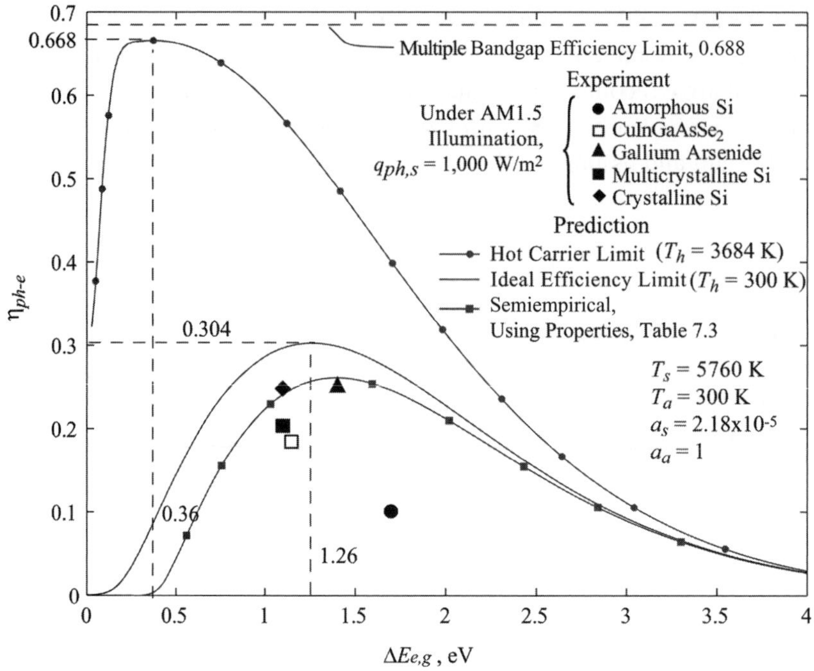

Figure 7.46. Variations of solar PV efficiency with respect to bandgap energy. Three different analyses are used, namely, ideal analysis with carriers at T_e, ideal with carriers at T_a, and semiempirical, infinite multiple bandgap. Some measured solar cell performances are also shown [229].

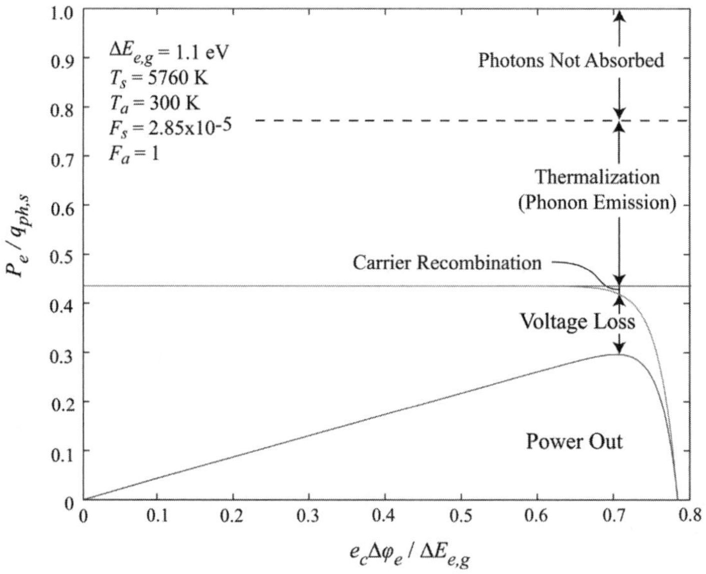

Figure 7.47. Distribution of incoming irradiation energy in PV energy conversion.

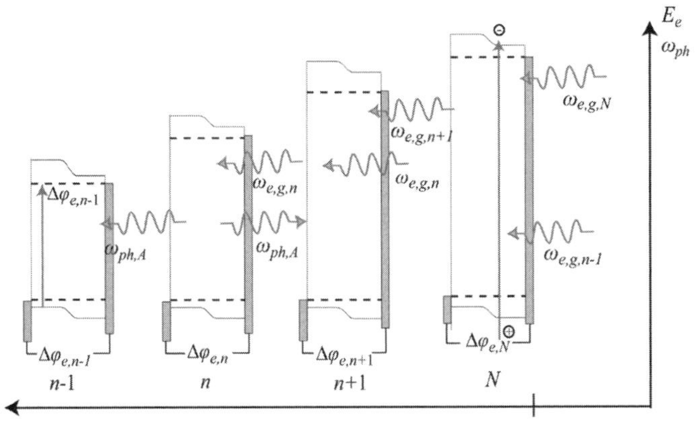

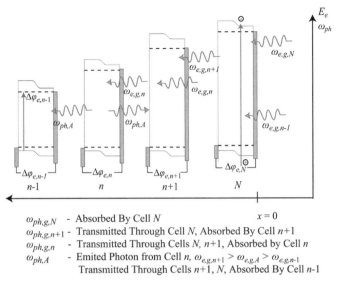

Figure 7.48. A rendering of a tandem multiple-bandgap PV solar cell, showing the continuously decreasing bandgap energy with emitted photon recycling.

has been shown to be equivalent to placing frequency-selective reflectors between cells, which reflect emitted photons with energy above the cell bandgap energy back to the cell [229]. With this assumption the efficiency calculation can be determined similarly to the single-bandgap cell with $T_h = T_a$ and the integral in the current equation taken from the bandgap energy of the given cell to the bandgap energy of the next cell for both absorbtion and emission; (7.243) becomes

$$\frac{j_e}{e_c} = \int_{\omega_{e,g,n}}^{\omega_{e,g,n+1}} \frac{a_s E_{b,s,\omega}}{\hbar\omega} \mathrm{d}\omega + \int_{\omega_{e,g,n}}^{\omega_{e,g,n+1}} \frac{a_s E_{b,a,\omega}}{\hbar\omega} \mathrm{d}\omega$$
$$- \int_{\omega_{e,g,n}}^{\omega_{e,g,n+1}} \frac{a_s E_{b,a,\omega}}{\hbar\omega} \exp(\frac{e_c \Delta\varphi_e}{k_B T_a}) \mathrm{d}\omega. \quad (7.245)$$

For a finite number of cells, at each cell, the power output is found by determining the maximum product of voltage and current. For the case of infinite tandem cells, $(j_{e,n}\Delta\varphi_{e,n})_{max}$ is determined for each photon frequency. For the total power out, the product of the maximum voltage and maximum current of each cell is integrated over all photon frequencies,

$$\frac{\dot{S}_{ph\text{-}e}}{A} = \frac{P_e}{A} = \int_0^\infty (\Delta\varphi_e j_e)_{max} \mathrm{d}\omega. \quad (7.246)$$

The efficiency limit for an infinite tandem cell with $T_s = 5760$ K, $T_a = 300$ K, $a_s = 2.85\times10^{-5}$ and $a_s = 1$ is 68.8%, similar to the limiting efficiency of a hot carrier solar cell.

7.14.3 Semiempirical Solar PV Efficiency

The preceding ideal analyses are very useful investigations because they require minimal information about the PV device. However, for a more realistic analysis, carrier transport must be considered. In the preceding analysis, the mobility of electrons were assumed to be infinite. Without this assumption, a PV solar cell will require an electric field and or a substantial concentration gradient to create a current. Also, recombination rates will include mechanisms other than radiative recombination. Here we analyze a *p–n* junction PV cell, adding to the analysis the general carrier transport given by the drift–diffusion equation (5.260),

$$j_e = e_c n_{e,c} \mu_e e_e + e_c D_e \frac{\mathrm{d}n_{e,c}}{\mathrm{d}x}, \qquad j_h = e_c n_h \mu_h e_e - e_c D_h \frac{\mathrm{d}n_h}{\mathrm{d}x}$$

$$j_{e,h} = j_e + j_h. \tag{7.247}$$

The electron diffusion coefficient (Table 3.6) is $D_e = k_B T \mu_e / e_c$, and is similar for holes. These equations are used along with the carrier continuity equation,

$$\frac{1}{e_c}\frac{\mathrm{d}j_e}{\mathrm{d}x} + \dot{n}_{e,rec} - \dot{n}_{ph\text{-}e,s} = 0, \qquad \frac{1}{e_c}\frac{\mathrm{d}j_h}{\mathrm{d}x} - \dot{n}_{h,rec} + \dot{n}_{ph\text{-}e,s} = 0. \tag{7.248}$$

To determine the current from these equations, several assumptions are made to simplify the analysis. All carriers immediately relax to the conduction band edge. The depletion region of the cell creates a well-defined boundary, where beyond the depletion region there is no electric field present (i.e., the cell is quasi-neutral). The generation rate is constant along the length of the cell. The carrier recombination is linearly dependent on the minority carrier concentration.

The recombination rate for conduction electrons in the *p*-type region is modeled as follows:

$$\dot{n}_{e,rec} = \frac{n_{e,c} - n_{e,i}}{\tau_{e,rec}}, \quad n_{e,i} = \frac{n_i^2}{n_{d,p}}, \tag{7.249}$$

$$\frac{1}{\tau_{e,rec}} = \frac{1}{\tau_{e\text{-}t}} + \frac{1}{\tau_{e,h\text{-}k}} + \frac{1}{\tau_{e,h\text{-}ph}}, \tag{7.250}$$

where we have used the dominant recombination mechanisms listed in Table 7.6.

Figure 7.49 shows the minority carrier lifetimes (kinetics) $\tau_{e\text{-}i}$, as functions of dopant concentrations for *p*-type Si. Note that the smallest $\tau_{e\text{-}i}$ dominates the recombination.

To find the current we first consider the minority carrier concentrations as functions of position $n_{e,c}(x)$ in the quasi-neutral regions of the cell. The following derivation is performed for electrons but is similarly applicable to holes. The dimensional parameters are those in Figure 7.45. Combining the drift–diffusion equation and the carrier continuity equation, considering no electric field, we obtain

$$\frac{\mathrm{d}^2 n_{e,c}}{\mathrm{d}x^2} - \frac{\dot{n}_{ph\text{-}e,s}}{D_e} + \frac{n_e - n_{e,i}}{L_e^2} = 0. \tag{7.251}$$

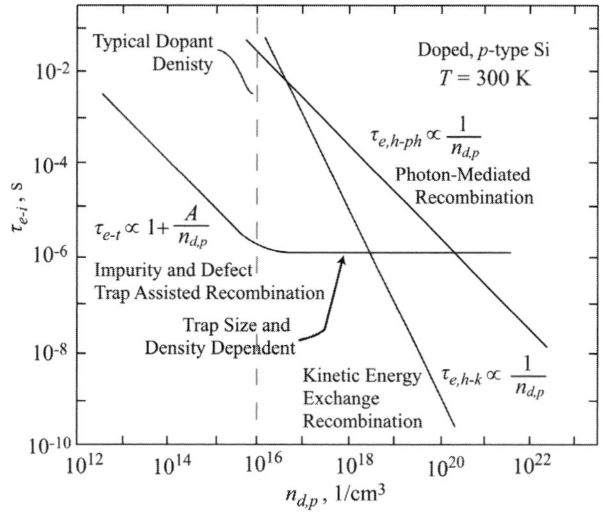

Figure 7.49. Variations of the measured, minority carrier lifetimes (kinetics) with respect to dopant concentration for p-type Si [128].

The diffusion length is introduced as $L_e = (D_e \tau_{e,rec})^{1/2}$. Then the solution is

$$n_{e,c}(x) = A \exp(\frac{x + l_n + l_d}{L_e}) + B \exp(\frac{-x + l_n + l_d}{L_e}) + \tau_{e,rec}\dot{n}_{ph\text{-}e,s} + n_{e,i}.$$

To solve for the constants, we consider the following boundary conditions. As $x \rightarrow \infty$, $n_e(x)$ is finite, therefore $A = 0$. The constant B is found by considering the carriers concentration at the edge of the depletion region $x = l_n + l_d$, defined as [128]

$$n_{e,c} - n_{e,i} = \frac{n_i^2}{n_{d,p}}[\exp(\frac{e_c \Delta\varphi_e}{k_B T_a}) - 1]. \tag{7.252}$$

Therefore, the conduction electron density in the p-type region is

$$n_{e,c}(x) = \tau_{e,rec}\dot{n}_{ph\text{-}e,s} + n_{e,i} + \{\frac{n_i^2}{n_{d,p}}[\exp(\frac{e_c \Delta\varphi_e}{k_B T_a}) - 1] - \tau_{e,rec}\dot{n}_{ph\text{-}e,s}\}\exp(\frac{-x + l_n + l_d}{L_e}). \tag{7.253}$$

The equations governing holes can be set up similarly. The carrier concentrations are then plugged into the drift–diffusion equation to determine the current. For steady state operation the total current across the cell is constant. Therefore, to determine the current, the electron and hole currents are summed at one position in the cell. Now, we define the electron current at the edge of the p-type region and depletion region $l_n + l_d$, as

$$j_e(l_n + l_d) = e_c L_e \dot{n}_{ph\text{-}e,s} - \frac{e_c n_i^2 D_e}{n_{d,p} L_e}[\exp(\frac{e_c \Delta\varphi_e}{k_B T_a}) - 1]. \tag{7.254}$$

Table 7.7. *Typical Si PV material properties [252].*

Quantity	Magnitude
D_e	40 cm^2/s
D_h	2 cm^2/s
L_e	14.0 μm
L_h	140 μm
$n_{d,p}$	10^{16} cm^{-3}
$n_{d,n}$	10^{19} cm^{-3}
$\tau_{e,rec}$	1×10^{-6} s
τ_h	5×10^{-6} s

The hole current is determined at the edge of the n-type region and depletion region, l_n,

$$j_h(l_n) = e_c L_h \dot{n}_{ph\text{-}e,s} - \frac{e_c n_i^2 D_h}{n_{d,n} L_h}[\exp(\frac{e_c \Delta\varphi_e}{k_B T_a}) - 1]. \qquad (7.255)$$

The total current $j_{e,h}$ determined at l_n is $j_h(l_n)$ + $j_e(l_n + l_d)$ plus the electron current generated in the depletion region, $j_{e,dep}$. The current density in the depletion region cannot be solved in the same manner as the other regions because an electric field is present. A simplification for this region can be made by allowing the current to be defined solely by the carrier continuity equation, neglecting recombination, $j_{e,dep} = e_c l_d \dot{n}_{ph\text{-}e,s}$. This simplification is made because the depletion region is relatively small and the electrons and holes are removed quickly by the electric field. From these equations the total current is

$$j_{e,h} = e_c \dot{n}_{ph-e,s}(L_e + L_h + l_d) - e_c n_i^2(\frac{D_e}{n_{d,p} L_e} + \frac{D_h}{n_{d,n} L_h})[\exp(\frac{e_c \Delta\varphi_e}{k_B T_a}) - 1]. \qquad (7.256)$$

The efficiency for this approach is plotted in Figure 7.46 as a function of bandgap energy using the material properties of Si listed in Table 7.7. Note that the efficiency is calculated assuming all solar irradiation above the bandgap energy is absorbed over $L_e + L_h + l_d$, equivalent to the width of the cell. Table 7.8 includes actual solar cell operating parameters along with some other defining characteristics. The fill factor F_F is the total power out $(\Delta\varphi_e j_e)_{max}$ divided by the product of the open circuit voltage $\Delta\varphi_o$ and short circuit current j_{sh}.

7.15 Size Effects

7.15.1 Enhanced Near-Field Radiative Heat Transfer

Thermally excited solid surface waves are due to thermal fluctuations of atoms adjacent to the surface. These surface waves influence the density of states of the emitted

Table 7.8. *Measured PV cell performance of some single-bandgap materials [47, 129].*

Material	$\eta_{ph\text{-}e}$	Band Gap	$\Delta E_{e,g}$, eV	l_p, μm	l_n, μm	$\Delta\varphi_o$, V	j_{sh}, mA/cm^2	F_F
Crystalline Si	0.247	indirect	1.1	300–500	0.5	0.706	42.2	0.828
Polycrystalline Si	0.203	indirect	1.1	300	0.5	0.664	33.7	0.809
Crystalline GaAs	0.251	direct	1.43	0.5	4	1.022	28.2	0.871
Polycrystalline GaAs	0.182	direct	1.43	–	–	0.994	23.0	0.797
Crystalline InP	0.219	direct	1.35	4.0	0.03	0.878	29.3	0.854
Amorphous (Nanocrystalline) Si	0.101	direct	1.7	0.01	0.02	0.539	24.4	0.766
($CuInGaSe_2$)	0.184	direct	1.1–1.2	5.0	0.3	0.669	35.7	0.770

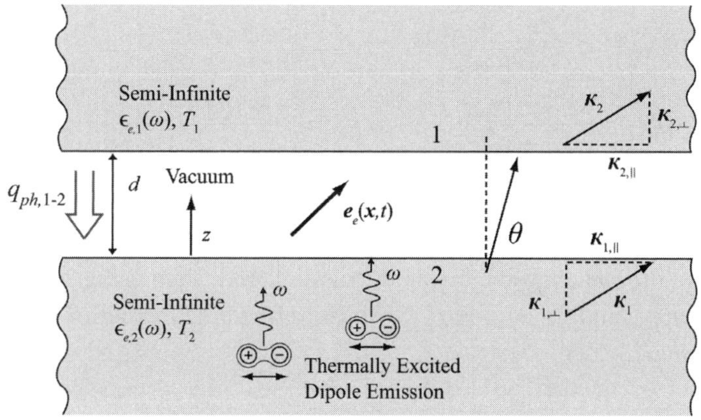

Figure 7.50. Thermally excited surface waves $\boldsymbol{e}_e(\boldsymbol{x}, t)$ and the near-field radiation exchange between two closely placed solid surfaces [114]. Here $\kappa_\parallel$ is the component of wave vector along the surface.

EM fields that have intensities that are orders of magnitude larger in the near field than in the far field [58, 164]. Also, the emitted EM waves are quasi-monochromatic (different than blackbody radiation) in the near field.

The EM waves propagating along the interface of two material with distinct EM properties (σ_e, ϵ_e, μ_e), decay exponentially perpendicular to the interface. The surface waves that are due to the coupling between the EM field and a resonant polarization oscillation in the solid are called the surface polaritons. The surface wave of a metal is due to charge density waves or plasmons, and is called the surface-plasmon polarization. For dielectric solids the surface wave is due to the optical phonons and phonon polaritons (which also exist within the solid) and is called the surface-phonon polariton.

In Section 7.2.2 we discussed the near-field surface emission and here we give a quantitative example for two surfaces at different temperatures separated by a vacuum gap of the order of nanometers. Here we examine the formulation of the thermally excited, near-field EM emission and review examples with radiation between two doped Si surfaces separated by a small gap.

Referring to Figure 7.50 for a medium 1 at temperature T, emitting to vacuum (with $T_2 = 0$), the spectral photon energy density (7.24) (for emission) is [164]

$$\frac{\mathrm{d}E_{ph,\omega}}{V\mathrm{d}\omega} = \frac{\epsilon_o}{2}\langle|\boldsymbol{e}_e|^2\rangle + \frac{\mu_o}{2}\langle|\boldsymbol{h}_e|^2\rangle$$

$$= \frac{\hbar\omega^3}{2\pi^2 u_{ph,o}^3[\exp(\frac{\hbar\omega}{k_B T}) - 1]} \int_0^{\omega/u_{ph,o}} \frac{\kappa \mathrm{d}\kappa}{\frac{\omega}{u_{ph,o}}|\kappa_{\perp,1}|} \frac{1 - |\rho_{r,12}^s|^2 + (1 - |\rho_{r,12}^p|^2)}{2}$$

propagating waves

$$+ [\int_{\omega/u_{ph,o}}^{\infty} \frac{4\kappa^3 \mathrm{d}\kappa}{\frac{\omega}{u_{ph,o}}\kappa_{\perp,1}} \frac{\mathrm{Im}(|\rho_{r,12}^s|) + \mathrm{Im}(|\rho_{r,12}^p|)}{2} e^{-2\mathrm{Im}(\kappa_{\perp,1},z)}], \quad (7.257)$$

evanescent waves

where

$$\kappa_{\perp,1} = \epsilon_{e,1}\mu_{e,1}\frac{\omega}{u_{ph,o}} - \kappa^2,$$

$$\rho_{r,12}^s = \frac{\kappa_{\perp,1} - \kappa_{\perp,2}}{\kappa_{\perp,1} + \kappa_{\perp,2}}, \qquad \rho_{r,12}^p = -\frac{\epsilon_{e,1}\mu_{\perp,2} + \epsilon_{e,2}\kappa_{\perp,1}}{\epsilon_{e,1}\kappa_{\perp,2} + \epsilon_{e,2}\kappa_{\perp,1}},$$

where in the s-polarization $\boldsymbol{e}_e$ is perpendicular to y–z plane and the p-polarization $\boldsymbol{e}_e$ lies in y–z plane. $\rho_{r,12}^s$ and $\rho_{r,12}^p$ are the Fresnel reflection factors.

The first integral gives the far-field emission, and when using $2\pi\kappa \mathrm{d}\kappa = \omega^2 \cos\theta \mathrm{d}\Omega/u_{ph,o}^2$, and for $\rho_{r,12}^s = \rho_{r,12}^p = 0$ (for absorbing medium 2), we have

$$\frac{\mathrm{d}E_{ph,\omega}}{\mathrm{d}\omega} = \frac{\mathrm{d}E_{b,\omega}}{\mathrm{d}\omega} = \frac{\hbar\omega^3}{\pi^2 u_{ph,o}^2} \frac{\mathrm{d}\omega}{e^{\hbar\omega/k_B T} - 1}, \quad (7.258)$$

which is the Planck blackbody spectral energy density (7.4). Note that spectral emissivity $\epsilon_{ph,\omega}$ (7.147) is given in terms of the Fresnel reflection factors as

$$\epsilon_{ph,\omega} = \frac{1 - |\rho_{r,12}^s|^2 + 1 - |\rho_{r,12}^p|^2}{2}, \quad (7.259)$$

which is 1/2 of the sum of the transmission factors for both polarizations.

The spectral energy flux from a dielectric medium 1 to another dielectric medium 2 (Figure 7.50) is given in terms of the exchange function $Z_{1\text{-}2}(\kappa_\parallel)$ as [114]

$$q_{ph,1\rightarrow 2,\omega} = \frac{1}{\pi^2}\frac{\hbar\omega}{\exp(\frac{\hbar\omega}{k_B T_1}) - 1}\int_0^{\infty} Z_{1\text{-}2}(\kappa_\parallel)\kappa_\parallel \mathrm{d}\omega \mathrm{d}\kappa_\parallel, \quad (7.260)$$

where $\kappa_\parallel$ is the surface-wave-vector component along the surface. This includes the propagating and evanescent waves (exponentially decaying). The net total heat flux is

$$q_{ph,1\text{-}2,\omega} = \frac{1}{\pi^2}\int_0^{\infty}\int_0^{\infty}[\frac{\hbar\omega}{\exp(\frac{\hbar\omega}{k_B T_1}) - 1} - \frac{\hbar\omega}{\exp(\frac{\hbar\omega}{k_B T_2}) - 1}] Z_{1\text{-}2}(\kappa_\parallel)\kappa_\parallel \mathrm{d}\omega \mathrm{d}\kappa_\parallel. \quad (7.261)$$

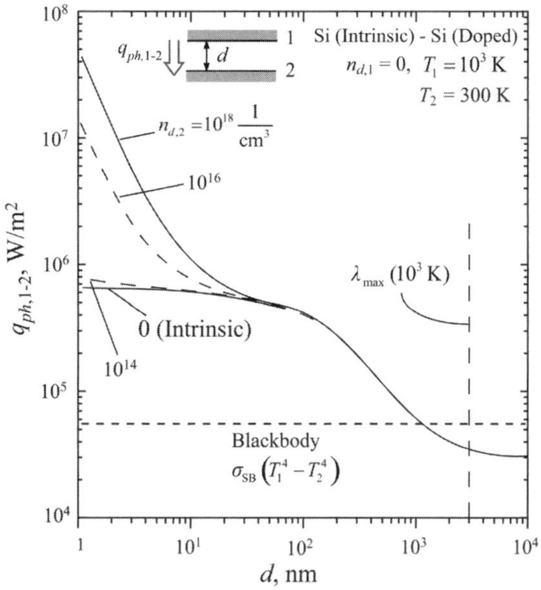

Figure 7.51. Variation of the net radiation heat transfer rate between two Si surfaces as functions of the separation distance. The temperature of each surface and the dopant concentration are also given [114].

The exchange function is a function of complex dielectric function ϵ_e, and for semiconductors the Drude model (3.49) is extended for each material as [357]

$$\epsilon_e(\omega) = \epsilon_{e,\Delta} - \frac{n_{e,c}e_c^2/\epsilon_o m_{e,e}}{\omega^2 + i\omega/\tau_e} - \frac{n_{h,c}e_c^2/\epsilon_o m_{h,e}}{\omega^2 + i\omega/\tau_h}$$

Drude model of dielectric function, (7.262)

where $\epsilon_{e,\Delta}$ is for interband (and lattice vibration) transition; the other terms are for the free-carrier conduction-band electron and valence band hole transitions [114]. The procedure for determining these parameter in the ϵ_e expression is given in [114]. Figure 7.51 shows the variation of the predicted net radiation heat transfer rate between two Si surfaces separated by a gap d.

One surface is at 1000 K (using $\lambda_{max}T$ = 2898 μm-K Section 7.1, gives λ_{max} = 3000 nm) and the other at 300 K (λ_{max} = 10,000 nm) [114]. One Si solid is intrinsic, and the other is doped. The results show that, for $d > 3000$ nm, the far-field behavior is approached, whereas for the doped Si the heat transfer radiation continues to increase with a decrease in the gap spacing. The blackbody radiation (with surface emissivity of unity) is also shown for reference. For doped Si, two different regimes are found, one influenced by doping and one independent of doping.

7.15.2 Photon Energy Confinement by Near-Field Optical Microscopy

Similar to the photon localization of Section 7.7, the electric field intensity in the near field of a probe tip (optical microscope) has been numerically determined

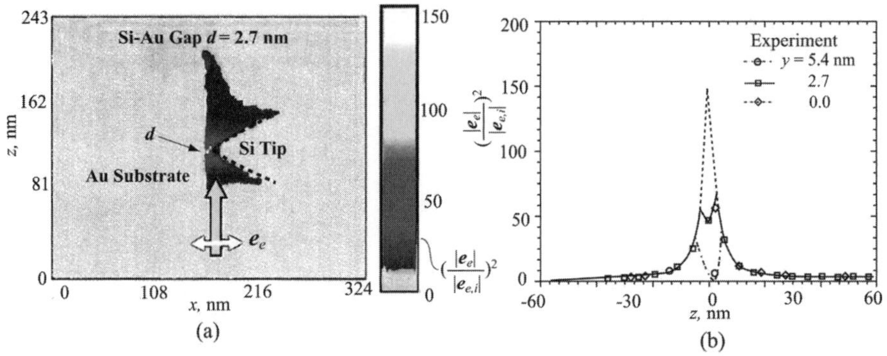

Figure 7.52. (a) and (b) Normalized electric field energy distribution underneath a Si probe tip irradiated with P-polarization femtosecond laser of $\lambda = 800$ nm in the proximity of the surface of Au film [69]. The tip apex/sample surface separation $d = 2.7$ nm.

using the classical Maxwell equations (Table 3.5) [69]. The probe (cylindrical cone with cone angle of 36°) is made of Si and the substrate is Au. Figures 7.52(a) and (b) show the probe-substrate and the predicted field energy enhancement (confinement) $(|\boldsymbol{e}_e|/|\boldsymbol{e}_{e,i}|)^2$, where $|\boldsymbol{e}_e|$ is the laser irradiation intensity through the probe (uniform entering the probe). The photon wavelength is 800 nm, with polarization parallel to the incident plane. The top, Au plate separation is $d = 2.7$ nm. Field enhancements as high as 150 appear on the Au surface. The lateral confinement of the high-field-enhancement regime is about 10 nm, allowing for high resolution probing or material alterations.

This photon confinement adjacent to a surface, and where used with sufficiently large, pulsed laser irradiation power, can increase the temperature of the surface. Predicted transient, local temperatures as high as 12,000 K have been reported [69] for metals, under electron, lattice thermal nonequilibrium. (Sections 5.19 and 5.20).

Radiation propagating through subwavelength aperture is concentrated (focused) in the near field of the aperture (near-field optical microscopy) [56, 161, 162]. Amongst the aperture geometries are the O- and H-apertures shown in Figures 7.52(a) and (b) [161], where the former has an analytical solution (to classical Maxwell equations) and the latter is solved numerically.

7.16 Problems

Problem 7.1

(a) Derive the expression for the photon Stefan–Boltzmann constant in (7.9).

(b) Starting from the definition of blackbody intensity, and using the Stefan–Boltzmann law, show that the Stefan–Boltzmann coefficient (but not constant) for a phonon (using the Debye model) is $\rho c_{v,p} u_{p,\mathrm{A}}/(16T^3)$.

Problem 7.2

(a) Derive Rayleigh–Jeans relation (7.10) from blackbody (Planck) distribution function (7.8).

(b) Show that $\mathrm{d}E_{b,\lambda}/\mathrm{d}\lambda$ (7.8) has a maximum at $\lambda_{max}T = 2898$ μm-K (we need to find the root of the transcendental equation resulting from differentiation). This peak relation is called the Wien displacement law (as T increases, λ_{max} is displaced).

Problem 7.3

Show that the Thomson cross-section area (for photon-free-electron elastic scattering) is $8\pi r_C^2/3$, where r_C is the classical electron radius and is equal to 6.652×10^{-29} m^2. Note that, as given by (7.81), this is also equal to $e_c^4/6\pi m_e^2\epsilon_o^2u_{ph,o}^4$, because $r_C = e_c^2/4\pi m_e\epsilon_o u_{ph,o}^2$.

Hint: Start from the energy of the free electron.

Problem 7.4

(a) Show that for an incident radiation intensity of $I_{ph,i}$ (energy per unit area irradiated is $\pi I_{ph,i}$) in a medium with an absorption coefficient σ_{ph}, the divergence (in one dimension, along x) of radiation heat flux q_{ph} is equal to $-\pi I_{ph,i}\sigma_{ph}\exp(-\sigma_{ph}x)$. In the positive direction, the radiation heat flux at $x = 0$ is from (7.32) as

$$q_{ph}(x=0) = 2\pi\int_0^{\pi/2} I_{ph,i}\cos\theta\sin\theta\mathrm{d}\theta = \pi I_{ph,i}.$$

Assume there is no emission and scattering. The divergence of $\boldsymbol{q}_{ph}$ appears in the macroscopic energy equation given in Table 1.1.

(b) Comment on how this term becomes a heat source in the macroscopic energy equation when radiation is absorbed.

Problem 7.5

Derive the expression for Thomson scattering cross-section area, (7.25), starting from a plane-polarized electric field (3.20):

$$\boldsymbol{e}_e = e_{e,o}\boldsymbol{s}_\alpha \sin\omega t.$$

Then use

$$\frac{\mathrm{d}A_{ph,s}}{\mathrm{d}\Omega} = \frac{1}{|\bar{\boldsymbol{s}}_e|}\frac{\mathrm{d}\bar{P}_e}{\mathrm{d}\Omega},\quad |\bar{\boldsymbol{s}}_e| = \frac{\epsilon_o u_{ph}e_{e,o}^2}{2},\quad \mathrm{d}\Omega = \mathrm{d}\cos\theta\mathrm{d}\phi,$$

where $|\bar{\boldsymbol{s}}_e|$ is the time-averaged Poynting vector (Table 3.5), and $\bar{P}_e$ is the time-averaged power per differential solid angle $\mathrm{d}\Omega$,

$$\bar{P}_e = \bar{\boldsymbol{s}}_e(r,\theta,\phi)\cdot\boldsymbol{s}_\alpha r^2\mathrm{d}\Omega = \frac{e_c 4e_{e,o}^2}{32\pi^2\epsilon_o u_{ph}^3 m_e^3}\sin^2\theta.$$

Problem 7.6

Plot the photon spectral absorption coefficient (Table 7.2) at $T = 290$ K, for the followings.

(a) Free-carrier absorption in Al [electrical conductivity of Al is given in Table A.1 and can be used to find $\langle\langle\tau_e\rangle\rangle$ using (5.146), and for metals m_e is used for the electron mass], use $\epsilon_e = 1$.

(b) Free-carrier absorption in doped Si, with $n_{e,c} = 10^{19}$ 1/m^3, and $m_{e,e} = 0.32 m_e$, (use σ_e from Figure 5.24 and $\epsilon_e = 1$).

(c) Direct-band absorption for GaAs (Table 5.1, at Γ point).

(d) Indirect-band absorption in Si with optical-phonon cut-off energy of 0.062 eV, determine the coupling constant $a_{ph\text{-}e\text{-}p,a}$ from the experimental results given in the figure. The Si lattice constant is listed in Table A.2. Use units of eV and cm.

Use photon frequency over the UV(ultraviolet), V(visible), and NIR(near-infrared) regimes, when appropriate. Compare with any data from the literature. The bandgap data are given in Table 5.1.

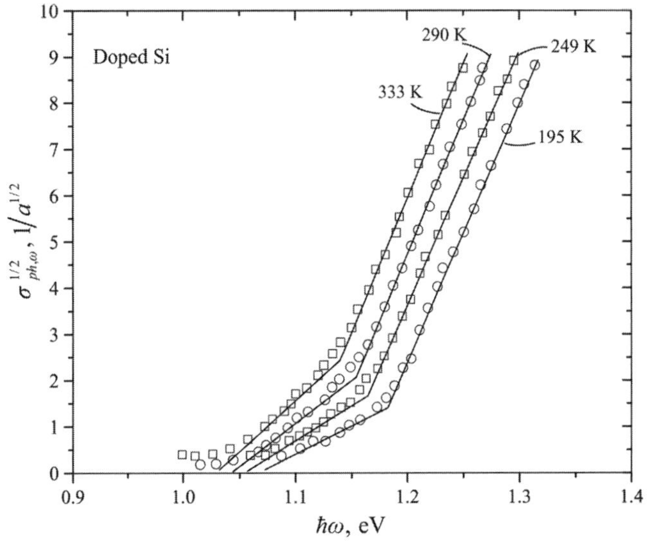

Problem 7.6. Spectral absorption coefficient of doped Si

Problem 7.7

(a) Using (7.154) to (7.156) and (7.160), plot the variations of (i) $\dot{\gamma}_{e\text{-}ph\text{-}p}$, (ii) $\dot{\gamma}_{e\text{-}ph}$, (iii) $\dot{\gamma}_{e\text{-}p}$, and (iv) $\eta_{e\text{-}ph}$ with respect to temperature for T from 0 to 2500 K, for 1% Yb^{3+}: Y_2O_3 with the following properties (melting temperature of yttria is 2705 K). Use the values $\dot{\gamma}_{e\text{-}ph\text{-}p}(T=0) = 952.38$ s^{-1}, $\hbar\omega_p = 0.2027$ eV, and $N_p = 6$.

(b) Explain the trend for each of the four (decay rates and efficiency) variables. Note that 0.2027 eV is rather large for a single phonon (even optical phonon).

Problem 7.8

(a) Using the blackbody emission, show that

$$n_b = \int_0^\infty f_{ph}^{\mathrm{o}}(\omega) D_{ph}(\omega) \mathrm{d}\omega = \frac{8\pi}{u_{ph}^3} \left(\frac{k_{\mathrm{B}} T}{h_{\mathrm{P}}}\right)^3 \int_0^\infty \frac{x^2}{e^x - 1} \mathrm{d}x$$

$$= 2.404 \times 8\pi \left(\frac{k_{\mathrm{B}}}{h_{\mathrm{P}} u_{ph}}\right)^3 T^3.$$

This shows that the number density of photons in the blackbody emission is proportional to T^3.

(b) Show that the total energy of photons per unit volume is $8\pi^5(k_B T)^4/15(h_P u_{ph})^3$. Note that $\int_0^\infty x^3(e^x - 1)^{-1}dx = \pi^4/15$.

(c) Evaluate this photon number density in $1/\text{cm}^3$, for $T = 10^2$, 10^3, and 10^4 K.

(d) Show that the average energy per photon $\langle E_b \rangle = E_b/Vn_b$ is $2.701 k_B T$. Evaluate this in eV, for $T = 10^2$, 10^3, and 10^4 K. Compare these with $\hbar\omega$ at the peak of the spectral emissive power of the spectra ($\lambda_{max} T = 2898$ μm-K) (continue to use eV for energy unit).

Problem 7.9

(a) Using (7.126), determine $I_{ph,\omega}/I_{ph,\omega}(x = 0)$, for $x = 1$ mm, $\lambda = 5$ μm (near infrared), and $\lambda = 0.5$ μm (green), at $T = 300$ K, and for (i) Cu (metal), (ii) Si (indirect-band semiconductor), (iii) GaAs (direct-band semiconductor), and (iv) SiO_2 (dielectric). Use available (general literature) $\sigma_{ph,\lambda}$ for Cu, Si, and GaAs. Use available κ_λ for quartz (SiO_2) and $\sigma_{ph,\lambda} = 4\pi\kappa_\lambda/\lambda$.

(b) Explain the mechanism of photon absorption for each of these materials.

Problem 7.10

The simple, dipole, spontaneous emission (classical dipole radiation) rate is (3.43) (discussed in Section 3.3.4 and derivation is Problem 7.23)

$$\dot{\gamma}_{ph,e,sp} = \frac{\omega^3 |\boldsymbol{\mu}_e|^2}{3\pi\epsilon_o \hbar u_{ph}^3},$$

where $\boldsymbol{\mu}_e$ is the average transition dipole moment.

The thermally stimulated emission is

$$\dot{\gamma}_{ph,e,st} \frac{\mathrm{d}E_{ph,\omega}}{V\mathrm{d}\omega} = \frac{\pi}{3\epsilon_o \hbar^2} |\boldsymbol{\mu}_e|^2 \frac{\mathrm{d}E_{ph,\omega}}{V\mathrm{d}\omega},$$

where $E_{ph,\omega}/V$ is given by (7.4).

(a) Show that

$$\frac{\dot{\gamma}_{ph,e,sp}}{\dot{\gamma}_{ph,e,st} \frac{\mathrm{d}E_{ph,\omega}}{V\mathrm{d}\omega}} = \exp\left(\frac{\hbar\omega}{k_B T}\right) - 1.$$

(b) For $T = 300$ K, show that for $\omega \gg 4.35 \times 10^{12}$ Hz, spontaneous emission dominates.

Problem 7.11

(a) Show that $\mathrm{d}I_{b,\omega}$ peaks at $\hbar\omega/k_B T = 2.821$. This is referred to as the most likely frequency for blackbody radiation at temperature T.

(b) Why are the peaks for $I_{b,\lambda}$ and $I_{b,\omega}$ different?

Problem 7.12

(a) Use Table 7.5 to calculate the fundamental angular frequency ω_n for CO_2 (C–O bond).

(b) Assume that the collision diameter of CO_2 listed in Table 6.4 can be used for its photon absorption cross-section area $\pi d^2/4$.

(c) Compare these with the measured resonance value of $A_{ph,\omega} = 1.4\times10^{-17}$ cm^2 at $\lambda = 4.23$ μm [Figure 7.16 (b)], and comment.

Problem 7.13

In a CO_2 laser, first there is a vibration energy interchange between N_2 (which are excited by electrons) and CO_2 (Figure 7.39). The $l = 1$ vibration of N_2 matches the level (0,0,1) vibration of CO_2, exciting the asymmetric vibration. Then there is a laser (photon emission) to symmetric mode (1,0,0). The presence of He in a CO_2 laser is to maintain the plasma discharge (provide the free electron).

(a) Calculate the natural (fundamental) frequency of CO_2, using Table 7.5 and (7.226), for the C=O bond.

(b) Then using the quantum numbers and energies given in Figure 7.39

$$E_{f,v}(l_1, l_2, l_3) = \hbar\omega_{f,1}(l_1 + \frac{1}{2}) + \hbar\omega_{f,2}(l_2 + \frac{1}{2}) + \hbar\omega_{f,3}(l_3 + \frac{1}{2}),$$

calculate $\omega_{f,1}$, $\omega_{f,2}$, and $\omega_{f,3}$.

(c) Compare the results of (a) and (b), and comment (including the role of rotational energy transitions).

Problem 7.14

(a) Photon scattering from packed particles results in a small photon mean free path $\lambda_{ph} = 1/\sigma_{ph}$. Consider spherical alumina (Al_2O_3) solid particles of diameter $d_s = 1$, 10, and 10^3 μm, surface emissivity of $\epsilon_{ph} = 0.5$, and the radiant conductivity k_{ph} relation [172]

$$k_{ph} = a_1\epsilon_{ph}\tan^{-1}(a_2\frac{k_s^{*a_3}}{\epsilon_{ph}} + a_4)(4d_s\sigma_{SB}T^3),$$

$$a_1 = 0.5756,\ a_2 = 1.5353,\ a_3 = 0.8011,$$

$$a_4 = 0.1843,\ k_s^* = \frac{k_s}{4d_s\sigma_{SB}T^3},$$

where k_s is the solid particle thermal conductivity, at $T = 500$ K, for this packed particle bed.

(b) Use (7.97) to find λ_{ph} for this medium and express this as multiples of d_s.

(c) The effective thermal conductivity of packed beds $\langle k\rangle$ depends on the solid and the fluid conductivities, k_s and k_f, porosity, particle geometry, and interparticle contact [172]. For a packed bed of spherical particles (porosity 0.38) of amorphous Al_2O_3 ($k_s = 36$ W/m-K), with air ($k_f = 0.0395$ W/m-K) as the fluid, use $\langle k\rangle = 10$ W/m-K [172], and compare with the preceding k_{ph}.

Problem 7.15

For the radiative (photon) assisted transition from state i to state f shown in the figure, the net electron–hole radiative recombination in volume V is given, similar to (7.132) and (7.133), as in [20] and [21]

$$\dot{\gamma}_{e\text{-}h\text{-}ph}(\mathrm{m}^3/\mathrm{s}) = \frac{tV}{32\pi^6\hbar^2}\int\int\sum_{s_c,s_v}|M_{f,i}|^2 F\frac{1-\cos(x)}{x^2}\mathrm{d}\kappa_{ph}\mathrm{d}\kappa_{e,i},$$

$$x = \frac{t}{\hbar}(E_{e,i} - E_{h,f} - \Delta E_{e,g} - \hbar\omega_{ph}),$$

$$F = f_{e,f}(1 - f_{e,i})(f_{ph}+1) - (1 - f_{e,f})f_{e,i}f_{ph},$$

where s_e and s_v are the spin states, E_{e,κ_e} is the electron kinetic energy in conduction band, and $E_{e,\kappa_e'}$ is for the hole kinetic energy in the valance band.

The matrix element is

$$|M_{f,i}|^2 = \frac{e_c^2\hbar^2}{2Vm_{e,e}^2\epsilon_o\epsilon_e(\omega)\hbar\omega}\sum_{s_c,s_v}\langle f|\mathbf{s}_{ph,\alpha}\cdot\mathbf{p}_e|i\rangle^2.$$

Show that the direction-averaged matrix element becomes

$$\sum_{s_c,s_v}|M_{f,i}|^2 = \frac{e_c^2}{6V\epsilon_o\epsilon_e(\omega)\hbar\omega}\frac{\hbar^2}{m_{e,e}^2}\sum_{s_c,s_v}\langle f|p_e|i\rangle^2.$$

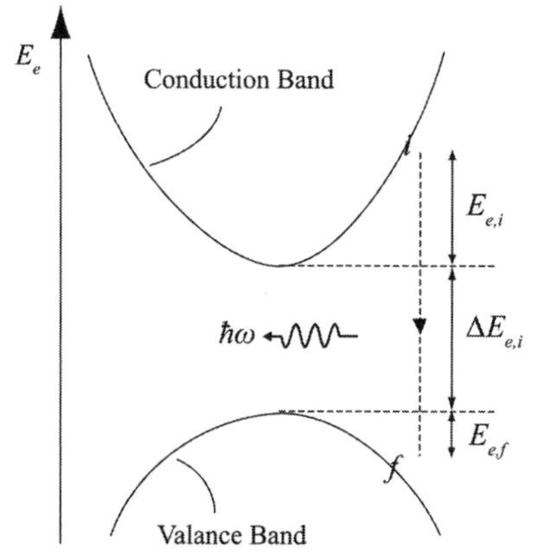

Problem 7.15. Electronic energy designations for radiative recombination

Problem 7.16

Determine the fundamental vibrational frequency of N–N (reported value is 2330 1/cm).

Problem 7.17

The crystalline calcium resonance transition can be modeled as a two-level system, with $\omega_{e,g} = 4.5\times10^{15}$ rad/s, and its radiative decay lifetimes is $\tau_{ph,e,sp} = 1/\dot{\gamma}_{ph,e,sp} = 4.5\times10^{-9}$ s. This decay is from 4^1P (B level) to 4^1S (A level), where $g_{e,B} = 3$, and $g_{e,A} = 1$.

(a) Find $\dot{\gamma}_{ph,a}$ (absorption rate) under equilibrium conditions.

(b) Find the stimulated emission rate $\dot{\gamma}_{ph,e,st}$ under equilibrium conditions.

(c) Find the absorption cross-section area $A_{ph,a}$ using $\int_\omega d\omega = 10^{12}$ rad/s.

(d) Using (3.43), find the transition electric dipole moment $|\boldsymbol{\mu}_e|$ for this transition. Compare this with Debye dipole moment D $= e_c r_\mathrm{B}/2.54$ (Section 1.5.2), where r_B is the Bohr radius. Is this transition dipole moment larger or smaller, and why?

Problem 7.18

The absorption cross-section area is temperature dependent; for example, empirical results show that [124]

$$A_{ph,a}(T) = A_{ph,a,\mathrm{o}} + a_{ph,a}(T_s - T_\mathrm{o}),$$

where for Yb^{+3}: ZBLANP laser cooling material, $A_{ph,o} = 3.70\times10^{-26}$ m^2, and $a_{ph,a} = 1.75\times10^{-28}$ m^2/K, for $T_\mathrm{o} = 301$ K. In laser cooling involving these materials, use $\lambda_{ph,i} = 1015$ nm, $\lambda_{ph.e} = 999.8$ nm, and a solid in form of a small fiber with $D = 250$ μm, $L = 14$ mm, $\epsilon_{ph} = 0.54$, $n_d = 2.42\times10^{20}$ 1/cm^3, and $Q_{ph,i} = 2.2$ W.

(a) Using the number of dopants per unit volume n_d, and $\alpha_{ph,\lambda,i} = \sigma_{ph,\lambda,i}L = A_{ph,a}n_d L$, calculate the absorptance $\alpha_{ph,\lambda,i}$, for target temperature $T_s = 300$ K.

(b) Using this absorptance, calculate the target temperature T_s, for $T_\infty = 300$ K and $\eta_{e\text{-}ph} = 0.98993$, and $\eta_{p\text{-}ph}$ given by (7.168).

(c) Repeat (a) and (b) with the new target temperature until convergence is reached.

Note that $A_{ph,a}$ is temperature dependent.

Problem 7.19

Show that under thermal equilibrium distribution and under a thermal source the population rate equation (7.37) leads to the two relation (7.40) and (7.42) between the Einstein coefficients.

Problem 7.20

Show that the angular-averaged transition dipole moment is

$$\langle\frac{|\mathbf{s}_\alpha \cdot \boldsymbol{\mu}_e|}{|\boldsymbol{\mu}_e|}\rangle = 3^{-1/2},$$

for an electric field with polarity $\mathbf{s}_\alpha$ and use $\boldsymbol{\mu}_e = |\boldsymbol{\mu}_e|\mathbf{s}_i$.

Note that

$$\mathbf{s}_\alpha \cdot \mathbf{s}_i = \cos\theta,$$

and that

$$\langle(\mathbf{s}_\alpha \cdot \mathbf{s}_i)^2\rangle = \frac{\int_0^{2\pi}\int_0^{\pi} \cos^2\theta \sin\theta \mathrm{d}\theta \mathrm{d}\phi}{\int_0^{2\pi}\int_0^{\pi} \sin\theta \mathrm{d}\theta \mathrm{d}\phi}.$$

Problem 7.21

Using (7.165) and Table 2.10, calculate the transition dipole moment for the following electronic transition in a hydrogen atom.

(a) $1s \rightarrow 2s$ (answer: $\boldsymbol{\mu}_e = 0$)

(b) $2s \rightarrow 2p_z$ (answer: $\boldsymbol{\mu}_e = -3e_c r_\mathrm{B} \mathbf{s}_z$).

(a) Starting from

$$\frac{\mathrm{d}A_{ph\text{-}e}}{\mathrm{d}\Omega} = 64\pi \frac{e_c^2 r_\mathrm{B}^2}{4\pi\epsilon_o \hbar u_{ph}} \sin\theta (\frac{E_{e,1}}{\hbar\omega})^4 \frac{e^{-4\gamma\cot^{-1}\gamma}}{1-e^{-2\pi\gamma}},$$

show that

$$A_{ph\text{-}e} = \frac{512\pi^2}{3} \frac{e_c^2 r_\mathrm{B}^2}{4\pi\epsilon_o \hbar u_{ph}} (\frac{E_{e,1}}{\hbar\omega})^4 \frac{e^{-4\gamma\cot^{-1}\gamma}}{1-e^{-2\pi\gamma}}.$$

Use $\mathrm{d}\Omega = \sin\theta \mathrm{d}\theta \mathrm{d}\phi$ and γ as defined in Table 7.2.

(b) Using the photoelectric absorption relation (also in Table 7.2) (more details are in Section E.3), plot the dimensional ionization cross-section area $A_{ph\text{-}e}$ as a function of incoming photon energy $\hbar\omega$ (in eV), for $\hbar\omega$ between $E_{e,1}$ (2.114) and $10E_{e,1}$.

(c) Using a circular cross-section geometry, comment on the radius of this absorption cross-section area and make comparisons with the Bohr radius.

Problem 7.23

Consider the derivation of (3.43) expressing the relation for the emission rate (or radiation decay rate) of an electric entity undergoing a transition dipole moment $\boldsymbol{\mu}_e$.

(a) Using the first order perturbation Hamiltonian (E.28), show that, on averaging over all directions, we have ($a_{e,\mathrm{o}} = e_{e,\mathrm{o}}/\omega_i$)

$$\langle f|\mathrm{H}'|i\rangle^2 = \frac{1}{12}\frac{e_c^2 e_{e,\mathrm{o}}^2}{m_e^2\omega_i^2}\langle f|\mathbf{p}_e|i\rangle^2,$$

where the average (in all directions) gives $3^{-1/2}$, which is squared here (Problem 7.20). Note, $\exp(i\boldsymbol{\kappa}\cdot\boldsymbol{x}) = 1 + i\boldsymbol{\kappa}\cdot\boldsymbol{x} + ... \simeq 1$, because the wavelength of EM radiation emitted is much larger than the typical size of the atom.

(b) Use $[\boldsymbol{x}, \mathrm{H}_\mathrm{o}] = -(i\hbar/m_e)\mathbf{p}_e$ (Problem 2.15) and using (7.135) to show that

$$|\langle f|\mathrm{H}'|i\rangle|^2 = \frac{1}{12}e_{e,\mathrm{o}}^2|\boldsymbol{\mu}_e|^2.$$

Note that $\langle f|i\rangle = 0$.

(c) Use the FGR along with (E.8), (7.2), and (7.3) to show that the stimulated emission rate is

$$\frac{\mathrm{d}E_{ph,\omega}}{V\mathrm{d}\omega}\dot{\gamma}_{ph,e,st} = \frac{\omega_{e,g}^3}{3\pi\epsilon_o\hbar u_{ph}^3}|\boldsymbol{\mu}_e|^2 f_{ph}^{\mathrm{o}}.$$

Note that from (3.38), $\epsilon_{\mathrm{o}} e_{e,\mathrm{o}}^2/2 = \hbar\omega_{e,g} f_{ph}^{\mathrm{o}} D_{b,\omega}$.

(d) Then using (7.39), (7.5), and (7.6), arrive at (3.43).

Problem 7.24

Using (7.29), (7.31), and the commutator relation, $[A, B] = AB - BA$ (Chapter 2 problem), derive the two relations given in (7.32).

Problem 7.25

For the $^2\mathrm{F}_{7/2} \rightarrow {}^2\mathrm{F}_{5/2}$ transition of Yb^{3+}, $\Delta E_{e,g} = 1.1998$ eV.

(a) estimate $|\boldsymbol{\mu}_e|$ using $|\boldsymbol{\mu}_e| = g_\mu e_c \Delta R_e$ [180]. Use $g_\mu = 1$, $E_{e,4f} = -19.86$ eV, $\langle R_{4f}\rangle = 0.379$ Å, and

$$\Delta E_{e,g} = A(\frac{1}{R_{e,f}^2} - \frac{1}{R_{e,i}^2}),\ \ A = E_{e,4f}\langle R_{4f}\rangle^2,$$

$$\frac{1}{R_{e,f}^2} + \frac{1}{R_{e,i}^2} = \frac{2}{\langle R_{4f}^2\rangle},\ \ \Delta R_e = R_{e,f} - R_{e,i}.$$

(b) Use this in (7.189) to calculate the spontaneous emission rate.

(c) Using (7.42), find the thermal equilibrium, stimulated emission rate. Note the units are explained in Section 7.4.1.

(d) Using (7.57) and (7.40), calculate the absorption cross-section area for $\int_\omega \mathrm{d}\omega = 10^{12}$ rad/s. Comment on the relation with any other cross-section area (e.g., fluid particle collision).

(e) Using (7.129), find the absorption coeffifient. Use $n_d = 2.8\times 10^{20}$ cm^{-3}.

Problem 7.26

For the case of small absorption (optically thin $\sigma_{ph,L} \ll 1$) and emission domination, consider the following.

(a) Show that the ERT (Table 7.1) under thermal equilibrium, along a path S, and under steady state and $\dot{s}_{ph,i} = 0$, becomes

$$I_{ph,\omega}(S) = \int_0^S \sigma_{ph,\omega} I_{b,\omega}\mathrm{d}S.$$

(b) Using the definition of the mean absorption coefficient,

$$\langle\sigma_{ph}\rangle = \frac{\int_0^\infty \sigma_{ph,\omega}E_{b,\omega}\mathrm{d}\omega}{\int_0^\infty E_{b,\omega}\mathrm{d}\omega} = \frac{\int_0^\infty \sigma_{ph,\omega}E_{b,\omega}\mathrm{d}\omega}{\sigma_{\mathrm{SB}}T^4},$$

which is called the Planck mean absorption coefficient, show that

$$I_{ph} = \int_0^S \langle\sigma_{ph}\rangle \frac{\sigma_{\mathrm{SB}}T^4(S)}{\pi}\mathrm{d}S.$$

(c) Using q_{ph}^{+} as the flux (per unit area) emerging from an isothermal slab made of material with $\langle\sigma_{ph}\rangle$, at T having a depth L, i.e.,

$$q_{ph}^{+} = 2\pi \int_0^{\pi/2} I_{ph}(\theta)\cos\theta \sin\theta \mathrm{d}\theta,$$

and using $S = L/\cos\theta$, show that this flux is

$$q_{ph}^{+} = 2\langle\sigma_{ph}\rangle\sigma_{\mathrm{SB}} L T^4.$$

This can be applied to emitted radiation from a gas or a solid slab that is optically thin.

APPENDIX A

Tables of Properties and Universal Constants

A.1 Periodic Table I

Table A.1. *Periodic table of bulk properties of elements.*

Key (example cell):

18 ← Group Number
Noble Elements ← Group Name

Entry in cell	Value (Ar)	Entry in cell	Value (Ar)
Element Abreviation	Ar	z Atomic Number	18
Element Name	Argon		
M(kg/kmole)	39.948	State	Gas
c_p(J/kg-K)	521	σ_e(1/Ω-cm)	–
ρ(kg/m³)	1.78	Δh_{sl}(J/kg)	4.898×10^6
k(W/m-K)	0.0177	Δh_{lg}(J/kg)	1.63×10^5
T_{sl}(K)	84	T_{lg}(K)	87

1 Alkali Metals (Except H)

Symbol	z	Name	M (kg/kmole)	State	c_p (J/kg-K)	σ_e (1/Ω-cm)	ρ (kg/m³)	Δh_{sl} (J/kg)	k (W/m-K)	Δh_{lg} (J/kg)	T_{sl} (K)	T_{lg} (K)
H	1	Hydrogen	1.008	Gas	14304	-	0.0899	1.16×10^5	0.1815	9.09×10^5	14	20
Li	3	Lithium	6.941	Solid	3582	1.07×10^5	534	4.32×10^5	84.8	2.134×10^7	454	1615
Na	11	Sodium	22.990	Solid	1228	2.11×10^5	970	1.13×10^5	142	4.263×10^6	371	1156
K	19	Potassium	39.098	Solid	757	1.39×10^5	862	6.013×10^4	102.5	2.052×10^6	337	1032
Rb	37	Rubidium	85.468	Solid	363	8.0×10^4	1530	2.64×10^4	58.2	8.1×10^5	313	961
Cs	55	Cesium	132.905	Solid	242	5.0×10^4	1930	1.64×10^4	35.9	4.96×10^5	302	944
Fr	87	Francium	[223]	Solid	-	-	-	-	15	-	300	879

2 Alkaline-Earth Metals

Symbol	z	Name	M (kg/kmole)	State	c_p (J/kg-K)	σ_e (1/Ω-cm)	ρ (kg/m³)	Δh_{sl} (J/kg)	k (W/m-K)	Δh_{lg} (J/kg)	T_{sl} (K)	T_{lg} (K)
Be	4	Berylium	9.012	Solid	1825	3.08×10^5	1850	1.354×10^6	201	3.245×10^7	1562	2745
Mg	12	Magnesium	24.305	Solid	1023	2.33×10^5	1740	3.68×10^5	156	5.242×10^6	923	1363
Ca	20	Calcium	40.078	Solid	647	2.78×10^5	1540	2.13×10^5	201	3.833×10^6	1115	1767
Sr	38	Strontium	87.620	Solid	301	4.7×10^4	2640	10.4×10^4	35.4	1.585×10^6	1042	1655
Ba	56	Barium	137.327	Solid	204	2.6×10^4	3620	5.7×10^4	18.4	1.1×10^6	1002	2170
Ra	88	Radium	[226]	Solid	119.5	-	5000	-	18.6	-	973	1900

Transition Metals (Groups 3–9)

Group	Symbol	z	Name	M (kg/kmole)	State	c_p (J/kg-K)	σ_e (1/Ω-cm)	ρ (kg/m³)	Δh_{sl} (J/kg)	k (W/m-K)	Δh_{lg} (J/kg)	T_{sl} (K)	T_{lg} (K)
3	Sc	21	Scandium	44.956	Solid	568	2.1×10^4	2990	3.13×10^5	15.8	6.989×10^6	1814	3109
3	Y	39	Yttrium	88.906	Solid	298	1.7×10^4	4470	1.28×10^5	17.2	4.435×10^6	1795	3611
4	Ti	22	Titanium	47.880	Solid	523	2.3×10^4	4500	3.24×10^5	21.9	8.791×10^6	1943	3562
4	Zr	40	Zirconium	91.224	Solid	278	2.4×10^4	6520	1.85×10^5	22.7	6.376×10^6	2128	4682
4	Hf	72	Hafnium	178.490	Solid	144	3.3×10^4	13300	1.35×10^5	23	3.211×10^6	2504	4876
5	V	23	Vanadium	50.942	Solid	489	5.0×10^4	6110	4.11×10^5	30.7	8.870×10^6	2183	3682
5	Nb	41	Niobium	92.906	Solid	265	6.9×10^4	8570	2.84×10^5	53.7	7.341×10^6	2742	5017
5	Ta	73	Tantalum	180.948	Solid	140	7.6×10^4	16400	1.74×10^5	57.5	4.211×10^6	3293	5731
6	Cr	24	Chromium	51.996	Solid	449	7.8×10^4	7150	3.256×10^5	93.9	6.622×10^6	2136	2945
6	Mo	42	Molybdenum	95.940	Solid	251	1.89×10^5	10200	4.17×10^5	138	6.175×10^6	2896	4912
6	W	74	Tungsten	183.850	Solid	132	1.89×10^5	19300	1.74×10^5	173	4.211×10^6	3695	5828
7	Mn	25	Manganese	54.931	Solid	479	7.2×10^3	7300	2.19×10^5	7.81	4.112×10^6	1519	2335
7	Tc	43	Technetium	[98]	Solid	243	7×10^4	11500	3.32×10^5	50.6	5.83×10^{10}	2430	4538
7	Re	75	Rhenium	186.207	Solid	137	5.4×10^4	20800	1.77×10^5	48	3.842×10^6	3459	5869
8	Fe	26	Iron	55.847	Solid	449	1.02×10^5	7870	2.47×10^5	80.4	6.259×10^6	1811	3135
8	Ru	44	Ruthenium	101.047	Solid	238	1.35×10^5	12100	2.56×10^5	117	5.838×10^6	2607	4423
8	Os	76	Osmium	190.200	Solid	130	1.10×10^4	22500	2.89×10^5	87.6	3.31×10^6	3306	5285
9	Co	27	Cobalt	58.993	Solid	421	1.72×10^5	8860	2.75×10^5	100	6.390×10^6	1768	3201
9	Rh	45	Rhodium	102.906	Solid	243	2.08×10^5	12400	2.65×10^5	150	4.798×10^6	2236	3970
9	Ir	77	Iridium	192.220	Solid	131	1.96×10^5	22500	2.13×10^5	147	3.185×10^6	2720	4701

Rare-Earth Metals

Symbol	z	Name	M (kg/kmole)	State	c_p (J/kg-K)	σ_e (1/Ω-cm)	ρ (kg/m³)	Δh_{sl} (J/kg)	k (W/m-K)	Δh_{lg} (J/kg)	T_{sl} (K)	T_{lg} (K)
La	57	Lanthanum	138.906	Solid	195	1.3×10^4	6150	4.46×10^4	13.4	2.978×10^6	1191	3737
Ce	58	Cerium	140.115	Solid	192	1.2×10^4	8160	3.9×10^4	11.3	2.955×10^6	1071	3716
Pr	59	Praseodymium	140.908	Solid	193	1.5×10^4	6770	4.89×10^4	12.5	2.105×10^6	1204	3793
Nd	60	Neodymium	144.240	Solid	190	1.7×10^4	7010	4.95×10^4	16.5	1.891×10^6	1294	3347
Pm	61	Promethium	[145]		7260	-	7220	-	17.9	-	1315	3273
Sm	62	Samarium	150.360	Solid	197	1.0×10^4	7520	5.73×10^4	13.3	1.107×10^6	1347	2067
Ac	89	Actinium	[227]	Solid	120	-	10000	6.3×10^4	12	1.75×10^6	1324	3473
Th	90	Thorium	232	Solid	113	6.6×10^6	11700	6.94×10^4	54	2.218×10^6	2028	5061
Pa	91	Protactinium	231	Solid	121	-	15370	-	47	6.48×10^4	1845	-
U	92	Uranium	238	Solid	116	3.9×10^4	19100	3.66×10^4	27.5	1.949×10^6	1408	4407
Np	93	Neptunium	237	Solid	20450	8.5×10^3	-	-	6.3	-	912	4173
Pu	94	Plutonium	[244]	Solid	138	7.0×10^3	19700	1.17×10^4	6.7	1.409×10^6	913	3503

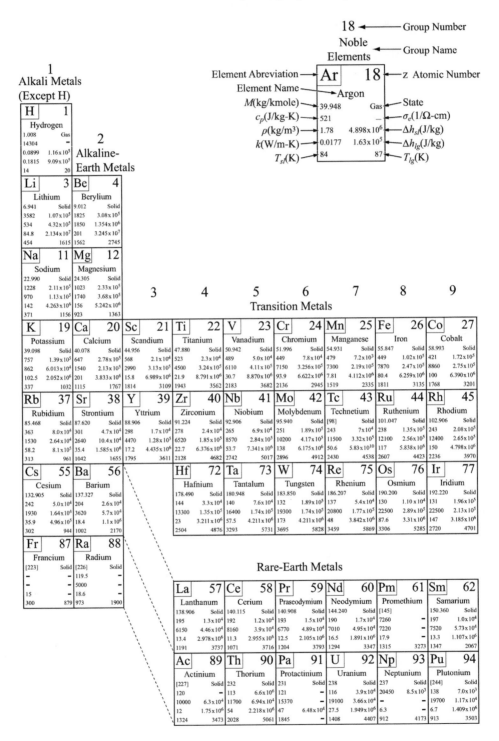

18 Noble Elements

He	2
Helium	
4.002	Gas
5197	-
0.179	2.1×10^{4}
0.152	2.1×10^{5}
3	4

13 Boron Group, 14 Carbon Group, 15 Pricogens, 16 Chalcogens, 17 Halogens

B	5	C	6	N	7	O	8	F	9	Ne	10
Boron		Carbon		Nitrogen		Oxygen		Fluorine		Neon	
10.811	Solid	12.011	Solid	14.007	Gas	15.999	Gas	18.998	Gas	20.180	Gas
1026	-	509	-	1039	-	919	-	828	-	1030	-
2340	1.933×10^{6}	3510	-	1.24	-	1.43	-	1.7	-	0.9	1.6×10^{5}
27.4	-	2300	-	0.026	3.99×10^{5}	0.2674	4.26×10^{5}	0.0279	3.44×10^{5}	0.0493	8.89×10^{4}
2365	4275	4100	4473	63	77	54	90	53	95	24	27

Al	13	Si	14	P	15	S	16	Cl	17	Ar	18
Aluminum		Silicon		Phosphorus		Sulfur		Chlorine		Argon	
26.982	Solid	28.086	Solid	30.974	Solid	32.066	Solid	35.453	Gas	39.948	Gas
897	3.77×10^{5}	678	-	791	-	732	-	479	-	521	-
2700	3.97×10^{5}	2330	1.802×10^{6}	2690	8.48×10^{4}	2070	5.34×10^{4}	3.21	1.8×10^{5}	1.78	4.898×10^{3}
237	1.0875×10^{7}	149	1.4×10^{7}	12.1	1.799×10^{6}	0.205	2.87×10^{5}	0.0089	5.76×10^{5}	0.0177	1.63×10^{5}
933	2793	1687	3540	317	553	388	718	172	239	84	87

10 11 12
Transition Metals

Ni	28	Cu	29	Zn	30	Ga	31	Ge	32	As	33	Se	34	Br	35	Kr	36
Nickel		Copper		Zinc		Gallium		Germanium		Arsenic		Selenium		Bromine		Krypton	
58.693	Solid	63.546	Solid	65.390	Solid	69.723	Solid	72.610	Solid	74.992	Solid	78.960	Solid	79.904	Liquid	83.800	Gas
444	1.43×10^{5}	385	5.95×10^{5}	388	1.69×10^{5}	371	6.7×10^{4}	310	-	343	-	339	-	226	-	248	-
8900	2.91×10^{5}	8960	2.07×10^{5}	7140	1.12×10^{5}	5910	7.97×10^{4}	5323	5.09×10^{5}	5727	-	4810	6.62×10^{4}	3120	6.6×10^{4}	3.73	1.9×10^{5}
90.9	6.309×10^{6}	401	4.726×10^{6}	116	1.767×10^{6}	40.6	3.688×10^{6}	60.2	4.557×10^{6}	50.2	1.7×10^{6}	2.04	1.2×10^{6}	0.122	1.9×10^{5}	0.0095	1.08×10^{5}
1728	3187	1358	2836	693	1180	303	2477	1211	3107	1090	888	494	958	266	332	116	120

Pd	46	Ag	47	Cd	48	In	49	Sn	50	Sb	51	Te	52	I	53	Xe	54
Palladium		Silver		Cadmium		Indium		Tin		Antimony		Tellurium		Iodine		Xenon	
106.420	Solid	107.868	Solid	112.411	Solid	114.820	Solid	118.710	Solid	121.757	Solid	127.600	Solid	126.904	Solid	131.290	Gas
251	9.5×10^{4}	235	6.21×10^{5}	232	1.38×10^{5}	233	1.14×10^{5}	228	9.1×10^{4}	207	2.4×10^{4}	197	-	145	-	158	-
12020	4.17×10^{5}	10500	1.02×10^{5}	8690	5.39×10^{4}	7310	2.86×10^{4}	7280	6.07×10^{4}	6680	1.66×10^{5}	6240	1.4×10^{6}	4953	-	5.89	1.7×10^{4}
71.8	5.567×10^{6}	429	2.323×10^{6}	96.9	8.86×10^{5}	81.8	2.019×10^{6}	66.8	2.495×10^{6}	24.4	-	2.35	8.95×10^{6}	0.449	3.29×10^{5}	0.0057	9.63×10^{4}
2896	4912	1235	2436	594	1040	430	2346	505	2876	904	1860	723	1261	387	457	161	165

Pt	78	Au	79	Hg	80	Tl	81	Pb	82	Bi	83	Po	84	At	85	Rn	86
Platinum		Gold		Mercury		Thallium		Lead		Bismuth		Polonium		Astatine		Radon	
195.080	Solid	196.967	Solid	200.590	Liquid	204.383	Solid	207.200	Solid	208.980	Solid	[209]	Solid	[210]	Solid	[222]	Gas
133	9.6×10^{4}	129	4.55×10^{5}	140	1.0×10^{4}	129	6.1×10^{4}	129	4.8×10^{4}	122	8.6×10^{3}	125	2.2×10^{4}	140	-	93.7	-
21500	1.15×10^{5}	19300	6.43×10^{4}	13534	1.14×10^{4}	11800	2.06×10^{4}	11300	2.32×10^{4}	9790	5.33×10^{4}	9320	5.98×10^{4}	-	-	9.73	1.2×10^{4}
71.6	2.612×10^{6}	318	1.701×10^{6}	8.3	2.96×10^{5}	46.1	8.06×10^{5}	35.3	8.58×10^{5}	7.92	7.3×10^{5}	20	-	1.7	-	0.0036	7.52×10^{4}
2042	4100	1337	3130	234	630	577	1746	600	2022	544	1837	527	1033	575	650	202	211

Rare-Earth Metals

Eu	63	Gd	64	Tb	65	Dy	66	Ho	67	Er	68	Tm	69	Yb	70	Lu	71
Europium		Gadolimium		Terbium		Dysprosium		Holmium		Erbium		Thulium		Ytterbium		Lutetium	
151.965	Solid	157.250	Solid	158.925	Solid	162.500	Solid	164.930	Solid	167.260	Solid	168.934	Solid	173.040	Solid	174.967	Solid
182	1.1×10^{4}	236	7.0×10^{3}	182	9.0×10^{3}	173	1.1×10^{4}	165	1.3×10^{4}	168	1.2×10^{4}	160	1.6×10^{4}	155	3.8×10^{4}	154	1.9×10^{4}
5240	6.06×10^{4}	7900	6.36×10^{4}	8230	6.79×10^{4}	8550	6.79×10^{4}	8800	1.03×10^{5}	9070	1.19×10^{5}	9320	9.9×10^{4}	6900	1.43×10^{4}	9840	1.26×10^{5}
13.9	9.44×10^{5}	10.5	2.286×10^{6}	11.1	1.42×10^{6}	10.7	2.083×10^{6}	16.2	1.461×10^{6}	14.5	1.563×10^{6}	16.9*	1.130×10^{6}	34.9*	7.45×10^{5}	16.4*	2.034×10^{6}
1095	1869	1586	3546	1629	3503	1685	2840	1747	2973	1802	3141	1818	2223	1092	1469	1936	3675

Am	95	Cm	96	Bk	97	Cf	98	Es	99	Fm	100	Md	101	No	102	Lr	103
Americium		Curium		Berkelium		Californium		Einsteinium		Fermium		Mendelevium		Nobelium		Lawrencium	
[243]	Solid	[247]	-	[247]	-	[251]	-	[252]	-	[257]	-	[258]	-	[259]	-	[260]	-
-	-	-	-	-	-	-	-	-	-	-	-	-	-	-	-	-	-
13670	-	13300	-	14790	-	-	-	-	-	-	-	-	-	-	-	-	-
10	-	10	-	10	-	10	-	10	-	10	-	10	-	10	-	10	-
1449	2284	1618	-	1323	-	1173	-	1133	-	1800	-	1100	-	1100	-	1900	-

A.1 Periodic Table II

Table A.2. *Periodic table of atomic-structure properties of elements.*

Key (example cell, *s* Block — Outer Orbit Group Designation):

Element Abreviation: Ar	Atomic Radius (Å): Filled Shell Configuration	1.88
	Tetrahedral Covalent Bond	–
	12-Coordinated Metal	–
Electronic Orbital Structure: [Ne]$3s^23p^6$	Grüneisen Parameter	2.83
First Ionization Energy(eV): 15.76	Oxidation States	–
Electronegativity: –	Isotopes (Three Most Abundant)	40,36
Crystal Structure: FCC	Lattice Constant(s) (Å, degrees)	5.26
Cohesive Energy(eV/atom): 0.080	Mean Debye Temperature(K) (LT = Low Temperature Value)	85

Structure	Lattice constants	Structure	Lattice constants
FCC - Face Centered Cubic	*a*	ORC - Orthorhombic	*a*/*b*/*c*
BCC - Body Centered Cubic	*a*	HEX - Hexagonal	*a*/*c*
SC - Simple Cubic	*a*	DIA - Diamond	*a*
CUB - Cubic	*a*	RHL - Rhombohedral	*a*/α
TET - Tetragonal	*a*/*c*	MCL - Monoclinic	*a*/*c*

Groups: 1 (*s* Block), 2 (*s* Block), 3, 4, 5, 6, 7, 8, 9 (*d* Block); *f* Block

Element	Group / Block	Filled Shell Radius	Tetrahedral Covalent Bond	12-Coordinated Metal	Electronic Orbital Structure	Grüneisen Parameter	First Ionization Energy (eV)	Oxidation States	Electronegativity	Isotopes	Crystal Structure	Lattice Constant(s)	Cohesive Energy (eV/atom)	Mean Debye Temperature (K)
H	1, *s* Block	2.08	–	–	$1s^1$	–	13.595	1	2.20	1,2	HEX	3.75/6.49	–	110
Li	1, *s* Block	0.68	–	1.56	[He]$2s^1$	1.06	5.39	1	0.98	7,6	BCC	3.49	1.63	400
Be	2, *s* Block	0.35	1.06	1.13	[He]$2s^2$	1.28	9.32	2	1.57	9	HEX	2.29/3.59	3.32	1000
Na	1, *s* Block	0.97	–	1.91	[Ne]$3s^1$	1.49	5.14	1	0.93	23	BCC	4.23	1.113	150
Mg	2, *s* Block	0.65	1.40	1.60	[Ne]$3s^2$	1.71	7.64	2	1.31	24,26,25	HEX	3.21/5.21	1.51	318
K	1, *s* Block	1.33	–	2.38	[Ar]$4s^1$	1.65	4.34	1	0.82	39,41	BCC	5.23	0.934	100
Ca	2, *s* Block	0.99	–	1.98	[Ar]$4s^2$	1.09	6.11	2	1.00	40,44,42	FCC	5.58	1.84	230
Sc	3, *d* Block	0.81	–	1.64	[Ar]$3d^14s^2$	–	6.56	3	1.36	45	HEX	3.31/5.28	3.90	359LT
Ti	4, *d* Block	0.68	–	1.46	[Ar]$3d^24s^2$	–	6.83	4	1.54	48,46,47	HEX	2.95/4.68	4.85	380
V	5, *d* Block	–	–	1.35	[Ar]$3d^34s^2$	1.36	6.74	5,3	1.63	51,50	BCC	3.02	5.31	390
Cr	6, *d* Block	–	–	1.28	[Ar]$3d^54s^1$	0.90	6.76	6,3,2	1.66	52,53,50	BCC	2.88	4.10	460
Mn	7, *d* Block	–	–	1.26	[Ar]$3d^54s^2$	–	7.43	7,6,4,2,3	1.55	55	CUB	8.89	2.92	2335
Fe	8, *d* Block	–	–	1.27	[Ar]$3d^64s^2$	1.88	7.90	2,3	1.83	56,54,57	BCC	2.87	4.28	400
Co	9, *d* Block	–	–	1.25	[Ar]$3d^74s^2$	2.37	7.86	2,3,4	1.88	59	HEX	2.51/4.07	4.39	385
Rb	1, *s* Block	1.48	–	2.55	[Kr]$5s^1$	–	4.18	1	0.82	85,87	BCC	5.59	0.852	56LT
Sr	2, *s* Block	1.13	–	2.15	[Kr]$5s^2$	–	5.69	2	0.95	88,86,87	FCC	6.08	1.72	147LT
Y	3, *d* Block	0.93	–	1.80	[Kr]$4d^15s^2$	–	6.5	3	1.22	89	HEX	3.65/5.73	4.37	256LT
Zr	4, *d* Block	0.80	–	1.60	[Kr]$4d^25s^2$	0.92	6.95	4	1.33	90,94,92	HEX	3.32/5.15	6.25	250
Nb	5, *d* Block	0.67	–	1.47	[Kr]$4d^45s^1$	–	6.77	5,3	1.6	93	BCC	3.30	7.57	275
Mo	6, *d* Block	–	–	1.40	[Kr]$4d^55s^1$	1.63	7.18	6,5,4,3,2	2.16	98,96,95	BCC	3.15	6.82	380
Tc	7, *d* Block	–	–	1.36	[Kr]$4d^55s^2$	–	7.28	7	1.90	93	HEX	2.74/4.39	6.85	–
Ru	8, *d* Block	–	–	1.34	[Kr]$4d^75s^1$	–	7.36	2,3,4,6,8	2.20	102,104,101	HEX	2.70/4.28	6.74	382LT
Rh	9, *d* Block	–	–	1.35	[Kr]$4d^85s^1$	–	7.46	2,3,4	2.28	103	FCC	3.80	5.75	350LT
Cs	1, *s* Block	1.67	–	2.73	[Xe]$6s^1$	–	3.89	1	0.79	1.64x10^4	BCC	6.05	0.804	40LT
Ba	2, *s* Block	1.35	–	2.24	[Xe]$6s^2$	–	5.21	2	0.89	5.7x10^4	BCC	5.02	1.90	110LT
Hf	4, *d* Block	–	–	1.58	[Xe]$4f^{14}5d^26s^2$	–	6.83	4	1.30	180,178,177	HEX	3.20/5.06	6.44	–
Ta	5, *d* Block	–	–	1.47	[Xe]$4f^{14}5d^36s^2$	1.66	7.88	5	1.50	181,180	BCC	3.31	8.10	225
W	6, *d* Block	–	–	1.41	[Xe]$4f^{14}5d^46s^2$	1.71	7.98	6,5,4,3,2	2.36	184,186,182	BCC	3.16	8.90	310
Re	7, *d* Block	–	–	1.38	[Xe]$4f^{14}5d^56s^2$	–	7.87	7,6,4,2,-1	1.90	187,185	HEX	2.76/4.46	8.03	416LT
Os	8, *d* Block	–	–	1.35	[Xe]$4f^{14}5d^66s^2$	–	8.7	2,3,4,6,8	2.20	192,190,189	HEX	2.74/4.33	8.17	400LT
Ir	9, *d* Block	–	–	1.36	[Xe]$4f^{14}5d^76s^2$	–	9.1	2,3,4,6	2.20	193,191	FCC	3.84	6.94	430
Fr	1, *s* Block	1.75	–	–	[Rn]$7s^1$	–	–	1	0.70	–	(BCC)	–	–	–
Ra	2, *s* Block	1.37	–	–	[Rn]$7s^2$	–	5.28	2	0.90	226	BCC	–	1.66	–
La	*f* Block	1.15	–	1.88	[Xe]$5d^16s^2$	–	5.77	3	1.10	138,139	HEX	3.75/6.07	4.47	132
Ce	*f* Block	1.10	1.71	1.82	[Xe]$4f^15d^16s^2$	–	5.54	3,4	1.12	140,142	FCC	5.16	4.32	139LT
Pr	*f* Block	–	–	1.84	[Xe]$4f^36s^2$	–	5.46	3,4	1.13	141	HEX	3.67/5.92	3.70	152LT
Nd	*f* Block	–	–	1.82	[Xe]$4f^46s^2$	–	5.53	3	1.14	142,143,146	HEX	3.66/5.91	3.40	157LT
Pm	*f* Block	–	–	1.81	[Xe]$4f^56s^2$	–	5.55	3	1.13	145	HEX	–	–	–
Sm	*f* Block	–	–	1.80	[Xe]$4f^66s^2$	–	5.64	3,2	1.17	152,154,147	RHL	9.00/23.1	2.14	166
Ac	*f* Block	1.11	–	–	[Rn]$6d^17s^2$	–	5.17	3	1.10	227	FCC	5.31	4.25	–
Th	*f* Block	0.99	–	1.80	[Rn]$6d^27s^2$	–	6.08	4	1.30	232	FCC	5.08	6.20	100
Pa	*f* Block	0.90	–	1.63	[Rn]$5f^26d^17s^2$	–	5.89	5,4	1.50	231	TET	3.92/3.23	–	–
U	*f* Block	0.83	–	1.56	[Rn]$5f^36d^17s^2$	–	6.19	6,5,4,3	1.38	238,235	ORC	2.85/5.86/4.95	5.55	210LT
Np	*f* Block	–	–	1.56	[Rn]$5f^46d^17s^2$	–	6.27	6,5,4,3	1.36	237	ORC	4.72/6.66/4.89	4.73	188LT
Pu	*f* Block	–	1.58	1.64	[Rn]$5f^67s^2$	–	6.06	6,5,4,3	1.28	244	MCL	–	3.60	150LT

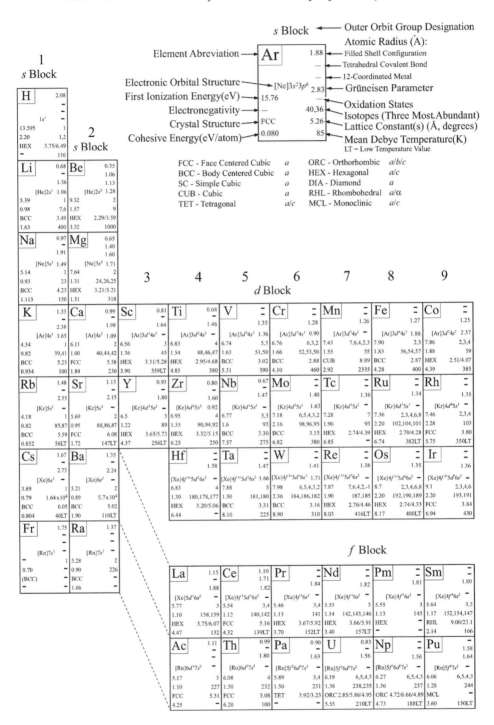

18
p Block
(Except He: *s* Block)

13 14 15 16 17
p Block

10 11 12
d Block

Group	Element				Configuration									
18	He	-	-	-	$1s^2$	-	24.58	-	-	4	HEX	3.57/5.83	-	26LT
13	B	0.23	0.88	0.98	[He]$2s^22p^1$	-	8.30	3	2.04	11,10	TET	8.73/5.03	5.81	1250
14	C	0.15	0.77	0.92	[He]$2s^22p^2$	0.90	11.26	4,2	2.55	12,13	DIA	3.57	7.37	1860
15	N	1.71	0.70	-	[He]$2s^22p^3$	-	14.54	3,5,4,2,-3	3.04	14,15	HEX	4.04/6.67	4.92	79LT
16	O	1.40	0.66	-	[He]$2s^22p^4$	-	13.61	-2	3.44	16,18	CUB	6.83	2.60	46LT
17	F	1.36	0.64	-	[He]$2s^22p^5$	-	17.42	-1	3.98	19	MCL	-	0.84	-
18	Ne	1.36	-	-	[He]$2s^22p^6$	2.76	21.56	-	-	20,22,21	FCC	8.89×10^4	0.020	63
13	Al	0.50	1.26	1.43	[Ne]$3s^23p^1$	2.41	5.98	3	1.61	27	FCC	4.05	3.39	394
14	Si	0.41	1.17	1.32	[Ne]$3s^23p^2$	0.44	8.15	4	1.90	28,29,30	DIA	5.43	4.63	645
15	P	2.12	1.10	-	[Ne]$3s^23p^3$	-	10.55	3,5,4,-3	2.19	31	CUB	7.17	3.43	-
16	S	1.84	1.04	-	[Ne]$3s^23p^4$	-	10.36	2,4,6,-2	2.58	32,34,33	ORC	10.47/24.49/12.87	2.85	-
17	Cl	1.81	0.99	-	[Ne]$3s^23p^5$	-	13.01	1,3,5,7,-1	3.16	35,37	ORC	6.24/8.26/4.48	1.40	-
18	Ar	1.88	-	-	[Ne]$3s^23p^6$	2.73	15.76	-	-	40,36	FCC	5.26	0.080	85
10	Ni	-	-	1.25	[Ar]$3d^84s^2$	2.11	7.63	2,3	1.91	58,60,62	FCC	3.52	4.44	375
11	Cu	-	1.35	1.28	[Ar]$3d^{10}4s^1$	2.14	7.72	2,1	1.90	63,65	FCC	3.61	3.49	315
12	Zn	0.74	1.31	1.39	[Ar]$3d^{10}4s^2$	2.75	9.39	2	1.65	64,66,68	HEX	2.66/4.94	1.35	234
13	Ga	0.62	1.26	1.41	[Ar]$3d^{10}4s^24p^1$		6.00	3	1.81	69,71	ORC	4.51/7.64/4.51	2.81	240
14	Ge	0.53	1.22	1.37	[Ar]$3d^{10}4s^24p^2$	0.77	7.88	4	2.01	74,72,70	DIA	5.86	3.85	360
15	As	2.22	1.18	1.39	[Ar]$3d^{10}4s^24p^3$	-	9.81	3,5,-3	2.18	75	RHL	4.16/54.2	2.96	285
16	Se	1.98	1.14	-	[Ar]$3d^{10}4s^24p^4$	-	9.75	2,4,6,-2	2.55	80,78,76	HEX	4.36/4.95	2.46	150LT
17	Br	1.95	1.11	-	[Ar]$3d^{10}4s^24p^5$	-	11.84	1,5,-1	2.96	79,81	ORC	6.67/8.72/4.48	1.22	-
18	Kr	2.00	-	-	[Ar]$3d^{10}4s^24p^6$	2.84	14.00	-	3.00	84,86,82	FCC	5.72	0.116	73LT
10	Pd	-	-	1.38	[Kr]$4d^{10}$	2.47	8.33	2,4	2.20	106,108,105	FCC	3.89	3.89	275
11	Ag	1.26	1.52	1.45	[Kr]$4d^{10}5s^1$	2.65	7.57	1	1.93	107,109	FCC	4.09	2.95	215
12	Cd	0.97	1.48	1.57	[Kr]$4d^{10}5s^2$	3.09	8.99	2	1.69	114,112,111	HEX	2.98/5.62	1.16	120
13	In	0.81	1.44	1.66	[Kr]$4d^{10}5s^25p^1$	2.75	5.78	3	1.78	115,113	TET	4.59/4.94	2.52	129
14	Sn	0.71	1.40	1.55	[Kr]$4d^{10}5s^25p^2$	-	7.34	4,2	1.96	120,118,116	TET	5.82/3.18	3.14	170
15	Sb	2.45	1.36	1.59	[Kr]$4d^{10}5s^25p^3$	-	8.64	3,5,-3	2.05	121,123	RHL	4.51/57.1	2.75	200
16	Te	2.21	1.32	-	[Kr]$4d^{10}5s^25p^4$	-	9.01	2,4,6,-2	2.10	130,128,126	HEX	4.45/5.92	2.19	139LT
17	I	2.16	1.28	-	[Kr]$4d^{10}5s^25p^5$	-	10.45	1,5,7,-1	2.66	127	ORC	7.27/9.79/4.79	1.11	-
18	Xe	2.17	-	-	[Kr]$4d^{10}5s^25p^6$	2.65	12.13	-	2.60	132,129,131	FCC	6.20	0.16	55LT
10	Pt	-	-	1.39	[Xe]$4f^{14}5d^96s^1$	2.89	8.96	2,4	2.28	195,194,196	FCC	3.92	5.84	230
11	Au	1.37	-	1.44	[Xe]$4f^{14}5d^{10}6s^1$	3.17	9.22	3,1	2.54	197	FCC	4.08	3.81	170
12	Hg	1.10	1.48	1.57	[Xe]$4f^{14}5d^{10}6s^2$	-	10.43	2,1	2.00	202,200,199	RHL	2.99/70.8	0.67	100
13	Tl	0.95	-	1.72	[Xe]$4f^{14}5d^{10}6s^26p^1$	2.36	6.11	3,1	1.62	205,203	HEX	3.46/5.53	1.88	96
14	Pb	0.84	-	1.75	[Xe]$4f^{14}5d^{10}6s^26p^2$	3.06	7.41	4,2	2.33	208,206,207	FCC	4.95	2.03	88
15	Bi	-	-	1.70	[Xe]$4f^{14}5d^{10}6s^26p^3$	-	7.29	3,5	2.02	209	RHL	4.75/57.2	2.18	120
16	Po	-	-	1.76	[Xe]$4f^{14}5d^{10}6s^26p^4$	-	8.43	4,2	2.00	210	SC	3.35	1.50	-
17	At	-	-	-	[Xe]$4f^{14}5d^{10}6s^26p^5$	-	-	1,3,5,7,-1	2.20	210	-	-	-	-
18	Rn	-	-	-	[Xe]$4f^{14}5d^{10}6s^26p^6$	-	10.74	-	-	222	(FCC)	-	0.202	-

f Block

Element				Configuration									
Eu	-	2.04	1.80	[Xe]$4f^76s^2$	-	5.67	3,2	1.20	153,151	BCC	4.61	1.86	107LT
Gd	-	-	1.80	[Xe]$4f^75d^16s^2$	-	6.16	3	1.20	158,160,156	HEX	3.64/5.78	4.14	176LT
Tb	-	-	1.78	[Xe]$4f^96s^2$	-	5.86	3,4	1.10	159	HEX	3.60/5.69	4.05	188LT
Dy	-	-	1.77	[Xe]$4f^{10}6s^2$	-	5.94	3	1.22	164,162,163	HEX	3.59/5.65	3.04	186LT
Ho	-	-	1.77	[Xe]$4f^{11}6s^2$	-	6.02	3	1.23	165	HEX	3.58/5.62	3.14	191LT
Er	-	-	1.76	[Xe]$4f^{12}6s^2$	-	6.11	3	1.24	166,168,167	HEX	3.56/5.59	3.29	195LT
Tm	-	-	1.75	[Xe]$4f^{13}6s^2$	-	6.18	3,2	1.25	169	HEX	3.54/5.56	2.42	200LT
Yb	-	1.94	1.81	[Xe]$4f^{14}6s^2$	-	6.25	3,2	1.10	174,172,173	FCC	5.49	1.60	118LT
Lu	-	-	-	[Xe]$4f^{14}5d^16s^2$	-	5.43	3	1.27	175,176	HEX	3.51/5.56	4.43	207LT
Am	-	-	1.81	[Rn]$5f^77s^2$	-	5.99	6,5,4,3	1.13	243	HEX	-	2.73	-
Cm	-	-	-	[Rn]$5f^76d^17s^2$	-	6.02	3	1.28	247	HEX	-	3.99	-
Bk	-	-	-	[Rn]$5f^97s^2$	-	6.23	4,3	1.30	247	HEX	-	-	-
Cf	-	-	-	[Rn]$5f^{10}7s^2$	-	6.30	3	1.30	251	HEX	-	-	-
Es	-	-	-	[Rn]$5f^{11}7s^2$	-	6.42	3	1.30	254	HEX	-	-	-
Fm	-	-	-	[Rn]$5f^{12}7s^2$	-	6.50	3	1.30	257	-	-	-	-
Md	-	-	-	[Rn]$5f^{13}7s^2$	-	6.58	3,2	1.30	256	-	-	-	-
No	-	-	-	[Rn]$5f^{14}7s^2$	-	6.65	3,2	1.30	254	-	-	-	-
Lr	-	-	-	[Rn]$5f^{14}6d^17s^2$	-	-	3	-	257	-	-	-	-

A.3 Fundamental Constants

Table A.3. *Fundamental constants, derived quantities, and units conversions [209, 210].*

Symbol	Magnitude and Units
Fundamental constants:	
c_o	speed of light in vacuum, 2.99792458×10^8 m/s
e_c	electron charge, 1.602×10^{-19} C (1 eV = 1.602×10^{-19} J where V = J/C)
G_N	Newton (gravitational) constant, 6.673×10^{-11} m^3/kg-s^2
h_P	Planck constant, 6.626×10^{-34} J-s = 4.136×10^{-15} eV-s
$\hbar$	$h_P/2\pi = 1.055 \times 10^{-34}$ J-s = 6.583×10^{-16} eV-s
k_B	Boltzmann constant, 1.381×10^{-23} J/K = 8.618×10^{-5} eV/K
m_e	electron mass, 9.109×10^{-31} kg
m_n	neutron mass, 1.675×10^{-27} kg
m_p	proton mass, 1.673×10^{-27} kg
N_A	Avogadro number, 6.022×10^{23} molecule/mole = 6.022×10^{26} molecule/kmole
μ_o	free-space magnetic permeability, $4\pi \times 10^{-7} = 1.257 \times 10^{-6}$ N-s^2/C^2

Derived quantities:

$\epsilon_o = \dfrac{1}{c_o^2 \mu_o} = 8.854 \times 10^{-12}$ C^2/N-m^2	free-space electric permittivity
$\mathrm{D} = e_c r_B/2.54 = 3.3356 \times 10^{-30}$ C-m	Debye (atomic dielectric dipole moment)
$N_L = \dfrac{\pi^2}{3}\dfrac{k_B T^2}{e_c^2} = 2.442 \times 10^{-8}$ W-Ω/K^2	Lorenz number
$r_C = \dfrac{e_c^2}{4\pi\epsilon_o m_e c_o^2} = 2.8179 \times 10^{-15}$ m	classical (Compton) electron radius
$r_B = \dfrac{4\pi\epsilon_o \hbar^2}{m_e e_c^2} = 5.292 \times 10^{-11}$ m	Bohr radius (atomic length unit)
$\dfrac{e_c^2}{4\pi\epsilon_o r_B} = 27.2114$ eV $= 4.35975 \times 10^{-18}$ J	hartree (atomic energy unit)
$\dfrac{e_c^2}{4\pi\epsilon_o r_B^2} = 8.2378 \times 10^{-8}$ N	atomic force unit
$R_g \equiv k_B N_A = 8.3145 \times 10^3$ J/kmole-K	universal gas constant
$\mathrm{Ry} = \dfrac{m_e e_c^4}{8\epsilon_o^2 c_o \hbar^3} = 1.0974 \times 10^5$ cm^{-1}	Rydberg ground-state energy of H
$\dfrac{1}{4\pi\epsilon_o} = 8.9876 \times 10^9$, Nm^2/C^2	Coulomb constant
$\alpha = \dfrac{e_c^2}{4\pi\epsilon_o \hbar c_o} = 7.29735 \times 10^{-3}$	fine-structure constant
$\sigma_{SB} \equiv \dfrac{\pi^2 k_B^4}{60\hbar^3 c_o^2} = 5.670 \times 10^{-8}$ W/m^2-K^4	Stefan–Boltzmann constant

Units conversions:

1 eV = 8.0655×10^4 cm^{-1} = 2.418×10^{14} Hz = 11,600 K = 1.602×10^{-19} J

1 cm^{-1} = 0.12398 meV = 2.998×10^{10} Hz

A.4 Unit Prefixes

Table A.4. *Unit prefixes.*

Factor	Prefix	Symbol	Factor	Prefix	Symbol
10^{24}	yotta	Y	10^{-24}	yocto	y
10^{21}	zetta	Z	10^{-21}	zepto	z
10^{18}	exa	E	10^{-18}	atto	a
10^{15}	peta	P	10^{-15}	femto	f
10^{12}	tera	T	10^{-12}	pico	p
10^{9}	giga	G	10^{-9}	nano	n
10^{6}	mega	M	10^{-6}	micro	μ
10^{3}	kilo	k	10^{-3}	milli	m
10^{2}	hecto	h	10^{-2}	centi	c
10^{1}	deka	da	10^{-1}	deci	d

One angstrom is $1\ \text{Å} = 10^{-10}$ m

APPENDIX B

Derivation of Green–Kubo Relation

The Green–Kubo (G–K) development of a time-correlation function expression for the thermal conductivity (the G–K approach) is based in classical statistical thermodynamics. Multiple methods can be used to arrive at the final result [361]. Similar approaches can be used to develop expressions for the self-diffusion coefficient, the shear viscosity, and the bulk viscosity. These are all transport coefficients that cannot be obtained by applying a perturbation to the system Hamiltonian, as can be done for some other properties (e.g., the electrical conductivity) for which there is a real force that drives the transport. Here, the method of Helfand [142] as outlined in [239], is presented step-by-step [232].

B.1 Perturbation Energy Equation and Fourier Transform

The G–K approach is valid in the case of small disturbances from equilibrium and for long times (i.e., the hydrodynamic limit). The key aspect of the derivation is the introduction of a microscopic description of a system to the solution of the macroscopic governing equation.

For the thermal conductivity, a canonical ensemble of particles (i.e., the NVT ensemble) is considered, and the energy equation (similar to the macroscopic energy equation of Table 1.1 for no net flow and no radiation heat transfer, with conduction heat flux given by the Fourier law, using the thermal conductivity tensor $\boldsymbol{K}$) is written as

$$nc_v \frac{\partial E'(\boldsymbol{x}, t)}{\partial t} = \nabla \cdot \boldsymbol{K} \cdot \nabla E'(\boldsymbol{x}, t), \tag{B.1}$$

where n is the number of particles per unit volume and c_v is the specific heat per particle.

Here the independent variable $E'(\boldsymbol{x}, t)$ is the deviation (perturbation) of the energy from its expectation value at position $\boldsymbol{x}$ at time t,

$$E'(\boldsymbol{x}, t) = E(\boldsymbol{x}, t) - \langle E(\boldsymbol{x}, t)\rangle, \tag{B.2}$$

where $E(\boldsymbol{x}, t)$ is the actual energy at that point and $\langle E(\boldsymbol{x}, t)\rangle$ is its expectation value. There is an initial condition $E'(\boldsymbol{x}, 0)$. The specific heat has units of J/K - particle, and is the number of particles per unit volume.

Define the Fourier transform of $E'(\boldsymbol{x}, t)$ as $L(\boldsymbol{\kappa}, t)$, such that

$$L(\boldsymbol{\kappa}, t) = \int \exp(i\boldsymbol{\kappa}\cdot\boldsymbol{x})E'(\boldsymbol{x}, t)\mathrm{d}\boldsymbol{x}. \tag{B.3}$$

Substituting this expression into the energy equation leads to

$$\frac{\mathrm{d}L(\boldsymbol{\kappa}, t)}{\mathrm{d}t} = -\frac{\kappa^2 k}{nc_v}L(\boldsymbol{\kappa}, t), \tag{B.4}$$

with initial condition $L(\boldsymbol{\kappa}, 0)$, which has a solution

$$L(\boldsymbol{\kappa}, t) = L(\boldsymbol{\kappa}, 0)\exp(-\frac{\kappa^2 kt}{nc_v}). \tag{B.5}$$

On a microscopic level, the system energy can be defined on a particle basis as

$$E'(\boldsymbol{x}, t) = \sum_j E'_j(t)\delta[\boldsymbol{x} - \boldsymbol{x}_j(t)], \tag{B.6}$$

where $E'_j(t)$ and $\boldsymbol{x}_j(t)$ are energy and position of particle j, and the sum is over the N particles in the system. The Fourier transform of (B.6) is

$$L(\boldsymbol{\kappa}, t) = \int \sum_j E'_j(t)\delta[\boldsymbol{x} - \boldsymbol{x}_j(t)]\exp(i\boldsymbol{\kappa}\cdot\boldsymbol{x})\mathrm{d}\boldsymbol{x} = \sum_j E'_j(t)\exp[i\boldsymbol{\kappa}\cdot\boldsymbol{x}_j(t)]. \tag{B.7}$$

Multiplying both sides of (B.5) by $L^*(\boldsymbol{\kappa}, 0)$, the complex conjugate of the initial condition, and using (B.7) gives

$$\sum_j E'_j(t)\exp[i\boldsymbol{\kappa}\cdot\boldsymbol{x}_j(t)]\sum_l E'_l(0)\exp[-i\boldsymbol{\kappa}\cdot\boldsymbol{x}_l(0)]$$

$$= \{\sum_j E'_j(0)\exp[i\boldsymbol{\kappa}\cdot\boldsymbol{x}_j(0)]\sum_l E'_l(0)\exp[-i\boldsymbol{\kappa}\cdot\boldsymbol{x}_l(0)]\}\exp(-\frac{k\kappa^2 t}{nc_v}). \tag{B.8}$$

Noting that $\sum_j a_j \sum_l b_l = \sum_j \sum_l a_j b_l$, and taking the ensemble average of both sides, we have

$$\langle\sum_j\sum_l E'_j(t)E'_l(0)\exp\{i\boldsymbol{\kappa}\cdot[\boldsymbol{x}_j(t) - \boldsymbol{x}_l(0)]\}\rangle$$

$$= \langle\sum_j\sum_l E'_j(0)E'_l(0)\exp\{i\boldsymbol{\kappa}\cdot[\boldsymbol{x}_j(0) - \boldsymbol{x}_l(0)]\}\rangle\exp(-\frac{k\kappa^2 t}{nc_v}). \tag{B.9}$$

B.2 Taylor Series Expansion and Ensemble Averaging

An expression for the thermal conductivity is found by expanding both sides of this equation as a Taylor series about $\kappa = 0$ and comparing the coefficients with second order. This procedure is justified by the assumption of small perturbations from equilibrium.

Without losing any generality, the remainder of the derivation is simplified by taking $\boldsymbol{\kappa}$ to be in the x direction [i.e., $\boldsymbol{\kappa} = (\kappa, 0, 0)$]. Consider the left-hand side of (B.9), expanded to second order in κ, with the derivatives evaluated at $\kappa = 0$:

$$\begin{aligned}&\langle \sum_j \sum_l E_j'(t)E_l'(0)\rangle + \kappa\langle \sum_j \sum_l E_j'(t)E_l'(0)i[x_j(t) - x_l(0)]\rangle \\ &+\frac{\kappa^2}{2}\langle \sum_j \sum_l E_j'(t)E_l'(0)i^2[x_j(t) - x_l(0)]^2\rangle. \end{aligned} \tag{B.10}$$

Before proceeding, a number of useful expressions related to the summations and ensemble averages to be evaluated are introduced. First, by conservation of energy,

$$\sum_j E_j'(0) = \sum_l E_l'(t). \tag{B.11}$$

This expression indicates that, although energy is exchanged between the particles, the total value is a constant. Second, the stationary condition for an equilibrium ensemble average states that, for some quantity $\phi(t)$ based on the particle positions and momenta, that

$$\langle \phi(t)\rangle = \langle \phi(0)\rangle. \tag{B.12}$$

Now, consider (B.10) term by term. The first term becomes

$$\begin{aligned}&\langle \sum_j \sum_l E_j'(t)E_l'(0)\rangle = \langle \sum_j E_j'(t)\sum_l E_l'(0)\rangle \\ &= \langle \sum_j E_j'(0)\sum_l E_l'(0)\rangle = \langle \sum_j \sum_l E_j'(0)E_l'(0)\rangle. \end{aligned} \tag{B.13}$$

Using the definition of the energy deviation, and dropping the time label, we have

$$\begin{aligned}&\langle \sum_j \sum_l E_j'(t)E_l'(0)\rangle = \langle \sum_j \sum_l (E_j - \langle E_j\rangle)(E_l - \langle E_l\rangle)\rangle \\ &= \langle \sum_j \sum_l (E_j E_l - E_j\langle E_j\rangle - E_l\langle E_j\rangle + \langle E_j\rangle\langle E_l\rangle)\rangle \\ &= \langle \xi^2 - 2\xi\langle \xi\rangle + \langle \xi\rangle^2\rangle = \langle (\xi - \langle \xi\rangle)^2\rangle = Nk_{\rm B}T^2 c_v, \end{aligned} \tag{B.14}$$

where ξ represents the total energy in the system [i.e., $\xi = \int E(\boldsymbol{x}, t)\mathrm{d}\boldsymbol{x}$] and $N = nV$ is the number of atoms in the system, where V is the volume of the system. The last step comes from the thermodynamic definition of temperature in the thermal fluctuation and canonical ensemble. (Table 2.4, where c_v contains all the degrees of freedom in atomic motion)

For the first term in (B.10),

$$
\begin{aligned}
& i\kappa\langle\sum_j\sum_l E_j'(t)E_l'(0)[x_j(t)-x_l(0)]\rangle \\
&= i\kappa\langle\sum_j\sum_l E_j'(t)E_l'(0)x_j(t)\rangle - i\kappa\langle\sum_j\sum_l E_j'(t)E_l'(0)x_l(0)\rangle \\
&= i\kappa\langle\sum_j E_j'(t)x_j(t)\sum_l E_l'(0)\rangle - i\kappa\langle\sum_j E_j'(t)\sum_l E_l'(0)x_l(0)\rangle \\
&= i\kappa\langle\sum_j E_j'(t)x_j(t)\sum_l E_l'(t)\rangle - i\kappa\langle\sum_j E_j'(0)\sum_l E_l'(0)x_l(0)\rangle \\
&= \kappa\langle\sum_j\sum_l E_j'(t)E_l'(t)x_j(t)\rangle - i\kappa\langle\sum_j\sum_l E_j'(0)E_l'(0)x_l(0)\rangle \\
&= i\kappa\langle\sum_j\sum_l E_j'(0)E_l'(0)x_j(0)\rangle - i\kappa\langle\sum_j\sum_l E_j'(0)E_l'(0)x_l(0)\rangle \\
&= 0.
\end{aligned}
\tag{B.15}
$$

For the ensemble average in the second-order term,

$$
\begin{aligned}
& \langle\sum_j\sum_l E_j'(t)E_l'(0)[x_j(t)-x_l(0)]^2\rangle \\
&= \langle\sum_j\sum_l E_j'(t)E_l'(0)[x_j(t)]^2\rangle - 2\langle\sum_j\sum_l E_j'(t)E_l'(0)x_j(t)x_l(0)\rangle \\
&\quad + \langle\sum_j\sum_l E_j'(t)E_l'(0)[x_l(0)]^2\rangle \\
&= \langle\sum_j E_j'(t)[x_j(t)]^2\sum_l E_l'(0)\rangle - 2\langle\sum_j E_j'(t)x_j(t)\sum_l E_l'(0)x_l(0)\rangle \\
&\quad + \langle\sum_j E_j'(t)\sum_l E_l'(0)[x_l(0)]^2\rangle \\
&= \langle\sum_j E_j'(t)[x_j(t)]^2\sum_l E_l'(t)\rangle - 2\langle\sum_j E_j'(t)x_j(t)\sum_l E_l'(0)x_l(0)\rangle \\
&\quad + \langle\sum_j E_j'(0)\sum_l E_l'(0)[x_l(0)]^2\rangle \\
&= \langle\sum_j\sum_l E_j'(t)E_l'(t)[x_j(t)]^2\rangle - 2\langle\sum_j E_j'(t)x_j(t)\sum_l E_l'(0)x_l(0)\rangle \\
&\quad + \langle\sum_j\sum_l E_j'(0)E_l'(0)[x_l(0)]^2\rangle.
\end{aligned}
\tag{B.16}
$$

If there is no correlation between the positions and energies of the particles, then

$$\langle \sum_j \sum_l E_j'(t) E_l'(t)[x_j(t)]^2 \rangle = \langle \sum_j [x_j(t)]^2 [E_j(t)]^2 \rangle$$
$$= \langle \sum_j E_j'(t) x_j(t) \sum_l E_l'(t) x_l(t) \rangle, \qquad \text{(B.17)}$$

so that the second-order term becomes

$$-\frac{\kappa^2}{2}[\langle \sum_j E_j'(t)x_j(t) \sum_l E_l'(t)x_l(t)\rangle - 2\langle \sum_j E_j'(t)x_j(t) \sum_l E_l'(0)x_l(0)\rangle$$
$$+ \langle \sum_j E_j'(0)x_j(0) \sum_l E_l'(0)x_l(0)\rangle]$$
$$= -\frac{\kappa^2}{2}\langle\{\sum_j [x_j(t)E_j'(t) - x_j(0)E_j'(0)]\}^2\rangle. \qquad \text{(B.18)}$$

The left-hand side of (B.9) is then

$$Nk_\text{B}T^2 c_v - \frac{\kappa^2}{2}\langle\{\sum_j [x_j(t)E_j'(t) - x_j(0)E_j'(0)]\}^2\rangle. \qquad \text{(B.19)}$$

Now consider the right-hand side of (B.9) expanded term by term about $\kappa = 0$. The zeroth-order term is

$$\langle \sum_j \sum_l E_j'(0) E_l'(0) \rangle, \qquad \text{(B.20)}$$

which is of the same form as the first term on the left-hand side, and thus equal to $Nk_\text{B}T^2 c_v$.

Let the ensemble average on the right-hand side of (B.9) be represented by A and the exponential by B. For the first-order term, the derivative

$$\frac{\partial}{\partial \kappa}(AB) = A\frac{\partial B}{\partial \kappa} + B\frac{\partial B}{\partial \kappa} \qquad \text{(B.21)}$$

is required. Letting $C = k_x t/nc_v$ (where k_x is the thermal conductivity in the x-direction), the required derivative is

$$-2AC\kappa \exp(-C\kappa^2) + \exp(-C\kappa^2)\frac{\partial A}{\partial \kappa}. \qquad \text{(B.22)}$$

For $\kappa = 0$, this expression reduces to $\partial A/\partial \kappa$, so that the first-order term is

$$i\kappa\langle \sum_j \sum_l E_j'(0)E_l'(0)[x_j(0) - x_l(0)]\rangle. \qquad \text{(B.23)}$$

This term can be evaluated in the same manner as the first-order expansion of the left-hand side, and is equal to zero [see (B.15)].

For the second-order term, the derivative

$$\frac{\partial^2}{\partial \kappa^2}(AB) = A\frac{\partial^2 B}{\partial \kappa^2} + 2\frac{\partial A}{\partial \kappa}\frac{\partial B}{\partial \kappa} + B\frac{\partial^2 A}{\partial \kappa^2} \qquad \text{(B.24)}$$

is required. As discussed for the first-order term, the derivative of the exponential goes to zero at $\kappa = 0$, so that the middle term above is zero. For the first term,

$$A\frac{\partial^2}{\partial\kappa^2}\exp(-C\kappa^2) = A\frac{\partial}{\partial\kappa}[-2C\kappa\exp(-C\kappa^2)]$$
$$= A[4C^2\kappa^2\exp(-C\kappa^2) - 2C\exp(-C\kappa^2)], \tag{B.25}$$

which, when evaluated at $\kappa = 0$, gives $2ac$, or

$$2\langle\sum_j\sum_l E_j'(0)E_l'(0)\rangle\frac{k_x t}{nc_v} = \frac{2Nk_B T^2 k_x t}{n}. \tag{B.26}$$

Note that $N/n = V$. For the last term in the second order derivative, noting that the exponential goes to unity at $\kappa = 0$,

$$\langle\sum_j\sum_l E_j'(0)E_l'(0)[x_j(0) - x_l(0)]^2\rangle. \tag{B.27}$$

Using the steps shown in (B.16) to (B.18), this term is found to be equal to zero. Multiplying the final result of (B.26) by $-\kappa^2/2$, the second-order term is thus

$$-Vk_B T^2 k_x\kappa^2 t, \tag{B.28}$$

and the expansion of the right-hand side of (B.9) is

$$Nk_B T^2 c_v - Vk_B T^2 k_x\kappa^2 t. \tag{B.29}$$

B.3 Thermal Conductivity Tensor

When the expansions of the two sides of (B.9), (B.19), and (B.29) are brought together, the zeroth-order terms are seen to be equal. Equating the coefficients of the second-order terms and solving for k_x (related to atomic motion in solid or fluid state) gives

$$k_x = \frac{1}{2k_B T^2 Vt}\langle\{\sum_j[E_j'(t)x_j(t) - E_j'(0)x_j(0)]\}^2\rangle. \tag{B.30}$$

The final step in the derivation is to transform the ensemble average/summation in (B.30) into the integral of a time correlation function. First, note that

$$\sum_j[E_j'(t)x_j(t) - E_j'(0)x_j(0)] = \int_0^t\frac{\mathrm{d}}{\mathrm{d}t}\sum_j[E_j'(t_1)x_j(t_1)]\mathrm{d}t_1. \tag{B.31}$$

This is done by noting that on the left-hand side the integral of the integrands are evaluated at $t = 0$ and $t = \infty$, so they can be written as time derivatives of an integral. Then because the integration, summation, and differentiation are linear, their order can be changed.

Defining the heat current in the x direction $\dot{w}_x$ as

$$\dot{w}_x(t) \equiv \frac{\mathrm{d}}{\mathrm{d}t}\sum_j[E_j'(t)x_j(t)], \tag{B.32}$$

the ensemble average in (B.30) can be written as

$$\begin{aligned}\langle\{\sum_j[E'_j(t)x_j(t)-E'_j(0)x_j(0)]\}^2\rangle &= \langle\int_0^t \dot{w}_x(t_1)\mathrm{d}t_1\int_0^t \dot{w}_x(t_2)\mathrm{d}t_2\rangle\\ &= \langle\int_0^t\int_0^t \dot{w}_x(t_1)\dot{w}_x(t_2)\mathrm{d}t_1\mathrm{d}t_2\rangle\\ &= \int_0^t\int_0^t \langle\dot{w}_x(t_1)\dot{w}_x(t_2)\rangle\mathrm{d}t_1\mathrm{d}t_2\\ &= \int_0^t\int_0^t \langle\dot{w}_x(t_2-t_1)\dot{w}_x(0)\rangle\mathrm{d}t_1\mathrm{d}t_2, \qquad \text{(B.33)}\end{aligned}$$

where the last step results from the stationary nature of the equilibrium ensemble. Using the identity

$$\int_0^t\int_0^t f(t_2-t_1)\mathrm{d}t_1\mathrm{d}t_2 \equiv 2t\int_0^t(1-\frac{\tau}{t})f(\tau)\mathrm{d}\tau \qquad \text{(B.34)}$$

leads to

$$k_x=\frac{2}{2Vk_\mathrm{B}T^2}\int_0^t(1-\frac{\tau}{t})\langle\dot{w}_x(\tau)\dot{w}_x(0)\rangle\mathrm{d}\tau. \qquad \text{(B.35)}$$

As t can be chosen to be arbitrarily large, τ/t in the integral can be ignored, and the thermal conductivity in the x direction can be written as

$$k_x=\frac{1}{k_\mathrm{B}T^2V}\int_0^\infty\langle\dot{w}_x(t)\dot{w}_x(0)\rangle\mathrm{d}t. \qquad \text{(B.36)}$$

To generalize the results, the heat current is expressed as a vectorial flux, i.e.,

$$\dot{\boldsymbol{w}}=\frac{\mathrm{d}}{\mathrm{d}t}\sum_j[E'_j(t)\boldsymbol{x}_j(t)], \qquad \text{(B.37)}$$

giving the thermal conductivity tensor

$$\boldsymbol{K}=\frac{1}{k_\mathrm{B}T^2V}\int_0^\infty\langle\dot{\boldsymbol{w}}(t)\dot{\boldsymbol{w}}(0)\rangle\mathrm{d}t. \qquad \text{(B.38)}$$

In a system with cubic isotropy, the scalar conductivity is given by

$$k=\frac{1}{k_\mathrm{B}T^2V}\int_0^\infty\frac{\langle\dot{\boldsymbol{w}}(t)\cdot\dot{\boldsymbol{w}}(0)\rangle}{3}\mathrm{d}t \quad \text{G–K relation for thermal conductivity.} \qquad \text{(B.39)}$$

For a two-body potential, the heat current vector can be expressed as

$$\dot{\boldsymbol{w}}=\sum_i E_i\boldsymbol{u}_i+\frac{1}{2}\sum_{i,j}(\boldsymbol{F}_{ij}\cdot\boldsymbol{u}_i)\boldsymbol{x}_{ij} \quad \text{heat current vector,} \qquad \text{(B.40)}$$

a form readily implemented in MD simulation.

B.4 Physical Interpretation of Heat Current Vector Autocorrelation

Autocorrelation is a measure of how a signature of a dynamic system matches a time-shifted version of itself (i.e., cross correlation of a signature or signal with itself). Developing a physical interpretation of the final result is useful. The argument of the derivative of the heat current can be taken as the energy center of mass of the system. It is a vector that indicates the direction of energy transfer in the system at an instant in time. How long this quantity stays correlated with itself is related to the thermal conductivity. In a material with a high thermal conductivity, the correlation will be long lasting. This can be alternatively stated as a system in which fluctuations from equilibrium dissipate slowly. For a material with a low thermal conductivity, the correlation will be short-lived. One interesting aspect of the G–K approach is that transport properties can be obtained from an equilibrium system. This is an important point for the thermal conductivity, as it is generally thought of in a nonequilibrium system with a temperature gradient. The G–K approach is a real-space formulation, unlike the $\boldsymbol{\kappa}$-space- or $\boldsymbol{p}$-space-based BTE.

APPENDIX C

Derivation of Minimum Phonon Conductivity Relations

C.1 Einstein Thermal Conductivity

The derivation presented in this section and the next is based on that given in [100, 52], with some of the mathematical steps given in more detail. The formulation of the Einstein thermal conductivity, $k_{\rm E}$, presented here is that given in [232], which is extended to arrive at the C–P high scatter limit, $k_{\rm CP}$, in Section C.2.

In the Einstein approach, the vibrational states do not correspond to phonons, but to the atoms themselves, which are assumed to be on a simple cubic lattice as shown in Figure C.1. As will be discussed, the choice of the crystal structure does not affect the final result. Each atom is treated as a set of three harmonic oscillators in mutually perpendicular directions. Although the atomic motions are taken to be independent, an atom is assumed to exchange energy with its first, second, and third nearest neighbors. The coupling is realized by modeling the atomic interactions as being a result of linear springs (with spring constant Γ) connecting the atoms. A given atom has 6 nearest neighbors at a distance of a, 12 second-nearest neighbors at a distance of $2^{1/2}a$, and 8 third-nearest neighbors at a distance of $3^{1/2}a$.

The derivation starts from the equation written in terms of the mean free path (4.109) ($k_p = nc_{v,p}u_{p,A}\lambda/3$). All the vibrational states are assumed to have the same angular frequency of oscillation, $\omega_{\rm E}$, which leads to a per particle specific heat of from (4.55), as

$$c_{v,p} = 3k_{\rm B}\frac{x_{\rm E}^2 e^{x_{\rm E}}}{(e^{x_{\rm E}}-1)^2}, \tag{C.1}$$

where x is defined as $\hbar\omega_{\rm E}/k_{\rm B}T \equiv T_{\rm E}/T$. The mean free path is taken as the distance between first nearest-neighbor atoms ($a = n^{-1/3}$), and the velocity is taken as $a/\tau_{\rm E}$ ($= n^{-1/3}/\tau_{\rm E}$), where $\tau_{\rm E}$ represents the time needed for energy to move from one

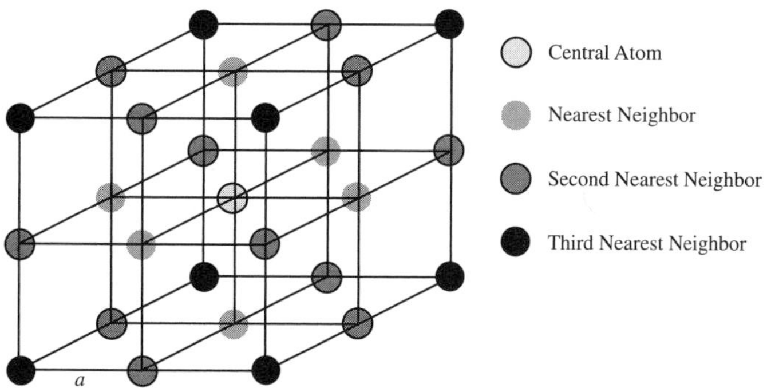

Figure C.1. The simple cubic crystal structure, with the first-, second- and third-nearest neighbors of the central atom shown.

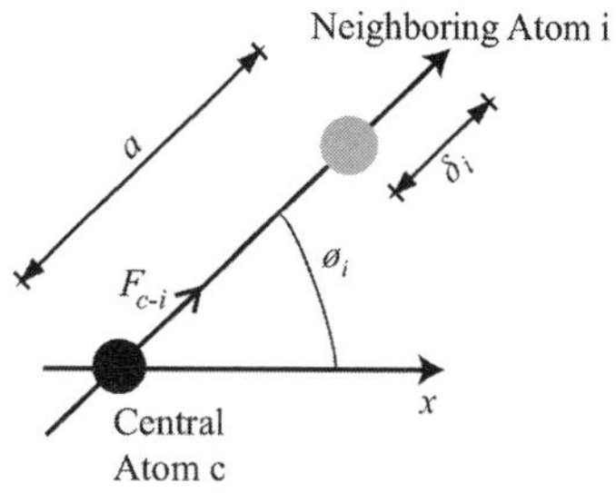

Figure C.2. Two neighboring atoms in the Einstein thermal conductivity formulation.

atom to another. Using these specifications leads to

$$k_{p,\mathrm{E}} = \frac{n^{-1/3} k_{\mathrm{B}}}{\tau_{\mathrm{E}}} \frac{x_{\mathrm{E}}^2 e^{x_{\mathrm{E}}}}{(e^{x_{\mathrm{E}}} - 1)^2}. \tag{C.2}$$

The value of $\tau_{p,\mathrm{E}}$ must be specified.

Consider a set of two atoms, as shown in Figure C.2. The equation of motion for the central atom in the x direction is desired. As such, consider a displacement of magnitude x ($x \ll a$) of the central atom in the x direction, which will cause the neighbor atom, labeled as atom i, to move along the line of action between it and the central atom by an amount δ_i ($\delta_i \ll a$). The angle formed by these two directions is ϕ_i. The change in the length of the spring is then $(\delta_i - x\cos\phi_i)$, and the force on the central atom in the x direction is

$$F_{c\text{-}i} = \Gamma(\delta_i - x\cos\phi_i)\cos\phi_i. \tag{C.3}$$

Summing over the contributions of all the neighbor atoms (26 in total) leads to the equation of motion

$$m\frac{\mathrm{d}^2 x}{\mathrm{d}t^2} = \sum_i \Gamma(\delta_i - x\cos\phi_i)\cos\phi_i, \tag{C.4}$$

where m is the mass of an atom. Multiplying through by $(\mathrm{d}x/\mathrm{d}t)\mathrm{d}t$ and integrating over time leads to

$$\int m\frac{\mathrm{d}^2x}{\mathrm{d}t^2}\frac{\mathrm{d}x}{\mathrm{d}t}\mathrm{d}t = -\int \Gamma x(\sum_i \cos^2\phi_i)\frac{\mathrm{d}x}{\mathrm{d}t}\mathrm{d}t + \int \Gamma(\sum_i \delta_i\cos\phi_i)\frac{\mathrm{d}x}{\mathrm{d}t}\mathrm{d}t$$

$$\int \frac{1}{2}m\mathrm{d}(\dot{x}^2) = -\int \frac{1}{2}\Gamma(\sum_i \cos^2\phi_i)\mathrm{d}(x^2) + \int \Gamma(\sum_i \delta_i\cos\phi_i)\frac{\mathrm{d}x}{\mathrm{d}t}\mathrm{d}t$$

$$\sum_i \Gamma\cos\phi_i\int \delta_i\frac{\mathrm{d}x}{\mathrm{d}t}\mathrm{d}t = \int \mathrm{d}[\frac{m}{2}(\frac{\mathrm{d}x}{\mathrm{d}t})^2 + \frac{\Gamma x^2}{2}\sum_i\cos^2\phi_i]. \quad \text{(C.5)}$$

The variable of integration of the last line of the right-hand side of (C.5) is taken as the energy associated with the central atom (kinetic and potential) so that the integral corresponds to the change of the energy of the atom over a specified time. The left-hand side of the last line of (C.5) represents the work done by the other atoms in the system to bring about this change in energy. Consider a time of one-half of the period of oscillation (π/ω_{E}), and define the summand of the right-hand side of the last line of (C.5) as η_i and the energy change as ΔE, such that the last line of (C.5) can be written as

$$\Delta E = \sum_i \eta_i. \quad \text{(C.6)}$$

Assume that the atomic motions are sinusoidal and uncorrelated so that

$$x = A\sin(\omega_{\mathrm{E}}t) \quad \text{(C.7)}$$

$$\delta_i = A_i\sin(\omega_{\mathrm{E}}t + \alpha_i), \quad \text{(C.8)}$$

where α_i is a random phase angle. This leads to

$$\begin{aligned}\eta_i &= \Gamma\cos\phi_i\int_0^{\pi/\omega_{\mathrm{E}}}\delta_i\frac{\mathrm{d}x}{\mathrm{d}t}\mathrm{d}t \\ &= \Gamma\cos\phi_i\int_0^{\pi/\omega_{\mathrm{E}}} A_i\sin(\omega_{\mathrm{E}}t+\alpha_i)A\omega_{\mathrm{E}}\cos(\omega_{\mathrm{E}}t)\,\mathrm{d}t \\ &= \Gamma\cos\phi_i AA_i\omega_{\mathrm{E}}\int_0^{\pi/\omega_{\mathrm{E}}}[\sin(\omega_{\mathrm{E}}t)\cos\alpha_i\cos(\omega_{\mathrm{E}}t) \\ &\quad + \cos(\omega_{\mathrm{E}}t)\sin\alpha_i\cos(\omega_{\mathrm{E}}t)]\,\mathrm{d}t.\end{aligned} \quad \text{(C.9)}$$

Consider the first term in the integral

$$\int_0^{\pi/\omega_{\mathrm{E}}}\sin(\omega_{\mathrm{E}}t)\cos(\omega_{\mathrm{E}}t)\mathrm{d}t = \int_{-\pi/2\omega_{\mathrm{E}}}^{\pi/2\omega_{\mathrm{E}}}\frac{1}{2}\sin(2\omega_{\mathrm{E}}t)\,\mathrm{d}t = 0. \quad \text{(C.10)}$$

Then, because of the odd nature of the sine function, the work done becomes

$$\begin{aligned}
\eta_i &= \Gamma \cos\phi_i\, AA_i\omega_{\rm E} \sin\alpha_i \int_0^{\pi/\omega_{\rm E}} \cos^2(\omega_{\rm E}t){\rm d}t \\
&= \Gamma \cos\phi_i\, AA_i\omega_{\rm E} \sin\alpha_i \int_0^{\pi/\omega_{\rm E}} [\frac{1}{2}+\frac{1}{2}\cos(2\omega_{\rm E}t)]{\rm d}t \\
&= \Gamma \cos\phi_i\, AA_i\omega_{\rm E} \sin\alpha_i .[\frac{t}{2}+\frac{1}{2\omega_{\rm E}}\sin(2\omega_{\rm E}t)]|_0^{\pi/\omega_{\rm E}} \\
&= \frac{\Gamma \cos\phi_i\, AA_i \sin\alpha_i \pi}{2}.
\end{aligned} \tag{C.11}$$

Because the phase angles α_i are random, the average value of the work done by the atoms is zero. As many atoms transfer energy to the central atom as those that obtain energy from it (as would be expected in an equilibrium system). However, the RMS fluctuation of the energy,

$$\overline{(\Delta E)^2} = \overline{(\sum_i \eta_i)^2}, \tag{C.12}$$

will not be zero. Because the α_i values are uncorrelated, $\overline{\eta_i\eta_j}=0$, so that

$$\overline{(\Delta E)^2} = \overline{\sum_i \eta_i^2} = \sum_i \overline{\eta_i^2}. \tag{C.13}$$

Thus, from (C.11),

$$\eta_i^2 = \frac{\pi^2\Gamma^2\cos^2\phi_i}{4} A^2 A_i^2 \sin^2\alpha_i, \tag{C.14}$$

and

$$\overline{\eta_i^2} = \frac{\pi^2\Gamma^2\cos^2\phi_i}{4} \overline{A^2}\, \overline{A_i^2}\, \overline{\sin^2\alpha_i}. \tag{C.15}$$

If all the atoms are identical, then $A = A_i$. For random α_i,

$$\overline{\sin^2\alpha_i} = \int_0^{2\pi} \sin^2\alpha {\rm d}\alpha = \frac{1}{2}, \tag{C.16}$$

so that

$$\overline{\eta_i^2} = \frac{\pi^2\Gamma^2\cos^2\phi_i}{4}\frac{\overline{A^2}^2}{2}, \tag{C.17}$$

and

$$\overline{(\Delta E)^2} = \sum_i \frac{\pi^2\Gamma^2\cos^2\phi_i}{4}\frac{\overline{A^2}^2}{2} = \frac{\pi^2\Gamma^2\overline{A^2}^2}{8}\sum \cos^2\phi_i. \tag{C.18}$$

As can be calculated, the sum in (C.18) is 26/3. Einstein estimated its value at 10, which leads to

$$(\overline{(\Delta E)^2})^{1/2} = (\frac{10}{8})^{1/2}\pi\Gamma\overline{A^2} \simeq 3.5\Gamma\overline{A^2} \quad \text{Einstein estimation.} \tag{C.19}$$

The size of the energy fluctuations can be compared to the total energy associated with each atom. The potential energy associated with an atom, as seen in the last line of (C.5), is†

$$\varphi = \frac{\Gamma x^2}{2} \sum_i \cos^2 \phi_i \simeq 5\Gamma x^2. \tag{C.20}$$

The average potential energy $\overline{\varphi}$ is then

$$\overline{\varphi} = 5\Gamma \overline{x^2}, \tag{C.21}$$

where, from (C.7),

$$x^2 = A^2 \sin^2(\omega_{\rm E} t), \tag{C.22}$$

so that

$$\overline{x^2} = \overline{A^2}\, \overline{\sin^2(\omega_{\rm E} t)} = \overline{A^2}\frac{1}{2}, \tag{C.23}$$

and

$$\overline{\varphi} = \frac{5\Gamma \overline{A^2}}{2} \quad \text{average potential energy.} \tag{C.24}$$

For a harmonic system in which the average kinetic and potential energies are equal (Section 2.5.3), the average total energy is then given by

$$\overline{E} = 5\Gamma \overline{A^2}. \tag{C.25}$$

This energy is of the same order as the energy fluctuations over a time equal to half of the atomic period of vibration, $\pi/\omega_{\rm E}$, seen in (C.19), suggesting that this is the appropriate time scale to use as $\tau_{p,{\rm E}}$ in the formulation of the thermal conductivity. This specification leads to

$$\frac{1}{\tau_{p,{\rm E}}} = \frac{\omega_{\rm E}}{\pi} = \frac{T_{\rm E} k_{\rm B}}{\pi \hbar}, \tag{C.26}$$

so that

$$k_{p,{\rm E}} = \frac{n^{-1/3} k_{\rm B}^2}{\pi \hbar} T_{\rm E} \frac{x_E^2 e^{x_{\rm E}}}{(e^{x_{\rm E}} - 1)^2} \quad \text{Einstein thermal conductivity,} \tag{C.27}$$

where $T_{\rm E}$ is the Einstein temperature, which can be found by fitting the Einstein specific heat model to experimental data.

Because of the approximate nature of the $\tau_{\rm E}$ specification, the geometry initially chosen for the crystal, the assumed form of the atomic displacements (which will not be sinusoidal), the treatment of the potential energy, and the approximation of the summation in (C.18) are of no great consequence.

† As mentioned, this is twice the potential energy associated with a single atom.

C.2 Cahill–Pohl Thermal Conductivity

The Cahill–Pohl (C–P) thermal conductivity model is a direct extension of the Einstein model. The derivation is that given in [232]. The idea is to include a range of frequencies, instead of the single frequency used by Einstein. The thermal conductivity is written as

$$k_p = \sum_i \int \frac{1}{3} D_{p,\alpha}(\omega) c_{v,p,\alpha} u_{p,\alpha} \lambda_{p,\alpha} \mathrm{d}\omega, \tag{C.28}$$

where the summation is over the vibrational polarizations (one longitudinal and two transverse), $D_{p,\alpha}(\omega)$ is the volumetric density of vibrational states of polarization α, $c_{v,p}$ is now the specific heat per degree of freedom (and not per particle, as used previously), and $u_{p,\alpha}$ is the low-frequency (long-wavelength) speed of sound for polarization α in the material.

Instead of assuming energy transfer to occur between only nearest-neighbor atoms, one assumes that energy is transferred between vibrating entities with frequency ω and size $\lambda/2$, both of which are variable. Assuming no dispersion of the vibrational waves,

$$\omega = u_p \kappa = u_p \frac{2\pi}{\lambda}, \tag{C.29}$$

where κ is the wave number and λ is the wavelength (taken as twice the distance over which the energy transfer occurs). Thus, for the mean free path in (C.28),

$$\lambda_{p,\alpha} = \frac{\lambda_\alpha}{2} = \frac{\pi u_{p,\alpha}}{\omega}. \tag{C.30}$$

With no dispersion, the Debye density of states for each polarization (4.26)

$$D_{p,\alpha}(\omega) = \frac{\omega^2}{2\pi^2 u_{p,\alpha}^3}, \tag{C.31}$$

the specific heat per mode is

$$c_{v,p} = k_{\mathrm{B}} \frac{x^2 e^x}{(e^x - 1)^2}, \tag{C.32}$$

and the maximum frequency for polarization α is found from (4.28), i.e.,

$$\omega_{\alpha,max} = (6\pi^2 n u_{p,\alpha}^3)^{1/3}, \tag{C.33}$$

which is taken as the upper limit of the integration in (C.28). These specifications lead to

$$\begin{aligned} k_{p,\text{C-P}} &= \sum_\alpha \frac{1}{3} \int_0^{\omega_{\alpha,max}} \frac{\omega^2}{2\pi^2 u_{p,\alpha}^3} k_{\mathrm{B}} \frac{x^2 e^x}{(e^x - 1)^2} u_{p,\alpha} \frac{\pi u_{p,\alpha}}{\omega} \mathrm{d}\omega \\ &= \sum_\alpha \frac{k_{\mathrm{B}}}{6\pi u_{p,\alpha}} \int_0^{x_{\alpha,max}} x \frac{k_{\mathrm{B}} T}{\hbar} \frac{x^2 e^x}{(e^x - 1)^2} \frac{k_{\mathrm{B}} T}{\hbar} \mathrm{d}x \\ &= \frac{k_{\mathrm{B}}^3 T^2}{6\pi\hbar^2} \sum_\alpha \frac{1}{u_{p,\alpha}} \int_0^{x_{\alpha,max}} \frac{x^3 e^x}{(e^x - 1)^2} \mathrm{d}x. \end{aligned} \tag{C.34}$$

A temperature $T_{D,\alpha}$ is found from (4.27)

$$T_{D,\alpha} = u_{p,i}\frac{\hbar}{k_B}(6\pi^2 n)^{1/3}, \tag{C.35}$$

(i.e., $T_{D,\alpha} = x_{i,max}T$, which is not the Debye temperature, as the vibrational polarizations have been differentiated), allowing the thermal conductivity to be written as

$$k_{p,\text{C-P}} = (\frac{\pi}{6})^{1/3} k_B n^{2/3} \sum_\alpha u_{p,\alpha}(\frac{T}{T_{D,\alpha}})^2 \int_0^{T_{D,\alpha}/T} \frac{x^3 e^x}{(e^x - 1)^2}\text{d}x$$

C–P thermal conductivity. (C.36)

Unlike the Einstein model, no fitting parameters are required. Both the number density and velocities (the speeds of sound) can be determined experimentally. It is important to note that the value of $k_{p,\text{CP}}$ depends only on the number density of atoms through the upper limit of the integrals. Also, the Einstein assumption of the time for energy transfer between neighboring entities being half the period of oscillation (i.e., $\tau_{p,\text{E}} = \pi/\omega_\text{E}$) is not necessary, but follows from the assumption of no dispersion and a mean free path equal to one-half of the wavelength, i.e.,

$$\tau_{p,\text{E}} = \frac{\lambda_p}{u_{p,A}} = \frac{\lambda/2}{\omega\lambda/2\pi} = \frac{\pi}{\omega_\text{E}}. \tag{C.37}$$

The key point that is taken from the Einstein model is the idea of energy transfer occurring between only neighboring vibrational entities. No long-range coherence is assumed to exist. It is for this reason that the C-P model gives reasonable agreement with the experimental thermal conductivities of amorphous materials, in which short-length-scale interactions are expected to dominate the thermal transport. From a phonon perspective, the C-P thermal conductivity can be interpreted as a model in which all phonons have a mean free path equal to one-half of their wavelength.

APPENDIX D

Derivation of Phonon Boundary Resistance

At a solid–solid (two semi-infinite media) interface (plane-parallel geometry), the net conduction heat flux q_k that is due to the difference of temperature between T_1 and T_2, assumed uniform in each (semi-infinite medium), for phonons on each side of the interface is [268, 322]

$$q_k = \frac{T_1 - T_2}{AR_{p,b}}, \tag{D.1}$$

where $R_{p,b}$ is the phonon boundary resistance. This resistance is determined by the number of phonons incident upon the interface, the energy carried by each phonon, and the probability of transmission across the interface.

The transmission probability τ_b depends on the side from which phonons arrive at the interface, angle of incidence, phonon frequency, phonon polarization, and T_1 and T_2. Using an analogy with blackbody radiation (7.8), we have

$$q_k = \sigma_{p,2} T_2^4 \tau_{b,2\text{-}1} - \sigma_{p,1} T_1^4 \tau_{b,1\text{-}2}, \tag{D.2}$$

where $\sigma_{p,i}$ is the phonon Stefan–Boltzmann constant, given by (Section 7.4)

$$\sigma_{p,i} = \frac{\rho_i c_{v_{p,i}} u_{p,i}}{16T^3}. \tag{D.3}$$

For $T_1 = T_2$, then $q_k = 0$, and

$$\sigma_{p,2}\tau_{b,2\text{-}1} = \sigma_{p,1}\tau_{b,1\text{-}2}, \tag{D.4}$$

i.e., under thermal equilibrium, the total number of phonons (sum over all phonon states, polarization α and frequency ω_p) leaving one side, is equal to the total number of phonons returning from the other side into that state (called the principle of detailed balance). From this, we define $q_{k,1\text{-}2}(T_1)$ and $q_{k,1\text{-}2}(T_2)$ for the incident phonons at temperatures T_1 and T_2, and write

$$q_k = q_{k,1\text{-}2}(T_1) - q_{k,1\text{-}2}(T_2). \tag{D.5}$$

Using the equilibrium distribution of thermal phonons, we have

$$q_{k,1\text{-}2}(T) = \sum_{\alpha} \int_0^{\infty} \int_0^{2\pi} \int_0^{\pi/2} \tau_{b,1\text{-}2}(\theta, \alpha, \omega_p) \times$$

$$\frac{D_{p,1}(\omega_p) f^{\rm o}_{p,1}(\omega_p, T_1)}{4\pi} u_{p,1,\alpha} \cos\theta \sin\theta {\rm d}\theta {\rm d}\phi \hbar\omega_p {\rm d}\omega_p. \qquad \text{(D.6)}$$

The second part inside the integral accounts for the number of phonons with mode α and frequency ω_p, at angles of incidence (θ, φ), speed $u_{p,1,\alpha}$, and energy $\hbar\omega_p$ that are incident from side 1, on the interface, per unit time. When using the Debye DOS, a cut-off frequency is used.

The transmission probability τ_b is given as a function of θ (angle between the phonon propagation direction and the normal to the interface), α, and ω_p, under the assumption that both sides of the interface are isotropic and there is no dependence on temperature. However, when all phonons are diffusely scattered at the interface, τ_b will be a function only of ω_p, and will be determined by the DOS and the principle of detailed balance subsequently given.

D.1 Diffuse Scattering

D.1.1 Transmission Probability for Diffuse Scattering

For diffuse scattering the RMS interface roughness $\langle \delta^2 \rangle^{1/2}$ is expected to be larger than the dominant phonon wavelength λ, which is estimated as $\lambda \simeq (T_{\rm D}/T)^a$, where a is the lattice spacing [268]. We examine (D.6) noting that the number of phonons of frequency ω_p leaving material 1, per unit area and per unit time, is

$$\sum_{\alpha} \int_0^{2\pi} \int_0^{\pi/2} \tau_{b,1\text{-}2}(\omega_p) \frac{D_{p,1}(\omega_p) f^{\rm o}_{p,1}(\omega_p, T_1)}{4\pi} u_{p,1,\alpha} \cos\theta \sin\theta {\rm d}\theta {\rm d}\phi. \qquad \text{(D.7)}$$

Because the transmission probability is independent of the incident angles, the angular integral result then is

$$\frac{1}{4} \sum_{\alpha} \tau_{b,1\text{-}2}(\omega_p) D_{p,1}(\omega_p) f^{\rm o}_{p,1}(\omega_p, T_1) u_{p,1,\alpha}. \qquad \text{(D.8)}$$

From the principle of detailed balance, in which the interface is in thermal equilibrium ($T_1 = T_2 = T$), we can write that

$$\frac{1}{4} \sum_{\alpha} \tau_{b,1\text{-}2}(\omega_p) D_{p,1}(\omega_p) f^{\rm o}_{p,1}(\omega_p, T_1) u_{p,1,\alpha}$$

$$= \frac{1}{4} \sum_{\alpha} \tau_{b,2\text{-}1}(\omega_p) D_{p,2}(\omega_p) f^{\rm o}_{p,2}(\omega_p, T_2) u_{p,2,\alpha}, \qquad \text{(D.9)}$$

i.e., the number of phonons leaving side 1 must equal the number of phonons leaving side 2. In addition, because in diffuse scattering a phonon loses the memory of where it comes from, it follows that

$$\tau_{b,1\text{-}2}(\omega_p) = \rho_{b,2\text{-}1}(\omega_p) = 1 - \tau_{b,2\text{-}1}(\omega_p), \quad \text{(D.10)}$$

where $\rho_{b,2\text{-}1}$ is the reflectivity. This equation states that the probability of reflection from one side must equal the probability of transmission from the other. Therefore, using (D.9), we have

$$\tau_{b,1\text{-}2}(\omega_p) \sum_\alpha D_{p,1}(\omega_p) f^{\text{o}}_{p,1}(\omega_p, T) u_{p,1,\alpha}$$

$$= [1 - \tau_{b,1\text{-}2}(\omega_p)] \sum_\alpha D_{p,2}(\omega_p) f^{\text{o}}_{p,2}(\omega_p, T) u_{p,2,\alpha}. \quad \text{(D.11)}$$

Solving for $\tau_{b,1\text{-}2}(\omega_p)$, we have

$$\tau_{b,1\text{-}2}(\omega_p) =$$

$$\frac{\sum_\alpha D_{p,2}(\omega_p) f^{\text{o}}_{p,2}(\omega_p, T) u_{p,2,\alpha}}{\sum_\alpha D_{p,1}(\omega_p) f^{\text{o}}_{p,1}(\omega_p, T) u_{p,1,\alpha} + \sum_\alpha D_{p,2}(\omega_p) f^{\text{o}}_{p,2}(\omega_p, T) u_{p,2,\alpha}}. \quad \text{(D.12)}$$

Using the Debye phonon DOS (4.26) and the phonon equilibrium distribution function (Table 1.2), (D.12) becomes

$$\tau_{b,1\text{-}2}(\omega_p) = 1 - \tau_{b,2\text{-}1} = \frac{\sum_\alpha u^{-2}_{p,2,\alpha}}{\sum_\alpha u^{-2}_{p,1,\alpha} + \sum_\alpha u^{-2}_{p,2,\alpha}} \quad \text{Debye transmissivity}, \quad \text{(D.13)}$$

i.e, the transmissivity is independent of frequency.

D.1.2 Net Heat Flow for Diffuse Scattering

For diffuse scattering of phonons, (D.1) and (D.5) become

$$\frac{1}{AR_{p,b}} = \frac{G_{p,b}}{A} = \frac{q_{k,1\text{-}2}}{T_1 - T_2} = \frac{\hbar \tau_{b,1\text{-}2}}{4(T_1 - T_2)} \sum_\alpha u_{p,1,\alpha} \times$$

$$\int_0^\infty D_{p,1}(\omega_p) [\frac{1}{\exp(\frac{\hbar\omega_p}{k_{\text{B}} T_1}) - 1} - \frac{1}{\exp(\frac{\hbar\omega_p}{k_{\text{B}} T_2}) - 1}] \omega_p \text{d}\omega_p, \quad \text{(D.14)}$$

where $G_{p,b}$ is the phonon boundary conductance.

Replacing $D_p(\omega_p)$ with the measured phonon DOS, this equation is numerically integrated to yield a value for q, which is substituted into (D.1) to calculate the phonon boundary resistance $R_{p,b}$.

Using the Debye density of states $D_{p,\text{D}}$, i.e.,

$$D_{p,1}(\omega_p) = D_{p,\text{D},1}(\omega_p) = \frac{\omega_p^2}{2\pi^2 u^3_{p,1,\alpha}}, \quad \text{(D.15)}$$

(D.14) becomes

$$\frac{1}{AR_{p,b}} = \frac{\hbar\tau_{b,1\text{-}2}}{8\pi^2(T_1 - T_2)} \sum_\alpha u_{p,1,\alpha}^{-2} \times \int_0^{\omega_D} [\frac{\omega_p^3}{\exp(\frac{\hbar\omega_p}{k_B T_1}) - 1} - \frac{\omega_p^3}{\exp(\frac{\hbar\omega_p}{k_B T_2}) - 1}] d\omega, \quad \text{(D.16)}$$

where $\omega_D = \omega_{D,1}$ is the Debye cut-off frequency of side 1.

Using the variable $z = \omega_p/\omega_D$ for normalization, the upper limit of the integral becomes equal to 1, and we have

$$\frac{1}{AR_{p,b}} = \frac{\hbar\tau_{b,1\text{-}2}\omega_D^4}{8\pi^2(T_1 - T_2)} \sum_\alpha u_{p,1,\alpha}^{-2} \times \int_0^1 [\frac{z^3}{\exp(zT_D/T_1) - 1} - \frac{z^3}{\exp(zT_D/T_2) - 1}] dz,$$

$$z = \frac{\omega_p}{\omega_D}, \quad \frac{T_D}{T} = \frac{\hbar\omega_D}{k_B T}. \quad \text{(D.17)}$$

Then for $T_1 \to T_2 \to T$, we can use a variable $x = zT_D/T_1 = zT_D/T_2$ and $T_D = T_D$, and (D.17) becomes

$$\frac{1}{AR_{p,b}} = \frac{\tau_{b,1\text{-}2} k_B^4 T_D^3}{8\pi^2\hbar^3} \sum_\alpha u_{p,1,\alpha}^{-2}$$
$$\times \frac{1}{T_D^3(T_1 - T_2)} [T_1^4 \int_0^{T_D/T_1} \frac{x^3}{\exp(x) - 1} dx - T_2^4 \int_0^{T_D/T_2} \frac{x^3}{\exp(x) - 1} dx]. \quad \text{(D.18)}$$

A dimensionless phonon boundary conductance can be given as

$$\frac{1}{(AR_{p,b})^*} \equiv \frac{1}{AR_{p,b}} \frac{8\pi^2\hbar^3}{k_B^4 T_D^3 \tau_{b,1\text{-}2} \sum_\alpha u_{p,1,\alpha}^2} = \frac{1}{T_D^3(T_1 - T_2)} \times$$
$$\left[T_1^4 \int_0^{T_D/T_1} \frac{x^3}{\exp(x) - 1} dx - T_2^4 \int_0^{T_D/T_2} \frac{x^3}{\exp(x) - 1} dx\right], \quad \text{(D.19)}$$

and its variation is shown in Figure D.1. The dimensionless phonon boundary conductance initially increases with a temperature, and reaches a plateau at $T/T_D = 1$. The maximum value is $1/(AR_{p,b})^* = 1/3$.

D.2 Specular Scattering

D.2.1 Transmission Probability for Specular Scattering

In specular scattering, phonons are treated as plane waves and the materials that form the interface are treated as continua. This applies to phonons with wavelengths larger than the interfacial roughness.

Because the longitudinal phonons are the quantized LA waves, the transmission probability will be given by the equivalent expression for acoustic waves [213]:

$$\tau_{b,1\text{-}2}(\theta_1) = \tau_{b,2-1}(\theta_2) = 4\frac{\frac{\rho_2 u_{p,2}}{\rho_1 u_{p,1}} \frac{\cos\theta_2}{\cos\theta_1}}{(\frac{\rho_2 u_{p,2}}{\rho_1 u_{p,1}} + \frac{\cos\theta_2}{\cos\theta_1})^2}, \quad \text{(D.20)}$$

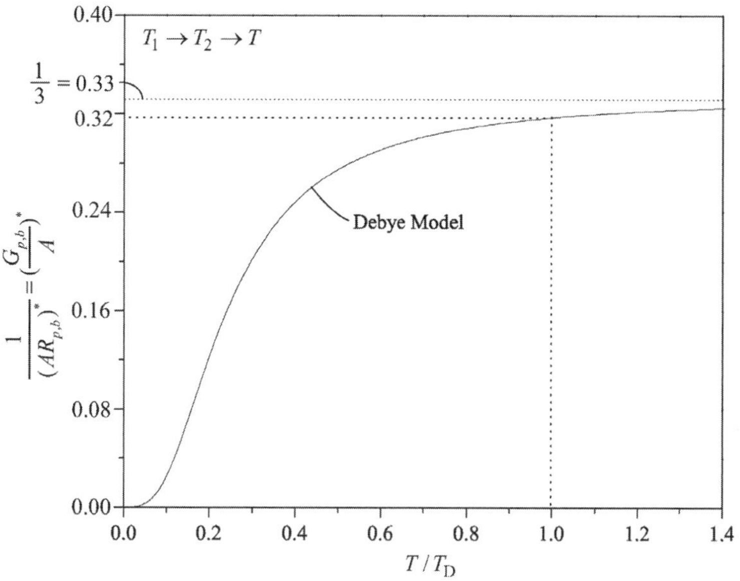

Figure D.1. Variations of dimensionless phonon boundary conductance with respect to dimensionless temperature. The high-temperature limit of 1/3 is also shown.

where θ_1 is the incident angle in side 1 and θ_2 is the refraction angle in side 2.

D.2.2 Net Heat Flow for Specular Scattering

For specular scattering of phonons, (D.1), (D.5), and (D.6) become

$$\frac{1}{AR_{p,b}} = \frac{\hbar}{4\pi^2(T_1 - T_2)} \sum_\alpha u_{p,1,\alpha}^{-2} \Gamma_{1,\alpha} \int_0^{\omega_D} [\frac{\omega_p^3}{\exp(\frac{\hbar\omega_p}{k_B T_1}) - 1} - \frac{\omega_p^3}{\exp(\frac{\hbar\omega_p}{k_B T_2}) - 1}] d\omega_p, \tag{D.21}$$

where $\omega_D = \omega_{D,1}$, and

$$\Gamma_{1,\alpha} = \int_0^{\pi/2} \tau_{b,1\text{-}2}(\theta, \alpha) \cos\theta \sin\theta d\theta, \tag{D.22}$$

is calculated numerically. Note that $\tau_{b,1\text{-}2}(\theta, \alpha)$ is assumed to be independent of ω_p for phonons with frequency less than $\omega_{D,1}$, and above that $\tau_{b,1\text{-}2}(\theta, \alpha)$ is zero.

Equation (D.21) can also be normalized with the variable $x = \hbar\omega_p/(k_B T)$, as performed in Section D.1.2. The upper limit of the integral then becomes $x = \hbar\omega_D/(k_B T)$, and for low T, $x \rightarrow \infty$. Therefore, for $[\hbar\omega_p/(k_B T)] \gg 1$, we have

$$\frac{1}{AR_{p,b}} = \frac{k_B^4}{4\pi^2\hbar^3(T_1 - T_2)} \sum_\alpha u_{p,1,\alpha}^{-2} \Gamma_{1,\alpha} \int_0^\infty [\frac{T_1^4 x^3}{e^x} - \frac{T_2^4 x^3}{e^x}] dx,$$

$$x = \frac{\hbar\omega_D}{k_B T} \text{ for low } T. \tag{D.23}$$

Note that, for normal incidence $\theta_1 = \theta_2 = 0^\circ$, (D.20) becomes

$$\tau_{b,1\text{-}2} = 4\frac{\rho_2 u_{p,2}\ \rho_1 u_{p,1}}{\rho_2 u_{p,2} + \rho_1 u_{p,1}}, \qquad \text{(D.24)}$$

and we have $\Gamma_{1,\alpha} = \tau_{b,1\text{-}2}/2$, i.e., (D.23) will differ from (D.17) only by the transmission probability.

APPENDIX E

Derivation of Fermi Golden Rule

The Fermi (or Fermi–Dirac) golden rule (3.27) allows for calculation of the transition probability rate between two eigenstates of a quantum system using the time-dependent perturbation theory. It can be derived from the time-dependent perturbation theory (the perturbation Hamiltonian, i.e., the scattering potential, is time dependent), under the assumption that the time of the measurement is much larger than the time needed for the transition.

It is the rate of gain of probability per unit time in the manifold of final eigenstate $|\psi_{\kappa'}\rangle$, which is equal to the rate of loss of probability per unit time from the initial eigenstate $|\psi_\kappa\rangle$.

A brief derivation of the FGR is given and more details can be found in [74, 219, 241]

E.1 Time-Dependent Perturbation

The general Hamiltonian of interest is of the form (5.88)

$$\mathrm{H} = \mathrm{H}_o + \mathrm{H}', \tag{E.1}$$

where H_o is a time independent Hamiltonian with a known solution ψ_κ, which is related to Ψ_κ through (2.68), i.e.,

$$\mathrm{H}_o \psi_\kappa = E(\kappa)\psi_\kappa \tag{E.2}$$

$$\Psi_\kappa^o = \psi_\kappa \exp[\frac{-i\,E(\kappa)t}{\hbar}], \tag{E.3}$$

and $E(\kappa)$ and ψ_κ are time-independent. Here H' causes time-dependent transitions between the states ψ_κ. The time-dependent Schrödinger equation is

$$i\hbar\frac{\partial \Psi}{\partial t} = \mathrm{H}\Psi, \tag{E.4}$$

using time-dependent coefficients $a_\kappa(t)$.

The approximation will involve expressing Ψ as an expansion of the time-independent eigenfunctions $\psi_\kappa \exp[-i\,E_e(\kappa)t/\hbar]$ of the unperturbed, time-independent system, i.e., (2.68)

$$\Psi = \sum_\kappa a_\kappa(t)\Psi_\kappa^{\circ} = \sum_\kappa a_\kappa(t)\psi_\kappa \exp[\frac{-iE(\kappa)t}{\hbar}]. \tag{E.5}$$

The time-dependent problem is solved when the coefficients $a_\kappa(t)$ are known, and for this we will develop time-rate equation, i.e., $\partial a_\kappa(t)/\partial t$.

Substituting (E.5) into (E.4), also using (E.2), we have,

$$\sum_\kappa i\hbar\frac{\partial a_\kappa(t)}{\partial t}\psi_\kappa \exp[\frac{-iE_e(\kappa)t}{\hbar}] + \sum_\kappa a_\kappa(t)E(\kappa)\psi_\kappa \exp[\frac{-iE(\kappa)t}{\hbar}]$$
$$= \sum_\kappa a_\kappa(t)(\mathrm{H_o} + \mathrm{H}')\psi_\kappa \exp[\frac{-iE(\kappa)t}{\hbar}]. \tag{E.6}$$

Now multiplying by $\psi_{\kappa'}^*$ and integrating over space, we have (using orthogonality of the eigenfunctions)

$$i\hbar\frac{\partial a_{\kappa'}(t)}{\partial t}\exp[\frac{-iE(\kappa')t}{\hbar}] = \sum_\kappa a_\kappa(t)\exp[\frac{-iE(\kappa)t}{\hbar}]\langle\psi_{\kappa'}|\mathrm{H}'|\psi_{\kappa'}\rangle. \tag{E.7}$$

The interaction matrix element is defined as

$$M_{\kappa',\kappa} = \langle\psi_{\kappa'}^*|\mathrm{H}'|\psi_\kappa\rangle, \quad \int \psi_{\kappa'}\mathrm{H}'\psi_\kappa dV \quad \text{interaction matrix element.} \tag{E.8}$$

Using this, (E.7) is rewritten as

$$\frac{\partial a_{\kappa'}(t)}{\partial t} = \frac{1}{i\hbar}\sum_\kappa M_{\kappa',\kappa}a_\kappa(t)\exp\{\frac{i[E(\kappa') - E(\kappa)]t}{\hbar}\}, \tag{E.9}$$

which is the equation for $a_\kappa'(t)$.

Next we apply the regular perturbation to find an approximation for $a_\kappa(t)$ appearing on the right-hand side of (E.9). Similar to the perturbation approximation used in Section 2.6.7, we have

$$\mathrm{H}' \to \epsilon\mathrm{H}'$$
$$a_{\kappa'} = a_{o,\kappa'} + \epsilon a_{1,\kappa'} + \epsilon^2 a_{2,\kappa'}. \tag{E.10}$$

Substituting these into (E.9), and collecting the terms with coefficients of the same power ϵ, we have

$$\frac{\partial a_{o,\kappa'}}{\partial t} = 0 \tag{E.11}$$
$$\frac{\partial a_{j+1,\kappa'}}{\partial t} = \frac{1}{i\hbar}\sum_\kappa M_{\kappa',\kappa}a_{j,\kappa}\exp\{\frac{i[E(\kappa') - E(\kappa)]t}{\hbar}\}.$$

To study the time evolution of the problem, we assume that the perturbation H′ is absent at time $t \le 0$ and starts at $t = 0$. With this assumption, the system is in a time-independent state up to $t = 0$. We assume that the system is in a single, well-defined state ψ_κ with

$$a_{o,\kappa}(t=0) = 1 \tag{E.12}$$

$$a_{o,\kappa'}(t=0) = \langle\kappa'|\kappa\rangle = 0 \text{ if } \kappa' \neq \kappa,$$

$$\text{single-state initial conditions.} \tag{E.13}$$

Integration of (E.9) to the first order term gives

$$a_{1,\kappa'}(t) = \frac{1}{i\hbar}\int_0^t M_{\kappa',\kappa}\exp\{\frac{i[E(\kappa') - E(\kappa)]t}{\hbar}\}\mathrm{d}t, \tag{E.14}$$

which satisfies initial conditions (E.12) and (E.13).

E.2 Transition Rate

A number of important problems in quantum mechanics involve a perturbation that has time dependence with a harmonic form. Examples include interactions of electrons with electromagnetic radiation (photons) and electrons in crystals interacting with phonons. In these cases, the time dependence of the perturbation potential is

$$\mathrm{H}' = \mathrm{H}'_x \exp(\mp i\omega t) \text{ space time-periodic behavior,} \tag{E.15}$$

where H'_x is spatial part (time independent) of the perturbation Hamiltonian. This makes the matrix element (E.8)

$$M_{\kappa',\kappa} = M_{o,\kappa',\kappa}\exp(\mp i\omega t), \quad M_{o,\kappa',\kappa} = \langle\psi_{\kappa'}|\mathrm{H}'_x|\psi_\kappa\rangle, \quad \text{time-periodic behavior.} \tag{E.16}$$

We now integrate (E.14), taking the constant of integration to be zero, as $a_{1,\kappa'} = 0$ for $t \rightarrow -\infty$ (no perturbation):

$$a_{1,\kappa'}(t) = -\frac{1}{i\hbar}M_{o,\kappa',\kappa}\frac{\exp\{\frac{i[E(\kappa')-E(\kappa)\mp\hbar\omega]t}{\hbar}\} - 1}{\frac{i[E(\kappa')-E(\kappa)\mp\hbar\omega]}{\hbar}}. \tag{E.17}$$

Equation (E.17) states that as $E(\kappa') - E(\kappa) \simeq \hbar\omega$ or $E(\kappa') - E(\kappa) \simeq -\hbar\omega$ (because of absorption or emission), the term to the right of $M_{o,\kappa',\kappa}$ will be very large. Because the probability of finding a state is related to $\psi^2_{\kappa'}$, then this probability is highest for the energies of transition (absorption/emission).

For simplicity, we define

$$\omega' = \frac{[E(\kappa') - E(\kappa) \mp \hbar\omega]}{\hbar}, \tag{E.18}$$

and then (E.17) becomes

$$a_{1,\kappa'}(t) = -\frac{1}{i\hbar}M_{o\kappa',\kappa}\exp(\frac{i\omega' t}{2})\frac{\sin(\omega' t/2)}{\omega' t/2}t. \tag{E.19}$$

Figure E.1 shows that the variation of the probability of transition $\kappa' \rightarrow \kappa'$, with absorption of $\hbar\omega$, due to the applied perturbation potential from $t = 0$ to $t = t_o$

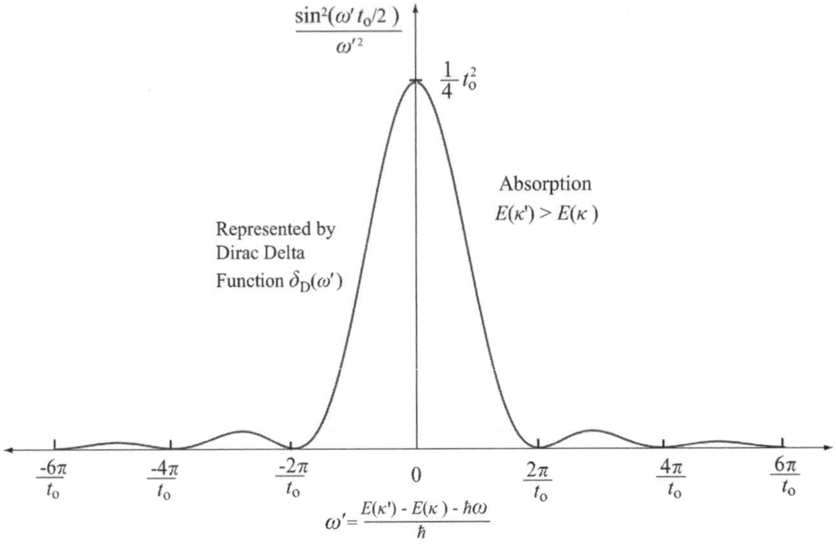

Figure E.1. Variations of the probability of finding the system in a state κ' after the perturbation has been applied at time t_o.

(Chapter 3 problem). The probability is $[\sin(\omega' t_o/2)/\omega']^2$. The probability is highest when $E(\boldsymbol{\kappa}') - E(\boldsymbol{\kappa}) \simeq \hbar\omega$ and the peak is proportional to t_o^2.

The transition rate is given by the transition probability per unit time, (5.91) and (5.93), i.e.,

$$\dot{\gamma}(\boldsymbol{\kappa}', \boldsymbol{\kappa}) = \lim_{t \to \infty} \frac{|a_{1,\boldsymbol{\kappa}'}(t)|^2}{t} = \lim_{t \to \infty} \frac{1}{t\hbar^2} |M_{o,\boldsymbol{\kappa}',\boldsymbol{\kappa}}|^2 [\frac{\sin(\omega' t/2)}{\omega' t/2}]^2 t^2. \tag{E.20}$$

Next we note that the integral over the infinite time domain is simplified as

$$\int_{-\infty}^{+\infty} \frac{\sin^2(\omega' t/2)}{(\omega' t/2)^2} \mathrm{d}x = \pi. \tag{E.21}$$

Now using this and the definition of the Dirac delta function δ_{D} (Glossary), we find the following equality:

$$\lim_{t \to \infty} [\frac{\sin(\omega' t/2)}{\omega' t/2}]^2 = \pi \delta_{\mathrm{D}}(\frac{\omega' t}{2}) = \frac{2\pi}{t} \delta_{\mathrm{D}}(\omega'). \tag{E.22}$$

Note that here we have used the relation

$$\delta_{\mathrm{D}}(ax) = \frac{1}{a} \delta_{\mathrm{D}}(x) \qquad a \text{ is a constant.} \tag{E.23}$$

Hence, transition rate (E.20) becomes

$$\dot{\gamma}(\boldsymbol{\kappa}', \boldsymbol{\kappa}) = \frac{1}{t\hbar^2} |M_{o,\boldsymbol{\kappa}',\boldsymbol{\kappa}}|^2 \frac{2\pi}{t} \delta_{\mathrm{D}}\{\frac{[E(\boldsymbol{\kappa}') - E(\boldsymbol{\kappa}) \mp \hbar\omega]}{\hbar}\} t^2$$

$$= \frac{2\pi}{\hbar} |M_{o\boldsymbol{\kappa}',\boldsymbol{\kappa}}|^2 \delta_{\mathrm{D}}[E(\boldsymbol{\kappa}') - E(\boldsymbol{\kappa}) \mp \hbar\omega] \text{ Fermi golden rule.} \tag{E.24}$$

This is the FGR (3.27).

E.3 Example: Photoelectric Transition Rate for Hydrogen

The hydrogen atom, in ground state when irradiated with photon of energy matching the ionization energy $\hbar\omega$ and amplitude (we will use the electric potential $\boldsymbol{a}_{e,\mathrm{o}}$, Table 3.5), ejects an electron (photoelectric effect) [110]. The transition probability rate is given by (E.24), i.e.,

$$\dot{\gamma}_{ph-e} = \frac{2\pi}{\hbar}|\langle f|\mathrm{H}'|i\rangle|^2\delta_{\mathrm{D}}(E_{e,f} - E_{e,i} - \hbar\omega), \tag{E.25}$$

where f is the final state, i is the initial state, and H' is the time-independent interaction Hamiltonian.

The photon is represented by an EM wave having vector potential $\boldsymbol{a}_e$ (Table 3.5, where $\boldsymbol{b}_e = \nabla \times \boldsymbol{a}_e$, $\nabla \cdot \boldsymbol{a}_e = 0$), i.e.,

$$\boldsymbol{a}_e(\boldsymbol{x}, t) = \boldsymbol{a}_{e,\mathrm{o}}\cos(\boldsymbol{\kappa} \cdot \boldsymbol{x} + \omega t). \tag{E.26}$$

Using this, the electron kinetic energy is (Chapter 5 problem)

$$\mathrm{H}'(t) = -\frac{1}{2m_e}[-e_c(\boldsymbol{p}_e \cdot \boldsymbol{a}_e + \boldsymbol{a}_e \cdot \boldsymbol{p}_e)] = \frac{e_c}{m_e}\boldsymbol{a}_e \cdot \boldsymbol{p}_e. \tag{E.27}$$

Then

$$\begin{aligned}\mathrm{H}'(t) &= \frac{e_e}{m_e}\cos(\boldsymbol{\kappa} \cdot \boldsymbol{x} - \omega t)\boldsymbol{a}_{e,\mathrm{o}} \cdot \boldsymbol{p}_e \\ &= \frac{e_c}{2m_e}[e^{i(\boldsymbol{\kappa}\cdot\boldsymbol{x}-\omega t)} + e^{-i(\boldsymbol{x}\cdot\boldsymbol{\kappa}-\omega t)}]\boldsymbol{a}_{e,\mathrm{o}} \cdot \boldsymbol{p}_e \\ &= \frac{e_c}{2m_e}e^{i\boldsymbol{\kappa}\cdot\boldsymbol{x}}\boldsymbol{a}_{e,\mathrm{o}} \cdot \boldsymbol{p}_e e^{-i\omega t} \\ &= \mathrm{H}'e^{-i\omega t},\end{aligned} \tag{E.28}$$

as the $\exp(i\omega t)$ corresponding to photon emission is neglected.

The ground-state energy is given by (2.114), i.e., $E_{e,1} = e_c^2/8\pi\epsilon_{\mathrm{o}}r_{\mathrm{B}}$, and its wave function (for hydrogen $z = 1$) is given in Table 2.10, i.e.,

$$|100\rangle = \frac{1}{\pi^{1/2}r_{\mathrm{B}}^{3/2}}e^{-x/r_{\mathrm{B}}}, \tag{E.29}$$

and we assume that the final state is $L^{-3/2}e^{i\boldsymbol{\kappa}_f\cdot\boldsymbol{x}}$. Then the matrix element is (using the Bloch plane wave and time-independent H')

$$\begin{aligned}\mathrm{M}_{\boldsymbol{\kappa}_f,\boldsymbol{\kappa}} &= \langle\boldsymbol{\kappa}_f|\frac{e_c}{2m_e u_{ph}}e^{i\boldsymbol{\kappa}\cdot\boldsymbol{x}}\boldsymbol{a}_{e,\mathrm{o}} \cdot \boldsymbol{p}_e|100\rangle \\ &= \frac{1}{L^{3/2}}\int e^{i\boldsymbol{\kappa}_f\cdot\boldsymbol{x}}\frac{e_c}{2m_e u_{ph}}e^{i\boldsymbol{\kappa}\cdot\boldsymbol{x}}\boldsymbol{a}_{e,\mathrm{o}} \cdot (-i\hbar\nabla)\frac{1}{\pi^{1/2}r_{\mathrm{B}}^3}e^{ix/r_{\mathrm{B}}}d\boldsymbol{x},\end{aligned} \tag{E.30}$$

where $L^3 = V$ is from the periodic, cubic box treatment of the electron (Section 2.6.6). Because $\lambda \gg r_{\mathrm{B}}$, we set $\exp(i\boldsymbol{\kappa} \cdot \boldsymbol{x}) = 1$.

The integral contains ∇, which is removed by integration by parts, i.e.,

$$\int e^{i\boldsymbol{\kappa}_f\cdot\boldsymbol{x}}\boldsymbol{a}_{e,\mathrm{o}}\cdot(i\hbar\nabla)e^{-x/r_\mathrm{B}}\mathrm{d}\boldsymbol{x}$$

$$= -\boldsymbol{a}_{e,\mathrm{o}}\cdot\boldsymbol{p}_f\int e^{i\boldsymbol{\kappa}_f\cdot\boldsymbol{r}}e^{-x/r_\mathrm{B}}\mathrm{d}\boldsymbol{x}. \tag{E.31}$$

Taking the z axis along $\boldsymbol{\kappa}_f$, the ϕ integration gives 2π and the θ integration (over $\sin\theta\mathrm{d}\theta$) gives

$$\int_0^\pi\int_0^{2\pi}\int_0^\infty e^{i\kappa_f\cdot s_z}e^{-z/r_\mathrm{B}}\mathrm{d}z\mathrm{d}\phi\sin\theta\mathrm{d}\theta = \frac{8\pi}{r_\mathrm{B}}\frac{1}{(r_\mathrm{B}^{-2}+\kappa_f^2)^2}. \tag{E.32}$$

Using this, we have

$$\dot{\gamma}_{ph\text{-}e} = \frac{2\pi}{\hbar}|\frac{1}{L^{3/2}}\frac{e_c}{2m_e u_{ph}}\frac{1}{\pi^{1/2}r_\mathrm{B}^{3/2}}\boldsymbol{a}_{e,\mathrm{o}}\cdot\boldsymbol{p}_e\frac{8\pi}{r_\mathrm{B}}\frac{1}{(r_\mathrm{B}^{-2}+\kappa_f^2)^2}|^2\delta_\mathrm{D}(E_{e,f}-E_{e,i}-\hbar\omega). \tag{E.33}$$

By taking the average over incidence angle, $\boldsymbol{a}_{e,\mathrm{o}}\cdot\boldsymbol{p}_f = a_{e,\mathrm{o}}p_f\cos\theta$, and noting that $\kappa_f = p_f/\hbar$, this total ionization (photoelectric transition) rate becomes

$$\dot{\gamma}_{ph\text{-}e} = \frac{4m_e p_f}{\pi r_\mathrm{B}\hbar^4}(\frac{e_c}{m_e u_{ph}})^2\frac{4\pi a_{e,\mathrm{o}}^2 p_f^2}{3}[\frac{1}{r_\mathrm{B}^{-2}+(\frac{p_f}{\hbar})^2}]^4$$

photoelectric transition rate for hydrogen atom, (E.34)

where we have used from (5.64) and (5.65) (not allowing for spin) that (Chapter 5 problem)

$$L^3\frac{D_e}{2} = \int\delta_\mathrm{D}(E_{e,f}-E_{e,i}-\hbar\omega) = (\frac{L}{2\pi})^3 4\pi\kappa_f\frac{m_e}{\hbar^2}, \tag{E.35}$$

and $\langle\cos^2\theta\rangle = \int\cos\theta\mathrm{d}\Omega = 1/3$, where $\mathrm{d}\Omega = \sin\theta\mathrm{d}\theta\mathrm{d}\phi$ (Table 7.1).

The cases of $\hbar\omega \gg E_{e,i}$ and $\hbar\omega - E_{e,i} \ll E_{e,i}$ have been examined in [125]. For the case $\hbar\omega - E_{e,i} \simeq E_{e,i}$ (near threshold, where the ejected electron has a very high velocity), we have for the total (integrated over the solid angle) absorption cross-section area

$$A_{ph\text{-}e} = \frac{512\pi^2}{3}\frac{r_\mathrm{B}^2 e_c^2}{4\pi\epsilon_\mathrm{o}\hbar c}(\frac{E_{e,1}}{\hbar\omega})^4\frac{e^{-4\gamma\cot^{-1}\gamma}}{1-e^{2\pi\gamma}},\quad \gamma = (\frac{\hbar\omega}{E_{e,1}}-1)^{-1/2}. \tag{E.36}$$

This shows a strong decay in $A_{ph\text{-}e}$ as $\hbar\omega/E_{e,1}$ increases. (Chapter 7 problem)

The transtion rate and the absorption (ionization) cross-section area are related through photon energy density (3.38), and is [110]

$$\frac{a_{e,\mathrm{o}}^2\omega A_{ph\text{-}e}}{8\pi c} = \hbar\omega\dot{\gamma}_{ph\text{-}e}. \tag{E.37}$$

Table E.1. *Examples of perturbation Hamiltonian.*

Interaction	Mechanism	Perturbation Hamiltonian
Electron–charge	piezoelectric scattering	$\mathrm{H}' = \frac{e_c e_{pz}}{\epsilon_o \epsilon_e} \boldsymbol{d} \cdot \boldsymbol{s}$
Electron–phonon	scattering by acoustic phonon	$\mathrm{H}' = \varphi_{d,\mathrm{A}} \nabla \cdot \boldsymbol{d}$
	scattering by optical phonon	$\mathrm{H}' = \varphi'_{d,\mathrm{O}}\, \boldsymbol{d} \cdot \boldsymbol{s}$
Photon–electron	absorption in gases	$\mathrm{H}' = -\lvert \boldsymbol{\mu}_e \rvert e_{e,\mathrm{o}} \cos(\omega t)$

$\boldsymbol{d}$	displacement vector, m
e_{pz}	piezoelectric constant, C/m^2
$\boldsymbol{s}$	unit vector
$\varphi_{d,\mathrm{A}}$	acoustic scattering potential, J
$\varphi'_{d,\mathrm{O}}$	optical scattering potential, J/m
$\boldsymbol{\mu}_e$	transition dipole moment, C-m

E.4 Examples of Perturbation Hamiltonian

Table E.1 gives examples of perturbation Hamiltonian H$'$ in the interaction matrix element (E.8). These are samples of scatterings/couplings discussed in energy transport and transformation kinetics. The phonon scatterings are discussed in Section 4.9.4. The electron scatterings are discussed in Section 5.15. The photon absorption, and its interaction matrix element, for solids and gases, are discussed in Section 7.8. Other photon couplings are discussed in Section 7.12. Examples of calculations of $\lvert \boldsymbol{\mu}_e \rvert$ and $\varphi'_{d,\mathrm{O}}$ are given in [180].

APPENDIX F

Derivation of Equilibrium, Particle Probability Distribution Functions

The Fermi–Dirac (fermion) and Bose–Einstein (boson) statistics include quantum effects and apply to interacting, indistinguishable particles (Table 1.2). The M–B statistics apply to noninteracting indistinguishable particles (classical particles), whose wave functions do not overlap and quantum effects vanish (footnote of Section 2.6.5). When the particle concentration is much less than the quantum limit, quantum effects will vanish and all the particles can be treated as classical particles. The ratio of the particle concentration and the quantum limit is called the quantum concentration (footnote of Section 6.1) n_q, which is defined by

$$n_q = \frac{N}{V}(\frac{2\pi\hbar^2}{mk_\mathrm{B}T})^{3/2}, \tag{F.1}$$

where N is the number of particles, V is the volume of the system, m is the mass of the particle, and T is the temperature. Therefore both fermions and bosons become the M–B statistics at high temperatures or low concentrations.

Derivation of the particle (including quantum) statistics distribution functions are given in [119, 239].

F.1 Partition Functions

For different ensembles [Section 2.5.1(A)], the partition function may have different forms. For a canonical ensemble, the partition function (2.27) is defined as

$$Z(N, V, T) = \sum_j e^{-E_j/k_\mathrm{B}T}, \tag{F.2}$$

where j designates the energy state of the system.

For a grand canonical ensemble (μVT), the partition function is defined as

$$Z(\mu, V, T) = \sum_{N=0}^{\infty} Z(N, \mu, V, T,) = \sum_{N=0}^{\infty}\sum_j e^{(E_j-\mu)/k_\mathrm{B}T}, \tag{F.3}$$

where N is the number of particles, and when it is a constant, then $\mu = 0$, and $Z(\mu, V, T)$ becomes $Z(N, V, T)$, relating (F.2) and (F.3).

The partition function (F.2) and (F.3) can be simplified when the Hamitonian (Table 2.5) is written as the sum of the various energy mechanisms and degrees of freedom, i.e.,

$$E = \mathrm{H} = \varphi + \mathrm{H}_e + \mathrm{H}_t + \mathrm{H}_v + \mathrm{H}_r + \cdots. \tag{F.4}$$

For an N-particle canonical ensemble, the energy of each particle can be expressed as E_j^k, where $k = a, b, c...$ designates the particle (here we assume that they are distinguishable) and j denotes the energy state of the system, the canonical partition function (F.2) becomes

$$\begin{aligned} Z(N, V, T) &= \sum_j e^{-E_j/k_\mathrm{B}T} = \sum_{m,n,l} e^{-(E_m^a+E_n^b+E_l^c+...)/k_\mathrm{B}T} \\ &= \sum_m e^{-E_m^a/k_\mathrm{B}T} \sum_n e^{-E_n^b/k_\mathrm{B}T} \sum_l e^{-E_l^c/k_\mathrm{B}T} ... \\ &= Z_a Z_b Z_c ..., \quad Z_a = \sum_m e^{-E_m^a/k_\mathrm{B}T}, \end{aligned} \tag{F.5}$$

where m, n, and l, are quantum numbers designating the energy state of each particle. (F.5) means the partition function of the system is the product of the partition functions of the subsystems. Then, for interacting, indistinguishable particles ($Z_a = Z_b = Z_c = ...$), we have

$$Z(N, V, T) = Z(V, T)^N \text{ interacting, indistinguishable particles.} \tag{F.6}$$

This reduces the N-body problem to a one-body problem.

For noninteracting, indistinguishable particles (no wave function overlaps), considering the degeneracy in the energy state of the system, we have [239]

$$Z(N, V, T) = \frac{Z(V, T)^N}{N!} \text{ noninteracting, indistinguishable particles.} \tag{F.7}$$

This also reduces the N-body problem to a one-body problem. For this case the number of available molecular states is much larger than the number of particles N. This is returned to the Boltzmann statistics and becomes increasingly valid at high temperatures.

Using relation (F.5) for the canonical ensemble, the partition function for a grand canonical ensemble is rewritten as

$$\begin{aligned} Z(\mu, V, T) &= \sum_{\{N_j\}} \prod_j e^{-N_j(E_j-\mu)/k_\mathrm{B}T} \\ &= \prod_j [\sum_{N_j} e^{-N_j(E_j-\mu)/k_\mathrm{B}T}] \\ &= \prod_j Z_j, \quad Z_j = \sum_{N_j} e^{-N_j(E_j-\mu)/k_\mathrm{B}T}. \end{aligned} \tag{F.8}$$

Here $\{N_j\}$ designates the configuration of the system, for example, the configuration $\{1, 2, ..\}$ means there is 1 particle in the first energy state, 2 particles in the second energy state, etc. (F.8) states that when an energy state rather than a particle is considered as a subsystem in a grand canonical ensemble, the partition function of the system may have the same relation with the partition functions of the subsystems as in a canonical ensemble [(F.5)].

The probability for a state j in the occupation representation is (2.20), i.e.,

$$P_j = \frac{e^{(E_j-\mu)/k_\mathrm{B}T}}{Z}. \tag{F.9}$$

This is valid for both the grand canonical ensemble and the canonical ensemble (for the canonical ensemble, $\mu = 0$). The mean equilibrium occupancy of particles i ($i = p, e, f, ph$) in the energy state j is therefore

$$f_{i,j}^{\circ} = \sum N_j P_j = -\frac{\partial \ln Z_j}{\partial \xi_j}, \quad \xi_j = (E_j - \mu)/k_B T. \tag{F.10}$$

The Fermi–Dirac (fermions), Bose–Einsten (bosons) and Maxwell–Boltzmann (M–B) distribution functions are directly derived in the subsequent sections using the statistics for a grand canonical ensemble, as there is no limit for the number of particles.

F.2 Fermi–Dirac Distribution Function f_e°

In the grand canonical ensemble (μVT) discussed in Sections 2.5.1 and F.1, the partition function (2.21) for a single-particle state j (of a multiparticle system) with energy E_j, for a quantum gas in a box, is

$$Z_j = \sum_{N_j} e^{-N_j(E_j-\mu)/k_\mathrm{B}T}, \tag{F.11}$$

where N_j is the number of particles in the jth state with energy E_j.

For fermions, a state is occupied by only a single particle or not occupied. This is called a multiplicity of two. Then (F.11) becomes

$$Z_j = \sum_{N_j=0}^{1} e^{-[E_j-\mu]N_j/k_\mathrm{B}T} = e^{-(E_j-\mu)/k_\mathrm{B}T} + 1. \tag{F.12}$$

Then the equilibrium occupation distribution (F.10) is

$$f_e^{\circ} = \frac{e^{-(E_j-\mu)/k_\mathrm{B}T}}{e^{-(E_j-\mu)/k_\mathrm{B}T}+1} = \frac{1}{e^{(E_j-\mu)/k_\mathrm{B}T}+1} \quad \text{fermion}. \tag{F.13}$$

F.3 Bose–Einstein Distribution Function f_p°, f_{ph}°

For bosons, there is no limit for the number of particles in a state j. Therefore we have

$$Z_j = \sum_{N_j=0}^{\infty} e^{-(E_j-\mu)N_j/k_\mathrm{B}T} = \frac{1}{1-e^{-(E_j-\mu)/k_\mathrm{B}T}}. \tag{F.14}$$

Thus the expected distribution (F.10) becomes

$$f_j^{\circ} = f_p^{\circ} = f_{ph}^{\circ} = \frac{e^{-(E_j-\mu)/k_{\rm B}T}}{1-e^{-(E_j-\mu)/k_{\rm B}T}}$$
$$= \frac{1}{e^{(E_j-\mu)/k_{\rm B}T}-1} \quad \text{boson.} \tag{F.15}$$

F.4 Maxwell–Boltzmann Distribution Function f_f°

For noninteracting, indistinguishable (this combination is also called distinguishable) particles (having degeneracy $N_j!$), using $\mu = 0$ and similar to (F.7), we have

$$Z_j = \sum_{N_j=0}^{\infty} \frac{e^{-N_j E_j/k_{\rm B}T}}{N_j!}$$
$$= \exp(e^{-E_j/k_{\rm B}T}). \tag{F.16}$$

Therefore, the expected distribution (F.10) becomes

$$f_f^{\circ} = \frac{1}{e^{E_j/k_{\rm B}T}} \quad \text{M–B.} \tag{F.17}$$

F.5 Comparison Among Fermion, Boson, and M–B Particles

The mean occupancy $f_i^{\circ}(E_{i,j}, \mu, T)$ $(i = p, e, f, ph)$ for an orbital with energy $E_{i,j}$ in a system with temperature T and chemical potential μ can be expressed in the generic form

$$f_i^{\circ} = \frac{1}{e^{(E_{i,j}-\mu)/k_BT}+\gamma}, \quad \gamma = 1 \text{ (fermion)},\ 0 \text{ (M-B)},\ -1 \text{ (boson)}. \tag{F.18}$$

When $(E_{i,j}-\mu)/k_BT \gg 1$, $f_i^{\circ} = 1/e^{(E_{i,j}-\mu)/k_BT}$, where the mean occupancy $f_i^{\circ} \ll 1$. The chemical potential is closely related to the quantum concentration (footnote, Section 6.1)

$$\frac{N}{V} = \int \frac{{\rm d}^3 p}{h_{\rm P}^3} \exp[-(\frac{p^2}{2m} - \mu)/k_{\rm B}T] = \lambda_{\rm dB}^{-3} e^{\mu/k_{\rm B}T},$$
$$\lambda_{\rm dB} = (\frac{2\pi\hbar^2}{mk_{\rm B}T})^{1/2} \quad \text{de Broglie wavelength.} \tag{F.19}$$

Then using (F.1) and (F.19), the chemical potential becomes

$$\mu = k_{\rm B}T \ln n_q \lambda_{\rm dB}^3. \tag{F.20}$$

The classical condition $(E_{i,j}-\mu)/k_BT \gg 1$ means that $\mu \ll E_{i,j} - k_BT$ or $\mu \ll 0$, and this leads to $n_q \ll 1$. Therefore both fermions and bosons can be treated as classical particles at high energies, where $f_i^{\rm o} \ll 1$. This is called the nondegenerate state.

Nomenclature

A boldface lowercase indicates a vector, and a boldface uppercase letter indicates a tensor; Roman letters are used when proper names are referred to.

a	interatomic spacing (lattice constant), m
a_s	speed of sound in fluids, m/s
$\boldsymbol{a}$	acceleration vector, m/s^2
$\boldsymbol{a}_e$	vector potential, A-kg/m
A	cross-sectional area, m^2
$\boldsymbol{b}_e$	magnetic field vector, V-s/m^2
c_o	speed of light in vacuum, 2.998×10^8 m/s
c_p	specific heat capacity at constant pressure, J/kg-K, or J/K
c_v	specific heat capacity at constant volume, J/kg-K, or J/K
D	Debye, units of electric dipole moment, 3.3356×10^{-30} C-m
D	diameter, m
$D_i(E)$	energy density of states, 1/m^3-eV, for carrier i
$D_i(\omega)$	frequency density of states, 1/m^3-rad/s, for carrier i
$\boldsymbol{d}$	displacement vector, m
E	energy, J
E_p	bulk modulus, Pa
$E_{\rm Y}$	Young modulus, Pa
e	specific energy, J/kg
e_c	electron charge, 1.602×10^{-19} C
$\boldsymbol{e}_e$	electric field intensity vector, V/m
Fo	Fourier number
$\boldsymbol{F}$	force vector, N
f	frequency, 1/s or Hz
f_i	probability distribution function for carrier i
g	degeneracy
$\boldsymbol{g}$	gravity vector, m/s^2
$G_{\rm N}$	Newton (gravitational) constant, 6.673×10^{11} m^3/kg-s^2

G	shear modulus, Pa
G_i	conductance of carrier i, W/K
H	Hamiltonian, J
h	enthalpy, J/kg or J/particle
h_P	Planck constant, 6.626×10^{-34} J-s
$\hbar$	$= h_\mathrm{P}/2\pi$
$\boldsymbol{h}_e$	magnetic field intensity, A/m
$I_{ph,\omega}$	spectral radiation intensity, W/m^2-(rad/s)
$\boldsymbol{j}_e$	electric current density vector, A/m^2
$\boldsymbol{j}_i$	flux vector for quantity i
$\mathbf{J}$	Jacobian matrix
$\boldsymbol{J}_i$	current vector of quantity i
k	thermal conductivity, W/m-K
k_B	Boltzmann constant, 1.381×10^{-23} J/K
Kn_l	Knudson number, λ_f/l
$\boldsymbol{K}$	thermal conductivity tensor, W/m-K
L	length, m
M	molecular weight, kg/kmole
$M_{i,j}$	energy interaction matrix element, J
m	mass, kg
m_e	electron mass, 9.109×10^{-31} kg
m_n	neutron mass, 1.675×10^{-27} kg
m_p	proton mass, 1.673×10^{-27} kg
m_λ	complex index of refraction $n_\lambda - i\kappa_\lambda$
n_i	number density of particle i, 1/m^3
n_λ	spectral index of refraction
$\dot{n}$	particle generation rate, 1/m^3-s
N	number of particle
N_A	Avogadro number, 6.022×10^{26} molecule/kmole
$N_{\mathrm{L,o}}$	Lorenz number, 2.44×10^{-8} W-Ω/K^2
p	pressure, Pa, or momentum, N-s
$\boldsymbol{p}$	momentum vector, N-s
$\boldsymbol{p}_e$	dipole moment vector, C-m
P	probability
Pe_D	Peclect number, $u_f D/\alpha_f$
Pr	Prandtl number, ν/α_f
Q	heat flow rate, W, or normal coordinate, m
$\boldsymbol{q}$	heat flux vector, W/m^2
q_i	charge of particle i
$\boldsymbol{r}$	radial location, m
$\boldsymbol{r}_\mathrm{B}$	Bohr radius, 5.292×10^{-11} m
$\boldsymbol{r}_\mathrm{C}$	Compton radius, 2.8179×10^{-15} m
R	resonance
R_g	universal gas constant, 8.3145×10^3 J/kmol-K

R_t	thermal resistance, K/W
Re_D	Reynolds number, $u_f D/\nu_f$
Ry	Rydberg constant, 1.0974×10^{-5} cm^{-1}, also unit of energy, 13.606 eV
$\dot{S}$	energy conversion rate, W
S	entropy, J/K, or quantum angular momentum spin J-s
$\dot{s}$	volumetric energy conversion rate, W/m^3
$\boldsymbol{s}$	unit vector
$\boldsymbol{s}_e$	Poynting vector, W/m^2
t	time, s
T	temperature, K
$\boldsymbol{u}$	velocity vector, m/s
$\boldsymbol{u}_f$	fluid velocity vector, m/s
V	volume, m^3
$\boldsymbol{\omega}$	heat current vector, W-m
$\boldsymbol{x}$	location, or position vector, m
x, y, z	Cartesian coordinate, m
z	atomic number
Z	partition function
z_e	number of conduction electrons per atom
Z_e	thermoelectric figure of merit, 1/K

Greek

α	diffusivity $= k/\rho c_p$, m^2/s, or fine-structure constant, 7.29735×10^{-3}
α_e	molecular polarization
α_{ph}	photon absorptivity
$\dot{\gamma}_{ad}, \dot{\gamma}_{de}$	adsorption and desorption rate coefficients, 1/m^3-s
$\dot{\gamma}_{i,j}$	interaction between carriers i and j, or transition rate between state i and j, 1/s
γ	specific heat capacities ratio, $c_{f,p}/c_{f,v}$
γ_{G}	Grüneisen constant (or parameter)
Γ	spring constant, N/m
$\mathit{\Gamma}$	gamma function
$\delta_{\mathrm{D}}(E)$	Dirac delta function, 1/J
δ_{ij}	Kronecker delta, $\delta_{ij}(i = j) = 1$, $\delta_{ij}(i \neq j) = 0$.
ϵ_e	electric permittivity (dielectric function)
ϵ_{o}	free-space electric permittivity, 8.854×10^{-12} A^2-s^2/N-m^2
ϵ_{ph}	emissivity
$\epsilon_{f,t}$	turbulent kinetic energy dissipation rate, m^2/s^3
η	efficiency
κ	wave number, $\kappa = 2\pi/\lambda$, 1/m
κ_λ	spectral index of extinction
κ_p	isothermal compressibility, Pa^{-1}
$\boldsymbol{\kappa}$	wave vector, 1/m
λ	wavelength, m
λ_i	mean free path of carrier i, m

μ	chemical potential, J/kg, or J/particle, or dynamical viscosity, Pa-s
μ_e	magnetic permeability, N/A^2
$\boldsymbol{\mu}_e$	transition dipole moment, C-m
μ_o	free-space magnetic permeability, 1.257×10^{-6} N/A^2
ν_P	Poisson ratio
ν_f	fluid kinematic viscosity, μ_f/ρ_f, m^2/s
ρ	density, kg/m^3
ρ_e	electrical resistivity, Ω-m
ρ_r	reflectivity
σ_i	absorption coefficient for carrier i, 1/m
σ_e	electrical conductivity, 1/Ω-m
σ_{SB}	Stefan–Boltzmann constant, 5.670×10^{-8} W/m^2-K^4
τ_a	atomic time scale, 2.4189×10^{-17} s
$\tau_{i\text{-}j}$	relaxation time from interaction between i and j, s
$\tau_{i,j}$	stress tensor component, Pa
τ_r	transmissivity
φ	potential energy, J
Ψ	time-dependent wave function, $m^{-3/2}$
ψ	time-independent wave function, $m^{-3/2}$
Ω	solid angle, sr
ω	angular frequency, $\omega = 2\pi f$, rad/s

Subscripts

A	acoustic
b	blackbody
B	Bohr, or Boltzmann
c	conduction
C	Coulombic, or Compton
C–P	Cahill–Pohl
dB	de Broglie
D	Debye, or Dirac
E	Einstein
e	electron, or electromagnetic, or emission
fm	free molecular
f	fluid particle, fluctuation
F	Fermi
g	group, or gap
G	Grüneisen
h	hole
i	incident, carrier
k	conduction
ku	surface convection
K	Kolmogoroff
l	laminar, or based on length l

L	longitudinal, or Lorenz
LJ	Lennard–Jones
m	molecular
MD	molecular dynamics
n	normal
N	Newton
o	reference value
O	optical
P	Planck, or Peltier, or Poisson
p	phonon, or potential energy, or constant pressure
ph	photon
r	radiation
sp	spontaneous
st	stimulated
S	Seebeck, or Slack
T	thermal
T	Thomson, or transverse
t	turbulent
tr	transition, or turbulent fluctuation
u	convection
v	valence
Y	Young
α	polarity, or diffusion
ω	frequency
μ	viscous

Superscripts

$'$	turbulent fluctuation
$*$	dimensionless (scaled)
o	equilibrium value
$\dagger$	complex conjugate, or Hermitian conjugate

Others

$\langle\ \rangle$	ensemble average, or spatial average
$-$	time average, or average
$\sim$	nearly
$\propto$	proportional
$O(\ )$	order of magnitude
Δ	difference
$\nabla\cdot$	divergence
∇	gradient
$\partial/\partial t$	time derivative

Abbreviations

AMM	acoustic mismatch model
BCC	body-centered cubic
BDC	1, 4-benzenedicarboxylate
B–K–S	van Beest–Krammer-van Staten
BTE	Boltzmann transport equation
BZBC	Brillouin zone boundary condition
C–P	Cahill–Pohl
DCF	displacement correlation function
DFT	density function theory
DMM	diffuse mismatch model
DOS	density of states
EF	exponential-fit
EM	electromagnetic
ERT	equation of radiative transfer
FAU	faujasite (a zeolite)
FCC	face-centered cubic
FD	first-dip
FGR	Fermi golden rule
fm	free-molecular
GGA	generalized gradient approximation
G–K	Green–Kubo
GTO	Gaussian-type orbital
HCACF	heat current autocorrelation function
HFSS	high-frequency structure simulator
L–J	Lennard–Jones
LA	longitudinal-acoustic
LAPW	linearized augmented plane wave
LCAO	linear combination of atomic orbitals
LDA	local density approximation
LTA	zeolite A (Linde Type A)
LO	longitudinal-optical

M–B	Maxwell–Boltzmann
MD	molecular dynamics
MHD	magnetohydrodynamics
MOF	metal-organic framework
MSD	mean-square displacemnt
MSRD	mean-square relative displacement
MWNT	multiwall nanotube
NIR	near-infrared regime
NMR	nuclear magnetic resonance
pdf	probability density function
PV	photovoltaic
QC	quadratic convergence
QD	quantum dot
QED	quantum electrodynamics
QW	quantum well
RDF	radial distribution function
RMS	root-mean-square
SC	simple cubic
SCF	self-consistent function
SI	*Systeme International* (system standardizing abbreviations of units)
SMRT	single-mode relaxation time
SOD	sodalite (a zeolite)
SOMO	singly occupied molecular orbital
S–W	Stillenger–Weber
TA	transverse-acoustic
TE	thermoelectric
TDDFT	time-dependent density function theory
TO	transverse-optical
W–K–B	Wentzel–Kramers–Brillouin
YAG	yttria–aluminum–garnet

Glossary

Adiabatic (Born–Oppenheimer) approximation: A technique used to decouple the motion of nuclei and electrons (i.e., separate the variables corresponding to the nuclear and electronic coordinates in the Schrödinger equation associated to the molecular Hamiltonian). It is based on the fact that typical electronic velocities far exceed those of nuclei.

Anti-Stokes emission: When the frequency of emitted photon is equal or larger than the exciting (absorbed) photon. The extra energy can be contributed from phonon absorption (lattice vibration) or from molecular kinetic energy.

Auger effect: One of two principal processes for relaxation of inner-shell electron vacancy in an excited or ionized atom. It is a two-electron process in which an electron makes a discrete transition from a less bound shell to the vacant but tightly bound electron shell. The energy gained in this process is transferred, by means of electron static interaction, to another bound electron, which then escapes from the atom (called the Auger electron).

Ballistic transport: The motion of electrons in ultrasmall (highly confined) regions in semiconductor structures at very high electric field with velocities much higher than their equilibrium thermal velocity. The ballistic electrons are not subjected to scattering, so they can move with ultrahigh velocity. The ballistic transport is determined by the electronic structure of the semiconductor and is different for different semiconductors and this allows ultrafast devices. Ballistic transport is also extended to other microscale energy carriers, indicating transport with no scattering.

Bandgap energy: The energy difference between the bottom of the conduction band and the top of the valence band in a semiconductor or an insulator.

Bloch theorem: This theorem states that the eigenfunction of the Schrödinger equation for a periodic potential is the product of a plane wave times a function that has the same periodicity as the periodic potential.

Bloch wave: A description of an electron in a crystal that takes account of the periodic potential of the positive ions. A free-electron wave, e.g., $\exp[i(\boldsymbol{\kappa} \cdot \boldsymbol{x})]$ is modulated near each atom of the lattice by a periodic potential function $a_{\kappa}(\boldsymbol{x})$.

The Bloch wave is thus represented by

$$\psi_\kappa(\boldsymbol{x}) = a_\kappa(\boldsymbol{x}) \exp[i(\boldsymbol{\kappa} \cdot \boldsymbol{x})].$$

Bloch showed that the periodic field of the crystal lattice neither scatters nor destroys the free-electron wave, but modulates it.

Bravais lattice: Periodic array in which the repeated units (single or multiatom) of a crystal are arranged. The atoms are presented as points and the infinite array of these discrete points appears exactly the same when viewed from any atom (translational symmetry).

Bose–Einstein condensate: A phase transition at low temperature and high density, corresponding to a phase-space density of

$$n_f \lambda_{\mathrm{dB}}^3 = 2.612,$$

where n_f is the fluid particle number density and λ_{dB} is the de Broglie wavelength. At this state, a gas of atoms undergoes transition to the Bose–Einstein condensate, which is the lowest accessible quantum state and a new form of matter.

Bremsstrahlung: German word for decelerating radiation. It describes photon emission (or absorption) when a free electron undergoes momentum (energy) change.

Brillouin scattering: Interaction of light in a medium with density variations that change its path. The density variations may be due to acoustic modes (traveling sound waves) or temperature gradient. When the medium is compressed, the index of refraction of light in the medium changes and its path bends. From quantum mechanics, Brillouin scattering is interaction of photons with acoustic phonons. The scattered light wavelength is changed slightly by a variable quantity known as the Brillouin shift.

Brillouin zone: Polyhedral zone in a crystal lattice used to define the frequency or energy of wave motion.

Chirality: The property of an object that cannot be superimposed on its mirror image.

Cohesive energy: The energy of a solid that is required for disassembling it into its constituent parts.

Compton scattering: An increase in the wavelength of a photon (X-ray and γ-ray regimes) due to absorption by free electrons (thus increasing the kinetic energy of the electrons).

Conduction band: A vacant or only partially occupied set of many closely spaced electronic levels resulting from an array of a large number of atoms forming a system in which the electrons can move freely or nearly so.

Cooperative processes: Summing of quanta of energy by the interaction of excited particles resulting in accumulation of excitation energy in one of them. For particles A and B the excitation energy can pass as $\ldots - A^* - A - B - \ldots \rightarrow - A - A - B^* - \ldots.$

Creation and annihilation operators: Creation $a^\dagger$ (annihilation a) operator is the operator in quantum field theory that increases (lowers) the number of particles in a given state by one (i.e., adjacent eigenfunctions of a particle wave function). In the context of the quantum harmonic oscillator, the ladder operators are the creation and annihilation operators, adding or subtracting fixed quanta of energy to the oscillator system. Creation/annihilation operators are different for bosons (integer spin) and fermions (half-integer spin). This is because their wave functions have different symmetry properties.

de Broglie wavelength: A wave associated with an electron (or other quasi-particles) in motion that gives it diffraction and interference characteristics. de Broglie suggested the wave nature of electrons with the momentum given by $\boldsymbol{p} = \hbar\boldsymbol{\kappa}$, where κ is the wave number ($2\pi\lambda$, where $\lambda = h_P/mu$ is the wavelength). The thermal de Broglie wavelength is

$$\lambda_{dB}(\text{thermal}) = (\frac{h_P^2}{2\pi m k_B T})^{1/2}.$$

Debye temperature: The temperature corresponding to the crystals highest normal mode of vibration (frequency f_D), i.e., the highest temperature that can be achieved that is due to a single normal vibration f_D and is given by

$$T_D = h_P f_D / k_B.$$

Degenerate energy state: Existing in two or more quantum states with the same energy level.

Density of states (DOS): the distribution of a fixed amount of energy among a number of identical particles depends on the density of available energy states. The probability that a given energy state will be occupied. The probability that a given energy state will be occupied is given by the distribution function, but if there are more available energy states in a given energy interval, will give greater weight to the probability for that energy interval, i.e.,

$$n_i(E_i) = D_i(E_i) f_i(E_i) \mathrm{d}E_i, \; i = p, e, f, i,$$

where n_i is the number of particles per unit volume with energy E_i to $E_i + \Delta E_i$, $D_i(E)$ is the energy DOS (number of energy states per unit volume in interval ΔE_i), and f_i is the probability distribution function for energy E_i.

The wave number DOS is given by the general expression for spherical energy surfaces as

$$D_i(\kappa)\mathrm{d}\kappa = \frac{4\pi\kappa^2}{(2\pi)^3}\mathrm{d}\kappa.$$

There is also the frequency DOS $Di(\omega)$. The relation between E_i (or ω) and κ is called the dispersion relation $E_i(\kappa)$, [or $\omega_i(\kappa)$], and depends on the particular carrier.

Dirac delta function: This function represents any point action, i.e., action that is highly localized in space and/or time. For interactions involving an energy match requirement, the energy Dirac delta function is defined as

$$\delta_D(E - E_i) = 0 \text{ for } E \neq E_i,$$

$$\text{and } \delta_D(E - E_i) \to \infty \text{ for } E = E_i,$$

$$\text{and } \int_0^\infty \delta_D(E - E_i)\mathrm{d}E = 1.$$

Note that $\delta_D(E)$ has the unit of inverse energy.† Also note that as given by (3.28) and (5.64), the energy integral of the energy Dirac delta function gives the DOS.

Dispersion: The frequency or mode dependence of the phase velocity in a medium. The chromatic dispersion of an optical medium is basically the frequency dependence of the phase velocity with which a wave propagates in the medium. More precisely, one defines dispersion of second and higher orders by means of the Taylor expansion of the wave number as a function of the angular frequency ω (around ω_o):

$$\kappa(\omega) = \kappa_o + \frac{\partial \kappa}{\partial \omega}(\omega - \omega_o) + \frac{1}{2}\frac{\partial^2 \kappa}{\partial \omega^2}(\omega - \omega_o)^2 + \frac{1}{6}\frac{\partial^3 \kappa}{\partial \omega^3}(\omega - \omega_o)^3,$$

where the zeroth-order term describes a common phase shift. The first-order term contains the inverse group velocity and describes an overall time delay without an effect on the pulse shape. The second-order term contains the second-order dispersion (group delay dispersion per unit length). The third-order term contains the third-order dispersion per unit length. The second-order dispersion is often specified in units of s^2/m, and is the derivative of the inverse group velocity with respect to the angular frequency. The normal dispersion refers to $\partial^2\kappa/\partial\omega^2 > 0$ (negative values are called anomalous).

Distinguishable and indistinguishable particles: Particles are considered to be indistinguishable if their wave packets overlap significantly. Two particles can be considered to be distinguishable if their separation is large compared with their de Broglie wavelength.

Dopant: In semiconductors a dopant results from introducing impurities into the intrinsic (pure) phase to change its thermoelectric properties. A lightly and moderately doped semiconductor is referred to as extrinsic. A semiconductor that is doped to such high levels that it acts more like a conductor than a semiconductor is called degenerate. For the group 14 semiconductors such as Si or Ge, the most

† The Fourier transfer of $F(\kappa) = 1$ is

$$f(x) = \int_{-\infty}^{\infty} F(\kappa)e^{i\kappa x}\mathrm{d}\kappa = \int_{-\infty}^{\infty} e^{i\kappa x}\mathrm{d}\kappa.$$

Taking the inverse, we have

$$1 = \frac{1}{2\pi}\int_{-\infty}^{\infty} f(x)e^{-i\kappa x}\mathrm{d}x.$$

Then $f(x)$ is zero when $x \neq 0$, and $f(x)$ tends to infinity for $x = 0$. Then, using the Dirac delta function, we have

$$\int_{-\infty}^{\infty} f(x)\delta_D(x - x_o)\mathrm{d}x = f(x_o).$$

Note that in the energy Dirac delta function $f(E)$ is equal to 1.

common dopants are group 13 or group 15 elements. Group 13 dopants are electron acceptors (because they are missing the fourth valence electron) and this creates holes; are called the p-type dopants. Group 15 dopants are electron donors, adding extra valence electrons which become unbonded (conduction electrons) from individual atoms and allow the compound to be electrically conductive, and are called the n-type dopants. Heavy (or high) doping refers to the order-of-1 dopant atoms per 10^4 of host atoms. This is often shown as n^+ for n-type doping or p^+ for p-type doping. Dopants are also added to dielectrics (such as oxides).

Drude electron transport model: This model assumes that the conduction electrons (a) do not interact with the cations (free-electron approximation), except for collisions, in which the velocity of the electron abruptly and randomly changes direction as a result of collision (relaxation time approximation), (b) maintain thermal equilibrium through collisions (classical statistics approximation), and (c) do not interact with each other (independent electron approximation).

Effective electron mass: The effective mass of a particle is its apparent mass in the semiclassical model of transport in a crystal. Electrons and holes in a crystal respond to electric and magnetic fields almost as if they were free particles in a vacuum, but with a different mass. This is a fraction of the free-electron mass m_e. Effective mass is defined by analogy with the Newton second law $\boldsymbol{F} = m\boldsymbol{a}$. Using quantum mechanics it can be shown (Chapter 5) that for an electron in an external electric field $\boldsymbol{e}_e$

$$\boldsymbol{a} = \frac{1}{\hbar^2}\frac{\mathrm{d}^2 E_e}{\mathrm{d}\kappa^2} e_c \boldsymbol{e}_e \equiv \frac{\boldsymbol{F}}{m_{e,e}}.$$

Also, in an external electric field, the electron would experience a force of $\boldsymbol{F} = m_{e,e} e_c \boldsymbol{e}_e$, where e_c is the charge. Then the electron effective mass $m_{e,e}$ becomes

$$m_{e,e} = \hbar^2 (\frac{\mathrm{d}^2 E_e}{\mathrm{d}\kappa^2})^{-1}.$$

Elastic collision (scattering): Scattering of an energy carrier by another energy carrier or by other entities (e.g., impurities, in solids) that leaves the energy of the carrier unchanged after the event. For example, impurity scattering and acoustic-phonon scattering of electrons in solids are generally treated as elastic scattering.

Electron bandgap: Under the periodic potential in crystals, electron energy gaps are formed by electrons with localized (nonpropagating) wave functions as opposed to those of electrons in allowed bands that have extended (propagating) wave functions. Bandgap energy is defined by use of this localization as the energy needed to remove an electron from a bond in the solid, enabling the electron to move freely through the solid to conduct electricity.

Electron volt: An amount of energy gained by an unbound electron when accelerated through a static potential of one volt.

Energy relaxation time: The time required for dissipating the carrier energy.

Ergodic hypothesis: This hypothesis states that the time-average behavior of an individual particle in a system is equal to the ensemble-average behavior of the system.

Evanescent wave: Evanescent means tending to vanish, and the intensity of evanescent waves decays exponentially (rather than sinusoidally) with distance from the interface at which they are formed (the wave number is purely imaginary). Evanescent waves are formed when sinusoidal waves are (internally) reflected at an interface at an angle greater than the critical angle so that total internal reflection occurs.

Exciton: The combination of an electron and a hole in an excited semiconductor crystal. An exciton consists of a single electron and a single hole bound by the Coulomb force.

Fermi energy or Fermi level: The energy of the highest occupied energy state of a system of fermions at $T = 0$ K. It is equal to the chemical potential of electrons in a solid (metals, semiconductors, or insulators) or in an electrolyte solution at $T = 0$ K. However, the chemical potential is temperature dependent. For conduction electrons, using a constant chemical potential energy results in negligible error.

Fermi golden rule: This rule states that the electronic transition rate depends on the strength of the coupling between the initial and final states of a system and also depends on the number of ways the transition can occur. The transition probability is also called the decay probability. A transition will proceed more rapidly if the coupling between the initial and final states is stronger. This coupling term is called the matrix element for the transition and comes from an alternative formulation of quantum mechanics in terms of matrices rather than the differential equations (i.e., Schrödinger approach). The matrix element can be placed in an integral form, in which the interaction that causes the transition is expressed as a potential φ that operates on the initial state wave function. The transition probability is proportional to the square of the integral of this interaction over the appropriate space.

Fermi surface: The surface of constant electron energy in $\boldsymbol{\kappa}$ (wave-number vector) space. The Fermi surface separates the unfilled orbitals from the filled orbitals at absolute zero. The electrical properties of the metal are determined by the shape of the Fermi surface, because the current is due to changes in the occupancy of states near the Fermi surfaces. The free-electron Fermi surfaces are developed from spheres of radius $\kappa_{\rm F}$ determined by the valence electron concentration.

Feynman diagram: This diagram visualizes quantum-electrodynamical interactions with lines and vertices (meeting of lines) and is used to derive the probability rate for the interactions (e.g., Fermi golden rule for weak interactions). For example, the single-phonon only (radiationless) electronic transition (absorption and emission) is shown as in the figure.

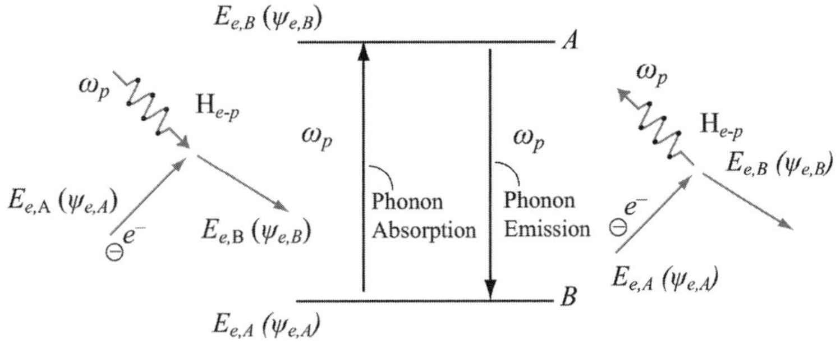

Feynman diagram for radiationless (phonon absorption/emission) two-level electronic transition. $H_{e\text{-}p}$ is the electron–phonon interaction Hamiltonian.

Field emission: Emission of electrons and ions by metals (filament) in the presence of an electric field.

Fluid particle: Charged or neutral atoms or molecules in a fluid state undergoing thermal motion (each particle undergoing random, translational, rotational, and vibrational motions representing their kinetic energy and temperature) with or without a net motion (under external force, such as gravitational field, pressure gradient, or electromagnetic field). The simplest treatment of fluid particle transport assumes particles in classical (Newtonian–Hamiltonian), translational thermal motion (Boltzmann theory of gases), subject to the statistical averaging.

Fluorescence: Visible photon emission that is due to X-rays or ultraviolet photon absorption. Fluorescence ends when the source is removed.

Group theory in quantum mechanics: In quantum-mechanical mathematical operations, matrices act on members of a vector space, and only certain members of the space that are symmetrical (with classified symmetry) can be created as described by the group theory theorems.

Grüneisen parameter: Is a dimensionless constant representing the dependence of vibrational frequencies of solid (shift in Debye frequency) on its volume; it is related to the bulk thermophysical properties through

$$\gamma_G = \frac{3\beta E_p}{\rho c_v},$$

where β is the linear thermal expansion coefficient and E_p is the isothermal bulk modulus. Table A.2 lists the experimental values of γ_G for elements. The Grüneisen relation can be written as

$$\frac{\partial p}{\partial T}|_V\, V = \gamma_G \frac{\partial E}{\partial T}|_V,$$

or when integrated as the Grüneisen equation of states for cubic solids it becomes,

$$p(T, V)V = \gamma_G[E(T, V) + G(V)],$$

where $G(V)$ is called the integration constant. An example of $G(V)$ is

$$G(V) = -A[(\frac{V_o}{V})^{m/3} - (\frac{V_o}{V})^{n/3}],$$

where A, m, and n are the L–J potential parameters (Table 2.1).

The modal Grüneisen constant (or parameter) is defined as

$$\gamma_{G,\kappa,\alpha} = -\frac{\partial \ln \omega_{\kappa,\alpha}}{\partial \ln V},$$

where α is the polarization. In the Debye approximation, can have

$$\gamma_{G,\kappa,\alpha} = \gamma_G = -\frac{\partial \ln \omega_D}{\partial \ln V} \quad \text{Debye approximation.}$$

γ_G is a measure of anharmonicity in a material.

Hamiltonian: Is a function describing the state of a mechanical system in terms of position and momentum (classical mechanics), and in quantum mechanics the Hamiltonian is an operator corresponding to the total energy of a system.

Inelastic collision: A collision between two particles in which part of their kinetic energy is transformed to another form of energy.

Inelastic scattering: The scattering of particle resulting from inelastic collision.

Ionic crystals: Crystals with at least two atoms in their lattice, which are ionized (with zero total charge). The binding is mostly electrostatic, isotropic, and strong. Ionic crystals are insulators (no conduction electron).

Kapitza resistance: In the presence of a heat current there is a discontinuity in temperature at an interface between two different materials, i.e., the interfaces possess a finite thermal resistance known as the Kapitza resistance.

Kinetics: The mechanism by which a physical or chemical change is affected. The microscale energy transformation kinetics describes mechanisms of the energy transition/conversion rate.

Kinetic theory: Particles in thermal motion involving interparticle collisions with the RMS of this thermal velocity being related to the temperature.

Kirchhoff law of radiation: Under local thermal equilibrium (the photon energy distribution function follows the boson distribution), the directional, spectral emissivity of a body is equal to its directional, spectral absorptivity at the same temperature.

Luminescence: Visible photon emission that is not due to thermal radiation emission; this includes phosphorescence and fluorescence.

Momentum relaxation time: The time required for randomizing the carrier momentum.

n- and p-type semiconductors: Addition of a small atomic percentage of dopant (electron donor and acceptor) atoms in the regular crystal lattice of semiconductor changes the weakly bound outer-shell electrons resulting in significant change in their electrical properties, producing n-type and p-type semiconductors. Group 16 impurity atoms with 5 valence electrons (donors) produce n-type semiconductors by contributing extra electrons. Group 13 impurity (dopant) atoms with 3 valence electrons (acceptors) produce p-type semiconductors by producing a hole (or electron deficiency). Thus the holes are the majority carriers, whereas electrons are the minority carriers in p-type semiconductors, and vice versa. The intrinsic (undoped) semiconductor is called i-type.

Normal displacement coordinates: Normal coordinates comprise a set of coordinates for a coupled system such that the equations of motion each involve one of these coordinates. In lattice displacement, the normal displacement coordinate is a Fourier transform of the lattice displacement and is a complex quantity variable.

Pauli exclusion principle: No two identical fermions may occupy the same quantum state simultaneously. For elements, no two electrons have the same four quantum numbers (n, l, m, and s, Section 2.6.6).

Particle probability distribution function: The probability that a particle is in an energy state E, and is the continuous generalization of discrete probability. There are three distributions, Maxwell–Boltzmann (classical), Bose–Einstein (boson), and Fermi–Dirac (fermion). The distribution functions scale E with $k_B T$ and show that the larger $E/k_B T$ has a lower probability.

Phonon drag on electron: This refers to the increase in the effective mass of conduction electrons due to its field interactions with the crystal lattice. It also refers to electron–phonon (lattice) inelastic scattering under lack of local thermal equilibrium between the two carriers. At low temperatures (around 0.2 T_D, where T_D is the Debye temperature), the nonequilibrium distribution of phonons plays a more significant role in electron–phonon scattering and in thermoelectricity this particularly affects the Seebeck coefficient, as phonons force the electrons to move under a temperature gradient.

Phonon gas: A crystal lattice at zero temperature lies in its ground state, and contains no phonons. According to thermodynamics, when the lattice is held at a nonzero temperature its energy is not constant, but fluctuates randomly about some mean value. These energy fluctuations are caused by random lattice vibrations, which can be viewed as a gas of phonons. Because these phonons are generated by the temperature of the lattice, they are sometimes referred to as thermal phonons. Unlike the atoms that make up an ordinary gas, thermal phonons can be created or destroyed by random energy fluctuations. Their behavior is similar to that of the photon gas produced by an electromagnetic cavity, wherein photons may be emitted or absorbed by the cavity walls. This similarity is not coincidental for it turns out that the electromagnetic field (boson) behaves like a set of harmonic oscillators. Both gases obey the Bose–Einstein statistics.

Phonon polariton: The propagating electromagnetic waves coupled to polar optical phonon. For example, the high-frequency dielectric properties are closely related to the optical-phonon modes at THz frequency.

Phosphor: A substance that exhibits the phenomenon of phosphorescence. Transition metal compounds and rare-earth compounds are among phosphors.

Phosphorescence: Visible photon emission that is due to X-ray or ultraviolet photon absorption. Phosphorescence continues for a period of time after the source is removed.

Plasma frequency: The natural frequency of oscillations of electrons in a plasma displaced relative to the ion background; the waves with frequency below the plasma frequency cannot propagate in the plasma.

Plasmon: A quasi-particle resulting from oscillation of plasma (free-electron gas). Plasmons can couple with photon to create plasma polariton.

Polariton: A quasi-particle resulting from coupling of light (photon) with an interacting resonance. Phonon-polariton results from coupling of infrared photon and optical phonon. Exciton-polaiton is coupling of visible light and exciton.

Polaron: A quasi-particle composed of an electron and its accompanying polarization field. In electron–phonon coupling, the electron moving through the crystal carries with it the lattice distortion (polarization) it causes.

Polar and nonpolar semiconductors: In nonpolar semiconductors the longitudinal-optical and the transverse-optical branches of phonon dispersion have the same frequency at the zone center Γ, and in polar semiconductors these two branches split (because of long-range dipolar interactions).

Polar molecules: Molecules in which there exists a permanent spatial separation of the centroid of positive and negative charge or dipole moment.

Potential energy: The energy in matter that is due to its position or the arrangement of its parts. The potential energy appears in gravitational potential, elastic potential, chemical (interatomic) potential, electrical potential, etc.

Primitive cell: A cell that when translated in space will fill the space without overlapping.

Pseudopotential: Effective potential for electrons in a crystal lattice that is calculated in the orthogonalized plane-wave method and in the pseudopotential method. This is a weak potential, because the electrons move rapidly past nuclei in the lattice.

Quantum Hamiltonian: The physical state (total energy) of a system, which may be characterized as a ray in an abstract Hilbert space (or, in the case of ensembles, as a trace class operator with trace 1). It generates the time evolution of quantum states.

Quasi-particle: A unit or quantum that has some of the properties of a particle, such as mass or momentum. Phonons and photons are treated as quasi-particles. It also refers to a particle that is a result of the renormalized self-interactions incorporated into it. For example, in the renormalization of all of the interactions that a single electron experiences in a periodic lattice, into self-energy, using an effective mass, the electron is treated as a quasi-particle.

Quantum mechanics: A physical theory that at very small distances produces results that are very different and much more accurate than the results of the classical mechanics. It is derived from a small set of basic principles and applies to at least three general types of phenomena that classical mechanics and classical electrodynamics cannot account for, namely quantization, wave–particle duality (interference of matter particles), and quantum entanglement.

Raman scattering: An inelastic scattering of a photon that creates or annihilates a phonon (and other molecular motion transitions), giving rise to the Stokes and anti-Stokes lines in the scattered spectrum.

Rattler: Loosely bound phonon scatterer (atoms) introduced into a host crystal to result in a short phonon mean free path (reducing lattice conductivity and also resulting in an amorphous phase behavior).

Rayleigh and Mie scatterings: The process by which small particles suspended in a medium of a different index of refraction scatter a portion of the incident radiation in all directions. In elastic scattering, no energy transformation results, only a change in the spatial distribution of the radiation; however, the particles also absorb radiation (inelastic scattering). This scattering varies as a function of the ratio of the particle diameter to the wavelength of the radiation (called the size parameter). When this size parameter is less than about one-tenth, the Rayleigh scattering occurs in which the scattering coefficient varies inversely as the fourth power of the wavelength.

Reciprocal lattice: Every crystal structure has two lattices associated with it, the direct lattice and the reciprocal lattice. A microscope image is a map of the direct lattice whereas the diffraction pattern is a map of the reciprocal lattice. The reciprocal lattice is essential in analytic studies of periodic structures, including the theory of crystal diffraction and the law of momentum conservation, in which the full translational symmetry of free space is reduced to that of a periodic potential.

Relaxation time: Average time between collisions.

Reststrahlen absorption and reflection: Reststrahlen, meaning residue rays in German, is the successive reflection of a broadband photon beam from the surface of a crystal to obtain a narrow-band beam. This is due to the interaction of electromagnetic waves with the vibrating dipoles (molecular vibrations in crystal, i.e., optical phonons).

Rydberg energy, formula, and constant: A unit of electron energy based on the ground-state energy of a hydrogen atom, one Ry energy is

$$\mathrm{Ry} = \frac{m_e e_c^4}{2(4\pi\epsilon_\mathrm{o})^2\hbar^2} = 13.607 \text{ eV}.$$

The Rydberg formula is

$$\frac{1}{\lambda} = \frac{m_e e_c^4}{8c_\mathrm{o}\epsilon_\mathrm{o}^2\hbar^3}\left(\frac{1}{n_f^2} - \frac{1}{n_i^2}\right) = \mathrm{Ry}\left(\frac{1}{n_f^2} - \frac{1}{n_i^2}\right),$$

where λ is the wavelength of photon emitted (in vacuum), for $n_f < n_i$, and the Rydberg constant is defined from the above relation as

$$\mathrm{Ry} = \frac{m_e e_c^4}{8c_\mathrm{o}\epsilon_\mathrm{o}^2\hbar^3} = 1.0974 \times 10^5 \text{ cm}^{-1}.$$

Semiconductors: Solid-state material in which, in contrast to metals and insulators, the electrical conductivity can be controlled by orders of magnitude by adding very small amounts of impurity elements (dopants) thus affecting the negatively charged electrons and positively charged holes. Their electrical conductivity is also sensitive to temperature, illumination, and magnetic field. These result from their interatomic bonds (mostly covalent) in which the valence band and the conduction band are separated by the energy gap. Semiconductor properties are displayed by the elements from the group 14 of the periodic table (Tables A.1 and A.2.), i.e., C in the form of diamond, Si, Ge and Sn and their compounds with elements from group 14, e.g., SiGe and SiC, as well as compounds from groups 13 and 15, e.g., GaAs, InP, or GaN, and groups 12 and 16, e.g., CdTe, ZnS.

Skutterudites: Complex minerals, $ReTm_4Pn_{12}$ (Re: rear earth, Tm: transition metals, Pn: pnicogen), with Re in a simple cubic structure of Tm, each in a tilted pnicogen octahedra.

Spherical, parabolic band: Idealized electron band structure in crystals, in which the energy surfaces are spherical and the energy variation with respect to momentum is parabolic (model band structure).

Spin: A quantum presentation of angular momentum in quantum electrodynamics. It obeys the commutation relation and is a property of all elementary particles (including electrons).

Spontaneous emission: Photon emission that is from perturbations that are due to the vacuum state energy fluctuations (including thermal motion), and the emitted photons are incoherent (no phase relationship).

Stimulated emission: Photon emission that is due to the initial presence of photons in the system and maintains phase coherence with those initial photons.

Stokes law: The frequency of luminescence excited by radiation (photoluminescence) is not higher than the frequency of the exciting radiation.

Surface plasmon: A collective excitation of the electrons at the interface between a conductor and an insulator. On a plane surface, they are nonradiative electromagnetic modes (they cannot be generated by a photon nor decay spontaneously into photons).

Thermophotovoltaic: A device that converts thermal radiation emission from a controlled heat source into electricity. The device is designed for maximum efficiency at the wavelength range of this source.

Thermionic emission: Emission of electrons or ions by metals (filament) heated to high temperatures; the charged particle is called a thermion.

Thomson scattering: In photons (electromagnetic wave) incident upon a charged particle, the electric and magnetic components of the wave exert a Lorentz force on the particle, setting it into motion. Because the wave is periodic in time, so is the motion of the particle. Thus the particle is accelerated and consequently emits radiation.

Tunneling: A quantum mechanical effect allowing transition through a classically forbidden energy state.

Umklapp processes: German word Umklapp for flipping over is the scattering process that conserves the phonon momentum with a wave-vector jump across the Brillouin zone.

Valence band: The highest energy continuum of energy levels in a semiconductor that is fully occupied by electrons at $T = 0$ K.

Wigner–Seitz primitive cell: In crystals (lattices), it is the region of space that is closer to that point than to any other lattice point.

Bibliography

[1] M. Abramowitz and I.A. Stegun. *Handbook of Mathematical Functions*. National Bureau of Standards, Applied Mathematics Series 55, 1972.

[2] S. Adachi. Model dielectric constants of GaP, GaAs, GaSb, InP, InAs, and InSb. *Physical Review B*, 35:7454–7463, 1987.

[3] M.P. Allen and D.J. Tildesley. *Computer Simulation of Liquids*. Clarendon, Oxford, 1989.

[4] P.S. Anderson, W.M. Kays, and R.J. Moffat. Experimental results for the transpired turbulent boundary layer in an adverse pressure gradient. *J. Fluid Mechanics*, 69:353–375, 1975.

[5] P.W. Anderson. Absence of diffusion in certain random lattices. *Physical Review*, 109:1492–1505, 1958.

[6] G. Arya, H.C. Chang, and E.J. McGinn. Molecular simulations of knudsen wall-slip: Effect of wall morphology. *Molecular Simulations*, 20:697–709, 2003.

[7] C.K. Asawa. Long-delayed fluorescence of Nd^{3+} in pure $LaCl_3$ and in $LaCl_3$ containing Ce^{3+}. *Physical Review*, 155:188–197, 1967.

[8] M. Asen-Palmer, E. Bartkowski, and E. Gmelin. Thermal conductivity of germanium crystals with different isotopic compositions. *Physical Review B*, 56:9431–9447, 1997.

[9] N.W. Ashcroft and N.D. Mermin. *Solid State Physics*. Saunder College Press, Philadelphia, PA, 1976.

[10] R. Astala, S.M. Auerbach, and P.A. Monson. Normal mode approach for predicting the mechanical properties of solids from first principles: Application to compressibility and thermal expansion of zeolites. *Physical Review B*, 71:014112 (1–15), 2005.

[11] F. Auzel. Multiphonon-assisted anti-stokes and stokes fluorescence of triply ionized rare-earth ions. *IEEE J. Quantum Electronics*, 35:115–122, 1999.

[12] R. Bairelein. *Thermal Physics*. Cambridge University Press, Cambridge, 1999.

[13] J. Bardeen. Electrical conductivity of metals. *J. Applied Physics*, 11:88–111, 1940.

[14] J.O. Barnes and J.A. Rayne. Lattice expansion of Bi_2Te_3 from 4.2 k to 600 k. *Physical Letters A*, 46:317–318, 1974.

[15] T.H.K. Barron, C.C. Huang, and A. Pasternak. Interatomic forces and lattice dynamics of α-quartz. *J. Physics C*, 9:3925–3940, 1976.

[16] G. Barrow. *The Structure of Molecules*. Benjamin, New York, 1963.

[17] B. Di Bartolo. *Optical Interaction in Solids*. Wiley, New York, 1968.

[18] M. Bartowiak and G.D. Mahan. Heat and electricity transport through interfaces, recent trends in thermoelectric materials. *Semiconductors and Semimetals*, 70:245–271, 2001.
[19] G.K. Batchelor. *The Theory of Homogeneous Turbulence*. Cambridge University Press, Cambridge, 1953.
[20] A.R. Beattie and A.M. White. An analytic approximation with a wide range of applicability for electron initiated auger transitions in narrow-gap semiconductors. *J. Applied Physics*, 79:802–813, 1996.
[21] A.R. Beattie and A.M. White. An analytic approximation with a wide range of applicability for electron initiated auger transitions in narrow-gap semiconductors. *Semiconductor Science and Technology*, 12:357–368, 1997.
[22] L.X. Benedict, E.L. Shirley, and R.B. Bohn. Theory of optical absorption in diamond, Si, Ge, and GaAs. *Physics Review B*, 57:R9385–R9387, 1998.
[23] G. Beni and P.M. Platzman. Temperature and polarization dependence of external x-ray absorption fine-structure spectra. *Physical Review B*, 14:1514–1518, 1976.
[24] R. Berman. *Thermal Conduction in Solids*. Clarendon, Oxford, 1974.
[25] R.M. Besancon. *Encyclopedia of Physics*. Second Edition, Van Nostrand Reinhold, New York, 1976.
[26] C.M. Bhandari and D.M. Rowe. *Thermal Conductivity in Semiconductors*. Wiley, New York, 1988.
[27] A.B. Bhatia. *Ultrasonic Absorption*. Dover, New York, 1985.
[28] J.P. Biersack and J.F. Ziegler. Refined universal potentials in atomic collisions. *Nuclear Instruments Methods*, 194:93–100, 1982.
[29] D. Bilc, S.D. Mahanti, and M.G. Kanatzidis. Electronic transport properties of PbTe and $AgPb_mSbTe_{2+m}$ systems. *Physical Review B*, 74:125202–125213, 2006.
[30] G.D. Billing. *Dynamics of Molecule Surface Interactions*. Wiley, New York, 2000.
[31] R.B. Bird, W.E. Stewant, and E.N. Lightfoot. *Transport Phenomena*. Second Edition, Wiley, 2001.
[32] L.A. Bisebergy and H.W. Moos. Multiphonon orbit-lattice relaxation of excited states of rare-earth ions in crystals. *Physical Review*, 174:429–438, 1968.
[33] P. Blaha, K. Schwarz, G. Madsen, D. Kvasnicka, and J. Luitz. *Wien2k User's Guide*. Vienna University of Technology, Getreidemarkt, 2001.
[34] N.P. Blake and H. Metiu. *Chemistry, Physics, and Material Science of Thermoelectric Materials: Beyond Bismuth Telluride*. M.G. Kanatzidis, et al., Editors, Kluwer Academic/Plenum, London, 2003.
[35] J. Blömer and A. E. Beylich. Molecular dynamic simulation of energy accomodation of internal and translational fluid degrees of freedom at gas-solid interfaces. *Surface Science*, 423:127–133, 1999.
[36] K. Blotekjaer. Transport equations for electrons in two-valley semiconductors. *IEEE Transactions on Electron Devices*, 17:38–47, 1970.
[37] L. Boltzmann. *Lecture Notes on Gas Theory, Translated from the German Editions (Part 1; 1896, and Part 2; 1898)*. Dover, New York, 1964.
[38] M. Born and E. Wolf. *Principles of Optics*. Sixth Edition, Pergamon, Elmsford, 1980.
[39] M.I. Boulos. Thermal plasma processing. *IEEE Transactions on Plasma Science*, 19:1078–1089, 1991.
[40] F.B. Bowden and W.R. Throssell. Adsorption of water vapor on solid surfaces. *Proceedings Royal Society London*, A209:297–308, 1951.
[41] S. Brandt and H.D. Dahmen. *The Picture Book of Quantum Mechanics*. Second Edition, Springer-Verlag, New York, 1995.

[42] P.W. Bridgman. The thermal conductivity of liquids under pressure. *Proceedings of American Academy of Arts and Science*, 99:141–169, 1923.

[43] H.W. Brinkman, W.J. Briels, and H. Verweij. Molecular dynamics simulations of yttria-stabilized zirconia. *Chemical Physics Letters*, 247:386–390, 1995.

[44] D.A. Broido and T.L. Reinecke. Effect of superlattice structure on the thermoelectric figure of merit. *Physical Review B*, 51:13797–13800, 1994.

[45] D.A. Broido, A. Ward, and N. Mingo. Lattice thermal conductivity of silicon from empirical interatomic potentials. *Physical Review B*, 72:014308-1-8, 2005.

[46] R.H. Bube. *Electrons in Solids: An Introductory Survey*. Third Edition, Academic Press, San Diego, 1992.

[47] R.H. Bube. *Photovoltaic Materials, In: Series on Properties of Semiconductor Materials - Vol.1*. Imperial College Press, New Jersey, 1998.

[48] R.A. Buckingham. The classical equation of state of gaseous helium, neon, and argon. *Proceedings Royal Society London, Series A*, 168:264–283, 1938.

[49] D.E. Burch. *Infrared Absorption by Carbon Dioxide, Water Vapor, and Minor Atmospheric Constituents*. US Department of Commerce Report, Washington, 1962.

[50] G.A. Burdick. Energy band structure of copper. *Physical Review*, 129:138–150, 1963.

[51] D.G. Cahill and S.K. Pohl. Heat flow and lattice vibrations in glasses. *Solid State Communications*, 70:6131–6140, 1992.

[52] D.G. Cahill, S.K. Watson, and R.O. Pohl. Lower limit to thermal conductivity of disordered crystals. *Physical Review B*, 119:1–9, 1992.

[53] J. Callaway. Model for lattice thermal conductivity at low temperatures. *Physical Review*, 113:1046–1051, 1959.

[54] H.B. Callen. *Thermodynamics and an Introduction to Thermostatistics*. Wiley, New York, 1985.

[55] V.P. Carey. *Statistical Thermodynamics and Microscale Thermophysics*. Cambridge University Press, Cambridge, 1999.

[56] V.P. Carey, G. Chen, C. Grigoropoulos, M. Kaviany, and A. Majumder. A review of heat transfer physics. *Nanoscale and Microscale Thermophysical Engineering*, 12:1-60, 2008.

[57] V.P. Carey and A.P. Wemhoff. Disjoining pressure effects in ultra-thin liquid films in micropassages-comparison of thermodynamic theory with predictions of molecular dynamics simulations. *ASME J. Heat Transfer*, 128:1276–1284, 2006.

[58] R. Carminati, P. Chantrenne, S. Dihaire, S. Gomex, N. Trannoy, and G. Tessier. *Microscale and Nanoscale Heat Transfer*. Springer, Berlin, 2007.

[59] S. Chandrasehkar. *Hydrodynamic and Hydromagnetic Stability*. Dover, New York, 1961.

[60] C.H. Chang and E. Pfender. Heat and momentum transport to particulates injected into low-pressure ($\tilde{8}0$ mbar) nonequilibrium plasmas. *IEEE Transactions on Plasma Science*, 18:958–967, 1990.

[61] C.W. Chang, D. Okawa, A. Majumdar, and A. Zettl. Solid-state thermal rectifier. *Science*, 314:1121–1124, 2006.

[62] J. Che, T. Cagin, W. Deng, and W.A. Goddard III. Thermal conductivity of diamond and related materials from molecular dynamics simulations. *J. Chemical Physics*, 113:6888–6900, 2000.

[63] J.R. Chelikowsky and M.L. Cohen. Non-local pseudopotential calculations for the electronic structure of eleven diamond and zinc-blende semiconductors. *Physical Review B*, 14:556–582, 1976.

[64] F.F. Chen. *Introduction to Plasma Physics and Controlled Fusion*. Plenum, New York, 1984.

[65] G. Chen. Thermal conductivity and ballistic-phonon transport in the cross-plane direction of superlattice. *Physical Review B*, 23:14598–14973, 1998.

[66] G. Chen. *Nanoscale Energy Transport and Conversion*. Oxford University Press, Oxford, 2005.

[67] G. Chen and A. Shakouri. Heat transfer in nanostructures for solid-state energy conversion. *ASME J. Heat Transfer*, 124:242–252, 2002.

[68] X. Chen and P. Han. On the thermodynamic derivation of the saha equation modified to a two-temperature plasma. *J. Physics D-Applied Physics*, 32:1711–1718, 1999.

[69] A. Chimmalgi, C.P. Grigoropoulos, and K. Komvopoulos. Surface nanostructuring by nano-/femtosecond laser-assisted scanning force microscopy. *J. Applied Physics*, 97:104319–1–12, 2005.

[70] F.C. Chou, J.R Lukes, X.G. Liang, K. Takahashi, and C.L. Tien. Molecular dynamics in microscale thermophysical engineering. *Annual Review of Heat Transfer*, 10:141–176, 1999.

[71] D.K. Christen and G.L. Pollack. Thermal conductivity of solid argon. *Physical Review B*, 12:3380–3391, 1975.

[72] J.D. Chung, A.J.H. McGaughey, and M. Kaviany. Role of phonon dispersion on lattice thermal conductivity modeling. *ASME J. Heat Transfer*, 126:376–380, 2004.

[73] W.T. Coffey, Yu.P. Kalmykov, and J.T. Waldron. *The Langevin Equation*. Second Edition, World Scientific, Hackensack, 2004.

[74] C. Cohen-Tannoudji, B. Diu, and F. Laloe. *Quantum Mechanics*. Wiley, New York, 2007.

[75] C. Cohen-Tannoudji, J. Dupont-Roc, and C. Grynberg. *Atom-Photon Interactions: Basic Processes and Applications*. Wiley-Interscience, Weinheim, 1998.

[76] E.U. Condon and G.H. Shortley. *The Theory of Atomic Spectra*. Cambridge Univeristy Press, London, 1935.

[77] R.M. Costescu, M.A. Wall, and D.G. Cahill. Thermal conductance of epitaxial interfaces. *Physical Review B*, 67:054302–1–5, 2003.

[78] S. Cottenier. *Density Functional Theory and the Family of (L)APW-Methods: A Step-by-Step Introduction*. http://www.wien2k.at/reg_user/textbooks/, 2004.

[79] E.G. Cravalho, C.L. Tien, and R.P. Caren. Effect of small spacings on radiative transfer between two dielectrics. *ASME J. Heat Transfer*, 89:351–358, 1967.

[80] P. Cross, J. Decius, and E. Wilson. *Molecular Vibrations*. McGraw-Hill, Totonto, 1955.

[81] S.J. Cyvin. Vibrational mean-square amplitude matrices. 1. secular equations involving mean-square amplitudes of vibration, and approximate computations of mean-square amplitude matrices. *Spectrochemica Acta*, 10:828–834, 1959.

[82] L.W. da Silva. *PhD Thesis: Integrated Micro Thermoelectric Cooler: Theory, Fabrication and Characterization*. University of Michigan, Ann Arbor, 2005.

[83] L.W. da Silva and M. Kaviany. Micro thermoelectric cooler: Interfacial effects on thermal and electrical transport. *International J. Heat Mass Transfer*, 47:2417–2435, 2004.

[84] G.G. de la Cruz and Yu.G. Gurvich. Electron and phonon thermal waves in semiconductors: An application to photothermal effects. *J. Applied Physics*, 80:1726–1730, 1996.

[85] P.J.U Debye. *The Collected Papers of Peter J.W. Debye*. OX Bow Press, Woodbridge, 1988.

[86] L.D. Deloach, S.A. Payne, L.L. Chase, L.K. Smith, W.L. Kway, and W.F. Krupke. Evaluation of absorption and emission of Yb^{3+} doped crystals for laser applications. *IEEE J. Quantum Electronics*, 29:1179–1191, 1993.

[87] G.H. Dieke and H.M. Crosswhite. The spectra of the doubly and triply ionized rare earths. *Applied Optics*, 2:675–686, 1963.

[88] M.J.F. Digonnet. *Rare Earth Doped Fiber Lasers and Amplifiers*. Marcel Dekker, New York, 1993.

[89] Y. Ding, A. Campargue, E. Bertseva, S. Tashkun, and V.I. Perevalov. Highly sensitive absorption spectroscopy of carbon dioxide by ICLAS-VeCSEL between 8800 and 9530 cm^{-1}. *J. Molecular Spectroscopy*, 231:117–123, 2005.

[90] G. Domingues, S. Volz, K. Joulian, and J.J. Greffet. Heat transfer between two nanoparticles through near field interactions. *Physical Review Letters*, 94:085901(1–4), 2005.

[91] M.T. Dove. *Introduction to Lattice Dynamics*. Cambridge University Press, Cambridge, 1993.

[92] M.S. Dresselhaus. *Solid State Physics, Part II: Optical Properties*. MIT, unpublished, 2005.

[93] S.J. Duclos, K. Brister, R.C. Haddon, A.R. Kortan, and F.A. Thiel. Effects of pressure and stress on C_{60} fullerite to 20 GPa. *Nature*, 351:380–382, 1991.

[94] J.S. Dugdale. *Electical Properties of Metals and Alloys*. Edward Arnold Publishers, London, 1977.

[95] W. Duley. CO_2 *Lasers: Effects and Applications*. Academic Press, London, 1976.

[96] E.R.G. Eckert and E. Pfender. Advances in plasma heat transfer. *Advances in Heat Transfer*, 4:229–310, 1967.

[97] B.C. Edwards, J.E. Anderson, R.I. Epstein, G.L. Mills, and A.J. Mord. Demonstration of a solid-state optical cooler: an approach to cryogenic refrigeration. *J. Applied Physics*, 86:6489–6493, 1999.

[98] E.K. Edwards. *Radiative Heat Transfer Notes*. Hemisphere, Washington, 1981.

[99] K. Eikema, W. Ubachs, W. Vassen, and H. Horgorvorst. Lamb shift measurement in the $1t^1s$ ground state of helium. *Physical Review A*, 55:1866–1884, 1997.

[100] A. Einstein. Elementare betrachtungen uber die thermische molekularbewegung in festen korpern. *Annalen der Physik*, 35:679–694, 1911.

[101] R.J. Elliolt and A.F. Gibson. *An Introduction to Solid State Physics and Its Applications*. Harper & Row, New York, 1974.

[102] R.I. Epstein, J.J. Brown, B.C. Edwards, and A. Gibbs. Measurements of optical refrigeration in ytterbium-doped crystals. *J. Applied Physics*, 90:4815–4819, 2001.

[103] R.I. Epstein, M.I. Buckwald, B.C. Edwards, T.R. Gosnell, and C.E. Mungan. Observation of laser-induced fluorescent cooling of a solid. *Nature*, 377:500–503, 1995.

[104] I.D. Feranchuk, A.A. Minkevich, and A.P. Ulyanenkov. Estimate of the Debye temperature for crystal, with polyatomic unit cell. *European Physics: J. Applied Physics*, 19:95–101, 2002.

[105] J. Fernandez, A. Mendioroz, A.J. Garcia, R. Balda, and J.L. Adam. Anti-stokes laser-induced internal cooling of Yb^{3+}-doped glasses. *Physical Review B*, 62:3213–3217, 2000.

[106] J. Fernandez, A. Mendioroz, A.J. Garcia, R. Balda, J.L. Adam, and M.A. Arriandiaga. On the origin of anti-stokes laser-induced cooling of Yb^{3+}-doped glass. *Optical Materials*, 16:173–179, 2001.

[107] C. Florea and K.A Winick. Ytterbium-doped glass waveguide laser fabricated by ion exchange. *IEEE J. Lightwave Technology*, 17:1593–1601, 1999.

[108] J.B. Foresman and A.E. Frisch. *Exploring Chemistry with Electronic Structure Methods*. Gaussian Inc., Pittsburgh, 1996.

[109] A.R. Forouhi and I. Bloomer. Optical properties of crystalline semiconductors and dielectrics. *Physical Review B*, 38:1865–1874, 1988.

[110] M. Fowler. http://galileo.phys.virginia.edu/classes/752.mf1i.spring03/PhotoelectricEffect.htm, 2004.

[111] A.J. Freeman and R.E. Watson. Theoretical investigation of some magnetic and spectroscopic properties of rare-earth ions. *Physical Review*, 127:2058–2075, 1962.

[112] D. Frenkel and B. Smit. *Understanding Molecular Simulation: From Algorithms to Applications*. Academic, San Diego, 1996.

[113] M.J. Frisch, G.W. Trucks, H.B. Schlegel, G.E. Scuseria, M.A. Robb, J.R. Cheeseman, J.A. Montgomery, T. Vreven, K.N. Kudin, J.C. Burant, J.M. Millam, S.S. Iyengar, J. Tomasi, V. Barone, B. Mennucci, M. Cossi, G. Scalmani, N. Rega, G.A. Petersson, H. Nakatsuji, M. Hada, M. Ehara, K. Toyota, R. Fukuda, J. Hasegawa, M. Ishida, T. Nakajima, Y. Honda, O. Kitao, H. Nakai, M. Klene, X. Li, J.E. Knox, H.P. Hratchian, J.B. Cross, C. Adamo, J. Jaramillo, R. Gomperts, R.E. Stratmann, O. Yazyev, A.J. Austin, R. Cammi, C. Pomelli, J.W. Ochterski, P.Y. Ayala, K. Morokuma, G.A. Voth, P. Salvador, J.J. Dannenberg, V.G. Zakrzewski, S. Dapprich, A.D. Daniels, M.C. Strain, O. Farkas, D.K. Malick, A.D. Rabuck, K. Raghavachari, J.B. Foresman, J.V. Ortiz, Q. Cui, A.G. Baboul, S. Clifford, J. Cioslowski, B.B. Stefanov, G. Liu, A. Liashenko, P. Piskorz, I. Komaromi, R.L. Martin, D.J. Fox, T. Keith, M.A. Al-Laham, C.Y. Peng, A. Nanayakkara, M. Challacombe, P.M.W. Gill, B. Johnson, W. Chen, M.W. Wong, C. Gonzalez, and J.A. Pople. *Gaussian 03, Revision C.02*. Gaussian, Inc., Wallingford CT, 2004.

[114] C.J. Fu and Z.M. Zhang. Nanoscale radiation heat transfer for silicon at different doping levels. *International J. Heat Mass Transfer*, 49:1703–1718, 2006.

[115] K. Fushinobu, A. Majumdar, and K. Hijikata. Heat generation and transport in submicron semiconductor devices. *ASME J. Heat Transfer*, 117:25–31, 1995.

[116] J.D. Gale. Gulp-a computer program for symmetry adapted simulation of solids. *J. Chemical Society, Faraday Transactions*, 93:629–637, 1997.

[117] J. Le Gall, M. Olivier, and J.J. Greffet. Experimental and theoretical study of reflection and coherent thermal emission by a SiC grating supporting a surface-phonon polariton. *Physical Review B*, 55:10105–10114, 1997.

[118] W.R. Gambill and J.H. Lienhard. An upper-bound for critical boiling heat fluxes. *ASME J. Heat Transfer*, 111:815–818, 1989.

[119] C. Garrod. *Statistical Mechanics and Thermodynamics*. Oxford University Press, Oxford, 1995.

[120] S. Gasiorowicz. *Quantum Physics*. Third Edition, Wiley, New York, 2003.

[121] T.H. Geballe and G.W. Hall. Seebeck effect in silicon. *Physical Review*, 98:940–947, 1955.

[122] T.H. Geballe and G.W. Hull. Isotopic and other types of thermal resistance in germanium. *Physical Review*, 110:773–775, 1958.

[123] H.J. Goldsmid. *Electronic Refrigeration*. Pion, London, 1986.

[124] T.R. Gosnell. Laser cooling of a solid by 65 k starting from room temperature. *Optics Letters*, 24:1041–1043, 1999.

[125] K. Gottfried and T.M. Yan. *Quantum Mechanics: Fundamentals*. Second Edition, Springer, New York, 2004.

[126] H.B. Gray. *Electrons and Chemical Bonding*. Benjamin, New York, 1964.

[127] R.B. Greegor and F. W. Lytle. Extended x-ray absorption fine structure determination of thermal disorder in Cu: comparison of theory and experiment. *Physical Review B*, 20:4902–4907, 1979.

[128] M.A. Green. *Solar Cells, Operating Principles, Technology, and Systems Applications*. Prentice-Hall, Englewood Cliffs, 1982.

[129] M.A. Green, K. Emery, D. King, S. Igari, and W. Warta. Solar cell efficiency tables, version 24. *Progress of Photovoltaics: Research and Application*, 12:365, 2004.

[130] J.S. Griffith. *The Theory of Transition Metal Ion*. Cambridge Univeristy Press, London, 1961.

[131] D.J. Griffiths. *Introduction to Quantum Mechanics*. Third Edition, Prentice-Hall, Upper Saddle River, 1995.

[132] E. Grüneisen and E.S. Goens. Analysis of metal crystals iii thermic expansion of zinc and cadmium. *Zeitschrift für Physik (Z. Phys.)*, 29:141–156, 1924.

[133] Y. Guissani and B. Guillot. A numerical investigation of the liquid-vapor co-existance curve of silica. *J. Chemical Physics*, 104:7633–7644, 1996.

[134] V.L. Gurevich. *Transport in Phonon Systems*. North-Holland, Amsterdam, 1986.

[135] R.A. Hamilton and J.E. Parrott. Variational calculation of the thermal conductivity of germanium. *Physical Review*, 178:1284–1292, 1969.

[136] T.W. Hansch and A.L. Schawlow. Cooling of gases by laser radiation. *Optics Communications*, 13:68–69, 1975.

[137] T.C. Harman, P.J. Taylor, D.L. Spears, and M.P. Walsh. Thermoelectric quantum-dot superlattices with high *ZT*. *J. Electronic Materials*, 29:L1–L4, 2000.

[138] M. Hayashi, A.M. Mebel, K.K. Liang, and S.H. Lin. *Ab initio* calculations of radiationless transitions between excited and ground singlet electronic states of ethylene. *J. Chemical Physics*, 108:2044–2055, 1998.

[139] E. Hecht. *Optics*. Addison Wesley, San Francisco, 2002.

[140] B. Heeg, M.D. Stone, A. Khizhnyak, G. Rumbles, G. Mills, and P.A. Debarber. Experimental demonstration of intracavity solid-state laser cooling of Yb^{3+}: ZrF_4-BaF_2-LaF_3-AlF_3-NaF glass. *Physical Review A*, 70:021401(1–4), 2004.

[141] P. Heino and E. Ristolainen. Thermal conduction at nano scale in some metals by md. *Microelectronics J.*, 34:773–777, 2003.

[142] E. Helfand. Transport coefficient from dissipation in canonical ensemble. *Physical Review*, 119:1–9, 1960.

[143] C. Herring. A new method for calculating wave functions in crystals. *Physical Review*, 57:1169–1177, 1940.

[144] C. Herring. Role of low-energy phonons in thermal conduction. *Physical Review*, 95:954–965, 1954.

[145] G. Herzberg. *Molecular Spectra and Molecular Structure II. Infrared and Raman Spectra of Polyatomic Molecules*. Van Nostrand, New York, 1945.

[146] G. Herzberg. *Molecular Spectra and Molecular Structure IV. Constants of Diatomic Molecules*. Van Nostrand, New York, 1979.

[147] R.C. Hilborn. Einstein coefficients, cross sections, f values, dipole moments, and all that. *American J. Physics*, 50:982–986, 1982.

[148] J.O. Hinze. *Turbulence*. Second Edition, McGraw-Hill, New York, 1975.

[149] C.Y. Ho and Y.S. Touloukian. *Thermophysical Properties of Matter, the TPRC Data Series*. IFI-Plenum, New York, 1970.

[150] K. Hoang, S.D. Mahanti, J. Androulakis, and M.G. Kanatzidis. Electronic structure of $AgPb_mSbTe_{m+2}$ compounds – implications on thermoelectric properties. *Materials Research Society Symposium Proceedings*, 0886:F05–06.5, 2006.

[151] M.G. Holland. Analysis of lattice thermal conductivity. *Physical Review*, 132:2461–2471, 1963.

[152] B.L. Huang and M. Kaviany. Structural metrics of high-temperature lattice conductivity. *Journal Applied Physics*, 100:123507–1–12, 2006.

[153] B.L. Huang and M. Kaviany. *Ab-Initio* and md predictions of electron and phonon transport in bismuth telluride. *Physical Review B*, 77:125209-1-19, 2008.

[154] B.L. Huang, A.J.H. Mcgaughey, and M. Kaviany. Thermal conductivity of metal-organic framework 5 (mof-5): Part 1. molecular dynamic simulations. *International J. Heat Mass Transfer*, 50:393–404, 2006.

[155] B.L. Huang, A.J.H. Mcgaughey, and M. Kaviany. Thermal conductivity of metal-organic framework 5 (mof-5): Part 2. measurement. *International J. Heat Mass Transfer*, 50:405–411, 2006.

[156] J.N. Israelachvili. *Intermolecular and Surface Forces: With Applications to Colloidal and Biological Systems*. Academic, San Diego, 1985.

[157] M.Y. Jaffrin. Shock structure in a partially ionized gas. *Physics Fluids*, 8:606–625, 1965.

[158] S.P. Jang and S.U.S. Choi. Role of brownian motion in the enhanced thermal conductivity of nanofluids. *Applied Physics Letters*, 84:431–4318, 2004.

[159] J. Javanainen. Light-pressure cooling of trapped ions in three dimensions. *Applied Physics A: Materials Science Processing*, 23:175–182, 1980.

[160] H.W. Jeon, H.P. Ha, D.B. Yun, and J.D. Shim. Electrical and thermoelectrical properties of undoped Bi_2Te_3-Sb_2Te_3 and Bi_2Te_3-Sb_2Te_3-Sb_2Se_3 single crystals. *J. Physics Chemistry of Solids*, 52:579–585, 1991.

[161] E.X. Jin and X. Xu. Radiating transfer through nanoscale apertures. *J. Quantitative Spectroscopy and Radiative Transfer*, 93:163–173, 2005.

[162] E.X. Jin and X. Xu. Obtaining subwavelength optical spots using nanoscale ridge apertures. *ASME J. Heat Transfer*, 129:37–43, 2007.

[163] D.D. Joseph. *Stability of Fluid Motion II*. Springer, Berlin, 1976.

[164] K. Joulian, J.P. Mulet, F. Marguier, R. Caminati, and J.J. Greffet. Surface electromagnetic waves thermally excited: Radiative heat transfer, coherence properties and casimir forces revisited in the near field. *Surface Science Reports*, 57:59–112, 2005.

[165] Y.S. Ju and K.E. Goodson. Phonon scattering in silicon films with thickness of order 100 nm. *Applied Physics Letters*, 74:3005–3007, 1999.

[166] B.R. Judd. The theory of spectra of europium salts. *Proceedings Royal Society London. Series A*, 228:120–128, 1955.

[167] C.L. Julian. Theory of conduction in rare-gas crystals. *Physical Review*, 137:A128–137, 1965.

[168] H. Kaburaki, J. Li, and S. Yip. Thermal conductivity of solid argon by classical molecular dynamics. *Materials Research Society Symposia Proceedings*, 503:503–508, 1998.

[169] M.I. Kaganov and I.M. Lifshits. *Quasiparticles*. V. Kissin Translator (from Russian), Mir, Moscow, 1979.

[170] B. Kallies and R. Meier. Electronic structure of 3d $[M(H_2O)_6]^{3+}$ ions from Sc^{III}: A quantum mechanical study based on dft computations and natural bond orbital analyses. *Inorganic Chemistry*, 40:3101–3112, 2001.

[171] M. Kaviany. Onset of thermal convection in a saturated porous medium. *International J. Heat Mass Transfer*, 27:2001–2010, 1984.

[172] M. Kaviany. *Principles of Heat Transfer in Porous Media*. Second Edition, Springer, New York, 1995.

[173] M. Kaviany. *Principles of Convective Heat Transfer*. Second Edition, Springer, New York, 2001.

[174] M. Kaviany. *Principles of Heat Transfer*. Wiley, New York, 2001.

[175] W.M. Kays and M.T. Crawford. *Convective Heat Mass Transfer*. Third Edition, McGraw-Hill, New York, 1993.

[176] P. Keblinski and D.G. Cahill. Comment on model for heat conduction in nano fluids. *Physical Review Letters*, 95:209401, 2005.
[177] E.H. Kennard. *Kinetic Theory of Gases*. McGraw-Hill, New York, 1938.
[178] M. Kerker. *The Scattering of Light*. Academic, New York, 1969.
[179] M. Kilo, R.A. Jackson, and G. Borchardt. Computer modelling of ion migration in zirconia. *Philosophical Magzine*, 83:3309–3325, 2003.
[180] J. Kim, A. Kapoor, and M. Kaviany. Material metrics for laser cooling of solids. *Physical Review B*, 77:115127-1-15, 2008.
[181] A. Kittel, W. Müller-Hirsch, J. Parisi, S.A. Biehs, D. Reddig, and M. Holthaus. Near-field heat transfer in a scanning thermal microscope. *Physical Review Letters*, 95:224301–1–4, 2005.
[182] C. Kittel. Ultrasonic resonance and properties of matter. *Reports on Progress in Physics*, 11:205–247, 1946.
[183] C. Kittel. *Introduction to Solid State Physics*. Seventh Edition, Wiley, New York, 1986.
[184] P.G. Klemens. The thermal conductivity of dielectric solids at low temperatures. *Proceedings Royal Society London, Series A*, 208:108–133, 1951.
[185] P.G. Klemens. The scattering of low-frequency lattice waves by static imperfections. *Proceedings Physical Society London, Series A*, 68:1113–1128, 1955.
[186] P.G. Klemens. Thermal conductivity and lattice vibrational modes. In: *Solid State Physics*, F. Seitz and D. Turnbull (eds.), Academic, New York, 1958, Vol. 7, p. 1.
[187] T. Koga, X. Sun, S.B. Cronin, and M.S. Dresselhaus. Carrier pocket engineering applied to design useful thermoelectric materials using GaAs/AlAs superlattices. *Applied Physics Letters*, 75:2438–2440, 1998.
[188] A. Konard, U. Herr, R. Tidecks, F. Kummer, and K. Samwer. Luminescence of bulk and nanocrystalline cubic yttria. *J. Applied Physics*, 90:3516–3523, 2001.
[189] M. Konuma. *Film Deposition by Plasma Techniques*. Springer-Verlag, Berlin, 1992.
[190] J. Korringa. On calculation of the energy of a bloch wave in a metal. *Physica*, 13:392–400, 1947.
[191] M. Kozlowski and J. Marciak-Kozlowska. Beyond the Fourier equation: quantum hyperbolic heat transport. *arXiv.org:cond-mat/0304052*, April 2003.
[192] G.J. Kramer, N.P. Farragher, B.W.H. van Beest, and R.A. van Saton. Interatomic force fields for silicas, aluminophosphates, and zeolites: Derivation based on *ab initio* calculations. *Physical Review B*, 43:5068–5080, 1991.
[193] G. Kresse and J. Furthmüler. Efficient iterative schemes for *ab initio* total-energy calculations using a plane-wave basis set. *Physical Review B*, 54:11169–11186, 1997.
[194] G. Kresse and J. Furthmüler. *VASP the Guide*. http://cms.mpi.univie.ac.at/vasp/vasp/vasp.html, 2005.
[195] T. Kunikiyo, M. Takenaka, and Y. Kamakura. A monte carlo simulation of anisotropic electron transport in silicon including full band structure and anisotropic impact-ionization model. *J. Applied Physics*, 75:297–312, 1994.
[196] A.J.C. Ladd, B. Moran, and W.G. Hoover. Lattice thermal conductivity: A comparison of molecular dynamics and anharmonic lattice dynamics. *J. Chemical Physics*, 34:5058–5064, 1986.
[197] J.F. Lancaster. *The Physics of Welding*. Pergamon, Oxford, 1986.
[198] L. Landau. On the thermodynamics of photoluminescence. *J. Physics (Moscow)*, 10:503–506, 1962.
[199] E.S. Landry, M.I. Hossein, and A.J.H. McGaughey. Complex superlattice unit cell designs for reduced thermal conductivity. *Physical Review B*, 77:184302–1–13, 2008.

[200] A.R. Leach. *Molecular Modelling Principles and Applications*. Addison-Wesley-Longman, Essex, 1996.

[201] S. Lee and P. von Allmen. Tight-binding modeling of thermoelectric properties of bismuth telluride. *Applied Physics Letters*, 88:221071–221073, 2006.

[202] S.M. Lee, D.G. Cahill, and R. Venkatasubramanian. Thermal conductivity of Si − Ge superlattices. *Applied Physics Letters*, 70:2957–2959, 1997.

[203] Y.H. Lee, R. Biswas, C.M. Soukoulis, C.Z. Wang, CT. Chan, and K.M. Ho. Molecular-dynamics simulation of thermal conductivity in amorphous silicon. *Physical Review B*, 43:6573–6580, 1991.

[204] G. Lei, J.E. Anderson, M.I. Buchwald, B.C. Edwards, R.I. Epstein, M.T. Murtagh, and G.H. Sigel. Spectroscopic evaluation of Yb^{3+}-doped glasses for optical refrigeration. *IEEE J. Quantum Electronics*, 34:1839–1845, 1998.

[205] I.N. Levine. *Physical Chemistry*. Third Edition, McGraw-Hill, New York, 1988.

[206] G.V. Lewis and C.R.A. Catlow. Potential models for ionic oxides. *J. Physics C: Solid State Physics*, 18:1149–1161, 1985.

[207] J. Li. *Ph.D. Thesis: Modeling microstrutural effects on deformation resistance and thermal conductivity*. Massachusetts Institute of Technology, Cambridge, 2000.

[208] J. Li, L. Porter, and S. Yip. Atomistic modeling of finite-temperature properties of crystalline β − SiC. 2. thermal conductivity and effects of point defects. *J. Nuclear Materials*, 255:139–152, 1998.

[209] M.D. Licker, editor. *Dictionary of Scientific and Technical Terms*. Sixth Edition, McGraw Hill, New York, 2003.

[210] D.R. Lide, editor. *CRC Handbook of Chemistry and Physics*. 82nd Edition, CRC Press, Boca Raton, 2001.

[211] S.H. Lin. Rate of interconversion of electronic and vibrational energy. *J. Chemical Physics*, 44:3759–3767, 1966.

[212] D. Lindley. *Boltzmann's Atoms*. Free Press, New York, 2001.

[213] W.A. Little. The transport of heat between dissimilar solids at low temperatures. *Canadian J. Physics*, 37:334–349, 1959.

[214] C.K. Loong, P. Vashishta, R.K. Kalia, W. Jin, M.H. Degani, D.G. Hinks, D.L. Price, J.D. Jorgensen, B. Dabrowski, A.W. Mitchell, D.R. Richards, and Y. Zheng. Phonon density of states and oxgen-isotope effect in $Ba_{1-x}K_xBiO_3$. *Physical Review B*, 45:8052–8064, 1992.

[215] P. Lorrain and D.R. Corson. *Electromagnetic Fields and Waves*. Second Edition, Freeman, San Francisco, 1970.

[216] R. Loudon. *The Quantum Theory of Light*. Third Edition, Oxford University Press, Oxford, 2000.

[217] S.K. Loyalka. Slip in thermal creep flow. *Physics of Fluids*, 14:21–24, 1971.

[218] V.A. Luchnikov, N.N. Medvedev, Y.I. Naberukhin, and V.N. Novikov. Inhomogeneity of the spatial distribution of vibrational modes in a computer model of amorphous argon. *Physical Review B*, 51:15569–15572, 1995.

[219] M. Lundstrom. *Fundamentals of Carrier Transport*. Second Edition, Cambridge University Press, Cambridge, 2000.

[220] X. Luo, M.D. Eisaman, and T.R. Gosnell. Laser cooling of a solid by 21 k starting from room temperature. *Optics Letters*, 23:639–641, 1998.

[221] O. Madelung, editor. *Semiconductors-Basic Data*. Springer, Berlin, 1996.

[222] G.K.H. Madsen and D.J. Singh. Boltztrap. a code for calculating band-structure dependent quantities. *Computer Physics Communication*, 175:67–71, 2006.

[223] G.D. Mahan. Good thermoelectrics. *Solid State Physics*, 51:82–152, 1997.

[224] G.D. Mahan and J.O. Sofo. Heat transfer in nanostructures for solid-state energy conversion. *Proceedings of National Academy of Science, USA*, 93:7436–7439, 1996.

[225] A. Maiti, G.D. Mahan, and S.T. Pantelides. Dynamical simulations of nonequilibrium processes - heat flow and the kapitza resistance across grain boundaries. *Solid State Communications*, 102:517–521, 1997.

[226] A. Majumdar. *Microscale energy transport in solids, In: Microscale Energy Transport*. Taylor & Francis, Washington, 1998.

[227] A. Majumdar and P. Reddy. Role of electron-phonon coupling in thermal conductance of metal-nonmetal interfaces. *Applied Physics Letter*, 84:4768–4771, 2004.

[228] J. Marciak-Kozlowska. Wave characteristic of femtosecond heat conduction in thin films. *Int. J. Thermophysics*, 14:593–398, 1993.

[229] A. Martí and G. Araújo. Limiting efficiencies for photovoltaic energy conversion in multigap systems. *Solar Energy Materials and Solar Cells*, 43:203–222, 1996.

[230] F.G. Mateo, J. Zuniga, A. Requena, and A. Hidalgo. Energy eigenvalues for lennard–jones potential using the hypervirial perturbative method. *J. Physics B: Atomic, Molecular and Optical Physics*, 23:2771–2781, 1990.

[231] S. Mazumder and A. Majumdar. Monte carlo study of phonon transport in solid thin films including dispersion and polarization. *ASME J. Heat Transfer*, 123:749–759, 2001.

[232] A.J.H. McGaughey. *Ph.D Thesis: Phonon Transport in Molecular Dynamics Simulations: Formulation and Thermal Conductivity Predictions*. University of Michigan, Ann Arbor, 2004.

[233] A.J.H. McGaughey and M. Kaviany. Quantitative validation of the boltzmann transport equation phonon thermal conductivity model under the single mode relaxation time approximation. *Physical Review B*, 69:94303–1–12, 2004.

[234] A.J.H. McGaughey and M. Kaviany. Thermal conductivity decomposition and analysis using molecular dynamics simulations: Part i. lennard–jones argon. *International J. Heat Mass Transfer*, 47:1783–1798, 2004.

[235] A.J.H. McGaughey and M. Kaviany. Thermal conductivity decomposition and analysis using molecular dynamics simulations: Part ii. complex silica structures. *International J. Heat Mass Transfer*, 47:1799–1816, 2004.

[236] A.J.H. McGaughey and M. Kaviany. Observation and analysis of phonon interactions in molecular dynamics simulations. *Physical Review B*, 71:184305–1–11, 2005.

[237] A.J.H. McGaughey and M. Kaviany. Phonon transport in molecular dynamics simulations: Formulation and thermal conductivity prediction. *Advances in Heat Transfer*, 37:169–225, 2005.

[238] A.R. McGurn, K.T. Christensen, F.M. Mueller, and A.A. Maradudin. Anderson localization in one-dimensional randomly disordered optical systems that are periodic on average. *Physical Review B*, 47:13120–13125, 1993.

[239] D.A. McQuarrie. *Statistical Mechanics*. University Science Books, Sausalito, 2000.

[240] D.A. McQuarrie and J.D. Simon. *Physical Chemistry: A molecular Approach*. University Science Books, Sausalito, 1997.

[241] E. Merzbacher. *Quantum Mechanics*. Third Edition, Wiley, New York, 1997.

[242] H.J. Metcalf and P. Van der Straten. *Laser Cooling and Trapping*. Springer, New York, 1999.

[243] R. Meyer, L.J. Lewis, S. Prakash, and P. Entel. Vibrational properties of nanoscale materials: from nanoparticles to nanocrystalline materials. *Physical Review B*, 68:104303(1–9), 2003.

[244] M.I. Mishchenko, L.D. Travis, and A.A. Lacis. *Scattering, Absorption, and Emission of Light by Small Particles*. Cambridge University Press, Cambridge, 2002.

[245] M.F. Modest. *Radiation Heat Transfer*. McGraw-Hill, New York, 1993.

[246] A.S. Monin and A.M. Yaglom. *Statistical Fluid Mechanics: Mechanics of Turbulence, Volume 2*. MIT Press, Cambridge, 1979.

[247] E. Montoya, F. Agullo-Rueda, S. Manotas, J. Garciá Solé, and L.E. Bausa. Electron-phonon coupling in Yb^{3+}: $LiNbO_3$ laser crystal. *J. Luminescence*, 94-95:701–705, 2001.

[248] M. Morrison, N. Lane, and T. Estle. *Quantum States of Atoms, Molecules, and Solids*. Prentice-Hall, Englewood Cliffs, 1976.

[249] J. Mostaghimi, G.Y. Zhao, and M.I. Boulos. The induction plasma chemical reactor: Part i. equilibrium model. *Plasma Chemistry and Plasma Processing*, 10:133–150, 1990.

[250] C.E. Mungan, M.I. Buchwald, B.C. Edwards, R.I. Epstein, and T.R. Gosnell. Internal laser cooling of Yb^{3+}-doped glass measured between 100 and 300 k. *Applied Physics Letters*, 71:1458–1460, 1997.

[251] A. Narayanasawmy, D.Z. Chen, and C. Chen. Near-field radiative transfer between two spheres. *IMECE 2006*, 15845, 2006.

[252] J. Nelson. *The Physics of Solar Cells*. Imperial College Press, New Jersey, 2003.

[253] D.J. Newman. Theory of lanthanide crystal fields. *Advances in Physics*, 20:197–256, 1971.

[254] G. Nilsson and G. Nelin. Phonon dispersion relations in Ge at 80 k. *Physical Review B*, 3:364–369, 1971.

[255] A.J. Nozik. Spectroscopy and hot electron relaxation dynamics in semiconductor quantum wells and quantum dots. *Annual Review of Physical Chemistry*, 52:193–231, 2001.

[256] B.H. O'Connor and T.M. Valentine. A neutron diffraction study of crystal structure of c-form of yttrium sesquioxide. *Acta Crystallographica, Section B*, B25:2140, 1969.

[257] L. Onsager. Reciprocal relationships in irreversible processes i. *Physical Review*, 37:405–426, 1931.

[258] C. Henkel K. Joulain P.-O. Chapuis, S. Volz, and J.-J Greffet. Effects of spatial dispersion in near-field radiative heat transfer between two parallel metallic surfaces. *Physical Review B*, 77:035431(1–9), 2008.

[259] J.P. Paolini. The bond order-bond length relationship. *J. Computational Chemistry*, 11:1160–1163, 1990.

[260] J.R. Pardo, J. Cernicharo, and E. Serabyn. Atmospheric transmission at microwaves (atm): An improved model for millimeter/submillimeter applications. *IEEE Transactions Antennas Propagation*, 49:1683–1694, 2001.

[261] S.P. Parker, J.C. Bellows, J.S. Gallagher, and A.H. Hervey. *Encyclopedia of Physics*. Second Edition, McGraw-Hill, New York, 1993.

[262] J.E. Parrott. High temperature thermal conductivity of semiconductor alloys. *Proceedings Physical Society*, 81:726–735, 1963.

[263] M.G. Paton and E.N. Maslen. A refinement of crystal structure of yttria. *Acta Crystallographica*, 19:307–310, 1965.

[264] L. Pauling. *The Nature of Chemical Bonds*. Third Edition, Cornell Universtiy Press, Ithaca, 1960.

[265] R.E. Peierls. On the kinetic theory of thermal conduction in crystals. *Annalen der Physik*, 3:1055–1101, 1929.

[266] H. Petschek and S. Byron. Approach to equilibrium ionization behind story shock waves in argon. *Annals of Physics*, 1:270–315, 1957.

[267] M. Peyrard. The design of thermal rectifier. *Europhysics Letters*, 76:49–55, 2006.

[268] P.E. Phelan. Application of diffuse mismatch theory to the prediction of thermal boundary resistance in thin-film high-t_c superconductors. *ASME J. Heat Transfer*, 120:37–43, 1998.

[269] W.D. Phillips. Laser cooling and trapping of neutral atoms (nobel lecture). *Review of Modern Physics*, 70:721–741, 1998.

[270] T. Plechacek, J. Navratil, J. Horak, and P. Lostak. Defect structure of Pb-doped Bi_2Te_3 single crystals. *Philosophical Magazine*, 84:2217–2227, 2004.

[271] R.C. Powell. *Physics of Solid-State Laser Materials*. Spinger-Verlag, New York, 1998.

[272] R.C. Powell, G.E. Venikouas, L. Xi, and J.K. Tyminski. Thermal effects on the optical spectra of $Al_2O_3 : Ti^{3+}$. *J. Chemical Physics*, 84:662–665, 1986.

[273] P. Pringsheim. Zwei bemerkungen uber den unterschied von lumineszenz und temperaturstrahlung. *Zeitshrift für Physik (Z. Phys.)*, 57:739–746, 1929.

[274] J. Puibasset and R.J.M. Pellanq. A grand canonical monte carlo simulation study of water adsorption in vycor-like hydrophilic mesoporous silica at different temperatures. *J. Physics: Condensed Matter*, 16:5329–5243, 2004.

[275] G. Racah. On the decomposition of tensors by contraction. *Reviews of Modern Physics*, 21:494–496, 1949.

[276] A.D. Rakic and M.L. Majewski. Modeling the optical dielectric function of GaAs and AlAs: Extension of adachi's model. *J. Applied Physics*, 80:5909–5914, 1996.

[277] S.C. Rand. Bright storage of light. *Optics & Photonics News*, May:32–37, 2003.

[278] Y.I. Ravich, B.A. Efimova, and V.I. Tamarchenko. Scattering of current carriers and transport phenomena in lead chalcogenides i. theory. *Physica Status Solidi (B)*, 43:11–33, 1971.

[279] A. Rayner, M.E.J. Friese, A.G. Truscott, N.R. Heckenberg, and H. Rubinsztein-Dunlop. Laser cooling of a solid from ambient temperature. *J. Modern Optics*, 48:103–114, 2001.

[280] S.M Raytov, Y.A. Kravtsov, and V.I. Tatarski. *Principles of Statistical Radio Physics*. Third Edition, Springer, New York, 1987.

[281] S.M. Redmond. *Ph.D Dissertation: Luminescence Instabilities and Non-Radiative Processes in Rare-Earth Systems*. University of Michigan, Ann Arbor, 2003.

[282] S.M. Redmond, G.L. Armstrong, H.Y. Chan, E. Mattson, A. Mock, B. Li, R. Potts, M. Cui, S.C. Rand, S.L. Oliveira, J. Marchal, T. Hinklin, and R.M. Laine. Electrical generation of stationary light in random scattering media. *J. Optical Society America B*, 21:214–222, 2004.

[283] E.D. Reed and H.W. Moos. Multiphonon relaxation of excited states of rare-earth ions in YVO_4, $YAsO_4$, and YPO_4. *Physical Review B*, 8:980–987, 1973.

[284] B.K. Ridley. The electron-phonon interaction in quasi-two-dimensional semiconductor quantum-well structures. *J. Physics C: Solid State Physics*, 15:5899–5977, 1982.

[285] L. A. Riseberg and H. W. Moos. Multiphonon orbit-lattice relaxation of excited states of rare-earth ions in crystals. *Physical Review*, 174:429–438, 1968.

[286] R.T. Ross and A.J. Nozik. Efficiency of hot-carrier solar energy converters. *J. Applied Physics*, 53:3813–3818, 1982.

[287] M. Roufosse and P.G. Klemens. Thermal-conductivity of complex dielectric crystals. *Physical Review B*, 7:5379–5386, 1973.

[288] D.M. Rowe. *CRC Handbook of Thermoelectrics*. CRC Press, Boca Raton, 1995.

[289] X.L. Ruan. *Ph.D Thesis: Fundamentals of Laser Cooling of Rare-Earth-Ion Doped Solids and Its Enhancement Using Nanopowders*. University of Michigan, Ann Arbor, 2006.

[290] X.L. Ruan, H. Bao, and M. Kaviany. Boundary-induced vibrational spectra broadening in nanostructures. *Physical Review B*, submitted, 2008.
[291] X.L. Ruan and M. Kaviany. Photon localization and electromagnetic field enhancement in laser irradiated, random porous media. *Nanoscale and Microscale Thermophysical Engineering*, 9:63–84, 2005.
[292] X.L. Ruan and M. Kaviany. Enhanced laser cooling of rare-earth-ion-doped nanocrystalline powders. *Physical Review B*, 73:155422–1–15, 2006.
[293] X.L. Ruan and M. Kaviany. *Ab initio* photon-electron and electron-vibration coupling calculations related to laser cooling of ion-doped solids. *J. Computational Theoretical Nanoscience*, 5:221–229, 2008.
[294] X.L. Ruan, S.C. Rand, and M. Kaviany. Entropy and efficiency in laser cooling of solids. *Physsical Review B*, 75:1–9, 2007.
[295] A.R. Ruffa. Statistical thermodynamics of insulators. *J. Chemical Physics*, 83:6405–64008, 1985.
[296] J.J. Sakurai. *Advanced Quantum Mechanics*. Benjamin/Cummings, Menlo Park, 1984.
[297] D.H. Sampson. *Radiative Contributions to Energy and Momentum Transport in Gases*. Wiley, New York, 1965.
[298] R.T. Sanderson. Chemical bonds and bond energy. *In Physical Chemistry Series*, 21:174–211, 1971.
[299] T. Sato. Spectral emissivity of silicon. *Japanese J. Physics*, 6:339–347, 1967.
[300] P.K. Schelling, S.R. Phillpot, and P. Keblinski. Comparison of atomic-level simulation methods for computing thermal conductivity. *Physical Review B*, 65:144306, 2002.
[301] K.W. Schlichting, N.P. Padture, and P.G. Klemens. Thermal conductivity of dense and porous yttria-stabilized zirconia. *J. Materials Science*, 36:3003–3010, 2001.
[302] D. Segal and A. Nitzan. Spin-boson thermal rectifier. *Physical Review Letters*, 94:034301–1–4, 2005.
[303] H. Septzler, C.G. Sammis, and R.J. O'Connell. Equation of state of sodium chloride ultrasonic measurements to 8 kbar and 800.deg. and static lattice theory. *J. Physics Chemistry of Solids*, 33:1727–1750, 1972.
[304] F. Sharipov and D. Kalempa. Velocity slip and temperature jump coefficients for gaseous mixtures. ii thermal slip coefficient. *Physics of Fluids*, 16:759–764, 2004.
[305] J.L. Shohet. *The Plasma State*. Academic, New York, 1971.
[306] R. Siegel and J. Howell. *Thermal Radiation Heat Transfer*. Taylor & Friends, New York, 2002.
[307] A.E. Siegman. *Lasers*. University Science Books, Sausalito, 1986.
[308] D.J. Singh and L. Nordström. *Planewaves, Pseudopotentials, and the LAPW Method*. Springer, New York, 2006.
[309] J. Singh. *Physics of Semiconductor and Their Heterostructures*. McGraw-Hill, New York, 1993.
[310] J. Singh. *Electronic and Optoelectronic Properties of Semiconductors Structures*. Cambridge University Press, Cambridge, 2003.
[311] D.B. Sirdeshmukh, P.G. Krishna, and K.G. Subhadra. Micro-macro property correlations in alkali halide crystals. *J. Materials Science*, 38:2001–2006, 2003.
[312] G.A. Slack. Effect of isotopes on low-temperature thermal conductivity. *Physical Review*, 105:829–831, 1957.
[313] G.A. Slack. *The Thermal Conductivity of Nonmetallic Crystals, Solid State Physics*. 34, 1-73, Academic Press, New York, 1979.
[314] G.A. Slack and C. Glassbrenner. In: Thermal conductivity of germanium from 3 k to 1020 k. *Physical Review*, 120:782–789, 1960.
[315] H. Smith and H.H. Jensen. *Transport Phenomena*. Clarendon, Oxford, 1989.

[316] K.C. Sood and M.K. Roy. Longitudinal and transverse parts of the correction term in the Callaway model for the phonon conductivity. *J. Physics: Condensed Matter*, 5:L245–L246, 1993.

[317] G.S. Springer. Heat transfer in rarefied gases. *Advances in Heat Transfer*, 7:163–218.

[318] G.P. Srivastava. *The Physics of Phonons*. Adam Hilger, Bristol, 1995.

[319] R. Stedman, L. Almqvist, and G. Nillson. Phonon-frequency distribution and heat capacities of aluminum and lead. *Physical Review*, 162:549–557, 1967.

[320] F.H. Stillinger and T.A. Weber. Computer simulation of local order in condensed phases of silicon. *Physical Review B*, 31:5262–5271, 1985.

[321] M.E. Striefler and G.R. Barsch. Lattice dynamics of alpha-quartz. *Physical Review B*, 12:4553–4566, 1975.

[322] E.T. Swartz and R.O. Pohl. Thermal boundary resistance. *Reviews of Modern Physics*, 61:605–658, 1989.

[323] L. Tablot, R.K. Cheng, R.W. Schefer, and D.R. Willis. Thermophoresis of particle in a heated boundary layer. *J. Fluid Mechanics*, 101:737–758, 1980.

[324] H. Tachikawa, T. Ichikawa, and H. Yoshida. Geometrical structure and electronic states of the hydrated titanium(iii) ion. an *ab initio* ci study. *J. American Chemical Society*, 112:982–987, 1990.

[325] C.L. Tien. Thermal radiation properties of gases. *Advances in Heat Transfer*, 5:253–324, 1968.

[326] C.L. Tien and J.H. Lienhard. *Statistical Thermodynamics*. Holt, Rinehart and Winston, New York, 1971.

[327] P. Tipler and R. A. Leewelly. *Modern Physics*. Third Edition, Freeman, New York, 2000.

[328] M.D. Tiwari and B.K. Agrawal. Analysis of the lattice thermal conductivity of germanium. *Physical Review B*, 4:3527–3532, 1971.

[329] Y. Touloukian. *Thermophysical Properties of Matter*. Plenum, New York, 1970.

[330] R.E. Trees. Configuration interaction in mn ii. *Physical Review*, 83:756–760, 1951.

[331] P.L.W. Tregenna-Piggott, S.P. Best, M.C.M. O'Brien, K.S. Knight, J.B. Forsyth, and J.R. Pilbrow. Cooperative Jahn-Teller effect in titanium alum. *J. American Chemical Society*, 119:3324–3332, 1997.

[332] M.C. Tsai and S. Kou. Heat transfer and fluid flow in welding arcs produced by sharpened and flat electrodes. *International J. Heat Mass Transfer*, 33:2089–2098, 1990.

[333] C. Uher. Skutterudites: Prospective novel thermoelectrics. *Semiconductors and Semimetals*, 69:140–253, 2001.

[334] B.W.H. van Beest, G.J. Krammer, and R.A. van Santen. Force field for silica and aluminophosphates based on *ab initio* calculations. *Physical Review Letters*, 64:1955–1958, 1990.

[335] B. van Zeghbroeck. http://ece-www.colorado.edu/~bart/book.

[336] S. Vavilov. Some remarks on the Stokes law. *J. Physics (Moscow)*, 9:68–73, 1945.

[337] S. Vavilov. Photoluminescence and thermodynamics. *J. Physics (Moscow)*, 10:499–501, 1946.

[338] F.J. Vesely. *University of Vienna*. http://homepage.univie.ac.at/franz.vesely/notes/md_neq/mdneq/node1.html, 2007.

[339] W.E. Vincenti and C.H. Kruger. *Introduction to Physical Gas Dynamics*. Wiley, New York, 1965.

[340] Y.A. Vlasov, M.A. Kaliteevski, and V.V. Nikolaev. Different regimes of light localization in a disordered photonic crystal. *Physical Review B*, 60:1555–1562, 1999.

[341] S. Volz, J.B. Saulnier, M. Lallemand, B. Perrin, P. Depondt, and M. Mareschal. Transient Fourier-law deviation by molecular dynamics in solid argon. *Physical Review B*, 54:340–347, 1996.

[342] S.G. Volz and G. Chen. Molecular-dynamics simulation of thermal conductivity of silicon crystals. *Physical Review B*, 61:2651–2656, 2000.

[343] H.M. Wastergaard. *Theory of Elasticity and Plasticity*. Dover, New York, 1952.

[344] D.P. White. The effect of ionization and displasive radiation on thermal conductivity of alumina. *J. Applied Physics*, 73:2254–2258, 1993.

[345] D.P. White. The effect of ionization and dissipative reaction on the thermal conductivity of alumina at low temperature. *J. Nuclear Materials*, 212-215:1069–1074, 2004.

[346] F. Widulle, T. Ruff, K. Konuma, I. Silier, M. Cordona, W. Kriegseis, and V.I. Ozhogin. Isotope effects in elemental semiconductors: a Raman study of silicon. *Solid State Communications*, 118:1–22, 2001.

[347] D.S. Wiersma, P. Bartolini, A. Lagendijk, and R. Righini. Localization of light in a disordered medium. *Nature*, 390:671–673, 1997.

[348] C. Willett. *Introduction to Gas Lasers: Population Inversion Mechanisms*. Pergamon, New York, 1974.

[349] E.B. Wilson, J.C. Decius, and P.C. Cross. *Molecular Vibrations: The Theory of Infrared and Raman Vibrational Spectra*. McGraw-Hill, 1955.

[350] D. Wolf, P. Keblinski, S.R. Philpot, and J. Eggebrecht. Exact method for the simulation of Coulombic systems by direct, pairwise 1/r summation. *J. Chemical Physics*, 110:8254–8282, 1999.

[351] T.O. Woodruff and H. Ehrenrich. Absorption of sound in insulators. *Physical Review*, 123:1553–1559, 1961.

[352] H.S. Yang, K.S. Hong, S.P. Feofilov, B.M. Tissue, R.S. Meltzer, and W.M. Dennis. Electron-phonon interaction in rare earth doped nanocrystals. *J. Luminescence*, 83–84:139–145, 1999.

[353] W.M. Yen, S. Shionoya, and H. Yamamoto. *Phosphor Handbook*. CRC Press, Boca Raton, 2002.

[354] R.C. Yu, N. Tea, M.B. Salamon, D. Lorents, and R. Malhotra. Thermal conductivity of single crystal C_{60}. *Physical Review Letters*, 68:2050–2053, 1992.

[355] R. Zallen and M.L. Slade. Influence of pressure and temperature on phonons in molecular chalcogenides: Crystalline As_4S_4 and S_4N_4. *Physical Review B*, 18:5775–5798, 1978.

[356] D.M. Zayachuk. The dominant mechanisms of charge-carrier scattering in lead telluride. *Semiconductors*, 31:173–176, 1997.

[357] Z.M. Zhang. *Nano/Microscale Heat Transfer*. McGraw-Hill, New York, 2007.

[358] Z.M. Zhang and M.I. Flik. Predicted absorption of $YBa_2Cu_3O_7$/YSZ/Si multilayer structures for infrared detectors. *IEEE Transactions on Applied Superconductivity*, 3:1604–1607, 1993.

[359] J.M. Ziman. *Electrons and Phonons: The Theory of Transport Phenomenon in Solids*. Oxford University Press, Oxford, 1960.

[360] J.M. Ziman. *Principles of the Theory of Solids*. Second Edition, Cambridge University Press, Cambridge, 1972.

[361] R. Zwanzig. Time-correlation functions and transport coefficients in statistical mechanics. *Annual Review of Physical Chemistry*, 16:67–102, 1965.

Index